Microbial Ecology
of the Oceans

Microbial Ecology of the Oceans

Second Edition

DAVID L. KIRCHMAN
College of Marine and Earth Studies, University of Delaware

A JOHN WILEY & SONS, INC., PUBLICATION

Copyright © 2008 by John Wiley & Sons, Inc. All rights reserved.

Published by John Wiley & Sons, Inc., Hoboken, New Jersey
Published simultaneously in Canada

Wiley-Blackwell is an imprint of John Wiley & Sons, formed by the merger of Wiley's global Scientific, Technical, and Medical business with Blackwell Publishing.

No part of this publication may be reproduced, stored in a retrieval system, or transmitted in any form or by any means, electronic, mechanical, photocopying, recording, scanning, or otherwise, except as permitted under Sections 107 or 108 of the 1976 United States Copyright Act, without either the prior written permission of the Publisher, or authorization through payment of the appropriate per-copy fee to the Copyright Clearance Center, Inc., 222 Rosewood Drive, Danvers, MA 01923, 978-750-8400, fax 978-750-4470, or on the web at www.copyright.com. Requests to the Publisher for permission should be addressed to the Permissions Department, John Wiley & Sons, Inc., 111 River Street, Hoboken, NJ 07030, 201-748-6011, fax 201-748-6008, or online at http://www.wiley.com/go/permission.

Limit of Liability/Disclaimer of Warranty: While the publisher and author have used their best efforts in preparing this book, they make no representations or warranties with respect to the accuracy or completeness of the contents of this book and specifically disclaim any implied warranties of merchantability or fitness for a particular purpose. No warranty may be created or extended by sales representatives or written sales materials. The advice and strategies contained herein may not be suitable for your situation. You should consult with a professional where appropriate. Neither the publisher nor author shall be liable for any loss of profit or any other commercial damages, including but not limited to special, incidental, consequential, or other damages.

For general information on our other products and services or for technical support, please contact our Customer Care Department within the United States at 800-762-2974, outside the United States at 317-572-3993 or fax 317-572-4002.

Wiley also publishes its books in variety of electronic formats. Some content that appears in print may not be available in electronic format. For more information about Wiley products, visit our web site at www.wiley.com.

Library of Congress Cataloging-in-Publication Data:

Kirchman, David L.
  Microbial ecology of the oceans / [edited by] David L. Kirchman. -- 2nd ed.
    p. cm.
  Includes bibliographical references and index.
  ISBN 978-0-470-04344-8 (pbk.)
1. Marine microbiology. 2. Marine ecology. 3. Carbon cycle
(Biogeochemistry) I. Title.
  QR106.M53 2008
  579'.177--dc22   2007051389

Printed in the United States of America

10 9 8 7 6 5 4 3 2 1

CONTENTS

PREFACE	xv
CONTRIBUTORS	xvii
1 INTRODUCTION AND OVERVIEW	1

David L. Kirchman

Eukaryotic Phytoplankton and Cyanobacteria	3
Photoheterotrophic Bacteria	5
Dissolved Organic Material	7
Heterotrophic Bacteria	10
Marine Archaea	13
Heterotrophic Protists	14
Nanoflagellates (2–20 μm)	14
Microzooplanktonic Protists (20–200 μm)	16
Dinoflagellates	16
Marine Fungi	16
Marine Viruses	17
N_2 Fixers	18
Nitrifiers and Other Chemolithotrophs	19
Denitrifiers	20
Concluding Remarks	21
Summary	22
Acknowledgments	22
References	23

2 UNDERSTANDING ROLES OF MICROBES IN MARINE PELAGIC FOOD WEBS: A BRIEF HISTORY 27

Evelyn Sherr and Barry Sherr

Introduction	27
Pre-1950s: The Early Years	28
1950–1974	29
1970s–1980s	32
Improvement in Methods	32
Bacterial Abundance	32
Bacterial Activity	33
Marine Heterotrophic Protists	34
The "Microbial Loop"	36
1990–Present: The Molecular Revolution	39
Summary	40
References	41

3 BACTERIAL AND ARCHAEAL COMMUNITY STRUCTURE AND ITS PATTERNS 45

Jed A. Fuhrman and Åke Hagström

Introduction	45
Major Groups of Prokaryotes in Seawater	47
"Classically" Culturable Bacteria	49
The *Roseobacter* Clade of Marine *Alphaproteobacteria*	50
Gammaproteobacteria	51
Bacteroidetes	52
Cyanobacteria	52
"Sea Water" Culturable Bacteria	55
SAR11 Cluster	55
Not-Yet-Cultured Bacteria	57
Marine Gammaproteobacterial Clusters	57
Actinobacteria	58
SAR116 Cluster	59
SAR202	59
Marine Group A	59
Marine Group B	59
Betaproteobacteria	59
Marine *Archaea*	60
Bacterioplankton Diversity	63
Species Concept	63

Microdiversity	64
Components of Diversity: Richness and Evenness	65
Community Structure: Description and Factors	67
Bottom-Up Control	68
Sideways Control	69
Top-Down Control	70
"Kill the Winner" Hypothesis	71
Temporal Variation (Days to Seasonal)	72
Short-Term Variation	72
Seasonal Variation	72
Spatial Variation	74
Microscale Patterns	74
Global Distribution	75
Latitudinal Gradient and Degree of Endemism	76
Patchiness and Large Eddies	77
Summary	79
References	80

4 GENOMICS AND METAGENOMICS OF MARINE PROKARYOTES   91

Mary Ann Moran

Introduction	91
The Basics of Prokaryotic Genomics	92
Genome Sequence and Assembly	92
Finding Genes	95
Finding Operons	96
Functional Annotation	96
Tame or Wild? Pure-Culture Genomics Versus Metagenomics	100
Genomics in Marine Microbial Ecology	103
The Ecology of Genome Composition	103
Reverse Biogeochemistry: Discovery of New Ecological Processes	104
Environmental Reductionism: New Details About Recognized Processes	106
Comparative Genomics and Metagenomics	107
Future Directions	122
Summary	125
Acknowledgments	125
References	125

5	PHOTOHETEROTROPHIC MARINE PROKARYOTES	131

Oded Béjà and Marcelino T. Suzuki

Introduction	131
Facultative Photoheterotrophy by Unicellular Cyanobacteria	132
Cyanobacteria as Facultative Heterotrophs	132
Uptake of Urea and DMSP	133
Uptake of Nucleosides and Amino Acids	134
Field Studies Using Light and Dark Incubations	135
Implications of Facultative Photoheterotrophy by Cyanobacteria	138
Marine AAnP Bacteria: Habitats and Diversity	139
Rediscovery of the Marine AAnP Bacteria	139
Diversity of AAnP Bacteria	139
Physiology of AAnP Bacteria	140
AAnP Bacterial Abundance and Ecological Significance	142
Proteorhodopsin-Containing Prokaryotes	143
Proteorhodopsin Genotypes and Taxonomic Distributions	144
Proteorhodopsin Spectral Tuning	145
Proteorhodopsin-Containing Prokaryotes: Abundance and Activity	146
Proteorhodopsin-Containing Prokaryotes: Ecological Significance	150
Summary	151
References	151

6	ECOLOGY AND DIVERSITY OF PICOEUKARYOTES	159

Alexandra Z. Worden and Fabrice Not

Introduction	159
Functional Roles, Classification, and Biological Traits	162
Photoautotrophs	163
Heterotrophs and Alternative Lifestyles	170
Environmental Diversity and Molecular Phylogenetics	172
Diversity of Uncultured Populations	174
Methodological Issues for envPCR Studies	178
Distribution, Abundance, and Activities	179
Methods for Quantifying Mixed Assemblages	180
Distribution, Abundance, and Activity of Mixed Picophytoplankton Assemblages	182
Quantifying Specific Picoeukaryote Populations	186
Methodological Challenges to Quantifying Specific Populations and Resolving Dynamics	190

	Mortality, Contributions to Microbial Food Webs, and Microbial Interactions	191
	Genomic Approaches to Picoeukaryote Ecology	193
	Integration of Picoeukaryotes to the Microbial Food Web: Research Directions	194
	Summary	195
	Acknowledgments	196
	References	196
7	**ORGANIC MATTER–BACTERIA INTERACTIONS IN SEAWATER**	**207**
	Toshi Nagata	
	Introduction	207
	Organic Matter Inventory and Fluxes	208
	DOM–Bacteria Interactions	211
	Labile Low-Molecular Weight (LMW) DOM	211
	Extracellular Hydrolytic Enzymes	215
	Polymeric DOM—Protein as a Model	217
	Refractory DOM	220
	POM–Bacteria Interactions	223
	POM Continuum	223
	POM Fluxes	223
	POM–Mineral Interactions	229
	Bacterial Community Structure and Utilization of Organic Matter	230
	Future Challenges	231
	Summary	232
	References	232
8	**PHYSIOLOGICAL STRUCTURE AND SINGLE-CELL ACTIVITY IN MARINE BACTERIOPLANKTON**	**243**
	Paul A. del Giorgio and Josep M. Gasol	
	Introduction	243
	Distribution of Physiological States in Bacterioplankton Assemblages	245
	The Concept of "Physiological Structure" of Bacterioplankton Assemblages	245
	Starvation, Dormancy, and Viability in Marine Bacterioplankton	246

Describing the Physiological Structure of Bacterioplankton	250
Single-Cell Properties and Methodological Approaches	250
Operational Categories of Single-Cell Activity	259
Regulation of Physiological Structure of Marine Bacterioplankton	260
Factors Influencing Physiological State of Bacterial Cells in Marine Ecosystems	261
Factors Influencing Loss and Persistence of Physiological Fractions	263
Distribution of Single-Cell Characteristics in Marine Bacterioplankton Assemblages	265
Distribution of Single-Cell Activity and Physiological States in Marine Bacterioplankton	265
Simultaneous Determination of Several Aspects of Single-Cell Activity and Physiology	270
Patterns in Distribution of Single-Cell Activity and Physiology Along Marine Gradients	271
Distribution of Activity and Growth Among Bacterial Size Classes	273
Distribution of Activity Across and Within Major Phylogenetic Groups	274
Dynamics of Single-Cell Activity and Physiological States	276
Ecological Implications of Patterns in Bacterioplankton Single-Cell Activity	279
Community Versus Individual Cell Growth and Metabolic Rates	280
Linking the Distribution of Single-Cell Parameters and the Bulk Assemblage Response	282
Ecological Role of Different Physiological Fractions	283
Concluding Remarks	284
Summary	285
Acknowledgments	285
References	285

9 HETEROTROPHIC BACTERIAL RESPIRATION 299

Carol Robinson

Introduction	299
Measurement of Bacterial Respiration and Production	301
Routine Measurement Techniques for Bacterial Respiration and Their Limitations	301
Routine Measurement Techniques for Bacterial Production and Their Limitations	304

Magnitude and Variability of Bacterial Respiration	304
Temporal Variability	308
Spatial Variability	309
Relationship Between Bacterial Respiration and Environmental and Ecological Factors	311
Bacterial Respiration as a Proportion of Community Respiration	315
Predicting Bacterial Respiration	317
Comparison Between Measurements and Predictions of Bacterial Respiration	319
Magnitude of Bacterial Respiration in Relation to Primary Production	321
Bacterial Respiration in a Changing Environment	324
Summary	326
Acknowledgments	327
References	327

10 RESOURCE CONTROL OF BACTERIAL DYNAMICS IN THE SEA   335

Matthew J. Church

Introduction	335
Growth in the Sea	336
Growth and Nutrient Uptake Kinetics	339
Approaches to Understanding Resource Control of Growth	343
Comparative Approaches	343
Experimental Approaches for Defining Limitation of Bacterial Growth	349
Limitation by Dissolved Organic Matter	351
Bacterial Growth on Bulk DOM Pools	353
Limitation by Specific DOM Compounds	354
Limitation by Inorganic Nutrients	361
Nitrogen	361
Phosphorus	364
Trace Nutrients	365
Temperature–DOM Interactions	366
Light	368
Resource Control of Specific Bacterial Populations in the Sea	369
Summary	371
Acknowledgments	371
References	371

11 PROTISTAN GRAZING ON MARINE BACTERIOPLANKTON — 383

Klaus Jürgens and Ramon Massana

Introduction	383
New Insights into Phylogenetic Organization	386
Functional Size Classes of Protists	390
Natural Assemblages of Marine Heterotrophic Nanoflagellates	391
Functional Ecology of Bacterivorous Flagellates	394
Living in a Dilute Environment	394
Using Culture Experiments to Infer the Ecological Role of HNF	397
Impact of Protistan Bacterivory on Marine Bacterioplankton	401
Search for the Perfect Method to Quantify Protistan Bacterivory	401
Rates of Protistan Bacterivory in the Sea	403
Balance of Bacterial Production and Protistan Grazing	404
Bottom-Up Versus Top-Down Control of Bacteria and Bacterivorous Protists	405
Ecological Functions of Bacterial Grazers	406
Grazing as a Shaping Force of Bacterial Assemblages	408
Bacterial Cell Size Determines Vulnerability Towards Grazers	408
Other Antipredator Traits of Prokaryotes	411
Grazing-Mediated Changes in Bacterial Community Composition	413
Molecular Tools for Protistan Ecology	414
Culturing Bias and Molecular Approaches	414
Global Distribution and Diversity of Marine Protists	420
Linking Diversity and Function for Uncultured Heterotrophic Flagellates	422
Summary	423
Acknowledgments	424
References	424

12 MARINE VIRUSES: COMMUNITY DYNAMICS, DIVERSITY AND IMPACT ON MICROBIAL PROCESSES — 443

Mya Breitbart, Mathias Middelboe, and Forest Rohwer

Introduction	443
Viruses and the Marine Microbial Food Web	444
Direct Counts and Viral Numbers	444

	Viral Production and Decay Rates	447
	Viral Decay and Rates of Production in Pelagic Systems	447
	Measurements of Viral Production in Marine Sediments	449
	General Rates of Viral Production	449
	Role of Viruses in Biogeochemical Cycling	450
	Impact of Viruses on Bacterial Diversity and Community Dynamics	452
	Marine Viral Diversity	457
	Methods for Examining Marine Viral Diversity	457
	Culture-Based Studies of Viral Diversity	458
	The Need for Culture-Independent Methods	459
	Culture-Independent Studies of Viral Diversity Using Transmission Electron Microscopy	460
	Whole-Genome Profiling of Viral Communities Based on Genome Size	461
	Studies of Viral Diversity Using Signature Genes	461
	Metagenomic Studies of Viral Diversity	462
	A Vision for the Future	466
	Summary	467
	References	468
13	**MOLECULAR ECOLOGICAL ASPECTS OF NITROGEN FIXATION IN THE MARINE ENVIRONMENT**	**481**
	Jonathan P. Zehr and Hans W. Paerl	
	Introduction	481
	Chemistry, Biochemistry, and Genetics of N_2 Fixation	482
	Genetics and Enzymology	483
	Evolution of N_2 Fixation	485
	Phylogeny of Nitrogenase	487
	Genomics of N_2 Fixation	487
	Diversity of N_2-Fixing Microorganisms	489
	Regulation in Diazotrophs	489
	Methods for Assessing Diazotroph Diversity, Gene Expression, and N_2 Fixation Activity	490
	Ecophysiological Aspects of N_2 Fixation	494
	Ecology of Diazotrophs in the Open Ocean	499
	Estuarine and Coastal Waters	505
	Benthic Habitats, Including Microbial Mats and Reefs	506
	Deep Water and Hydrothermal Vents	507
	Summary	508
	Acknowledgments	509
	References	509

| 14 | NITROGEN CYCLING IN SEDIMENTS | 527 |

Bo Thamdrup and Tage Dalsgaard

Introduction	527
Inputs	531
Transformations	532
Microbes and Microbial Processes	532
Processes Involving Mn and Fe	548
Nitrogen Budgets	550
Benthic Budgets	550
Oceanic Budgets	552
Summary	554
References	555

INDEX   569

PREFACE

It has been nearly 10 years since work started on the first edition of this book. Ten years is a long time for just about any field of science, but especially for a fast-moving one such as marine microbial ecology. Here, finally, is the second edition.

This book is more than just a revision of the first edition which was published back in 2000. Some chapters from that edition are not repeated here, because work in those areas has slowed and the basic principles covered before have not changed. However, those topics and principles remain as important and as valid today as 10 years ago, and readers should hang onto the 2000 book (or get it if they do not have it already); much of it is still relevant. Other chapters of this book have titles similar to those in the first edition, but even in these cases, the chapters have been substantially rewritten, often by authors who have different perspectives on the topics covered previously. Finally, several chapters discuss microbes and biogeochemical processes that we were just beginning to learn about 10 years ago, and still others that we did not even know existed back then.

What remains the same is the intended audience: advanced undergraduates, beginning graduate students, and colleagues from other fields wishing to learn about microbes and the processes they mediate in marine systems. This book, aided by the first edition, is meant to be as close to a textbook as a multi-authored book can be.

I wish to thank several people who helped to get this book published. First and most importantly, I thank the chapter authors for agreeing to work on this project and also for looking over another chapter (or two) in the book. Each chapter was reviewed by another chapter author and an outsider not connected to the book. I especially want to thank Jens Boenigk, Hugh Ducklow, Pete Conway, Stefan Hulth, Rick Keil, Karin Lochte, Alison Murray, Jack Middelburg, Jarone Pinhassi, Thomas Reinthaler, Janice Thompson, Daniel Vaulot, Tracy Villareal, and Peter Williams. The anonymous reviewers (and those I've forgotten to mention by name—sorry) also deserve thanks. I greatly appreciated Dave Karl's support during a critical junction of this project, and I acknowledge the help of Karen

Chambers, Thom Moore, and others at Wiley. Finally, I thank the readers of the first edition of this book. This second edition would not have come about if not for your positive comments and feedback.

DAVID KIRCHMAN

Lewes, Delaware

CONTRIBUTORS

ODED BÉJÀ   Faculty of Biology, Technion–Israel Institute of Technology, Haifa, Israel [beja@techunix.technion.ac.il]

MYA BREITBART   College of Marine Science, University of South Florida, St. Petersburg, FL 33701, U.S.A. [mya@marine.usf.edu]

MATTHEW J. CHURCH   Department of Oceanography, University of Hawaii, Honolulu, HI 96822, U.S.A. [mjchurch@hawaii.edu]

TAGE DALSGAARD   National Environmental Research Institute, University of Aarhus, DK-8600 Silkeborg, Denmark [tda@dmu.dk]

JED A. FUHRMAN   Department of Biological Sciences, University of Southern California, Los Angeles, CA 90089, U.S.A. [fuhrman@usc.edu]

JOSEP M. GASOL   Institut de Ciències del Mar, CMIMA (CSIC), Passeig Marítim de la Barceloneta 37–49, 08003 Barcelona, Catalunya, Spain [pepgasol@icm.csic.es]

PAUL A. DEL GIORGIO   Département des Sciences Biologiques, Université du Québec à Montréal, CP 8888, succ. Centre Ville, Montréal, Québec, Canada H3C 3P8 [del_giorgio.paul@uqam.ca]

ÅKE HAGSTRÖM   Kalmar University, Sweden [ake.hagstrom@hik.se]

KLAUS JÜRGENS   Leibniz Institute for Baltic Sea Research, 18119 Rostock, Germany [klaus.juergens@io-warnemuende.de]

DAVID L. KIRCHMAN   College of Marine and Earth Studies, University of Delaware, Lewes, DE 19958, U.S.A. [kirchman@udel.edu]

RAMON MASSANA   Institut de Ciències del Mar, CMIMA (CSIC), Passeig Marítim de la Barceloneta 37–49, 08003 Barcelona, Catalunya, Spain [ramonm@icm.csic.es]

MATHIAS MIDDELBOE   Marine Biological Laboratory, University of Copenhagen, DK-3000 Helsingør, Denmark [mmiddelboe@bio.ku.dk]

MARY ANN MORAN   Department of Marine Sciences, University of Georgia, Athens, GA 30602-3636, U.S.A. [mmoran@uga.edu]

TOSHI NAGATA   Ocean Research Institute, The University of Tokyo, Tokyo, Japan [nagata@ori.u-tokyo.ac.jp]

FABRICE NOT   Evolution du Plancton et PaleOceans Laboratory, CNRS, Université Paris 06, UMR7144, Station Biologique de Roscoff, 29682, Roscoff, Cedex BP 74, France [not@sb-roscoff.fr]

HANS W. PAERL   Institute of Marine Sciences, University of North Carolina at Chapel Hill, Morehead City, NC 28557, U.S.A. [hans_paerl@unc.edu]

CAROL ROBINSON   School of Environmental Sciences, University of East Anglia, Norwich NR4 7TJ, U.K. [carol.robinson@uea.ac.uk]

FOREST ROHWER   Department of Biology, San Diego State University, San Diego, CA 92182, U.S.A. [frohwer@gmail.com]

EVELYN SHERR   College of Oceanic and Atmospheric Sciences, Oregon State University, Corvallis, OR 97331-5503, U.S.A. [sherre@coas.oregonstate.edu]

BARRY SHERR   College of Oceanic and Atmospheric Sciences, Oregon State University, Corvallis, OR 97331-5503, U.S.A. [sherrb@coas.oregonstate.edu]

MARCELINO T. SUZUKI   Chesapeake Biological Laboratory, University of Maryland, Center for Environmental Science, Solomons, MD 20688, U.S.A. [suzuki@cbl.umces.edu]

BO THAMDRUP   Institute of Biology, University of Southern Denmark, DK-5230 Odense M, Denmark [bot@biology.sdu.dk]

ALEXANDRA Z. WORDEN   Monterey Bay Aquarium Research Institute, Moss Landing, CA 95039, U.S.A. [azworden@mbari.org]

JONATHAN P. ZEHR   Ocean Sciences Department, University of California, Santa Cruz, CA 95064, U.S.A. [zehrj@pmc.ucsc.edu]

INTRODUCTION AND OVERVIEW

DAVID L. KIRCHMAN
College of Marine and Earth Studies, University of Delaware, Lewes, DE 19958, U.S.A.

Marine microbes are capable of flourishing in all oceanic habitats, from several kilometers below the seafloor to the top millimeter of the ocean surface. They thrive in environmental conditions where other organisms cannot, ranging from supercooled brine channels of Arctic ice floes to near-boiling waters of hydrothermal vents. Consequently, marine microbes are the most numerous group of organisms on the planet. In addition to being abundant, the many different types of marine microbes carry out many different types of metabolism. As a consequence of this diversity, marine microbes are involved in virtually all geochemical reactions occurring in the oceans.

Many of these microbes, the ecological interactions among them, and the biogeochemical processes they mediate are the topics covered by this book.

What is marine microbial ecology? A complicated answer is given in the first edition of this book (Kirchman and Williams 2000). A simple answer is that it is the study of the ecology of microbes in marine systems. "Microbes" includes all organisms smaller than about 100 μm, which can be seen only with a microscope. These organisms include bacteria, archaea, and protists (single-celled eukaryotes). Chapter 12 examines the ecological roles of viruses and phages, things that arguably are not living and thus are not microbes. Colleagues outside the field sometimes assume that "microbe" and "microorganisms" refer only to bacteria, even just heterotrophic bacteria. Certainly these microbes are quite abundant and ecologically important in the oceans, and readers will see several chapters

---

Microbial Ecology of the Oceans, Second Edition. Edited by David L. Kirchman
Copyright © 2008 John Wiley & Sons, Inc.

about them. But there is more to microbial ecology than just the study of heterotrophic bacteria.

The purpose of this chapter is to provide an overview of the book and of some important marine microbes and the parts of biogeochemical processes they mediate. The summary by Sherr and Sherr (2000) remains relevant today, and you are urged to read it. This chapter will take a complementary approach. In fact, much of the entire first edition of this book remains relevant today, and readers are urged to look it over. Table 1.1 summarizes some of the functional groups of microbes discussed here and in the book as a whole.

TABLE 1.1 Functional Groups of Microbes in the Oceans Discussed in this Book

Functional Group	Function	Type of Microbe	Discussed in Chapters
Primary producers	Fix CO_2 to produce organic material using light energy	Eukaryotes and cyanobacteria	2, 3, 5, 6, 13, 14
Photoheterotrophs	Use organic material, aided by light energy	Cyanobacteria and other bacteria[a]	3, 5, 10
Heterotrophic prokaryotes	Mineralize and oxidize dissolved organic matter (DOM) to produce biomass and inorganic byproducts	Bacteria and archaea	2–5, 7–11, 13, 14
Grazers	Control prey populations and release dissolved material	Eukaryotes	2, 6, 8, 11
Viruses	Control prey populations, release dissolved material, and mediate genetic exchange	Not applicable	3, 8, 12
N_2 fixers (diazotrophs)	Reduce N_2 to ammonium	Cyanobacteria[b]	3, 13, 14
Nitrifiers	Oxidization of ammonium to nitrate	Bacteria and archaea	3, 4, 14
Denitrifiers[c]	Release of N_2 or N_2O during oxidation of ammonium or reduction of nitrate	Bacteria and archaea	14

[a]Many protists are mixotrophs (see Chapters 6 and 11) and some eukaryotic phytoplankton are capable of using DOM, but heterotrophic bacteria and archaea usually dominate DOM fluxes.

[b]Some heterotrophic bacteria and archaea are capable of fixing N_2, but cyanobacteria dominate N_2, fixation in the oceans.

[c]The term "denitrification" is often reserved for the production of N_2 or N_2O by dissimilatory nitrate reduction. Here the anammox reaction (oxidation of ammonium) is included because it too results in the loss of N as N_2 gas from the system. See Chapter 14 for more on these definitions.

EUKARYOTIC PHYTOPLANKTON AND CYANOBACTERIA

A starting point for the carbon cycle is carbon fixation or the transformation of CO_2 to a "fixed," nongaseous form—organic carbon (Fig. 1.1). Unlike terrestrial ecosystems, carbon fixation in marine systems is nearly exclusively by free-floating microbial "plants." The exceptions include a few near-shore environments such as salt marshes and mangrove stands where higher-plant production dominates, and in some shallow marine environments where much of the primary production can be by benthic algae (Behringer and Butler 2006; Gattuso et al. 2006; Segal et al. 2006). Aquatic ecologists use the term "phytoplankton" (or algae) rather than "plants," but in fact there are some important similarities between plants on land and phytoplankton in lakes and the oceans. (Algae are found in terrestrial environments, but here I use "land plants" to mean larger, higher plants, which dominate terrestrial primary production.) Both land plants and phytoplankton are autotrophs (CO_2 is their carbon source) and are the main primary producers in their respective ecosystems, using the same mechanism for fixing CO_2, the Calvin–Benson–Bassham cycle. Both have chlorophyll a in reaction centers where light energy is converted to chemical energy. However, unlike land plants, in marine phytoplankton the main pigments absorbing light energy, "the light-harvesting pigments," are not chlorophylls. An example of these other pigments include the carotenoids, one being fucoxanthin, which is abundant in diatoms. For this reason, many phytoplankton are not green, because their dominant light-harvesting pigments absorb light with wavelengths (color) that differ from that absorbed by land plants.

The most important difference, however, between land plants and phytoplankton is the most obvious one: land plants, such as California coastal redwoods and giant sequoias that tower 100 m above the ground, are among the largest creatures on the

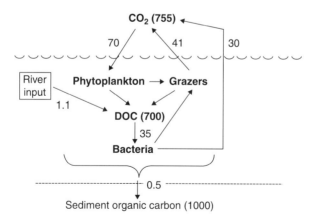

Figure 1.1 The role of microbes in the oceanic carbon cycle. The numbers in parentheses are standing stocks of carbon with units of pgC (1 pgC = 10^{15} gC). The other numbers are rates with units of pgC per year. The numbers are from Hedges and Oades (1997) and from estimates of the average fraction of primary production routed through dissolved organic carbon (DOC) to bacteria (Chapter 9), assuming a growth efficiency of 15 percent.

planet, while phytoplankton are among the smallest, some as tiny as a micrometer or less (10^{-6} m). This difference in size has many implications for how marine and indeed all aquatic ecosystems are structured. By "structured," I mean the size and number of organisms, biochemical composition, phylogenetic diversity, growth rates and net changes in population sizes, and trophic interactions (who is eating whom). Size matters for many of the processes discussed in this book.

Another huge difference between terrestrial plants and phytoplankton is that the latter includes cyanobacteria, in addition to eukaryotic algae. Cyanobacteria are discussed in Chapters 3 and 5. Especially in the nutrient-poor, oligotrophic oceans, cyanobacteria can account for a high fraction (nearly 90 percent) of primary production and of total phytoplankton biomass. Estimates for cyanobacteria may decrease as we learn more about small eukaryotic phytoplankton (see Chapter 6), but cyanobacteria will undoubtedly remain important in the oceans. Two groups of cyanobacteria are especially large contributors to primary production and phytoplankton biomass: *Synechococcus* and *Prochlorococcus* (Table 1.2). The cells in both cyanobacterial groups are small (1 μm in diameter or less), smaller than eukaryotic phytoplankton. Phylogenetically, cyanobacteria are bacteria; they lack a nucleus, and their cell wall and membranes are like those of Gram-negative bacteria (Hoiczyk and Hansel 2000). Functionally, however, both *Synechococcus* and *Prochlorococcus* are members of the phytoplankton community because they are mainly photoautotrophic and use light energy to fix CO_2 by similar mechanisms (e.g., both have the Calvin–Benson cycle) as found in eukaryotic phytoplankton.

Cyanobacteria and Blue–Green Algae

The old term for cyanobacteria is "blue–green algae," which hints at the main pigments of some of these microbes. The green is due to chlorophyll a, while the blueish tinge comes from phycocyanin. When isolated from other pigments, the striking blue color of phycocyanin emerges. Marine *Synechococcus*, one of the main cyanobacteria found in the ocean, has phycocyanin, but this microbe in pure cultures is blood-red due to phycoerythrin.

TABLE 1.2 Comparison of the Two Major Coccoid Cyanobacterial Genera Found in the Oceans

Property	*Synechococcus*	*Prochlorococcus*
Size (diameter)	0.9 μm	0.7 μm
Chlorophyll a	Yes	Yes, but modified
Chlorophyll b	No	Yes, but modified
Phycobilins[a]	Yes	Variable
Distribution[b]	Cosmopolitan	Oceanic gyres
N_2 fixation	Some species	No

[a]Phycobilins are the major light-harvesting pigments in cyanobacteria.
[b]Neither group of cyanobacteria is found in cold, high-latitude oceans. *Synechococcus* can be abundant in coastal waters, unlike *Prochlorococcus*.

Microbiologists knew about cyanobacteria for over a century, but the high abundances of oceanic *Synechococcus* and *Prochlorococcus* were discovered only around 1977 and 1986, respectively (Chisholm et al. 1988; Waterbury et al. 1979). Since then, we have learned much about these organisms. In contrast to most other marine bacteria, marine cyanobacteria have convenient markers, their pigments, for studying them. These unique pigments make it possible to examine *Prochlorococcus* by flow cytometry and *Synechococcus* by microscopy as well as by flow cytometry (see Chapter 6 for a description of flow cytometry). Again unlike many other marine bacteria, ecologically relevant representatives from the two cyanobacterial groups can be isolated and grown alone in pure culture in the laboratory. Consequently, we have learned much about the physiology and biochemistry of these microbes, and the genomes of several of them have been completely sequenced (Dufresne et al. 2003; Palenik et al. 2007; Rocap et al. 2003).

The eukaryotic members of the phytoplankton community are also important in many oceanic waters. The large species (10–100 μm, which is large in the microbial world) are relatively easy to identify because they have distinctive shapes and sizes, in addition to their distinctive light-harvesting pigments. Large phytoplankton, such as diatoms and coccolithophorids, have been studied for years, and are well known for their importance in coastal waters, especially in spring when they form dense blooms. But often, much smaller eukaryotic phytoplankton species are much more abundant and dominate phytoplankton biomass, along with the cyanobacteria. These small eukaryotes are only slightly larger than bacteria and are members of the "picoplankton" community, which includes all microbes 2 μm or less in size (see Chapter 2). Unlike large phytoplankton, the picophytoplankton are hard to identify by traditional methods, because of their size and lack of distinguishing features. In Chapter 6, Worden and Not discuss these important microbes and the use of molecular tools and other methods for examining them.

PHOTOHETEROTROPHIC BACTERIA

All microbes, not just eukaryotic phytoplankton and cyanobacteria, are affected by light directly or indirectly in the surface layer of the oceans (Moran and Zepp 2000). An example of an indirect effect is the photochemical modification of dissolved organic matter (DOM) used by heterotrophic bacteria. The direct effects include the damage of microbial DNA by short-wavelength light, especially in the ultraviolet region (200–400 nm). These types of light effects are well known.

Less well known is the use of light by phototrophic bacteria to drive adenosine triphosphate (ATP) synthesis while also obtaining energy from other sources, most prominently the oxidation of organic material (Table 1.3). Chapter 5 discusses photoheterotrophic bacteria in the oceans, including cyanobacteria, proteorhodopsin-bearing bacteria and aerobic anoxygenic phototrophic bacteria (AAP bacteria, or AAnP in Chapter 5). Except for cyanobacteria, these bacteria are not autotrophic and do not contribute to primary production. We think that they can synthesis ATP via both phototrophic and heterotrophic mechanisms, hence making them

TABLE 1.3 Phototrophic Microbes in the Oceans

Microbial Group	Pigment in Energy Production[a]	CO_2 Fixation?
Eukaryotic phytoplankton	Chlorophyll a	Yes
Cyanobacteria	Chlorophyll a	Yes
AAP bacteria[b]	Bacteriochlorophyll a	No
Several	Proteorhodopsin	No

[a]The pigment most directly involved in ATP production. Chlorophyll a and bacteriochlorophyll a are the main pigments in the reaction centers in the respective microbes. Other pigments, called "accessory pigments," such as fucoxanthin for eukaryotic diatoms and phycoerythrin for *Synechococcus*, often absorb more light energy than the energy-production pigment.

[b]Aerobic anoxygenic phototrophic bacteria. The abbreviation "AAnP" is used in Chapter 5 and by some authors of studies published in the primary literature.

photoheterotrophic, but little else is known about them. We probably know more about the diversity of genes encoding proteorhodopsin and of the diagnostic gene (*pufM*) for AAP bacteria than we do about their ecophysiology and biogeochemical roles in the oceans. For instance, the number and diversity of proteorhodopsin genes in the oceans suggest that bacteria bearing them are abundant, but how proteorhodopsin benefits these bacteria is not entirely clear; light has no effect on the growth of *Pelagibacter ubique*, a cultured marine bacterium from the SAR11 clade with proteorhodopsin (Giovannoni et al. 2005), whereas light does stimulate the growth of another marine heterotrophic bacterium in pure culture experiments (Gomez-Consarnau et al. 2007). One hypothesis is that photoheterotrophic bacteria should be most abundant in oligotrophic oceans, where light energy can augment that gained from heterotrophy and the oxidation of organic material. However, AAP bacteria can be quite abundant in eutrophic estuaries and coastal waters (Cottrell et al. 2006; Schwalbach and Fuhrman 2005; Sieracki et al. 2006; Waidner and Kirchman 2007). We know little about even basic parameters of photoheterotrophic bacteria.

One measure of the importance of photoheterotrophic microbes is the effect of light on bacterial production, as measured by leucine incorporation. This effect varies among environments (Pakulski et al. 2007). In some cases, visible light can inhibit leucine incorporation, whereas in other systems, it stimulates incorporation rates. One factor hypothesized to explain this variable effect is the abundance of *Prochlorococcus* (Church et al. 2004). The role of *Prochlorococcus* in DOM fluxes has been demonstrated by flow cytometric studies of methionine assimilation (Zubkov et al. 2003; Zubkov and Tarran 2005). More recent flow cytometric studies confirmed that *Prochlorococcus* was responsible over 75 percent of light-stimulated leucine incorporation in the North Atlantic Ocean, but *Synechococcus* and probably other bacterial groups also accounted for substantial light-dependent leucine incorporation (Michelou et al. 2007). While per cell rates can be quite high for *Prochlorococcus* and *Synechococcus*, their abundance is not as high as other microbial groups, such as heterotrophic bacteria and photoheterotrophic bacteria other than cyanobacteria. Consequently, heterotrophic bacteria and photoheterotrophic bacteria other than cyanobacteria account for most of leucine incorporation in the oceans regardless of light intensities.

Chapter 5 calls cyanobacteria "operational heterotrophs" because they appear to use relatively few organic compounds in experiments with laboratory cultures. However, we really do not know which DOM compounds are used by cyanobacteria in the oceans. These microbes are probably mainly photoautotrophic, but they could still contribute substantially to the uptake of some DOM components.

DISSOLVED ORGANIC MATERIAL

This chapter is focused on organisms and viruses, but DOM is the "800-pound gorilla" in the ocean, making it hard to ignore: it is the largest pool of organic carbon in the ocean and one of the largest in the biosphere. Concentrations of DOM are not measured directly, but rather the key components are examined, given here in decreasing concentration and order of our understanding: dissolved organic carbon (DOC), dissolved organic nitrogen (DON), and dissolved organic phosphorus (DOP). The amount of C in the DOC pool is nearly equivalent to the carbon in atmospheric CO_2 (Hedges and Oades 1997), but concentrations of individual DOM components are very low, often nanomolar (10^{-9} mol/L). The book edited by D. Hansell and C. Carlson provides an excellent overview of DOM in the oceans (Hansell and Carlson 2002) and another book discusses DOM–microbe interactions in several aquatic environments (Findlay and Sinsabaugh 2003). Chapter 7 also discusses DOM extensively.

The large size of the DOC pool is perhaps the most obvious reason why it is included in oceanic carbon budgets and climate change models, but the flux through this pool is also quite large and in many ways is a more important parameter. By "flux," I mean the rate (usually expressed as $mgC/m^2/d$ or $mmol-C/m^2/d$) at which DOM components are produced or utilized. Chapter 7 discusses what we know about the fluxes of different components of the DOM pool. In spite of most oceanic DOC being old and refractory, the rest of the DOC pool is sufficiently labile (used readily by bacteria) that overall DOC fluxes are usually equivalent to about 50 percent of primary production, and in some marine systems, the fraction is even higher (Chapter 9). Some DOM comes directly from phytoplankton, for example via release of small compounds or the sloughing off of large polymers associated with the outside of phytoplankton cells. However, most of the DOM appears to be produced by grazers (Nagata 2000), but every organism (and virus) in the ocean contributes to the production of DOM. Organic carbon from terrestrial sources can be a substantial fraction of total DOM in lakes and estuaries (Cauwet 2002).

The high fluxes of DOM help explain many features of biological communities in marine water columns. One important feature is the retention of material in the surface layer of the oceans. Because most of the DOC used by bacteria is respired and because of the small size of microbes, as much as 90 percent of primary production is mineralized by grazers and heterotrophic bacteria in the upper surface layer of the open oceans and only 10 percent or less sinks out to deeper waters. These percentages vary greatly, especially in coastal waters where less primary production is mineralized in the surface layer and more is exported to deep waters or horizontally to less productive waters. Figure 1.2 illustrates the fates of primary production and the possible relationships between DOM fluxes and export. At the risk of oversimplifying complex relationships, the figure suggests that any carbon going into the microbial food webs is less likely to end up in larger organisms and higher trophic levels or to sink out of the surface layer, leading to the prediction that there should be an inverse relationship between DOM fluxes and export out of the surface layer. Any organic carbon going into the dissolved pool has to pass through too many transfers among trophic levels before it reaches organisms large enough to produce large particles that sink out of the surface layer.

> **"Mineralization"** is the transformation of organic material back to its inorganic starting material ("minerals"), most importantly CO_2, NH_4^+ and PO_4^{3-}.

There are some data supporting the inverse relationship between DOM fluxes and export (Cho et al. 2001). Cho et al. (2001) examined bacterial biomass production in the Yellow Sea, off the coast of South Korea, and calculated the ratio of bacterial production to primary production, which is a useful index of the relative size of the DOM flux. Heterotrophic bacteria are the main users of DOM in the oceans, as discussed below, although respiration by bacteria (oxidation of organic carbon back to CO_2) is usually much larger than biomass production by bacteria (Chapters 8 and 9). Cho and colleagues found an inverse relationship between an index of organic carbon export (the f-ratio—see text box) and the ratio of bacterial production to primary production (Fig. 1.3), consistent with the negative relationship between export and the phytoplankton–DOC–bacteria pathway implied by Figure 1.2.

While photochemical and other abiotic reactions can modify and even mineralize DOM (especially from terrestrial sources) back to CO_2 and other inorganic compounds (e.g., NH_4^+), the dominant reactions affecting marine DOM are catalyzed by microbes. Phytoplankton, grazers, and viral lysis all contribute to DOM production, and heterotrophic bacteria are the main users of DOM. Microbes thus set what goes into the DOM pool and what remains.

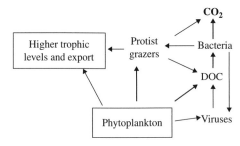

Figure 1.2 Main trophic interactions and the fate of primary production in the oceans. "DOC" is dissolved organic carbon. "Higher trophic levels" refers to organisms much larger than microbes, such as zooplankton and fish. "Export" is the sinking of organic material out of the surface layer.

DISSOLVED ORGANIC MATERIAL

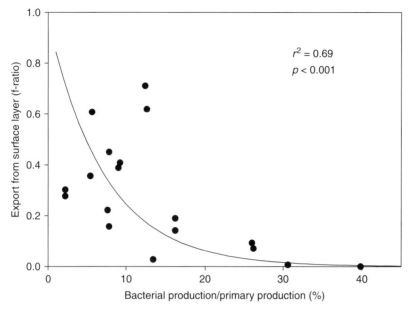

Figure 1.3 An example of the relationship between export and the ratio of bacterial production to primary production. Export was estimated from the f-ratio, which is the ratio of nitrate uptake to total N assimilation. Export and the f-ratio are equal at steady state. Data taken from Cho et al. (2001).

The f-Ratio, New Production, and Export

These three interrelated terms are used to describe relationships among nutrient sources, nutrient uptake, and export of material out of the surface mixed layer. The f-ratio is simply the ratio of nitrate uptake to total N use. New production refers to primary production supported by nutrients "new" to the surface layer (Dugdale and Goering 1967). The archetypal new nutrient is nitrate, because it was thought that its only source was from nitrification in deep waters, whereas ammonium was released by grazers and microbes in surface waters. If so, then the upward flux of nutrients would bring in nitrate but not ammonium, hence making nitrate the "new" nutrient for the surface layer. At steady state, nitrate uptake should equal export of N. While they have their uses, these terms have several problems, and they seem to be falling out of use. One problem is that nitrification in euphotic zone could introduce nitrate to the surface layer that would not be new (Yool et al. 2007), and N_2 fixation would produce ammonium that would be new, not regenerated. Whereas biological oceanographers once thought that N_2 fixation rates were slow enough to be ignored, we now think that these rates are high relative to total N uptake and regeneration (Chapter 13).

HETEROTROPHIC BACTERIA

More chapters of this book touch on heterotrophic bacteria than on any other group of microbes (Table 1.1), for good reason. After viruses, heterotrophic bacteria are the most abundant organisms in the ocean and the entire biosphere. Their biomass can be substantial as well (see below). More importantly, heterotrophic bacteria dominate DOM assimilation in the oceans, although DOM can be used by several types of microbes, including cyanobacteria (see above), eukaryotic phytoplankton (Mulholland et al. 2002), protists, and perhaps even larger eukaryotes. The main reason for the dominance by bacteria is size, more specifically the relationship between surface area and volume. Large organisms have more surface area than small organisms and thus more transport systems in membranes for taking up dissolved compounds from the surrounding water. However, large organisms have an even more biomass or volume to support, the reason being surface area (SA) increases only as the square of the radius (SA $= 4\pi r^2$) whereas volume (V) increases as the cube ($V = 4/3\pi r^3$). Consequently, the smallest microbes, bacteria, have the largest surface area-to-volume ratio of all microbes and thus are able to out-compete all other microbes for dissolved compounds. Figure 1.4 gives a rough idea about the sizes of the major microbial groups in the oceans and in addition shows how abundance increases with decreasing size.

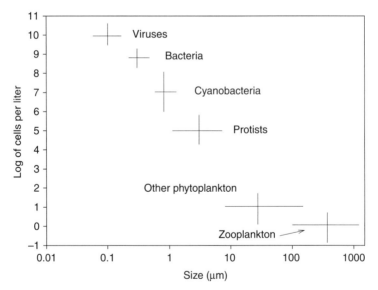

Figure 1.4 Cell abundance and size of major microbial groups in surface waters of aquatic habitats. These estimates are provided as a rough guide to the size and abundance of these microbes. Here "bacteria" refers to heterotrophic bacteria and "zooplankton" to zooplankton such copepods. Abundances can deviate greatly from the numbers provided here, as can sizes. Some protists can be as large as 50 μm and some eukaryotic phytoplankton are as small as 1 μm.

Chapter 9 concentrates on the most important fate of the DOC taken up by bacteria: respiration. Most of the organic carbon taken up by bacteria is respired to CO_2, and the associated organic nitrogen and phosphorus and the other elements also may or may not, depending on the need for these elements, be mineralized back to NH_4^+, PO_4^{3-}, and so on (Kirchman 2000). Here I will make only a few points about biomass and growth of these microbes to complement Chapter 9. The reader is urged to review Ducklow (2000).

First, although heterotrophic bacteria are small, each having as little as 12 fgC (1.2×10^{-14} gC) per cell in the open ocean (Fukuda et al. 1998), their sheer numbers add up such that the total biomass of heterotrophic bacteria (gC of bacterial cell mass per liter) can be quite large, sometimes rivaling that of phytoplankton, especially in oligotrophic waters. (As discussed in Chapter 3, microbes in the other prokaryotic domain, archaea, are only abundant in deep waters (>500 m), where their biomass is roughly equal to that of bacteria.) Estimates of bacterial biomass depend on conversion factors, such as the amount of carbon per cell or biovolume (Ducklow 2000), but these problems do not affect the general trends. Unlike terrestrial systems and eutrophic waters, the biomass of primary producers is smaller than that of all heterotrophs (not just bacteria) in oligotrophic marine systems such as the open ocean gyres (Gasol et al. 1997). Chapter 10 discusses how bacterial biomass alone becomes a larger portion of total biomass in oligotrophic systems where phytoplankton biomass is low.

This large bacterial biomass sets some limits on growth rates for bacteria. Most heterotrophic bacteria grow rapidly in laboratory pure cultures, with doubling times as short as about 30 minutes. (Doubling time is the time needed for a cell to divide and the population to double.) However, in most natural aquatic systems, bacterial generation times are longer than a day, and sometimes are even weeks long in oligotrophic oceans, at least when calculated for the entire community. These generation times are set by the concentrations and supply of labile DOM and other factors, as discussed in Chapter 10.

Community Growth Rates

The growth rates and generation time discussed here are for the entire bacterial assemblage. One common method for estimating these rates in aquatic ecosystems is to divide cell or biomass production by cell numbers of bacterial biomass, making sure that the units agree such that the end result has units of per time (e.g., per day). This approach and any other one that lumps all bacteria together yields a composite estimate, not necessarily an average rate. It is quite likely that some microbes grow much faster than this composite estimate, whereas others may grow more slowly or not at all (see Chapter 8).

Low bacterial growth rates are consistent with current estimates of bacterial biomass (again, for the entire bacterial community). Biomass (B_b) is connected to growth rates (μ, roughly the inverse of generation time) via bacterial production (BP):

$$\mu_b = BP/B_b \tag{1}$$

A similar equation can be written for phytoplankton biomass (B_p), growth rates (μ_b) and primary production (PP). Since bacterial respiration (R) must be less than primary production over sufficiently long time and space scales, an upper bound for bacterial production is then

$$\text{BP} < \text{PP}\left(\frac{\text{BGE}}{1 - \text{BGE}}\right) \qquad (2)$$

since respiration and BP are linked by the bacterial growth efficiency (BGE) (see Chapters 8 and 9). Combining Equation (1), the corresponding equation for phytoplankton growth rates, and Equation (2) yields

$$\frac{\mu_b}{\mu_p} < \frac{B_p}{B_b}\left(\frac{\text{BGE}}{1 - \text{BGE}}\right) \qquad (3)$$

As illustrated in Figure 1.5, Equation (3) indicates that bacterial growth rates cannot exceed phytoplankton growth rates, except for low ratios of bacterial to phytoplankton biomass or high bacterial growth efficiencies, or some combination of both. However, even if the growth efficiency is high (50 percent), much higher than the best estimate for the oceans (approximately 15 percent), and unless bacterial biomass is substantially less than that of phytoplankton, bacterial growth rates have to be slower than that for phytoplankton. Figure 1.6 gives some data from the

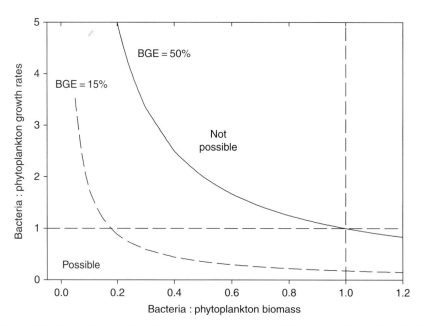

Figure 1.5 Theoretical relationship among growth rates and biomass levels for phytoplankton and heterotrophic bacteria. BGE is bacterial growth efficiency. The horizontal and vertical dashed lines indicate equal rates or biomass levels for phytoplankton and bacteria.

MARINE ARCHAEA

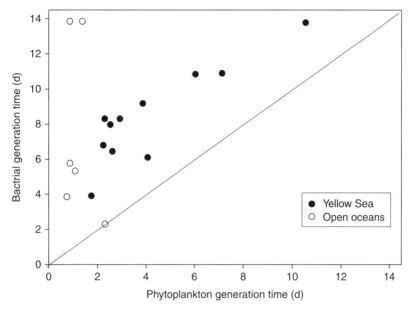

Figure 1.6 Example of generation times for bacteria and phytoplankton. Data are community generation times averaged over the euphotic in the Yellow Sea (Cho et al. 2001) and select open oceans (Ducklow 2000). The diagonal line is the 1 : 1 line.

Yellow Sea indicating that this is the case. Although Cho et al. (2001) found a positive correlation between bacterial and phytoplankton growth rates, bacteria grew more slowly by about twofold on average than phytoplankton, with generation times ranging from about 4 to nearly 14 days (Fig. 1.6). Bacteria also grow more slowly than phytoplankton in most of the open oceans examined to date (Ducklow 2000).

MARINE ARCHAEA

Archaea share some rather superficial similarities with bacteria, the other major domain of prokaryotes. Bacteria and archaea cannot be distinguished by simple microscopy, and cells in both microbial groups do not have a nucleus and lack the organelles found in eukaryotes. The old term for these microbes, archaebacteria, hints at this commonality. However, in many respects, archaea are as different from bacteria as both prokaryotic groups are from eukaryotes. In fact, archaea may be phylogenetically more closely related to eukaryotes than to bacteria. Of the many differences between bacteria and archaea, perhaps the most important one is the composition of the cell wall and membrane. Archaea do not have the peptidoglycan found in bacterial cell walls, and the archaeal membrane has unique lipids not found in either bacteria or eukaryotes.

Oceanic archaea had been discovered in the early 1990s, but their abundance and biogeochemical role in the oceanic water column were unknown until recently

(Ingalls et al. 2006; Karner et al. 2001), as discussed in Chapter 3. These studies found that archaea are everywhere, but may be particular abundant, relative to bacteria, in the deep ocean. Even more surprising has been the evidence that these oceanic archaea are chemolithoautotrophs that oxidize ammonium (chemolithotrophy) to drive CO_2 fixation (chemoautotrophy). Natural-abundance ^{14}C data and incorporation of $^{13}CO_2$ into archaeal lipids provided some of the first clues that marine archaea are autotrophs and derive their biomass C from CO_2 rather than from organic material (Pearson et al. 2001; Wuchter et al. 2003). Subsequent microautoradiographic studies demonstrated fixation of $^{14}CO_2$ by archaea from the deep North Atlantic Ocean, albeit in decompressed samples (Herndl et al. 2005). Ingalls et al. (2006) suggested that over 80 percent of archaeal carbon is from CO_2 assimilated via chemoautotrophy in 670 m water of the North Pacific gyre.

We still know very little about marine archaea, about chemolithoautotrophy, and about the many other microbes and processes in the deep ocean. Like most biological oceanographers, microbial ecologists have focused on the surface layer and the euphotic zone where sunlight is available for primary producers. The water masses below the euphotic zone—the "dark ocean"—are greatly undersampled, especially considering their vastness. Over 75 percent of the ocean is below 1000 m, making the dark ocean the largest habitat in the biosphere.

HETEROTROPHIC PROTISTS

Marine protists, which are single-celled eukaryotes, are very diverse, in both form and function. Some are members of the phytoplankton community and contribute substantially to primary production and phytoplankton biomass, especially in oligotrophic oceanic regimes. Others are heterotrophic and eat other microbes by phagotrophy; they engulf other microbes into food vacuoles, where the prey is digested. Grazing is discussed in Chapter 11. Still other eukaryotic microbes probably are mixotrophic (Adolf et al. 2006), carrying out both light-driven photosynthesis (like a plant) and grazing (like an animal). Some protists used to be called protozoa, but this term, which refers to the lowest division of the animal kingdom, is misleading because many of these microbes are mixotrophic, if not strict autotrophs.

As discussed in Chapter 6, we are beginning to understand the phylogenetic diversity of eukaryotic picoplankton and other protists, but we know little about their ecological roles in the oceans. However, we do know some things about the role of several protist groups that are distinguishable by size and other characteristics observable by simple microscopy (Sherr and Sherr 2000). Figure 1.7 gives just a few examples of the many shapes taken on by protists in the oceans, and Chapter 11 provides several other examples.

Nanoflagellates (2–20 μm)

Microscopy studies revealed that the oceans harbor a large number of small eukaryotic cells without any chlorophyll, bearing one or more flagella. These

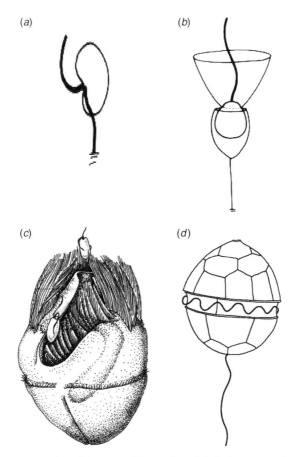

Figure 1.7 Some examples of heterotrophic protists. (*a*) *Cafeteria roenbergensis* Fenchel and Patterson 1988; (*b*) *Salpingoeca infusorium* Kent 1880; (*c*) *Strombidium capitatum*, including small flagellated prey; and (*d*) *Stoeckeria algicida*. *Cafeteria* and *Salpingoeca* are nanoflagellates (about 5 and 20 μm, respectively), whereas *Strombidium* is a ciliate and *Stoeckeria* is a dinoflagellate. (*a*) and (*b*) are from Sherr and Sherr (2000), (*c*) is modified from (Montagnes 1996), and (*d*) was provided by Hae Jin Jeong (see Jeong et al. 2005).

microbes have been called "colorless microflagellates," "heterotrophic nanoplankton" (HNAN), or simply "flagellates." As with all such groups, nanoflagellates are taxonomically diverse, with probably equally diverse ecological and biogeochemical roles. Given the cell size of nanoflagellates, they probably graze on heterotrophic bacteria, cyanobacteria, and picoeukaryotes because these potential prey have roughly the same size. As discussed in Chapter 11, similar-sized prey are likely eaten by the same predator, although a predator may choose among prey based on "taste" or chemical properties of the prey surface (Strom 2000).

Microzooplanktonic Protists (20–200 μm)

Microbial ecologists first focused on one group of microzooplankton protists, tintinnids, a type of ciliate with elaborate houses (loricae) surrounding them (Sherr and Sherr 2000). However, improved methods for studying microzooplankton protists led to the discovery of many "aloricate" or "naked" ciliates. These microbes can be abundant in the microzooplankton community and are quite important as grazers of phytoplankton (herbivory) and predators of flagellates that in turn feed on bacteria, cyanobacteria, and picoeukaryotes in the oceans. Ciliates are critical links in the microbial food chain and are likely important prey for carnivorous zooplankton. Some ciliates, such as *Laboea spiralis*, are mixotrophic and carry out photosynthesis with the help of chloroplasts from phytoplankton ingested (but not completely digested) by the ciliate (McManus and Fuhrman 1986). Still others, such as *Mesodinium rubrum*, are strictly autotrophic and can form large blooms in coastal waters (Sherr and Sherr 2000). The term "ciliate" gives a general idea about the appearance and mode of locomotion for these protists, but their ecological roles in the oceans vary greatly.

Dinoflagellates

These fascinating microbes vary greatly in size and shape and in their roles in estuaries, coastal waters, and the oceans. The size of dinoflagellates ranges from unarmored dinoflagellates, such as *Gymnodinium*, which are 8–15 μm, to *Noctiluca* which are 200–2000 μm. A "typical" dinoflagellate is armored by cellulose plates (theca) with two grooves to accommodate the two flagellates that are used for locomotion. Many dinoflagellates, such as *Gymnodinium*, are pigmented photoautotrophs and are members of the phytoplankton community. However, about 50 percent of dinoflagellates, including *Noctiluca*, are colorless and probably are heterotrophic. Along with ciliates, heterotrophic dinoflagellates are important members of the microzooplankton community, and can account for a large fraction of grazing on phytoplankton (Strom et al. 2007). Some heterotrophic dinoflagellates feed on microbial prey by phagocytosis, but others use feeding tubes to suck out the cytoplasm of their prey, and still other dinoflagellates capture their prey with a feeding veil. Dinoflagellates include several pathogens of higher marine life and humans. Red tides are well known to be caused by dinoflagellates such as *Gonyaulax polyhedra* and *Gymnodinium splendens*, which secrete potent neurotoxins called saxotoxins. Less well understood are the dinoflagellates, such as *Pfiesteria piscicida*, that are potential fish pathogens. Some are even parasites. *Amoebophrya*, for example, is a genus of dinoflagellates that parasitize other dinoflagellates (Coats and Park 2002). These topics and the original literature are discussed by Sherr and Sherr (2000).

Marine Fungi

Filamentous fungi and yeasts are found in the oceans, but these heterotrophic eukaryotes do not appear to be abundant and are not ecologically important in aquatic ecosystems, except as decomposers of dry detritus from vascular plants in coastal regions (Newell 2003). Fungi and yeasts live on DOM or particulate organic detritus, and thus would have to compete with heterotrophic bacteria and archaea in the oceans. The low abundance of fungi and yeasts indicate that they do not fare well in this competition, probably because they are larger than prokaryotes and thus have much higher surface area-to-volume ratios, putting them at a disadvantage in transporting dissolved compounds. Filamentous fungi do well in the standing dead (above the water line) plants in salt marshes, because their hyphae are able to penetrate and propagate up the dry, dead plant, something that prokaryotes cannot do. A similar reason explains why fungi are so important in the carbon cycle of terrestrial systems.

MARINE VIRUSES

In addition to grazing by protists, lysis by viruses is the other main type of mortality inflicted upon marine microbes. As pointed out earlier and emphasized in Chapter 12, viruses are the most abundant biological entity in the biosphere, and even though they are small, much smaller than bacteria, their biomass in the oceans is substantial. There are probably viruses for every micro- and macroorganism on the planet. Since viral abundance follows host abundance, viruses for heterotrophic bacteria (bacteriophages or just phages) are probably the most common type in the oceans. However, viruses also attack eukaryotic phytoplankton, and there is some evidence of viruses stopping phytoplankton blooms from growing even further (Suttle 2005). Viruses may be especially critical in controlling blooms of harmful algae in coastal waters (Lawrence and Suttle 2004).

Chapter 12 mentions that viral lysis and grazing are about equal in causing bacterial mortality, but in fact the percentages probably vary greatly. We expect viral lysis to be the dominant form of bacterial mortality in anaerobic environments such as sediments (Fischer et al. 2006) where few protists can grow. In contrast, viral lysis may be lower in oligotrophic environments where bacterial abundance is low (Strom 2000). In support of the latter hypothesis, Strom (2000) pointed out that estimates of grazing rates matched bacterial production in oligotrophic systems, but not so in eutrophic environments, where bacterial growth exceeded grazing, necessitating another form of bacterial mortality, such as viral lysis. Very few studies have examined both grazing and viral lysis, in part because the methods for both are difficult and time-consuming.

Viruses and protists have similar but also different impacts on bacteria and other microbial prey. Chapter 12 points out that the release of DOM from cells lysed by viruses probably leads to higher growth rates for the surviving bacteria. Like viral lysis, grazing also can release potentially limiting organic carbon and inorganic nutrients, but any released DOM is probably more altered, due to enzymatic attack in food vacuoles, than the DOM released by viruses. However, unlike viruses, grazers may have a direct negative effect on bacterial growth rates. Since grazing rates vary with prey size and since bacterial cells must get bigger before dividing, grazers should tend to graze more heavily on rapidly growing cells (see Chapter 11). This mechanism could explain why growth rates are often higher in treatments that minimize grazing (e.g., dilution and filtration), although artificial breakage of cells and other DOM contamination during sample preparation would have the same effect.

Laboratory experiments provide more insights into how grazers and protists affect their microbial prey. In the absence of larger organisms and other predators, a protist grazer will reduce prey numbers to vanishingly low levels, although some bacteria can form filaments and flocs that are resistant to grazing (Chapter 11). In contrast, viruses can apparently keep bacterial abundance constant over time (Wilcox and Fuhrman 1994), but they cannot reduce their host numbers to low levels, unlike a grazer feeding on bacteria. The difference is because microbes develop resistance against viruses. A field study of *Synechococcus* provides a good example of this coexistence (Waterbury and Valois 1993). In the end, mortality via viral lysis or grazing matches bacterial growth most of the time, since bacterial abundance varies relatively little over time and space in the oceans.

N_2 FIXERS

Just as the carbon cycle starts with the fixation of CO_2 to organic C, so too can the nitrogen cycle be said to start with the fixation of N_2 to ammonium (Fig. 1.8). There the similarity between carbon and nitrogen fixation ends. Unlike carbon fixation, N_2 fixation is carried out only by prokaryotes and in the oceans mainly by some cyanobacteria. Fixation by heterotrophic bacteria is thought to be rare in the oceans because the supply of labile DOM is too low to support the high expense of N_2 reduction to ammonium. Once N_2 is fixed, there are several similarities between the exchange of C and N among trophic levels and between fixed inorganic

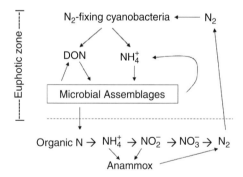

Figure 1.8 The nitrogen cycle in the oceans. DON is dissolved organic nitrogen. Nitrification, which is oxidization of ammonium to nitrite (NO_2^-) and eventually to nitrate (NO_3^-), requires oxygen, whereas dissimilatory nitrate reduction (reduction of NO_3^- to N_2) occurs only when oxygen is very low. Likewise, the anammox reaction also occurs only when oxygen is absent.

NITRIFIERS AND OTHER CHEMOLITHOTROPHS

and organic compounds. Similar to C, N is transferred to the rest of the marine biota by a variety of mechanisms, ranging from the direct release of ammonium and DON to grazing on diazotrophs by protist grazers and zooplankton. Chapter 13 discusses the genetics, microbiology, and biogeochemistry of N_2 fixation in the oceans.

> **Definition of Diazotrophs**
>
> N_2 fixers are also called diazotrophs. The prefix "diazo-" is derived from "di" (two) and "azote," from the Greek word ("no life") used by Antoine Lavoisier (1743–1794) for the gas we now call nitrogen. "Troph" is from the Greek word *trophē* for food.

Chapter 13 also discusses possible limits of N_2 fixation and why diazotrophs are not more abundant than they are. Two elements other than N are thought to potentially control N_2 fixation and microbial production in some oceans. Chemical oceanographers have always argued that phosphorus is the ultimate limiting element in the oceans, and now biological oceanographers have recognized the importance of P limitation in regimes such as the Mediterranean Sea (Thingstad et al. 2005) and the North Pacific Gyre, arguably the largest biome on the planet (Karl 1999). The second element, iron, has also been examined extensively by both oceanographers and microbial ecologists. The role of Fe in limiting primary production in high nutrient–low chlorophyll (HNLC) oceans is now well established (de Baar et al. 2005). The impact of low Fe and P concentrations in the oceans on N_2 fixation is being intensively examined (Chapter 13). Chapter 10 also discusses the role of these two elements in regulating the growth of the total bacterial community.

NITRIFIERS AND OTHER CHEMOLITHOTROPHS

We have assumed that oceanic bacteria gain energy either from light, which fuels the metabolism of cyanobacteria and drives the synthesis of organic material from CO_2, or from the oxidation of organic material back to CO_2, which defines what heterotrophic bacteria do. But already we have seen that some of these bacteria may really be photoheterotrophic and fuel ATP synthesis with light energy as well as by organic carbon oxidation. Another possible mechanism for ATP synthesis is by chemolithotrophy, which is the oxidation of reduced inorganic compounds coupled with the reduction of an electron acceptor, which is often oxygen. We have encountered one form of chemolithotrophy, the oxidation of ammonium. Other inorganic compounds used by chemolithotrophic bacteria include various reduced sulfur compounds, carbon monoxide, and methane. Ammonium oxidizers were discovered decades ago, but microbial ecologists thought bacteria carrying out this reaction were not abundant in the surface ocean because of slow growth rates (Ward 2000). Other forms of chemolithotrophy had been found previously only close to sources of reduced inorganic compounds, such as anoxic sediments. These bacteria were not thought to be common in the surface ocean, because concentrations of reduced inorganic compounds are very low in the oceans.

Genomic and experimental data are changing this picture. In addition to the work on ammonium oxidation by archaea in the deep ocean, new data now suggest that marine bacteria may be mixotrophic, harvesting energy from whatever reduced compound, organic or inorganic, they can obtain (Moran et al. 2004). While concentrations of reduced inorganic compounds are very low in the oceans, so too are concentrations of labile DOM components. Chemolithotrophy is one of several fascinating findings from genomic studies of marine microbes, as discussed in Chapter 4.

DENITRIFIERS

The microbes and processes discussed in this book are found mainly in the upper surface layer of the oceans, but the organisms discussed in Chapter 14 are exceptions. Chapter 14 focuses on N-cycle reactions and associated microbes that are found mainly in sediments and anoxic water columns, oxygen-deficient zones in the eastern tropical Pacific and the Arabian Sea, and in basins such as the Black Sea and the Cariaco Basin. Oxygen supply and concentrations are key to understanding these microbes. While factors such as DOM quantity and quality, grazing, and viral lysis all affect the activity and composition of microbial communities, oxygen has probably an even more dramatic impact. The lack of oxygen excludes nearly all eukaryotes from anoxic systems, the exceptions being some protists, and there are large shifts at the division (phylum) level in bacteria as well. The shift is due in part to the rise in organisms that use inorganic sulfur compounds in ATP synthesis, the most important being the sulfate-reducing bacteria. Many of these bacteria are *Deltaproteobacteria* (Llobet-Brossa et al. 2002; Ravenschlag et al. 2000), which are common in anoxic systems but not in oxic waters. Although sulfate as an electron acceptor (analogous to oxygen) is not energetically favorable (Table 1.4), the high concentrations of sulfate ensure the dominance of sulfate reduction in the mineralization of organic carbon in anoxic marine systems. After oxygen is depleted, sulfate reduction often accounts for a large fraction (20 percent > 50 percent) of organic carbon oxidation in salt marshes, estuaries, and shallow coastal sediments (Table 1.4). In contrast to sulfate, nitrate respiration does not oxidize much organic carbon in most ecosystems, because nitrate concentrations are too low, even though it is an energetically-favorable electron acceptor (Table 1.4).

These forms of anaerobic respiration with nitrate and sulfate have been well known for several years. More recent was the discovery of ANaerobic AMMonium OXidation (anammox), as discussed in Chapter 14. In contrast to nitrification, in which ammonium is oxidized aerobically, in anammox, ammonium is oxidized using nitrite instead of oxygen as the electron acceptor. Because it produces N_2 gas, this process could be called a form of denitrification. Chapter 14, however, reserves that term for heterotrophic dissimilatory nitrate reduction when the endproducts are N_2 or N_2O. Table 1.5 summarizes the differences between these two mechanisms by which the oceans lose N.

CONCLUDING REMARKS

TABLE 1.4 Summary of Electron Acceptors Used by Bacteria and Archaea in Marine Environments[a]

Electron Acceptor	Energetic Yield (kJ/mol)[b]	% of Total Organic Carbon Oxidation[c]	Comments
Oxygen	−3190	33	100% in most water columns
Nitrate	−3030	<10	Higher in polluted waters
Mn(IV)	−3090	<10	90–100% in Skagerrak and Panama Basin
Fe(III)	−1410	23	Needs replenishment of Fe oxides
Sulfate	−380	31	Less important in deep waters
CO_2	−350	<10	More important in freshwaters

[a] The electron acceptor accounting for all or nearly all organic carbon oxidation is oxygen in the open ocean and in most coastal waters. The percentages given above are for marine systems where oxygen supply can be insufficient.
[b] These changes in Gibbs free energy ($\Delta G^{\circ\prime}$) represent the oxidation of 1 mol glucose, and were taken from Nealson and Saffarini (1994). The authors did the calculations assuming that glucose was the electron donor, although glucose is not oxidized by organisms using metal oxides, sulfate, and CO_2 as electron acceptors.
[c] Values are averages taken from a wide range of systems summarized by Canfield et al. (2005).

TABLE 1.5 Comparison of the Two Processes Producing N Gases (N_2 and N_2O)[a]

	Dissimilatory NO_3^- Reduction	Anammox[b]
Starting N compound	NO_3^-	NH_4^+ and NO_2^-
Electron donor	Organic carbon	NH_4^+
Electron acceptor	NO_3^-	NO_2^-
N endproduct	N_2 or N_2O[c]	N_2
Energetic yield (kJ/mol)	−631[d]	−357
Carbon source	Organic carbon	CO_2
Growth rates	Fast	Slow
Phylogenetic distribution	Many species	*Planctomycetes*
Unique biomarkers	None	Ladderane (a lipid)

[a] Information mainly from Canfield et al. (2005). See also Chapter 14.
[b] Anaerobic oxidation of ammonium.
[c] Dissimilatory NO_3^- reduction can also produce NH_4^+ and NO_2^- but when these nongaseous N endproducts are produced, it is not denitrification.
[d] Energy yield per mole of nitrate, assuming glucose is the electron donor and N_2 is the N endproduct.

CONCLUDING REMARKS

The subheadings in this chapter refer to the various functional groups of microbes (and viruses and DOM) listed in Table 1.1, but these subheadings could have as easily been the associated biogeochemical processes. This combination of both organisms and processes is what separates marine microbial ecology from marine microbiology and biogeochemistry. The first chapter in the first edition of this book (Kirchman and Williams 2000) argued that the most important advances in microbial ecology were made initially by scientists who focused on biogeochemical processes without being bothered by their inability to identify the microbes. Although much remains to be done, this identification problem and the "great plate count anomaly" have been solved by the application of 16S rRNA-based technology and novel cultivation approaches. Application of this technology made up the next great wave of microbial ecology. We are now in the middle of still another wave, really a tsunami of genomic sequence data from isolated microbes and metagenomic studies of oceanic environments. The challenge facing microbial ecologists is to use these data to understand not only the organisms but the biogeochemical processes as well. It is the integration of these disciplines and approaches that makes microbial ecology so exciting today and so important for understanding current problems in the oceans and for predicting what the future holds for the entire biosphere.

SUMMARY

1. Marine microbial ecology is an important field of science because of the abundance and diversity of marine microbes, but also because these microbes are essential in mediating various biochemical cycles and other processes that control greenhouse gases.
2. Marine microbes include all organisms that smaller than about 100 μm, including cyanobacteria, heterotrophic bacteria, archaea, eukaryotic phytoplankton, and a diverse array of autotrophic, heterotrophic and mixotrophic protists. Microbial ecologists study all of these, as well as viruses, the most numerous biological entities in the oceans.
3. Microbial ecologists have discovered new organisms (e.g., coccoid cyanobacteria and archaea) carrying out well-known reactions (N_2 fixation and nitrification), as well as entirely novel microbes and processes (the anammox reaction) contributing to important biogeochemical fluxes (N loss).
4. Microbial ecology is being changed by genomics and the huge increase in sequence data from isolated microbes and metagenomic studies of oceanic habitats. Microbial ecologists now face the challenge of using these data to understand microbes and biogeochemical processes important in the ocean.

ACKNOWLEDGMENTS

I would like to thank the following colleagues who commented on this chapter and offered information and feedback: Ev and Barry Sherr, Suzanne Strom, Bo Thramdrup, and Alex Worden. I also thank David Montagues and Hae Jin Jeong for their drawings of some protists. The writing of this chapter was supported by a grant from the National Science Foundation (MCB-0453993).

REFERENCES

Adolf, J. E., D. K. Stoecker, and L. W. Harding, Jr. 2006. The balance of autotrophy and heterotrophy during mixotrophic growth of *Karlodinium micrum* (Dinophyceae). *J. Plankton Res.* 28: 737–751.

Behringer, D. C., and M. J. Butler. 2006. Stable isotope analysis of production and trophic relationships in a tropical marine hard-bottom community. *Oecologia* 148: 334–341.

Canfield, D. E., B. Thamdrup, and E. Kristensen. 2005. *Aquatic Geomicrobiology*. Elsevier Academic Press.

Cauwet, G. 2002. DOM in the coastal zone. In D. A. Hansell and C. A. Carlson (eds.), *Biogeochemistry of Marine Dissolved Organic Matter*. Academic Press, pp. 579–609.

Chisholm, S. W., R. J. Olson, E. R. Zettler, R. Goericke, J. B. Waterbury, and N. A. Welschmeyer. 1988. A novel free-living prochlorophyte abundant in the oceanic euphotic zone. *Nature* 334: 340–343.

Cho, B. C., M. G. Park, J. H. Shim, and D. H. Choi. 2001. Sea-surface temperature and f-ratio explain large variability in the ratio of bacterial production to primary production in the Yellow Sea. *Mar. Ecol. Prog. Ser.* 216: 31–41.

Church, M. J., H. W. Ducklow, and D. M. Karl. 2004. Light dependence of [^{3}H]leucine incorporation in the oligotrophic North Pacific Ocean. *Appl. Environ. Microbiol.* 70: 4079–4087.

Coats, D. W., and M. G. Park. 2002. Parasitism of photosynthetic dinoflagellates by three strains of *Amoebophrya* (Dinophyta): Parasite survival, infectivity, generation time, and host specificity. *J. Phycol.* 38: 520–528.

Cottrell, M. T., A. Mannino, and D. L. Kirchman. 2006. Aerobic anoxygenic phototrophic bacteria in the Mid-Atlantic Bight and the North Pacific Gyre. *Appl. Environ. Microbiol.* 72: 557–564.

de Baar, H. J. W., P. W. Boyd, K. H. Coale, et al. 2005. Synthesis of iron fertilization experiments: From the Iron Age in the Age of Enlightenment. *J. Geophys. Res. (Oceans)* 110: C09S16.

Ducklow, H. 2000. Bacterial production and biomass in the oceans. In D. L. Kirchman (ed.), *Microbial Ecology of the Oceans*, 1st edn. Wiley-Liss, pp. 85–120.

Dufresne, A., M. Salanoubat, F. Partensky, et al. 2003. Genome sequence of the cyanobacterium *Prochlorococcus marinus* SS120, a nearly minimal oxyphototrophic genome. *Proc. Natl. Acad. Sci. USA* 100: 10020–10025.

Dugdale, R. C., and J. J. Goering. 1967. Uptake of new and regenerated forms of nitrogen in primary productivity. *Limnol. Oceanogr.* 12: 196–206.

Findlay, S. E. G., and R. L. Sinsabaugh (eds.). 2003. *Aquatic Ecosystems: Interactivity of Dissolved Organic Matter*. Academic Press.

Fischer, U. R., C. Wieltschnig, A. K. T. Kirschner, and B. Velimirov. 2006. Contribution of virus-induced lysis and protozoan grazing to benthic bacterial mortality estimated simultaneously in microcosms. *Environ. Microbiol.* 8: 1394–1407.

Fukuda, R., H. Ogawa, T. Nagata, and I. Koike. 1998. Direct determination of carbon and nitrogen contents of natural bacterial assemblages in marine environments. *Appl. Environ. Microbiol.* 64: 3352–3358.

Gasol, J. M., P. A. del Giorgio, and C. M. Duarte. 1997. Biomass distribution in marine planktonic communities. *Limnol. Oceanogr.* 42: 1353–1363.

Gattuso, J. P., B. Gentili, C. M. Duarte, J. A. Kleypas, J. J. Middelburg, and D. Antoine. 2006. Light availability in the coastal ocean: Impact on the distribution of benthic photosynthetic organisms and their contribution to primary production. *Biogeosciences* 3: 489–513.

Giovannoni, S. J., L. Bibbs, J. C. Cho, et al. 2005. Proteorhodopsin in the ubiquitous marine bacterium SAR11. *Nature* 438: 82–85.

Gomez-Consarnau, L., J. M. Gonzalez, M. Coll-Llado, P. Gourdon, T. Pascher, R. Neutze, C. Pedros-Alio, and J. Pinhassi. 2007. Light stimulates growth of proteorhodopsin-containing marine Flavobacteria. *Nature* 445: 210–213.

Hansell, D. A., and C. A. Carlson (eds.). 2002. *Biogeochemistry of Marine Dissolved Organic Matter*. Academic Press.

Hedges, J. I., and J. M. Oades. 1997. Comparative organic geochemistries of soils and marine sediments. *Org. Geochem.* 27: 319–361.

Herndl, G. J., T. Reinthaler, E. Teira, H. Van Aken, C. Veth, A. Pernthaler, and J. Pernthaler. 2005. Contribution of archaea to total prokaryotic production in the deep Atlantic Ocean. *Appl. Environ. Microbiol.* 71: 2303–2309.

Hoiczyk, E., and A. Hansel. 2000. Cyanobacterial cell walls: News from an unusual prokaryotic envelope. *J. Bacteriol.* 182: 1191–1199.

Ingalls, A. E., S. R. Shah, R. L. Hansman, L. I. Aluwihare, G. M. Santos, E. R. M. Druffel, and A. Pearson. 2006. Quantifying archaeal community autotrophy in the mesopelagic ocean using natural radiocarbon. *Proc. Natl. Acad. Sci. USA* 103: 6442–6447.

Jeong, H. J., J. S. Kim, J. Y. Park, et al. 2005. *Stoeckeria algicida* n. gen., n. sp (Dinophyceae) from the coastal waters off Southern Korea: Morphology and small subunit ribosomal DNA gene sequence. *J. Eukaryot. Microbiol.* 52: 382–390.

Karl, D. M. 1999. A sea of change: Biogeochemical variability in the North Pacific subtropical gyre. *Ecosystems* 2: 181–214.

Karner, M. B., E. F. DeLong, and D. M. Karl. 2001. Archaeal dominance in the mesopelagic zone of the Pacific Ocean. *Nature* 409: 507–510.

Kirchman, D. L. 2000. Uptake and regeneration of inorganic nutrients by marine heterotrophic bacteria. In D. L. Kirchman (ed.), *Microbial Ecology of the Oceans*, 1st edn. Wiley-Liss, pp. 261–288.

Kirchman, D. L., and P. J. L. B. Williams. 2000. Introduction. In D. L. Kirchman (ed.), *Microbial Ecology of the Oceans*, 1st edn. Wiley-Liss, pp. 1–11.

Lawrence, J. E., and C. A. Suttle. 2004. Effect of viral infection on sinking rates of *Heterosigma akashiwo* and its implications for bloom termination. *Aquat. Microb. Ecol.* 37: 1–7.

Llobet-Brossa, E., R. Rabus, M. E. Bottcher, et al. 2002. Community structure and activity of sulfate-reducing bacteria in an intertidal surface sediment: a multi-method approach. *Aquat. Microb. Ecol.* 29: 211–226.

McManus, G. B., and J. A. Fuhrman. 1986. Photosynthetic pigments in the ciliate *Laboea strobila* from Long Island Sound, USA. *J. Plankt. Res.* 8: 317–327.

Michelou, V. K., M. T. Cottrell, and D. L. Kirchman. 2007. Light-stimulated bacterial production and amino acid assimilation by cyanobacteria and other microbes in the North Atlantic Ocean. *Appl. Environ. Microbiol.* 73: 5539–5546.

Montagnes, D. J. S. 1996. Growth responses of planktonic ciliates in the genera *Strobilidium* and *Strombidium*. *Mar. Ecol. Prog. Ser.* 130: 241–254.

Moran, M. A., and R. G. Zepp. 2000. UV radiation effects on microbes and microbial processes. In D. L. Kirchman (ed.), *Microbial Ecology of the Oceans*, 1st edn. Wiley-Liss, pp. 201–228.

Moran, M. A., A. Buchan, J. M. Gonzalez, et al. 2004. Genome sequence of *Silicibacter pomeroyi* reveals adaptations to the marine environment. *Nature* 432: 910–913.

Mulholland, M. R., C. J. Gobler, and C. Lee. 2002. Peptide hydrolysis, amino acid oxidation, and nitrogen uptake in communities seasonally dominated by *Aureococcus anophagefferens*. *Limnol. Oceanogr.* 47: 1094–1108.

Nagata, T. 2000. Production mechanisms of dissolved organic matter. In D. L. Kirchman (ed.), *Microbial Ecology of the Oceans*, 1st edn. Wiley-Liss, pp. 121–152.

Nealson, K. H., and D. Saffarini. 1994. Iron and manganese in anaerobic respiration: Environmental significance, physiology, and regulation. *Annu. Rev. Microbiol.* 48: 311–343.

Newell, S. Y. 2003. Fungal content and activities in standing-decaying leaf blades of plants of the Georgia Coastal Ecosystems research area. *Aquat. Microb. Ecol.* 32: 95–103.

Pakulski, J. D., A. Baldwin, A. L. Dean, et al. 2007. Responses of heterotrophic bacteria to solar irradiance in the eastern Pacific Ocean. *Aquat. Microb. Ecol.* 47: 153–162.

Pearson, A., A. P. Mcnichol, B. C. Benitez-Nelson, J. M. Hayes, and T. I. Eglinton. 2001. Origins of lipid biomarkers in Santa Monica Basin surface sediment: A case study using compound-specific delta C-14 analysis. *Geochem. Cosmochim. Acta* 65: 3123–3137.

Ravenschlag, K., K. Sahm, C. Knoblauch, B. B. Jørgensen, and R. Amann. 2000. Community structure, cellular rRNA content, and activity of sulfate-reducing bacteria in marine Arctic sediments. *Appl. Environ. Microbiol.* 66: 3592–3602.

Rocap, G., F. W. Larimer, J. Lamerdin, et al. 2003. Genome divergence in two *Prochlorococcus* ecotypes reflects oceanic niche differentiation. *Nature* 424: 1042–1047.

Schwalbach, M. S., and J. A. Fuhrman. 2005. Wide-ranging abundances of aerobic anoxygenic phototrophic bacteria in the world ocean revealed by epifluorescence microscopy and quantitative PCR. *Limnol. Oceanogr.* 50: 620–628.

Segal, R. D., A. M. Waite, and D. P. Hamilton. 2006. Transition from planktonic to benthic algal dominance along a salinity gradient. *Hydrobiologia* 556: 119–135.

Sherr, E. B., and B. F. Sherr. 2000. Marine microbes: An overview. In D. L. Kirchman (ed.), *Microbial Ecology of the Oceans*, 1st edn. Wiley-Liss, pp. 13–46.

Sieracki, M. E., I. C. Gilg, E. C. Thier, N. J. Poulton, and R. Goericke. 2006. Distribution of planktonic aerobic anoxygenic photoheterotrophic bacteria in the northwest Atlantic. *Limnol. Oceanogr.* 51: 38–46.

Strom, S. L. 2000. Bacterivory: Interactions between bacteria and their grazers. In D. L. Kirchman (ed.), *Microbial Ecology of the Oceans*, 1st edn. Wiley-Liss, pp. 351–386.

Strom, S. L., E. L. Macri, and M. B. Olson. 2007. Microzooplankton grazing in the coastal Gulf of Alaska: Variations in top-down control of phytoplankton. *Limnol. Oceanogr.* 52.

Suttle, C. A. 2005. Viruses in the sea. *Nature* 437: 356–361.

Thingstad, T. F., M. D. Krom, R. F. C. Mantoura, et al. 2005. Nature of phosphorus limitation in the ultraoligotrophic eastern Mediterranean. *Science* 309: 1068–1071.

Waidner, L. A., and D. L. Kirchman. 2007. Aerobic anoxygenic phototrophic bacteria attached to particles in turbid waters of the Delaware and Chesapeake estuaries. *Appl. Environ. Microbiol.* 73: 3936–3944.

Ward, B. B. 2000. Nitrification and the marine nitrogen cycle. In D. L. Kirchman (ed.), *Microbial Ecology of the Oceans*, 1st edn. Wiley-Liss, pp. 427–453.

Waterbury, J. B., and F. W. Valois. 1993. Resistance to co-occurring phages enables marine *Synechococcus* communities to coexist with cyanophages abundant in seawater. *Appl. Environ. Microbiol.* 59: 3393–3399.

Waterbury, J. B., S. W. Watson, R. R. L. Guillard, and L. E. Brand. 1979. Widespread occurrence of a unicellular, marine, planktonic, cyanobacterium. *Nature* 277: 293–294.

Wilcox, R. M., and J. A. Fuhrman. 1994. Bacterial viruses in coastal seawater: Lytic rather than lysogenic production. *Mar. Ecol. Prog. Ser.* 114: 35–45.

Wuchter, C., S. Schouten, H. T. S. Boschker, and J. S. S. Damste. 2003. Bicarbonate uptake by marine Crenarchaeota. *FEMS Microb. Let.* 219: 203–207.

Yool, A., A. P. Martin, C. Fernandez, and D. R. Clark. 2007. The significance of nitrification for oceanic new production. *Nature* 447: 999–1002.

Zubkov, M. V., and G. A. Tarran. 2005. Amino acid uptake of *Prochlorococcus* spp. in surface waters across the South Atlantic subtropical front. *Aquat. Microb. Ecol.* 40: 241–249.

Zubkov, M. V., B. M. Fuchs, G. A. Tarran, P. H. Burkill, and R. Amann. 2003. High rate of uptake of organic nitrogen compounds by *Prochlorococcus* cyanobacteria as a key to their dominance in oligotrophic oceanic waters. *Appl. Environ. Microbiol.* 69: 1299–1304.

UNDERSTANDING ROLES OF MICROBES IN MARINE PELAGIC FOOD WEBS: A BRIEF HISTORY

EVELYN SHERR and BARRY SHERR

College of Oceanic and Atmospheric Sciences, Oregon State University, Corvallis, OR 97331-5503, U.S.A.

INTRODUCTION

> The sea harbors an extensive population of bacteria, varying greatly in numbers and in the variety of their activities
> —Waksman 1934

> ... microorganisms are widely distributed in sea water and on the ocean floor, where they influence chemical, physiochemical, geological, and biological conditions
> —ZoBell 1946

Progress in understanding of microbes in ocean systems has depended both on the research focus of individual scientists, and on the development of novel approaches to study of microbes in the sea. A brief history of the study of ecological roles of microbes in the sea shows that many of the current research thrusts were of interest early on in the development of the field. Often, however, progress on specific facets of marine microbial ecology was impeded by lack of instrumentation and methods to address the problems. A great deal of what we know about marine microbes has only been discovered since the mid-1970s. New methodology, including molecular genetic approaches and

Microbial Ecology of the Oceans, Second Edition. Edited by David L. Kirchman
Copyright © 2008 John Wiley & Sons, Inc.

major oceanographic expeditions in the past few decades, have revolutionized our understanding of the vital roles microbes play in marine ecosystems.

Every aquatic microbiologist who was working during the 1970s and 1980s has a unique perspective of the development of the field at that time. Our own perspective is influenced by our years at the University of Georgia Marine Institute on Sapelo Island, where L. R. Pomeroy and colleagues did research that led to novel concepts about the role of microbes in marine ecosystems. We are also grateful for the perspectives of several of our colleagues who have authored highly influential papers.

PRE-1950s: THE EARLY YEARS

In the first half of the 20th century, scientists were aware that a variety of bacteria existed in the sea, and that their activities were likely important in biogeochemical cycling in marine ecosystems. Methods for studying marine bacteria were mainly drawn from clinical microbiology. For example, assessment of the abundance of bacteria in seawater was done by plating diluted water samples onto nutrient agar plates, and then counting the number of bacterial colonies that grew up on the agar per unit volume of original sample. The presumption was that each colony grew from a single cell; thus, the number of "culturable" bacteria per milliliter of seawater could be estimated.

The development of marine microbiology in the first part of the 20th century was greatly influenced by Selman Waksman (1888–1973), a soil microbiologist who developed a Department of Marine Bacteriology at the Woods Hole Oceanographic Institution, and by Claude ZoBell (1904–1989), a research scientist at Scripps Institution of Oceanography (Ehrlich 2000; McGraw 2004). Waksman wrote a series of papers on the biogeochemical activities of marine bacteria in the 1930s, focused mainly on the role of bacteria in decomposing organic matter (Waksman 1934; Waksman and Renn 1936). In 1952, Waksman received a Nobel Prize for his work in soil microbiology, which led to discovery of a number of new antibiotics. ZoBell (1946) compiled existing knowledge in the field in *Marine Microbiology: A Monograph on Hydrobacteriology*. Research carried out by ZoBell and his students demonstrated that microbial life was adapted to living in all parts of the ocean, including abyssal depths and deep in ocean sediments (McGraw 2004). As a PhD student in ZoBell's laboratory, Richard Morita carried out much of the field work on long ocean cruises, and wondered how bacteria in the deep ocean survived with very little substrate. Morita later focused his long research career at Oregon State University on starvation-survival mechanisms of marine bacteria and, in particular, on how marine microbes persisted at cold temperatures (Morita 1997). ZoBell's own research was on the roles that marine bacteria played in geochemical processes in the biosphere, including the deep subsurface. Deep-rock microbiology in the 1940s and 1950s was largely focused on oil deposits. ZoBell published extensively on petroleum microbiology, arguably laying the groundwork for today's growing field of subsurface microbiology.

It is humbling to realize that many research themes that still occupy microbial oceanographers were either formally investigated or anticipated by Waksman, ZoBell, and other early researchers cited in their publications. These themes included

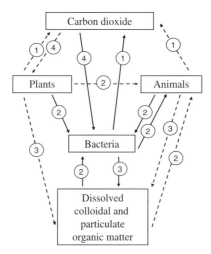

Figure 2.1 Diagram redrawn from Figure 10 in Chapter X of ZoBell (1946), depicting the carbon cycle in the sea. In ZoBell's diagram the solid lines represented processes in which bacteria were thought to participate, and the dashed lines processes in which bacteria did not participate. The number codes represented specific biological processes: (1) respiration, (2) nutrition, (3) decomposition, and (4) carbon dioxide fixation.

the role of bacteria in elemental cycling in the sea; the problem of decomposition of organic matter, especially dissolved organic matter, by marine bacteria; the idea that there were fractions of bacterial cells that are live/active versus dead/inactive; the potential for allelopathic interactions between microbes; viral lysis as a source of bacterial mortality; the fact that some marine bacteria were adapted to oligotrophic conditions in the open ocean; and the role of bacteria in benthic systems.

Both Waksman (1934) and ZoBell (1946) presented conceptual diagrams that showed bacteria as having central roles in biogeochemical cycles of marine ecosystems. ZoBell's box diagram of carbon flows in pelagic ecosystems clearly presaged modern versions of such models (Fig. 2.1). Unfortunately, the methodologies and data needed to confirm the inferred importance of bacteria in marine ecosystems and biogeochemical cycles were not available at the time. In fact, it would take another 30 years before the central place of bacteria in marine food webs hypothesized by these early workers was given serious consideration.

1950–1974

In the years after World War II, oceanography thrived along with other natural sciences, and research in the field of marine microbial ecology expanded in both Europe and in North America. The concept of ecosystems—how they were structured and how they functioned—was a crucial step in the progress of microbial oceanography. The brothers Eugene and Howard Odum collaborated on a book, *Fundamentals of Ecology*, first published in 1953, which became a standard text focused on the structure of ecosystems.

The backstory of one now widely cited paper, Pomeroy (1974), is an example of how interactions between scientific disciplines can result in new concepts. Eugene Odum, a professor at the University of Georgia in Athens, established an Institute of Ecology on campus, and the University of Georgia Marine Institute on Sapelo Island, a barrier island on the Georgia coast that is still reachable only by boat. In 1954, as a young PhD, Lawrence Pomeroy was recruited as one of the first group of scientists to staff the Marine Institute. Research at the Marine Institute focused on the coastal salt marsh estuaries that surrounded Sapelo Island (Odum and Smalley 1959; Odum and de la Cruz 1963). John Teal (1962) carried out a pioneering study of energy flow through this salt marsh ecosystem. Southeastern salt marshes are composed of vast green swaths of the salt-tolerant cordgrass, *Spartina alterniflora*. The grasses grow quickly in the coastal Georgia heat, producing an amount of biomass per square meter of marsh comparable to that produced in an Iowan corn field. In the fall, the marsh grass turns yellow and dies back. The twice daily tide that floods the marsh carries off much of the grass leaves and stems to the tidal creeks and rivers. Dead *Spartina* leaves are colonized by marine fungi and bacteria, and then decomposed by the microbes to detrital fragments on the marsh surface and in the estuarine water. In the process, microbial biomass enriches the particles with organic nitrogen and phosphorus, making the leaf fragments a more nutritious food for marsh animals. The consensus from the research carried out by Odum, Teal, and colleagues at the Marine Institute was that the food web of salt marsh estuaries was largely based on this microbially enriched detritus. The "detritus food web" concept was influential in Pomeroy's consideration of the quantitative importance of microbial degradation in the cycling of organic matter in pelagic marine systems.

Pomeroy was joined at the Marine Institute by a postdoctoral colleague also interested in elemental cycles in aquatic ecosystems, Robert E. Johannes. Johannes (1965) carried out experiments with bacterivorous flagellates cultured from estuarine water that showed very high biomass-specific rates of phosphorus excretion, much greater than the biomass-specific rates of phosphorus excretion determined for copepods and other zooplankton. This was dramatic evidence that small organisms in the plankton had a much higher "rate of living" than larger plankton. This finding in part led Pomeroy and Johannes to collaborate on experiments with plankton to demonstrate that whole-water respiration rates routinely measured as disappearance of oxygen in dark bottles were mainly due to activity of microbes rather than to metazoans such as copepods. The colleagues compared rates of respiration in whole seawater and in seawater passed through a No. 2 plankton net with a mesh size of 366 μm). They found that the organisms that passed through the plankton net were responsible for virtually all of the oxygen decrease measured during their experiments (Pomeroy and Johannes 1968). Their results indicated that on a per-volume basis, the smaller organisms in the plankton, mainly microbes, had a 10-fold greater rate of respiration compared with larger plankton.

By the early 1970s, Pomeroy had concluded that the prevailing concept of the structure and functioning of marine food webs was inadequate. This idea of marine food webs was formalized in John Steele's 1974 monograph on modeling: *The Structure of Marine Ecosystems*. Steele's simplified diagram of marine food webs relegated heterotrophic microbes to a "bacteria" compartment in the benthos responsible for decomposing fecal material. The model had no formal role at all for bacteria or heterotrophic protists in the plankton. Steele (1974) wrote "The phytoplankton of the open sea is eaten nearly as fast as it is produced, so that effectively all plant production goes through the herbivores." From his and Johannes' remineralization and respiration experiments, Pomeroy knew that heterotrophic microbes played a much bigger role in planktonic food webs than Steele's model suggested. Pomeroy was able to integrate the ecosystem theory promoted by the Odums with the concept of detritus-based food webs developed at the Marine Institute on Sapelo Island to formulate a new view of the role of microbes in the sea.

Around this time, John Bardach, the editor of the journal *BioScience*, was seeking review papers with wide appeal to promote readership. He asked his friends, one of whom was Robert Johannes, for names of potential authors. Johannes suggested Pomeroy. Pomeroy obliged with a paper summarizing evidence that supported his new view that microbes were central to the functioning of marine ecosystems. According to Pomeroy, his manuscript was sent out to two external referees. One never responded, and the other said the paper was nonsense and should be rejected. However, Bardach liked the review and published it anyway. Although not widely cited at first, by the end of the 1980s, Pomeroy's (1974) paper in *BioSciences*, "The ocean's food web: A changing paradigm," was regarded as a pivotal description of new concepts about how the bulk of elemental flows in planktonic food webs flowed through small planktonic organisms, including small phytoplankton, marine bacteria, and their protist grazers (Fig. 2.2).

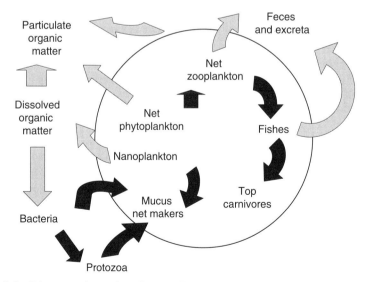

Figure 2.2 Diagram redrawn from Pomeroy's 1974 *BioScience* paper. Classic pelagic food web components are inside the circle, black arrows show transfer of materials and energy between these components of the food web. Microbial components added by Pomeroy to the classic food web are shown outside the circle.

1970s–1980s

There was great ferment in marine microbiology during the 1970s and 1980s, stimulated by application of newly available methodology. In the early 1980s, a critical mass of work on marine microbes sparked two symposia on marine microbiology sponsored by NATO (Fasham 1984; Hobbie and Willams 1984) that stimulated the publication of influential review papers summarizing the new understanding of the roles of microbes in the sea (Williams 1979; Azam et al. 1983).

IMPROVEMENT IN METHODS

Bacterial Abundance

The late 1970s and early 1980s saw major advances in methods to quantify the abundance of marine bacteria. Direct count assays based on epifluorescence microscopy were introduced, which allowed easy visualization of bacterial cells, which are difficult to detect using regular transmitted-light microscopy. The procedures involved staining cells with a fluorochrome such as blue-light-excited, green-fluorescing acridine orange, filtering the cells down onto a membrane filter, and then examining the preparation using an epifluorescence microscope. In this type of microscopy, a series of optical filters selects a specific segment of light from a broad spectrum lamp to excite fluorochromes present in cells at the surface (hence "epi-") of the stained preparation, and then allows only specific wavelengths of emitted (fluoresced) light to pass up to the observer's eye at the objective lens. The marine microbiologist E. J. Ferguson Wood (1955) had previously suggested the use of acridine orange and fluorescence microscopy to visualize microbes, based on a paper by a soil microbiologist (Strugger 1948). However, Wood used a transmitted light microscope fitted with special optical filters to look at bacteria suspended in water. It was only after true epifluorescence microscopes became readily available and protocols based on settling cells onto membrane filters were suggested (Zimmerman and Meyer-Reil 1974; Hobbie et al. 1977) that this approach became widely used. Porter and Feig (1980) introduced the UV-excited, blue-fluorescing DNA stain DAPI as an alternate fluorochrome for bacterial counts, which had the advantage of less background interference from the filter surface compared with acridine orange.

The direct count methods revealed that the abundances of bacterial cells in seawater were orders of magnitude greater than counts made from bacterial colonies on agar plates had indicated. Instead of hundreds to thousands of bacterial cells per milliliter of seawater, in fact there were hundreds of thousands to millions of bacteria per milliliter. This discrepancy was dubbed "the great plate count anomaly" (Staley and Konopka 1985). Once the abundance of bacteria in seawater could be easily quantified, researchers could estimate bacterial growth rates by monitoring the rate of increase over time of bacterial numbers in seawater samples.

Bacterial Activity

More sophisticated approaches to quantifying bacterial activity were soon developed. Hobbie et al. (1968) were among the first to demonstrate, using radiolabeled amino acids, that marine bacteria were able to rapidly assimilate the low-molecular-weight organic substrates present in seawater. Assaying the rate at which bacterial cells incorporated radiolabeled substrates into biomolecules was an obvious approach to measuring the rate of bacterial biomass production. Organic substrates such as sugars and amino acids, while readily assimilated by bacteria, were also respired. Fuhrman and Azam (1982) suggested using tritium-labeled thymidine (TdR), a nucleotide that is incorporated into DNA with little respiratory loss. Empirical conversion factors of amount of radiolabeled thymidine incorporated per number of bacterial cells produced allowed estimation of the rate of bacterial biomass production. Kirchman et al. (1985) subsequently developed a method based on incorporation of radiolabeled leucine, a common amino acid in protein, to quantify the rate of protein production by marine bacteria. Since protein is a large and relatively constant proportion of bacterial cell biomass, the rate of leucine incorporation into bacterial protein could be directly converted into rate of production of bacterial biomass, without the need for empirically determined conversion factors. These approaches were quickly adopted by the community of marine microbiologists, and soon there was a growing data set of bacterial productivity in various parts of the world ocean (Ducklow and Carlson 1992; Ducklow 2000).

Reviews by Peter Williams (1979, 1981) focused on the importance of microbes in a variety of marine systems based on respiration rates in various size fractions of seawater. Williams (1979) reported results from his own work in an experiment in a British Columbia, Canada, fjord, in which a very large plastic bag was filled and suspended in the water column. Enclosure, or "mesocosm," experiments such as this allowed researchers to track changes in a planktonic ecosystem over time within a single, isolated water mass. Comparison of rates of respiration in water

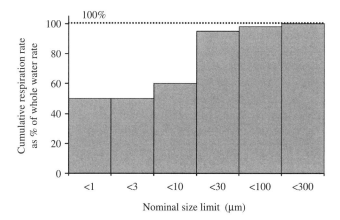

Figure 2.3 Distribution of plankton respiratory activity with size (oxygen consumption per unit time, measured for sequentially smaller size fractions in a mesocosm (from data of Williams 1981).

TABLE 2.1 Comparison of Biomass and Living Surface Area of Various Groups of Plankton in the CEPEX Experimental Enclosure CEE-2

Planktonic Group	Biomass (μg dry wt/L)[a]	Surface Area (cm^2/L)[a]
Bacterioplankton	26 (4.6%)	24.6 (69%)
Protozoa	9.2 (1.7%)	0.3 (0.7%)
Phytoplankton	310 (56%)	10.7 (30%)
Zooplankton	206 (37%)	0.3 (0.9%)
Total	551	35.9

Source: From Williams (1981), Table 5.
[a]Percentage of total in parenthesis.

collected from the mesocosm bag that was passed through sequentially finer filters demonstrated that organisms smaller than 30 μm accounted for almost all of the respiratory activity (Fig. 2.3). Williams also noted that although the biomass of bacteria was smaller than that of phytoplankton or zooplankton, when the surface areas of the various categories of planktonic organisms were compared, bacteria were overwhelmingly dominant (Table 2.1). This interesting idea, which Williams (1984) credited to Scripps microbiologist Farooq Azam, highlights a major reason why marine bacteria are so important in material and energy fluxes in marine ecosystems; because of their large surface area, these small but abundant organisms have the greatest probability of encountering, and interacting with, chemical substances in seawater. In his 1981 review, Williams presented a simple compartment diagram that showed bacteria consuming dissolved organic matter produced from phytoplankton exudates (and also from zooplankton feeding processes, which he charmingly termed "munchates") and presumably regenerating inorganic nitrogen and phosphorus nutrients to further fuel phytoplankton growth.

During the same period, Holgar Jannasch and A. Aristides Yayanos developed methods of studying bacterial abundance and activity in the deep ocean, which involved the use of special sampling and incubation chambers that allowed microbes to be kept at the high pressure of the depths at which water samples were collected during measurements of activity rates (Jannasch and Wirsen 1977; Jannasch 1984; Yayanos 1986).

MARINE HETEROTROPHIC PROTISTS

Russian scientists have had a long history of study of marine microbes. One of the most influential Russian microbial ecologists, Yuri Sorokin, was broadly interested in the roles of heterotrophic microbes in the sea, including bacteria and heterotrophic protists. He was one of the first marine microbial ecologists who attempted to quantify the abundance and biomass of colorless, heterotrophic flagellates as well as ciliates in the open ocean (Sorokin 1981).

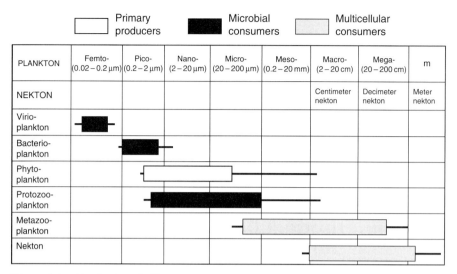

Figure 2.4 Distribution of different taxonomic–trophic compartments of plankton in a spectrum of size fractions, with a comparison of size range of nekton. Based on Figure 1 of Sieburth et al. (1978).

In the west, John Sieburth, a professor at the University of Rhode Island, was also interested in the whole spectrum of marine microbes, from viruses to bacteria to protists. Sieburth authored two books that served to greatly spark interest in both marine bacteria and in the wide diversity of unicellular eukaryotes in the sea: *Microbial Seascapes* (1975), a compendium of scanning electron micrographs of marine microbes, and *Sea Microbes* (1979), an encyclopedic narrative of what was then known about marine microbes, with a strong emphasis on protists. He and colleagues (Sieburth et al. 1978) formalized terms that were used to describe various size categories of marine organisms (Fig. 2.4). Their terms pico- (0.2–20 μm), nano- (2–20 μm), and micro- (20–200 μm) are now routinely used by aquatic microbial ecologists in referring to size classes of microorganisms. Davis and Sieburth (1982) advanced the study of protists in marine ecosystems by adapting the acridine orange staining and epifluorescence microscopy method used to count bacteria to enumerate nonpigmented, presumably bacterivorous, flagellates in seawater. A number of methods based on epifluorescence microscopy to determine in situ abundance, growth rates, and rates of prey ingestion of these small, colorless protists soon followed (Caron 1983; Sherr and Sherr 1983; McManus and Fuhrman 1986; Sherr et al. 1987).

A leading light of marine protozoology from the mid 1960s to the present has been the Danish scientist, Tom Fenchel. A benthic ecologist, Fenchel focused his research on how organisms lived in their natural habitats (Fenchel 1967, 1977), and was particularly interested in bioenergetics of protists. A four-paper series on the ecology of marine heterotrophic flagellates (Fenchel 1982a–d) was extraordinarily important in focusing attention on the role of marine bacterivores in the sea.

THE "MICROBIAL LOOP"

At the NATO Advanced Research Institute Flows of Energy and Material in Marine Ecosystems in Caiscais, Portugal, in 1981, a group of marine microbial ecologists drew together emerging information about microbial abundance and activity in the sea, which resulted in the paper of Azam et al. (1983). These authors suggested that the microbial components of pelagic food webs formed a separate entity that they termed the "microbial loop." They segregated the "classic food web" of larger phytoplankton to zooplankton to fish and the "microbial loop" that began with heterotrophic bacteria consuming dissolved organic matter (Fig. 2.5). The size spectrum of the various components of their food web was based on the terminology of Sieburth et al. (1978). Some of the phytoplankton in their diagram were too small to be consumed by copepods; thus, herbivorous protists were suggested to be an important pathway from small phytoplankton to multicellular zooplankton such as copepods. We now know that phagotrophic protists are significant grazers of all size classes of phytoplankton, as discussed in Chapter 11.

The idea of a "microbial loop" in marine food webs was reinforced by another paper in the same year by Hugh Ducklow (1983). Ducklow reviewed the basis for the importance of heterotrophic microbes—bacteria and protists—in marine ecosystems. He also presented a simple block diagram (Fig. 2.6) which clearly identified the microbial side of the food web outlined in Pomeroy (1974) as a three-step pathway from heterotrophic bacteria to bacterivorous protists (mainly flagellates), to larger protists (mainly ciliates) that consumed the bacterivores. Ducklow's figure stressed that a multistep pathway between heterotrophic microbes, and not simply degradation of organic matter by bacteria, was necessary for regeneration of inorganic nutrients for phytoplankton production. He also demonstrated that the "microbial

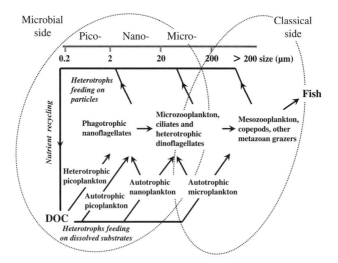

Figure 2.5 Conceptual diagram of the "microbial loop" idea of Azam et al. (1983), based on the presentation of the relation of the microbial and classical components of pelagic food webs in Figure 8.1 in Fenchel (1987). DOC, dissolved organic carbon.

THE "MICROBIAL LOOP"

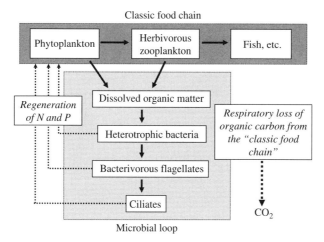

Figure 2.6 Simplified box model diagram of the microbial loop concept. Redrawn from Figure 1 of Ducklow (1983).

loop" was mainly a sink for organic carbon, with virtually all of the organic matter that flowed through the loop being lost to the food web as respiratory carbon dioxide. Ducklow's diagram neatly captured all of the previous concepts and data concerning the roles of bacteria and protists in marine food webs up to that time. This model emphasized the role of the microbial loop in regeneration of macronutrients, particularly nitrogen and phosphorus, which allowed further phytoplankton growth. When oceanographers considered the "microbial loop" concept, they often cited Azam et al. (1983), but visualized the diagram of Ducklow (1983).

However, Ducklow's simplified "microbial loop" diagram (Fig. 2.6) did leave out the smaller phytoplankton as components of the microbial food web depicted in the diagrams of Pomeroy (Fig. 2.2) and Azam et al. (Fig. 2.5). During the 1980s, there was a growing body of data on the large fraction of phytoplankton production consumed by planktonic grazers smaller than 200 μm (the microzooplankton: mainly ciliates and phagotrophic dinoflagellates), based on the dilution assay protocol of Landry and Hassett (1982). At the same time, pico- and nanosized phytoplankton including coccoid cyanobacteria (Waterbury et al. 1979) and small eukaryotes (Murphy and Haugen 1985; Shapiro and Guillard 1986) were found to comprise a large proportion of phytoplankton biomass in the open ocean; these cells are too small to be effectively grazed by copepods. Sherr and Sherr (1988) reintroduced the small phytoplankton and herbivorous protist components of the Azam et al. (1983) paper in an expanded conceptual diagram in which the microbial loop as depicted by Ducklow (1984) was embedded in an overall microbial food web (Fig. 2.7). In this conceptual model, the multicellular components of pelagic food webs were supported both by "classic" large phytoplankton and by the heterotrophic protist component of the "microbial" side. The model shown in Figure 2.7 has been updated by including the potential for viral lysis of bacteria and phytoplankton to affect carbon flows in marine food webs (Fuhrman and Suttle 1993). Pico- and nanosized phytoplankton are discussed in Chapter 6 and viruses in Chapter 12.

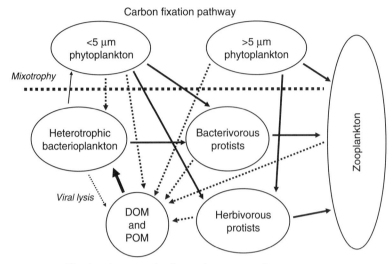

Figure 2.7 Diagram of complex trophic interactions within the microbial food web, which is separated here into phytoplankton and "microbial loop" (i.e., bacteria and heterotrophic protist compartments). Solid arrows represent trophic links; dotted arrows represent regenerative pathways. The dashed line between phytoplankton and bacteria represents mixotrophy by phytoflagellates. In this conceptual model, production of <5 μm sized phytoplankton is accessible to zooplankton only after being "repackaged" into larger protist cells. This figure has been redrawn from Figure 2 of Sherr and Sherr (1988), and updated by relabeling the original "ciliate" box as "herbivorous protists," which includes both ciliates and phagotrophic dinoflagellates, and by adding a dotted arrow between bacteria and dissolved organic matter (DOM) to indicate the mortality process of viral lysis. POM, particulate organic matter.

Joel Goldman (1984) further elaborated the microbial loop concept by proposing that an important source of regenerated ammonium and phosphate was the "spinning wheel" of microbial activity associated with suspended organic aggregates. Goldman and his student and colleague David Caron subsequently evaluated the potential for nitrogen and phosphorous regeneration by protists fed prey with differing C:N and C:P ratios (Caron and Goldman 1988), and wrote a key review of the major contribution of phagotrophic protists to nutrient regeneration in the sea (Caron and Goldman 1990).

The revised ideas about the role of heterotrophic microbes—bacteria and phagotrophic protists—and the importance of smaller phytoplankton in marine food webs had a great influence in the sampling and experimental design of large oceanographic field projects carried out in the world ocean during the 1980s and 1990s. The Subarctic Pacific Ecosystem Research (or SUPER) program (1984–1987), led by Charles B. Miller at Oregon State University, investigated the cause of uniformly low phytoplankton stocks year-round, despite high nutrient concentrations, in the subarctic Pacific Ocean. Experiments that evaluated rates of recycling of nitrogen and grazing by phagotrophic protists on phytoplankton lead to a combined iron-limitation, micrograzer/recycled ammonium hypothesis to explain the observation of constantly low phytoplankton biomass in this region (Miller et al. 1991). Subsequently, during the decade-long Joint Ocean Flux Study (JGOFS) starting in 1989, multinational efforts to understand time-varying fluxes of carbon, nitrogen, and other biogenic elements in major regions of the world ocean included measurement of the biomass and activity of heterotrophic bacteria and of phagotrophic protists (Landry 2002; Ducklow et al. 2004).

The data gained from these large programs, as well as from other studies carried out in other marine systems during the same period, were incorporated into mathematical models of marine ecosystems, which demonstrated that the relative magnitude of material flow through small phytoplankton and heterotrophic microbes is fundamental to the fates of organic matter in the sea, that is, how much is respired/regenerated, stored, or exported. The first influential model that incorporated microbial processes into elemental flows in pelagic ecosystems was that of Pace et al. (1984). Subsequent nitrogen- or carbon-based models in which microbes were of central importance were developed by Michael Fasham, Hugh Ducklow, Louis Legendre, and their colleagues (Fasham et al. 1990; Legendre and Lefevre 1995; Ducklow et al. 2004).

While this history of the development of marine microbial ecology has focused on understanding the role of heterotrophic microbes in pelagic food webs, study of microbes in benthic systems has proceeded as well. Much of this effort has involved the study of mediation of specific biogeochemical processes by microbes (Fenchel et al. 1998). However, work on chemosynthetic processes and symbiotic relationships in hydrothermal vent and methane seep environments has led to spectacular new discoveries about marine microbes in benthic systems (Jannasch and Wirsen 1979; Karl 1995; Boetius et al. 2000; Michaelis et al. 2002; Van Dover et al. 2003).

1990–PRESENT: THE MOLECULAR REVOLUTION

Up until the 1990s, heterotrophic bacteria in the sea were lumped together in what microbial ecologists acknowledged was a "black box" in terms of species diversity. The "plate count anomaly," that is, the difference in abundance between the number of bacterial cells that could be grown up as colonies on agar plates and the number of bacterial cells actually enumerated in seawater by epifluorescence microscopy, suggested that a very large portion of bacterial diversity was composed of strains that had not yet been brought into culture, and was therefore unknown. The bacterial isolates that were obtained for use in laboratory experimentation were typically microbes that grew well on surfaces or at high substrate concentration.

The groundwork for the current transformation of our understanding of microbial diversity in the sea was laid by Carl Woese and his colleague George Fox with a 1977 publication in the *Proceedings of the National Academy of Sciences*. Woese and Fox used differences in gene sequences of evolutionarily conserved regions of the bacterial genome that coded for the 16S portion of the ribosome to show that bacteria, or prokaryotes, actually were split into two quite separate domains of life: the true bacteria, or *Bacteria*, and the *Archaea*. Their completely unexpected finding was stiffly resisted by traditional microbiologists for years (Morell 1997). However, a powerful new approach to getting inside the "black box" of microbial diversity was launched. Molecular biologist Norman Pace and colleagues developed methods for cloning specific gene sequences from DNA extracted from environmental samples (Pace et al. 1986; Olson et al. 1986). Stephen Giovannoni, who worked in Norman Pace's laboratory, led a research team in the first description of prokaryotic diversity in seawater based on 16S ribosomal DNA gene sequences (Giovannoni et al. 1990). This research team discovered novel prokaryotic gene sequences that were unrelated to those of any previously cultured marine bacteria. Among these sequences were those of an uncultured bacterial "phylotype" that appeared to comprise a large fraction of all the gene sequences isolated from samples collected in the Sargasso Sea. This gene clone was labeled SAR11. The phylogenetic subgroup, or clade, of *Bacteria* with close affiliation to the initial SAR11 16S rRNA gene sequence has subsequently been found to be widely distributed and abundant in all regions of the world ocean (Giovannoni and Rappé 2000) (see also Chapter 3).

The molecular biology revolution has swept through the various fields of aquatic microbial ecology. Studies on diversity and distribution of phylotypes of both marine prokaryotes and eukaryotes, and on the distribution of specific functional genes, are now a large component of research in this field. Spectacular discoveries have been made, which will be highlighted in subsequent chapters. This new wave of research is only at the beginning, though, with exciting prospects for the future, including study of how microbial genes function and how gene products (specific proteins produced) affect biogeochemical processes in the sea.

SUMMARY

1. Roles of heterotrophic microbes in marine plankton were mainly speculative before the 1950s, because methods for accurately assessing bacterial abundance and activity in situ were not yet available, and only a few scientists, notably Selman Waksman and Claude ZoBell, had research programs focused on marine microbes.

2. The paper of L. R. Pomeroy in *BioScience* in 1974 marked a renewed interest in elucidating the importance of heterotrophic microbes, including heterotrophic bacteria and phagotrophic protists, in pelagic food webs.

3. In the 1970s, new methods of enumerating heterotrophic microbes, and quantifying rates of microbial activity, led to discoveries that bacteria and heterotrophic protists were much more abundant in the sea than had previously been recognized, and were major consumers of phytoplankton production.

4. International workshops on marine microbes in the early 1980s compiled growing data on marine bacteria and protists, and resulted in a number of influential review papers, including the "microbial loop" paper of Azam et al. (1983).

5. Beginning in the 1990s, molecular genetic approaches to understanding phylogenetic diversity and gene function of bacteria and protists have yielded exciting new information about microbes in the sea. Molecular biology approaches continue to be a major theme in the field of marine microbiology.

REFERENCES

Azam, F., T. Fenchel, J. G. Field, J. S. Gray, L. A. Meyer-Reil, and F. Thingstad. 1983. The ecological role of water-column microbes in the sea. *Mar. Ecol. Prog. Ser.* 10: 257–263.

Boetius, A., K. Ravenschlag, C. J. Schubert, et al. 2000. A marine microbial consortium apparently mediating anaerobic oxidation of methane. *Nature* 407: 623–626.

Caron, D. A. 1983. Technique for enumeration of heterotrophic and phototrophic nanoplankton, using epifluorescence microscopy, and comparison with other procedures. *Appl. Environ. Microbiol.* 46: 491–498.

Caron, D. A., and J. C. Goldman. 1988. Dynamics of protistan carbon and nutrient cycling. *J. Protozool.* 35: 247–249.

Caron, D. A., and J. C. Goldman. 1990. Nutrient regeneration. In G. M. Capriulo (ed.), *Ecology of Marine Protozoa*. Oxford University Press, pp. 283–306.

Davis, P. G., and J. McN. Sieburth. 1982. Differentiation of phototrophic and heterotrophic nanoplankton populations in marine waters by epifluorescence microscopy. *Ann. Inst. Oceanogr.* 58(Suppl): 249–259.

Ducklow, H. W. 1983. Production and fate of bacteria in the oceans. *BioScience* 33: 494–501.

Ducklow, H. W., and C. A. Carlson. 1992. Oceanic bacterial production. *Adv. Microb. Ecol.* 12: 113–181.

Ducklow, H. W. 2000. Bacterial production and biomass in the oceans. In D. L. Kirchman (ed.), *Microbial Ecology of the Ocean*, 1st edn. Wiley-Liss, pp. 85–120.

Ducklow, H. W., D. K. Steinberg, and K. O. Buesseler. 2004. Upper ocean carbon export and the biological pump. Special Issue—JGOFS. *Oceanography* 14: 50–58.

Ehrlich, H. L. 2000. ZoBell and his contributions to the geosciences. In C. R. Bell, M. Brylinsky, and P. Johnson-Green (eds.), *Microbial Biosystems: New Frontiers. Proceedings of the 8th International Symposium on Microbiology Ecology, 9–14 August 1998*. Atlantic Canada Society for Microbial Ecology, Halifax (Nova Scotia), Canada, pp. 57–62.

Fasham, M. J. R. (ed.). 1984. *Flows of Energy and Materials in Marine Ecosystems: Theory and Practice. NATO Conference Series IV: Marine Sciences*, Vol. 13. Plenum Press.

Fasham, M. J. R., H. W. Ducklow, and D. S. McKelvie. 1990. A nitrogen-based model of plankton dynamics in the oceanic mixed layer. *J. Mar. Res.* 48: 591–639.

Fenchel, T. 1967. The ecology of marine microbenthos. I. The quantitative importance of ciliates as compared with metazoans in various types of sediments. *Ophelia* 4: 121–137.

Fenchel, T. 1977. The significance of bacterivorous protozoa in the microbial community of detrital particles. In J. Cairns (ed.), *Aquatic Microbial Communities*. Garland, pp. 529–544.

Fenchel, T. 1982a. Ecology of heterotrophic flagellates. I. Some important forms and their functional morphology. *Mar. Ecol. Prog. Ser.* 8: 211–223.

Fenchel, T. 1982b. Ecology of heterotrophic microflagellates. II. Bioenergetics and growth. *Mar. Ecol. Prog. Ser.* 8: 225–231.

Fenchel, T. 1982c. Ecology of heterotrophic flagellates. III. Adaptations to heterogeneous environments. *Mar. Ecol. Prog. Ser.* 9: 25–33.

Fenchel, T. 1982d. Ecology of heterotrophic flagellates. IV. Quantitative occurrence and importance as bacterial consumers. *Mar. Ecol. Prog. Ser.* 9: 35–42.

Fenchel, T. 1987. *Ecology of Protozoa: The Biology of Free-Living Phagotrophic Protists*. Science Tech Publishers. MA.

Fenchel, T., G. M. King, and T. H. Blackburn. 1998. *Bacterial Biogeochemistry: An Ecophysiological Approach*. Academic Press.

Fuhrman, J. A., and F. Azam. 1982. Thymidine incorporation as a measure of heterotrophic bacterioplankton production in marine surface waters: evaluation and field results. *Mar. Biol.* 66: 109–120.

Fuhrman, J. A., and C. A. Suttle. 1993. Viruses in marine planktonic systems. *Oceanography* 6: 51–63.

Giovannoni, S., and M. Rappé. 2000. Evolution, diversity, and molecular ecology of marine prokaryotes. In D. L. Kirchman (ed.), *Microbial Ecology of the Oceans*, 1st edn. Wiley-Liss, pp. 47–86.

Giovannoni, S. J., T. B. Britschgi, C. L. Moyer, and K. G. Field. 1990. Genetic diversity in Sargasso Sea bacterioplankton. *Nature* 345: 60–63.

Goldman, J. C. 1984. Conceptual role for microaggregates in pelagic waters. *Bull. Mar. Sci.* 35: 462–476.

Hobbie, J. E., and P. J. leB. Williams (eds.). 1984. *Heterotrophic Activity in the Sea. NATO Conference Series IV: Marine Sciences*, Vol. 14. Plenum Press.

Hobbie, J. E., C. C. Crawford, and K. L. Webb. 1968. Amino acid flux in an estuary. *Science* 159: 1963–1964.

Hobbie, J. E., R. J. Daley, and S. Jasper. 1977. Use of Nuclepore filters for counting bacteria by fluorescence microscopy. *Appl. Environ. Microbiol.* 33: 1225–1228.

Jannasch, W. W. 1984. Aspects of measuring bacterial activity in the deep ocean. In J. E. Hobbie, and P. J. leB. Williams (eds.), *Heterotrophic Activity in the Sea*. Plenum Press, pp. 505–522.

Jannasch, H. W., and C. O. Wirsen. 1977. Retrieval of concentrated and undecompressed microbial populations from the deep sea. *Appl. Environ. Microbiol.* 33: 642–646.

Jannasch, H. W., and C. O. Wirsen. 1979. Chemosynthetic primary production at East Pacific sea floor spreading centers. *BioScience* 29: 529–598.

Johannes, E. E. 1965. Influence of marine protozoa on nutrient regeneration. *Limnol. Oceanogr.* 10: 434–442.

Karl, D. M. (ed.). 1995. *The Microbiology of Deep-Sea Hydrothermal Vents*. CRC Press.

Kirchman, D. L., E. K'Nees, and R. E. Hodson. 1985. Leucine incorporation and its potential as a measure of protein synthesis by bacteria in natural aquatic systems. *Appl. Environ. Microbiol.* 49: 599–607.

Landry, M. R. 2002. Integrating classical and microbial food web concepts: Evolving views from the open-ocean tropical Pacific. *Hydrobiologia* 480: 29–39.

Landry, M. R., and R. P. Hassett. 1982. Estimating the grazing impact of marine microzooplankton. *Mar. Biol.* 67: 283–288.

REFERENCES

Legendre, L., and J. LeFevre. 1995. Microbial food webs and the export of biogenic carbon in oceans. *Aquat. Microb. Ecol.* 9: 69–77.

McGraw, D. J. 2004. Claude ZoBell and the foundations of marine microbiology (1933–1939). In S. Morcos and G. Wright (eds.), *Ocean Sciences Bridging the Millennia: A Spectrum of Historical Accounts, Proceedings of International Congress on the History of Oceanography (ICHO) VI.* China Ocean Press, pp. 45–64.

McManus, G. B., and J. A. Fuhrman. 1986. Bacterivory in seawater studied with the use of inert fluorescent particles. *Limnol. Oceanogr.* 31: 420–426.

Michaelis, W., R. Seifert, K. Nauhaus, et al. 2002. Microbial reefs in the Black Sea fueled by anaerobic oxidation of methane. *Science* 297: 1013–1015.

Miller, C. B., B. W. Frost, P. A. Wheeler, M. R. Landry, N. Welschmeyer, and T. M. Powell, 1991. Ecological dynamics in the subarctic Pacific, a possibly iron-limited ecosystem. *Limnol. Oceanogr.* 36: 1600–1615.

Morell, V. 1997. Microbiology's scarred revolutionary. *Science* 276: 699–702.

Morita, R. Y. 1997. *Bacteria in Oligotrophic Environments: Starvation-Survival Lifestyle.* Chapman & Hall.

Murphy, L. S., and E. M. Haugen. 1985. The distribution and abundance of phototrophic ultraplankton in the north Atlantic. *Limnol. Oceanogr.* 30: 47–58.

Odum, E. P., and A. A. de la Cruz. 1963. Detritus as a major component of ecosystems. *AIBS Bull* (later BioScience) 13: 39–40.

Odum, E. P., and H. T. Odum. 1959. *Fundamentals of Ecology.* W. B. Saunders, Philadelphia.

Odum, E. P., and A. E. Smalley. 1959. Comparison of population energy flow of an herbivorous and a deposit-feeding invertebrate in a salt marsh ecosystem. *Proc. Natl. Acad. Sci. USA* 45: 617–622.

Olson, G. J., D. L. Lane, S. J. Giovanonni, N. R. Pace, and D. A. Stahl. 1986. Microbial ecology and evolution: A ribosomal RNA approach. *Annu. Rev. Microbiol.* 40: 337–366.

Pace, M. L., J. E. Glasser, and L. R. Pomeroy. 1984. A simulation analysis of continental shelf food webs. *Mar. Biol.* 82: 47–63.

Pace, N. R., D. A. Stahl, D. L. Lane, and G. L. Olsen. 1986. The analysis of natural microbial populations by rRNA sequences. *Adv. Microbiol.* 40: 337–365.

Pomeroy, L. R. 1974. The ocean's food web: A changing paradigm. *BioScience* 24: 499–504.

Pomeroy, L. R., and R. E. Johannes. 1968. Occurrence and respiration of the ultraplankton in the upper 500 meters of the ocean. *Deep-Sea Res.* 15: 381–391.

Porter, K. G., and Y. S. Feig. 1980. The use of DAPI for identifying and counting aquatic microflora. *Limnol. Oceanogr.* 25: 943–948.

Shapiro, L. P., and R. R. L. Guillard. 1986. Physiology and ecology of the marine eukaryotic ultraplankton. In T. Platt and W. K. W. Li (eds.), *Photosynthesis in the Sea*, Vol. 214. Can. J. Fish. Aquat. Sci. pp. 371–389.

Sherr, E. B., and B. F. Sherr. 1983. Double-staining epifluorescence technique to assess frequency of dividing cells and bacterivory in natural populations of heterotrophic microprotozoa. *Appl. Environ. Microbiol.* 46: 1388–1393.

Sherr, E. B., and B. F. Sherr 1988. Role of microbes in pelagic food webs: A revised concept. *Limnol. Oceanogr.* 33: 1225–1227.

Sherr, B. F., E. B. Sherr, and R. D. Fallon. 1987. Use of monodispersed, fluorescently labeled bacteria to estimate in situ protozoan bacterivory. *Appl. Environ. Microbiol.* 53: 958–965.

Sieburth, J. McN. 1975. *Microbial Seascapes: A Pictorial Essay on Marine Microorganisms and their Environments.* University Park Press.

Sieburth, J. McN. 1979. *Sea Microbes.* Oxford University Press.

Sieburth, J. McN. 1981. Protozoan bacterivory in pelagic marine waters. In J. E., Hobbie, and P. J. leB. Williams (eds.), *Heterotrophic Activity in the Sea.* Plenum Press, pp. 405–444.

Sieburth, J. McN, V. Smetacek, and J. Lenz. 1978. Pelagic ecosystem structure: heterotrophic compartments of the plankton and their relationship to plankton size fractions. *Limnol. Oceanogr.* 23: 1256–1263.

Sorokin, Y. I. 1981. Microheterotrophic organisms in marine ecosystems. In A. R. Longhurst, (ed.), *Analysis of Marine Ecosystems.* Academic Press, pp. 293–332.

Staley, J. T., and A. Konopka. 1985. Measurement of in situ activities of nonphotosynthetic microorganisms in aquatic and terrestrial habitats. *Annu. Rev. Microbiol.* 39: 321–346.

Steele, J. H. 1974. *Structure of Marine Ecosystems.* Harvard University Press.

Strugger, S. 1948. Fluorescent microscope examination of bacteria in soil. *Can. J. Res.* 26: 188–193.

Teal, J. M. 1962. Energy flow in the salt marsh ecosystem of Georgia. *Ecology* 43: 614–624.

Van Dover, C. L., P. Aharon, J. M. Bernhard, et al. 2003. Blake Ridge methane seep: characterization of a soft-sediment, chemosynthetically based ecosystem. *Deep-Sea Res.* I 50: 281–300.

Waksman, S. A. 1934. The role of bacteria in the cycle of life in the sea. *Sci. Monthly* 38: 35–49.

Waksman, S. A., and C. E. Renn. 1936. Decomposition of organic matter in sea water by bacteria. III. Factors influencing the rate of decomposition. *Biol. Bull.* 70: 472–483.

Waterbury, J. B., S. W. Watson, R. R. Guillard, and L. E. Brand. 1979. Widespread occurrence of a unicellular, marine, planktonic, cyanobacterium. *Nature* 277: 293–294.

Williams, P. J. leB. 1981. Incorporation of heterotrophic processes into the classical paradigm of the microbial food web. *Kieler Meersforsch. Suppl.* 5: 1–28.

Williams, P. J. leB. 1984. Bacterial production in the marine food chain: The Emperor's new suit of clothes? In M. J. R. Fasham (ed.), *Flows of Energy and Materials in Marine Ecosystems.* Plenum Press, pp. 271–344.

Woese, C. R., and G. E. Fox. 1977. Phylogenetic structure of the prokaryotic domain: The primary kingdoms. *Proc. Natl. Acad. Sci. USA* 74: 5088–5090.

Wood, E. J. F. 1955. Fluorescence microscopy in marine microbiology. *J. Cons. Int. Expl. Mer.* 21: 6–7.

Yayanos, A. A. 1986. Evolutional and ecological implications of the properties of deep-sea barophilic bacteria. *Proc. Natl. Acad. Sci. USA* 83: 9542–9546.

Zimmerman, R., and L.-A. Meyer-Reil. 1974. A new method for fluorescence staining of bacterial populations on membrane filters. *Kieler Meeresforsch.* 30: 24–27.

ZoBell, C. E. 1946. *Marine Microbiology: A Monograph on Hydrobacteriology.* Chronica Botanica.

BACTERIAL AND ARCHAEAL COMMUNITY STRUCTURE AND ITS PATTERNS

JED A. FUHRMAN
Department of Biological Sciences, University of Southern California, Los Angeles, CA 90089, U.S.A.

ÅKE HAGSTRÖM
Kalmar University, Sweden

INTRODUCTION

Community structure is the characterization of the types of organisms present in an environment and the relative proportions of those types. It is of interest for its own sake, as a central object of study in the broad field of ecology, and it also can yield much information on the functional aspects of the microbial community, including biogeochemical processes of interest to biological oceanographers. Community structure can be defined by functional types (e.g., photosynthetic bacteria or nitrogen-fixing organisms), but more commonly it is defined by the taxonomic or phylogenetic identification of the organisms. The latter can be at different levels of resolution, from domain or kingdom down to clonal strains, that is, strains recently derived from a single parent cell. This phylogenetic characterization of bacteria and archaea in seawater will be the primary focus of this chapter, connected to function when that is possible. We will not only discuss what types of bacteria and archaea are found in seawater, but also how they are distributed over time, with geographic location, and with depth.

Microbial Ecology of the Oceans, Second Edition. Edited by David L. Kirchman
Copyright © 2008 John Wiley & Sons, Inc.

Early studies of marine bacteria, starting in the late 19th century, often focused on isolating pure cultures, and cultures were needed to identify the bacteria (Certes 1884; Frankland and Frankland 1894). But half a century later, it became widely recognized that only a tiny fraction of the total bacteria present in seawater are cultivable on standard microbiological media (e.g., Jannasch and Jones 1959). When epifluorescence microscopy became available to easily count the total bacterial abundance in seawater (typically 10^9/L) (Ferguson and Rublee 1976; Hobbie et al. 1977), marine microbial ecologists learned that only about 0.1 percent of the observable bacteria were readily cultivable on standard rich growth media (Ferguson et al. 1984). This problem was not unique to seawater, and the difference has been called the "great plate anomaly" (Staley and Konopka 1985). In any case, with much less than 1 percent of the bacteria being able to be studied and identified by classical cultivation, it became obvious that this approach was not suitable to characterize overall bacterial community structure, even though it may identify some members of the community. Alternatives were needed.

Such alternatives came in the 1980s, when bacterial community structure studies experienced a renaissance following the development of molecular phylogeny. This permits classification of organisms according to evolutionary relatedness as shown by sequence similarity of molecules such as ribosomal RNA (Woese 1987). While originally developed with pure cultures, molecular phylogeny could be applied in natural settings after it was recognized that nucleic acid could be extracted directly from organisms collected from their natural environment, and the genes of interest could be cloned and sequenced (Olsen et al. 1986; Pace et al. 1986). Thus one could obtain genetic sequence data without the need to culture the organism, and even if that organism had never been seen before, the sequence shows how it is related to other organisms and provides ways to at least tentatively name it. This approach was first applied with 16S rRNA genes, universally present and sufficiently conserved to permit comparisons among all organisms. Such studies have been done widely with marine microorganisms, and the ensuing results have revolutionized our ideas about what bacteria are present in seawater and how they are distributed. Associated research has also yielded information about the ecological roles of different bacteria.

Culture-independent genetic analysis of microbial communities is based upon extraction of DNA directly from biomass, and then determination of gene sequences from that DNA. Such genes can be used to identify the microorganisms that were present in the sample, even for organisms never seen before, because well-conserved genes such as that for 16S rRNA can place any organisms on a universal "tree of life." This approach can also be used to determine specific capabilities as indicated by the presence of functional genes. Originally, most such work used the polymerase chain reaction (PCR) to look for particular genes, but recent metagenomic studies (see Chapter 4) randomly sequence portions of the genomes and potentially can show the presence of capabilities not otherwise known from that environment. PCR and the different metagenomic approaches each have advantages and disadvantages.

MAJOR GROUPS OF PROKARYOTES IN SEAWATER

The large majority of known marine bacteria and archaea tend to occur in about a dozen generally broad phylogenetic divisions whose relationships can be visualized in phylogenetic trees (Fig. 3.1), with some of these divisions arguably broader than animal phyla (Giovannoni and Rappé 2000; Giovannoni and Stingl 2005). This outline of marine bacterioplankton, which is based on only 16S rDNA phylogeny, can be considered in the context of about 20 recognized *culturable* and well-studied major divisions of *Bacteria* depicted in the most current version of the tree of life by Ciccarelli et al. (2006). ("Divisions" are also called kingdoms or phyla.) Their analysis is based on a new computational approach, which focuses on 31 "universally" occurring genes with indisputable orthology in 191 species with largely annotated genomes. As more whole-genome sequences of marine bacteria (currently >150, as described in Chapter 4) become available for analysis according to a similar multilocus approach, a revised representation of the bacterial divisions might be expected. However, the main lines in Figure 3.1 are likely to remain similar. It is also important to recognize that there are many more (>50 total) major bacterial divisions, including groups within which cultured members are not yet available, and many marine organisms fall into this latter category. Since the major divisions are ancient, it is important to recognize that shared membership in a broad division does not necessarily imply strong functional similarities between modern organisms, although it is tempting to think this is the case. While relatedness as based upon analysis of "housekeeping" or "core" genes such as 16S rRNA might sometimes imply many shared characteristics, evolution of microbes has been shown to sometimes include some dramatic changes in functionality via gene loss or horizontal genetic exchange of a variety of functional genes. Current studies are starting to show a remarkable amount of practical functional diversity and niche partitioning within some surprisingly similar clusters of closely related organisms, such as *Prochlorococcus*. One of the biggest challenges in marine microbial diversity studies today is to understand when it is safe to conceptually lump groups of organisms functionally or ecologically, and when we should split them.

> **Phylogenetic trees** are attempts to represent evolutionary relationships among organisms, and most often are comparisons of homologous gene sequences (individual or sometimes multiple genes) where the genes are used to represent the organisms that possess them. Trees are calculated based upon a variety of models and algorithms. In radially drawn trees (e.g., Figs. 3.1 and 3.2), the evolutionary distance can be read directly as distance along the tree joining organisms, but in trees where all organisms are at the end of horizontal branches (Fig. 3.3), the vertical distances connecting organisms are ignored.

Therefore, while we will describe characteristics of groups of organisms from what is known now about various members of the groups, it needs to be borne in mind that the entire group may well not share those characteristics.

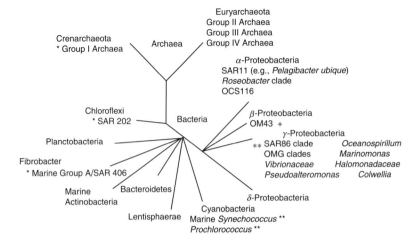
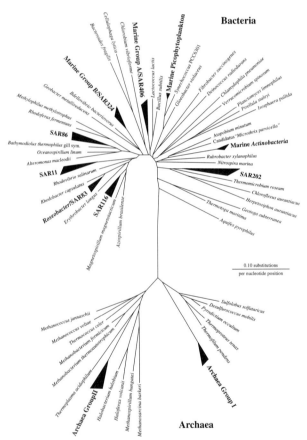

Figure 3.1 Schematic illustration of the phylogeny of the major *Archaea* and *Bacteria* clades, showing only the major marine groups (top) and also in the context of other bacteria (bottom). In the top diagram, groups with a single asterisk are mostly found in the mesopelagic and surface waters during polar winters (deep mixing), those with two asterisks are mostly in the photic zone, and those with a + are mostly coastal. Others seem to be ubiquitous in seawater. The top diagram is modified from Giovannoni and Stingl (2005) and the bottom diagram is from Giovannoni and Rappé (2000).

"CLASSICALLY" CULTURABLE BACTERIA

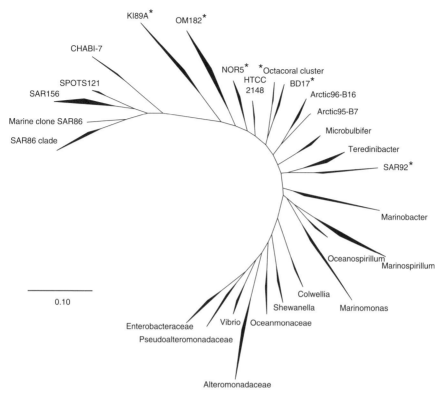

Figure 3.2 Unrooted Bayesian tree of marine *Gammaproteobacteria* and close relatives, from 16S rRNA. Groups with Latin names have formally named cultures. An * indicates that a member of the clade has been isolated by dilution culture but not yet given a formal name. Others are not yet cultivated. Courtesy of Michael Schwalbach.

"CLASSICALLY" CULTURABLE BACTERIA

In the past, "bacteria" was often used to mean "colony-forming organisms," that is, bacteria that could multiply and form large aggregations on a solid substrate, usually agar, fortified with a general broth or a more specific composition of nutrients. It was assumed that, given the right conditions and the right substrate, all marine bacteria could form colonies and thus be isolated. When this ambition failed, it was argued that the fundamental needs (chemical, nutritional, and physical) of marine bacteria were unclear, which made it difficult to provide a culturing media that would be suitable for all microorganisms sampled from a heterogeneous environment. It was also realized that rich media such as "marine broth" excluded oligotrophic or facultative oligotrophic cells adapted to nutrient poor environments, due to factors such as nutrient stress (Button et al. 1993). Furthermore, since colony formation is an active process—flagellae need to be shut down, adhesive components must be manufactured, and so on—the assumption of total culturability of this sort appears to have

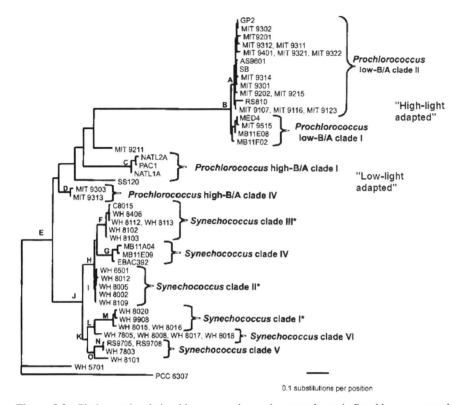

Figure 3.3 Phylogenetic relationships among the marine cyanobacteria *Prochlorococcus* and *Synechococcus* as demonstrated by 16S–23S rRNA intergenic spacers. Modified from Rocap et al. (2002).

been fundamentally wrong. Instead, the behavior of aggregated (colony-forming) growth should be seen as the characteristics of eutrophic species that may be considered "r-ecotypes" or "bloomers," that is, a species with a relatively unstable population size that fluctuates widely, while seldom approaching its carrying capacity in nature. These bacteria are adapted to growth in nutrient-rich environments and generally have large cell size, achieve high cell density in cultures, and are able to grow with short generation times. Some of the prominent readily culturable types from seawater are described below.

The *Roseobacter* Clade of Marine *Alphaproteobacteria*

While most marine bacterial groups found to be common by culture-independent methods either have not yet been cultured or require special cultivation techniques (see below), a broad group of *Alphaproteobacteria*, within the *Roseobacter* lineage, possesses many members commonly found by conventional marine culture surveys (on relatively rich media) and also numerically important in culture-independent assays (Gonzalez and Moran 1997). They occur readily in a broad variety of marine environments (including plankton, sediments, sea ice, animal surfaces, etc.), and typically constitute about 20 percent of communities in coastal waters, about 15 percent in mixed layer ocean communities, and <1 percent at depths below a few hundred meters (reviewed by Buchan et al. 2005). While some members of the genus *Roseobacter* contain the pigment bacteriochlorophyll a and are aerobic anoxygenic phototrophs (see Chapter 5), many or most members of this broad clade (currently at 17 genera) do not possess this property. This group has been found to be particularly important in transformations of sulfur compounds such as dimethylsulfonylpropionate, an abundant osmolyte produced by many phytoplankton and implicated in climate regulation via dimethylsulfide gas (Gonzalez et al. 2000, 2003; Buchan et al. 2005). Several members of this clade have had their genomes fully sequenced, the first being *Silicibacter pomeroyi*, which has interesting capabilities, such as the oxidation of carbon monoxide and reduced sulfur compounds (Moran et al. 2004); more extensive results are described in Chapter 4.

Gammaproteobacteria

Perhaps the best known culturable *Gammaproteobacteria* are the vibrios, which are generally rapidly growing polar-flagellated curved rods, readily isolated from marine waters and sediments, and aquaculture settings worldwide, often in association with eukaryotes. Their potential for explosively rapid growth, with doubling times in rich culture measured in tens of minutes, means they are readily represented in cultivation-based surveys. They include *Vibrio cholerae*, some strains of which are the causative agent of the human disease cholera (Colwell 1996), *V. vulnificus* (implicated in wound infection), fish pathogens such as *V. anguilarum*, and many non-pathogenic species. A review on the biodiversity of *Vibrio* by Thompson et al. (2004) describes this large group of organisms whose best-studied members have a bias towards attacking other living cells. Vibrios often show intricate behavior, such as *V. fischeri* cells living symbiotically in the light organs of squid or flashlight fish, and thereby generating bioluminescence. Cell-to-cell communication, or "quorum sensing," likely plays important roles for these organisms. Such bacteria can sense the density of specific cells by signaling molecules or pheromones (e.g., *N*-acylhomoserine lactones). This has been documented for most pathogenic *Vibrio* and many other pathogens and biofilm- or multicell-forming organisms (Waters and Bassler 2005; Camilli and Bassler 2006). Quorum sensing involves regulation of the expression of virulence or other genes (e.g., for light production), and these recent reports suggest not only that it is a widespread phenomenon, but also that the chemical communication within a species may be intercepted or interfered with by other species.

There are also a variety of other readily culturable marine *Gammaproteobacteria*, including *Alteromonas*, *Pseudoalteromonas*, *Marinomonas*, *Shewanella*, *Glaciecola Oceanospirillum*, *Colwellia*, and a few others (Giovannoni and Rappé 2000). They have a tendency for rapid potential growth rates, well suited for a feast-and-famine lifestyle, such as on organic surfaces or in the guts of fish. *Shewanella*, in particular, has a large versatility of potential inorganic electron acceptors in addition to oxygen. A study of close relatives of one species, *Alteromonas macleodii*, has shown remarkable microdiversity and geographic partitioning of different subtypes (García-Martínez et al. 2002). Phylogenetic relationships among the *Gammaproteobacteria* are shown in Figure 3.2.

Bacteroidetes

Bacteria belonging to the *Bacteroidetes* (previously known as *Cytophaga–Flavobacteria–Bacteroides*) group are increasingly recognized as important members of the marine bacterioplankton. These bacteria are highly diverse and have been found in a wide range of habitats (e.g., rumen, human colon, lakes, pelagic oceans, sediments, hydrothermal vents, and sea ice). However, little is known about the diversity of the *Bacteroidetes* members that are most commonly found in the sea. According to some fluorescence in situ hybridization (FISH) analyses (see Chapter 6 for a description of the FISH method), this cluster may account for as much as half of all bacterioplankton cells potentially hybridized in a seawater sample (Cottrell and Kirchman 2000a). However, in contrast, few clone libraries of 16S rRNA genes from seawater possess such a high proportion of *Bacteroidetes* representatives. This might be explained by the apparent bias against this cluster by primers commonly used in clone library building (Weaver et al. 2003), but the relatively low representation of these organisms in metagenomic studies so far suggests that perhaps the FISH results are not broadly representative. Commonly, *Bacteroidetes* are thought to be taking part in organic material degradation, and have been reported repeatedly with the natural occurrence of algal blooms (Riemann et al. 2000). A member of this group (*Dokdonia* sp. strain MED134) has also recently been reported to use the light-driven proton-pumping pigment proteorhodopsin to capture light energy in a way that directly benefits growth, apparently making the organism functionally photoheterotrophic (Gomez-Consarnau et al. 2007); it is not known how broadly this property occurs in the group, but this could have significance for growth efficiency, energy budgets, and marine carbon cycling in general.

- **Autotrophic** organisms use inorganic carbon (CO_2) as a C source, heterotrophic ones use organic carbon.

- **Photoautotrophs** use light energy and fix CO_2, while **photoheterotrophs** use light energy but require organic carbon.

- **Chemoautotrophs** and **chemoheterotrophs** use chemical energy, with most chemoautotrophs performing inorganic oxidation reactions, such as oxidation of ammonia to nitrite or of sulfide to sulfate, and chemoheterotrophs oxidizing organic matter for energy.

Cyanobacteria

Marine cyanobacteria are probably the best understood group of marine bacteria, in terms of genetics and ecology, and many members not only have been cultured, but now also have had their genomes completely sequenced. All the known members of this group possess chlorophyll a and perform oxygenic photosynthesis (and no known members of any other bacterial or archaeal group can do so); why this is so is one of the most intriguing questions of microbial evolution. (See Chapter 5 for definitions of oxygenic and anoxygenic photosynthesis.) These organisms are common in all but polar waters, and often are among the dominant phytoplankton in the tropics and subtropics (Partensky et al. 1999; Landry and Kirchman 2002). By far the most abundant cyanobacteria are in the genus *Prochlorococcus*, first discovered by flow cytometry (Chisholm et al. 1988), with abundances on the order of 10^5 cells/mL in some warm oligotrophic environments (Partensky et al. 1999). *Prochlorococcus* has small cells (diameter about 0.6 μm) that until the advent of flow cytometry had been seen by epifluorescence microscopy (with DNA stains) for many years but not recognized as photosynthetic because its pigments (divinylchlorophyll a and b) do not fluoresce brightly enough to be seen easily via epifluorescence microscopy. *Synechococcus*, a close relative of *Prochlorococcus* (Fig. 3.3), but with slightly larger cells (about 0.9 μm) had been found to be common in seawater as early as 1979 (Johnson and Sieburth 1979; Waterbury et al. 1979), but is much more easily detected because its phycoerythrin accessory pigment autofluoresces bright yellow–orange under blue or green excitation.

Although closely related phylogenetically and generally similar in size, most *Prochlorococcus* and *Synechococcus* are thought to have different ecological strategies, with *Prochlorococcus* occurring more in oligotrophic water and *Synechococcus* typically more abundant in higher nutrient or dynamic situations. *Synechococcus* has been described as more of a generalist and better able to take advantage of fluctuating environments (Rocap et al. 2002; Dufresne et al. 2005; Palenik et al. 2003). *Prochlorococcus* occurs in multiple ecotypes, with the most pronounced difference between high-light-adapted versus low-light-adapted varieties that partition with depth and environmental gradients (Rocap et al. 2002; Johnson et al. 2006) (Fig. 3.4). The high-light *Prochlorococcus* ecotype has the smallest genome known from any oxygenic photosynthetic organism, with one strain having a genome of around 1.7 million base pairs. Low-light-adapted strains can have genomes close to 2.4 million base pairs, comparable to some *Synechococcus* strains (Rocap et al. 2002; Palenik et al. 2003). *Prochlorococcus* also has capabilities to compete successfully against other bacterioplankton for uptake of amino acids, as shown experimentally (Zubkov et al. 2003), and this probably contributes to its high abundance in oligotrophic waters. As such, it might be considered mixotrophic or photoheterotrophic (partly autotrophic and partly heterotrophic) to some extent. Chapters 5 and 10 discuss photoheterotrophy by cyanobacteria and other microbes.

In addition to *Synechococcus* and *Prochlorococcus*, other cyanobacteria are known to have particular significance in marine ecosystems due to their roles in fixation of atmospheric nitrogen. The colony-forming *Trichodesmium*, with tuft- or puff-shaped colonies often visible to the naked eye, lives generally in tropical and subtropical waters, and, although usually rare, it can have intense episodic local blooms and is thought to contribute significantly to the global nitrogen budget (Capone et al. 1997, 2005). It also has interesting capabilities such as utilization of

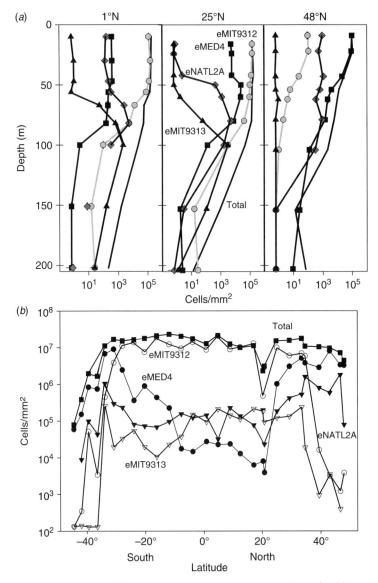

Figure 3.4 Distribution of different *Prochlorococcus* ecotypes as determined by quantitative PCR along a mid-Atlantic transect from the UK to the Falkland Islands. (*a*) Depth variations at particular sites. (*b*) Variations of depth-integrated abundance with latitude along the transect. The solid black line represents the total. See Figure 3.5 for phylogenetic relationships of the ecotypes. Modified from Johnson et al. (2006).

phosphonates (Dyhrman et al. 2006). There are also nitrogen-fixing symbiotic *Richelia* cyanobacteria that live within diatoms. More recently, unicellular cyanobacteria a few micrometers in diameter have been reported to be significant in open ocean nitrogen fixation (Montoya et al. 2004). In the Baltic Sea proper (brackish), nitrogen fixation contributes significantly to the nitrogen budget (Larsson et al. 2001). Annual blooms of *Nodularia spumigena* and *Aphanizomenon* sp. cover large Baltic Sea areas and are responsible for the majority of the nitrogen fixation, though unicellular cyanobacteria are also believed to participate (Wasmund et al. 2005). Chapter 13 discusses nitrogen fixation in more detail.

"SEA WATER" CULTURABLE BACTERIA

The dilution-to-extinction culturing method for the isolation of oligotrophic bacteria in sea water was developed by Button and colleagues (Button et al. 1993) and later modified for high throughput by Rappé, Connon and others (Connon and Giovannoni 2002; Rappé et al. 2002). It is particularly suited to the culture of oligotrophic bacteria. Species adapted to the oligotrophic conditions in the free water might be considered to be K-ecotypes, or "crumb pickers." The population growth of such species results in a relatively stable population size that fluctuates near what appears to be a "carrying capacity," without exceeding it (Whittaker 1975). However, the typical yield of *Pelagibacter ubique* in a sea water culture (see below) is lower than for a mixed bacteria inoculum in the corresponding sea water (Rappé et al. 2002). This could indicate that the utilizable fraction of the dissolved organic matter is different for the different cultures or that the oligotrophs are able to sense the density and thus control their abundance (Ammerman et al. 1984).

For oligotrophic bacteria competing for scarce resources, dispersed single-cell growth would allow optimal utilization of the available substrate, and colony formation may not be an adaptive trait (Simu and Hagström 2004). Thus, although several significant marine bacterial groups are represented by organisms culturable by conventional means (e.g., colonies on a plate), there are reasons to think that these do not include many prominent oligotrophs. Oligotrophic organisms adapted to utilize low concentrations of nutrients are characterized by their slow growth rate, small size and often by inability to grow in environments with high levels of nutrients.

SAR11 Cluster

The SAR11 cluster, a distinct branch within the *Alphaproteobacteria*, is the first novel major group of marine bacteria discovered by the cloning and sequencing of rRNA genes (Giovannoni et al. 1990). It is collectively highly abundant, reportedly constituting about 33 percent of many euphotic zone communities and 25 percent of mesopelagic communities (Morris et al. 2002). The group is highly diverse, with different subclades found in different depths and habitat types (Fig. 3.5) (Field et al. 1997; Garciá-Martinez and Rodríguez-Valera 2000; Brown and Fuhrman 2005; Pommier et al. 2005). Although for many years recalcitrant to cultivation, a few strains have been grown, with considerable difficulty, in dilution culture (Rappé et al. 2002; Simu and Hagström 2004). One strain, *P. ubique*, has been maintained in long-term culture, and its genome has been fully sequenced (Giovannoni et al. 2005b). Its genome, at 1.3 million base pairs, is reportedly the smallest

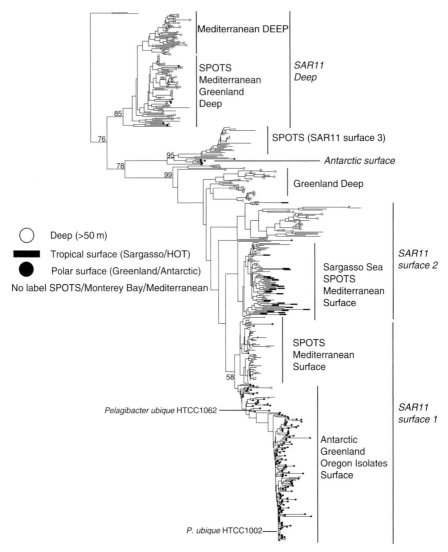

Figure 3.5 Diversity of members of the SAR11 clade as determined from sequence variation of the 16S–23S rRNA ITS spacer region. Note the groupings with depth and habitat type (SPOTS refers to San Pedro Ocean Time Series, off the Southern California coast, HOT to Hawaii Ocean Time-series). *Pelagibacter ubique* is a cultivated member of the clade. Modified from color version in Brown and Fuhrman (2005).

and encodes the smallest number of predicted open reading frames known for a free-living microorganism. This genome codes for complete biosynthetic pathways for all 20 amino acids and most cofactors, and has remarkably few extragenetic elements (no pseudogenes, introns, transposons, extrachromosomal elements, or integrins), and the shortest intergenic spacers yet observed for any cell. Interestingly,

P. ubique possesses the pigment proteorhodopsin that can act as a light-driven proton pump and has been thought to potentially be important in supplying energy to cells, but the organisms in culture as studied to date show identical growth rates and yield in light and dark conditions (Giovannoni et al. 2005a). Whether this pigment is uniform throughout the group (in surface-dwelling clades, one would guess), and its particular benefits for growth of SAR11 cluster members, are not yet known.

NOT-YET-CULTURED BACTERIA

Marine Gammaproteobacterial Clusters

A few divisions within the *Gammaproteobacteria* have been found to be common in marine clone libraries. Originally detected in clones from near Hawaii (Schmidt et al. 1991), California, and Bermuda (Fuhrman et al. 1993), many clones related broadly to the clone SAR86 within the *Gammaproteobacteria* have since been found worldwide. The group was originally named after SAR86 by Mullins et al. (1995) when relatively few members were known, but a recent evaluation of hundreds of clones indicated that what had been lumped into one very broad group (Fig. 3.1) is actually made up of subsets, each containing further subdivisions (Fig. 3.2).

As has been seen with other groups described above, the subclusters vary with environmental parameters, implying significant niche specialization with depth, seasons, nutrients, and so on. Isolates of one group, called OMG (oligotrophic marine *Gammaproteobacteria*) have been cultivated (Cho and Giovannoni 2004), and they have been found to be obligate oligotrophs, unable to grow on rich media, and with very small cells. Interestingly, a metagenomic DNA fragment from one of the as-yet uncultivated subclades related to SAR86 was discovered to have a gene coding for what was named proteorhodopsin (Béjà et al. 2000), the first report of a pigment like this in a bacterium; similar pigments had been known from archaea. That report sparked considerable interest in this potentially important mechanism for capturing light energy in marine ecosystems (discussed in Chapter 5). Subsequent to its discovery in the *Gammaproteobacteria*, the gene was also found in *P. ubique* (the SAR11 cluster in the *Alphaproteobacteria*), in *Dokodinoa* (*Flavobacteria*, which is in the *Bacteroidetes*) as described above, in a marine euryarchaeon (see below), and to be remarkably abundant in a metagenomic library from the Sargasso Sea (Venter et al. 2004). So what was initially called "proteorhodopsin" because it was originally found in a proteobacterium turns out to be broadly present in bacteria and archaea, and its name might be confusing (but no worse than the related pigment called "bacteriorhodopsin," found so far only in archaea!). See Chapter 5 for more on proteorhodopsin.

There are a few hints about other metabolic properties of these organisms. Recently, Todd et al. (2007) found a *dddD* gene, part of the pathway of dimethylsulfide production, in the cultivated HTCC2207 within the SAR92 group. This suggests the group contains members which produce dimethylsulfide gas.

Actinobacteria

The marine actinobacterial group was first reported by Fuhrman et al. (1993) as the third most common group they found (after SAR11 and cyanobacteria) among their 16S rRNA clones from off Southern California and Bermuda. They are routinely found in clone libraries worldwide (Fig. 3.6) but may not be as common as *Actinobacteria* in terrestrial and freshwater environments. Although in the same major division, it is not clear how the marine forms identified from clone libraries relate to the better-studied cultivable *Actinomycetes*. The ecology of cultivable *Actinomycetes* in marine and littoral habitats has been reviewed by Goodfellow and Haynes (1984). Whereas it is still debated whether or not *Actinomycetes* are part of indigenous marine microflora (Goodfellow and Williams 1983), observations of this group decrease with increasing distance from land (Okazaki and Okami 1972). Little is know of the roles of these organisms in seawater. The known cultivable relatives have been reported to decompose diverse large molecules (e.g., agar, alginates,

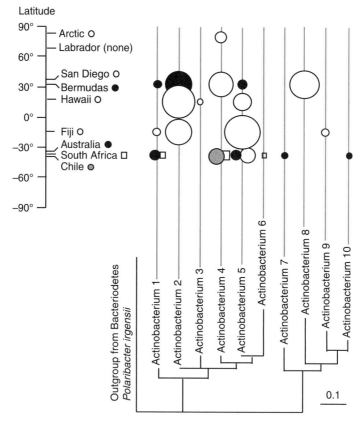

Figure 3.6 Geographical distribution of two clusters of *Actinobacteria* based on 16S rRNA clones organized according to their abundance (size of circles) and presence in clone libraries from nine locations world wide. Modified from Pommier et al. (2007).

laminrarin, cellulose, chitin, and oil and other hydrocarbons) and have been implicated in the degradation of wood and petroleum hydrocarbons in marine habitats (Goodfellow and Williams 1983). Whether the noncultivable marine *Actinobacteria* are particularly adept as digesting similar polymers is not yet known. Kirchman et al. (2005) noted in a study of the Delaware River estuary by FISH that *Actinobacteria* were especially abundant in the freshwater part of the estuary, and that they were particularly active in the uptake of thymidine and leucine, implying that they contributed significantly to bacterial production.

SAR116 Cluster

Like the SAR11 cluster, clones related to SAR116 in the *Alphaproteobacteria* are commonly found in clone libraries, especially in the euphotic zone (but not as commonly as SAR11 relatives). This group has several subclades that differ by up to about 10 percent in 16S rRNA sequence positions. There are no known cultures and their physiology is not known.

SAR202

The SAR202 cluster is deeply branched within the "green non-sulfur" (GNS) bacterial division. This group has been reported to be maximal near the lower boundary of the deep chlorophyll maximum layer in the Sargasso Sea (Giovannoni et al. 1996), and was one of the first groups reported to show strong depth stratification. It was also reported from 3000 m in the northwest Pacific and 1000 m in the Atlantic near Bermuda (Fuhrman and Davis 1997); hence it appears widespread in the deep sea. Subsequently, it has been found in many clone libraries.

Marine Group A

Marine Group A is apparently a major division of the *Bacteria*. It was first found in marine subsurface waters (100 and 500 m depths) (Fuhrman et al. 1993), and Gordon and Giovannoni (1996) showed that it is strongly stratified in the water column, peaking below the deep chlorophyll maximum. The group seasonally oscillates in surface waters of the Sargasso Sea, correlated to chlorophyll a concentrations (Gordon and Giovannoni 1996). Its physiology is not known.

Marine Group B

Marine Group B in the *Deltaproteobacteria* has been found primarily in deeper waters (500–3000 m) (Fuhrman and Davis 1997; Wright et al. 1997). Its physiology is unknown.

Betaproteobacteria

The *Betaproteobacteria* are a large group (perhaps 75 genera) with a global presence. In terms of metabolic, morphological, and ecological characteristics, they are heterogeneous (Cavalier-Smith 2002). They contain some purple nonsulfur phototrophs, various chemolithotrophs, some methylotrophs, many chemoorganotrophs, some nitrogen-fixing bacteria, and some pathogens. While less is known of *Betaproteobacteria* in the marine environment, their morphologies can vary from rods or cocci to spiral and sheathed cells. Novel betaproteobacterial isolates have been obtained from surface seawater in many locations using the dilution-to-extinction approach. Phylogenetic analysis of 16S rRNA gene sequences have frequently placed these isolates within a clade composed of largely uncultured environmental clones known as the OM43 clade (Uphoff et al. 2001; Morris et al. 2006). This clade has been reported as potentially significant, with populations comprising 5–10 percent of coastal seawater picoplankton worldwide.

MARINE *ARCHAEA*

Probably the most unexpected discovery in the study of marine bacterial community structure involves organisms that are not strictly *Bacteria*. Carl Woese and his colleagues' early foray into molecular phylogeny based upon ribosomal RNA sequences found many organisms that had been thought to be bacteria actually are as distant evolutionarily from the bacteria as are the eukaryotes. These cultured organisms consisted of methanogens (strict anaerobes) and certain extreme thermophiles and halophiles, which clustered together broadly in a phylogenetic tree (Fig. 3.1). Woese originally named the group *Archaebacteria* (Woese and Fox 1977), and later renamed them *Archaea* (Woese et al. 1990) to make it clearer that these are not just another type of bacteria, but something wholly different. Thus, life has been split into three domains: the *Archaea*, the *Bacteria*, and *Eukarya*, roughly equally distant evolutionarily from each other. Initially, all organisms in the *Archaea* were thought to be "extremophiles," based upon evidence from cultured organisms. However, the first study to apply "universal" PCR primers to the cloning and sequencing of rRNA genes from seawater found that a majority of random clones from 500 m depth, and some others from 100 m depth, belonged to the *Archaea*, not the *Bacteria* (Fuhrman et al. 1992). Those authors concluded from the clone results and other DNA hybridization data that archaea might be quite abundant in the deep sea, which later was shown to be true. Many microbiologists were skeptical of this report, because the closest cultured relatives (although not really close at all; Fig. 3.1) to these initially reported marine archaea were extreme thermophiles (in the division *Crenarchaeota*), but the organisms were living at refrigerator-like temperatures in aerobic water of normal ocean salinity. Later the same year, DeLong (1992) reported in another cloning and sequencing study that near-surface coastal waters contained a very small percentage (<2 percent) of archaea in two groups, one being the same as that reported by Fuhrman et al. (1992), and the other in the *Euryarchaeota* (the second major archaeal division). DeLong (1992) used PCR primers that target archaea only; hence his study could find archaea even if very rare. Note that archaea are often thought to be part of the "bacterial" community, and are in fact indistinguishable from bacteria by methods such as standard epifluorescence microscopy.

Early information on the abundance of marine archaea suggested that *Crenarchaeota* are particularly abundant in the deep sea; for example, they make up about one-third of random 16S rRNA clone libraries at 1000–3000 m depths (Fuhrman and Davis 1997) and 20–30 percent of rRNA below 100 m depth, as estimated by probe hybridization (Massana et al. 1997). FISH has confirmed these patterns, showing that *Crenarchaeota* are typically only a few percentages of the prokaryotes in the euphotic zone, but average about 20–40 percent at depths below a few hundred meters (Fuhrman and Ouverney 1998; DeLong et al. 1999; Karner et al. 2001; Teira et al. 2004; Herndl et al. 2005). The distribution of bacteria and archaea in the North Atlantic is shown in Figure 3.7. *Euryarchaeota* are typically only 5–20 percent in deep or surface waters most of the time, yet relatively more abundant than *Crenarchaeota* near the surface, where on rare occasions they may exceed 30 percent of the total cell counts (Fig. 3.7). There are also reports of relatively high abundance of archaea part of the year in Antarctic surface waters (Massana et al. 1998), which may be in part due to deep mixing bringing them to the surface.

Information on the metabolism of various members of the marine archaea is fragmentary. Investigations by an approach that combines FISH and microautoradiography (STARFISH, also abbreviated as MAR-FISH and Micro-FISH) indicated that over half of the archaea take up free amino acids from seawater (Ouverney and Fuhrman 2000; Teira et al. 2004, 2006a), indicating heterotrophic capability. (See Chapter 5 for more on microautoradiography.) However, reports based upon isotope composition of archaeal lipids indicated an autotrophic (i.e., inorganic) source of carbon (Pearson et al. 2001), and were confirmed by an elegant uptake experiment with [^{13}C]bicarbonate (Wuchter et al. 2003). The autotrophic mechanism was unclear, but, using a metagenomic approach (see Chapter 4), Venter et al. (2004) and Schleper et al. (2005) found genes for ammonia oxidation in environmental *Crenarchaeota*; in particular, crenarchaeal genes were linked to a gene for ammonia oxidation (ammonium monooxygenase or *amoA*). Konneke et al. (2005) reported the first isolation, from a marine aquarium, of a member of the marine *Crenarchaeota*, which turned out to be an autotroph that could use aerobic oxidation of ammonia to nitrite as its sole source of energy. Francis et al. (2005) reported that archaeal *amoA* genes are remarkably widespread and diverse in a variety of marine environments

Therefore current evidence suggests that the marine *Crenarchaeota* possess both heterotrophic and autotrophic capabilities, and it is not known if these capabilities are routinely present in the same individuals. Ingalls et al. (2006) investigated natural ^{14}C abundance in crenarchaeal lipids collected near Hawaii at 670 m depth, and concluded that about 83 percent of the lipid was derived from autotrophic metabolism (i.e., derived from local inorganic carbon) and about 17 percent from heterotrophic metabolism (i.e., from surface-derived organic matter); it is not known to what extent this applies to other depths and locations. Furthermore, Wuchter et al. (2006) reported that archaeal abundance, and not that of bacteria, correlated well with ammonium oxidation in mesocosms, and that at depths to 1000 m

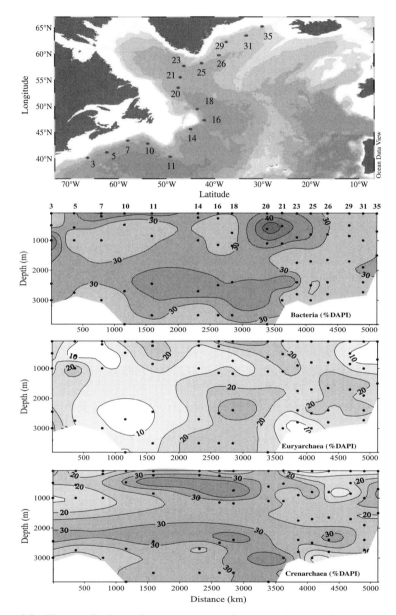

Figure 3.7 The contributions of *Bacteria*, *Euryarchaeota*, and *Crenarchaeota* to the total prokaryotic abundance, as measured by the percentage of total DAPI-stainable cells that are detectable by FISH with group-specific probes, determined on a transect in the North Atlantic Ocean shown in the top panel. Note that below several hundred meters, the *Archaea* typically exceed the *Bacteria* in total abundance. From Teira et al. (2006b).

the abundance of archaeal *amoA* is much higher than that of bacterial *amoA* genes. Hallam et al. (2006) reported that the genome of *Cenarchaeum symbiosum*, a crenarchaeal symbiont present in certain sponges, also suggest autotrophic metabolism and ammonia-oxidizing capabilities. These results all point to the likely significance of *Crenarchaeota* in the biogeochemically critical process of ammonia oxidation, and the data suggest that archaea might be more important than bacteria in this process. But the data so far only suggest that the archaea perform this first step in the nitrification process, with bacteria completing the process by oxidizing nitrite (NO_2^-) to nitrate (NO_3^-).

As for the metabolic capabilities of *Euryarchaeota*, little is known in general. However, Frigaard et al. (2006) found that individuals in the euphotic zone can possess a gene for proteorhodopsin, which might be used to capture light energy for metabolic functions (see Chapter 5). The presence of such similar genes in *Bacteria* and *Euryarchaeota* suggest "lateral" or "horizontal" gene transfer, that is, transfer between different kinds of organisms.

BACTERIOPLANKTON DIVERSITY

Species Concept

For most eukaryotic organisms, including plants, animal, fungi and some protists, a species is defined as an interbreeding population. But bacteria don't "breed"—they do not have "sex," at least not as their main mechanism to exchange DNA. Instead they acquire DNA fragments for recombination in a variety of ways, including genes from taxonomically distant bacteria, that is, horizontal gene transfer. Thus the traditional biological species concept serves poorly in relation to bacterial speciation. Empirically, a definition for bacterial species has been based on DNA–DNA cross-hybridization compared with 16S rDNA sequences for bacterial species showing similar phenotypes. Accordingly, organisms with 97 percent or less similarity in the 16S rRNA gene or less than 70 percent DNA–DNA hybridization have been proposed to belong to different species (Stackebrandt and Göbel 1994).

Today, this delineation threshold has matured to the point that it is described in major microbiological textbooks and has found use in medical and environmental applications (Madigan and Martinko 2004). It is commonly (mis)interpreted as suggesting its converse, namely, that organisms with more than 97 percent sequence similarity are necessarily in the same species. However, this definition lacks theoretical foundation and, more importantly, it does not seem to adequately describe ecologically relevant units of the microbial community (Rossello-Mora and Amann 2001; Keswani and Whitman 2001). For instance, with a 97 percent or greater similarity threshold, several subclades of phenotypically distinct cyanobacteria form a single species (Carr and Mann 1994; West et al. 2001; Casamayor et al. 2002; Rodriguez et al. 2005). Some organisms with more than 97 percent rRNA similarity have surprisingly unrelated genomes and plasmid-encoded properties. Recently, multilocus sequence typing (MLST) has been proposed as a more robust method to identify boundaries between bacterial species (Gevers et al. 2005). However, neither the DNA–DNA hybridization used to "calibrate" the 16S rRNA method nor the MLST approach are applicable to environmental studies where the communities contain a large fraction of bacteria that cannot be isolated on rich media.

As an alternative, Cohan (2001, 2002) has proposed the concept of "ecotype" defined as populations of cells that show genetic cohesion and are ecologically distinct. In Cohan's model, neutral mutations accumulate in the genomes of the close relatives of an ancestral type until a periodic selection event (also called "selective sweep") occurs. Selective sweeps occur when an adaptive mutant develops a significantly higher fitness than the other cells of the population and "purges" them out from the environment, insuring genetic cohesion. This concept is appealing in phylogenetic studies because it predicts that ecotypes are revealed in phylogenetic trees made from environmental gene clones as distinct clusters of highly related sequences. Such clusters have been reported for several members of marine bacterial communities, such as *Prochlorococcus* spp. (Moore et al. 1998; Rocap et al. 2002) or *Alteromonas macleodii* (Lopez-Lopez et al. 2005). This theory has now been applied to a broad-scale study of bacterioplankton diversity in the marine environment (Acinas et al. 2004), and seems to be consistent with the recurrent occurrence of closely related sequences or clusters in environmental clone libraries (Brown and Fuhrman 2005; Fuhrman et al. 2006).

Nevertheless, virtually all of the current information on bacterial distribution is based on the occurrence of ribotypes (i.e., single 16S rRNA gene sequences). or clones that are regarded as proxies for specific taxonomic units. Thus, while the legitimacy of the ribotype concept can be questioned, we have elected to use this information to elucidate patterns of distribution as long as no other means are available for defining an operational taxonomic unit (OTU). Therefore, an important disclaimer for the following discussion is to underline the current uncertainty regarding the delineation of bacterial species.

Microdiversity

The clusters of close relatives mentioned above exhibit what has been termed "microdiversity" (Garcia-Martinez and Rodriguez-Valera 2000). An example can be seen in Figure 3.5, which illustrates the fine-scale diversity of several members of the SAR11 cluster, and similarly in Figure 3.3 for several *Prochlorococcus*, both based on sequences of the highly variable spacer region between 16S and 23S rRNA genes (the ITS region). Because we do not have a great deal of information on all these organisms, it is not clear what portion of this microdiversity relates to functional differences and what portion may be neutral mutations that are unrelated to function. One recent study of microdiversity of *Vibrio splendidus* suggested that numerous fine-scale differences in genotypes of individuals are largely neutral (Thompson et al. 2005). It is tempting to think that each end cluster is a coherent species with shared functionality (true ecotype), but it is not certain if this is fully correct. We know most about the cyanobacteria, because many members have been cultivated and various adaptations to light and nutrients known; cultures have also facilitated whole-genome analysis. Recent genomic evidence suggests that in *Prochlorococcus*, significant functionally important differences among very close relatives (more than 99 percent sequence identity of 16S rRNA) tend to occur in "genomic islands" that possess genes coding for particular adaptations such as specific nutrient utilization (Coleman et al. 2006). Nevertheless, these differences are between adjacent end clusters (e.g., MED4 and MIT 9312 in the two different high-light clades), and at this time, we do not know the extent of functional variation between members of the same end cluster. We do know that some highly host-specific viruses infect certain members of these terminal clusters and not others (Sullivan et al. 2003), so at least there are differences with respect to this important part of the organisms' ecological niches.

Some observed microdiversity has been seen to correspond to depth and biogeographic distributions, and two of the most common marine bacterial clades provide good examples. The cyanobacterium *Prochlorococcus* occurs in different subclades that are very closely related to each other (Fig. 3.3), and their distribution has been measured over a mid-Atlantic transect (from off Ireland to near the Falkland Islands) by quantitative PCR methods. There are clear patterns of changing relative abundance with depth and latitude (Fig. 3.4). Another example is the SAR11 clade (Fig. 3.5). A few different patterns are evident. First, as with *Prochlorococcus*, some SAR11 clusters are widespread, occurring in distant locations and over a wide depth range. More commonly, clusters tend to be associated primarily either with the euphotic zone or with deep waters. There are also clusters that occur predominantly in similar environments far apart, such as the polar clade found in the surface waters of the Antarctic and near Greenland, or the surface clade found in the Mediterranean or off California. Interestingly, the cultivated SAR11 occurs in the predominantly polar clade, even though isolated off the Oregon coast. One SAR11 subcluster so far has been found in only one location, off Southern California, but it is important to remember that the global sampling is so sparse that this group could well occur elsewhere.

Components of Diversity: Richness and Evenness

The view that bacterioplankton diversity in the sea may be immense is probably the foremost image at present (Giovannoni et al. 1990). Certainly the plentiful input of 16S rRNA gene sequences over the last 15 years has lent support to this image. However, as Giovannoni and Rappé (2000) emphasized, already 10 years after the first marine bacterioplankton clone library was reported, a relatively small number of major marine bacterioplankton clades account for 80 percent of all marine bacterial ribotypes recovered from the sea. This seeming inconsistency can be explained. Since marine bacterial abundance is about 10^9 cells/L, one cannot hope to directly detect the complete diversity of any sample. It should be noted that the clades identified by Giovannoni and Rappé are very broad, and each could be as diverse as the insects. On the other hand, ribotypes represented by unique sequences can be detected recurrently between years at specific localities, as described later in the chapter. This would indicate that while the species richness may be huge, the number of specific niches may be limited in a particular environment.

The recorded richness will be limited by the sensitivity of the assay used, as is true of all methods in species-rich systems. The observed richness using Automated rRNA Intergenic Spacer Analysis (ARISA—see the text box on DNA fingerprinting methods) in a study of marine coastal communities ranged from 34 to 129 OTUs per sample (Fuhrman et al. 2006). While such fingerprinting techniques represent today's state of the art in rapid whole-community analysis, the outcome depends on the sensitivity, and the result certainly does not represent the total species richness in these samples, but rather a standardized measure of richness; ARISA generally detects OTUs that each represent more than about 0.1 percent of the total, or about $10^6/L$ (Hewson and Fuhrman 2004). In a similar fashion, the clone libraries of 16S rRNA genes only show the "dominant" species even if a thousand clones are

---

DNA Fingerprinting Methods

Fingerprinting methods give a snapshot of the entire microbial community at once, with the ability to tentatively identify different components. The most commonly used versions include terminal restriction fragment polymorphism (T-RFLP) (Avaniss-Aghajani et al. 1994; Liu et al. 1997; Osborn et al. 2000) and amplified ribosomal intergenic spacer analysis (ARISA) (Fisher and Triplett 1999; Brown et al. 2005). In ARISA, variations in the spacer length between the 16S and 23S rRNA genes yield the different-sized products. Typically, several hundred different taxa can theoretically be distinguished. In these methods, it is possible for multiple taxa to have the same length of detected product, so identification of a particular peak on the basis of database information is not definitive. Nevertheless, clone libraries from the environment in question can be used to find the most likely identification of peaks in the fingerprints (Gonzalez et al. 2000; Brown et al. 2005).

Another type of fingerprint analysis is denaturing gradient gel electrophoresis (DGGE), which separates similar-length nucleic acid molecules, such as PCR products, on the basis of small differences in the sequences (Muyzer et al. 1993; Muyzer and Smalla 1998). Typically, extracted DNA from an environmental sample is amplified with bacterial primers, and the products separated by DGGE. The dominant organisms are represented by different bands in such an analysis. It is also possible to use probes to characterize individual bands, or the bands may be excised, cloned, and sequenced for phylogenetic analysis.

Fingerprints might be chosen in preference to cloning methods if one wishes to analyze and compare composition of many samples, because fingerprinting detects even relatively rare organisms at low cost, yet also is highly amenable to statistical comparison (especially ARISA and TRFLP). But cloning has higher phylogenetic resolution, and large clone libraries (thousands of clones) can show organisms too rare to appear in fingerprints.

analyzed. Even with massive metagenomic studies (see Chapter 4), relatively few 16S rRNA or other phylogenetically comparable genes are reported; with about 1.6×10^9 bases sequenced from the Sargasso Sea, Venter et al. (2004) reported 643 16S rRNA sequences that were different at the over 99 percent similarity level, out of about 1400 total 16S rRNA sequences.

Bacterioplankton biodiversity consists of two components: richness, or taxonomic diversity, and evenness, defined as the distribution of individuals among taxa. Because richness is a parameter that describes an extreme (the maximum number of taxa), it is theoretically unknowable on the basis of finite samples. To find a single bacterium based on a 16S rDNA sequence identity directly in a liter of water, at least 5×10^9 clones would have to be sequenced, a task that appears impossible today even with the highest-capacity equipment. Richness can only be estimated with models that are still highly uncertain. One such model, based on a lognormal species abundance curve (Curtis et al. 2002), estimated 160 species per milliliter of seawater and >2 million species in the world ocean. The other component of biodiversity, evenness, describes the relative abundances or proportions of taxa occurring in a community. Even communities tend to have taxa at similar abundances while uneven ones have some taxa much more abundant than others. Because random sampling yields proportions that are unbiased, evenness of the more abundant taxa can be estimated from small samples with reasonable precision.

COMMUNITY STRUCTURE: DESCRIPTION AND FACTORS

A prominent idea among microbiologists is that because free-living marine planktonic microorganisms are very small and have large population sizes, they should be cosmopolitan, endemism should be rare, and consequently their global diversity should be low (Fenchel and Finlay 2004). The famous statement of Baas-Becking 1934 that "everything is everywhere, but the environment selects" has led to the notion that biogeographical patterns of marine bacterioplankton would be fuzzy, weak or absent. However, despite admittedly very wide dispersal of many microbes, considerable recent evidence disputes the suggestion that all microbes are universally distributed (Martiny et al. 2006). Typically, 16S rRNA gene clone libraries of bacterioplankton have few dominant species, which could indicate competition for a narrow range of readily used substrates. Also common is a large number of infrequent species, that is, a long tail on the species rank correlation curve (Pedrós-Alió 2006). An example of such community composition curves is presented by Pommier et al. (2007) from 16S rRNA gene libraries collected at nine distant localities spread around the world (Fig. 3.8). The species rank correlation curves from these localities all show a strong left-skewed distribution that deviates from the lognormal. This species distribution could result from a strong niche assembly through competition for common resources or from high turnover of rare species in the tail of the distribution (Wilson and Lundberg 2004).

Bottom-Up Control

The size of a bacterial population (i.e., members of a particular species—bearing in mind the problems with defining species) is controlled by the balance of growth and death rates. If growth exceeds death, then the population increases, and if death exceeds growth, then it decreases. We often think of bacterial populations as being in a sort of quasisteady state, where growth is approximately equal to death in the long term and communities have a sort of long-term stability, but we do not know if this really holds true. However, it follows that if it does not hold true, then, given the observed near-steady state in bacterial abundance, new species would have to emerge frequently, balanced by extinctions and the disappearance of other species.

"Bottom-up" effects on the bacterial community have been inferred by experimental manipulations, and a few recent examples will be given. Bacterioplankton community

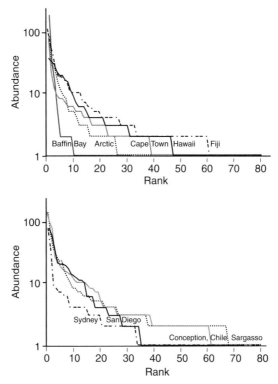

Figure 3.8 Rank-abundance distributions of bacterioplankton "species" at nine localities, as evaluated by 16S rRNA clone libraries with a "same species" similarity cutoff of greater then 98 percent sequence identity. All were strongly left-skewed, suggesting that the local bacterioplankton communities were open communities strongly influenced by high immigration rates and rapid turnover and/or highly competitive such that most species share a common set of resource and have very narrow niches. Modified from Pommier et al. (2007).

dynamics as a result of inorganic nutrient additions were followed in mesocosm experiment experiments with seawater from the northwest Mediterranean Sea. Using DGGE of PCR-amplified 16S rDNA, a change in the community structure compared with a control was observed within a few days of nutrient manipulations (Schafer et al. 2001). However, the control also changed considerably compared with the in situ community during the experimental period, presumably due to containment and the initial removal of the size fraction larger then 200 μm, thereby removing part of the predatory community. Such containment artifacts often complicate the interpretation of field experiments with marine microbes.

In culture experiments based on seawater from the northwestern Sargasso Sea, the combined addition of glucose and inorganic nutrients markedly changed the bacterial community measured as changing peak patterns by length heterogeneity PCR (LH-PCR) (Carlson et al. 2002). Changes were observed within 2 days of experimental initiation and corresponded to a response of increased production.

By adding protein to a mesocosm based on Southern California Bight water, species specific responses of culturable bacteria have also been observed by use of the DNA–DNA hybridization technique (Pinhassi et al. 1999). Four out of the five species with the most pronounced response to the protein addition belonged to the *Bacteroidetes* group. The increase in specific bacteria coincided with a massive increase in protease activity, suggesting a direct link between dominant bacteria and favorable traits in the manipulated environment. The link between protein uptake and *Bacteroidetes* is consistent with observations of Cottrell and Kirchman (2000b) who used a microautoradiography–FISH method to show that this group is particularly important in protein utilization.

"Bottom-up" control might be expected to lead to competition between bacteria for the same limiting resources. Such competition between bacteria is classically studied in controlled laboratory conditions, and difficult to isolate from other effects in the field (Korona et al. 1994). It is clear that many marine bacteria compete for specific limiting nutrients, (e.g., P, N, and Fe), and the relatively limited variety of readily available forms of these elements makes competition inevitable. Sometimes this is mediated with molecules such as siderophores (Weaver et al. 2003).

Chapter 10 discusses other aspects of bottom-up control of microbial communities in the oceans.

Sideways Control

We talk of both "bottom-up" and "top-down" controls on populations, where "bottom-up" refers to resources and "top-down" refers to predators (including parasites and viruses). For consistency in the metaphor, one could also think of "sideways" control where bacterial populations are influenced by competition for resource, negative interactions such as allelopathy (where an organism releases chemicals to attack others), or syntrophy (cross-feeding) (Fig. 3.9). With marine bacteria, there is relatively modest specific information on the relative importance of the different possible control mechanisms on community structure. We do know that all the potential mechanisms operate to some extent.

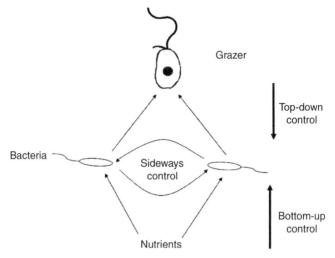

Figure 3.9 Schematic diagram of controls on bacterial community structure, including sideways control between different bacteria that could be negative (allelopathy) or positive (e.g., syntrophy or cross-feeding).

A special case of sideways control is that of bacteria preying on other cells, whether prokaryotic or eukaryotic. Direct bacterial predation is known to occur. *Bdellovibrio* and similar organisms invade their bacterial prey and digest them in order to grow and reproduce by multiple fission; this is the best studied example of the variety of bacteria known to prey on other bacteria by numerous mechanisms (Yair et al. 2003). Also, allelopathy, or antagonistic chemical interactions, between marine bacteria is relatively little studied, but has been shown to occur and be potentially important in structuring bacterial community composition (Long and Azam 2001a; Long et al. 2003). The potential for bacterioplankton to attack zooplankton and fish was touched upon in relation to the vibrios, but in addition bacteria have been observed frequently to prey on algae (Mayali and Azam 2004). This suggests that the presence of a suitable prey could determine the succession of important components of the bacterioplankton community. Finally, bacteria may also interact positively with each other (between species) by processes like syntrophy or cross-feeding of vitamins or other nutrients. Comparative genomics based on the whole-genome sequencing of a large number of bacterioplankton ribotypes that currently is under way (see Chapter 4) will allow possibilities for more thorough investigations of this subject.

Top-Down Control

Grazing can affect community composition, apparently due to selectivity in grazing organisms that preferentially remove particular taxa and less so others (Šimek et al. 1999; Jürgens and Matz 2002; Hahn and Höfle 2001; Jardillier et al. 2005). These effects could be due to factors such as prey size, where bigger bacteria tend to be grazed more efficiently in part because the grazers are more likely to encounter larger prey (Fenchel 1986), or where filter-feeding grazers have an effective minimum prey size that includes some bacteria but not others. Selective effects could also be from active processes, such as tactile or chemical-based acceptance or rejection of certain prey types. Also, some bacteria may be indigestible and survive passage through a grazer. It has even been suggested that some bacteria "prefer" to be ingested by an animal because their ideal habitat is the animal gut—this has been posited as a possible explanation for light production in some *Vibrio* species, where the light attracts fish to eat bacteria-laden particles.

There are relatively few comparisons of top-down and bottom-up controls. One continuous culture study with lake water noted some pronounced effects of different limiting nutrients and also different flagellate grazing pressure. There were also interactions between nutrient and grazing controls, but interestingly there were only moderate overall genotypic changes in the bacteria induced by nutrients and separately by protist grazing. This is comparable to the modest effects on composition sometimes seen for viruses, such as by Matz and Jürgens (2003) and Schwalbach et al. (2004). Clearly this is an area where considerably more research is needed. Chapter 11 also discusses grazing and top-down control of bacterial communities.

"Kill the Winner" Hypothesis

Viruses (Chapter 12) have interesting properties that are thought to give them particular influence on bacterial community structure. Because virus infection is host-density-dependent and also thought to be largely species-specific, viruses might have the effect of preferentially infecting organisms that become abundant (through competitive exclusion), because abundant organisms are more susceptible to an epidemic infection, while rare organisms are not. This idea was first proposed when it became apparent that viruses are important potential killers of bacteria (Bratbak et al. 1992; Fuhrman and Suttle 1993), and has been more recently mathematically modeled and termed the "kill the winner" hypothesis (Thingstad and Lignell 1997; Thingstad 2000). It has even spawned secondary hypotheses about cascading effects, relating to "killing the killer of the winner" (Miki and Yamamura 2005).

Such viral effects on community composition are consistent with laboratory experiments (Levin et al. 1977) and some field data (Hennes et al. 1995). However, evidence to support the hypotheses is relatively sparse, and some experiments suggest that viruses may not consistently have a dramatic short-term effect on overall community composition, but they do have distinct effects on particular taxa (Schwalbach et al. 2004; Winter et al. 2004; Hewson and Fuhrman 2006). Also, bacterial taxonomic richness has been reported to be negatively correlated with virus abundance, which is not consistent with expectation from the "kill the winner" hypothesis" (Winter et al. 2005). On the other hand, covariance has been observed between bacterial and viral community structure (Hewson et al. 2006c). It needs to be realized that the effects in nature are expected to occur over long

TEMPORAL VARIATION (DAYS TO SEASONAL)

Short-Term Variation

Under steady-state conditions often implicit in many simple conceptual models, bacterial communities would not be expected to change. While there may indeed be some environments that approximate a steady state for some time period, in reality the abiotic and biotic marine environments often change, sometimes abruptly (e.g., by a change in weather). On what time scales might we expect bacterial communities to change? With euphotic zone average bacterial doubling times typically on the order of a few days, faster in eutrophic areas, slower in oligotrophic (Fuhrman and Azam 1982; Ducklow 2000), it is reasonable to expect that the community composition could change on this time scale or longer. This is basically what has been observed.

In an early test of short-time-scale variations, Lee and Fuhrman (1991) found little change in bacterial composition of an enclosed 20 L plankton sample over 48 hours as measured by DNA–DNA hybridization. Similarly, Acinas et al. (1997) found little change in field samples from near-surface and deep chlorophyll maximum depths from the northwestern Mediterranean over 48 hours by community fingerprinting. More recent studies have sometimes found subtle changes over days; Hewson et al. (2006b) reported that drifter studies showed 10–20 percent change in community composition per day in subtropical oligotrophic waters (by ARISA, as evaluated by presence–absence of taxa or relative proportions). However, Riemann and Middelboe (2002) found remarkable stability over 2 months in bacterial communities of coastal Denmark as measured by DGGE. Differences in the extent of variation could be due to methodological sensitivities, but probably reflect differences in different environments. Some environments are no doubt more stable than others.

Seasonal Variation

If bacteria are adapted to recurring physical and metabolic conditions in the environment, we would expect temporal succession of bacterioplankton. In a "premolecular" study of seasonal variation, Waterbury and Valois (1993) focused on just one type of bacterium, the cyanobacterium *Synechococcus*, as it varied over the year in coastal waters of Woods Hole, Massachusetts. They found a very strong summer-high, winter-low pattern, but with considerable month-to-month variation. More recently, studies on seasonal variation have been conducted using a diverse set of molecular methods that either detect changes in the overall community structure or quantify and follow changes of specific bacterioplankton taxa. From these studies, some general patterns on marine bacterial community structure seem deducible: the marine bacterioplankton community changes over the seasonal cycle, but the rate of change is relative slow with time scales of weeks to months. Another conclusion is that bacterioplankton communities reoccur in annual patterns. To illustrate these conclusions on seasonal population dynamics, we have chosen studies using four different methods. It is worthwhile to observe that in these studies it has been taken for granted that the taxonomic markers would be stable over long periods of time.

Using hybridization of specific oligonucleotide 16S rDNA probes, it can be shown that single specific bacterioplankton species reoccur over time. The presence of *Vibrio angularium* isolated in December was detected in December during two consecutive years, but not in other months, in coastal waters of the Southern California Bight (Rehnstam et al. 1993). Another example of seasonal variation was observed in the assemblage of uncultured archaea in coastal Antarctic waters (Murray et al. 1998). Highest abundance, as quantified by rRNA oligonucleotide hybridization, was observed during austral winter, while archaeal rRNA almost disappeared during summer and autumn. Seasonal succession was followed in the Baltic Sea using whole DNA–DNA hybridization of specific bacterioplankton isolates and environmental DNA (Pinhassi and Hagström 2000). The dominant bacteria persisted for weeks to months, and a clear seasonal pattern could be deduced, with a dominant spring and a summer community clearly separated in time while a group of bacteria in lower number existed throughout the growth season. The *Cytophaga–Flavobacteria–Bacteroides* group dominated during spring, when nutrient concentrations are high, while during summer, with stratified, nutrient depleted surface water, *Alphaproteobacteria* dominated the community.

A gradual change of the *Alphaproteobacteria* and *Bacteroidetes* assemblage was also detected by analysis of DGGE fingerprints in a seasonal cycle study in Blanes Bay, Mediterranean Sea (Schauer et al. 2003). The changing intensity of the bands revealed that single populations maintained high concentrations on a weekly to monthly time scale. While the contribution of major phylogenetic groups remained relatively constant over time, distinct phylotypes within the groups replaced each other between seasons. The structure of the bacterial community in the northwestern Mediterranean has also been examined on a seasonal scale by use of capillary electrophoresis–single-strand conformation polymorphism (CE-SSCP), which gives a community fingerprinting based on PCR-amplified 16S rDNA (Ghiglione et al. 2005), similar to DGGE.

Showing a truly "seasonal" pattern, however, needs data from several years at the same site. Such studies are rare, but those available from marine environments indicate reoccurrence of communities on a seasonal scale. Multi-annual studies in the North Atlantic (the Bermuda Atlantic Times Series (BATS) site) have shown that specific kinds of bacteria reappear during certain seasons (Morris et al. 2005). For example, the relative abundance of SAR11 rRNA increased in surface waters from March to September during three consecutive years. The wide range of abiotic and biotic factors that control the composition of the community indicates a strong forcing by the environment and the need to study the entire active bacterioplankton community. A study using ARISA found that the bacterial community structure off the southern California coast varies in a repeatable and predictable temporal pattern over several

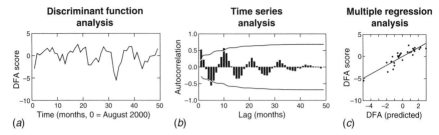

Figure 3.10 Demonstration of repeatable and predictable bacterial community composition at the San Pedro Ocean Time Series (California). Samples collected monthly from 5 m were studied by ARISA, and the relative abundance of individual OTU (taxa) were followed and analyzed by discriminant function analysis (DFA), autocorrelation, and multiple regression against environmental parameters (e.g., temperature, nutrients, chlorophyll, bacteria, and viruses). (*a*) DFA scores, a measure of community composition, as repeating seasonally. (*b*) Autcorrelation analysis shows that the community composition is significantly correlated from month to month (indicating relatively slow changes), but gets less similar to the point of being negatively correlated in the opposite season, and then correlated again in the same season the following year. (*c*) Prediction of community composition, as measured by DFA score, with the environmental parameters. DFA focuses on only a portion of the OTUs (a majority, but not all), and these showed the patterns described above (Fuhrman et al. 2006).

years, with the community predictable not only by time of year but also on the basis of environmental parameters such as temperature, and chlorophyll (Fig. 3.10) (Brown et al. 2005; Fuhrman et al. 2006). These three multi-annual studies all show that structure and dominance of specific species are maintained on a weekly to monthly scale. They also suggest most bacteria have well-defined niches.

One point not yet clear in the studies mentioned above is the extent of genetic variation over time within what appears to be a single bacterial species or OTU. It is quite possible that what appears to be a near-constant presence of an organism might actually consist of a succession of variants, such as what one might expect from the "kill the winner" hypothesis. In a recent report from the Global Ocean Sampling, a large metagenomics study, Rusch et al. (2007) demonstrated that there seems to exist "a considerable and in many cases conflicting variation among related organisms." This variation may in fact represent the succession of variants assumed in the "kill the winner" hypothesis. In this scenario, the environmental conditions might favor long-term presence of bacteria occupying a particular niche, but viruses may knock down the abundance of the dominant strain while resistant conspecifics might increase in abundance to fill the niche. This situation might repeat itself several times, and would be difficult to detect by current approaches to measuring diversity.

SPATIAL VARIATION
Microscale Patterns

As illustrated by Azam (1998), a drop of water consists of a network of organic matter and living organisms in a seemingly chaotic and complex relationship focused in "hot spots." In this environment, bacteria need various biogeochemical abilities and behaviors in order to deal with the diverse nutrient pools. Evidence to support this view was presented by DeLong et al. (1993) and Acinas et al. (1997, 1999), who found the phylogenetic diversity of the particle-attached bacteria assemblage to be substantially different from that of the free-living bacteria. This spatial distribution also results in microscale patches of bacterial activity based on single-cell activity, as illustrated by Seymour et al. (2004). The spatial scale was investigated by Long and Azam (2001b); they found a significant variation in the bacterial community composition at the microscale when they examined diversity in 1 μL water samples. But, interestingly, microscale variations essentially vanished when samples as large as liters were compared. Little variation was also found by Hewson et al. (2006b), where 20 L ocean samples were collected within about a kilometer and compared by ARISA. The fact that the many microscale variations tend to average out to very similar communities when millions of individual microliters are homogenized and analyzed in bulk is a nice illustration of how population statistics works. In the following, we turn to large-scale patterns, but it is important to remember that these are made up of microorganisms distributed on the microscale.

Global Distribution

As discussed above, the total bacterioplankton species richness cannot easily be observed. Even so, patterns of bacterioplankton assemblages and the degree of cosmopolitanism can be deduced from studies of bacterial ribotype abundance. This was demonstrated by Pommier et al. (2007) in a global sampling campaign. The results of their clone library-based study showed generally similar composition of major taxonomic groups and divisions at the local community level (Fig. 3.11), but this refers to broad groups that can be diverse as insects or angiosperms. Still, only two of the ribotypes showed a ubiquitous distribution, that is, they were represented at all localities. The first ribotype was an uncultured alphaproteobacterium previously found in the San Pedro Channel, California (Brown et al. 2005), Antarctica (Prabagaran et al. 2007), and the North Sea (Zubkov et al. 2003), which confirmed its ubiquitous distribution. The second was closely related to *Pelagibacter ubique*, a member of the alphaproteobacterial SAR11 clade (Morris et al. 2002).

Marine bacterioplankton 16S rRNA sequences stored in GenBank can also provide a view of the global distribution, since sample location can be found for many of the sequences (Pommier et al. 2005). Based on data from GenBank in 2004, analyzed liberally to consider identity based upon greater than 97 percent sequence identity, it can be seen that less then 10 percent of all entered bacterioplankton 16S rRNA sequences show a ubiquitous geographic distribution. In the alphaproteobacterial group, two members of the SAR11 cluster, *P. ubique*, and the uncultured alphaproteobacterium NHF49-1 and one member of the *Roseobacter* group (an unidentified isolate) were reported all over the globe. From the *Gammaproteobacteria*, one member of the SAR86 cluster, the uncultured gammaproteobacterial OM10 and four isolated strains (*Alteromonas macleodii*, *Pseudoalteromonas haloplanktis*, *Marinobacter aquaeoiei*), and one unidentified gammaproteobacterium SWAT9

76 BACTERIAL AND ARCHAEAL COMMUNITY STRUCTURE AND ITS PATTERNS

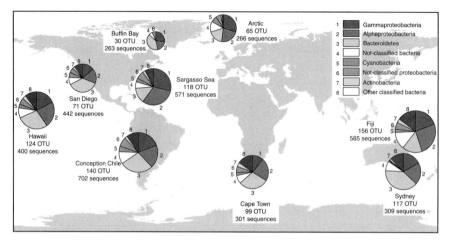

Figure 3.11 Distribution of major bacterioplankton groups recorded from 16S rRNA clone libraries sampled at 5 m depth from nine localities worldwide. Local numbers of distinct OTUs and numbers of sequenced clones are indicated besides major taxonomic group distribution pies (specified in the key at top right) for each locality. The size of the pie is proportional to the total number of OTUs sampled in each locality. Modified from Pommier et al. (2007).

were also found from the tropics to the poles. Moreover, one uncultured deltaproteobacterium, SAR 324, one strain of the cyanobacterium *Synechococcus*, and one isolated *Bacteroidetes* bacterium were also found to be distributed ubiquitously. Noticeably, no evidence for cosmopolitan distribution for any members of the *Firmicutes* group was found. These results can be compared to the representation of ubiquitous groups presented in Figure 1 of Giovannoni and Stingl (2005).

Remarkably, and despite the frequent sampling in the temperate regions, eight ribotypes were found to be confined to the polar regions of the globe, with no occurrence in the temperate or tropical areas. All sequences included in these ribotypes were obtained by culture-independent techniques (e.g., cloning or direct amplification of environmental samples). Three of them were *Gammaproteobacteria*, two *Betaproteobacteria*, one from the *Deltaproteobacteria*, one *Bacteroidetes*, and one unidentified marine bacterium showing 93 percent similarity to a *Roseobacter* species.

Latitudinal Gradient and Degree of Endemism

Most organisms show latitudinal gradients in species diversity (Magurran and Henderson 2003; Hillebrand 2004; Magurran 2004), and species richness generally increases from the poles towards the equator. This ubiquitous pattern lacks a general explanation. Productivity, environmental continuity, species–area relationships, and evolutionary history are some frequently suggested mechanisms (Hillebrand 2004), none of which would unambiguously predict marine bacterioplankton diversity gradients. Currently, we can find three datasets that could address a latitudinal gradient of marine bacterioplankton, showing higher species richness at low latitudes. In the first, a transect from the Bering Strait to Antarctic, Baldwin and colleagues reported that they did not find increasing numbers of OTU towards low latitudes (Baldwin et al. 2005). However, their TRFLP approach detected few taxa per sample, and a large uncertainty was noted due to the presence of eukaryotic cells in the sampling protocol. In contrast, from nine clone libraries collected worldwide, Pommier and colleagues demonstrated the presence of a latitudinal gradient in ribotype diversity (Pommier et al. 2007). Similarly, Fuhrman et al. (in press) demonstrated a marked latitudinal gradient in a large dataset (>100 samples) based on ARISA screening of the bacterioplankton community. These authors reported OTU richness is significantly correlated to temperature and not productivity, supporting the hypothesis that richness is influenced by the kinetics of metabolism.

The observed patterns of ribotype richness in local communities may result from several, not necessarily mutually exclusive, ecological and evolutionary processes. Bacterioplankton communities, with a large number of small individuals unbound by migration barriers, are open at small to intermediate scales, and immigrants can exploit niches found ubiquitously. At larger scales, wide-ranging oligotrophic regions separate local-resource-rich parts of the water (i.e., "hot spots"). Accordingly, there is room for local idiosyncrasies in niche packing and OTU composition. Furthermore as discussed above, the "kill the winner" hypothesis, that is, the frequency-dependent viral predation of temporally abundant genotypes (or species), has been suggested to explain local coexistence and high diversity by allowing growth of rare types due to competitive release (Thingstad 2000). High diversity of local endemics can also reflect the accumulation of neutral mutations within clusters of diverging ecotypes (Cohan 2001). Additionally, genome streamlining, which is the selective process of minimizing genome size by optimizing the functional genes/replication ratio, might favor a few, highly competitive types in a wide range of environments. This process, which has already been proposed for the two marine cosmopolitans *Synechoccocus* (Carr and Mann 1994) and *Pelagibacter ubique* (Giovannoni et al. 2005b), may be widespread.

Patchiness and Large Eddies

The large-scale patterns described above give an idea about the global distribution of various bacterial taxa in the sea, comparing sites that tend to be thousands of kilometers apart, or comparing depths in the stratified ocean. The question then arises about smaller scales, especially in the horizontal direction, where ocean mixing is most likely to occur. Is the bacterial community composition uniform on regional scales of hundreds of kilometers at a particular depth or isopycnal (depth of same density)? Do bacterial communities occur in coherent patches, and, if so, what is their scale? This problem has only begun to be addressed. Acinas et al. (1997) reported little community change over scales less than a few kilometers in the Mediterranean. Riemann and Middelboe (2002) reported changes measurable by DGGE in microbial composition over a transect crossing frontal waters between the Baltic and North Seas. Suzuki et al. (2001) used a quantitative PCR method to map bacterial composition changes in the vicinity of an upwelling plume off Monterey California, and reported clear patterns of variation on scales of several kilometers in this environment known for strong horizontal gradients.

Recent studies by Hewson and colleagues (Hewson et al. 2006a; b) used whole-community fingerprinting to compare bacterial communities in near-surface, and meso- and bathypelagic depths of the open sea on scales ranging to thousands of kilometers. They reported that in most surface waters, community similarity was very close for samples collected within about 2 km, but sampling locations tens of kilometers apart had relatively low similarity that was about the same as for locations hundreds of kilometers apart (Fig. 3.12). This might be interpreted as meaning that coherent community patches in the open sea have a scale of a few to <50 km, presumed to occur because, on this scale, mixing (e.g., by eddies) predominates over biological interactions that might change community composition (Hewson et al. 2006b). Interestingly, patch composition can vary to the same extent whether the patches are next to each other or hundreds to thousands of kilometers apart. An

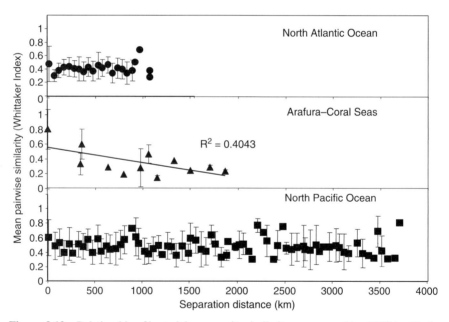

Figure 3.12 Relationship of bacterial community similarity as measured by ARISA with distance between samples. The North Atlantic samples are primarily from north and east of South America, the Arafura–Coral Sea from north and east of Australia, and the North Pacific samples from near Hawaii towards Japan. For each pair of samples, the relative abundance (proportions) of the same OTUs is compared, and the composite similarity totaling all OTUs is shown here, on a scale ranging from 0 (completely different communities) to 1 (identical communities). A decreasing similarity with distance is evident only for the Arafura Sea (Indian Ocean) to Coral Sea (Pacific Ocean) region, which includes a narrow and shallow strait between Australia and Papua–New Guinea; this result may represent a mixing curve between different end-member communities in the two seas. From Hewson et al. (2006b).

exception was for samples collected across the shallow and narrow Torres Strait between Australia and Papua New Guinea, which separates the Coral Sea in the Pacific from the Arafura Sea, a finger of the Indian Ocean. Those samples had decreasing similarity with increasing distance, as if they represented a mixing curve between distinct communities in the two oceans (Fig. 3.12); this apparent mixing pattern resembled that across the Baltic/North Sea front observed by Riemann and Middelboe (2002). Thus, as one might expect, patterns differ according to the situation.

While it might be expected that deep-sea bacteria, with much slower growth rates than those in the euphotic zone (with mean doubling times of weeks to months versus days), would have a more uniform horizontal distribution pattern over space, measurements suggest that they are patchy as well. Samples collected at 500–3000 m show patchiness that appears to move with the currents (Hewson et al. 2006a). Therefore, samples collected at the same geographic location only weeks apart can have a very different microbial community (presumably because patches are moving past the location), while samples hundreds of kilometers apart, sampled over a period of days, might have extremely similar communities. The likely source for this patchiness is variations in the rain of organic matter from the overlying surface waters, where mesoscale patches in features such as chlorophyll are clearly observed by satellite, and where bacterial communities also have patchy distribution as described in the previous paragraph.

SUMMARY

1. Marine planktonic bacteria and archaea are highly diverse.
2. The majority of marine prokaryotes can be found in about a dozen major phylogenetic divisions (phyla or kingdoms), which are generally found worldwide.
3. Major groups have subdivisions that often show distribution patterns such as stratification with depth or variation by region or habitat.
4. Microdiversity, or the presence of large numbers of closely related taxa, is extremely common, and there are several examples where close relatives have divergent phenotypes indicating niche specialization.
5. "Drawing the line" between close relatives to define prokaryotic species (or their equivalent) is one of the biggest challenges today, and it hampers but does not prevent investigation of diversity patterns.
6. Factors that control community composition include nutrient type and availability, physical and chemical conditions, predation, viruses, competition, symbiosis, and allelopathy. The relative importance of these mechanisms varies in time and space, but how is not currently known.
7. Limited data available to date indicate that time scales for variation of community composition in a given water mass are such that most communities change

relatively little in a day, somewhat in a week, and substantially over months in euphotic zone waters. Annually repeating patterns are also evident.

8. Spatial variation of community composition in the euphotic zone can be substantial on the microscale (micrometers scale), but tends to average out over centimeter and meter scales, and then increases again on scale of a few to tens of kilometers. The limited data from the deep sea also suggest substantial patchiness on scales of tens of kilometers.

REFERENCES

Acinas S. G., F. Rodriguez-Valera, and C. Pedrós-Alió. 1997. Spatial and temporal variation in marine bacterioplankton diversity as shown by RFLP fingerprinting of PCR amplified 16S rDNA. *FEMS Microbiol. Ecol.* 24: 27–40.

Acinas S. G., J. Anton, and F. Rodriguez-Valera. 1999. Diversity of free-living and attached bacteria in offshore western Mediterranean waters as depicted by analysis of genes encoding 16S rRNA. *Appl. Environ. Microbiol.* 65: 514–522.

Acinas, S. G., V. Klepac-Ceraj, D. E. Hunt, C. Pharino, I. Ceraj, D. L. Distel, and M. F. Polz. 2004. Fine-scale phylogenetic architecture of a complex bacterial community. *Nature* 430: 551–554.

Ammerman, J. W., J. A. Fuhrman, Å., Hagström, and F. Azam. 1984. Growth characteristics of marine bacteria growth in "seawater cultures" (unsupplemented particle-free seawater). *Mar. Ecol. Prog. Ser.* 18: 31–39.

Avaniss-Aghajani, E., K. Jones, D. Chapman, and C. Brunk. 1994. A molecular technique for identification of bacteria using small subunit ribosomal RNA sequences. *Biotechniques* 17: 144–149.

Azam, F. 1998. Microbial control of oceanic carbon flux: the plot thickens. *Science* 280: 694–696.

Baas-Becking, L. G. M. 1934. *Geobiologie of inleiding tot de milieukunde*. W.P. Van Stockum & Zoon.

Baldwin, A. J., J. A. Moss, J. D. Pakulski, P. Catala, F. Joux, and W. H. Jeffrey. 2005. Microbial diversity in a Pacific Ocean transect from the Arctic to Antarctic circles. *Aquat. Microb. Ecol* 102: 41–91.

Béjà, O., L. Aravind, E. V. Koonin, et al. 2000. Bacterial rhodopsin: Evidence for a new type of phototrophy in the sea. *Science* 289: 1902–1906.

Bratbak, G., M. Heldal, T. F. Thingstad, B. Riemann, and O. H. Haslund. 1992. Incorporation of viruses into the budget of microbial C-transfer. A first approach. *Mar. Ecol. Prog. Ser.* 83: 273–280.

Brown, M. V., and J. A. Fuhrman. 2005. Marine bacterial microdiversity as revealed by internal transcribed spacer analysis. *Aquat. Microb. Ecol.* 41: 15–23.

Brown, M. V., M. S. Schwalbach, I. Hewson, and J. A. Fuhrman. 2005. Coupling 16S-ITS rDNA clone libraries and ARISA to show marine microbial diversity: Development and application to a time series. *Environ. Microbiol.* 7: 1466–1479.

Buchan, A., J. M. Gonzalez, and M. A. Moran. 2005. Overview of the marine *Roseobacter* lineage. *Appl. Environ. Microbiol.* 71: 5665–5677.

REFERENCES

Button, D. K., F. Schuts, P. Quang, R. Martin, and B. R. Robertson. 1993. Viability and isolation of marine bacteria by dilution culture: Theory, procedures, and initial results. *Appl. Environ. Microbiol.* 59: 881–891.

Camilli, A., and B. L. Bassler. 2006. Bacterial small-molecule signaling pathways. *Science* 311: 1113–1116.

Capone, D. G., J. P. Zehr, H. W. Paerl, B. Bergman, and E. J. Carpenter. 1997. *Trichodesmium*, a globally significant marine cyanobacterium. *Science* 276: 1221–1229.

Capone, D. G., J. A. Burns, J. P. Montoya, A. Subramaniam, C. Mahaffey, T. Gunderson, A. F. Michaels, and E. J. Carpenter. 2005. Nitrogen fixation by *Trichodesmium* spp.: An important source of new nitrogen to the tropical and subtropical North Atlantic Ocean. *Global Biogeochem. Cycles* 19: GB2024, doi:10.1029/2004GB002331.

Carlson, C. A., S. J. Giovannoni, D. A. Hansell, S. J. Goldberg, R. Parsons, M. P. Otero, K. Vergin, and B. R. Wheeler. 2002. Effect of nutrient amendments on bacterioplankton production, community structure, and DOC utilization in the northwestern Sargasso Sea. *Aquat. Microb. Ecol.* 30: 19–36.

Carr, N. G., and N. H. Mann. 1994. The oceanic cyanobacterial picoplankton. In D. G. Bryant (ed.), *The Molecular Biology of Cyanobacteria*. Kluwer, pp. 27–48.

Casamayor E. O., C. Pedrós-Alió, G. Muyzer, and R. Amann. 2002. Microheterogeneity in 16S ribosomal DNA-defined bacterial populations from a stratified planktonic environment is related to temporal changes and to ecological adaptations. *Appl. Environ. Microbiol.* 68: 1706–1714.

Cavalier-Smith, T. 2002. The neomuran origin of archaebacteria, the negibacterial root of the universal tree and bacterial megaclassification. *Int. J. Syst. Evol. Microbiol.* 52: 7–76.

Certes, A. 1884. On the culture, free from known sources of contamination, from waters and from sediments brought back by the expeditions of the *Travailleur* and the *Talisman;* 1882–1883. *Seances Acad. Sci.* 98: 690–693.

Chisholm, S. W., R. J. Olson, E. R. Zettler, J. Waterbury, R. Goericke, and N. Welschmeyer. 1988. A novel free-living prochlorophyte abundant in the oceanic euphotic zone. *Nature* 334: 340–343.

Cho, J. C., and S. J. Giovannoni. 2004. Cultivation and growth characteristics of a diverse group of oligotrophic marine *Gammaproteobacteria*. *Appl. Environ. Microbiol.* 70: 432–440.

Ciccarelli, F. D., T. Doerks, C. von Mering, C. J. Creevey, B. Snel, and P. Bork. 2006. Toward automatic reconstruction of a highly resolved tree of life. *Science* 311: 1283–1287.

Cohan, F. M. 2001. Bacterial species and speciation. *Syst. Biol.* 50: 513–524.

Cohan, F. M. 2002. What are bacterial species? *Annu Rev Microbiol.* 56: 457–487.

Colwell, R. R. 1996. Global climate and infectious disease: The cholera paradigm. *Science* 274: 2025–2031.

Coleman, M. L., M. B. Sullivan, A. C. Martiny, C. Steglich, K. Barry, E. F. DeLong, and S. W. Chisholm. 2006. Genomic islands and the ecology and evolution of *Prochlorococcus*. *Science* 311: 1768–1770.

Connon, S. A., and S. J. Giovannoni. 2002. High-throughput methods for culturing microorganisms in very-low-nutrient media yield diverse new marine isolates. *Appl. Environ. Microbiol.* 68: 3878–3885.

Cottrell, M. T., and D. L. Kirchman. 2000a. Community composition of marine bacterioplankton determined by 16S rRNA gene clone libraries and fluorescence in situ hybridization. *Appl. Environ. Microbiol.* 66: 5116–5122.

Cottrell, M. T., and D. L. Kirchman. 2000b. Natural assemblages of marine proteobacteria and members of the *Cytophaga–Flavobacter* cluster consuming low- and high-molecular-weight dissolved organic matter. *Appl. Environ. Microbiol.* 66: 1692–1697.

Curtis, T. P., W. T. Sloan, and J. W. Scannell. 2002. Estimating prokaryotic diversity and its limits. *Proc. Natl. Acad. Sci. USA* 99: 10494–10499.

Delong, E. F. 1992. Archaea in coastal marine environments. *Proc. Nat. Acad. Sci. USA* 89: 5685–5689.

DeLong, E. F., D. G. Franks, and A. A. Alldredge. 1993. Phylogenetic diversity of aggregate-attached vs. free-living marine bacterial assemblages. *Limnol. Oceanogr.* 38: 924–934.

DeLong, E. F., L. T. Taylor, T. L. Marsh, and C. M. Preston. 1999. Visualization and enumeration of marine planktonic archaea and bacteria by using polyribonucleotide probes and fluorescent in situ hybridization. *Appl. Environ. Microbiol.* 65: 5554–5563.

Ducklow, H. W. 2000. Bacterial production and biomass in the oceans. In D. L. Kirchman (ed.), *Microbial Ecology of the Oceans*, 1st edn. Wiley-Liss, pp. 85–120.

Dufresne, A., L. Garczarek, and F. Partensky. 2005. Accelerated evolution associated with genome reduction in a free-living prokaryote. *Genome Biol.* 6: doi: 10.1186/gb-2005-6-2-r14.

Dyhrman, S. T., P. D. Chappell, S. T. Haley, J. W. Moffett, E. D. Orchard, J. B. Waterbury, and E. A. Webb. 2006. Phosphonate utilization by the globally important marine diazotroph *Trichodesmium*. *Nature* 439: 68–71.

Fenchel, T. 1986. Protozoan filter feeding. *Prog. Protistol.* 1: 65–113.

Fenchel, T., and B. J. Finlay. 2004. The ubiquity of small species: Patterns of local and global diversity. *BioScience* 54: 777–784.

Ferguson, R. L., and P. Rublee. 1976. Contribution of bacteria to standing crop of coastal plankton. *Limnol. Oceanogr.* 21: 141–145.

Ferguson, R. L., E. N. Buckley, and A. V. Palumbo. 1984. Response of marine bacterioplankton to differential filtration and confinement. *Appl. Environ. Microbiol.* 47: 49–55.

Field, K. G., D. Gordon, T. Wright, M. Rappé, E. Urbach, K. Vergin, and S. J. Giovannoni. 1997. Diversity and depth-specific distribution of SAR11 cluster rRNA genes from marine planktonic bacteria. *Appl. Environ. Microbiol.* 63: 63–70.

Fisher, M. M., and E. W. Triplett. 1999. Automated approach for ribosomal intergenic spacer analysis of microbial diversity and its application to freshwater bacterial communities. *Appl. Environ. Microbiol.* 65: 4630–4636.

Francis, C. A., K. J. Roberts, J. M. Beman, A. E. Santoro, and B. B. Oakley. 2005. Ubiquity and diversity of ammonia-oxidizing archaea in water columns and sediments of the ocean. *Proc. Natl. Acad. Sci. USA* 102: 14683–14688.

Frankland, P., and P. Frankland. 1894. *Micro-Organisms in Water; Their Significance, Identification and Removal*. Longmans Green.

Frigaard, N. U., A. Martinez, T. J. Mincer, and E. F. DeLong. 2006. Proteorhodopsin lateral gene transfer between marine planktonic Bacteria and Archaea. *Nature* 439: 847–850.

REFERENCES

Fuhrman, J. A., and F. Azam. 1982. Thymidine incorporation as a measure of heterotrophic bacterioplankton production in marine surface waters: Evaluation and field results. *Mar. Biol.* 66: 109–120.

Fuhrman, J. A., and A. A. Davis. 1997. Widespread Archaea and novel Bacteria from the deep sea as shown by 16S rRNA gene sequences. *Mar. Ecol. Prog. Ser.* 150: 275–285.

Fuhrman, J. A., and C. C. Ouverney. 1998. Marine microbial diversity studied via 16S rRNA sequences: cloning results from coastal waters and counting of native archaea with fluorescent single cell probes. *Aquat. Ecol.* 32: 3–15.

Fuhrman, J. A., and C. A. Suttle. 1993. Viruses in marine planktonic systems. *Oceanography* 6: 51–63.

Fuhrman, J. A., K. McCallum, and A. A. Davis. 1992. Novel major archaebacterial group from marine plankton. *Nature* 356: 148–149.

Fuhrman, J. A., K. McCallum, and A. A. Davis. 1993. Phylogenetic diversity of subsurface marine microbial communities from the Atlantic and Pacific Oceans. *Appl. Environ. Microbiol.* 59: 1294–1302. [Erratum 1995; 61: 4517].

Fuhrman, J. A., S. H. Lee, Y. Masuchi, A. A. Davis, and R. M. Wilcox. 1994. Characterization of marine prokaryotic communities via DNA and RNA. *Microb. Ecol.* 28: 133–145.

Fuhrman, J. A., I. Hewson, M. S. Schwalbach, J. A. Steele, M. V. Brown, and S. Naeem. 2006. Annually reoccurring bacterial communities are predictable from ocean conditions. *Proc. Natl. Acad. Sci. USA* 103: 13104–13109.

Fuhrman, J. A., J. A. Steele, I. Hewson, M. S. Schwalbach, M. V. Brown, J. L. Green, and J. H. Brown. A latitudinal diversity gradient in planktonic marine bacteria. *Proc. Natl. Acad. USA*, in press.

García-Martínez, J., and F. Rodríguez-Valera. 2000. Microdiversity of uncultured marine prokaryotes: the SAR11 cluster and the marine Archaea of Group I. *Mol. Ecol.* 9: 935–948.

García-Martínez, J., S. G. Acinas, R. Massana, and F. Rodríguez-Valera. 2002. Prevalence and microdiversity of *Alteromonas macleodii*-like microorganisms in different oceanic regions. *Environ. Microbiol.* 4: 42–50.

Gevers, D., F. M. Cohan, J. G. Lawrence, B. G. Spratt, T. Coenye, E. J. Feil, E. Stackebrandt, Y. V. de Peer, P. Vandamme, F. L. Thompson, and J. Swings. 2005. Re-evaluating prokaryotic species. *Nat. Rev. Microbiol.* 3: 733–739.

Ghiglione, J.-F., M. Larcher, and P. Lebaron. 2005. Spatial and temporal scales of variation in bacterioplankton community structure in the NW Mediterranean Sea. *Aquat. Microb. Ecol.* 40: 229–240.

Giovannoni, S. J., and M. Rappé. 2000. Evolution, diversity, and molecular ecology of marine prokaryotes. In Kirchman D. L. (ed.), *Microbial Ecology of the Oceans*, 1st edn. Wiley-Liss, pp. 47–84.

Giovannoni, S. J., and U. Stingl. 2005. Molecular diversity and ecology of microbial plankton. *Nature* 437: 343–348.

Giovannoni, S. J., and T. B. Britschgi, C. L. Moyer, and K. G. Field. 1990. Genetic diversity in Sargasso Sea bacterioplankton. *Nature* 345: 60–63.

Giovannoni, S. J., M. S. Rappé, K. L. Vergin, and N. L. Adair. 1996. 16S rRNA genes reveal stratified open ocean bacterioplankton populations related to the green non-sulfur bacteria. *Proc. Nat. Acad. Sci. USA* 93: 7979–7984.

Giovannoni, S. J., L. Bibbs, J. C. Cho, et al. 2005a. Proteorhodopsin in the ubiquitous marine bacterium SAR11. *Nature* 438: 82–85.

Giovannoni, S. J., H. J. Tripp, S. Givan, et al. 2005b. Genome streamlining in a cosmopolitan oceanic bacterium. *Science* 309: 1242–1245.

Gomez-Consarnau, L., J. M. Gonzalez, M. Coll-Llado, P. Gourdon, T. Pascher, R. Neutze, C. Pedrós-Alió, and J. Pinhassi. 2007. Light stimulates growth of proteorhodopsin-containing marine Flavobacteria. *Nature* 445: 210–213.

Gonzalez, J. M., and M. A. Moran. 1997. Numerical dominance of a group of marine bacteria in the alpha-subclass of the class Proteobacteria in coastal seawater. *Appl. Environ. Microbiol.* 63: 4237–4242.

Gonzalez, J. M., R. Simo, R. Massana, J. S. Covert, E. O. Casamayor, C. Pedrós-Alió, and M. A. Moran. 2000. Bacterial community structure associated with a dimethylsulfoniopropionate-producing North Atlantic algal bloom. *Appl. Environ. Microbiol.* 66: 4237–4246.

Gonzalez, J. M., J. S. Covert, W. B. Whitman, et al. 2003. *Silicibacter pomeroyi* sp. nov. and *Roseovarius nubinhibens* sp. nov., dimethylsulfoniopropionate-demethylating bacteria from marine environments. *Int. J. Syst. Evol. Microbiol.* 53: 1261–1269.

Goodfellow, M., and J. A. Haynes. 1984. *Actinomycetes* in marine sediments. In L. Ortiz-Ortiz, J. F. Bojalil, and V. Yakoleff (eds.), *Biological, Biochemical, and Biomedical Aspects of Actinomycetes*. Academic Press, pp. 453–472.

Goodfellow, M., and S. T. Williams. 1983. Ecology of *Actinomycetes*. *Annu. Rev. Microbiol.* 37: 189–216.

Gordon, D. A., and S. J. Giovannoni. 1996. Detection of stratified microbial populations related to *Chlorobium* and *Fibrobacter* species in the Atlantic and Pacific Oceans. *Appl. Environ. Microbiol.* 62: 1171–1177.

Hahn, M. W., and M. G. Höfle. 2001. Grazing of protozoa and its effect on populations of aquatic bacteria. *FEMS Microbiol. Ecol.* 35: 113–121.

Hallam, S. J., T. J. Mincer, C. Schleper, C. M. Preston, K. Roberts, P. M. Richardson, and E. F. DeLong. 2006. Pathways of carbon assimilation and ammonia oxidation suggested by environmental genomic analyses of marine Crenarchaeota. *PLoS Biology* 4: e95.

Hennes, K. P., C. A. Suttle, and A. M. Chan. 1995. Fluorescently labeled virus probes show that natural virus populations can control the structure of marine microbial communities. *Appl. Environ. Microbiol.* 61: 3623–3627.

Herndl, G. J., T. Reinthaler, E. Teira, H. van Aken, C. Veth, A. Pernthaler, and J. Pernthaler. 2005. Contribution of Archaea to total prokaryotic production in the deep Atlantic Ocean. *Appl. Environ. Microbiol.* 71: 2303–2309.

Hewson, I., and J. A. Fuhrman. 2004. Richness and diversity of bacterioplankton species along an estuarine gradient in Moreton Bay, Australia. *Appl. Environ. Microbiol.* 70: 3425–3433.

Hewson, I., and J. A. Fuhrman. 2006. Viral impacts upon marine bacterioplankton assemblage structure. *J. Mar. Biol. Assoc. UK* 86: 577–589.

Hewson, I., J. A. Steele, M. V. Brown, and J. A. Fuhrman. 2006a. Remarkable heterogeneity in meso- and bathypelagic bacterioplankton community composition. *Limnol. Oceanogr.* 51: 1274–1283.

Hewson, I., J. A. Steele, D. G. Capone, and J. A. Fuhrman. 2006b. Temporal and spatial scales of variation in bacterioplankton assemblages of oligotrophic surface waters. *Mar. Ecol. Prog. Ser.* 311: 67–77.

REFERENCES

Hewson, I., D. M. Winget, K. E. Williamson, J. A. Fuhrman, and K. E. Wommack. 2006c. Viral and bacterial assemblage covariance in oligotrophic waters of the West Florida Shelf (Gulf of Mexico). *J. Mar. Biol. Assoc. UK* 86: 591–603.

Hillebrand, H. 2004. Strength, slope and variability of marine latitudinal gradients. *Mar. Ecol. Prog. Ser.* 273: 251–267.

Hobbie, J. E., R. J. Daley, and S. Jasper. 1977. Use of nuclepore filters for counting bacteria by fluorescence microscopy. *Appl. Environ. Microbiol.* 33: 1225–1228.

Ingalls, A. E., S. R. Shah, R. L. Hansman, L. I. Aluwihare, G. M. Santos, E. R. M. Druffel, and A. Pearson. 2006. Quantifying archaeal community autotrophy in the mesopelagic ocean using natural radiocarbon. *Proc. Natl. Acad. Sci. USA* 103: 6442–6447.

Jannasch, H. W., and G. E. Jones. 1959. Bacterial populations in sea water as determined by different methods of enumeration. *Limnol. Oceanogr.* 4: 128–139.

Jardillier, L., Y. Bettarel, M. Richardot, C. Bardot, C. Amblard, T. Sime-Ngando, and D. Debroas. 2005. Effects of viruses and predators on prokaryotic community composition. *Microb. Ecol.* 50: 557–569.

Johnson, P. W., and J. M. Sieburth. 1979. Chroococcoid cyanobacteria in the sea—Ubiquitous and diverse phototropic biomass. *Limnol. Oceanogr.* 24: 928–935.

Johnson, Z. I., E. R. Zinser, A. Coe, N. P. McNulty, E. M. Woodward, and S. W. Chisholm. 2006. Niche partitioning among *Prochlorococcus* ecotypes along ocean-scale environmental gradients. *Science* 311: 1737–1740.

Jürgens, K., and C. Matz. 2002. Predation as a shaping force for the phenotypic and genotypic composition of planktonic bacteria. *Antonie Van Leeuwenhoek* 81: 413–434.

Karner, M. B., E. F. DeLong, and D. M. Karl. 2001. Archaeal dominance in the mesopelagic zone of the Pacific Ocean. *Nature* 409: 507–510.

Kirchman, D. L., A. I. Dittel, R. R. Malmstrom, and M. T. Cottrell. 2005. Biogeography of major bacterial groups in the Delaware Estuary. *Limnol. Oceanogr.* 50: 1697–1706.

Keswani, J., and W. B. Whitman. 2001. Relationship of 16S rRNA sequence similarity to DNA hybridization in prokaryotes. *Int. J. Syst. Evol. Microbiol.* 51: 667–678.

Konneke, M., A. E. Bernhard, J. R. de la Torre, C. B. Walker, J. B. Waterbury, and D. A. Stahl. 2005. Isolation of an autotrophic ammonia-oxidizing marine archaeon. *Nature* 437: 543–546.

Korona, R., C. H. Nakatsu, L. J. Forney, and R. E. Lenski. 1994. Evidence for multiple adaptive peaks from populations of bacteria evolving in a structured habitat. *Proc. Natl. Acad. Sci. USA* 91: 9037–9041.

Landry, M. R., and D. L. Kirchman. 2002. Microbial community structure and variability in the tropical Pacific. *Deep-Sea Res. Part II.* 49: 2669–2693.

Larsson, U., S. Hajdu J., Walve, and R. Elmgren. 2001. Estimating Baltic nitrogen fixation from the summer increase in upper mixed layer total nitrogen. *Limnol. Oceanogr.* 46: 811–820.

Lee, S., and J. A. Fuhrman. 1991. Spatial and temporal variation of natural bacterioplankton assemblages studied by total genomic DNA cross-hybridization. *Limnol. Oceanogr.* 36: 1277–1287.

Levin, B. R., F. M. Stewart, and L. Chao. 1977. Resource-limited growth, competition, and predation: A model and experimental studies with bacteria and bacteriophage. *Am. Nat.* 111: 3–24.

Liu, W. T., T. L. Marsh, H. Cheng, and L. J. Forney. 1997. Characterization of microbial diversity by determining terminal restriction fragment length polymorphisms of genes encoding 16S rRNA. *Appl. Environ. Microbiol.* 63: 4516–4522.

Long, R. A., and F. Azam. 2001a. Antagonistic interactions among marine pelagic bacteria. *Appl. Environ. Microbiol.* 67: 4975–4983.

Long, R. A., and F. Azam. 2001b. Microscale patchiness of bacterioplankton assemblage richness in seawater. *Aquat. Microb. Ecol.* 26: 103–113.

Long, R. A., A. Qureshi, D. J. Faulkner, and F. Azam. 2003. 2-n-Pentyl-4-quinolinol produced by a marine *Alteromonas* sp. and its potential ecological and biogeochemical roles. *Appl. Environ. Microbiol.* 69: 568–576.

Lopez-Lopez, A., S. G. Bartual, L. Stal, O. Onyshchenko, and F. Rodriguez-Valera. 2005. Genetic analysis of housekeeping genes reveals a deep-sea ecotype of *Alteromonas macleodii* in the Mediterranean Sea. *Environ. Microbiol.* 7: 649–659.

Madigan, M. T., and J. M. Martinko. 2006. *Brock Biology of Microorganisms*. Pearson.

Magurran, A. E. (2004). *Measuring Biological Diversity*. Blackwell Publishing.

Magurran, A. E., and P. A. Henderson. 2003. Explaining the excess of rare species abundance distributions. *Nat. Rev. Microbiol.* 422: 714–716.

Martiny, J. B. H., B. J. M. Bohannan, J. H. Brown, et al. 2006. Microbial biogeography: putting microorganisms on the map. *Nat. Rev. Microbiol.* 4: 102–112.

Massana, R., A. E. Murray, C. M. Preston, and E. F. DeLong. 1997. Vertical distribution and phylogenetic characterization of marine planktonic Archaea in the Santa Barbara Channel. *Appl. Environ. Microbiol.* 63: 50–56.

Massana, R., L. J. Taylor, A. E. Murray, K. Y. Wu, W. H. Jeffrey, and E. F. DeLong. 1998. Vertical distribution and temporal variation of marine planktonic archaea in the Gerlache Strait, Antarctica, during early spring. *Limnol. Oceanogr.* 43: 607–617.

Matz, C., and K. Jürgens. 2003. Interaction of nutrient limitation and protozoan grazing determines the phenotypic structure of a bacterial community. *Microb. Ecol.* 45: 384–398.

Mayali, X., and F. Azam. 2004. Algicidal bacteria in the sea and their impact on algal blooms. *J. Eukaryot. Microbiol.* 51: 139–144.

Miki, T., and N. Yamamura. 2005. Intraguild predation reduces bacterial species richness and loosens the viral loop in aquatic systems: "kill the killer of the winner" hypothesis. *Aquat. Microb. Ecol.* 40: 1–12.

Montoya, J. P., C. M. Holl, J. P. Zehr, A. Hansen, T. A. Villareal, and D. G. Capone. 2004. High rates of N_2 fixation by unicellular diazotrophs in the oligotrophic Pacific Ocean. *Nature* 430: 1027–1031.

Moore, L. R., G. Rocap, and S. W. Chisholm. 1998. Physiology and molecular phylogeny of coexisting *Prochlorococcus* ecotypes. *Nature* 393: 464–467.

Moran, M. A., A. Buchan, J. M. Gonzalez, et al. 2004. Genome sequence of *Silicibacter pomeroyi* reveals adaptations to the marine environment. *Nature* 432: 910–913.

Morris, R. M., M. S. Rappé, S. A. Connon, K. L. Vergin, W. A. Siebold, C. A. Carlson, and S. J. Giovannoni. 2002. SAR11 clade dominates ocean surface bacterioplankton communities. *Nature* 420: 806–810.

Morris, R. M., K. L. Vergin, J. C. Cho, M. S. Rappé, C. A. Carlson, and S. J. Giovannoni. 2005. Temporal and spatial response of bacterioplankton lineages to annual convective overturn at the Bermuda Atlantic Time-series Study site. *Limnol. Oceanogr.* 50: 1687–1696.

Morris, R. M., K. Longnecker, and S. J. Giovannoni. 2006. *Pirellula* and OM43 are among the dominant lineages identified in an Oregon coast diatom bloom. *Environ. Microbiol.* 8: 1361–1370.

Mullins, T. D., T. B. Britschgi, R. L. Krest, and S. J. Giovannoni. 1995. Genetic comparisons reveal the same unknown bacterial lineages in Atlantic and Pacific bacterioplankton communities. *Limnol. Oceanogr.* 40: 148–158.

Murray A. E., C. M. Preston, R. Massana, L. T. Taylor, A. Blakis, K. Wu, and E. F. DeLong. 1998. Seasonal and spatial variability of bacterial and archaeal assemblages in the coastal waters near Anvers Island, Antarctica. *Appl. Environ. Microbiol.* 64: 2585–2595.

Muyzer G., and K. Smalla. 1998. Application of denaturing gradient gel electrophoresis (DGGE) and temperature gradient gel electrophoresis (TGGE) in microbial ecology. *Antonie Van Leeuwenhoek Int. J. Gen. Mol. Microbiol.* 73: 127–141.

Muyzer G., E. D Waal, and A. G. Uitterlinden. 1993. Profiling of complex microbial populations by denaturing gradient gel electrophoresis analysis of polymerase chain reaction-amplified genes coding for 16S rRNA. *Appl. Environ. Microbiol.* 59: 695–700.

Okazaki, T., and Y. Okami. 1972. Studies on microorganisms. II. *Actinomycetes* in Sagami Bay and their antibiotic substances. *J. Antibiot. (Tokyo)* 25: 461–466.

Olsen, G. J., D. L. Lane, S. J. Giovannoni, and N. R. Pace. 1986. Microbial ecology and evolution: A ribosomal RNA approach. *Annu. Rev. Microbiol.* 40: 337–365.

Osborn, A. M., E. R. B. Moore, and K. N. Timmis. 2000. An evaluation of terminal-restriction fragment length polymorphism (T-RFLP) analysis for the study of microbial community structure and dynamics. *Environ. Microbiol.* 2: 39–50.

Ouverney, C. C., and J. A. Fuhrman. 2000. Marine planktonic Archaea take up amino acids. *Appl. Environ. Microbiol.* 66: 4829–4833.

Pace, N. R., D. A. Stahl, D. L. Lane, and G. J. Olsen. 1986. The analysis of natural microbial populations by rRNA sequences. *Adv. Microbiol. Ecol.* 9: 1–55.

Palenik, B., B. Brahamsha, F. W. Larimer, et al. 2003. The genome of a motile marine *Synechococcus*. *Nature* 424: 1037–1042.

Partensky, F., W. R. Hess, and D. Vaulot. 1999. *Prochlorococcus*, a marine photosynthetic prokaryote of global significance. *Microbiol. Mol. Biol. Rev.* 63: 106–127.

Pearson, A., A. P. McNichol, B. C. Benitez-Nelson, J. M. Hayes, and T. I. Eglinton. 2001. Origins of lipid biomarkers in Santa Monica Basin surface sediment: A case study using compound-specific delta C-14 analysis. *Geochim. Cosmochim. Acta* 65: 3123–3137.

Pedrós-Alió, C. 2006. Marine microbial diversity: can it be determined? *Trends Microbiol.* 14: 257–263.

Pinhassi, J., and Å. Hagström. 2000. Seasonal succession in marine bacterioplankton. *Aquat. Microb. Ecol.* 21: 245–256.

Pinhassi, J., F. Azam, J. Hemphala, R. A. Long, J. Martinez, U. L. Zweifel, and Å. Hagström. 1999. Coupling between bacterioplankton species composition, population dynamics, and organic matter degradation. *Aquat. Microb. Ecol.* 17: 13–26.

Pommier, T., K. Boström, B. Canbäck, P. Lundberg, L. Riemann, K. Simu, A. Tunlid, and Å. Hagström. 2007. Global patterns of diversity and community structure in marine bacterioplankton. *Mol. Ecol.* 16: 867–880.

Pommier, T., J. Pinhassi, and Å. Hagström. 2005. Biogeography analysis of ribosomal RNA clusters from marine bacterioplankton. *Aquat. Microb. Ecol.* 41: 79–89.

Prabagaran, S. R., R. Manorama, D. Delille, and S. Shivaji. 2007. Predominance of *Roseobacter*, *Sulfitobacter*, *Glaciecola* and *Psychrobacter* in seawater collected off Ushuaia, Argentina, Sub-Antarctica. *FEMS Microbiol. Ecol.* 59: 342–355.

Rappé, M. S., S. A. Connon, K. L. Vergin, and S. J. Giovannoni. 2002. Cultivation of the ubiquitous SAR11 marine bacterioplankton clade. *Nature* 418: 630–633.

Rehnstam, A. S., S. Backman, D. C. Smith, F. Azam, and Å. Hagström. 1993. Bloom of sequence-specific culturable bacteria in the sea. *FEMS Microbiol. Ecol.* 102: 161–166.

Riemann, L., and M. Middelboe. 2002. Stability of bacterial and viral community compositions in Danish coastal waters as depicted by DNA fingerprinting techniques. *Aquat. Microb. Ecol.* 27: 219–232.

Riemann, L., G. F. Steward, and F. Azam. 2000. Dynamics of bacterial community composition and activity during a mesocosm diatom bloom. *Appl. Environ. Microbiol.* 66: 578–587.

Rocap, G., D. L. Distel, J. B. Waterbury, and S. W. Chisholm. 2002. Resolution of *Prochlorococcus* and *Synechococcus* ecotypes by using 16S-23S ribosomal DNA internal transcribed spacer sequences. *Appl. Environ. Microbiol.* 68: 1180–1191.

Rodriguez, F., E. Derelle, L. Guillou, F. Le Gall, D. Vaulot, and H. Moreau. 2005. Ecotype diversity in the marine picoeukaryote *Ostreococcus* (Chlorophyta, Prasinophyceae). *Environ. Microbiol.* 7: 853–859.

Rossello-Mora, R., and R. Amann. 2001. The species concept for prokaryotes. *FEMS Microbiol. Rev.* 25: 39–67.

Rusch, D. B., A. L. Halpern, G. Sutton, et al. 2007. The Sorcerer II Global Ocean Sampling Expedition: Northwest Atlantic through Eastern Tropical Pacific. *PLoS Biol* 5: 398–431.

Schafer, H., L. Bernard, C. Courties, et al. 2001. Microbial community dynamics in Mediterranean nutrient-enriched seawater mesocosms: changes in the genetic diversity of bacterial populations. *FEMS Microbiol. Ecol.* 34: 243–253.

Schauer, M., V. Balague, C. Pedrós-Alió, and R. Massana. 2003. Seasonal changes in the taxonomic composition of bacterioplankton in a coastal oligotrophic system. *Aquat. Microb. Ecol.* 31: 163–174.

Schleper, C., G. Jurgens, and M. Jonuscheit. 2005. Genomic studies of uncultivated archaea. *Nat. Rev. Microbiol.* 3: 479–488.

Schmidt, T. M., E. F. DeLong, and N. R. Pace. 1991. Analysis of a marine picoplankton community by 16S rRNA gene cloning and sequencing. *J. Bacteriol.* 173: 4371–4378.

Schwalbach, M. S., I. Hewson, and J. A. Fuhrman. 2004. Viral effects on bacterial community composition in marine plankton microcosms. *Aquat. Microb. Ecol.* 34: 117–127.

Seymour, J. R., J. G. Mitchell, and L. Seuront. 2004. Microscale heterogeneity in the activity of coastal bacterioplankton communities. *Aquat. Microb. Ecol.* 35: 1–16.

Šimek, K., P. Kojecká, J. Nedoma, P. Hartman, J. Vrba, and J. R. Dolan. 1999. Shifts in bacterial community composition associated with different microzooplankton size fractions in a eutrophic reservoir. *Limnol. Oceanogr.* 44: 1634–1644.

Simu, K., and Å. Hagström. 2004. Oligotrophic bacterioplankton with a novel single-cell life strategy. *Appl. Environ. Microbiol.* 70: 2445–2451.

Stackebrandt, E., and B. M. Göbel. 1994. Taxonomic note: A place for DNA–DNA reassociation and 16S rRNA sequence analysis in the present species definition in bacteriology. *Int. J. Syst. Bacteriol.* 44: 846–849.

Staley, J. T., and A. Konopka. 1985. Measurement of in situ activities of nonphotosynthetic microorganisms in aquatic and terrestrial habitats. *Annu. Rev. Microbiol.* 39: 321–346.

Sullivan, M. B., J. B. Waterbury, and S. W. Chisholm. 2003. Cyanophages infecting the oceanic cyanobacterium *Prochlorococcus*. *Nature* 424: 1047–1051.

Suzuki, M. T., C. M. Preston, F. P. Chavez, and E. F. DeLong. 2001. Quantitative mapping of bacterioplankton populations in seawater: Field tests across an upwelling plume in Monterey Bay. *Aquat. Microb. Ecol.* 24: 117–127.

Teira, E., T. Reinthaler, A. Pernthaler, J. Pernthaler, and G. J. Herndl. 2004. Combining catalyzed reporter deposition–fluorescence in situ hybridization and microautoradiography to detect substrate utilization by bacteria and archaea in the deep ocean. *Appl. Environ. Microbiol.* 70: 4411–4414.

Teira, E., H. van Aken, C. Veth, and G. J. Herndl. 2006a. Archaeal uptake of enantiomeric amino acids in the meso- and bathypelagic waters of the North Atlantic. *Limnol. Oceanogr.* 51: 60–69.

Teira, E., P. Lebaron, H. van Aken, and G. J. Herndl. 2006b. Distribution and activity of Bacteria and Archaea in the deep water masses of the North Atlantic. *Limnol. Oceanogr.* 51: 2131–2144.

Thingstad, T. F. 2000. Elements of a theory for the mechanisms controlling abundance, diversity, and biogeochemical role of lytic bacterial viruses in aquatic systems. *Limnol. Oceanogr.* 45: 1320–1328.

Thingstad, T. F., and R. Lignell. 1997. Theoretical models for the control of bacterial growth rate, abundance, diversity and carbon demand. *Aquat. Microb. Ecol.* 13: 19–27.

Thompson, F. L., T. Iida, and J. Swings. 2004. Biodiversity of vibrios. *Microbiol. Mol. Biol. Rev.* 68: 403–431.

Thompson, J. R., S. Pacocha, C. Pharino, V. Klepac-Ceraj, D. E. Hunt, J. Benoit, R. Sarma-Rupavtarm, D. L. Distel, and M. F. Polz. 2005. Genotypic diversity within a natural coastal bacterioplankton population. *Science* 307: 1311–1313.

Todd, J. D., R. Rogers, Y. G. Li, et al. 2007. Structural and regulatory genes required to make the gas dimethyl sulfide in bacteria. *Science* 315: 666–669.

Trebesius, K., R. Amann, W. Ludwig, K. Muhlegger, and K. H. Schleifer. 1994. Identification of whole fixed bacterial cells with nonradioactive 23S rRNA-targeted polynucleotide probes. *Appl. Environ. Microbiol.* 60: 3228–3235.

Uphoff, H. U., A. Felske, W. Fehr, and I. Wagner-Dobler. 2001. The microbial diversity in picoplankton enrichment cultures: A molecular screening of marine isolates. *FEMS Microbiol. Ecol.* 35: 249–258.

Venter, J. C., K. Remington, J. F. Heidelberg, et al. 2004. Environmental genome shotgun sequencing of the Sargasso Sea. *Science* 304: 66–74.

Wasmund, N., G. Nausch, B. Schneider, K. Nagel, and M. Voss. 2005. Comparison of nitrogen fixation rates determined with different methods: a study in the Baltic Proper. *Mar. Ecol. Prog. Ser.* 297: 23–31.

Waterbury, J. B., and F. W. Valois. 1993. Resistance to co-occurring phages enables marine *Synechococcus* communities to coexist with cyanophages abundant in seawater. *Appl. Environ. Microbiol.* 59: 3393–3399.

Waterbury, J. B., S. W. Watson, R. L. L. Guillard, and L. E. Brand. 1979. Widespread occurrence of a unicellular, marine, planktonic cyanobacterium. *Nature* 227: 293–294.

Waters, C. M., and B. L. Bassler. 2005. Quorum sensing: Cell-to-cell communication in bacteria. *Annu. Rev. Cell Dev. Biol.* 21: 319–346.

Weaver, R. S., D. L. Kirchman, and D. A. Hutchins. 2003. Utilization of iron/organic ligand complexes by marine bacterioplankton. *Aquat. Microb. Ecol.* 31: 227–239.

West, N. J., W. A. Schonhuber, N. J. Fuller, R. I. Amann, R. Rippka, A. F. Post, and D. J. Scanlan. 2001. Closely related *Prochlorococcus* genotypes show remarkably different depth distributions in two oceanic regions as revealed by in situ hybridization using 16S rRNA-targeted oligonucleotides. *Microbiology* 147: 1731–1744.

Whittaker, R. H. 1975. Community stability. In R. H. Whittaker (ed.), *Communities and Ecosystems*. Macmillan, pp. 42–53.

Wilson, W. G., and P. Lundberg. 2004. Biodiversity and the Lotka–Volterra theory of species interactions: Open systems and the distribution of logarithmic densities. *Proc. Roy. Soc. Lond. Ser. B* 271: 1977–1984.

Winter, C., A. Smit, G. J. Herndl, and M. G. Weinbauer. 2004. Impact of virioplankton on archaeal and bacterial community richness as assessed in seawater batch cultures. *Appl. Environ. Microbiol.* 70: 804–813.

Winter, C., A. Smit, G. J. Herndl, and M. G. Weinbauer. 2005. Linking bacterial richness with viral abundance and prokaryotic activity. *Limnol. Oceanogr.* 50: 968–977.

Woese, C. R. 1987. Bacterial evolution. *Microbiol. Rev.* 51: 221–271.

Woese, C. R., and G. E. Fox. 1977. Phylogenetic structure of the prokaryotic domain: the primary kingdoms. *Proc. Natl. Acad. Sci. USA* 74: 5088–5090.

Woese, C. R., O. Kandler, and M. L. Wheelis. 1990. Towards a natural system of organisms: Proposal for the domains, Archaea, Bacteria, and Eukarya. *Proc. Natl. Acad. Sci. USA* 87: 4576–4579.

Wright, T. D., K. L. Vergin, P. W. Boyd, and S. J. Giovannoni. 1997. A novel delta-subdivision proteobacterial lineage from the lower ocean surface layer. *Appl. Environ. Microbiol.* 63: 1441–1448.

Wuchter, C., S. Schouten, H. T. S. Boschker, and J. S. S. Damste. 2003. Bicarbonate uptake by marine Crenarchaeota. *FEMS Microbiol. Lett.* 219: 203–207.

Wuchter, C., B. Abbas, M. J. L. Coolen, et al. 2006. Archaeal nitrification in the ocean. *Proc. Natl. Acad. Sci. USA* 103: 12317–12322.

Yair, S., D. Yaacov, K. Susan, and E. Jurkevitch. 2003. Small eats big: Ecology and diversity of *Bdellovibrio* and like organisms, and their dynamics in predator–prey interactions. *Agronomie* 23: 433–439.

Zubkov, M. V., B. M. Fuchs, G. A. Tarran, P. H. Burkill, and R. Amann. 2003. High rate of uptake of organic nitrogen compounds by *Prochlorococcus* cyanobacteria as a key to their dominance in oligotrophic oceanic waters. *Appl. Environ. Microbiol.* 69: 1299–1304.

GENOMICS AND METAGENOMICS OF MARINE PROKARYOTES

MARY ANN MORAN

Department of Marine Sciences, University of Georgia, Athens, GA 30602-3636, U.S.A.

INTRODUCTION

Earliest glimpses of the abundance (Hobbie et al. 1977) and metabolism (Williams and Askew 1978) of marine prokaryotes led to the realization that most biogeochemical properties of the ocean are determined by the activities of bacterioplankton (Azam and Worden 2004). However, progress in understanding those activities has often been limited by the availability of tools and instruments to measure them. Genomics provides a new set of tools to microbial ecologists by providing access to the genes that underlie the metabolic activities of bacterioplankton. In concert with development of culture-independent molecular techniques (see Chapter 3) and advances in culturing methods (Connon and Giovannoni 2002), the field of genomics is unraveling the biology of those previously cryptic organisms that govern ocean processes.

Genomics is the study of an organism's complete genetic material. Bioinformatics is the process of identifying and interpreting the biologically important information stored in the genetic material. For marine microbial ecologists, genomics and bioinformatics advance understanding in two key ways. First, they generate testable hypotheses about novel microbial processes in the ocean based on the discovery of

---

Microbial Ecology of the Oceans, Second Edition. Edited by David L. Kirchman
Copyright © 2008 John Wiley & Sons, Inc.

genes for previously uncharacterized or unsuspected metabolic activities. One of the most dramatic examples of "reverse biogeochemistry" was the discovery of a gene from an uncultured marine bacterium encoding a light-driven proton pump for the generation of adenosine triphosphate (ATP) (Béjà et al. 2000). This fundamentally changed our understanding of how energy from light is used in the ocean (see below and Chapter 5).

The second major advantage of genomics and bioinformatics is the accumulation of abundant biological details which can provide insight into the workings of marine ecosystems. This "environmental reductionism" approach relies on cellular biochemistry as the foundation for successfully understanding and predicting ocean-scale biogeochemical processes (Doney et al. 2004). For example, comparative analysis of *Prochloroccocus* genomes has revealed the genetic basis for how this dominant oceanic primary producer acclimates to light levels, uses nitrogen, and acquires iron in an iron-limited environment (Rocap et al. 2003).

This chapter focuses on prokaryotic genomics, with a particular emphasis on marine bacterioplankton. We first look generally at the basics of genomes and genomic analysis, and then see how genomic data can address ecological and biogeochemical questions in marine microbial ecology from both reverse biogeochemistry and environmental reductionism viewpoints.

THE BASICS OF PROKARYOTIC GENOMICS

Genome Sequence and Assembly

The DNA sequence of a cultured prokaryote is typically obtained by a process referred to as "whole-genome shotgun" (WGS) sequencing. DNA is extracted from the organism and fragmented into small pieces, which are randomly selected for sequencing. Random sequencing is typically stopped when 8 to 10 times the number of bases in the genome have been sequenced (i.e., $8\times$ to $10\times$ coverage). Computer programs then re-assemble the sequenced fragments, essentially searching for overlapping regions of identical sequence that can be joined together to form

---

Complete or Draft?

Many marine prokaryotic genomes are now being sequenced only to draft status, rather than fully closing the genome. In this case, the sequencing process stops after the automated assembly step, thereby eliminating the expensive procedure of fitting the contigs together and closing gaps. The downside of this approach is that some genes may be missing, but we do not know which ones, and the final gene order is not fully known. The upside is that many genomes can be sequenced quickly and cheaply. For ecological questions that do not require knowing all genes in every genome, the added power of many genome sequences might compensate for the loss of information about specific ones, but debate on this issue continues.

growing regions of contiguous DNA. When the assembly programs have pieced together as much sequence as possible, a human-intensive process of ordering the larger contiguous pieces, or contigs, begins. This typically involves designing individual PCR primers to link the contigs and fill in any sequencing gaps in the assembled genome, finally resulting in a closed genome.

The completed microbial genome sequences confirm that prokaryotic DNA is typically organized into circular chromosomes as well as extrachromosomal plasmids that replicate independently of the chromosome. Traditionally, the permanent prokaryotic genome was assumed to be carried on a single chromosome, and plasmids were viewed as transient gene repositories that could be largely ignored. For marine microbial ecologists, this perspective has two drawbacks. First, it is becoming evident that many prokaryotes carry more than one large DNA molecule. Whether these represent the splitting of a parent chromosome into two or more parts or the accumulation of smaller plasmids into a larger plasmid (i.e., a megaplasmid) is not always clear (Bentley and Parkhill 2004), but a significant percentage of the genome can be housed on these additional replicons (i.e., independently replicating DNA molecules). Members of the *Alphaproteobacteria*, a group very well represented in marine bacterioplankton communities, have a particular propensity for arranging genomes into more than one large molecule (Jumas-Bilak et al. 1998; Moran et al. 2004), as do some *Gammaproteobacteria* (Okada et al. 2005). Complex genome structure is therefore likely to be a prominent feature of marine prokaryotes. Of 34 fully sequenced marine bacterial and archaeal genomes, 35 percnt have more than one replicon (Table 4.1).

A second drawback to ignoring genes on nonchromosomal replicons is that ecologically interesting genes could be overlooked. While plasmid-borne genes may be somewhat transient over an evolutionary time frame, they are informative over an ecological time frame and can reflect selection pressures of the organism's environment (Sobecky 1999). To the extent that the plasmid-borne gene pool is more dynamic and opportunistic in an ecological sense, it is an essential part of the microbial ecologist's definition of a genome.

When is a Chromosome a Chromosome?

Designation of a second (or even third) replicon as a chromosome rather than a plasmid is often based on the presence of a ribosomal RNA operon (Bentley and Parkhill 2004) or other essential genes. But there are no hard and fast rules, and size often matters. A small (128 kb) replicon in the marine bacterium *Silicibacter* sp. TM1040 has been designated a plasmid even though it contains an rRNA operon (Moran et al. 2007). A large replicon in the Antarctic bacterium *Pseudoalteromonas haloplanktis* TAC125 has been designated a chromosome even though it has a plasmid-like replication system (Médigue et al. 2005).

TABLE 4.1 Complete Marine Bacterial and Archaeal Genomes as of November 2006[a]

Organism	Genome Size (Mb)	No. of replicons	No. of ORFs	No. of rRNA operons	GC %
Aeropyrum pernix K1	1.67	1	1841	1	56
Archaeoglobus fulgidus DSM4304	2.18	1	2420	1	48
Colwellia psychrerythraea 34H	5.37	1	4910	9	38
Desulfotalea psychrophila LSv54	3.66	3	3234	7	46
Geobacillus kaustophilus HTA426	3.59	2	3540	9	52
Gramella forsetii	3.80	1	3585	3	37
Idiomarina loihiensis L2TR	2.84	1	2628	4	47
Jannaschia sp. CCS1	4.40	2	4283	1	62
Methanocaldococcus jannaschii DSM 2661	1.74	3	1786	2	31
Methanococcus maripaludis S2	1.66	1	1722	3	33
Methanopyrus kandleri AV19	1.69	1	1687	1	61
Methanosarcina acetivorans C2A	5.75	1	4540	3	42
Nanoarchaeum equitans Kin4-M	0.49	1	536	1	31
Oceanobacillus iheyensis HTE831	3.63	1	3500	7	35
Pelagibacter ubique HTCC106	1.31	1	1354	1	29
Photobacterium profundum SS9	6.40	3	5491	15	41
Prochlorococcus marinus CCMP1375	1.75	1	1882	1	36
Prochlorococcus marinus MED4	1.66	1	1713	1	30
Prochlorococcus marinus MIT9313	2.41	1	2265	1	50
Pseudoalteromonas haloplanktis TAC125	3.85	2	3487	9	40
Pyrobaculum aerophilum IM2	2.22	1	2605	1	51
Pyrococcus abyssi GE5	1.77	2	1898	1	44
Pyrococcus furiosus DSM3638	1.91	1	2125	1	40
Pyrococcus horikoshii OT3	1.74	1	1955	1	41
Rhodopirellula baltica SH1	7.15	1	7325	1	55
Silicibacter pomeroyi DSS-3	4.60	2	4284	3	64
Synechococcus sp. WH8102	2.43	1	2517	2	59
Synechococcus sp. CC9902	2.235	1	2304	2	54
Thermotoga maritima MSB8	1.86	1	1858	1	46
Vibrio fischeri ES114	4.28	3	3802	12	38
Vibrio parahaemolyticus RIMD 2210633	5.16	3	4832	11	45
Vibrio vulnificus CMCP6	5.13	3	4488	8	46
Vibrio vulnificus YJ016	5.26	3	5024	9	46

[a]Shaded cells indicate genomes detailed in Table 4.2.
ORF, open reading frame.

Finding Genes

The information encoded in the genome that controls how the DNA is interpreted by the cell (i.e., the "punctuation" of the genome) is fairly conserved across prokaryotes and is therefore useful for identifying gene locations. Two sets of nucleotides with a predictable sequence, referred to as the −35 and −10 regions or the promoter sequence, are typically found just before the beginning of a gene. These nucleotides indicate to RNA polymerase that transcription should begin downstream (at position +1) (Fig. 4.1). Because the nucleotide composition at the promoter could occur by chance only once in 70 million nucleotides (Krane and Raymer 2003), its occurrence indicates with high likelihood that a gene is encoded immediately downstream. When resultant mRNA is translated into protein, a ribosome binding site (RBS; also called a Shine–Dalgarno sequence) on the transcript acts as the recognition site for the ribosome, and this is reflected in the DNA template with a sequence approximating 5′-AGGAGGT-3′ (Fig. 4.1). Downstream of this site, a start codon for protein translation (typically ATG, coding for methionine) marks the beginning of protein translation. At the end of the gene, a stop codon (TAA, TGA, or TAG) indicates where translation stops. Using these typical gene features (promoter sequences, RBSs, start codons, and stop codons), along with information on sequence similarity to protein coding regions that are already known, gene-finding software automatically identifies open reading frames (ORFs) in prokaryotic genomes. These are the likely locations of the genes.

Finding genes is easier in prokaryotes than eukaryotes because there is little space between genes and there are no interruptions by noncoding sequence within genes. Typically, 85–88 percent of bacterial and archaeal genomes are within coding regions (Krane and Raymer 2003). Among marine genomes, the spacing between genes (i.e., the intergenic DNA region) varies from a median of only three bases for *Pelagibacter ubique* to 78 for *Silicibacter pomeroyi* and 137 for *Photobacterium profundum* (Giovannoni et al. 2005a).

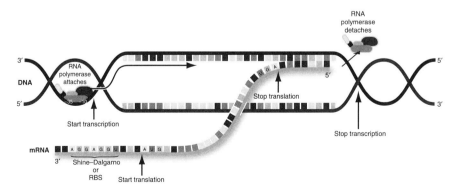

Figure 4.1 Features of prokaryotic genes leave signatures in the genome that are used to identify the protein-encoding regions.

- An **open reading frame** (ORF) is a sequence of nucleotides that begins with an initiation codon (usually methionine, coded by ATG), ends with a stop codon, and has the potential to encode a protein.
- A **gene** is a sequence of nucleotides that encodes a protein or RNA product. Protein-encoding genes are always ORFs, but not all ORFs turn out to be genes.

Based on such gene-finding analyses, we have learned that the number of genes harbored by marine prokaryotes varies widely. Among free-living marine prokaryotes sequenced to date, the SAR11 clade member *Pelagibacter ubique* sports the smallest genome (1.3 Mbp) and the fewest putative genes at 1354. The *Planctomycete Rhodopirellula baltica* has the largest genome (7.1 Mbp) and fivefold more putative genes than *P. ubique* at 7325 (Table 4.1). If we consider the symbiotic marine archaeon *Nanoarchaeum equitans* (with only 536 putative reading frames), variation across marine prokaryotic taxa in predicted gene number increases to more than 13-fold. This wide range in gene number must reflect fundamental differences in the ecological roles and strategies of marine prokaryotic taxa, despite the fact that the ocean appears fairly homogeneous from a bulk perspective. Most of these differences are unexplored.

Finding Operons

When prokaryotic proteins must interact to carry out a cellular process, the genes that encode them are often adjacent to one another on the chromosome and form an operon. Within the operon, genes are all translated at the same time under the direction of a single promoter. Identification of operons in prokaryotic genomes therefore provides information about which genes are expressed together in the cell. These genes might, for example, carry out a multistep metabolic pathway, encode two processes that must be co-regulated, or synthesize a protein consisting of multiple subunits. Identifying operons can also help determine gene function, since neighboring genes of known function provide hints about an unidentified gene. Identification of operons usually involves analysis of the size of the intergenic regions (genes that are within an operon typically have smaller spaces between genes), similarity of function (genes that are within an operon usually have related functions), and conservation of gene organization (genes that are within operons in one species are likely to be so in others) (Westover et al. 2005). The importance of operon structure in bacteria is evident in the *S. pomeroyi* genome, where almost half the genes (about 43 percent) are predicted to be part of an operon. These 611 predicted operons have an average length of three genes, but range from just two (315 operons) to eleven (1 operon).

Functional Annotation

Because the genome sequence by itself tells us little about an organism's biology, annotation is used to infer protein function encoded by each gene. In autoannotation, computer algorithms first accumulate evidence for function by comparing the sequence of an ORF to existing gene and protein databases. If a good match to the new sequence is found among characterized proteins, this suggests that the two have evolved from a common ancestor and consequently may have similar functions. Autoannotion therefore ascribes a likely function for the encoded protein (i.e., makes a gene call) based on the function assigned to similar sequences in the database. In manual annotation, human examination of a broader suite of evidence is then used to reject, accept, or adjust the automated gene call.

For genes that encode proteins (as opposed to RNAs), annotation is typically carried out on the amino acid translation of the nucleic acid sequence. Because more than one codon (or triplet of nucleotides) codes for a particular amino acid, a considerable variation at the nucleotide sequence level could translate into the same amino acid sequence. Thus, comparing amino acid sequences is a more reliable predictor of similarity of function between the proteins encoded by two sequences. Two similar proteins in different organisms that are thought to have the same biological function are defined as orthologs.

- Two genes are **homologs** if amino acid sequence similarity indicates they are descended from a common ancestor. They can be either orthologs or paralogs.
- Two genes are **orthologs** if they are descended from a common ancestor without any intervening gene duplication. In this case, the function is likely to be the same. By definition, orthologs cannot occur within the same genome. They are usually identified using the reciprocal best-hit criterion: each gene must be most similar to the other when all genes in two genomes are compared.
- Two genes are **paralogs** if they are descended from two different copies of a gene that arose by gene duplication. In this case, the function is likely to be different. Paralogs can occur within the same genome.

Gene annotation is best viewed as an approximation to be improved over time, rather than a definitive identification of protein function. The ability to correctly annotate genes, either with automated or manual approaches, is hampered by the fact that the rate of generating new gene sequences has greatly outpaced the rate of laboratory experiments designed to assign function. Because prokaryotic genes evolve from other genes, and because new genes can incorporate domains from several unrelated genes (Abascal and Valencia 2003), all or part of a gene sequence may be similar to another gene but have a distinctly different function. Indeed, many annotations of prokaryotic gene sequences are based on amino acid sequence identities of only 30 percent (Devos and Valencia 2001), leaving significant room for departure from the experimentally characterized function. For genes encoding common proteins that mediate conserved metabolic properties (e.g., tricarboxylic acid (TCA) cycle, amino acid biosynthesis, ribosome proteins, or DNA repair), identifications of orthologs is generally robust and not likely to lead us astray. But for unique or niche-specific genes in marine prokaryotic genomes, many either will not be assigned a function or will be assigned to a protein family with similar biochemical mechanisms but different substrates and products. For example, DNA sequences retrieved from seawater with more than 60 percent amino acid identity to sequences of purported bacterial cellulases (CelM) were determined experimentally to have peptidase activity instead (Cottrell et al. 2005). A gene encoding demethylation of dimethylsulfoniopropionate (DMSP), a key step in the marine sulfur cycle, was initially autoannotated as functioning in glycine metabolism (Howard et al. 2006).

Thus, annotation, by and large, involves making an educated guess at protein function based on the evidence available at the time. Gene products are typically placed into three categories: hypothetical proteins are those that have no significant sequence similarity to another protein and about which nothing can be inferred; conserved hypothetical proteins are those that have sequence similarity to other hypothetical proteins, and therefore are found in multiple organisms even if their function is unknown; assigned genes are those that have been designated a function, although this designation may be very general (e.g., putative oxidoreductase) or very specific (e.g., assimilatory nitrate reductase large subunit, EC:1.7.1.4).

Tools to Annotate a Gene

- GenBank searches with BLASTp: searches the largest sequence database for similar proteins (www.ncbi.nlm.nih.gov/blast/)
- COG: assigns function based on similarity to established clusters of evolutionarily related proteins (www.ncbi.nlm.nih.gov/COG/)
- KEGG: reconstructs metabolic pathways on clickable maps that place individual gene annotations in a physiological framework (www.genome.ad.jp/kegg/kegg2.html)
- Pfam: finds common protein domains (i.e., regions within proteins that have defined functions) using a curated collection of sequences from >7000 protein families (www.sanger.ac.uk/Software/Pfam/)
- TIGRFAM: finds members of sets of homologous proteins that are conserved with respect to function since their last common ancestor (i.e., "equivalogs") (www.tigr.org/TIGRFAMs/)
- SEED: classifies genes into conserved cellular subsystems curated by expert annotators (www.theseed.org/wiki/Home_of_the_SEED)

Assigned genes can be further classified into role categories that group genes encoding similar cellular functions (Riley 1993). For example, the "transport and binding" role category harbors a diversity of genes responsible for selectively moving compounds into or out of the cell, while the "signal transduction" category harbors genes for transmitting information about external conditions to the inside of the cell and mediating the cell's response. The number or type of genes in these categories may give hints about the cell's physiology or ecology. The high percentage of cell envelope genes in the Antarctic bacterium *Pseudoalteromonas haloplanktis* may reflect special adaptations to maintain membrane function in cold temperatures (Médigue et al. 2005), while the low percentage of signal transduction genes in *Prochlorococcus marinus* may reflect the reduced need of an autotroph to sense and respond to extracellular compounds (Table 4.2).

TABLE 4.2 Role Category Distribution for Genes in Four Marine Bacteria Representing Diverse Taxa

Gene Role Category	*S. pomeroyi* % of 4284 genes	*P. haloplanktis* % of 3487 genes	*P. marinus* % of 1713 genes	*R. baltica* % of 7325 genes
Amino acid biosynthesis	2.8	3.1	4.4	1.3
Biosynthesis of cofactors, prosthetic groups, and carriers	3.2	3.2	5.4	1.5
Cell envelope	5.5	8.9	5.3	4.8
Cellular processes	3.8	7.7	2.6	2.3
Central intermediary metabolism	3.6	4.8	3.9	3.2
Disrupted reading frame	0.5	0	0	0
DNA metabolism	2.2	4.3	3.2	2.0
Energy metabolism	10.2	11.0	13.4	4.8
Fatty acid and phospholipid metabolism	2.7	3.0	1.4	1.1
Hypothetical proteins	8.2	9.5	17.2	51.8
Hypothetical proteins, conserved	15.2	8.1	15.1	5.4
Mobile and extrachromosomal element functions	1.0	1.1	0	1.2
Pathogen responses	0	0	0	0
Protein fate	3.1	7.1	4.0	2.7
Protein synthesis	3.0	4.4	10.0	2.0
Purines, pyrimidines, nucleosides, and nucleotides	1.5	1.7	2.9	0.9
Regulatory functions	7.8	7.3	5.6	3.5
Signal transduction	1.6	0.9	0.1	0.6
Transcription	1.0	1.7	1.8	1.3
Transport and binding proteins	12.1	9.5	3.3	3.4
Unclassified	0	1.4	10.4	5.7
Unknown function	16.9	11.3	4.7	5.1
Viral functions	0	0	0	0

Currently, most genomes are analyzed using a "horizontal" annotation strategy, in which genes are annotated one at a time in a systematic progression through the genome. However, the growing database of prokaryotic genome sequences now makes a "vertical" annotation strategy feasible. In this case, a small set of functionally related genes are annotated simultaneously across many genomes by a single annotator. For example, the genes involved in osmolyte uptake might be manually annotated by one person for all available marine prokaryotic genomes. This approach has the potential to improve the accuracy of annotation, because it produces uniform gene identifications and terms, and allows annotators to concentrate on those gene sets for which they are experts (Overbeek et al. 2005).

Starting Points for Gene Annotation

Unfortunately, there are no definitive rules for identifying orthologs in prokaryotic genomes, because many pieces of information need to be weighed simultaneously. Manual annotation relies on amino acid sequence similarity, gene neighborhoods, taxonomic relatedness, and experimental evidence. But some very rough guidelines are as follows:

- Genes are potential orthologs if genome-to-genome comparisons indicate that they are reciprocal best hits (RBH) with a BLASTp expect value (E) $<10^{-5}$.
- Genes are potential orthologs if gene-to-gene comparisons indicate a BLASTp E-value of $<10^{-40}$ and a sequence identity of >40 percent for unrelated taxa, or an E-value of $<10^{-60}$ and a sequence identity of >60 percent for related taxa.

The E-value is a statistical parameter that represents how often the comparison of a sequence to a database would result in the same match purely by chance. The lower the E-value, the more significant the match. The E-value is influenced not only by the similarity of the sequences, but also the sequence length and the database size.

TAME OR WILD? PURE-CULTURE GENOMICS VERSUS METAGENOMICS

With less than one in every 10,000 prokaryotes in seawater amenable to culturing (Amann et al. 1995), it is obvious that pure-culture genomics is inadequate for accessing the marine microbial gene pool. Metagenomics (also referred to as "environmental genomics" or "community genomics") addresses this predicament by analyzing mixed genomes from natural microbial communities using DNA extracted directly from the environment (Fig. 4.2). Whereas pure-culture genomics is based on sequences from all the genes in one organism, metagenomics is based on sequences from some of the genes in many organims, regardless of whether or not they can be cultured.

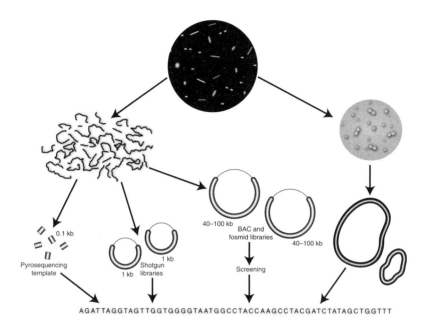

Figure 4.2 The two main approaches in marine microbial genomics are based on obtaining genome sequences of relevant cultured organisms (pure-culture genomics; right side) and on directly retrieving and sequencing fragments of genomes from natural communities (metagenomics; left side).

A number of different metagenomic techniques have been used successfully for marine microbial communities, with the size of the genomic fragment retrieved from community DNA being the primary technical difference among them. Fragments containing as many as 100 contiguous genes are obtained by cloning microbial DNA into bacterial artificial chromosome (BAC) vectors (Béjà et al. 2000); fragments containing as many as 40 contiguous genes are obtained by cloning DNA into fosmid vectors (DeLong et al. 2006); fragments containing one or two genes are obtained by WGS sequencing (Venter et al. 2004); and fragments containing about one-tenth of a gene are obtained by vector-free pyrosequencing (Edwards et al. 2006) (Fig. 4.2).

Although it may first seem like a trivial difference, the size of the fragment is fundamental to the nature of the information that can be extracted from metagenomic data, which in turn determines the type of questions that can be addressed. In the case of BAC and fosmid cloning, the large size of the insert provides access to adjacent genes, operon structure, and metabolic pathways; it permits taxonomic assignment of functional genes for those fragments that also contain ribosomal RNA genes or other taxonomic markers; and it allows screening (using PCR or expression; see below) to focus sequencing efforts on those fragments most likely to yield valuable information. On the other hand, in the case of WGS or pyrosequencing approaches, the small size of the fragment allows high-throughput sampling of many sequences, and the methods require little or no pre-sequencing investments in library stocking and screening. Theoretically, the large- versus small- insert methods converge if the small fragments can be assembled to produce large genome regions or even complete genomes. This is possible if the community DNA is sequenced deeply enough to produce overlapping sequences from contiguous regions of individual genomes. In reality, the high diversity of most natural microbial systems has precluded assembly of complete genomes or even genome regions from most small-fragment metagenomic libraries (DeLong 2005). The only exception among marine metagenomic studies is the partial genome assemblies obtained from some of the dominant members (e.g., *Prochlorococcus* and SAR11) of a deeply sequenced (i.e., millions of sequencing reads) marine planktonic community in the Sargasso Sea (Venter et al. 2004).

Large-fragment metagenomics is enabling a growing list of discoveries in marine microbial ecology. For example, genome fragments cloned into BAC or fosmid vectors have revealed novel transport proteins in an uncultured sponge symbiont (Fieseler et al. 2006), and the basis for cold adaptation in proteins from Antarctic bacteria (Grzymski et al. 2006). Typically, the genomic contents of large-fragment libraries are surveyed by "end sequencing," in which each end of the fragment is sequenced inward from a primer annealing to the fosmid or BAC vector. These end sequence surveys provide an initial picture of the captured genomic material and help identify fragments worthy of complete sequencing.

But while large pieces of ordered genome regions are critical for some questions, they are by no means necessary for all. For example, discovering an overabundance of selenocysteine-containing proteins in the Sargasso Sea (Zhang et al. 2005) only required sequences large enough to definitively identify individual genes. Likewise, investigation of the abundance and diversity of genes harbored in three disparate microbial communities was carried out with unassembled sequence reads (referred to as "environmental genome tags," EGTs; Tringe et al. 2005). Accordingly, for some ecological questions, the amount of sequence data in a meta-genomic library may be more important than the size of the original fragment that housed the DNA. The WGS sequencing libraries constructed from surface waters of the Sargasso Sea (Venter et al. 2004) and along a transect from Nova Scotia to the Galapagos Islands as part of the Global Ocean Sampling project (Rusch et al. 2007) are the largest marine metagenomic libraries available to date, harboring 1600 Mbp and 7500 Mbp of sequence data. The larger the metagenomic database, the more representative it will be of the source community.

Ideally, the ecological question at hand should dictate the type of genomic data that is collected. To learn about the biology of a recently isolated novel marine prokaryote, for example, a pure-culture genomics approach is an obvious next step. But to discover novel metabolic processes in the ocean, a large WGS metagenomic library is a better tactic to use. However, because metagenomic datasets are expensive to obtain and serve as public resources long after their initial publication, they will be used by the microbial ecology community for many questions outside the scope of the original study. Thus the North Pacific Subtropical Gyre (Station Aloha) metagenomic dataset initially collected to investigate stratification of microbial genes in the ocean (DeLong et al. 2006) was subsequently used to quantify DMSP

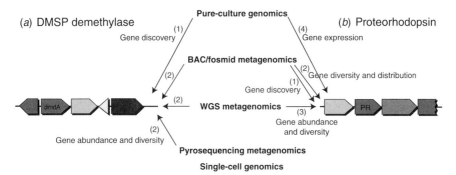

Figure 4.3 Current knowledge of the biogeochemically relevant genes for DMSP demethylation (*a*) and light harvesting via proteorhodopsin (*b*) is based on a variety of genomic methods. Single-cell genomics is likely to contribute significantly in future studies. In (*a*): (1, 2) Howard et al. (2006). In (*b*): (1) Béjà et al. (2000); (2) Béjà et al. (2001) and Sabehi et al. (2003); (3) Venter et al. (2004); (4) Giovannoni et al. (2005b).

demethylase genes (Howard et al. 2006), while the Sargasso Sea data (Venter et al. 2004) was used to search for genes encoding methylotrophy in marine bacterioplankton (Kalyuzhnaya et al. 2005). Often, a range of genomic data types can shed light on a central ecological question, each emphasizing a different aspect of that question (Fig. 4.3).

Metagenomic datasets are assuming an increasingly prominent place in marine microbial ecology. The first marine metagenomic libraries, now a decade old, were used for sequencing genome fragments from uncultured *Crenarchaeota* (Stein et al. 1996) and *Planctomycetales* (Vergin et al. 1998) in the North Pacific. The largest single marine metagenomic library is the Sargasso Sea dataset (Venter et al. 2004), composed of sequences from four different sites near Bermuda sampled in February and May of 2003. The first depth-resolved marine metagenomic library is from Station Aloha (DeLong et al. 2006). The best spatially resolved marine metagenomic library is from the Global Ocean Sampling of 48 surface samples from the eastern United States to the Gulf of Mexico, Panama, and the Galapagos Islands (Rusch et al. 2007). Finally, the first marine metagenomic libraries of DNA-based (Breitbart et al. 2002) and RNA-based virus communities (Culley et al. 2006) are from two coastal Pacific sites.

GENOMICS IN MARINE MICROBIAL ECOLOGY

The Ecology of Genome Composition

Genes are added to prokaryotic genomes over evolutionary time through duplication events, lateral transfer of genes, and acquisition of plasmid and bacteriophage DNA (Frank et al. 2002; Lerat et al. 2005). But the energetic costs of maintaining genes (including replication, repair, and regulation) act to counterbalance the potential benefits of genome expansion, so that most acquired genes are lost (Lawrence et al. 2001). Thus prokaryotic genomes are in flux over evolutionary time, and may not be finely tuned for the exact conditions in which an organism is currently found. Yet, with this caveat in mind, the composition of genes in a genome can nonetheless provide insights into a marine prokaryote's ecological strategy and biogeochemical role.

Reverse Biogeochemistry: Discovery of New Ecological Processes

How marine microbes obtain carbon and energy forms the basis of the role that they play in ocean biogeochemistry. Genome science has brought new insights to bear on this topic through the discovery of unsuspected genes indicating previously unknown activities, essentially allowing us to learn about biogeochemical processes going upward from the genes rather than downward from the processes. For example, a genome fragment from a member of the uncultured SAR86 bacteria was found to encode a rhodopsin-like protein, leading to the discovery of light-driven proton pumping in the membranes of marine bacterioplankton (Béjà et al. 2000). This amazing finding uncovered a formerly unrecognized mechanism for harvesting light energy in ocean surface waters (see Chapter 5). Genomic data revealed that the "proteorhodopsin" protein is widespread in the ocean (Béjà et al. 2001; Venter et al. 2004) and harbored in SAR11 (Giovannoni et al. 2005b) and other major bacterioplankton taxa (see Table 4.6). Both metagenomics (Bèjá et al. 2001; Venter et al. 2004; Sabehi et al. 2005; Frigaard et al. 2006) and pure-culture genomics (Giovannoni et al. 2005b, Table 4.6) have contributed significantly to our understanding of this process in the ocean.

Another example of a reverse biogeochemistry discovery comes from the growing evidence that the elusive marine crenarchaeotes (the "Group I.1A" *Crenarchaeota*), which account for as much as 20 percent of the prokaryotes in the ocean, live as autotrophs, using ammonia produced from the decomposition of organic matter at depth as the source of reducing power to fuel carbon fixation. Recent isolation of a free-living ammonia-oxidizing marine *Crenarchaeota* (Könneke et al. 2005) now provides the opportunity for detailed metabolic studies of a representative of this group, but it was metagenomic techniques that provided the first information about the pathways for carbon assimilation and ammonia oxidation. Initial evidence of ammonia oxidation came from an archaeal genome fragment assembled from the Sargasso Sea dataset (Venter et al. 2004). The presence of an ammonium monoxygenase gene in a genome assembly of archaeal origin suggested that marine *Archaea* participate in nitrification, a process previously thought to be carried out only by *Bacteria*. Subsequently, archaeal genome fragments from a fosmid library containing DNA from the uncultured marine sponge symbiont *Cenarchaeum symbiosum* provided information on mechanisms of ammonia oxidation (Hallam et al. 2006). Evidence for carbon fixation by the 3-hydroxypropionate cycle and for an oxidative TCA cycle suggested that these nitrifying prokaryotes can grow using either autotrophy or mixotrophy (Hallam et al. 2006).

TABLE 4.3 Key to Functional Genes in Tables 4.4–4.6

Nitrogen Cycle Genes

narG   nitrate reductase, α subunit
nirS   nitrite reductase
nirK   nitrite reductase
norB   nitric oxide reductase, large subunit
noxZ   nitrous oxide reductase
amoA   ammonia monooxygenase, subunit A
hao   hydroxylamine oxidoreductase
nifH   nitrogenase, azoferredoxin

Methylotrophy Genes

mtdB   methylene tetrahydromethanopterin dehydrogenase
mch   methenyltetrahydromethanopterin cyclohydrolase
fhcD   formylmethanofuran–tetrahydromethanopterin formyltransferase
fae   formaldehyde-activating enzyme

Oxidative Genes

xsc   sulfoacetaldehyde acetyltransferase (taurine degradation)
aphA   acetylpolyamine aminohydrolase (polyamine degradation)
dmdA   DMSP demethylase (DMSP degradation)
soxB   sulfur oxidation B protein (inorganic sulfur oxidation)
dmgdh dimethylglycine dehydrogenase (glycine betaine degradation)
chiC   group I chitinase
pcaH   protocatechuate 3,4-dioxygenase (aromatic ring cleavage)
boxA   benzoyl CoA oxygenase (aromatic ring cleavage)
nahH   catechol dioxygenase (aromatic ring cleavage)
coxL   carbon monoxide dehydrogenase, large subunit

Photoheterotrophy Genes

pufM   photosynthetic reaction center gene
cbbM/L RuBisCo
prho   proteorhodopsin

Nutrient Acquisition Genes

amt   ammonium transporter
napA   nitrate permease
narB   nitrate reductase
nasA   assimilatory nitrate reductase
nirA   nitrite reductase
nirB   nitrite reductase
ureC   urease
phnC   phosphonate transport (ABC-type transporter component)
phoA   alkaline phosphatase
pstS   phosphate uptake (ABC-type transporter component)

While the evolutionary trajectory of a gene cannot be ascertained from a single snapshot of genome composition, it is likely that many genes carried by marine prokaryotes are indeed of ecological value to them. Thus genes may give us hints as to what conditions the microbe expects to encounter in the ocean. For example, the presence of virulence (*vir*) genes in genomes of cultured members of the *Roseobacter* clade suggests that DNA or proteins might be directly transferred to eukaryotic plankton (Moran et al. 2007). If borne out by experimental evidence, this gives new insights into the nature of physical coupling between autotrophic and heterotrophic plankton. Likewise, the presence of 110 sulfatase genes in the genome of marine planctomycete *Rhodopirellula baltica*, a number two orders of magnitude higher than the next closest (i.e., six sulfatases in *Pseudomonas aeruginosa*), suggests that *R. baltica* acquires carbon from sulfur-containing glycopolymers abundant in algae and marine snow (Glöckner et al. 2003). Not surprisingly, the biggest opportunities for discovery of new or unexpected microbial activities come from genomes of uncultured prokaryotes, particularly those without close relatives in culture. Thus metagenomic studies typically reveal more novel processes or metabolisms in marine environments than pure culture genomic studies do.

Environmental Reductionism: New Details About Recognized Processes

Genomic science provides a powerful way to rapidly expand what we know about recognized biogeochemical processes, including which organisms are capable of mediating a given function, the diversity of enzymes that carry out the function, and where in the marine environment the function could occur. Thus we can collect detailed "reductionist" information on the underpinnings of a bulk process that is collectively measured for the bacterioplankton community. Such studies can focus on specific pathways (e.g., nitrogen fixation and carbon monoxide oxidation) or on specific taxa (e.g., *Prochlorococcus* or *Roseobacter*), using genomic and metagenomic techniques to obtain new details about recognized processes and interactions in the ocean.

The genome of the marine bacterium *Gramella forsetii*, for example, includes many genes for glycoside hydrolases, peptidases, and polymer binding via cell surface complexes (Bauer et al. 2006). These genes suggest the genetic foundation for frequent observations (Kirchman 2002) that marine *Bacteroidetes* specialize in the mineralization of high-molecular-weight organic matter in the sea (Bauer et al. 2006). The genome of deep-sea bacterium *Photobacterium profundum* contains genes for chitin and cellulose degradation that are upregulated under high pressure. These genes suggest that polymer-rich particles exported from ocean surface waters are important substrates for deep-sea bacteria (Vezzi et al. 2005).

Insights into how prokaryotic cells may adapt to substrate limitation emerged from analysis of the *Colwellia psychrerythraea* genome. This Arctic sediment bacterium produces polyhydroxyalkanoate (PHA) and polyamides, two storage polymers functioning as carbon and nitrogen reserves for the cell. Such high intracellular reserves can assist the bacterium in surviving cold-imposed limitations on carbon and nutrient uptake (Methé et al. 2005). Another cold-loving bacterium, *P. haloplanktis*, has abundant mechanisms for dealing with damaging reactive oxygen species, a significant challenge in cold environments because of the increased solubility of oxygen in seawater (Médigue et al. 2005).

Analyses of marine cyanophage genomes have provided details about bacterial–viral interactions in the ocean. Homologs of the gene for the cyanobacterial photosystem II core reaction center protein D1 (*psbA*) were found in several cyanophage genomes (Lindell et al. 2005). The phage-encoded genes are transcribed during infection, presumably to bolster host photosynthesis by replacing the non-functional host *psbA* genes and maintaining energy generation until phage replication is complete (Lindell et al. 2005). While the importance of phage in controlling marine cyanobacterial populations has certainly been recognized previously (Suttle and Chan 1994), genomic studies are now providing insights into the mechanisms by which they do so. In another example, the genome sequence of roseophage SIO1 suggests that the phage makes use of a host transcriptional activator (*phoB*) to transcribe its own genes for scavenging phosphorus from host DNA (Rohwer et al. 2000). Thus nucleotides from degraded host DNA are recycled by the phage.

While these studies used sequence similarity to known proteins as the basis for identifying genes of ecological interest in prokaryotic and viral genomes, another strategy is to screen microbial genomes based on observed phenotype. In the earliest marine application of functional screening, random genome fragments from seawater were searched for the ability to produce enzymes capable of degrading chitin, an important structural polysaccharide synthesized by marine invertebrates and plankton (Cottrell et al. 1999). Environmental DNA fragments (2–10 kb) were cloned into lambda phage, and 230,000 clones were screened for expression of chitin-relevant proteins using fluorogenic reporter compounds. Although marine environmental libraries had been screened previously using sequence homology to known genes (Schmidt et al. 1991; Stein et al. 1996; Vergin et al. 1998), this study was the first example of sequence-independent expression-based screening of wild marine DNA. Functional screens have since been used to identify novel polyketides, modular proteins commonly involved in cell–cell communication and defense, in prokaryotic symbionts of a marine sponge (Schirmer et al. 2005). The use of heterologous expression systems to search through genomes of uncultured microbes holds enormous promise for discovering new versions of prokaryotic enzymes encoding a variety of known functions (Schloss and Handelsman 2003; Robertson et al. 2004; Uchiyama et al. 2005).

A **functional screen** is a test to identify individuals possessing a phenotype of interest. To apply this method to marine microbes, genome or community genome fragments can be inserted into *Escherichia coli* cells, and the recombinant *E. coli* then screened for phenotypes of interest, such as chitinase activity (Cottrell et al. 1999; LeCleir et al. 2006) or nitrilase activity (Robertson et al. 2004).

Comparative Genomics and Metagenomics

Many advances in marine microbial genomics have been based on a single organism or a single ecosystem, such as identifying SAR11 strategies for life in low-nutrient seawater (Giovannoni et al. 2005a) or obtaining the first functional gene sequences from an uncultivated candidate phylum (Fieseler et al. 2006). A complementary approach addresses multiorganism or multiecosystem questions using comparative genomic methodologies to search for patterns across sequence datasets. The feasibility of comparative analyses hinges on the availability of sequence data, but growing numbers of microbial sequences are providing opportunities to investigate patterns across microbial groups (i.e., taxonomic patterns) or across space and time in the environment (i.e., biogeographical and temporal patterns). Such analyses hinge on the development of databases to integrate available marine genomic/metagenomic data and accompanying environmental data into a single data repository (e.g., MegX and CAMERA: Lombardot et al. 2006; Seshadri et al. 2007).

The first comparative analyses of marine metagenomic datasets have led to insights into environment-specific fine-tuning of gene inventories. In a comparison of metagenomic libraries from three environments (two of which were marine), gene stoichiometries in unassembled genome fragments suggested functions that may be characteristic of each environment (Tringe et al. 2005). Potential orthologs for proteorhodopsins, organic osmolyte transporters, and sodium export genes were abundant in genomes of ocean surface-water bacterioplankton (the Sargasso Sea), for example, but not in genomes from a soil microbial community. Hypothetical genes (those whose biological function is not known) were often the most-overrepresented genes in a given environment (Tringe et al. 2005), emphasizing the importance of increased efforts in gene and protein discovery and characterization.

Another comparative metagenomic analysis evaluated genomic fragments from seven depths at Station Aloha in the North Pacific Subtropical Gyre (DeLong et al. 2006). Surface-water samples had a higher representation of genes associated with photosynthesis, proteorhodopsins, chemotaxis, and DNA photorepair, while deep-water samples had relatively more genes for protein degradation, small-organic-molecule degradation, and polysaccharide and antibiotic production. The high abundance of cyanophage genes in the photic zone community DNA, presumably from viral particles replicating inside host cells, indicated the potential for viruses to significantly affect primary production (DeLong et al. 2006).

Comparative genomic analyses of genomes of cultured marine prokaryotes have led to insights into the basis for phenotypic diversity among closely related strains. Genome comparisons of two *Prochlorococcus* strains identified regions where most of the strain-specific genes were concentrated. These "genomic islands" appeared in the same positions in both strains, as well as in uncultured *Prochlorococcus* populations from the Sargasso Sea (Fig. 4.4), indicating they are hotspots for genetic recombination and niche differentiation (Coleman et al. 2006). In cultured marine roseobacters, the strain-specific genes are instead concentrated in small (approximately 100 kb) plasmids (Moran et al. 2007). Many of these plasmid genes have highest sequence similarities to genes in different taxa (Fig. 4.4), including *Deltaproteobacteria*, *Betaproteobacteria*, *Cyanobacteria*, and *Actinobacteria*.

Patterns in gene occurrence among cultured prokaryotes can also generate hypotheses about differences in ecological strategy and biogeochemical potential across major taxa. For example, genome comparisons of representatives of the SAR11

GENOMICS IN MARINE MICROBIAL ECOLOGY

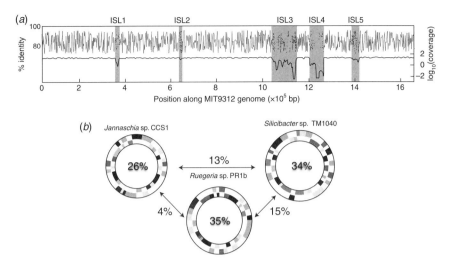

Figure 4.4 Comparative genomics has provided insights into the basis for ecological diversity among related prokaryotic strains. (*a*) Genomic islands in *Prochlorococcus* strain MIT9312 (indicated by "ISL") are evident as regions of low coverage when Sargasso Sea metagenomic sequences are aligned to the MIT9312 genome (from Coleman et al. 2006. *Science* 311: 1768–1770. Reprinted with permission from AAAS). (*b*) Small *Roseobacter* plasmids contain many unique genes. Double-ended arrows indicate how many genes are shared among the plasmids; numbers inside the plasmids indicate how many genes are species-specific based on absence of an ortholog in 11 other *Roseobacter* group genomes (Moran et al. 2007).

and *Roseobacter* clades, two of the most abundant heterotrophic bacterioplankton groups in the ocean, suggest that the former's ecological strategy is to make very efficient use of dilute compounds in oligotrophic regions of the ocean, while the latter's is to take advantage of high-nutrient microzones associated with particles and plankton (Moran et al. 2004, 2007; Giovannoni et al. 2005b).

As the database of marine prokaryotic genome sequences grows to include more of the numerically important marine taxa (see Chapter 3), comparative genomic studies across major groups will shape our understanding of what defines and differentiates each one. In a simple approach to cross-taxa analysis, the available marine bacterioplankton genomes (as of June 2006) were searched for orthologs to selected biogeochemically relevant genes including those involved in nitrogen cycling (e.g., *narG*, *nirS*, *norB*, *nosZ*, *amo*, and *nifH*), sulfur cycling (e.g., *soxB*, *dmdA*, and *xsc*), carbon metabolism (e.g., *chiA*, *pcaH*, *boxA*, *coxM*, *mtdB*, and *fhcD*), and nutrient acquisition (e.g., *amt*, *napA*, *narB*, *nirA*, *phnC*, and *phoA*). With the important caveat that genes were identified by autoannotation (using cut-off values of $E < 10^{-40}$ and a percentage similarity >40 percent), patterns that emerge from the analysis can be used to generate hypotheses about each group's role in the ocean (Tables 4.3–4.6). For example, chitin degradation may be dominated by *Gammaproteobacteria* in the *Vibrionales*, since *chiC* orthologs were found only in this group, while carbon monoxide oxidation (via *coxL*) may be dominated by

TABLE 4.4 Comparative Genomic Analysis of Biogeochemically Relevant Genes in Marine Bacterioplankton: Nitrogen Cycling and Methylotrophy[a]

	narG	nirS	nlrK	norB	nosZ	amoA	hao	nifH	mtdB	mch	fhcD	fae
Alphaproteobacteria												
Roseobacter												
Silicibacter pomeroyi		x		x	x		x					
Silicibacter sp. TM1040												
Jannaschia sp. CCS1												
Sulfitobacter EE36												
Sulfitobacter NAS4.1												
Roseovarius sp. 217			x	x	x							
Roseovarius nubinhibens												
Roseobacter sp. MED193	x											
Loktanella vestfoldensis SKA53												
Oceanicola batensis												
Oceanicola granulosus												
Rhodobacterales bacterium HTCC2654			x	x	x							
SAR11												
Pelagibacter ubique HTCC1002												
Pelagibacter ubique HTCC1062												
Rhizobacterales												
Oceanicaulis alexandrii	x											
Nitrobacter sp. Nb 311A									x	x	x	
Aurantimonas sp. SI85-9A1												x

Species						
Other alphas						
Erythrobacter NAP1						
Erythrobacter litoralis HTCC2594						
Parvularcula bermudensis						
Sphingomonas SKA58						
Sphingopyxis alaskensis				x		
Magnetococcus MC1					x	x
Gammaproteobacteria						
Oceanospirilliaceae						
Reinekea sp. MED297			x			
Oceanospirillum MED92	x					
Marinomonas sp. MED121						
Oceanobacter sp. RED65						
Alteromonadales						
Idiomarina baltica	x					
Idiomarina loihiensis						
Pseudoalteromonas haloplanktis	x					
Pseudoalteromonas tunicata D2						
Alteromonas macleodii		x				
Psychromonas sp. CNPT3				x		
Shewanella frigidimarina	x			x		
Vibrionales						
Vibrio angustum					x	
Vibrio sp. MED222					x	
Vibrio alginolyticus 12G01					x	

(*Continued*)

TABLE 4.4 Continued

	narG	nirS	nlrK	norB	nosZ	amoA	hao	nlfH	mtdB	mch	fhcD	fae
Vibrio splendidus 12B01							x					
Photobacterium SKA34												
Photobacterium profundum 3TCK				x	x		x					
Photobacterium profundum SS9				x	x		x					
Other gammaproteobacteria												
Gammaproteobacterium KT71												
Nitrococcus mobilis Nb231	x											
Gammaproteobacterium HTCC2207												
Bacteroidetes												
Flavobacteria												
Polaribacter irgensii 23-P												
Flavobacteria sp. MED0217												
Cellulophaga sp. MED134												
Flavobacteria bacterium BBFL7												
Flavobacteriales bacterium HTCC2170				x								
Robiginitalea biformata HTCC2501				x								
Croceibacter atlanticus												
Tenacibaculum sp. MED152												

Planctomycetes
Blastopirellula marina DSM3645T x x
Rhodopirellula baltica x x x

Actinobacteria
Janibacter HTCC2649 x
Agreia sp. PHSC20c1

Cyanobacteria
Prochlorococcus marinus str. MIT9211
Prochlorococcus MED4
Prochlorococcus 9312
Prochlorococcus 9313
Prochlorococcus CCMP1375
Synechococcus sp. RS9917
Synechococcus sp. WH5701
Synechococcus sp. WH7805
Synechococcus WH8102

Firmicutes
Bacillus sp. NRRL B-14911
Oceanobacillus iheyensis

[a]See Table 4.3 for gene key.

TABLE 4.5 Comparative Genomic Analysis of Biogeochemically Relevant Genes in Marine Bacterioplankton: Taurine, Polyamine, DMSP, Sulfur Oxidation, Glycine Betaine, Chitin, Aromatic Monomers, and Carbon Monoxide Oxidation.[a]

	xsc	aphA	dmdA	soxB	dmgdh	chlC	pcaH	boxA	nahH	coxL
Alphaproteobacteria										
Roseobacter										
Silicibacter pomeroyi	x	x	x	x	x		x	x		x
Silicibacter sp. TM1040	x	x	x		x		x			x
Jannaschia sp. CCS1	x		x		x		x			x
Sulfitobacter EE36				x	x		x			x
Sulfitobacter NAS4.1				x	x		x			x
Roseovarius sp. 217	x	x	x	x	x		x			x
Roseovarius nubinhibens	x	x	x	x	x		x			
Roseobacter sp. MED193	x	x	x	x	x		x			x
Loktanella vestfoldensis SKA53					x		x			x
Oceanicola batensis					x		x			x
Oceanicola granulosus					x					x
Rhodobacterales bacterium HTCC2654				x	x			x		x
SAR11										
Pelagibacter ubique HTCC1002	x		x		x					
Pelagibacter ubique HTCC1062	x		x		x					
Rhizobacterales										
Oceanicaulis alexandril										
Nitrobacter sp. Nb 311A					x		x			x
Aurantimonas sp. SI85-9A1										x

Other alphas
Erythrobacter NAP1   x
Erythrobacter litoralis HTCC2594
Parvularcula bermudensis
Sphingomonas SKA58
Sphingopyxis alaskensis
Magnetococcus MC1   x

Gammaproteobacteria
Oceanospirilliaceae
Reinekea sp. MED297
Oceanospirillum MED92   x
Marinomonas sp. MED121   x
Oceanobacter sp. RED65

Alteromonadales
Idiomarina baltica
Idiomarina loihiensis
Pseudoalteromonas haloplanktis   x
Pseudoalteromonas tunicata D2
Alteromonas macleodii
Psychromonas sp. CNPT3
Shewanella frigidimarina

Vibrionales
Vibrio angustum   x
Vibrio sp. MED222   x
Vibrio alginolyticus 12G01   x

(*Continued*)

TABLE 4.5 Continued

	xsc	aphA	dmdA	soxB	dmgdh	chlC	pcaH	boxA	nahH	coxL
Vibrio splendidus 12B01	x									
Photobacterium SKA34						x				
Photobacterium profundum 3TCK	x					x				
Photobacterium profundum SS9										
Other gammaproteobacteria										
Gammaproteobacterium KT71				x						
Nitrococcus mobilis Nb231										
Gammaproteobacterium HTCC2207			x		x				x	
Bacteroidetes										
Flavobacteria										
Polaribacter irgensii 23-P										
Flavobacteria sp. MED0217										
Cellulophaga sp. MED134										
Flavobacteria bacterium BBFL7										
Flavobacteriales bacterium HTCC2170										
Robiginitalea biformata HTCC2501										
Croceibacter atlanticus										
Tenacibaculum sp. MED152										

Planctomycetes
Blastopirellula marina DSM3645T
Rhodopirellula baltica   x

Actinobacteria
Janibacter HTCC2649
Agreia sp. PHSC20c1   x

Cyanobacteria
Prochlorococcus marinus str. MIT9211
Prochlorococcus MED4
Prochlorococcus 9312
Prochlorococcus 9313
Prochlorococcus CCMP1375   x
Synechococcus sp. RS9917
Synechococcus sp. WH5701
Synechococcus sp. WH7805
Synechococcus WH8102

Firmicutes
Bacillus sp. NRRL B-14911
Oceanobacillus iheyensis

[a]See Table 4.3 for gene key.

TABLE 4.6 Comparative Genomic Analysis of Biogeochemically Relevant Genes in Marine Bacterioplankton: Phototrophy and Nutrient Acquisition[a]

	pufM	cbbML	prho	amt	napA	narB	nasA	nirA	nirB	ureC	phnC	phoA	pstS
Alphaproteobacteria													
Roseobacter													
Silicibacter pomeroyi				x	x					x	x		x
Silicibacter sp. TM1040				x						x	x		x
Jannaschia sp. CCS1	x		x					x			x	x	x
Sulfitobacter EE36				x		x		x	x		x	x	x
Sulfitobacter NAS4.1				x							x		x
Roseovarius sp. 217	x			x	x		x			x	x		x
Roseovarius nubinhibens				x			x				x		x
Roseobacter sp. MED193				x			x			x	x		x
Loktanella vestfoldensis SKA53	x			x						x	x		x
Oceanicola batensis				x						x			x
Oceanicola granulosus				x						x			x
Rhodobacterales bacterium HTCC2654					x		x			x	x		x
SAR11													
Pelagibacter ubique HTCC1002			x	x									
Pelagibacter ubique HTCC1062			x	x									x
Rhizobacterales													
Oceanicaulis alexandrii				x									
Nitrobacter sp. Nb 311A		x					x						x
Aurantimonas sp. SI85-9A1		x		x						x	x		x

Taxon	C1	C2	C3	C4	C5	C6	C7	C8
Other alphas								
Erythrobacter NAP1		x					x	x
Erythrobacter litoralis HTCC2594			x				x	x
Parvularcula bermudensis			x		x	x		
Sphingomonas SKA58			x		x	x	x	x
Sphingopyxis alaskensis			x		x		x	x
Magnetococcus MC1			x		x	x	x	x
Gammaproteobacteria								
Oceanospirilliaceae								
Reinekea sp. MED297			x	x	x	x	x	x
Oceanospirillum MED92			x		x	x	x	x
Marinomonas sp. MED121			x		x	x	x	x
Oceanobacter sp. RED65			x		x	x	x	x
Alteromonadales								
Idiomarina baltica			x				x	x
Idiomarina loihiensis			x				x	x
Pseudoalteromonas haloplanktis			x		x		x	x
Pseudoalteromonas tunicata D2			x		x		x	x
Alteromonas macleodii			x	x	x		x	x
Psychromonas sp. CNPT3			x				x	x
Shewanella frigidimarina			x	x	x		x	x
Vibrionales								
Vibrio angustum	x		x		x	x	x	x
Vibrio sp. MED222			x		x	x	x	x
Vibrio alginolyticus 12G01				x	x		x	x
Vibrio splendidus 12B01	x		x	x	x	x	x	x
Photobacterium SKA34			x	x	x		x	x

(*Continued*)

TABLE 4.6 Continued

	pufM	cbbML	prho	amt	napA	narB	nasA	nirA	nirB	ureC	phnC	phoA	pstS
Photobacterium profundum 3TCK	x			x	x		x		x	x	x		x
Photobacterium profundum SS9				x	x		x		x				x
Other gammaproteobacteria													
Gammaproteobacterium KT71				x						x			x
Nitrococcus mobilis Nb231		x		x									x
Gammaproteobacterium HTCC2207			x	x			x		x	x			x
Bacteroidetes													
Flavobacteria													
Polaribacter irgensii 23-P			x	x									
Flavobacteria sp. MED0217				x									
Cellulophaga sp. MED134				x									
Flavobacteria bacterium BBFL7													
Flavobacteriales bacterium HTCC2170				x									x
Robiginitalea biformata HTCC2501				x									
Croceibacter atlanticus													
Tenacibaculum sp. MED152			x	x									
Planctomycetes													
Blastopirellula marina DSM3645T				x	x				x		x		
Rhodopirellula baltica				x	x				x		x		x

Actinobacteria
Janibacter HTCC2649
Agreia sp. PHSC20c1

Cyanobacteria
Prochlorococcus marinus str. MIT 9211
Prochlorococcus MED4
Prochlorococcus 9312
Prochlorococcus 9313
Prochlorococcus CCMP1375
Synechococcus sp. RS9917
Synechococcus sp. WH5701
Synechococcus sp. WH7805
Synechococcus WH8102

Firmicutes
Bacillus sp. NRRL B-14911
Oceanobacillus iheyensis

[a]See Table 4.3 for gene key.

Alphaproteobacteria in the *Roseobacter* group. On the other hand, proteorhodopsin orthologs are widely distributed taxonomically among marine bacteria, and the ability to assimilate phosphonates (carbon-bonded organic phosphorus encoded by *phoA* or as C–P encoded by *phnC*) is quite common. Whether these trends hold up as more (and more ecologically relevant) genomes are sequenced remains to be seen, but this exercise demonstrates the power of comparative genome analysis for addressing biological characteristics and ecological roles that define marine prokaryotic taxa.

FUTURE DIRECTIONS

Several facets of marine genomics that are now on the horizon promise increasingly powerful tools for microbial ecologists. The most significant of these may be developments in DNA sequencing methodology that will dramatically decrease sequencing costs. These new approaches include pyrosequencing (Margulies et al. 2005) and nanopore sequencing, among others (Service 2006), and many are expected to have the advantage of both high accuracy and long reads. A massive decrease in sequencing costs promises an equally massive increase in access to genomes of cultured and uncultured marine microbes.

The two main avenues for marine environmental genomics that now exist, pure-culture genomics and metagenomics (Fig. 4.1), will soon be joined by a third: single-cell sequencing. The ability to sequence the genome of a single cell is made possible by "multiple-displacement amplification" (MDA; Dean et al. 2002). In this amplification method, random oligonucleotide primers initiate DNA synthesis from the original DNA template, and then secondary priming of these products leads to gradual amplification of the genome (Fig. 4.5). MDA produces microgram quantities of DNA from femtogram quantities of starting template. Whole-genome shotgun sequencing of a single cell of a cultured *Prochlorococcus* has been demonstrated using MDA amplification of the genome (Zhang et al. 2006). This approach has the potential to make available genome sequences of uncultured marine prokaryotes, as well, provided that the individual cells can be captured from the environment. MDA is also useful in metagenomic studies for which sufficient high-quality DNA cannot be readily obtained (Yokouchi et al. 2006; Abulencia et al. 2006).

Functional genomics is the process of converting information represented by genomic DNA into an understanding of gene functions and effects; in other words, using the genome sequence as a starting point from which to understand expression, function, and regulation of genes and proteins. Environmental proteomics or "metaproteomics" is a type of functional genomics that analyzes the collective proteins synthesized by a microbial community. In the first application of this method, two-dimensional gel electrophoresis was used to separate dominant proteins from Chesapeake Bay bacterioplankton communities based on differences in charge and molecular weight (Kan et al. 2005). Mass spectrometry was unable to identify most of 41 selected proteins, however, largely because the existing protein databases do not yet contain sequences with sufficiently high amino acid similarity (estimated to be a minimum of 97 percent identity) to produce a significant match to the environmental proteins. Proteins translated from the Sargasso Sea metagenomic

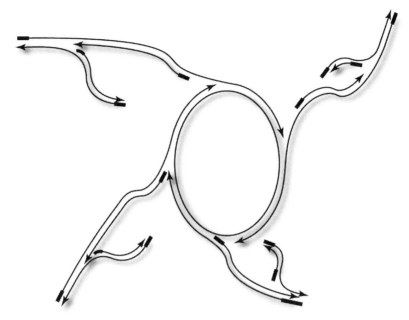

Figure 4.5 Multiple-displacement amplification (MDA) allows amplification of small amounts of starting DNA (e.g., from single cells or sparse communities) for genomic or metagenomic sequencing. The DNA polymerase used in MDA can synthesize over 10,000 nucleotides without dissociating from the template (in contrast to standard DNA polymerases, which dissociate readily). Random oligonucleotide primers initiate synthesis, and replicated strands are displaced by subsequent synthesis from the same template. Priming and elongation on the replicated strands results in a cascade of replication that amplifies the original template. (from Dean et al. 2002. PNAS 99: 5261–5266. Copyright 2002. National Academy of Sciences U.S.A.)

database (Venter et al. 2004) fared better in matching the Chesapeake Bay proteins (Kan et al. 2005).

Environmental transcriptomics or "metatranscriptomics" is another functional genomics technique that analyzes the collective genes transcribed by a microbial community, essentially providing a way to focus on the genes being expressed under particular ecological conditions (Poretsky et al. 2008). In the first application of this method, mRNAs were retrieved directly from a coastal bacterioplankton community and sequenced after reverse transcription and cloning (Fig. 4.6). The transcripts indicated that oxidation of sulfur (via the *soxA* gene), assimilation of C1 compounds (via *fdh1B*), and acquisition of nitrogen from polyamine metabolism (via *aphA*) were ongoing in the microbial community at the time of sample collection (Poretsky et al. 2005). Both environmental proteomics and transcriptomics provide a means of exploring cellular-level controls on the activity of natural microbial communities without bias toward the expected.

Finally, environmental microarrays based on genomic and metagenomic data are developing functional genomics tools for microbial ecology. Early use of this approach

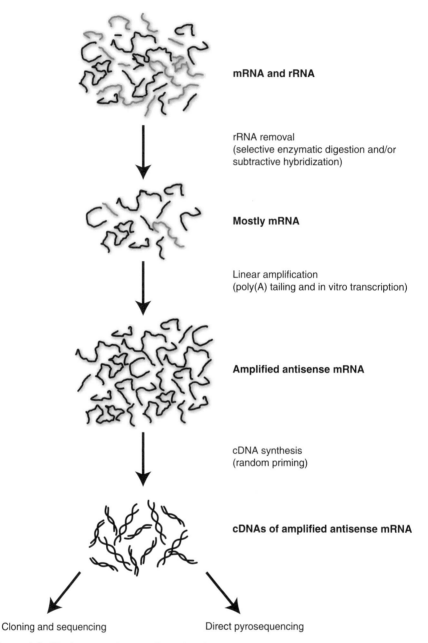

Figure 4.6 Environmental transcriptomics focuses sequencing efforts on the genes in a natural microbial community that are being transcribed at the time of sample collection. The protocol is from Poretsky et al. (2008).

focused on DNA in marine environments, including analysis of inventories of nitrogen cycle genes in estuarine sediments (Taroncher-Oldenburg 2003) and taxonomic characterization of marine bacterioplankton communities from arrayed 16S rDNA oligonucleotides (Peplies et al. 2004). Current use is focusing on mRNAs, using dynamics of the transcript pool to investigate spatial and temporal patterns in community gene expression.

While genomic techniques have initiated a revolution in how we can query microbes in the ocean, their value is ultimately tied to how much we know about the function of genes and proteins (Oremland et al. 2005) and how well we can measure rates of key biogeochemical processes that are intimately linked to the microbial gene pool.

SUMMARY

1. Marine microbial genomics and metagenomics allow the discovery of new ecological processes in the ocean starting from a gene sequence (i.e., "reverse biogeochemistry").
2. Marine microbial genomics and metagenomics allow accumulation of details about the genetic underpinnings of microbially mediated processes in the ocean (i.e., "environmental reductionism").
3. Templates for current genomic approaches can be relevant cultured microbes (pure-culture genomics) or DNA fragments (ranging in size from 0.1 to 100 kb) retrieved directly from natural communities (metagenomics). In the future, single-cell genomics will provide a third avenue for accessing relevant gene sequences from marine environments.
4. Developing genomic approaches that will continue to expand our understanding of microbial activity in the ocean include environmental proteomics, environmental transcriptomics, and environmental microarrays.

ACKNOWLEDGMENTS

This work was supported by funds from the Gordon and Betty Moore Foundation and the National Science Foundation. I thank C. English for graphics assistance; S. Sun for bioinformatics assistance; and D. Kirchman, R. Poretsky, and two reviewers for comments on an earlier draft.

REFERENCES

Abascal, F., and A. Valencia. 2003. Automatic annotation of protein function based on family identification. *Proteins: Struct. Funct. Genet.* 53: 683–692.

Abulencia, C. B., D. L. Wyborski, J. A. Garcia, et al. 2006. Environmental whole-genome amplification to access microbial populations in contaminated sediments. *Appl. Environ. Microbiol.* 72: 3291–3301.

Amann, R. I., W. Ludwig, and K. H. Schleifer. 1995. Phylogenetic identification and in situ detection of individual microbial cells without cultivation. *Microbiol. Rev.* 59: 143–169.

Azam, F., and A. Z. Worden. 2004. Microbes, molecules, and marine ecosystems. *Science* 303: 1622–1624.

Bauer, M., M. Kube, H. Teeling, et al. 2006. Whole genome analysis of the marine Bacteriodetes '*Gramella forsetii*' reveals adaptations to degradation of polymeric organic matter. *Environ. Microbiol.* 8: 2201–2213. doi: 10.1111/j.1462-2920.2006.01152x.

Béjà, O., E. V. Koonin, M. T. Suzuki, et al. 2000. Bacterial rhodopsin: Evidence for a new type of phototrophy in the sea. *Science* 289: 1902–1906.

Béjà, O., E. N. Spudich, J. L. Spudich, M. Leclerc, and E. F. DeLong. 2001. Proteorhodopsin phototrophy in the ocean. *Nature* 411: 786–789.

Bentley, S. D., and J. Parkhill. 2004. Comparative genomic structure of prokaryotes. *Annu. Rev. Genet.* 38: 771–791.

Breitbart, M., P. Salamon, B. Anderson, et al. 2002. Genomic analysis of uncultured marine viral communities. *Proc. Natl. Acad. Sci. USA* 99: 14250–14255.

Coleman, M. L., M. B. Sullivan, A. C. Martiny, et al. 2006. Genomic islands and the ecology and evolution of *Prochlorococcus*. *Science* 311: 1768–1770.

Connon, S. A., and S. J. Giovannoni. 2002. High-throughput methods for culturing microorganisms in very-low-nutrient media yield diverse new marine isolates. *Appl. Environ. Microbiol.* 68: 3878–3885.

Cottrell, M. T., J. A. Moore, and D. L. Kirchman. 1999. Chitinases from uncultured marine microorganisms. *Appl. Environ. Microbiol.* 65: 2553–2557.

Cottrell, M. T., L. Yu, and D. L. Kirchman. 2005. Sequence and expression analysis of *Cytophaga*-like hydrolases in western Arctic metagenomic library and the Sargasso Sea. *Appl. Environ. Microbiol.* 71: 8506–8513.

Culley, A. I., A. S. Lang, and C. A. Suttle. 2006. Metagenomic analysis of coastal RNA virus communities. *Science* 312: 1795–1798.

Dean, F. B., S. Hosono, L. Fang, et al. 2002. Comprehensive human genome amplification using multiple displacement amplification. *Proc. Natl. Acad. Sci. USA* 99: 5261–5266.

DeLong, E. F., 2005. Microbial community genomics in the ocean. *Nat. Rev. Microbiol.* 3: 459–469.

DeLong, E. F., C. M. Preston, T. Mincer, et al. 2006. Community genomics among stratified microbial assemblages in the ocean's interior. *Science* 311: 496–503.

Devos, D., and A. Valencia. 2001. Intrinsic errors in genome annotation. *Trends Genet.* 17: 429–431.

Doney, S. C., M. R. Abbott, J. J. Cullen, D. M. Karl, and L. Rothstein. 2004. From genes to ecosystems: The ocean's new frontier. *Front. Ecol. Environ.* 2: 457–466.

Edwards, R. A., B. Rodriguez-Brito, L. Wegley, et al. 2006. Using pyrosequencing to shed light on deep mine microbial ecology. *BMC Genomics* 7: 57.

Fieseler, L. A., Quaiser, C. Schleper, and U. Hentschel. 2006. Analysis of the first genome fragment from the marine sponge-associated, novel candidate phylum *Poribacteria* by environmental genomics. *Environ. Microbiol.* 8: 612–624.

Frank, A. C., H. Amiri, and S. G. Andersson. 2002. Genome deterioration: Loss of repeated sequences and accumulation of junk DNA. *Genetica* 115: 1–12.

Frigaard, N.-U., A. Martinez, T. J. Mincer, and E. F. DeLong. 2006. Proteorhodopsin lateral gene transfer between marine planktonic Bacteria and Archaea. *Nature* 439: 847–850.

Giovannoni, S. J., H. J. Tripp, S. Givan, et al. 2005a. Genome streamlining in a cosmopolitan oceanic bacterium. *Science* 309: 1242–1245.

Giovannoni, S. J., L. Bibbs, J. Cho, et al. 2005b. Proteorhodopsin in the ubiquitous marine bacterium SAR11. *Nature* 438: 82–85.

Glöckner, F. O., M. Kube, M. Bauer, et al. 2003. Complete genome sequence of the marine planctomycete *Pirellula* sp. strain 1. *Proc. Natl. Acad. Sci. USA* 100: 8298–8303.

Grzymski, J. J., B. J. Carter, E. F. DeLong, et al. 2006. Comparative genomics of DNA fragments from six Antarctic marine planktonic bacteria. *Appl. Environ. Microbiol.* 72: 1532–1541.

Hallam S. J., T. J. Mincer, C. Schleper, et al. 2006. Pathways of carbon assimilation and ammonia oxidation suggested by environmental genomic analyses of marine Crenarchaeota. *PLoS Biol.* 4: e95.

Hobbie, J. E., R. J. Daley, and S. Jasper. 1977. Use of Nucleopore filters for counting bacteria by fluorescence microscopy. *Appl. Environ. Microbiol.* 33: 1225–1228.

Howard, E. C., J. R. Henriksen, A. Buchan, et al. 2006. Bacterial taxa that limit sulfur flux from the ocean. *Science* 314: 649–653.

Jumas-Bilak, E., S. Michaux-Charachon, G. Bourg, M. Ramuz, and A. Allardet-Servent. 1998. Unconventional genomic organization in the alpha subgroup of the *Proteobacteria*. *J. Bacteriol.* 180: 2749–2755.

Kalyuzhnaya, M. G., O. Nercessian, A. Lapidus, and L. Chistoserdova. 2005. Fishing for biodiversity: Novel methanopterin-linked C-1 transfer genes deduced from the Sargasso Sea metagenome. *Environ. Microbiol.* 7: 1090–1916.

Kan, J., T. E. Hanson, J. M. Ginter, K. Wang, and F. Chen. 2005. Metaproteomic analysis of Chesapeake Bay microbial communities. *Saline Syst.* 1: 7.

Kirchman, D. L., 2002. The ecology of *Cytophaga–Flavobacteria* in aquatic environments. *FEMS Microbiol. Ecol.* 39: 91–100.

Könneke, M., A. E. Bernhard, J. R. De la Torre, et al. 2005. Isolation of an autotrophic ammonia-oxidizing marine archaeon. *Nature* 437: 543–546.

Krane, D. E., and M. L. Raymer. 2003. *Fundamental Concepts of Bioinformatics*. Benjamin Cummings.

Lawrence, J. G., R. W. Hendrix, and S. Casjens. 2001. Where are the pseudogenes in bacterial genomes? *Trends Microbiol.* 9: 535–540.

LeCleir, G. R., A. Buchan, J. Maurer, M. A. Moran, and J. T. Hollibaugh. 2007. Comparison of chitinolytic enzymes from an alkaline hypersaline lake and an estuary. *Environ. Microbiol.* 9: 197–205.

Lerat, E., V. Daubin, H. Ochman, and N. A. Moran. 2005. Evolutionary origins of genomic repertoires in bacteria. *PLoS Biol.* 3: e130.

Lindell, D., J. D. Jaffe, Z. I. Johnson, G. M. Church, and S. W. Chisholm. 2005. Photosynthesis genes in marine viruses yield proteins during host infection. *Nature* 438: 86–89.

Lombardot, T., R. Kottmann, H., Pfeffer, et al. 2006. Megx.net—database resource for marine ecological genomics. *Nucleic Acids Res.* 34: D390–D393.

Margulies, M., M. Egholm, W. E. Altman, et al. 2005. Genome sequencing in microfabricated high-density picolitre reactors. *Nature* 437: 376–380.

Médigue, C., E. Krin, G. Pascal, et al. 2005. Coping with cold: The genome of the versatile marine Antarctica bacterium *Pseudoalteromonas haloplanktis* TAC125. *Genome Res.* 15: 1325–1335.

Methé, B. A., K. E. Nelson, J. W. Deming, et al. 2005. The psychrophilic lifestyle as revealed by the genome sequence of *Colwellia psychrerythraea* 34H through genomic and proteomic analyses. *Proc. Natl. Acad. Sci. USA.* 102: 10913–10918.

Moran, M. A., A. Buchan, J. M., González et al. 2004. Genome sequence of *Silicibacter pomeroyi* reveals adaptations to the marine environment. *Nature* 432: 910–913.

Moran, M. A., R. Belas, M. A. Schell, et al. 2007. Ecological genomics of marine roseobacters. *Appl. Environ. Microbiol.* 73: 4559–4569.

Okada, K., T. Iida, K. Kita-Tsukamoto, and T. Honda. 2005. Vibrios commonly possess two chromosomes. *J. Bacteriol.* 187: 752–757.

Oremland, R. S., D. G. Capone, J. F. Stolz and J. Fuhrman. 2005. Whither or wither geomicrobiology in the era of "community metagenomics." *Nat. Rev. Microbiol.* 3: 572–578.

Overbeek, R., T. Begley, R. M. Butler, et al. 2005. The subsystems approach to genome annotation and its use in the project to annotate 1000 genomes. *Nucleic Acids Res.* 33: 5691–5702.

Peplies, J., S. C. K. Lau, J. Pernthaler, R. Amann, and F. O. Glöckner. 2004. Application and validation of DNA microarrays for the16S rRNA-based analysis of marine bacterioplankton. *Environ. Microbiol.* 6: 638–645.

Poretsky, R. S., N. Bano, A. Buchan, et al. 2005. Analysis of microbial gene transcripts in environmental samples. *Appl. Environ. Microbiol.* 71: 4121–4126.

Poretsky, R. S., N. Bano, A. Buchan, J. T. Hollibaugh and M. A. Moran. 2008. Environmental transcriptomics: A method for exploring community-level gene expression in natural samples. In Kowalchuk, G. A., F. J. de Bruijn, I. M. Head, A. D. Akkermans, J. D. van Elsas, (eds.), *Molecular Microbial Ecology Manual.* doi: 10.1007/978-1-4020-2177-0.

Riley, M. 1993. Functions of the gene products of *Escherichia coli*. *Microbiol. Rev.* 57: 862–952.

Robertson, D. E., J. A. Chaplin, G. DeSantis, et al. 2004. Exploring nitrilase sequence space for enantioselective catalysis. *Appl. Environ. Microbiol.* 70: 2429–2436.

Rocap, G., F. W. Larimer, J. Lamerdin, et al. 2003. Genome divergence in two *Prochlorococcus* ecotypes reflects oceanic niche differentiation. *Nature* 424: 1042–1047.

Rohwer, F., A. Segall, G. Steward, et al. 2000. The complete genomic sequence of the marine phage Roseophage SIO1 shares homology with nonmarine phages. *Limnol. Oceanogr.* 45: 408–418.

Rusch, D., A. L. Halpern, G. Sutton, et al. 2007. The Sorcerer II Global Ocean sampling expedition: Northwest Atlantic through Eastern Tropical Pacific. *PLoS Biol.* 5: e77.

Sabehi, G., R. Massana, J. P. Bielawski, et al. 2003. Novel proteorhodopsin variants from the Mediterranean and Red Seas. *Environ. Microbiol.* 5: 842–849.

Sabehi, G., A. Loy, K. H. Jung, et al. 2005. New insights into metabolic properties of marine bacteria encoding proteorhodopsins. *PLoS Biol.* 3: e273.

Schirmer, A., R., Gadkari, C. D. Reeves, et al. 2005. Metagenomic analysis reveals diverse polyketide synthase gene clusters in microorganisms associated with the marine sponge *Discodermia dissoluta*. *Appl. Environ. Microbiol.* 71: 4840–4849.

REFERENCES

Schloss, P. D., and J. Handelsman. 2003. Biotechnological prospects from metagenomics *Curr. Opin. Biotechnol.* 14: 303–310.

Schmidt, T. M., E. F. DeLong, and N. R. Pace. 1991. Analysis of a marine picoplankton community by 16S rRNA gene cloning and sequencing. *J. Bacteriol.* 173: 4371–4378.

Service, R. F. 2006. The race for the $1000 genome. *Science* 311: 1544–1546.

Seshadri R., S. A. Kravitz, L. Smarr, P. Gilna, and M. Frazier. 2007. CAMERA: A community resource for metagenomics. *PLoS Biol.* 5: e75.

Sobecky, P. A. 1999. Plasmid ecology of marine sediment microbial communities. *Hydrobiologia* 401: 9–18.

Stein, J. L., T. L. Marsh, K. Y. Wu, H. Shizuya, and E. F. DeLong. 1996. Characterization of uncultivated prokaryotes: Isolation and analysis of a 40-kilobase-pair genome fragment from a planktonic marine archaeon. *J. Bacteriol.* 178: 591–599.

Suttle, C. A., and A. M. Chan. 1994. Dynamics and distribution of cyanophages and their effect on marine *Synechococcus* spp. *Appl. Environ. Microbiol.* 60: 3167–3174.

Taroncher-Oldenburg, G., E. M. Griner, C. A. Francis, and B. B. Ward. 2003. Oligonucleotide microarray for the study of functional gene diversity in the nitrogen cycle in the environment. *Appl. Environ. Microbiol.* 69: 1159–1171.

Tringe, S. G., C. von Mering, A. Kobayashi, et al. 2005. Comparative metagenomics of microbial communities. *Science* 308: 554–557.

Venter, J. C., K. Remington, J. F. Heidelberg, et al. 2004. Environmental genome shotgun sequencing of the Sargasso Sea. *Science* 304: 66–74.

Vergin, K. L., E. Urbach, J. L. Stein, et al. 1998. Screening of a fosmid library of marine environmental genomic DNA fragments reveals four clones related to members of the order Planctomycetales. *Appl. Environ. Microbiol.* 64: 3075–3078.

Vezzi, A., S. Campanaro, M. D'Angelo, et al. 2005. Life at depth: *Photobacterium profundum* genome sequence and expression analysis. *Science* 307: 1459–1461.

Uchiyama, T., T. Abe, T. Ikemura, and K. Watanabe. 2005. Substrate-induced gene-expression screening of environmental metagenome libraries for isolation of catabolic genes. *Nat. Biotechnol.* 23: 88–93.

Westover, B. P., J. D. Buhler, J. L. Sonnenburg, and J. I. Gordon. 2005. Operon prediction without a training set. *Bioinformatics* 21: 880–888.

Williams, P. J., Le, B., and C. Askew. 1968. A method of measuring the mineralization by micro-organisms of organic compounds in sea water. *Deep-Sea Res.* 15: 365–378.

Yokouchi, H., Y. Fukuoka, D. Mukoyama, et al. 2006. Whole-metagenome amplification of a microbial community associated with scleractinian coral by multiple displacement amplification using φ29 polymerase. *Environ. Microbiol.* 8: 1155–1163.

Zhang, K., A. C. Martiny, N. B. Reppas, et al. 2006. Sequencing genomes from single cells by polymerase cloning. *Nat. Biotechnol.* 24: 680–686.

Zhang, Y., D. E. Fomenko, and V. N. Gladyshev. 2005. The microbial selenoproteome of the Sargasso Sea. *Genome Biol.* 6: R37.

5

PHOTOHETEROTROPHIC MARINE PROKARYOTES

ODED BÉJÀ
Faculty of Biology, Technion–Israel Institute of Technology, Haifa, Israel

MARCELINO T. SUZUKI
Chesapeake Biological Laboratory, University of Maryland, Center for Environmental Science, Solomons, MD 20688, U.S.A.

INTRODUCTION

Photoheterotrophs are microorganisms that use light as an energy source while exploiting organic compounds as their carbon and energy source. In the marine environment, these microorganisms are currently divided into three groups. The first group includes photoautotrophic cyanobacteria such as *Prochlorococcus* species. These cyanobacteria have been suggested to act as facultative photoheterotrophs in oligotrophic ecosystems. A second group is composed of aerobic anoxygenic phototrophic (AAnP) bacteria. This group was recently "rediscovered" in marine systems and accounts in some oceanic regimes for as much as 20 percent of total bacteria in the euphotic zone. Last is a collection of newly discovered and divergent bacterial groups that use bacterial rhodopsins as their light-harvesting pigments.

This chapter will discuss and focus on the occurrence, diversity, and insights into the significance of these photoheterotrophic organisms in the marine environment.

Microbial Ecology of the Oceans, Second Edition. Edited by David L. Kirchman
Copyright © 2008 John Wiley & Sons, Inc.

> **Phototrophs** These organisms use light energy for the production of cellular energy. They can be further divided as *photoautotrophs*, which obtain energy from light and synthesize cellular biomass predominantly from inorganic molecules, and *photoheterotrophs*, which obtain energy from light and synthesize cellular biomass predominantly from organic molecules.
>
> **Anoxygenic Phototrophs** The metabolism of these microbes does not produce molecular oxygen (O_2). There are two types: *anaerobic anoxygenic phototrophs*, which thrive in anoxic environments and in many cases use reduced inorganic compounds to produce cellular biomass from carbon dioxide (CO_2), and *aerobic anoxygenic phototrophs*, which thrive in oxic environments and, so far, have not been shown to produce significant amounts of cellular biomass from carbon dioxide.

FACULTATIVE PHOTOHETEROTROPHY BY UNICELLULAR CYANOBACTERIA

Cyanobacteria as Facultative Heterotrophs

Cyanobacteria in the genera *Synechococcus* and *Prochlorococcus* are now recognized as main contributors to the biomass and primary production of marine planktonic systems, particular in the oligotrophic regions of the Ocean (Johnson and Sieburth 1979; Waterbury et al. 1979; Chisholm et al. 1988). Their physiology and ecology have also been the subjects of a number of reviews (Waterbury et al. 1986; Carr and Mann 1994; Partensky et al. 1999; Scanlan and West 2002). Furthermore, several organisms in this group have been subject of whole-genome sequencing and analysis, which helped to delineate the genetic basis of metabolic capability and ecotype differentiation among these microbes (Dufresne et al. 2003; Palenik et al. 2003, 2006; Rocap et al. 2003).

Interestingly, even prior to the discovery of large abundances of these organisms in the oceans (Johnson and Sieburth 1979; Waterbury et al. 1979; Chisholm et al. 1988), the possibility was raised that they could be facultative heterotrophs. This idea likely resulted from observations as early as the 1930s indicating that cultures of heterocystous nitrogen-fixing cyanobacteria were able to growth in the dark with sugars (Allison et al. 1937; Khoja and Whitton 1971). (Heterocysts and nitrogen fixation are discussed in Chapter 13.) In an early study, Rippka (1972) evaluated growth on glucose by unicellular cyanobacteria with photosynthesis (more specifically, electron transfer from photosystem II) inhibited by 3(3,4-dichlorophenyl)-1,1-dimethylurea (DCMU). Her results indicated that the vast majority of unicellular cyanobacteria, and more significantly, all *Synechococcus* strains tested were unable to grow under these conditions. Although subsequent tests did indicate that three marine strains grew heterotrophically with glycerol (Rippka et al. 1983), none of 19 phycoerythrin-containing strains typical of coastal and oceanic regions was capable of photoheterotrophic growth (Waterbury et al. 1986). (Phycoerythrin is the dominant light-harvesting pigment in marine cyanobacteria; see also Chapter 1). This type of experiment has not been done with *Prochlorococcus* strains. Based on these results, it is currently accepted that *Prochlorococcus* and marine *Synechococcus* depend on light for most of their energy generation, and do not *require* organic compounds for biomass generation, and therefore, in the strict sense, they are photoautotrophs.

However, it is now clear that both *Synechococcus* and *Prochlorococcus* are capable of taking up carbon, sulfur, and (more significantly) nitrogen from organic compounds, and of using these organic-derived elements to produce cellular biomass. This is particularly relevant, since uptake of many of these compounds, such as nucleosides and amino acids, have been used to measure growth and production of heterotrophic bacterioplankton in aquatic systems. Uptake of these types of compounds by *Synechococcus* and *Prochlorococcus* leads to an overestimation of heterotrophic bacterioplankton growth and production, and thus *Synechococcus* and *Prochlorococcus* can be considered *operationally heterotrophic*.

Uptake of Urea and DMSP

Uptake of the organic molecule urea by unicellular cyanobacteria has long been recognized (McLachlan and Gorham 1962), and later studies confirmed this uptake by both marine *Synechoccocus* and *Prochlorococcus* (Waterbury et al. 1986; Willey and Waterbury 1989; Rippka et al. 2000). Urease-coding genes have also been retrieved from strains in both genera (Collier et al. 1999; Palinska et al. 2000). Since urease activity mainly results in the production of ammonia (NH_3) and carbon dioxide (CO_2), urea uptake, while significant for nitrogen budgets of these unicellular cyanobacteria, should not affect the measurement of heterotrophic activity by methods based on either incorporation of radiolabeled compounds or oxygen (O_2) consumption.

Marine *Synechococcus* and *Prochlorococcus* have also shown to take up dimethylsulfoniopropionate (DMSP), a small sulfur-containing organic molecule that is mainly produced by marine phytoplankton and is believed to function as an osmolite (Stefels 2000), antioxidant (Sunda et al. 2002), or a deterrent of grazing (Wolfe et al. 1997). DMSP transformations in the ocean have received considerable attention, since dimethylsulfide (DMS), a product of DMSP breakdown, affects cloud nucleation, and thus potentially influences albedo, and ultimately the global climate (Charlson et al. 1987; Andreae 1990). In a study employing microautoradiography and epifluorescence microscopy, Malmstrom et al. (2005) measured the contribution of *Synechococcus* populations in the Gulf of Mexico and northwest Atlantic Ocean to total [^{35}S]DMSP uptake and found that this uptake was greater than expected from the abundance of these organisms. They also found these organisms contributed significantly to the assimilation of [^{35}S]methanethiol (MeSH), and that cultured strains did not produce DMS from DMSP, indicating that DMSP uptake likely leads to uptake into biomass. In marine bacterioplankton populations, the main fate of the radiolabeled sulfur from [^{35}S]DMSP (Kiene et al. 1999) appears to be sulfur amino acids, in particular methionine, incorporated into proteins. Although the synthesis of methionine from MeSH does involve the incorporation of carbon molecules, this reaction is likely more significant to cellular energy budgets. Despite the fact that sulfate is a major ion in sea water, uptake of the reduced sulfur in MeSH requires less energy than sulfate use (Kiene and Linn 2000). In addition, the degradation from DMSP to MeSH might generate some cellular energy. As in the case of urea, DMSP uptake likely does not significantly affect measurements or heterotrophic bacterioplankton activity.

> **Microautoradiography—Examining Microbial Activity at the Single-Cell Level**
>
> One technique that has been important in examining potential photoheterotrophy is microautoradiography. In this approach, radioactive molecules added to a water sample are taken up by microbes, which are then collected onto a filter. The cells are fixed with formaldehyde and transferred to a photographic emulsion. After development, silver grains form in the emulsion around cells that have taken up the radioactive material. These grains and the cells can be observed and enumerated by epifluorescence microscopy.

Uptake of Nucleosides and Amino Acids

In contrast to urea and DMSP, the uptake of nucleosides and amino acids by *Synechococcus* and *Prochlorococcus* does affect measurements of heterotrophic bacterioplankton activity, and while not directly involved in cellular energy generation, could affect cellular energy budgets in a similar fashion to DMSP incorporation. Incorporation of reduced forms of carbon, sulfur, and nitrogen in oxic environments such as the open ocean requires less energy than incorporation of more prevalent oxidized forms. In a early report of the capacity of unicellular cyanobacteria to use amino acids as nitrogen sources, McLachlan and Gorham (1962) observed growth of a non-axenic *Microcystis aeruginosa* strain on a number of amino acids, and although this growth was lesser than with nitrate, these results suggested that amino acids could at least partially support this organism's growth.

Photoheterotrophy by Synechococcus Studies in the 1980s and 1990s demonstrated uptake of a selected number of nucleosides and amino acids by axenic marine *Synechoccocus* strains, although these results are confounded by the different levels of tracer additions. Incorporation of nucleosides by *Synechococcus* appears to vary between different nucleosides. Cuhel and Waterbury (1984) measured significant incorporation of radiolabeled adenine added at nanomolar levels by the marine *Synechoccocus* strain WH7803, but not of glucose or methylthymidine, supporting an earlier report by Fuhrman and Azam (1982) that marine *Synechoccocus* were not labeled during tritiated thymidine incorporation assays designed to measure heterotrophic bacterioplankton production.

Similarly, uptake of amino acids by *Synechococcus* strains also appear to be dependent on the specific amino acids tested. Willey and Waterbury (1989) showed chemotactic response by the motile *Synechoccocus* strain WH8113 to the amino acids β-alanine and glycine above nanomolar concentrations, but no responses to any of the sugars or other amino acids tested. This strain was also shown to take up micromolar levels of radiolabeled β-alanine and glycine (Willey and Waterbury 1989). In addition, Paerl (1991) found uptake of a mixture of amino acids by synechococcus strains WH7803 and WH8101 at subnanomolar levels, and he also found that the extent of uptake was light dependent.

Interestingly, incorporation of the amino acid leucine by *Synechococcus* is less clear, and this is significant, considering that this amino acid is commonly used in the measurement of heterotrophic bacterioplankton production (Kirchman et al. 1985). In the description of the leucine incorporation method, Kirchman et al. (1985) reported no incorporation of leucine at nanomolar concentrations by *Synechococcus*, in mixed bacterioplankton communities, which they attributed to competition for leucine by other bacterioplankton. In contrast, Paerl (1991) showed light-dependent uptake of mixed amino acids at subnanomolar levels by autofluorescent picoplankton in mixed bacterioplankton assemblages. Leucine incorporation was used by others to measure protein synthesis by synechococcus strain WH7803, albeit at micromolar additions (Kramer 1990). The apparently discrepancy between the results of Kirchman and colleagues and the uptake of mixed amino acids reported by Paerl might be explained by a lower ability of *Synechococcus* to uptake leucine at nanomolar and subnanomolar concentrations in comparison with other amino acids.

Photoheterotrophy by Prochlorococcus Even more remarkable than these observations for *Synechococcus* is the fact that it now appears that uptake of macromolecule precursors by *Prochlorococcus* is substantial. Uptake of radiolabeled amino acids—added at nanomolar concentrations—by cells sorted using flow cytometry has been measured in several oceanic and coastal regions (Zubkov et al. 2003, 2004, 2006; Zubkov and Tarran 2005; Michelou et al. 2007). The results of these studies are compelling, and indicate that the relative levels of incorporation of methionine and leucine by *Prochlorococcus*, compared with other bacterioplankton cells, roughly correspond to their relative abundances, meaning that these organisms can efficiently compete with heterotrophic bacterioplankton cells for amino acid uptake. In contrast, uptake of amino acids by *Synechoccocus* in the same experiments was much less, and the relative contribution of *Synechoccocus* to the entire amino acid uptake was significantly lower than its relative abundance (Zubkov and Tarran 2005).

Field Studies Using Light and Dark Incubations

The contribution of *Prochlorococcus* and *Synechococcus* to amino acid and nucleoside uptake has also been inferred from measurements of irradiance effects on the uptake of radiolabeled leucine (Morán et al. 2001; Church et al. 2004) and thymidine (Church et al. 2006). Figure 5.1 represents one example of these measurements taken

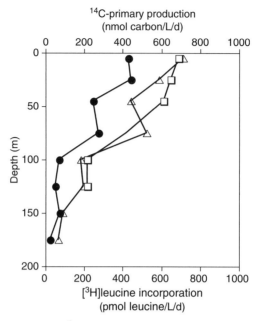

Figure 5.1 A depth profile of [^{3}H]leucine incorporation in light (open triangles) and dark (closed circles) incubations, and photoautotrophic production, estimated from $^{14}CO_2$ uptake (open squares) at Station ALOHA in May 2002. Adapted from Church et al. (2004).

at the North Pacific gyre. The results of these measurements indicate that in the vast majority of cases, leucine incorporation with light is higher than incorporation in dark controls (Table 5.1). Morán et al. (2001) observed higher leucine incorporation in darkness for samples from the Mediterranean Sea. However, the same group measured leucine incorporation at a range of irradiance levels, and concluded that dark stimulation of leucine uptake was likely an artifact. Increases of leucine uptake with light levels, similar to those in photosynthesis versus irradiance curves (PI curves), were also reported for samples from the North Pacific station ALOHA (Church et al. 2004).

In contrast, a series of experiments at station ALOHA showed that while leucine incorporation was stimulated by light, particularly in the mid-euphotic zone, the incorporation of tritiated thymidine does not seem to be higher in light incubations than in dark incubations (Church et al. 2006). Based on the lack of previous evidence of thymidine incorporation (at nanomolar levels) by cyanobacteria, and the absence of nucleoside transporters or thymidine kinases in the genomes of *Synechoccocus* and *Prochlorococcus*, Church and colleagues attributed the discrepancy between light effects on leucine but not thymidine incorporation to these cyanobacteria. These authors also argued that alternative explanations to light stimulation of leucine incorporation based on increased heterotrophic bacterioplankton production, resulting either from a tight coupling of heterotrophic bacterioplankton to

TABLE 5.1 Examples of Studies Examining The Effect of Light on Leucine Incorporation

Location	Depth (m)	Comments	Percent Stimulation[a]	Reference
NW Mediterranean, Open sea	5		118	Morán et al. (2001)
NW Mediterranean, Open sea	40		81	Morán et al. (2001)
NW Mediterranean, Shelf Break	5		58	Morán et al. (2001)
NW Mediterranean, Shelf Break	40		−65[b]	Morán et al. (2001)
NW Mediterranean, Shelf Break	5	Night	31	Morán et al. (2001)
NW Mediterranean, Shelf Break	40	Night	245	Morán et al. (2001)
Average for Mediterranean			61 ± 27	Morán et al. (2001)
N Pacific Gyre	5–25	In situ	27–62	Church et al. (2004)
N Pacific Gyre	75	In situ	20–114	Church et al. (2004)
Average for N Pacific Gyre	0–25	15 cruises	30 ± 20	Church et al. (2006)
Average for N Pacific Gyre	45–125	15 cruises	70 ± 20	Church et al. (2006)
North Atlantic Florida-Azores	5	12 stations	35 ± 52	Michelou et al. (2007)
North Atlantic Azores-Iceland		Three depths at 21 stations	6.2 ± 23	Michelou et al. (2007)

[a] Ratio of the difference between high and low irradiance values relative to the low irradiance value.
[b] The negative number indicates inhibition by light.

photosynthetic production or from the photolabilization of dissolved organic matter, would have led to enhanced uptake of both leucine and thymidine, which was not observed.

One interesting point not considered in this discussion was the possibility that other photoheterotrophic microbes might also have a differential capability of uptake of leucine (or other amino acids) and thymidine, which could partially explain differences between leucine and thymidine uptake under light and dark conditions. We searched for thymidine kinases in the genomes of photoheterotrophic organisms with complete genomes, including *Pelagibacter ubique*, HTCC1062 (Giovannoni et al. 2005b), which is a member of the ubiquitous and numerically abundant SAR11 group (see Chapter 3), *Congregibacter litoralis* KT71 which is a member of the gammaproteobacterial OM60/NOR5 group (Fuchs et al. 2007), and the *Bacteroidetes* strains MED134 and MED152 (Gómez-Consarnau et al. 2007).

Interestingly, while the genomes of strains KT71, MED132 and MED134 all contain thymidine kinase, the genome of *P. ubique* HTCC1062, much like *Prochlorococcus* and *Synechococcus*, does not. But the *P. ubique* genome does include genes likely involved in leucine transport. Since SAR11 is a major bacterioplankton group in oligotrophic regions of the ocean, enhanced growth by photoheterotrophic SAR11 in light versus dark incubations could at least partially explain differences between leucine and thymidine uptake in dark and light incubations.

Implications of Facultative Photoheterotrophy by Cyanobacteria

Uptake of dissolved organic matter by *Synechococcus* and *Prochlorococcus* could have significant implications for the microbial ecology of the oceans, although currently a quantitative assessment is lacking. As pointed out above, measurement of secondary production by heterotrophic bacterioplankton will be overestimated from measurements based on radiotracer-labeled precursor incorporation, in particular leucine incorporation, when these measurements are taken under in situ light conditions. Since the vast majority of past measurements have used dark incubations, this effect has been somewhat mitigated. However, as it will be discussed in the following sections, dark incubations might in turn underestimate biomass production and growth efficiency by photoheterotrophic bacterioplankton.

A second effect of organic matter uptake by *Synechococcus* and *Prochlorococcus* is competition with heterotrophic bacterioplankton for dissolved organic matter (DOM), particularly amino acids, nucleosides, and methylated sulfur compounds. In nutrient-limited open ocean regions, competition for amino acids and nucleosides is likely more relevant for nitrogen pools and transformations, and the uptake of amino acids will affect estimates of regenerated production (sensu Eppley and Peterson 1979) in these systems, although usually it has been assumed that ammonium is the main nitrogen source supporting regenerated production. In terms of the energy and carbon budgets of the ocean, the energy saved by the use of reduced organic compounds by these primarily photoautotrophic organisms would result in a higher overall growth efficiency by *Synechococcus* and *Prochlorococcus* in the euphotic zone, as less daytime photoautotrophically fixed carbon would be needed to be oxidized at nighttime for energy production and maintenance, and would be available for biomass production. Since current methods for estimating primary production use long (24-hour) incubations, this effect is likely accounted for. This higher efficiency might also allow these organisms to inhabit deeper strata of the water column, where light radiation is limiting. Finally, uptake of nitrogen-rich organic matter, added to the energy savings from using reduced sulfur, could also lead to an excess of carbohydrate carbon fixed during oxygenic photosynthesis, or glycolate produced by photorespiration (Fogg 1983). In turn, this carbon could be exudated and be available as energy sources for other bacterioplankton (Lau and Armbrust 2006; Lau et al. 2007). Thus of measurements of primary production in the particulate phase would underestimate the total carbon fixed during photosynthesis that is available for secondary production by heterotrophic bacterioplankton.

MARINE AAnP BACTERIA: HABITATS AND DIVERSITY

Rediscovery of the Marine AAnP Bacteria

AAnP bacteria were discovered as members of marine bacterioplankton more than two decades ago (Shiba et al. 1979, 1991). Besides growing photoheterotrophically, using light as an additional source of energy, these bacteria require oxygen for both growth and synthesis of their photosynthetic apparatus. In addition, AAnP bacteria have other features that distinguish them from anaerobic anoxygenic phototrophs: they contain relatively low amounts of photosynthetic units per cell and have very high concentrations of carotenoid pigments (Yurkov and Beatty 1998; Rathgeber et al. 2004). Although the first strains were isolated from diverse niches, including seawater samples, AAnP bacteria had been thought to be restricted to benthic niches such as seaweeds or beach sand samples, and were not considered to play a major role in the open ocean plankton. This view was dramatically changed when AAnP bacteria were discovered in the open ocean using infrared fast-repetition-rate (IRFRR) fluorometry (Kolber et al. 2000, 2001). Based on these biophysical measurements and thus bypassing the need to culture these bacteria, Kolber and co-workers suggested that AAnP bacteria account for up to 5 percent of surface ocean photosynthetic electron transport and compose up to 11 percent of the total microbial community in the euphotic zone.

Diversity of AAnP Bacteria

Until recently, diversity of AAnP bacteria was studied using classical cultivation methods and numerous strains have been isolated since 1987 (see the summary in Rathgeber et al. 2004). Most of these marine AAnP isolates belong to two orders of the *Alphaproteobacteria*, the *Rhodobacterales* (particularly *Roseobacter* and relatives) and the *Sphingomonadales* (particularly *Erythrobacter*). Kolber et al. (2001) suggested that cultured *Erythrobacter* species represent the predominant AAnPs in the upper ocean. However, since very few sequences related to *Erythrobacter* species have been reported in rDNA clone libraries constructed from marine plankton, while *Roseobacter*, on the other hand, is a well-represented group in different marine environments (Buchan et al. 2005), it appears that *Erythrobacter* species may not represent the predominant AAnP bacteria in the upper ocean. In fact, studies using fluorescence in situ hybridization (FISH) found that the abundance of these *Erythrobacter* species is very low, at least in the North Atlantic Ocean (Cottrell et al. 2006).

Because of this discrepancy between cultured AAnP bacteria and genes in environmental rRNA clone libraries, the diversity of oceanic AAnP bacteria has also been directly evaluated using molecular tools targeting genes other than for 16S rRNA (Béjà et al. 2002; Oz et al. 2005; Schwalbach and Fuhrman 2005; Waidner and Kirchman 2005; Yutin et al. 2005; Du et al. 2006). AAnP bacteria in environmental BAC and fosmid libraries (see Chapter 4) or in environmental samples have been successfully identified using known photosynthetic reaction center genes (e.g., *pufM*) or other markers that are known to be present on anoxygenic photosynthetic superoperons, (i.e., different *bch* genes; Oz et al. 2005) and that are involved in bacteriochlorophyll biosynthetic pathways. Diverse AAnP bacterial signals were also detected in large marine metagenomic libraries from the Atlantic and Pacific Oceans (Venter et al. 2004; DeLong et al. 2006), further increasing estimates of AAnP bacterial diversity. Based on data gathered from all these efforts, several wider targeting PCR primers targeting AAnP bacterial genes were also designed and successfully used to recover additional AAnP bacterial diversity in a variety of environments (Yutin et al. 2005).

> **Bacteriochlorophylls** These photosynthetic pigments, which are similar to the chlorophylls of cyanobacteria and higher plants, contain four pyrrole rings, an associated magnesium ion, and a number of side radicals. The main differences between bacteriochlorophylls and chlorophylls are in these side radicals.
>
> **Photosynthetic Reaction Center** This term is used to describe a complex of membrane proteins and pigments associated with light harvesting and electron transfer during photosynthesis.

Figure 5.2 illustrates AAnP bacterial diversity uncovered by these molecular approaches. While *Roseobacter*-like *pufM* genes are readily found in marine environments, *Erythrobacter*-like sequences are absent. In addition, *Rhodobacter*-like *pufM* sequences are also present in marine environments, despite the fact that these niches are oxic; *Rhodobacter* isolates are classic freshwater anaerobic anoxygenic phototrophic bacteria. Surprisingly, in addition to *Alphaproteobacteria*-like AAnP bacteria, a number of *pufM* sequences related to *Gammaproteobacteria* were also retrieved from different environments (Fig. 5.2).

Recent results from The Gordon and Betty Moore Foundation microbial genome sequencing project initiative (http://www.moore.org/microgenome/), which is an intensive sequencing effort currently targeting marine bacteria, have already led to several surprising discoveries related to marine AAnP bacteria. It was found that the gammaproteobacterium KT-71 isolate (Eilers et al. 2000), which belongs to the widespread OM60 clade (Rappé et al. 1997), is similar to *Gammaproteobacteria*-like marine AAnP bacteria previously identified only using molecular tools (Fuchs et al. 2007). Bacteriochlorophyll-based phototrophy has been also observed in other members of the OM60 clade (Cho et al. 2007), including strains HTCC 2080, HTCC 2148, and HTCC 2246, indicating that phototrophy might be common in this clade. Together, these findings show that AAnP bacterial diversity is not restricted to a single bacterial group and that aerobic anoxygenic photosynthesis is employed by different bacterial taxa in marine environments.

Physiology of AAnP Bacteria

The physiology of cultured AAnP bacteria, particularly those related to *Erythrobacter*, has been thoroughly reviewed (Yurkov and Beatty 1998; Rathgeber

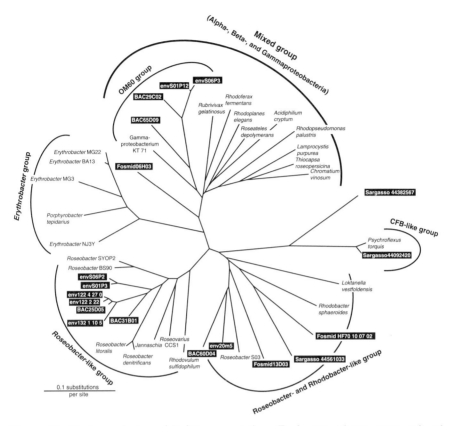

Figure 5.2 Phylogenetic tree of PufM representatives. Environmental sequences retrieved from PCR amplicons and BAC and fosmid libraries, as well as from the Sargasso Sea metagenome (Venter et al. 2004), are marked white on a black background. Evolutionary distances were determined using neighbor-joining (NJ) analysis with the test version 4.0b10 of PAUP* (Swofford 2002).

et al. 2004). The following remarkable characteristics of AAnP bacterial physiology are of significance to the marine environments:

- Efficient photoinduced electron transfer only occurs under aerobic conditions (Okamura et al. 1985; Garcia et al. 1994; Yurkov et al. 1995) and energy production is achieved via cyclic photophosphorylation.
- Bacteriochlorophyll (BChl$a$) absorbs in the near-infrared (800–880 nm) in cell suspensions and in the far-red (770 nm) in methanol extracts (Yurkov and Beatty 1998), and its synthesis appears to be highly inhibited by light.
- All cultured AAnP species produce diverse carotenoid compounds, and the ratio of bacteriochlorophyll to carotenoids (1 : 8–1 : 10) is low compared with anaerobic anoxygenic phototrophs (Yurkov and Beatty 1998). However, most carotenoids do not appear to be involved in light harvesting (Yurkov et al. 1994a).

- Like their anaerobic counterparts, AAnP bacteria have versatile carbon metabolism, but, unlike anaerobes, they lack ribulose-1,5-bisphosphate carboxylase (RuBisCO) and are not capable of autotrophic growth. CO_2 fixation does occur, but has been attributed to PEP carboxylase (Yurkov et al. 1994b) or other anaplerotic reactions (Shiba 1984) that do not support autotrophic growth.

Anaplerotic Reactions These reactions replenish components in the tricarboxylic acid (TCA) cycle. An example is the synthesis of oxaloacetate from pyruvate and CO_2. This replenishment is necessary, as oxaloacetate is used for protein biosynthesis as well as being in the TCA cycle. Lack of oxaloacetate would reduce the efficiency of the TCA cycle.

AAnP Bacterial Abundance and Ecological Significance

The relative abundance and importance of AAnP bacteria to the flow of energy and carbon in the ocean is still controversial. Using infrared epifluorescence microscopy as well as real-time polymerase chain reaction (PCR), Schwalbach and Fuhrman (2005) suggested that AAnP bacteria represent a small portion of total prokaryotic cells in the upper ocean (up to 2.2 percent). The discrepancy between their results and previous higher estimates is partly explained by the inclusion of phycoerythrin-containing *Synechococcus* spp. in infrared epifluorescence microscopy counts. Furthermore, based on BChl*a* concentrations, Goericke (2002) suggested that the contribution of BChl*a*-driven anoxygenic bacterial photosynthesis in the ocean to energy productions is likely substantially smaller than the previously suggested 5–10 percent global average (Kolber et al. 2000, 2001). Other recent studies report different abundance estimates for AAnP bacteria ranging from up to 16 percent of total prokaryotes in the Atlantic Ocean to 5 percent or less in the Pacific Ocean (Cottrell et al. 2006) and an average 2.3 percent (range 0.8–9.0 percent) in the North Western Atlantic (Sieracki et al. 2006). The latter study also measured biovolumes, and estimated that the contribution by AAnP bacteria to total bacterial biomass was higher (2–13 percent) than suggested by abundance alone, as the average cell volume of AAnP bacteria is significantly larger than the average (Sieracki et al. 2006). Finally, Yutin and colleagues evaluated the frequency of three genes associated with anoxygenic phototrophy (*pufM*, *pufL*, and *bchX*) in a global survey of environmental genomes from marine microorganisms (Rusch et al. 2007). Yutin et al. (2007) found that these genes were present in about 1–10 percent of bacterial genomes in samples taken from the Northeast Atlantic to the Eastern Tropical Pacific. Interestingly, in most samples, the most prevalent *pufM* genes originated from four unidentified groups. The degree of uncertainty associated with all these measurements, such as the detection limits of infrared epifluorescence microscopy, the universality of real-time PCR assays, and the distribution of AAnP lifestyle among typical marine bacterioplankton (i.e., the *Roseobacter* group) makes it is difficult to accurately estimate the contribution by AAnP bacteria to carbon cycling in the ocean.

Irrespective of these uncertainties, it is safe to say that AAnP bacterioplankton are present in significant numbers, comparable to or higher than for instance those of *Synechococcus* spp. and are actively synthesizing BChl*a* and other components of the photosynthetic machinery; expression (mRNA synthesis) of AAnP reaction center genes, particularly *Roseobacter*-like *pufM*, has been detected in the Pacific Ocean (Béjà et al. 2002) and in the Mediterranean Sea (Béjà, unpublished results). Due to their shear abundance, it is unquestionable that these organisms contribute to energy budgets of marine systems.

More significantly, the capability of AAnP bacteria likely confers a competitive advantage to these bacterioplankton in dilute and substrate-limited planktonic environments, thus increasing their roles in the cycling of DOM and major elements. However, since it does appears that AAnP bacteria might be more abundant in coastal rather than in oligotrophic regions (Sieracki et al. 2006), this competitive advantage might not be true in the open ocean. Future studies employing genomics, novel cultivation strategies, and epifluorescence microscopy analyses will hopefully aid in further assessing the role of AAnP phototrophy in the ocean.

PROTEORHODOPSIN-CONTAINING PROKARYOTES

Proteorhodopsins are retinal-binding integral membrane proteins recently found via environmental genomics surveys (Béjà et al. 2000; de la Torre et al. 2003; Venter et al. 2004). They are homologs of bacteriorhodopsins that were discovered in extremely halophilic archaea more than 30 years ago (Oesterhelt and Stoeckenius 1971). ("Bacterio" was used in the term for these rhodopsins before it was realized that the *Archaea* form a domain of life separate from the *Bacteria*, as discussed in Chapter 3.) The finding of microbial rhodopsin other than the classic archaeal bacteriorhodopsins came as a complete surprise. Not only were they in the *Bacteria* domain, but they were also found to exist in diverse and abundant marine bacterial groups.

Many bacteriorhodopsins and proteorhodopsins function as light-driven proton pumps capable of generating a chemiosmotic membrane potential from light energy. This membrane potential could be harnessed for adenosine triphosphate (ATP) synthesis or to drive active transport. More generally, rhodopsins are divided into two distinct protein families. One family consists of the visual rhodopsins, found in visual systems throughout the animal kingdom, which are photosensory pigments. The second family is composed of different microbial rhodopsins found in *Archaea*, *Bacteria* and lower *Eukarya*, that function as light-driven proton pumps, chloride ion pumps (halorhodopsins), or photosensory receptors (for detailed reviews on rhodopsins see Spudich et al. 2000; Spudich and Jung 2005). The two protein families share identical tertiary structures characterized by seven transmembrane α-helices that form a pocket in which retinal is covalently linked, as a protonated Schiff base, to a lysine in the seventh transmembrane helix (helix G).

The primary sequences of these different rhodopsins are not significantly similar, and may in fact have different origins.

In this section, the diversity and distribution of proteorhodopsins from marine bacterioplankton will be described and discussed.

> **Rhodopsins** These are light-sensitive proteins (opsins) with seven transmembrane domains and an associated retinal molecule. Rhodopsins change in conformation under light energy, and this can be translated as a signal in sensory rhodopsin systems, or the transmembrane transport of protons or ions, which can drive ATP synthesis.

Proteorhodopsin Genotypes and Taxonomic Distributions

Proteorhodopsins were first detected in 2000 in a marine environmental BAC library screened for 16S rRNA-containing clones (Béjà et al. 2000). The first proteorhodopsin was found on a genomic fragment originating from the uncultured SAR86 group II (Suzuki et al. 2001a). This marine bacterial group belongs to the *Gammaproteobacteria* and is widespread in marine plankton. Since its initial detection, numerous other proteorhodopsins have been found using molecular biology techniques (Fig. 5.3). While the initial surveys to report on proteorhodopsin diversity mainly relied on polymerase chain reaction (PCR) amplification and therefore were unable to determine the origin of detected genes, further screening of BAC and fosmid libraries also revealed the presence of proteorhodopsins in *Alphaproteobacteria* (de la Torre et al. 2003).

The diversity of proteorhodopsin increased substantially during the analysis of the first massive shotgun library from marine plankton (Venter et al. 2004). Almost 800 new proteorhodopsins from different families were detected in Sargasso Sea plankton samples. While this analysis could not definitively link these new proteorhodopsins to specific bacterial groups, this enormous and previously uncovered diversity was used to design new general proteorhodopsin primers that were used to screen different BAC libraries from the Pacific Ocean and the Mediterranean Sea. Proteorhodopsins have also been identified in the SAR86-I (Sabehi et al. 2005) and in marine *Archaea* (marine *Euryarchaeota* Group II; Frigaard et al. 2006).

Full genomic sequencing studies have uncovered yet additional bacterioplankton groups containing proteorhodopsins. A cultured representative, *Pelagibacter ubique*, of the cosmopolitan planktonic alphaproteobacterial SAR11 group (see Chapter 3), was shown to contain a proteorhodopsin gene (Giovannoni et al. 2005a, b). The marine microbial genomic sequencing initiative from the Gordon and Betty Moore Foundation revealed the presence of proteorhodopsins in other marine bacterial groups. Proteorhodopsins were found not only in cultured *Gammaproteobacteria* (*Vibrio angustum* S14; *Photobacterium* sp. SKA34; and SAR92 clade strain HTCC2207) and *Alphaproteobacteria* (additional *P. ubique*

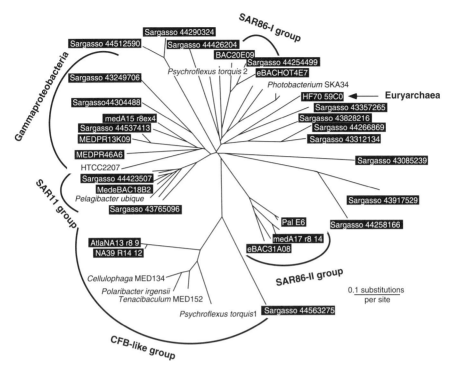

Figure 5.3 Phylogenetic tree of proteorhodopsin representatives. Environmental sequences retrieved from PCR amplicons and BAC and fosmid libraries, as well as from the Sargasso Sea metagenome (Venter et al. 2004), are marked white on a black background. Evolutionary distances were determined using neighbor-joining (NJ) analysis with the test version 4.0b10 of PAUP* (Swofford 2002).

strains) but also in the *Bacteroidetes* (*Dokdonia* sp. MED134; *Psychroflexus torquis* ATCC 700755; *Polaribacter* sp. MED152; and *Polaribacter irgensii* 23-P). While not all environmental proteorhodopsin genes detected so far can be assigned a taxonomic group, a conservative estimate based on BAC libraries predicts that at least 14 percent of the microorganisms in the photic zone contain proteorhodopsin (Sabehi et al. 2005).

Proteorhodopsin Spectral Tuning

Two related SAR86 proteorhodopsin subgroups that absorbed light with absorption maxima at 527 nm (green) and 490 nm (blue) were found in the Pacific Ocean. This spectral change (37 nm shift) results from a single amino acid change at position 105, where leucine in green light-absorbing proteorhodopsins (GPRs) is replaced by glutamine in blue light-absorbing proteorhodopsin (BPRs) (Man et al. 2003). The distribution of these different light-absorbing proteorhodopsin pigments is depth-dependent, with GPRs mainly at the surface and BPRs at deeper (75 m) waters, consistent with the light available at these depths (Béjà et al. 2001). In oligotrophic oceanic waters, most of the light energy is distributed in the blue–green, with maximal intensity usually in the blue (Kirk 1983). This energy peak is maintained over depth, even though the total energy decreases with depth. At the surface, the energy peak is broad, with a half-bandwidth between 400 and 650 nm, while in water below 5–10 m, all light except for blue light is absorbed or scattered, and the peak narrows and is shifted towards blue.

Due to the limited diversity retrieved by early PCR primers targeting proteorhodopsin, these initial observations of spectral tuning were restricted to the SAR86 proteorhodopsins. Based on new proteorhodopsin sequences retrieved in BAC libraries (Sabehi et al. 2005), on metagenomic libraries from the Sargasso Sea (Venter et al. 2004), and on the use of degenerate proteorhodopsin primers (Sabehi et al. 2007), we now know that this spectral tuning occurs in different proteorhodopsin families, including gammaproteobacterial proteorhodopsins and alphaproteobacterial proteorhodopsins (including members of the SAR11), as well as in other proteorhodopsin families. Interestingly, in most cases, the amino acid at position 105 in the different proteorhodopsin pigments is restricted to leucine and glutamine. Other amino acids are also observed (methionine and threonine), but are present in a small number of sequences.

Proteorhodopsin-Containing Prokaryotes: Abundance and Activity

Unlike chlorophyll and bacteriochlorophyll, proteorhodopsin does not fluoresce, and thus proteorhodopsin-containing microbes cannot be easily quantified using standard techniques such as epifluorescence microscopy or flow cytometry. Enumeration of the different clades containing proteorhodopsin (i.e., SAR86-II or SAR11) by FISH or other techniques could in theory be used to infer the contribution of proteorhodopsin-containing microbes, but currently it is not clear how widely proteorhodopsin is distributed among members of these clades. Thus, currently, relatively little is known regarding the abundance of proteorhodopsin-containing microbes in the oceans. Of even more significance for microbial oceanography, the activity of proteorhodopsin-containing prokaryotes as well as the contribution of proteorhodopsin-based energy production to the overall metabolic balance of these organisms is still very poorly understood. However, we do know a few things about proteorhodopsin in the oceans.

Proteorhodopsin molecules in marine bacterioplankton samples have been directly examined using laser flash-photolysis induced absorbance (Béjà et al. 2001). Basically, this technique measures absorbance at different wavelengths after a very fast (nanoseconds) laser pulse. The duration of the photochemical reaction cycle and absorbance spectrum are then used to identify these molecules. Using this technique, Béjà et al. (2000) were able to detect proteorhodopsin molecules in cell membranes prepared from large (700 L) cell concentrates from Monterey Bay. These proteorhodopsins had photochemical characteristics similar to those of the

SAR86-II proteorhodopsins expressed in *Escherichia coli* (Béjà et al. 2000), indicating that photoactive proteorhodopsins are present in marine plankton.

While proteorhodopsin measurements at the protein level are the most direct assessments of proteorhodopsin-containing microbial activity, low sensitivity precludes laser flash-photolysis techniques to be used regularly with marine plankton samples. Quantitative real-time PCR (qPCR) is an alternative technique that can be used for examining proteorhodopsin genes as well proteorhodopsin gene expression based on mRNA quantification. Although qPCR quantification is not free of problems, over the past few years we have used this technique to gather information regarding the spatial and temporal distribution of proteorhodopsin-containing prokaryotes in the ocean, based on quantification of proteorhodopsin genes. We also have examined the activity of these organisms by measuring expression of proteorhodopsins in the ocean based on proteorhodopsin mRNA levels.

Gene regulation at the transcriptional level is a critical requirement for the use of mRNA as a measurement of gene expression. So far, no study has directly tested whether proteorhodopsin is transcriptionally regulated, but some evidence exists that this might be true for at least some marine proteorhodopsin-containing organisms. Transcriptional regulation is well established for the bacteriorhodopsin system of *Halobacterium* sp. NRC1 (Baliga et al. 2001), with a relatively well-conserved region upstream of genes involved in bacteriorhodopsin and retinal synthesis. Interestingly, a homologous region exists upstream of the SAR86-II proteorhodopsin gene (Baliga et al. 2001), and a non-coding region with a putatively homologous region also appears upstream of the proteorhodopsin of *P. ubique* (S. Dassarma and M. Suzuki, unpublished observations), suggesting that these proteorhodopsin genes might be transcriptionally regulated.

SAR86-II in Monterey Bay A qPCR assay targeting SAR86-II (Suzuki et al. 2001a) proteorhodopsin genes was designed based on a limited dataset existing in 2000. While it appears from the vastly expanded proteorhodopsin datasets that this assay does not measures all SAR86-II-like genes, the data offer some insights regarding the distribution and activity of these yet-to-be-cultured proteorhodopsin-containing organisms. SAR86-II proteorhodopsin gene copy numbers were measured in DNA extracts collected during a mapping cruise in Monterey Bay (Suzuki et al. 2001a), and SAR86-II PR and SAR86 (all subgroups) rRNA gene abundances were measured as a percentage of bacterial rRNA genes (Fig. 5.4). Interestingly, both the general distribution and, more notably, the percentages are remarkably similar, suggesting that a majority of SAR86 contain proteorhodopsin genes, and that SAR86-II was likely the prevalent SAR86 subgroup in Monterey Bay at the sampling time.

The same qPCR assay was also used to follow the seasonal and spatial dynamics of SAR86-II proteorhodopsin genes at Station 4B in mid-Monterey Bay (Fig. 5.5). In vertical profiles down to 200 m depth, except for the January 2002 sample, proteorhodopsin-containing SAR86-II showed a typical phototroph response, with proteorhodopsin gene (and likely cell) concentrations higher at the surface and decreasing with depth. Proteorhodopsin mRNA peaked between 25 and 40 m depths, and thus

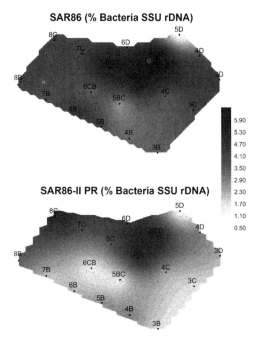

Figure 5.4 Mapping of SAR86 16S rRNA genes and SAR86-II proteorhodopsin genes in surface waters of the Monterey Bay (April 2000). Gene percentages measured by real time PCR (TaqMan) are relative to bacterial rRNA genes. Station coordinates are described in Suzuki et al. (2001b).

activity levels based on mRNA-to-gene ratios for proteorhodopsin were higher at these mid-depths (Suzuki et al., unpublished results). SAR86-II proteorhodopsin genes and SAR86 rRNA (all subgroups) genes covaried remarkably, indicating that SAR86-II was a prevalent SAR86 subgroup in Monterey Bay year-round (Suzuki et al., unpublished results).

Proteorhodopsin in the North Pacific Central Gyre One of our students conducted a study on proteorhodopsin gene expression at Station ALOHA in the North Pacific Central Gyre (Shi 2005). Using optimized reverse transcriptase (RT)-PCR protocols, a proteorhodopsin cDNA library was constructed using highly degenerate primers (Sabehi et al. 2005) designed based on an extended dataset that included sequences from the Sargasso Sea shotgun library (Venter et al. 2004). Among these clones, many could be identified as putative SAR11 clones based on a Bayesian phylogenetic analysis, as well as on high similarity to proteorhodopsin genes in Sargasso Sea contigs, and BACs containing syntenic genes highly similar to those of *P. ubique*.

A second qPCR assay was designed to the most common SAR11 gene type and mRNA levels over three diel cycles at Station ALOHA in December 2002 (Shi 2005). Even though advective processes complicated the interpretation of gene expression

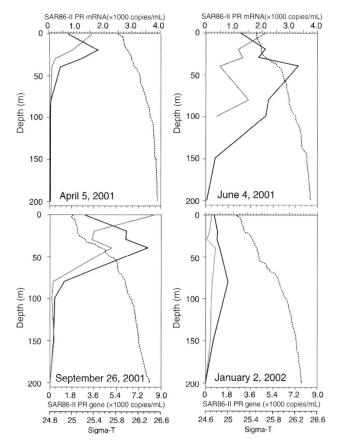

Figure 5.5 Vertical profiles of SAR86-II proteorhodopsin genes (gray lines), SAR86-II proteorhodopsin mRNA (full lines) and density (Sigma-T, dotted lines), in different seasons at station 4B (36.775°N, 122.016°W). Proteorhodopsin genes were measured by real-time PCR (TaqMan). Proteorhodopsin mRNAs were measured as cDNAs reverse-transcribed using random primers and measured by real-time PCR (TaqMan).

results based on proteorhodopsin mRNA-to-gene ratios, two main lines of evidence indicated that proteorhodopsin expression responded to light. First, proteorhodopsin expression was highest in early morning and late afternoon samples, decreasing towards mid-day, with minima at midnight. Second, superimposed on these diel patterns, there were responses associated with internal tides where shallowing of pycnocline surfaces coincided with relative decreases in gene copies and relative increases in mRNA copies (M.A. McManus, unpublished results). On the other hand, deepening of pycnocline surfaces was associated with increases in gene copies and relative decreases in mRNA copies (M.A. McManus, unpublished results), suggesting a depth-dependent response in gene expression similar to that observed for the SAR86-II proteorhodopsin in Monterey Bay. The fact that these responses also occurred at night suggests that the SAR11 proteorhodopsin is expressed at some level during the entire photoperiod, in agreement with the observation that proteorhodopsin is produced in the dark by the cultivated *P. ubique* strain HTCC1062 (Giovannoni et al. 2005a).

Proteorhodopsin-Containing Prokaryotes: Ecological Significance

It is unlikely that proteorhodopsin light-driven metabolism could support autotrophic CO_2 fixation. None of the data obtained so far from different proteorhodopsin-containing genomic fragments has produced any evidence for autotrophy. Although proton pumping by proteorhodopsin is capable of producing energy in the form of ATP, proteorhodopsin-containing microorganisms likely do not produce the reducing power necessary for carbon fixation. Thus, it is likely that proteorhodopsin can only support a photoheterotrophic lifestyle. In fact, even the energetic advantage of proteorhodopsin-based phototrophy in not well established. Growth experiments performed with *P. ubique* HTCC1062 failed to detect difference in yields between cells grown under light condition (Giovannoni et al. 2005a). Also, in field experiments, bacterial groups thought to harbor proteorhodopsin did not increase in abundance in the light compared with dark incubations (Schwalbach and Fuhrman 2005). While these studies failed to find light-enhanced growth in proteorhodopsin-containing isolates and natural microbial communities, light does significantly enhance both growth rate and yield in proteorhodopsin-expressing marine *Bacteroidetes* (Gómez-Consarnau et al. 2007). Whether cells living in the oligotrophic ocean might obtain significant amounts of energy from proteorhodopsin-mediated light metabolism remains to be seen.

Proteorhodopsins from cultivated microorganisms, except for *Pelagibacter*, are not closely related to the majority of proteorhodopsins detected in environmental samples, and thus environmental genomic techniques are still required to evaluate proteorhodopsin diversity in the environment. Continued analysis of syntenic sequences in large insert clones or shotgun contigs might also help to determine which microbes harbor proteorhodopsin and to explore other putative metabolic capabilities of proteorhodopsin-carrying organisms. In one such analysis (Sabehi et al. 2005), a proteorhodopsin-bearing clone was found to harbor an entire *dsr* operon containing the genes for both subunits of a reverse siroheme sulfite reductase *(dsrAB)*, which is typically used by chemolithotrophic or anaerobic phototrophic bacteria for exploiting reduced sulfur compounds as electron donors. The close relationship of the reverse sulfite reductase from a proteorhodopsin-carrying clone to the enzyme of the phototrophic (bacteriochlorophyll-based) gammaproteobacterium *Allochromatium vinosum* suggests the existence of anoxygenic phototrophs that might exploit light for energy generation not only by its bacteriochlorophyll-containing photosystem, but also by proteorhodopsin. The *dsr* genes might alternatively originate from an oxidizer of reduced sulfur compounds. Reduced inorganic sulfur compounds are not readily found in oxic marine environments. Dimethyl sulfide (DMS), on the other hand, is present in oxic marine surface waters, and marine bacteria containing both proteorhodopsin and the *dsr* genes might be involved in degradation of this compound, which plays a key role in cloud nucleation and might control transport of sulfur from oceanic to terrestrial systems. Alternatively, as suggested by Bryant and Frigaard (2006), putative reverse dissimilatory sulfite reductase and related enzymes could provide reducing equivalents for autotrophic growth by oxidizing reduced sulfur compounds, thus leading to proteorhodopsin-based photolithoautotrophic growth. Finally, a pathway frequently found in proteorhodopsin-containing BACs is involved in direct retinal biosynthesis from carotenes (Sabehi et al. 2005; Martinez et al. 2007). It appears that different proteorhodopsin-carrying bacteria can therefore produce this essential chromophore necessary to self-maintain a light-harvesting system based on proteorhodopsin.

SUMMARY

1. *Synechococcus* and *Prochlorococcus* are photoautotrophs that are capable of taking up DOM and compete with other nonautotrophic bacterioplankton for these compounds.
2. AAnP and proteorhodopsin-carrying bacteria are hypothesized to be abundant and important components of the oligotrophic oceanic bacterial community, because their potential use of light and DOM would give them an advantage over their heterotrophic counterparts.
3. The large numbers of proteorhodopsin genes detected in metagenomic libraries strongly suggests an important role for these pigments in photoheterotrophy.
4. AAnP bacteria are readily detected in these genomic surveys, although to a lesser extent, but appear to be abundant in some oceanic waters based on other data.
5. Both types of putative photoheterotrophs appear to be active as AAnP photosynthetic reaction center genes, and proteorhodopsin gene transcripts are observed in marine RNA samples.
6. However, the significance of these processes to overall ocean biogeochemistry is still unknown, as there is little direct evidence for photoheterotrophy for AAnP bacteria and no direct evidence of photoheterotrophy exists for proteorhodopsin-carrying bacteria.

REFERENCES

Allison, F. E., S. R. Hoover, and H. J. Morris. 1937. Physiological studies with the nitrogen-fixing alga. *Nostoc muscorum. Bot. Gaz.* 98: 433–463.

Andreae, M. O. 1990. Ocean–atmosphere interactions in the global biogeochemical sulfur cycle. *Mar. Chem.* 30: 1–29.

Baliga, N. S., S. P. Kennedy, W. V. Ng, L. Hood, and S. DasSarma. 2001. Genomic and genetic dissection of an archaeal regulon. *Proc. Natl. Acad. Sci. USA* 98: 2521–2525.

Béjà, O., L. Aravind, E. V. Koonin, et al. 2000. Bacterial rhodopsin: Evidence for a new type of phototrophy in the sea. *Science* 289: 1902–1906.

Béjà, O., E. N. Spudich, J. L. Spudich, M. Leclerc, and E. F. DeLong. 2001. Proteorhodopsin phototrophy in the ocean. *Nature* 411: 786–789.

Béjà, O., M. T. Suzuki, J. F. Heidelberg, W. C. Nelson, C. M. Preston, T. Hamada, J. A. Eisen, C. M. Fraser, and E. F. DeLong. 2002. Unsuspected diversity among marine aerobic anoxygenic phototrophs. *Nature* 415: 630–633.

Bryant, D. A., and N. U. Frigaard. 2006. Prokaryotic photosynthesis and phototrophy illuminated. *Trends Microbiol.* 14: 488–496.

Buchan, A., J. M. González, and M. A. Moran. 2005. Overview of the marine *Roseobacter* lineage. *Appl. Environ. Microbiol.* 71: 5665–5677.

Carr, N. G., and N. H. Mann. 1994. The oceanic cyanobacterial picoplankton. In D. A. Bryant (ed.), *The Molecular Biology of Cyanobacteria*. Kluwer, pp. 27–48.

Charlson, R. J., J. E. Lovelock, M. O. Andreae, and S. G. Warren. 1987. Oceanic phytoplankton, atmospheric sulfur, cloud albedo and climate. *Nature* 326: 655–661.

Chisholm, S. W., R. J. Olson, E. R. Zettler, R. Goerick, J. B. Waterbury, and N. A. Welschmeyer. 1988. A novel free-living prochlorophyte abundant in the oceanic euphotic zone. *Nature* 334: 340–343.

Cho, J.-C, M. D. Stapels, R. M. Morris, K. L. Vergin, M. S. Schwalbach, S. A. Givan, D. F. Barofsky, and S. J. Giovannoni. 2007. Polyphyletic photosynthetic reaction center genes in oligotrophic marine Gammaproteobacteria. *Environ. Microbiol.* 1456–1463.

Church, M. J., H. W. Ducklow, and D. M. Karl. 2004. Light dependence of [^{3}H]leucine incorporation in the oligotrophic North Pacific ocean. *Appl. Environ. Microbiol.* 70: 4079–4087.

Church, M. J., H. W. Ducklow, R. M. Letelier, and D. M. Karl. 2006. Temporal and vertical dynamics in picoplankton photoheterotrophic production in the subtropical North Pacific Ocean. *Aquat. Microb. Ecol.* 45: 41–53.

Collier, J. L., B. Brahamsha, and B. Palenik. 1999. The marine cyanobacterium *Synechococcus* sp. WH7805 requires urease (urea amidohydrolase, EC 3.5.1.5) to utilize urea as a nitrogen source: Molecular-genetic and biochemical analysis of the enzyme. *Microbiology* 145: 447–459.

Cottrell, M. T., A. Mannino, and D. L. Kirchman. 2006. Aerobic anoxygenic phototrophic bacteria in the Mid-Atlantic Bight and the North Pacific Gyre. *Appl. Environ. Microbiol.* 72: 557–564.

Cuhel, R. L., and J. B. Waterbury. 1984. Biochemical composition and short term nutrient incorporation patterns in a unicellular marine cyanobacterium, *Synechococcus* (WH7803). *Limnol. Oceanogr.* 29: 370–374.

de la Torre, J. R., L. Christianson, O. Béjà, M. T. Suzuki, D. Karl, J. F. Heidelberg, and E. F. DeLong. 2003. Proteorhodopsin genes are widely distributed among divergent bacterial taxa. *Proc. Natl. Acad. Sci. USA* 100: 12830–12835.

DeLong, E. F., C. M. Preston, T. Mincer, et al. 2006. Community genomics among stratified microbial assemblages in the ocean's interior. *Science* 311: 496–503.

Du, H., N. Jiao, Y. Hu, and Y. Zeng. 2006. Real-time PCR for quantification of aerobic anoxygenic phototrophic bacteria based on *pufM* gene in marine environment. *J. Exp. Mar. Biol. Ecol.* 329: 113–121.

Dufresne, A., M. Salanoubat, F. Partensky. 2003. Genome sequence of the cyanobacterium *Prochlorococcus marinus* SS120, a nearly minimal oxyphototrophic genome. *Proc. Natl. Acad. Sci. USA* 100: 10020–10025.

REFERENCES

Eilers, H., J. Pernthaler, F. O. Glockner, and R. Amann. 2000. Culturability and in situ abundance of pelagic bacteria from the North Sea. *Appl. Environ. Microbiol.* 66: 3044–3051.

Eppley, R. W., and B. J. Peterson. 1979. Particulate organic matter flux and planktonic new production in the deep ocean. *Nature* 282: 677–680.

Fogg, G. E. 1983. The ecological significance of extracellular products of phytoplankton photosysnthesis. *Bot. Mar.* 26: 3–14.

Frigaard, N.-U., A. Martinez, T. J. Mincer, and E. F. DeLong. 2006. Proteorhodopsin lateral gene transfer between marine planktonic Bacteria and Archaea. *Nature* 439: 847–850.

Fuchs, B. M., S. Spring, H. Teeling, et al. 2007. Characterization of a marine gammaproteobacterium capable of aerobic anoxygenic photosynthesis. *Proc. Natl. Acad. Sci. USA* 104: 2891–2896.

Fuhrman, J. A., and F. Azam. 1982. Thymidine incorporation as a measure of heterotrophic bacterioplankton production in marine surface waters: Evaluation and field results. *Mar. Biol.* 66: 109–120.

Garcia, D., P. Richaud, J. Breton, and A. Vermeglio. 1994. Structure and function of the tetraheme cytochrome associated to the reaction centers of *Roseobacter denitrificans*. *Biochimie* 76: 666–673.

Giovannoni, S. J., L. Bibbs, J.-C. Cho, et al. 2005a. Proteorhodopsin in the ubiquitous marine bacterium SAR11. *Nature* 438: 82–85.

Giovannoni, S. J., H. J. Tripp, S. Givan, et al. 2005b. Genome streamlining in a cosmopolitan oceanic bacterium. *Science* 309: 1242–1245.

Goericke, R. 2002. Bacteriochlorophyll *a* in the ocean: Is anoxygenic bacterial photosynthesis important? *Limnol. Oceanogr.* 47: 290–295.

Gómez-Consarnau, L., J. M. González, M. Coll-Lladó, P. Gourdon, T. Pascher, R. Neutze, C. Pedrós-Alió, and J. Pinhassi. 2007. Light stimulates growth of proteorhodopsin-containing marine Flavobacteria. *Nature* 445: 210–213.

Johnson, P. W., and J. M. Sieburth. 1979. Chroococcoid cyanobacteria in the sea: A ubiquitous and diverse phototrophic biomass. *Limnol. Oceanogr.* 24: 928–935.

Khoja, T., B. A. Whitton. 1971. Heterotrophic growth of blue–green algae. *Arch. Mikrobiol.* 79: 280.

Kiene, R. P., and L. J. Linn. 2000. Distribution and turnover of dissolved DMSP and its relationship with bacterial production and dimethylsulfide in the Gulf of Mexico. *Limnol. Oceanogr.* 45: 849–861.

Kiene, R. P., L. J. Linn, J. Gonzalez, M. A. Moran, and J. A. Bruton. 1999. Dimethylsulfoniopropionate and methanethiol are important precursors of methionine and protein-sulfur in marine bacterioplankton. *Appl. Environ. Microbiol.* 65: 4549–4558.

Kirchman, D. L., E. K'nees, and R. Hodson. 1985. Leucine incorporation and its potential as a measure of protein synthesis by bacteria in natural aquatic systems. *Appl. Environ. Microbiol.* 49: 599–607.

Kirk, J. T. O. 1983. The nature of the underwater light field. *Light and Photosynthesis in Aquatic Ecosystems*. Cambridge University Press, pp. 104–134.

Kolber, Z. S., C. L. Van Dover, R. A. Niderman, and P. G. Falkowski. 2000. Bacterial photosynthesis in surface waters of the open ocean. *Nature* 407: 177–179.

Kolber, Z. S., F. G. Plumley, A. S. Lang, et al. 2001. Contribution of aerobic photoheterotrophic bacteria to the carbon cycle in the ocean. *Science* 292: 2492–2495.

Kramer, J. G. 1990. The effect of irradiance and specific inhibitors on protein and nucleic-acid synthesis in the marine cyanobacterium *Synechococcus* sp. WH-7803. *Arch. Microbiol.* 154: 280–285.

Lau, W. W. Y., and E. V. Armbrust. 2006. Detection of glycolate oxidase gene *glcD* diversity among cultured and environmental marine bacteria. *Environ. Microbiol.* 8: 1688–1702.

Lau, W. W. Y., R. G. Keil, and E. V. Armbrust. 2007. Succession and diel transcriptional response of the glycolate-utilizing component of the bacterial community during a spring phytoplankton bloom. *Appl. Environ. Microbiol.* 73: 2440–2450.

McLachlan, J., and P. R. Gorham. 1962. Effects of pH and nitrogen sources on growth of *Microcystis aeruginosa* Kutz. *Can J Microbiol.* 8: 1–11.

Malmstrom, R. R., R. P. Kiene, M. Vila, and D. L. Kirchman. 2005. Dimethylsulfoniopropionate (DMSP) assimilation by *Synechococcus* in the Gulf of Mexico and northwest Atlantic Ocean. *Limnol. Oceanogr.* 50: 1924–1931.

Man, D., W. Wang, G. Sabehi, L. Aravind, A. F. Post, R. Massana, E. N. Spudich, J. L. Spudich, and O. Béjà. 2003. Diversification and spectral tuning in marine proteorhodopsins. *EMBO J* 22: 1725–1731.

Martinez, A., A. S. Bradley, J. Waldbauer, R. E. Summons, and E. F. DeLong. 2007. Proteorhodopsin photosystem gene expression enables photophosphorylation in heterologous host. *Proc. Natl. Acad. Sci. USA* 104: 5590–5595.

Michelou, V. K., M. T. Cottrell, and D. L. Kirchman. 2007. Light-stimulated bacterial production and amino acid assimilation by cyanobacteria and other microbes in the North Altantic Ocean. *Appl. Environ. Microbiol.* 73: 5539–5546.

Morán. X. A., R. Massana, and J. M. Gasol. 2001. Light conditions affect the measurement of oceanic bacterial production via leucine uptake. *Appl. Environ. Microbiol.* 67: 3795–3801.

Oesterhelt, D., and W. Stoeckenius. 1971. Rhodopsin-like protein from the purple membrane of *Halobacterium halobium*. *Nat. New Biol.* 233: 149–152.

Okamura, K., K. Takamiya, and M. Nishimura. 1985. Photosynthetic electron transfer system is inoperative in anaerobic cells of *Erythrobacter* species strain OCh114. *Arch. Microbiol.* 142: 12–17.

Oz, A., G. Sabehi, M. Koblízek, R. Massana, and O. Béjà. 2005. *Roseobacter*-like bacteria in Red and Mediterranean Sea aerobic anoxygenic photosynthetic populations. *Appl. Environ. Microbiol.* 71: 344–353.

Paerl, H. W. 1991. Ecophysiological and trophic implications of light-stimulated amino acid utilization in marine picoplankton. *Appl. Environ. Microbiol.* 57: 473–479.

Palenik, B., B. Brahamsha, F. W. Larimer, et al. 2003. The genome of a motile marine *Synechococcus*. *Nature* 424: 1037–1042.

Palenik, B., Q. Ren, C. L. Dupont, et al. 2006. Genome sequence of *Synechococcus* CC9311: Insights into adaptation to a coastal environment. *Proc. Natl. Acad. Sci. USA* 103: 13555–13559.

Palinska, K. A., T. Jahns, R. Rippka, and N. Tandeau De Marsac. 2000. *Prochlorococcus marinus* strain PCC 9511, a picoplanktonic cyanobacterium, synthesizes the smallest urease. *Microbiology* 146: 3099–3107.

Partensky, F., W. R. Hess, and D. Vaulot. 1999. *Prochlorococcus*, a marine photosynthetic prokaryote of global significance. *Microbiol. Mol. Biol. Rev.* 63: 106–127.

Rappé. M. S., P. F. Kemp, and S. J. Giovannoni. 1997. Phylogenetic diversity of marine coastal picoplankton 16S rRNA genes cloned from the continental shelf off Cape Hatteras, North Carolina. *Limnol. Oceanogr.* 42: 811–826.

Rathgeber, C., J. T. Beatty, and V. Yurkov. 2004. Aerobic phototrophic bacteria: New evidence for the diversity, ecological importance and applied potential of this previously overlooked group. *Photosynth. Res.* 81: 113–128.

Rippka, R. 1972. Photoheterotrophy and chemoheterotrophy among unicellular blue–green-algae. *Arch. Mikrobiol.* 87: 93.

Rippka, R., J. Deruelles, J. B. Waterbury, M. Herdman, and R. Y. Stanier. 1983. Generic assignments, strain histories and properties of pure cultures of cyanobacteria. *J. Gen. Microbiol.* 111: 1–61.

Rippka, R., T. Coursin, W. Hess, et al. 2000. *Prochlorococcus marinus* Chisholm et al. 1992 subsp. pastoris subsp. nov. strain PCC 9511, the first axenic chlorophyll a(2)/b(2)-containing cyanobacterium (Oxyphotobacteria). *Int. J. Syst. Evol. Microbiol.* 50: 1833–1847.

Rocap, G., F. W. Larimer, J. Lamerdin, et al. 2003. Genome divergence in two *Prochlorococcus* ecotypes reflects oceanic niche differentiation. *Nature* 424: 1042–1047.

Rusch, D. B., A. L. Halpern, K. B. Heidelberg, et al. 2007. The Sorcerer II Global Ocean Sampling expedition: I, The northwest Atlantic through the eastern tropical Pacific. *PLoS Biol.* 5: e77.

Sabehi, G., O. Béjà, M. T. Suzuki, C. M. Preston, and E. F. DeLong. 2004. Different SAR86 subgroups harbour divergent proteorhodopsins. *Environ. Microbiol.* 6: 903–910.

Sabehi, G., A. Loy, K. H. Jung, R. Partha, J. L. Spudich, T. Isaacson, J. Hirschberg, M. Wagner, and O. Béjà. 2005. New insights into metabolic properties of marine bacteria encoding proteorhodopsins. *PLoS Biol.* 3: e273.

Sabehi, G., B. C. Kirkup, M. Rosenberg, N. Stambler, M. F. Polz, and O. Béjà. 2007. Adaptation and spectral tuning in divergent marine proteorhodopsins from the eastern Mediterranean and the Sargasso Seas. *ISME J* 1: 48–55.

Scanlan, D. J., and N. J. West. 2002. Molecular ecology of the marine cyanobacterial genera *Prochlorococcus* and *Synechococcus*. *FEMS Microbiol. Ecol.* 40: 1–12.

Schwalbach, M. S., and J. A. Fuhrman. 2005. Wide-ranging abundances of aerobic anoxygenic phototrophic bacteria in the world ocean revealed by epifluorescence microscopy and quantitative PCR. *Limnol. Oceanogr.* 50: 620–628.

Shi, Y. 2005. Measurement of in situ expression of proteorhodopsin genes at the North Pacific Central Gyre Station ALOHA. MSc. Thesis, University of Maryland, College Park.

Shiba, T. 1984. Utilization of light energy by the strictly aerobic bacterium *Erythrobacter* sp. OCh114. *J. Gen. Appl. Microbiol.* 30: 239–244.

Shiba, T., U. Simidu, and N. Taga, 1979. Distribution of aerobic bacteria which contain bacteriochlorophyll *a*. *Appl. Environ. Microbiol.* 38: 43–48.

Shiba, T., Y. Shioi, K. Takamiya, D. C. Sutton, and C. R. Wilkinson. 1991. Distribution and physiology of aerobic bacteria containing bacteriochlorophyll *a* on the east and west coast of Australia. *Appl. Environ. Microbiol.* 57: 295–300.

Sieracki, M. E., I. C. Gilg, E. C. Thier, N. J. Poulton, and R. Goericke. 2006. Distribution of planktonic aerobic anoxygenic photoheterotrophic bacteria in the northwest Atlantic. *Limnol. Oceanogr.* 51: 38–46.

Spudich, J. L., and K. H. Jung. 2005. Microbial rhodopsins: Phylogenetic and functional diversity. In W. R. Briggs, and J. L. Spudich (eds.), *Handbook of Photosensory Receptors*. Wiley-VCH, pp. 1–24.

Spudich, J. L., C. S. Yang, K. H. Jung, and E. N. Spudich. 2000. Retinylidene proteins: Structures and functions from Archaea to humans. *Annu. Rev. Cell Dev. Biol.* 16: 365–392.

Stefels, J. 2000. Physiological aspects of the production and conversion of DMSP in marine algae and higher plants. *J. Sea Res.* 43: 183–197.

Sunda, W., D. J. Kieber, R. P. Kiene, and S. Huntsman. 2002. An antioxidant function for DMSP and DMS in marine algae. *Nature* 418: 317–320.

Suzuki, M. T., O. Béjà, L. T. Taylor, and E. F. DeLong. 2001a. Phylogenetic analysis of ribosomal RNA operons from uncultivated coastal marine bacterioplankton. *Environ. Microbiol.* 3: 323–331.

Suzuki, M. T., C. M. Preston, F. P. Chavez, and E. F. DeLong. 2001b. Quantitative mapping of bacterioplankton populations in seawater: Field tests across an upwelling plume in Monterey Bay. *Aquat. Microbiol. Ecol.* 24: 117–127.

Swofford, D. L. 2002. *PAUP*. Phylogenetic Analysis Using Parsimony (*And Other Methods)*. Sinauer Associates.

Venter, J. C., K. Remington, J. Heidelberg, et al. 2004. Environmental genome shotgun sequencing of the Sargasso Sea. *Science* 304: 66–74.

Waidner, L. A., and D. L. Kirchman. 2005. Aerobic anoxygenic photosynthesis genes and operons in uncultured bacteria in the Delaware River. *Environ. Microbiol.* 7: 1896–1908.

Waterbury, J. B., S. W. Watson, R. R. L. Guillard, and L. E. Brand. 1979. Widespread occurrence of a unicellular, marine, planktonic, cyanobacterium. *Nature* 277: 293–294.

Waterbury, J. B., S. W. Watson, F. W. Valois, and D. G. Franks. 1986. Biological and ecological characterization of the marine unicellular cyanobacterium *Synechococcus*. *Can. Bull. Fish. Aquat. Sci.* 214: 159–204.

Willey, J. M., and J. B. Waterbury. 1989. Chemotaxis toward nitrogenous compounds by swimming strains of marine *Synechococcus* spp. *Appl. Environ. Microbiol.* 55: 1888–1894.

Wolfe, G. V., M. Steinke, and G. O. Kirst. 1997. Grazing-activated chemical defense in a unicellular marine alga. *Nature* 387: 894–897.

Yurkov, V. V., and J. T. Beatty. 1998. Aerobic anoxygenic phototrophic bacteria. *Microbiol. Mol. Biol. Rev.* 62: 695–724.

Yurkov, V., N. Gad'on, A. Angerhofer, and G. Drews. 1994a. Light-harvesting complexes of aerobic bacteriochlorophyll-containing bacteria *Roseococcus thiosulfatophilus*, RB3 and *Erythromicrobium ramosum*, E5 and the transfer of excitation energy from carotenoids to bacteriochlorophyll. *Z. Naturforsch.* 49(c): 579–586.

Yurkov, V. V., E. Stackebrandt, A. Holmes, et al. 1994b. Phylogenetic positions of novel aerobic, bacteriochlorophyll *a*-containing bacteria and description of *Roseococcus thiosulfatophilus* gen. nov., sp. nov., *Erythromicrobium ramosum* gen. nov., sp. nov., and *Erythrobacter litoralis* sp. nov. *Int. J. Syst. Bacteriol.* 44: 427–434.

Yurkov, V., B. Schoepp, and A. Vermeglio. 1995. Electron transfer carriers in obligately aerobic photosynthetic bacteria from genera *Roseococcus* and *Erythromicrobium*. In P. Matthis (ed.), *Photosynthesis: From Light to Biosphere*. Kluwer, pp. 543–546.

Yutin, N., M. T. Suzuki, and O. Béjà. 2005. Novel primers reveal a wider diversity among marine aerobic anoxygenic phototrophs. *Appl. Env. Microbiol.* 71: 8958–8962.

Yutin, N., M. T. Suzuki, H. Teeling, M. Weber, J. C. Venter, D. B. Rusch, O. Béjà. 2007. Assessing diversity and biogeography of aerobic anoxygenic phototrophic bacteria in surface waters of the Atlantic and Pacific Oceans using the Global Ocean Sampling expedition metagenomes. *Environ. Microbiol.* 9: 1464–1475.

Zubkov, M. V., and G. A. Tarran. 2005. Amino acid uptake of *Prochlorococcus* spp. in surface waters across the South Atlantic Subtropical Front. *Aquat. Microb. Ecol.* 40: 241–249.

Zubkov, M. V., B. M. Fuchs, G. A. Tarran, P. H. Burkill, and R. Amann. 2003. High rate of uptake of organic nitrogen compounds by *Prochlorococcus* cyanobacteria as a key to their dominance in oligotrophic oceanic waters. *Appl. Environ. Microbiol.* 69: 1299–1304.

Zubkov, M. V., G. A. Tarran, and B. M. Fuchs. 2004. Depth related amino acid uptake by *Prochlorococcus* cyanobacteria in the Southern Atlantic tropical gyre. *FEMS Microbiol. Ecol.* 50: 153–161.

Zubkov, M. V., G. A. Tarran, and P. H. Burkill. 2006. Bacterioplankton of low and high DNA content in the suboxic waters of the Arabian Sea and the Gulf of Oman: Abundance and amino acid uptake. *Aquat. Microb. Ecol.* 43: 23–32.

ECOLOGY AND DIVERSITY OF PICOEUKARYOTES

ALEXANDRA Z. WORDEN

Monterey Bay Aquarium Research Institute, Moss Landing, CA 95039, U.S.A.

FABRICE NOT

Evolution du Plancton et Paleoceans Laboratory, CNRS, Université Paris 06, UMR7144, Station Biologique de Roscoff, 29682, Roscoff, France

INTRODUCTION

Microbial, single-celled eukaryotes (protists) play a wide range of ecological roles in marine environments. Like all eukaryotes, protists are the evolutionary product of one or several endosymbiosis events (Fig. 6.1). Thus, in contrast to bacteria and archaea, they have a nucleus and other organelles such as a mitochondrion and chloroplasts (if photosynthetic). Modern protistan communities are composed of several high-level taxa (taxa defined as a formal grouping of organisms), which result from the specifics of endosymbiotic event(s), including the particular organisms involved in the event(s) as well as subsequent evolutionary processes (Fig. 6.1). Photosynthetic protists are often referred to as phytoplankton or algae, although algae are not exclusively protistan, since there are also multicellular algae. Heterotrophic protists are generally known as protozoa (single-celled animals), but the fact that some may also have phototrophic capabilities blurs the distinction between trophic modes. Picoeukaryotes are the smallest protistan size class; they are highly diverse and are found throughout the world's oceans.

Microbial Ecology of the Oceans, Second Edition. Edited by David L. Kirchman
Copyright © 2008 John Wiley & Sons, Inc.

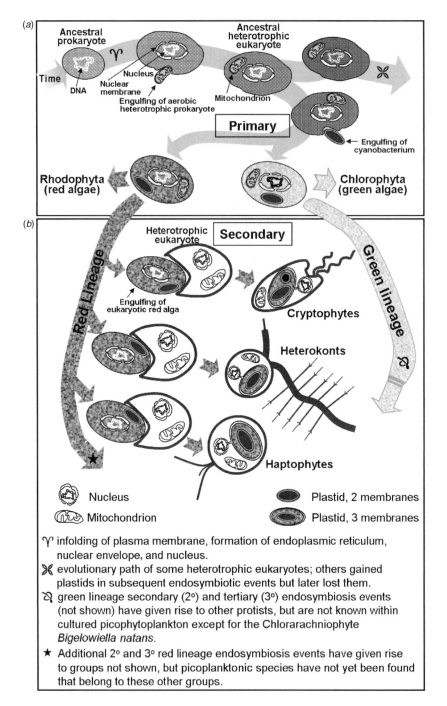

Figure 6.1

INTRODUCTION

> **Protist Metabolisms** Phagotrophy, in which prey are ingested or engulfed, and digested in food vacuoles, is the primary heterotrophic mode attributed to picoeukaryotes, and includes bacterivory (predation on bacteria, or more loosely consumption of small microbes). Mixotrophs can derive metabolic energy from both photosynthesis and heterotrophy. In eukaryotic phytoplankton the energetic contribution by mixotrophy varies across taxa. Many are purely photoautotrophic, although mixotrophy is now thought to be more common than early work indicated (Sanders et al. 2001).

To start our exploration of picoeukaryotes, we must ask the questions: What is the picosize fraction? And, how does size classification relate to organism ecology? Unlike marine prokaryotes, for which one has (at least for the most part) a general sense of organism size, the cell diameter of different protists ranges from less than one to several hundred micrometers (μm). Thus, it has become standard practice to group microbial eukaryotes according to a series of size fractions. These fractions are based on cell diameter, with the prefix pico- indicating $0.2-2.0$ μm cell diameter; nano- indicating $2.0-20$ μm; and micro- indicating $20-200$ μm (Sieburth et al. 1978; see also Chapter 2). Picoeukaryotes belong to the smallest of these size fractions ($0.2-2$ μm), but in current practice size range is most often defined as either $0.2-2.0$ μm or $0.2-3.0$ μm, depending on the filter pore size used by different research groups.

It may seem arbitrary to lump together organisms solely on the basis of size. For example, the grouping "picoplankton" includes eukaryotes and prokaryotes. However, it is helpful when considering the various life strategies and the competition processes that go on among different microbial populations. Moreover, due to their tiny size, all picoplankton, whether eukaryotic or prokaryotic, have low Reynolds numbers (Re), indicating that their movement is dominated by viscous forces, rather than inertial forces (Aris 1989). Because of their low Re picoplankton do not sink through the water column as individual cells; they can sink only if packaged into larger material (e.g., through predation, fecal pellet, and marine snow formation). The viscosity of seawater also has evolutionary consequences for the energetics and mechanisms of directional movement by motile microbes, such as the structure and placement of flagellar apparatuses. A similarity in cell size also puts different species of photosynthetic picoeukaryotes and photosynthetic bacteria under similar constraints in terms of the ratio of cell surface area to

Figure 6.1 Cartoon of the endosymbiotic events thought to have led to the development of eukaryotes and their diversification. Only endosymbiosis events relevant to cultured picophytoeukaryotes are depicted. (*a*) Initially, an aerobic heterotrophic prokaryote was engulfed, forming the mitochondrion; subsequently, in the development of photosynthetic organisms, a cyanobacterium was engulfed in what is often referred to as the primary endosymbiosis event. (*b*) Later endosymbioses events involved engulfment of eukaryotic organisms (i.e., organisms that had already undergone primary endosymbiosis), which are commonly referred to secondary endosymbiosis events. The differences we see in extant divisions (e.g., pigmentation) depend at least partially on the alga engulfed at the time of the secondary event. Tertiary endosymbiosis events (not shown) have resulted in microbes such as modern day dinoflagellates. (*b*) is conceptually based on a cartoon by Delwiche (1999).

volume. This ratio impacts the efficiency of nutrient acquisition (since nutrient transporters are located on the surface of cells), as well as packaging of photosynthetic pigments inside the cell (Raven 1986, 1998). These factors are critical to success in oligotrophic environments such as the open ocean. Overall, such size-based considerations are important for integrating organism dynamics into food webs and global biogeochemical cycles, including the carbon cycle and the flux to the deep ocean.

In this chapter, we explore the diversity and ecology of free-living marine picoeukaryotes. Picoeukaryotes have been studied for many years, with high profile publications dating back to the 1950s (Butcher 1952; Knight-Jones and Walne 1951). However, their tremendous diversity has been revealed only recently. The primary topic of this chapter is picoeukaryotes found in the euphotic zone. This bias does not necessarily reflect a lack of their importance in deeper waters, but simply a lack of data on picoeukaryote distribution, diversity, and physiology in such environments. To start, we focus on functional roles of cultured picoeukaryotes, which can be grown and studied in the laboratory, then move on to environmental diversity, molecular phylogenetics, and potential links with functional diversity. Next, we investigate the distribution of picoeukaryotes and methods for studying dynamics and activities of individual populations. The final section addresses genomic approaches for developing and testing hypotheses on picoeukaryote ecology. We hope to highlight topics in critical need of research for development of predictive microbial food web models and understanding of global biogeochemical cycles.

FUNCTIONAL ROLES, CLASSIFICATION, AND BIOLOGICAL TRAITS

Functional diversity or functional roles can be defined broadly according to different trophic modes. Picoeukaryotes are known to live by photoautotrophy and heterotrophy. Other trophic modes have been hypothesized as well, but since they have not been demonstrated (e.g., in cultured organisms), they are only given brief attention here.

An important aspect for the ecologist to consider is that this level of categorization of "functional diversity" is broad, and can be misleading if it engenders the idea of functional redundancy. Functional redundancy is the concept that two or more species play the same role in the ecosystem. Essentially, when species are identified as being similar or functionally redundant within a system (e.g., photoautotrophs), the loss of one species is viewed as having relatively little effect on ecosystem processes. This view is detrimental to understanding ecosystem responses to short- or long-term perturbations (see also Azam and Worden 2004; Fuhrman et al. 2002; Ward 2005; Weithoff 2003). The ocean lends itself to development of microbial specialization and niche differentiation due to the complexity and variability of physicochemical factors across environmental gradients. If a particular phytoplankter relies solely on ammonium as a nitrogen source while another can utilize other nitrogen sources, we should be careful in evaluating their redundancy; they do not necessarily perform photosynthesis well under the same conditions, and if one species is lost, but then conditions change such that the other is no longer successful, the function (photosynthesis in this case) could be lost. Other factors, such as enzyme kinetics or temperature tolerance ranges, vary among species, leading potentially to selection and subsequent consequences for food web interactions. For example, if not all photosynthetic picoplankton are grazed in the same way, at the same rate, or by the same grazer, then the loss (from the system) of one picophytoplankter could affect ecosystem processes via alteration of specific resources available to populations within other trophic levels.

These considerations, whether of differences in resource utilization or prey susceptibility, are important when considering scenarios resulting from climate change, ocean acidification, or other perturbations. Lastly, the interplay between top-down forces (e.g., predation) and bottom-up forces (e.g., resource availability) on different populations (Raven et al. 2005; Worden and Binder 2003) is an important aspect of ecosystem dynamics. Since many picoeukaryotes have not yet been cultured, knowledge of their biology and physiology is still too limited to tease apart resource-based controls or specific causes of mortality. Furthermore, the data reported to date on distributions, both spatial and temporal, indicate that there is much to learn.

Photoautotrophs

Picophytoeukaryotes might be considered functionally akin to the picocyanobacteria *Prochlorococcus* and *Synechococcus* (see below and Chapters 1, 3, and 5), because they are all photosynthetic picoplankton. However, picoeukaryotes have some different resource requirements (e.g., they may not be capable of synthesizing vitamins) and the "top-down" controls vary (e.g., the extent of viral lysis and predation). Picophytoeukaryotes are a diverse assemblage of species, with different photosynthetic pigment composition and light-harvesting capabilities, as well as different ultrastructures, physiologies, and environmental distributions. Currently, picophytoeukaryotes from three divisions (a classification rank equivalent to phylum) are considered ecologically important in marine systems. These divisions, Chlorophyta, Heterokonta, and Haptophyta, are not marine protistan per se, but also include nonmarine protists and, in some cases, multicellular organisms.

The division Chlorophyta (green algae) is evolutionarily at the base of the green lineage that gave rise to green plants, the dominant form of plant life in terrestrial environments (Lewis and McCourt 2004). The class Prasinophyceae (from the Greek *prasinos*, meaning green), is ancient within the Chlorophyta, and contains some of the smallest known marine picoeukaryotes. Many of the common extant picophytoeukaryotes (e.g., *Bathycoccus*, *Micromonas* and *Ostreococcus*), belong to the order Mamiellales within the prasinophytes (Fig. 6.2*a*, *b*). HPLC is often used to categorize different prasinophytes based on their pigment composition (Table 6.1).

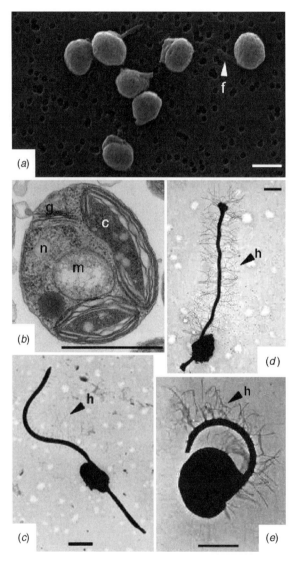

Figure 6.2 Images of various picoeukaryotes. (*a*) Scanning environmental micrograph of *Micromonas* RCC299; f, flagella. (*b*) Transmission electron micrograph (TEM) of *Ostreococcus*; c, chloroplast; m, mitochondrion; n, nucleus; g, Golgi apparatus. (*c*) TEM whole mount of *Bolidomonas mediterranea*; h, flagellar hairs. (*d*) TEM whole mount of *Picophagus flagellatus*; h, flagellar hairs. (*e*) TEM whole mount of the Bicosoecid *Symbiomonas scintillans*; h, flagellar hairs. Scale bars represent 1 μm. (*a*, *b*) Courtesy of Augustin Engman, Rory Welsh, and Alexandra Worden. (*c*) From Guillou et al. (1999a), reproduced with permission of the *Journal of Phycology*. (*d*, *e*) Guillou et al. (1999b), reprinted with permission of Elsevier.

TABLE 6.1 Characteristic Chlorophylls from Typical Picoeukaryotic Taxa

Taxa	Representative	Chl a	Chl b	MgDVP	Chl c_1	Chl c_2	Chl c_3	Chl c_2-MGDG	Chl c_3-like
Prasinophyceae	*Ostreococcus* sp. (surface)	+	+	+					
Prasinophyceae	*Ostreococcus* sp. (deep)	+	+	+					+
Prasinophyceae	*Micromonas* sp. (clade D Slapeta et al. 2006)	+	+	+					
Prasinophyceae	*Micromonas pusilla* (clade B, Slapeta et al. 2006)	+	+	+					+
Prasinophyceae	*Bathycoccus prasinos*	+	+	+					
Prasinophyceae	*Pycnococcus provasolii*	+	+	+					
Bolidophyceae	*Bolidomonas pacifica*	+				+	+		
Pinguiophyceae	*Pinguiochrysis pyriformis*	+		+		+	+		
Pelagophyceae	*Pelagococcus subviridis*	+		+		+	+		
Pelagophyceae	*Aureococcus anophagefferens*	+			+				
Cryptophyceae	*Hillea marina*	+		+		+			
Prymnesiophyceae	*Chrysochromulina leadbeasteri*	+		+		+	+	+	
Prymnesiophyceae	*Imantonia rotunda*	+		+	+	+	+	+	
Trebouxiophyceae	*Picochlorum* sp.	+	+						
Eustigmatophyceae	*Nannochloropsis granulata*	+							

Abbreviations: Chl a, chlorophyll a; Chl b, chlorophyll b; Chl c_1, chlorophyll c_1; Chl c_2, chlorophyll c_2; Chl c_3, chlorophyll c_3; Chl c_3-like, unknown chlorophyll c_3 (CS-170); MgDVP, magnesium 2,4-divinyl-pheoporphyrin a_5 monomethyl ester; MGDG, monogalactosyldiacylglyceride ester. In addition to the pigments given in the table, some Prymnesiophyceae have monovinylchlorophyll c_3, and Cryptophyceae have phycobilins. See Table 6.2 for carotenoids. Compiled from Zapata et al. (2004) (*Imantonia rotunda*), Zapata (2005) (Pelagophyceae, Cryptophyceae, Trebouxiophyceae, Pinguiophyceae, Zapata et al. (2001) (*Chrysochromulina*), Guillou et al. (1999b), and F. Rodriguez (personal communication) (Bolidophyceae), Latasa et al. (2004) (Prasinophyceae), Kawachi et al. (2002a, b) (Pinguiophyceae), Henley et al. (2004) (Trebouxiophyceae), and Karlson et al. (1996) (Eustigmatophyceae).

TABLE 6.2 Characteristic Carotenoids from Typical Picoeukaryotic Taxa

Taxa	F	4k-HF	19'-HF	Dd	Dt	19'-BF	Al	U	V-der	N	P	V	Mal	UnZ1	UnZ2	An	Z	L	Dh	UnM1	ε	α	β
Prasinophyceae								+	+	+	+	+	+	+	+	+	+	+	+	+			+
Prasinophyceae								+	+	+	+	+	+	+	+		+	+	+	+		+	+
Prasinophyceae								+	+	+	+	+	+	+	+	+	+	+	+	+		+	+
Prasinophyceae								+	+	+	+	+	+	+	+	+	+	+	+				+
Prasinophyceae								+	+	+	+	+	+	+	+		+	+	+				+
Prasinophyceae										+	+							+	+			+	+
Bolidophyceae	+			+	+																		+
Pinguiophyceae	+						+					+					+						+
Pelagophyceae	+			+	+	+																	+
Pelagophyceae	+			+	+	+																	+
Cryptophyceae							+														+	+	
Prymnesiophyceae	+	+	+																		+	+	+
Prymnesiophyceae	+	+	+			+										+	+	+				+	+
Trebouxiophyceae										+		+					+	+					+
Eustigmatophyceae												+				+	+						+

Abbreviations: Al, alloxanthin; An, antheraxanthin; Dd, diadinoxanthin; Dt, diatoxanthin; Dh, dihydrolutein; F, fucoxanthin; L, lutein; N, neoxanthin; MAL, micromonal; P, Prasinoxanthin; UnZ1, unidentified Z1; UnZ2, unidentified Z2; UnM1, unidentified M1; U, uriolide; V-der, violaxanthin derivative; V, violaxanthin; Z, zeaxanthin; α, α-carotene (β,ε-carotene); β, β-carotene (β,β-carotene); ε, ε-carotene (ε,ε-carotene); 4k-HF, 4-keto-19'-hexanoyloxyfucoxanthin; 19'-BF, 19'-butanoyloxyfucoxanthin; 19'-HF, 19'-hexanoyloxyfucoxanthin.

In addition to the pigments given in the table, Eustigmatophyceae have canthaxanthin and vaucheriaxanthin-like pigments, Cryptophyceae have monadoxanthin and crocoxanthin, and some Prymnesiophyceae have an unknown fucoxanthin-like pigment. See Table 6.1 for chlorophylls and phycobilins and for references.

> **High-performance liquid chromatography (HPLC)** is a chromatographic technique used to identify photosynthetic pigments present in microbial communities. An organic solvent is used to extract pigments from cells, then run through a retention column under high pressure, and pigments are separated according to their retention time. Based on the pigment ratios found in cultured phytoplankton, and analysis using specialized software (e.g., CHEMTAX), the taxonomic composition can be inferred for natural assemblages.

Micromonas pusilla (Fig. 6.2a) was described as "highly abundant" as far back as 1951 (although it was known as *Chromulina pusilla* at that time) by electron microscopy techniques (Knight-Jones and Walne 1951). Like many eukaryotic pelagic phytoplankton, it is a "minimalist" unicell, having one chloroplast, one mitochondrion, and one Golgi apparatus. *Micromonas* has a flagellum and is motile, with an estimated swimming speed of 100 μm/s (i.e., 75 body lengths/s) and is strongly phototactic (Crawford 1992; Throndsen 1973). Its physiology has been investigated with respect to nutrient uptake, carbon concentration, and light response (Cochlan and Harrison 1991; Durand et al. 2002; Iglesias-Rodriguez et al. 1998). *M. pusilla* has now been shown to be composed of phylogenetically distinct lineages (Fig. 6.3), suggesting that it is not a single species, but rather several (Worden 2006). The evolutionary distances between these lineages are pronounced when data from several molecular markers (e.g., several different genes) is analyzed (Slapeta et al. 2006). *Bathycoccus prasinos*, another Mamiellales, has an exterior covered by overlapping nonmineralized scales (Eikrem and Throndsen 1990), a feature of most larger (nonpico) prasinophytes, but which both *Micromonas* and *Ostreococcus* lack. In 1994, the smallest of the Mamiellales, actually the smallest free-living eukaryote known on the planet (as small as 0.7–0.8 μm diameter), *Ostreococcus tauri* (Fig. 6.2b), was discovered (Courties et al. 1994). *Ostreococcus* and *Bathycoccus* have the same limited number of organelles as *Micromonas*. Different *Ostreococcus* strains have distinct light-regulated growth optima, and these differences correspond to the grouping of strains within phylogenetic clades (Rodriguez et al. 2005). This indicates that some types may be better adapted for growth in deep water than others, which appear adapted for growth at the surface. Like many very small picophytoplankton (e.g., the cyanobacteria *Prochlorococcus* and most *Synechococcus*), *Ostreococcus* is not motile.

Several other prasinophytes fall within the picoplanktonic size class, including *Pycnococcus*, which belongs to the order Pseudoscourfieldiales. *Picochlorum* (e.g., the former *Nanochlorum eucaryotum*, now *P. eucaryotum* as well as *P. atomus*) is a purely marine Chlorophyta genus (Henley et al. 2004). Overall, though, data on environmental abundances indicate that the Mamiellales are one of the more ecologically important picophytoeukaryote groups, especially in coastal waters. *Micromonas* also appear to be important in polar waters (Not et al. 2005).

The Heterokonta (meaning having flagella of unequal length) is a major lineage also referred to as "stramenopiles," from the Greek for "straw-haired," referring to hairs or "mastigonems" (Fig. 6.2c–e) that occur on their flagella (Gray et al. 2004). Some picotaxa in this lineage have been cultured, but many have yet to be cultured. Unlike the Chlorophyta, which is composed mostly of

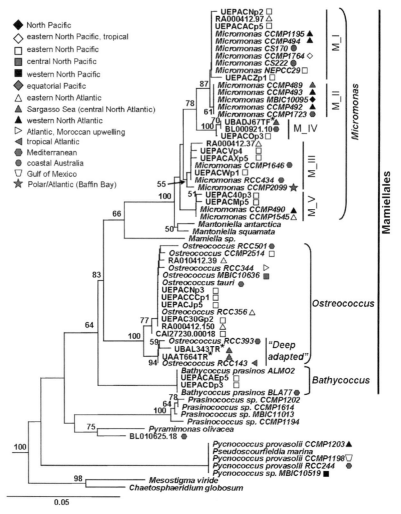

Figure 6.3 A neighbor-joining distance tree of selected prasinophyte clades containing picoeukaryotes. The tree is based on 429 nucleotide positions of the 18S rRNA gene. Cultured strains are italicized and all others are from envPCR studies (Guillou et al. 2004; Worden 2006), except those with an asterisk, which are from a shotgun clone library (Venter et al. 2004; Worden 2006). Symbols to the right of tree branches indicate, first, that the sequence is from either a cultured picoeukaryote or a picosize-fractionated envPCR study and, second, the general oceanic region of isolation or DNA collection (see the key on the figure). Numbers at branch nodes represent bootstrap support (percentage of 100 replicates); support is shown when over 50. The F84 model was used in Phylip (Felsenstien 2005) with γ-distributed rates ($\alpha = 1.4635$) and $T_i/T_v = 1.7091$ as computed in Model-Test (Posada and Crandall 1998). Note that the limited number of nucleotide positions and type of analysis used for this tree are meant to show clade structure only, and not overall evolutionary relationships or distances.

photosynthetic organisms, the Heterokonta includes phototrophs, heterotrophs, and mixotrophs. Heterokonts are products of a past secondary endosymbiosis event (Fig. 6.1), which adds greater complexity for intracellular trafficking and may influence the extent of cell size reduction possible due to presence of extra membrane layers.

Pelagomonas calceolata, *Pelagococcus* sp., and *Aureococcus anophagefferens* are photosynthetic picoeukaryotic heterokonts that belong to the class Pelagophyceae. Pelagophytes have a distinct pigment complement compared with other heterokonts and phytoplankton from other divisions (Table 6.1). *Aureococcus* uses a broad resource base and demonstrates how differences in nitrogen acquisition capabilities influence competitive dominance under certain environmental conditions (Sieracki et al. 2004; Taylor et al. 2006). Dissolved organic nitrogen (DON) was not considered a primary nitrogen source for phytoplankton blooms, although some eukaryotic phytoplankton species were thought to have the enzymatic capability to utilize it (Berg et al. 2003). Nevertheless, *Aureococcus* has been hypothesized to form brown tides using DON (Berg et al. 1997; Lomas et al. 2001). Subsequently, axenic cultures of *Aureococcus* were shown to utilize high-molecular-weight (HMW) DON and sustain growth on it (Berg et al. 2002). Comparing *Aureococcus* and its nitrogen utilization capabilities with those of the prasinophyte *Ostreococcus* demonstrates how organisms lumped together at a broad functional level (e.g., photoautotrophs) often reside in different niches and will be subject to different long- and short-term ecological constraints. Based on genome analysis, *Ostreococcus* seems to be optimized for use of inorganic nitrogen. Two *Ostreococcus* strains (one isolated from the coastal Pacific and one from a Mediterranean lagoon, both having sequenced genomes) have four genes each encoding ammonium transporters, in addition to genes for nitrate and nitrite utilization (Derelle et al. 2006; Worden, unpublished data). However, the genetic basis for HMW DON use in the laboratory has not yet been demonstrated for *Ostreococcus*.

Other described picophytoeukaryotic Heterokonta include *Bolidomonas* (Bolidophyceae), *Nannochloropsis* (Eustigmatophyceae), *Pinguiochrysis pyriformis* (Pinguiophyceae), and the Parmales order (putatively Chrysophyceae). *Bolidomonas* (Fig. 6.2c) belongs to a sister group of the diatoms, but has no siliceous structures and swims rapidly, at 800 body lengths/s (Guillou et al. 1999b). Some *Nannochloropsis* strains biosynthesize nonhydrolysable macromolecules (algaenans) that may significantly contribute to the pool of organic matter trapped in marine sediments (Gelin et al. 1996). The Pinguiophyceae are unusual in that they produce large amounts of omega-3 fatty acids (polyunsaturated fatty acids), molecules used in medicine and of high commercial value (Kawachi et al. 2002a, b). *P. pyriformis* was isolated from tropical western Pacific Ocean surface waters, but has rarely been observed elsewhere. The Parmales order, for example, *Tetraparma pelagica*, are uncultured picoeukaryotes reported in the North Pacific, Antarctic (Booth and Marchant 1987), subtropical and tropical waters (Bravo-Sierra and Hernandez-Becerril 2003). Parmales cells are covered with distinctive siliceous plates, which show up in the sediment record and are used for micropaleontological work (Booth and Marchant 1987). Because there are no cultured representatives, formal ultrastructural analyses, physiological studies, and phylogenetic analyses have not been performed. However, given their distribution (cold and tropical waters), the composition of their cell wall, and their record in marine sediments, they may be important to biogeochemical cycles (or have been in past times). Haptophyta is the third major division containing photosynthetic picoeukaryotes. This division is also a product of secondary endosymbiosis (Fig. 6.1), but is thought to be more ancient than the Heterokonta algae (Fujiwara et al. 2001). The distinguishing feature of haptophytes is the haptonema (from the Greek *hapsis*, touch, and *nema*, thread), which appears similar to a flagellum but differs in structural features and use. The haptonema is now vestigial in some haptophytes. The Haptophyta (sometimes also referred to as Prymnesiophyta) includes the coccolithophores, which have an exoskeleton of calcareous plates and only a vestigial haptonema. The coccolithophore *Emiliania huxleyi* is a microphytoplankter that forms massive blooms that can be detected and identified from space. There are few cultured pico-Haptophyta. *Imantonia rotunda* and *Phaeocystis cordata* (Vaulot et al. 2004) are the two notable examples, with some strains of the former being about 2.5–3 µm and the latter being 3–3.5 µm. Evidence is emerging from culture-independent environmental polymerase chain reaction (envPCR) surveys that there are uncultured picoplankton (Fig. 6.4) within this division (Moon-Van Der Staay et al. 2000) and that these populations have chloroplasts (Worden, unpublished data). Nevertheless, it is unclear whether they are photoautotrophic or mixotrophic, since both trophic modes occur in larger prymnesiophytes (e.g., *Phaeocystis* and *Prymnesium parvum*, respectively). Some are even thought to be strictly heterotrophic, such as the polar species *Balaniger balticus* (Jordan and Chamberlain 1997).

> **Environmental PCR (envPCR)** is used in combination with clone library construction to investigate the composition of uncultured microbial communities. Generally, total DNA is extracted from seawater, the target gene (e.g., the 18S rRNA gene) is PCR-amplified, the PCR products are ligated into a vector (e.g., a bacterial plasmid), and the vector is placed in a bacterial host (e.g., *Escherichia coli*). When the host bacteria are plated, each colony carries one version of the target gene from an uncultured microbe. Individual colonies are then picked randomly, the plasmids purified, and the inserts sequenced.

The division Cryptophyta (Fig. 6.1) contains one cultured picoplanktonic species. *Hillea marina* (Butcher 1952) was first isolated from shellfish tanks in North Wales and has rarely been noted otherwise (Chang 1983; Novarino 2003). Cryptophytes have a unique orange color derived from their phycobiliproteins (photosynthetic pigments; Table 6.1. Other picocryptophytes have been shown in envPCR surveys (Fuller et al. 2006b; Not et al. 2007; Romari and Vaulot 2004). The paucity of cultured picophytoeukaryotes from this and other divisions is a major restriction on physiological and ecological research on photosynthetic picoeukaryotes. We also know there are lineages for which there is not a single cultured representative (e.g., Not, Valentin et al. 2007), these will be discussed in later sections.

Heterotrophs and Alternative Lifestyles

Picoeukaryotic heterotrophs are peculiar, since they are similar in size to their prey, or only slightly larger. There are two formally described cultured strains of marine

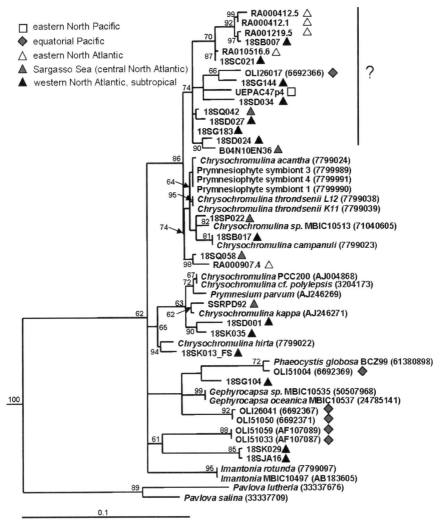

Figure 6.4 A preliminary quartet puzzling tree of prymnesiophytes focusing on novel sequences from several picosize-fractionated environmental samples. The tree is based on 530 nucleotide positions of the 18S rRNA gene. Font and symbols have the same meanings as in Figure 6.3. Numbers at branch nodes represent percentage support and are shown only when over 60 percent. 10,000 quartet puzzling steps were used and the TN model of substitution (Schmidt et al. 2002) was employed. Note that the limited number of nucleotide positions and type of analysis used for this tree are meant to show clade structure only, and not overall evolutionary relationships or distances.

heterotrophic picoeukaryotes: *Picophagus flagellatus* (Fig. 6.2*d*) and *Symbiomonas scintillans* (Fig. 6.2*e*). *P. flagellatus* has been placed at the base of the Chrysophyceae/Synurophyceae lineages (Guillou et al. 1999a), a slightly different position than the unsupported placement in the partial-length analysis shown here (Fig. 6.5). Laboratory results indicate that these phagotrophs graze selectively (on different prey species of similar size) and that there are variations in the efficiency with which different prey are assimilated (Guillou et al. 2001). *S. scintillans* contains an endosymbiotic bacterium (Guillou et al. 1999a), which is interesting considering the comparable sizes of the host and symbiont cells. Lastly, although not truly picoplanktonic, very small (2×5 μm) prasinophytes belonging to the *Pedinomonas* genus have been observed as symbionts of larger organisms such as dinoflagellates (Sweeney 1976) or Polycystinea (Cachon and Caram 1979). Moreover, symbionts related to the haptophyte genus *Chrysochromulina* (composed of numerous small species; see Fig. 6.4) have also been found in Foraminifera and Radiolaria.

Culture-independent environmental 18S rRNA gene surveys show that there are many other heterotrophic, potentially mixotrophic, or parasitic picoeukaryotes, which are not yet in culture (discussed in detail below). The ecological roles of such "organisms" are currently being explored, for example, the derivation of unique environmental gene sequences belonging to the Alveolates that may correspond to uncultured parasitic picoeukaryotes (see below).

ENVIRONMENTAL DIVERSITY AND MOLECULAR PHYLOGENETICS

The tremendous diversity of marine picoeukaryotes was discovered in large part by targeting the 18S rRNA gene. In particular, generation of envPCR clone libraries from picosize-fractionated water samples has revealed many unknown organisms since 2001, when the first studies using this approach were published (Diez et al. 2001; Lopez-Garcia et al. 2001b; Moon-Van Der Staay et al. 2001). Most of these organisms have not been cultured, or in some cases may be in culture but have not been sequenced (and therefore cannot be connected with envPCR data). The diversity discovered in envPCR surveys is much greater than expected based on paradigms of the time. High bacterial diversity (Giovannoni et al. 1990) had been explained by the fact that heterotrophic bacteria have to use many different compounds in order to grow, and are challenged by both viruses and grazers (see Chapter 3). Hence, specialization and speciation occurred frequently, especially for organisms with short generation times. In the case of eukaryotic microbes, their roles had been considered to be more simplistic, with less specialization, a view that clearly overlooked their widely varying physiologies and ecological strategies (Vaulot et al. 2002).

Interpretation of diversity data presents several challenges for the marine microbial ecologist. Foremost are questions of whether information on function can be derived from phylogenetic placement and how this diversity relates to the species concept. These are basic issues for data interpretation of all microbial molecular diversity studies, whether they are of prokaryotes or of picoeukaryotes. The traditional definition of "species" (known as "the biological species concept") relies on factors related to sexual reproduction as well as phenotypic characterization; both are difficult to assess for most marine microbes, and potentially are not the most relevant features

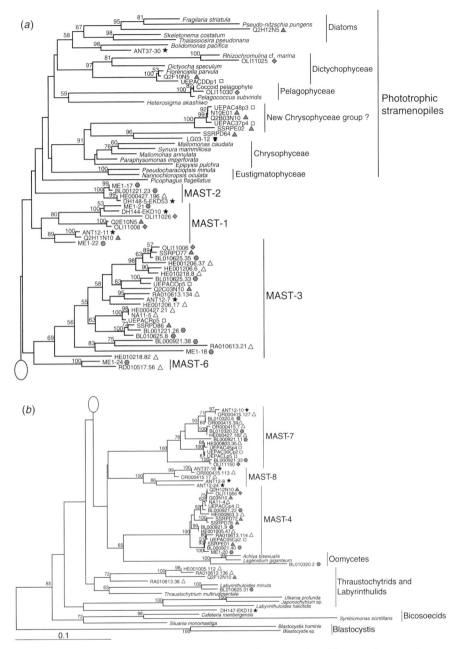

Figure 6.5 Picoeukaryote phylogeny within the Stramenopiles. The tree is a maximum-likelihood tree using GTR $+ I + \gamma$, as selected by Model-Test (Posada and Crandall 1998), and computed in Phylip modules (Felsenstien 2005). In this case, $I = 0.2010$ and $\gamma = 0.4303$. Numbers at nodes represent statistical support greater than 50 of 100 replicates. Note the MAST clades as well as the novel picochrysophytes. In addition, note that the Chrysophyceae, which are designated as "Phototrophic stramenopiles," actually contain both heterotrophic, mixotrophic and photosynthetic organisms; the actual trophic mode of the novel picochrysophytes is still unknown. See Chapter 11 for another version of this tree.

for describing microbial species. Thus, in microbiology, the species concept is still being debated, but is considered to be important for investigating ecological questions, since it is the common tool for assigning specific functions to a living entity. In the case of eukaryotes, sex is a defining feature of the species concept. Although sex-related genes are present in the genomes of *Micromonas* and *Ostreococcus*, sexual events have not been observed in these or other picoeukaryotes (Derelle et al. 2006; Farinas et al. 2006). Diatoms (larger phytoplankton) do undergo sexual events, but not under standard culture conditions. Thus, the environmental triggers for sexual reproduction remain unknown. Further discussion is not within the scope of this chapter, but, when dealing with microbial diversity, one has to keep in mind difficulties in defining species, and recognize that the term is frequently used as a descriptor without a precise definition (see, e.g., Finlay 2004; Rossello-Mora and Amann 2001).

Some investigators focus purely on molecular phylogenetic relationships among sequences obtained from envPCR studies and from cultured organisms, while others seek to relate environmentally derived sequences to organism ecology. Environmental sequences sometimes fall within well-known groups or phylogenetic clades, and can therefore be assigned a tentative functional role, such as having a photoautotrophic lifestyle. For example, it can be inferred that the environmental prasinophyte sequences (Fig. 6.3) are derived from photoautotrophs. But, for phylotypes in other groups, it is more difficult to assess ecological function.

> The term "**phylotype**" is often used to refer to a group of highly related sequences and avoids the quagmire involved in making a species determination. "**Clade**" generally refers to a broader group of sequences that are highly related and supported by statistical analysis. Likewise, the designation of an operational taxonomic unit (OTU) is used to group organisms without designating a formal taxonomic rank and without requiring phylogenetic analysis. In marine envPCR surveys, various specific sequence identity thresholds have been used to define OTUs.

Diversity of Uncultured Populations

With respect to picoeukaryotic sequences from clearly photoautotrophic organisms, there is excellent coverage of environmental prasinophyte sequences (Fig. 6.3). This is partially due to the greater number of diversity studies conducted in coastal zones, where they are common (see below). Phylogenetic analysis of sequences from envPCR surveys shows statistically supported clades that are distinct from clades containing cultured strains. This is the case for *Micromonas* (Fig. 6.3) and other prasinophytes (Guillou et al. 2004; Worden 2006). Overall, the different clades likely represent different ecotypes (niche differentiation) or even species whether composed primarily of cultured strains, a mixture, or only sequences from uncultured strains. Molecular surveys often reveal presence of picoeukaryotes not previously observed in an area, such as the first report of *Ostreococcus* in the Sargasso Sea (Worden 2006), which was found in shotgun clone sequences from a study on the prokaryotic community (Venter et al. 2004). Although the sequences were recovered from surface waters, they fell within a clade (four sequences; Fig. 6.3) previously identified as "deep-adapted," not in the "surface-adapted" clade (Rodriguez et al. 2005). Hence, the deeply mixing water column at the time of study is hypothesized to favor strains that can succeed in low light, although also exposed to surface light levels (Worden 2006).

Diverse haptophyte sequences have also been recovered from microbes passing through small-pore-size filters (1–3 µm; "picosize-fractionated") in samples from oligotrophic environments, and are unlike sequences from cultured haptophytes (Fig. 6.4) (Moon-Van Der Staay et al. 2000; Not et al. 2007). Flow-cytometric sorting (see below) from one of these oligotrophic sites allowed picophytoeukaryotes to be separated and viewed under the microscope, revealing cells approximately 1.5 µm × 2 µm with a characteristic prymnesiophyte chloroplast arrangement (Worden, unpublished data). Furthermore, fluorescence in situ hybridization (FISH) studies targeted to specifically label prymnesiophytes showed pico-sized cells harboring a characteristic red fluorescence from chlorophyll (see Fig. 6.8c, d). Thus, it has been hypothesized that these novel haptophyte sequences are from photosynthetic picoeukaryotes.

Identifying Microbes by Microscopy

In microbial ecology, fluorescence in situ hybridization (FISH) generally refers to whole cell in situ hybridization (as opposed to hybridization to extracted nucleic acids). Cells that have been preserved and collected on a filter are permeabilized, dehydrated, and hybridized with an rRNA-targeted probe. Using an epifluorescence microscope, the number of cells hybridized by each taxonomic probe is counted. Because ribosome content can be low, and consequently the probe fluorescence low, tyramide signal amplification (TSA-FISH, also referred to as CARD-FISH) is often used, in which an enzymatic reaction is used to amplify the fluorescence signal.

Heterokonta (or stramenopiles) is another division for which many sequences have been recovered from the picosize fraction that appear to be derived from uncultured organisms of unknown ecological function (Fig. 6.5). Novel Chrysophyceae have recently been identified in Arabian Sea picoplankton clone libraries (Fuller et al. 2006a). In this work, plastid-encoded 16S rRNA genes were amplified, cloned, and sequenced, rather than the nuclear-encoded 18S rRNA gene. This was done to intentionally skew the results to capture more diverse photosynthetic picoeukaryotes, by excluding picoeukaryotes that never had, or have completely lost, their plastids. Chrysophyceae from two of the open ocean sites in the Arabian Sea study had no close cultured representative. Although more research is necessary, these newly unveiled sequences could correspond to picoeukaryotes such as the Parmales. This approach is akin to earlier studies that used unfractionated samples and explored the diversity of the Rubisco large-subunit gene (*rbc*L), again with the aim of focusing on photosynthetic organisms (e.g., Paul et al. 2000). Some studies have targeted even narrower eukaryotic groups (although not focused on picoeukaryotes), by constructing libraries using group-specific primers designed for focussing on the diversity of a specific taxon (e.g., Bass and Cavalier-Smith 2004).

Other heterokont sequences within picoeukaryote libraries revealed 12 very distantly related groups (Fig. 6.5) of uncultured marine stramenopiles (MAST) (Massana et al. 2004b, 2006). MAST-2, -6, -9, -10, -11, and -12 were reported sporadically in envPCR studies, and thought to be minor components of marine picoeukaryotic assemblages. Among novel stramenopile clades, MAST-2 rRNA gene (rDNA) sequences appeared infrequency in clone libraries, but were widespread, coming from many different types of environments. Sequences belonging to MAST-9 and MAST-12 were found mostly in anoxic environments, and may represent anaerobic or anoxia-tolerant organisms (Massana et al. 2004b). The MAST-8 clade is relatively frequent in two of the open-sea systems analyzed to date, but is absent from the other sites. Finally, the clades MAST-1, -3, -4, and -7 are widely distributed in pelagic environments. They have been found in almost all open-ocean and coastal samples, but have almost never (with one exception) been retrieved from anoxic environments (Massana et al. 2004b). The commonality between the MAST clades is that they do not cluster with clearly photoautotrophic protists. Furthermore, the large phylogenetic distances between MAST clusters indicate they probably each represent populations with very different physiological and ecological roles (see also below). Chapter 11 discusses these microbes, and also includes more details about how they are studied.

One of the most dramatic discoveries to emerge from marine envPCR studies is the many novel alveolate sequences within picosize-fractionated libraries. These 18S rRNA gene sequences are unlike known cultured organisms, and are referred to as novel alveolates (NA). The name Alveolata (meaning "hollow or having hollow spaces," from the Latin *alveus*), refers to the fact that alveolates have small membrane-bound cavities (called *alveoli*) under their cell surfaces. Larger alveolates include dinoflagellates, such as the toxic red-tide-causing *Karenia brevis* and the bioluminescent dinoflagellate *Lingulodinium polyhedron*. Sequences in the novel picofractionated groups are not only diverse (Groisillier et al. 2006; Lopez-Garcia et al. 2001a; Moon-Van Der Staay et al. 2001), but also distributed through the deep ocean, as first shown in Antarctic waters (Lopez-Garcia et al. 2001a). Since the Antarctic study, analysis of a nonpolar water depth profile (surface to 3000 m) showed similar trends (Not et al. 2007). At this site, in the Sargasso Sea, 65 percent of sequences recovered from the euphotic zone belonged to the NA (51 out of 78), representing 55 percent of the OTUs using a threshold of 98 percent sequence identity. Similarly, 67 percent of the rDNA sequences fell within the alveolates at 500 m depth and 68 percent at 3000 m depth (Not et al. 2007). These NA form two main groups, NAI and NAII, each of which is composed of many clades, some clades having sequences from a broad range of environments (e.g., anoxic, deep-sea, and open-ocean waters), while others appear to be of a more limited environmental range (Groisillier et al. 2006). Although life strategies and trophic modes of the affiliated organisms (from which the sequences are derived) are not yet clear, several hypotheses exist to explain the distributions of these sequences. First, the high level of phylogenetic diversity encountered within these groups and seemingly specific environmental distributions of some groups are thought to indicate different life strategies and resource specialization. Sequences such as those in the NAII, which have tentatively been classified as Syndiniales, and are phylogenetically related to the parasitic genus *Amoebophrya*, have been hypothesized to represent parasitic picoeukaryotes (Groisillier et al. 2006). Recently, small unicellular parasites (5 μm length) have been isolated from copepods, and these also branch close to the NAII (Skovgaard et al. 2005).

Sometimes, sequences found in picosize-fractionated samples are from organisms of quite uncertain taxonomic affiliation and are not necessarily picoplanktonic. For example, environmental sequences that were initially difficult to place were subsequently found to be from a cultured heterotrophic protist, about 2–6 μm × 4–10 μm in size, named *Telonema*. Together with other gene sequences (HSP90, α-tubulin, and β-tubulin) and ultrastructural data, 18S rDNA data have now been used to refine the placement of the Telonemia taxon and indicate that may constitute a newly described, ancient eukaryotic phylum (Klaveness et al. 2005; Shalchian-Tabrizi et al. 2006). The picobiliphytes are another phylogenetically distinct taxon recently identified in clone libraries from picosize-fractionated water samples (Romari and Vaulot 2004), and may represent a new first-rank taxon (Not, Valentin et al. 2007). They do not seem to be in culture yet, but, based on observation of natural samples with specific FISH probes, they have a fluorescence typical of phycobiliproteins and are quite large, with a cell size of approximately 2 μm × 6 μm (Not, Valentin et al. 2007). Another study, conducted in subtropical waters showed that the two subgroups within the picobiliphytes average 4.1 ± 1.0 μm × 3.5 ± 0.8 μm and 3.5 ± 0.9 μm × 3.0 ± 0.9 μm, respectively. Consequently these authors recommended use of the name biliphytes, so as to not provide confusion with respect to size of these organisms. So far, they do not appear to be particularly abundant, although only a few samples have been quantified. Their highest abundances were reported in warm waters (<24 °C; Cuvelier et al. 2008). However, picobiliphytes appear to be widespread: sequences have been recovered from a number of different sites, specifically the English Channel (Romari and Vaulot 2004), the Sargasso Sea (Not et al. 2007), the North Sea, the Mediterranean Sea, and the Canada Basin of the Arctic Ocean (Not, Valentin et al. 2007), as well as the Straits of Florida (Cuvelier et al. 2008).

Other interesting groups of sequences have been recovered from 18S rDNA envPCR surveys using picosize fractionation. Numerous sequences have been recovered that are affiliated with larger planktonic species, such as radiolarians *sensu lato*. The environmental sequences from the picofraction have been helpful for refining the molecular phylogeny of radiolarians, a long-debated topic (Lopez-Garcia et al. 2002; Nikolaev et al. 2004; Not et al. 2007; Yuasa et al. 2006). Similar sequences have been found in the euphotic zone of the South China Sea (Yuan et al. 2004), the Mediterranean Sea (Marie et al. 2006), and the Sargasso Sea, where they are also well represented in deep waters (Not et al. 2007). However, cultured representatives do not exist, nor have the corresponding cells been identified or quantified in the environment.

Taken together, these findings call attention to the fact that each new study seems to reveal new populations of picoeukaryotes, reflecting low sequence coverage of picoeukaryote diversity, as well as that of larger microbial eukaryotes, particularly in the open ocean. At a broader level, sequences from envPCR studies have, in some cases, strongly impacted molecular phylogenetic schema previously based solely on cultured organisms, sometimes calling for revision of accepted taxonomy (Baldauf 2003). In addition to restructuring the molecular phylogeny of the Radiolaria (Not et al. 2007), the phylogeny of Prasinophyceae has been noticeably reworked. Indeed, input of new environmental sequences changed the taxonomy of this algal class, emphasizing the need to revise its internal phylogeny (Guillou et al. 2004; Lewis and McCourt 2004). Newly placed organisms such as *Telonema* and the picobiliphytes also restructure relationships and help identify new taxa, such as the new phylum Telonemia. However, classification guidelines require morphological and ultrastructural description, along with molecular phylogenetic characterization, making this task only partially feasible, given that a number of clades are not yet represented in culture.

Methodological Issues for envPCR Studies

Several methodological issues should be considered before drawing inferences from picoeukaryote gene libraries. First and foremost is the fact that generation of these libraries includes a size-fractionation step. Therefore, it is possible that some of these sequences are derived from larger organisms, which could have broken during the prefiltration step (Massana et al. 2004; Romari and Vaulot 2004; Worden 2006). For example, choanoflagellate-related sequences are occasionally found in picosize-fractionated clone libraries, although composing a minor fraction of the total gene sequences (Diez et al. 2001; Lovejoy et al. 2006; Not et al. 2007; Romari and Vaulot 2004). Another possibility is that certain life-stages of larger organisms (e.g., swarmers) could be small enough to pass through the prefilter, but because of their minute size, simply are not yet recognized as belonging to a larger species. Size fractionation only allows us to infer that the sequence-linked organism was probably smaller than the prefilter pore size (typically 2–3 μm), but this is not definitive. Cell size must be confirmed microscopically, either with specific FISH probes or via knowledge of cultured representatives. Some protists may be able to "squeeze" through a filter pore size of smaller diameter than their own cell size, or they may be much smaller in one dimension than another (e.g., *Telonema* or the picobiliphytes). In at least one study, diatoms were used to "control" for passage of larger organisms through the prefilter (Moon-Van der Staay et al. 2001). Since no diatom sequences were recovered, the authors concluded that sequences that were retrieved were picoeukaryotic. Diatoms may not be the best indicators, since they are less fragile than many other organisms, but at least in this way there is some evaluation for broken material. In other cases, for instance when ciliate-like sequences were recovered (ciliates generally being nano- or microplanktonic in size), it was proposed that the sequences were derived from broken cells (e.g., Massana et al. 2004; Romari and Vaulot 2004). Thus, it is particularly important to design follow-up studies to determine cell size of uncultured organisms first identified in size-fractionated libraries, as was done for the picobiliphytes and MAST clades.

Potential PCR biases do not invalidate the approach, but simply raise a warning against making conclusions concerning absence or abundance of a particular organism from a particular habitat (Wintzingerode et al. 1997). First, there may be biases with respect to DNA extraction protocols working more or less efficiently on different organisms. In addition, some DNA templates appear to amplify by PCR or clone more easily than others. Furthermore, primers designated as "universal primers" can be less than universal (see, e.g., Moon-Van der Staay et al. 2001; Stoeck et al. 2006). Thus, PCR and cloning do not necessarily reflect environmental abundance of the organism, and so relative organism abundance cannot be inferred reliably from abundance in clone libraries. This last point, however, is a topic of debate in the research community, and some do feel it valid to deduce relative abundance from such data.

After generation of the actual nucleotide sequences, these are considered from an evolutionary perspective, and their taxonomic placement determined. The most sensitive method for doing this is phylogenetic analysis (see Figs. 6.3–6.6), which is covered in detail in several books (e.g., Hillis et al. 1996; Salemi and Vandamme 2003). One of the apparent issues is the use of inappropriate phylogenetic analyses and inference of, for example, novel kingdoms based on analysis of environmental samples (for discussion, see, e.g., Berney et al. 2004; Cavalier-Smith 2004). Several factors can play into this, including: (1) the inclusion of chimeric sequences, (2) reliance on partial or single-stranded sequence, (3) limited databases, and (4) the long-branch attraction phenomenon. The inclusion of chimeric sequences is an artifact produced during the PCR reaction causing the seamless joining of gene sequences from two different organisms. For example, the chimeric sequence DH147-EKD10 falls in a basal position in the tree (Fig. 6.5), and likely compromises the statistical strength of the various other clades. The reliance on partial sequence (where only a small section of a gene is sequenced) or single-stranded sequence (where sequencing errors or simple misreading are overlooked) may produce errors when deriving major phylogenetic delineations. The effect of a limited database may include, for example, selecting too few taxa to be used in a particular analysis, as well as the fact that the 18S rRNA genes of all described eukaryotes have not yet been sequenced. The long-branch attraction phenomenon (Philippe et al. 2000; Van de Peer et al. 1996) causes the artifact that fast-evolving sequences are attracted to one another. From an ecological perspective, it is important to be aware of target populations that do not have ascribed ecological functions, and gene libraries help us to identify such "unknown" organisms. Thus, if a lineage defined as "novel" is simply a large cluster of sequences belonging to a known phylum (and not necessarily a new phylum), it is still important to explore the ecological significance of a large group of distinct organisms. However, in terms of phylogenetic distinctions, it is important to maintain rigor in the analyses and recognize the limitations imposed by the database and methods. Several envPCR surveys of marine microbial eukaryotic sequences inferred discovery of novel eukaryotic kingdoms that subsequent analyses (reanalyzing the sequences) showed to belong to previously identified kingdoms (Berney et al. 2004; Cavalier-Smith 2004).

DISTRIBUTION, ABUNDANCE, AND ACTIVITIES

General trends describing picoeukaryotic assemblages (composites of multiple taxa) are slowly becoming established. Two of the primary methods used to establish these trends are flow cytometry (FCM) and HPLC, with the former generally being used on photosynthetic populations, but having the potential to measure heterotrophic populations and the latter used for pigmented populations only.

Molecular approaches such as FISH and Q-PCR can be adapted to target very specific groups or can be adapted to target broad assemblages (e.g., all eukaryotes) through use of general (i.e., universal) eukaryotic probes that simultaneously target photosynthetic and heterotrophic eukaryotes.

> **Flow cytometers** collect information on individual particles as they pass through a laminar flow system and past a laser interrogation point. Data include light scatter and a variety of fluorescence signals (either natural fluorescence from pigments or from staining with fluorescent dyes or probes) emitted by that particle (see Chisholm et al. 1986; Marie et al. 2005; Olson et al. 1991; Shapiro 2003). Generally, a laser exciting at 488 nm is used, although others, such as UV-emitting lasers, are also used (Binder et al. 1996; Monger and Landry 1993). The scatter of light in the forward direction (forward-angle light scatter, FALS; Fig. 6.6) is an approximate indicator of cell size. The pigment chlorophyll emits a red fluorescence when excited, whereas phycobiliproteins, including those of many marine *Synechococcus*, as well as Cryptophytes, emit orange fluorescence. Cluster analysis of similarities in characteristics such as FALS and fluorescence allows definition and enumeration of discrete populations (Fig. 6.6). FCM can also be used to sort populations or individual cells from natural microbial communities.

Methods for Quantifying Mixed Assemblages

In the case of picoeukaryotes, a factor for FCM analysis is determining which populations are truly of mean diameter less than $2-3$ μm versus belonging to a larger size class. This is not trivial, since light scatter is only an approximate indicator of size. Some researchers have used synthetic beads of a known size to establish cell size. However, beads, which are perfectly spherical, scatter light more efficiently than most organic particles. This makes them appear larger, relative to most natural cells, than they are in actuality. Consequently, natural cells that are larger than 2 μm diameter beads can fall "below" the beads and be included in a count of cells smaller than $2-3$ μm causing an overestimate of the picoeukaryotic fraction. Preservation using paraformaldehyde or glutaraldehyde (as is standard practice, allowing sample analysis at a later date) can cause cells to shrink, but for picoeukaryotes the amount of shrinkage has not been systematically documented. One way to perform a fairly rigorous size calibration of the instrument is to use a series of cells of known size determined on a Coulter Multisizer (Durand 2001; Shalapyonok et al. 2001; Worden et al. 2004), which estimates volume via displacement of an electrolytic solution, or by measuring cell diameter microscopically, to define an analysis zone representing cells smaller than 2 μm. This method can also be used to calibrate forward-angle light scatter (FALS) or FALS time of flight

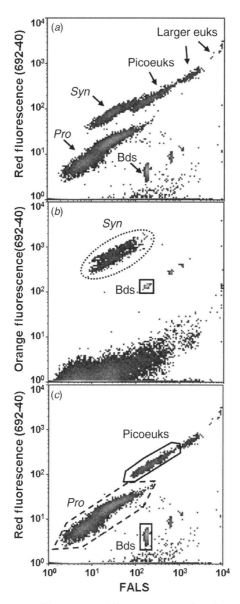

Figure 6.6 Two-parameter histograms of flow-cytometry data from a Gulf Stream DCM sample collected in March 2005. (*a*) Ungated forward-angle light scatter (FALS) versus red fluorescence (chlorophyll) data showing all picophytoplankton populations (*Pro*, *Prochlorococcus*; *Syn*, *Synechococcus*; Picoeuks, Picoeukaryotes) in the sample as well as 0.75 μm synthetic polystyrene beads (Bds, beads). Beads are added to samples for standardization. (*b*) *Synechococcus* is quantified based on its orange fluorescence (phycobilins) and FALS signature. (*c*) *Synechococcus* have been "gated-out" (i.e., removed), based on the *Synechococcus* definition performed in (*b*) so that *Prochlorococcus* and picoeukaryotes can be quantified without interference from *Synechococcus* populations. Note that the minor bead populations, slightly greater in FALS and red fluorescence, are not used for normalization; these populations are doublet and triplet bead populations caused by beads sticking together.

(the length of time a particle is in the laser beam) to yield cell size information (Shapiro 2003). This again is not an absolute measure of size, since different phytoplankton cells will scatter light differently, but nevertheless is extremely valuable, especially if the calibration is performed with a range of different genera that are also represented in the natural samples. Then mean diameter of individual environmental populations can be calculated, as well as the biomass (using a biomass conversion factor) in each of those populations. For example, five different picophytoplankton cultures (the picocyanobacteria *Prochlorococcus* and *Synechococcus*, as well as the picoeukaryotes *Pelagomonas* and two *Ostreococcus* strains) had an average of 238 ± 7 fg C/μm^3 (Worden et al. 2004). This value was then combined with flow estimated diameter information from natural populations, to calculate the biomass contributed by each of the natural populations. Thus, not only is abundance measured but biomass is also estimated. When exploring the ecological contributions of picoplankton, biomass is an important parameter to quantify. Generally, although picoeukaryotes are less abundant than the picocyanobacteria, they are larger in terms of cell size and cell biomass (Li 1994; Shalapyonok et al. 2001; Worden et al. 2004). Using the above approaches natural picophytoeukaryote populations in the coastal Pacific contained 530 ± 180 fg C/cell on average, while *Synechococcus* contained 82 fg C/cell and *Prochlorococcus* 39 fg C/cell (Worden et al. 2004). Flow cytometry can also be used to enumerate heterotrophs using a calibrated instrument and the stain LysoTracker Green, which stains the acidic food vacuole of protists (Rose et al. 2004). With this stain, cells must be alive, limiting its application to fresh (unfixed) samples.

HPLC can also be used to estimate the biomass and composition of picophytoplankton populations. For analysis of picoeukaryotes, sea water samples must be size-fractionated and the pigments present quantified by HPLC. Then assessment of each algal taxon's contribution (in proportions of total chlorophyll *a*) can be performed by iterative methods, regression analyses, multivariate statistical analyses, or factorization matrix procedures (Mackey et al. 1996; Zapata 2005). HPLC can provide information on the class level of organisms (Table 6.1), but generally cannot discriminate at greater resolution levels (e.g., genera) in field samples. Other limitations of HPLC-based methods include the large volume of water needed to analyze organisms that are low in abundance.

Distribution, Abundance, and Activity of Mixed Picophytoplankton Assemblages

Picophytoplankton dominate in oligotrophic waters, comprising over 50 percent of total biomass and production, whereas in nutrient-rich (eutrophic) waters, their contribution is lower (Agawin et al. 2000). Picophytoeukaryotes are typically much less abundant in the open ocean than *Prochlorococcus*, and similar in magnitude or lower than *Synechococcus*. Studies in the Arabian Sea and the Pacific and Atlantic Oceans found picoeukaryote abundance to be roughly similar to, or slightly lower than, *Synechococcus*. These microbes co-vary with *Synechococcus*, while they do not with *Prochlorococcus* (Fig. 6.7*a*, *b*), which can frequently be up to two orders of magnitude more abundant than the picoeukaryotes or *Synechococcus* (Binder et al.

DISTRIBUTION, ABUNDANCE, AND ACTIVITIES

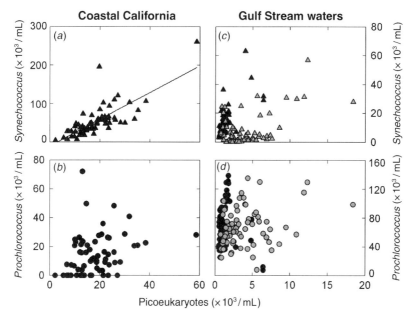

Figure 6.7 Abundance relationships between picophytoplankton populations quantified by flow cytometry. (*a, b*) Relationships between picoeukaryote abundance and (*a*) *Synechococcus* (▲), or (*b*) *Prochlorococcus* (•) at a coastal Pacific Ocean Site sampled at the surface on weekly basis over approximately 66 weeks (data have been replotted from Worden et al. 2004). Note there is a statistically significant positive correlation between picoeukaryote and *Synechococcus* abundance ($r = 0.8$, $p < 0.001$, Spearman rank). (*c, d*) show relationships between picoeukaryotes and *Synechococcus* (*c*) or *Prochlorococcus* (*d*) for a time series conducted in the Gulf Stream on periodic cruises from March 2005 to December 2005 (Hilton, Worden, and colleagues, unpublished data). In this case, the symbol shape indicates the same as above, but data from two depths are shown: the surface (5 m, black) and the DCM (gray). In these waters, no statistically significant relationship was found between the different picophytoplankton populations.

1996; Blanchot et al. 2001; Blanchot and Rodier 1996; Brown et al. 1999; Campbell and Vaulot 1993; Landry et al. 1996; Partensky et al. 1999; Zubkov et al. 1998). Covariance of picophytoeukaryotes with *Synechococcus* is not always the case. For example, in the Straits of Florida (the upstream source waters of the Gulf Stream), picoeukaryotes do not covary with either *Synechococcus* or *Prochlorococcus* (Fig. 6.7*c, d*), indicating potential differences in the composition and resources base of populations in this region from those encountered in the previous studies. Exceptions to covariance between picoeukaryotes and *Synechococcus* are not surprising, given that current knowledge is based on relatively low spatial and temporal coverage of picoeukaryote dynamics.

In general, picoeukaryote abundance is approximately 1000 cells/mL in many oligotrophic waters, and in coastal regimes averages around 5000 cells/mL, periodically ranging up to 20,000 cells/mL or more (see Figs. 6.7, 6.9, and 6.10). In the South Pacific at a longitude of 170°W, picoeukaryotes were more abundant moving northwards, particularly in the upper 80 m of the water column (DiTullio et al. 2003). In addition, prasinophyte-specific pigments also increased northwards, indicating the picoeukaryotes may have been prasinophytes. However, pigment samples were not size-fractionated, so it is unclear whether the increase in prasinophytes truly reflects picophytoeukaryote community composition, since larger species of prasinophytes also exist.

Based on HPLC, pelagophytes, which are represented by several marine strains in culture, appeared to be abundant in this same South Pacific region. Pelagophytes have also been reported to be abundant in other Pacific studies (Vaillancourt et al. 2003), as well as the Sargasso Sea (Goericke 1998). In another study, picoeukaryotes comprised 50–90 percent of the eukaryotic phytoplankton in both Pacific (Station ALOHA) and Sargasso Sea waters, and many picopelagophytes and picoprymnesiophytes were identified using electron microscopy (Anderson et al. 1999). Picophytoeukaryotes (<2 or <3 μm) have been found to be more abundant with higher nitrate concentrations, although these concentrations are still lower than eutrophic nutrient levels (Blanchot and Rodier 1996; Matsumoto et al. 2004; Shalapyonok et al. 2001). To summarize, two general trends appear to be picoeukaryote abundance covariance with *Synechococcus* (especially in surface waters), and increasing picophytoeukaryotes abundance with depth and a pronounced deep chlorophyll maximum (DCM).

Other important terms to understanding picoeukaryote ecology are biomass and activity. In the North Atlantic, FCM was used to demonstrate that picoeukaryotes formed a large portion of the primary producer biomass in this size fraction. However, when only looking at abundance they appeared much less important than the picocyanobacteria (Li 1994; Li et al. 1992). Later, eukaryotes mostly in the 2–4 μm size range, but including picoeukaryotes, were shown to comprise a significant portion of the phytoplankton biomass in the Sargasso Sea using size-calibrated FCM (Durand 2001). In an Atlantic transect from 50°N to 50°S, the abundance of picoeukaryotes was low at all latitudes, but still constituted a substantial part of total picophytoplankton biomass at most latitudes (Marañón et al. 2001; Zubkov et al. 1998). In combination with a recent equatorial Atlantic study where picophytoplankton contributed over 60 percent of chlorophyll *a* (Chl*a*) biomass and primary production (Pérez et al. 2005), this suggests that picoeukaryotes are extremely important to both biomass and productivity in this region. These levels appear similar in other water masses. In the Arabian Sea, picophytoeukaryotes (defined as <3 μm) contributed 29–60 percent of the picophytoplankton biomass averaged over four transects, and 18–33 percent of the total phytoplankton biomass in this region (Shalapyonok et al. 2001). Picoeukaryotes (<2 μm) comprised on average only 24 percent of picophytoplankton cells at a coastal Pacific Ocean site, but were responsible for 76 percent of the picophytoplankton biomass production (Worden et al. 2004). In that study, picophytoeukaryotes grew rapidly (0.71–1.29/day) and were subject to heavy grazing pressure—in which case, their carbon could be channeled to the "biological pump," depending on the types of organisms grazing on picoeukaryotes and the efficiency, including the number of food chain links, with which their carbon is "moved" to other trophic levels. The biological pump brings carbon dioxide, other gases, and nutrients to the bottom of the ocean in the form of sinking organic matter, including dead (large) phytoplankton and fecal pellets. In the equatorial Pacific, growth rates were 0.53 ± 0.18/day for *Prochlorococcus*, 0.42 ± 13/day for picoeukaryotes (<2 μm), and 0.56 ± 0.21/day for *Synechococcus*, with production contributions by each group respectively being 57, 33, and 10 percent of the picoplankton total (André et al. 1999). Recently, zooplankton were shown to be responsible for moving large amounts of picophytoplankton carbon to the deep sea (Richardson and Jackson 2007). For understanding primary production and carbon flux, it is important that growth rates of individual populations be estimated in a greater range of environments.

Primary production estimates from analysis of size-fractionated pigments after incubation with $^{14}CO_2$ also show the importance of picophytoeukaryotes in a variety of marine environments. These Chl *a*-based picoeukaryote growth rates are valuable indications of gross activity, representing a composite of presumably different growth rates of the different members within this size fraction. Some studies have estimated growth rates of groups such as prymnesiophytes and pelagophytes to be, for example, 0.5/day in the equatorial Pacific (Latasa et al. 1997) and $0.2-0.6$/day in Sargasso Sea surface waters (Goericke 1998). In these studies, size-fractionated rates for different groups were not reported, so primary production by, for example, picoplanktonic prymnesiophytes versus larger size fractions remains uncharacterized. Latasa et al. (1997) showed that prymnesiophytes growth was balanced by grazing mortality. Brown et al. (2002) found that picophytoeukaryote growth was often, but not always, balanced by grazing in the Arabian Sea. This is important, because it indicates that large proportions of picoeukaryotic primary production are transferred to higher trophic levels (e.g., heterotrophic protists).

Relative contributions of different taxa derived from HPLC-based abundances, in combination with flow-cytometry data, can be used to infer the composition of eukaryotic picophytoplankton. In oligotrophic Pacific waters, this approach was used to infer that eukaryotes smaller than 3 μm were prasinophytes and pelagophytes (Vaillancourt et al. 2003). We still know little about the dynamics of picophytoplanktonic prymnesiophytes or other taxonomic groups (e.g., prasinophytes), and much less about subtaxa within these broad groups. Furthermore, problems with chlorophyll-based estimates can also be significant. For instance, the possibility that unknown population of eukaryotes could be a source of Chl b_1 (a marker for *Prochlorococcus*) in upwelled waters is a potential problem, especially given that picoeukaryote diversity is extensive and still largely unknown (Mackey et al. 2002). Combining HPLC and flow-cytometry data, equatorial Pacific picoeukaryotes were deduced in one study to belong to the Haptophyta (Mackey et al. 2002). However, the authors of that study emphasized the lack of data from size-fractionated samples to support this conclusion and the need for more of this type of analysis to truly determine whether their conclusions were correct (Mackey et al. 2002). Unfortunately, just as with bacteria and archaea, picophytoeukaryotes are difficult to identify unambiguously, since they lack morphologically distinct features (Mackey et al. 2002; Simon et al. 1994).

Quantifying Specific Picoeukaryote Populations

Currently, data on distribution are not well connected to abundance and activity data. As outlined above, HPLC or flow cytometry reveals distribution and abundance (and, depending on the experiments performed, activity as well) at broad levels (e.g., photosynthetic, or whole eukaryotic taxa such as prymnesiophytes), but these methods do not discriminate at levels of greater ecological relevance. Individual population dynamics need to be tracked at more refined levels, such as genus, phylotype, or ecotype level, in order to understand picophytoplankton dynamics and contributions to primary production, as opposed to tracking a composite of different taxa—a fact recognized decades ago (Talling 1984). As mentioned above, the problem with focusing on composite populations (based on broad functional categorization, e.g., photoautotrophy) is that the data represent a "fictitious average" (Talling 1984), as if different species produce at the same rate and live under the same controls. This does not allow the system to be modeled in a predictive manner with respect to transitions through time, or perturbation. FISH (Fig. 6.8), quantitative PCR (Q-PCR), and dot-blot hybridization techniques make more specific population measurements possible.

A useful approach for quantifying specific populations is Q-PCR, an indirect method for quantitatively measuring amounts of DNA (or cDNA) derived from a particular gene sequence. Here, the level of specificity in Q-PCR assays is based on the probe design (and experimental goals). In a coastal Pacific study, a genus-level probe for *Ostreococcus* showed that concentrations ranged from a surface average of less than 5000 cells/mL to its greatest maximum during the 2-year study, of 3.2×10^5 cells/mL at the DCM (Countway and Caron 2006). Probes for several Mamiellales genera, as well as broader groups, were used in other studies, showing that in the Mediterranean, *Bathycoccus* was the most abundant among the Mamiellales and had a clear maximum at the DCM (Marie et al. 2006; Zhu et al. 2005).

FISH is another method used to quantify specific picoeukaryote types (Fig. 6.8). Application of rRNA-targeted probes, specific for a variety of prasinophyte genera, over a time series conducted in the English Channel (Fig. 6.9) showed *Micromonas* (Fig. 6.8b) dominance year-round (Not et al. 2004). Another study carried out in the summer, in the Norwegian and Barents Seas (northern Europe), also demonstrated the dominance of this genus at coastal and polar front stations (Not et al. 2005). Molecular phylogenetic studies on the genus *Micromonas* have identified distinct clades (Slapeta et al. 2006; Worden 2006). Several of these *Micromonas* clades were detected by FISH at a time-series station in the English Channel, and varied in their relative contribution throughout the year (Foulon et al., in preparation). *Micromonas* clade M_V was a minority, whereas clade M_I/II and clade M_III were the most abundant, with clade M_III being dominant in the summer. Less is known about the open ocean at this clade-specific level. In terms of distribution, Prasinophyceae have also been documented in oligotrophic open-ocean waters by HPLC studies, and *Micromonas* and *Bathycoccus* have been isolated from the Sargasso Sea, as has *Ostreococcus* from the tropical Atlantic. When compared across water masses representing a large nutrient gradient, the

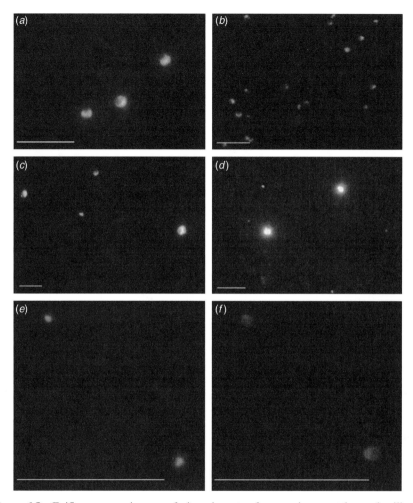

Figure 6.8 Epifluorescence images of picoeukaryotes from environmental samples illuminated via FISH. (*a, b*) Surface samples from the English Channel hybridized with a Chlorophyta-specific probe (CHLO02) and a *Micromonas* genus-specific probe (MICRO01), respectively, using TSA-FISH. The cytoplasm of hybridized cells has green fluorescence (FITC dye). (*c, d*) Florida Straits samples from 5 m and 85 m depth, respectively, hybridized with the Haptophyta-specific probe (PRYM02) using the TSA-FISH technique and FITC. The natural red/orange fluorescence of the cyanobacteria can be seen. The sample shown in (*d*) has been counterstained with the DNA-specific dye DAPI (blue) to allow visualization of heterotrophic bacteria. (*e, f*) A surface sample from the Blanes Bay (Mediterranean Sea) hybridized with the probe NS4 (Massana et al. 2004), specific for MAST-4 cells. In contrast to other images shown, the FISH method used to illuminate MAST cells in (*f*) did not involve signal amplification, but rather used a monolabeled probe linked to the fluorescent dye CY3 (red). The DAPI counterstaining from the same microscopic field is shown in (*e*). Scale bars represent 10 μm, except in (*e, f*), for which they represent 20 μm. (*a*) Courtesy of F. Not. (*b*) Courtesy of E. Foulon. (*c, d*) Courtesy of J. A. Hilton, F. Not, and A. Z. Worden. (*e, f*) Courtesy of R. Massana. (See insert for color representation.)

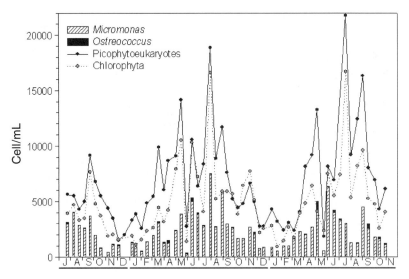

Figure 6.9 Picoeukaryote population dynamics at a coastal English Channel site over a multiyear time series, sampled bimonthly. The solid line (black circles) represents the picophytoeukaryote abundance measured by flow cytometry. The dotted line (gray circles) shows Chlorophyta abundances obtained using TSA-FISH and the CHLO02 probe. Stacked bars show specific abundances of *Micromonas* and *Ostreococcus*, using the specific probes MICRO01 and OSTREO01, respectively. Modified from Not et al. (2004) with additional information provided by N. Simon.

Chlorophyta contribution to the total picoeukaryote cells decreased from about 90 percent at eutrophic sampling sites down to about 20 percent in more oligotrophic waters (Fig. 6.10). The same pattern was observed for the contribution of *Micromonas*, which decreased from about 45 percent to 5 percent across the same gradient (Fig. 6.10). In the Gulf Stream-forming waters (Straits of Florida), FISH probes specific to an assortment of groups showed that prymnesiophytes were the most abundant of the enumerated picoeukaryote groups (Fig. 6.8c, d). However, based on the proportion of cells enumerated by these probes relative to those counted by general eukaryotic probes and flow cytometry, a large fraction of this assemblage has not yet been classified at the group or genus level (Hilton et al., in preparation).

The FISH-based abundance of prymnesiophytes in the size fraction below 2–3 μm, in combination with results from 18S rDNA clone libraries, suggests that the most ecologically relevant picoprymnesiophytes have not yet been cultured (Figs. 6.4 and 6.8 c, d). Pigment-based studies indicate that prymnesiophytes are important members of oceanic communities, for example, in the equatorial Pacific, where prymnesiophyte-specific carotenoids are abundant and most of Chl *a* is in the fraction below 2 μm (Bidigare and Ondrusek 1996). However, DNA dot-blot hybridizations on equatorial Pacific samples indicated that prymnesiophytes formed a small portion of picoeukaryotes, at least at the site investigated (Moon-Van der Staay et al. 2000).

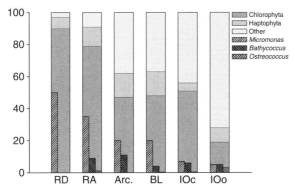

Figure 6.10 Relative abundances of major picophytoeukaryotic taxa from coastal to oceanic sites, estimated by TSA-FISH. The contributions of each taxonomic group and total picoeukaryotes were quantified using 3 μm-prefiltered samples. Division level data are shown in stack bars (solid gray shades) and genus level in grouped bars (patterned). Grouped bars represent genera belonging to the Chlorophyta and are shown inside the corresponding bar. Data correspond to averages of time-series data at a single depth for coastal samples (RD, RA, BL), or ocean transects including multiple euphotic zone depths (Arc., IOc, IOo). A rough nutrient gradient from eutrophic (RD) to oligotrophic (IOo) is represented. Sampling site codes are as follows: RD, Roscoff Dourduff, Estuarine station (48°38′N, 3°51′W), average of 1-year time series; RA, Roscoff ASTAN, English Channel (48°46′N, 3°57′W), average of 3-year time series; Arc., Arctic, Norwegian and Barents Seas (two transects, 70°30′N, 20°04′E to 73°59 N, 19°14′E and 76°19′N, 23°45′E to 76°20′N, 3°59′E), average from all stations, all depths sampled; BL, Blanes, coastal Mediterranean Sea (41°40′N, 2°48′E), average of 1-year time series; IOc, Indian Ocean Coast, coastal waters off South Africa (transect from 35°03′S, 23°44′E to 35°50′S, 32°02′E), average of three stations, all depth-sampled; IOo, Indian Ocean open, open ocean waters from the subtropical Indian Ocean (transect from 34°21′S, 37°41′E to 17°11′S, 83°40′E), average of 8 stations, all depth-sampled. Data for *Bathycoccus* and *Ostreococcus* are not available for the RD site. Data shown are from Not et al. (2004, 2005, in preparation) and Zhu et al. (2005).

More recently, the same approach showed that picoprymnesiophytes were abundant in Arabian Sea surface waters (Fuller et al. 2006b). Thus, the conditions under which prymnesiophytes form a significant fraction of picoeukaryotes in open-ocean ecosystems remains to be determined.

More studies on the dynamics of specific taxa within the picoeukaryotic fraction are needed, particularly in open ocean environments, where the dominant primary producers are generally picoplanktonic. One of the most recent studies takes a less labor intensive route by using dot-blot hybridizations and specific probes against plastid 16S rDNA amplicons generated by PCR. Although the probes were still at a fairly general level, the resolution of the data over the water column and through the Arabian Sea transect is excellent. The regions investigated ranged from oligotrophic to upwelling zones. In the oligotrophic waters, picoprymnesiophytes were more abundant in surface waters, whereas picochrysophytes were abundant throughout the water column (Fuller et al. 2006b). Overall picophytoplankton averaged

1000/mL and ranged up to 30,000/mL in the upwelling zone, where picocryptophytes were abundant. Again, picoprymnesiophytes and picochrysophytes were most abundant. A potential problem with use of data based on 16S rDNA plastid sequences for inferring ecological role as a primary producer is that the plastid can be maintained in a colorless form (no longer photosynthetically active), and some (ancestral-like) bacteria might harbor similar gene sequences. Nonetheless, on combining results from all the techniques discussed, picophytoeukaryotes appear to be predominantly composed of prasinophytes (order Mamiellales) in coastal waters. The open ocean seems to be dominated by prymnesiophytes and potentially chrysophytes or pelagophytes, although the data are not sufficient to draw firm conclusions on picophytoeukaryotic communities in open oceans. These trends are exemplified in Fig. 6.10. These FISH data taken from sites over time show that many coastal communities are dominated by Chlorophyta, particularly by *Micromonas*, whereas the category "others" (unknown) dominates at oligotrophic sites (Fig. 6.10).

Other picoeukaryotes have also been quantified using FISH. This has allowed the abundance, distribution, and ecological roles of uncultured MAST clades to be explored (see also Chapter 11). The clades MAST-1A, -1B, -1C, -2, and -4 have been investigated in terms of abundances and distribution (Massana et al. 2006). They are distributed worldwide, with the exception of MAST-4, which has not been found in polar waters. Taken together, these clades can contribute up to 35 percent of the heterotrophic flagellate population (e.g., in the Southern Ocean). MAST-4 cells appear to be the most abundant of those probed, ranging up to 500 cells/mL (in the Pacific Ocean) and on average compose 9 percent of the total heterotrophic flagellate population (Massana et al. 2006). Distribution and abundance of other open ocean clades (MAST-3 and MAST-7) have yet to be reported. Functional characterization of individual MAST clades also involved specific FISH probes, in combination with enrichment experiments (Massana 2002, 2004b, 2006). Thus far, MAST-1B, -1C, and -4 are reported to be heterotrophic bacterivores, based on microscopic observation (Massana et al. 2006). Other groups, such as MAST-1A and -2, are presumed to be bacterivores because they grow in the dark, but food vacuoles containing prey have not yet been observed. The size of MAST organisms varies; MAST-4 is approximately 2 μm in diameter, whereas others are larger; for example, MAST-1 is approximately 4–8 μm (Massana et al. 2006).

Methodological Challenges to Quantifying Specific Populations and Resolving Dynamics

A limiting factor for establishing the accuracy of both FISH and Q-PCR probes is the "completeness" of the sequence database. Probe and primer design relies on the representation of diversity within DNA sequence databases (e.g., GenBank). As the database continues to grow, it often reveals sequence discrepancies with earlier published work. Even recently designed probes end up having discrepancies with the database due to the deposition of new environmental sequences. This makes it prudent to check frequently the probe's true coverage with public databases. For example, in the case of a specific *Micromonas* Q-PCR probe used in an environmental survey (Marie et al. 2006), mismatches can be identified when compared with a group of envPCR sequences that were not available at the time of probe design (Fig. 6.3). The new environmental sequences revealed an uncultured *Micromonas* clade present in the Mediterranean Sea, the Sargasso Sea, and the Pacific Ocean. Such mismatches may result in an underestimation of the abundance of this genus. Increased sequence data on natural diversity will aid future investigations focused at more refined levels (e.g., phylotypes or ecotypes).

Other problems lie in the ability to design a specific probe given restrictions on probe length and placement on the molecule targeted etc. In this case the use of genes with more variability than the 18S rRNA gene may alleviate such design limitations. Furthermore, the number of 18S rRNA gene copies in eukaryotic genomes can vary greatly (from 1 to over 1.2×10^4) between species (see, e.g., Zhu et al. 2005). Unless the gene copy number is known, or unless a dilution series with known numbers of cells is conducted for calibration, Q-PCR quantification of specific populations reflects only relative change within the probed species, but not absolute numerical differences. Previously, issues of sensitivity (in terms of signal intensity) could prevent detection of target cells, but both TSA-FISH and Q-PCR methods largely overcome this problem. However, lack of cultured species hinders appropriate testing during the development of new primers or probes.

The hardest challenges for assessing picoeukaryotic diversity at this point are perhaps less due to methodological limitations than to the challenge of sampling a vast environment such as the world's oceans at ecologically relevant scales. In oceanography, we are often left with a snapshot at broad spatial or temporal resolution from which organism distribution, ecology and system dynamics are deduced. Several remote systems have been developed for "observing" microbial populations at high temporal (and potentially spatial) resolution that can be deployed on buoys. For example, the environmental sample processor (ESP) is an electromechanical-fluidic system that collects a water sample, concentrates microorganisms therein, and conducts an autonomous DNA array or antibody-based application to enable remote and near real-time detection of target organisms (e.g., Goffredi et al. 2006; Scholin et al. 2006). The ESP can also archive samples for nucleic acid analyses, microscopy, and other types of analyses after the instrument is recovered. In coastal Atlantic waters, an automated submersible flow cytometer has been deployed that allows quantification of individual cells based on their pigment and scatter signals (Olson et al. 2003). Combined with tools for specifically targeting picoeukaryotes, these systems should improve temporal and spatial resolution given the deployment of numerous buoys carrying multiple sensors.

Mortality, Contributions to Microbial Food Webs, and Microbial Interactions

Estimating contributions to microbial food webs relies on knowledge of the fate of the organisms under consideration. Two important causes of mortality (relevant to fate) are predation or "grazing" (consumption by predators) and viral lysis. Both of these topics are covered in more depth in Chapters 11 and 12.

Research on mortality of picophytoeukaryotes is largely driven by interest in the fate of primary production (carbon flow in the environment). Much less research has addressed the fate of heterotrophic populations. As emphasized earlier, photosynthetic eukaryotes lend themselves to measurement because of their pigments, making them suitable for HPLC or flow-cytometric measurements. They also tend to be more abundant than heterotrophic eukaryotes, making quantification easier. Observation and sizing of picoeukaryotes from various water types (coastal Mediterranean, open Indian Ocean, Antarctica and the Barents Sea), using the DNA-specific dye DAPI, showed on average that 80 percent of the cells were 3 μm or smaller and, within this size fraction, 80 percent of cells were photosynthetic, that is, contained chloroplasts (see Chapter 11). Significant levels of picophytoeukaryote mortality by predation were observed at a coastal Pacific site, where 79 percent of the picophytoplanktonic production consumed was from picophytoeukaryotes, while only 21 percent was from *Synechococcus* and *Prochlorococcu*s (Worden et al. 2004). Differential grazing rates on different members of the picophytoplankton community indicate that prey selection occurs regularly (e.g., Worden and Binder 2003; Worden et al. 2004; see also Jezbera et al. 2006). What remains unknown are the mechanisms and "rules" of selectivity—which prey are preferred, and why?

Microbial mortality due to viral lysis is covered in depth in Chapter 12, and a series of publications have discussed their potential influence on ecosystem dynamics and the microbial loop (Fuhrman 2000). Viruses infecting *Micromonas* are probably the best studied of picoeukaryote viruses. Several types have been shown to infect *Micromonas*, including an RNA virus, belonging to the family Reoviridae (Attoui et al. 2006), and the dsDNA virus Prasinovirus (Brussaard 2004). Mortality due to viral lysis was estimated to be equivalent to turnover rates of 9–25 percent of the *Micromonas* standing stock in a mesocosm containing Norwegian Fjord water (Evans et al. 2003). Similarly, a study of *Micromonas* and viral dynamics in the Mediterranean showed that the viruses were not able to terminate host blooms and that viral infection was not responsible for the observed decline in *Micromonas* abundance (Zingone et al. 1999). These field experiments correspond to laboratory results demonstrating that lytic viruses accounted for 2–10 percent of *Micromonas* turnover and coexisted stably with their hosts (Cottrell and Suttle 1995). Several prymnesioviruses have been identified, all of which are apparently DNA viruses, but their host ranges are not known (Brussaard 2004). In all likelihood there are particular conditions, and issues of host frequency and specificity, influencing the degree to which viruses regulate picoeukaryotic populations but these have not yet been characterized.

Causes and rates of picoeukaryote mortality require more experimental work, in both the laboratory and the field, since fate is an important aspect of how picoeukaryotes contribute to the ecosystem. The greatest challenge to the understanding of microbial contributions and ecology is the fact that individual adaptive strategies and real-time behavior are critical to food web dynamics (see Azam and Worden 2004). As we have discussed, the environment and organisms within it contribute to different selective pressures on individual populations comprising the "picoeukaryote group." Methods to improve upon bulk activity measurements include the combination of microautoradiography with FISH (see Chapters 3 and 5), but have primarily been used on prokaryotes (Cottrell and Kirchman 2000; Overney and Fuhrman 1999). It should also be possible to apply methods such as the double in situ hybridization technique (used in freshwater environments) to target specific prey types inside the food vacuoles of specific grazers (Jezbera et al. 2005; Medina-Sanchez et al. 2005).

GENOMIC APPROACHES TO PICOEUKARYOTE ECOLOGY

Complete genome sequencing of cultured isolates and environmental genomic approaches yield unique insights into the biology and ecology of picoeukaryotes. Genomic data facilitate hypothesis development and testing, revealing in depth information about the physiological capabilities that an organism harbors. Such information is helpful for understanding the cellular organization and evolution of these very small protists. Environmental genomic approaches are powerful because we can learn about processes in the absence of prescribed hunting (for a discussion, see Worden et al. 2006). Here we touch only lightly on genomics, as more detail (but with a prokaryotic focus) is provided in Chapter 4.

Ostreococcus tauri was the first environmentally relevant picoeukaryote to undergo complete genome sequencing (Derelle et al. 2006). This genome has revealed several surprises, including greater genome complexity than expected, with some chromosomes being structurally very different than others. In addition, the gene complement indicates that *Ostreococcus* is well poised for ammonium updake based on transporter numbers and types. Based on genome analysis, *O. tauri*, the coastal Pacific *Ostreococcus* strain (CCE9901), and two *Micromonas* strains have several types of ammonium transporters, some closely related to those expected for plant-like eukaryotes and others more akin to bacterial/archaeal ammonium transporters (Worden, unpublished work). Furthermore, the light-harvesting complex structure of Mamiellales has been revised based on data from the *O. tauri* genome sequence and subsequent research tailored by the genome findings (Six et al. 2005). Genomes of other picoeukaryotic genera have also recently been sequenced, including the brown-tide-forming *Aureococcus anophagefferens* and the two *Micromonas* strains. The *Micromonas* strains come from phylogenetically-distinct clades, M_II and M_V (Fig. 6.3). A suite of interesting findings are emerging from these studies. Nonphotosynthetic picoeukaryotes have not yet been sequenced, in part because several of the environmentally relevant groups (e.g., MAST and NA) have yet to be cultured. Those that have been cultured (e.g., *P. flagellatus* and *S. scintillans*) are not yet axenic, which is preferable for complete genome sequencing (otherwise subtraction of contaminating material via bioinformatic approaches is necessary). Eukaryotic genomes also require sequencing of mRNAs, generally in the form of expressed sequence tag (EST) libraries, in order to model protein encoding sequences. This is necessary for several reasons, including the fact that the intron–exon organization, which is common in eukaryotes but rare in prokaryotes, makes gene calling more complicated than in bacteria.

> **Expressed sequence tags (ESTs)** are derived from mRNAs (reverse transcribed into cDNA) and used to provide a profile of gene expression at a particular time or condition. They do not require prior knowledge of the sequence of the expressed gene to be generated because the mRNA polyA+ tail can be used.

A variety of approaches can be taken to conduct environmental genomic studies, such as shotgun or BAC sequencing (for reviews, see DeLong and Karl 2005; Ward 2006; Worden et al. 2006). Large environmental genomic datasets, such as one from the Sargasso Sea (Venter et al. 2004), that are focused on prokaryotes, also contain picoeukaryote sequences, since the size-fractionation step (e.g., <0.8 μm) only eliminates DNA from larger (or particle-attached) organisms. Both *Micromonas* and *Ostreococcus* 18S rDNA sequences were found at GOS Sargasso Sea Station 3 (Fig. 6.3), but not at the other GOS Sargasso Sea stations (see Worden 2006). Interestingly, based on SEAWiFS satellite imagery and in situ fluorescence measurements, Station 3 had elevated chlorophyll relative to other stations. If so, this could indicate that possibly *Micromonas* and *Ostreococcus* were part of a more active (blooming) phytoplankton community. Another valuable aspect of such shotgun data is that retrieved *Micromonas* sequences fell within an uncultured clade (Fig. 6.3). This is important, because often primers are designed based on sequences from cultures, but as for prokaryotes, cultured picoeukaryotes will not reveal everything we need to know about environmental populations. Environmental genomic databases can be queried for eukaryotic genes, which to date have focused on predominantly prokaryotic size fractions (e.g., <0.8 μm). For example, in a recent study a BAC clone containing a DNA fragment from the novel alveolate group I was sequenced, and shown to contain multiple repeats of the rRNA operon (Massana et al. 2008). Environmental genomic studies intentionally focused on sequencing genomic material from protists will be highly beneficial. Uncultured discrete populations can be sequenced, although this is more challenging. For example, uncultured populations of *Bathycoccus prasinos* and picoprymnesiophytes, that were flow sorted apart at sea, and then underwent whole genome amplification, have now been sequenced (Worden and colleagues, unpubl.).

INTEGRATION OF PICOEUKARYOTES TO THE MICROBIAL FOOD WEB: RESEARCH DIRECTIONS

As we have illustrated, picoeukaryotes play a variety of roles in the world's oceans, but many questions remain about the ecology and controls of picophytoeukaryote populations. For instance, to what extent are they competing for the same resources, or do their resource needs overlap with picocyanobacteria? Since picophytoeukaryotes are grazed, we can hypothesize that they are important to higher trophic levels, but there is little quantitative data on transfer of carbon to higher trophic levels. This is the case for bacterioplankton and cyanobacteria as well, even though picophytoplankton play a significant role in the biological pump. Furthermore, the specific consumers, whether heterotrophic nanoflagellates (HNAN), ciliates, or larger (e.g., copepod nauplii), are still poorly known (see also Chapter 11). Other unresolved questions include how picosized prey are selected by consumers, to what extent different prey are consumed at different rates, as well as whether different prey are assimilated with the same efficiency, leading to relatively greater respiration of their carbon (as opposed to incorporation into biomass). There are also a number of broader questions. Do picoalveolate sequences come from parasitic organisms, and, if so, how important is parasitism in marine microbial food webs? How extensive is the impact of MAST grazing activities and what is the fate of MAST populations themselves? What is the ecological impact of viral mortality on picoeukaryote populations? These are fertile and critically important areas for research. Rapid advancements in molecular and genomic approaches open new research directions that can used to address these questions. Then, it will be necessary to integrate a mechanistic understanding of picoeukaryote populations, and their interactions and activities, into microbial food webs at ecologically relevant scales.

SUMMARY

1. Picoeukaryotes are unicellular organisms $2-3$ μm or less in diameter. They have a range of lifestyle modes: from photosynthesis (picophytoeukaryotes) to phagotrophy. Other potential roles include parasitic, mixotrophic, and symbiotic relationships.
2. Picoeukaryotes are diverse at broad taxonomic levels (e.g., Stramenopiles and Alveolates) and at more refined levels, such as a single genus being composed of several distinct lineages (e.g., *Micromonas*).
3. The dominant picophytoeukaryotes in coastal regions are often Prasinophyceae, specifically those of the order Mamiellales. Individual activities and abundance of the genera (*Bathycoccus*, *Micromonas*, and *Ostreococcus*) composing this order, and subgroups within them, are still poorly characterized. Even fewer data are available on which picoeukaryote species dominate in the open ocean, but prymnesiophytes and pelagophytes are likely candidates.
4. Picophytoeukaryotes can contribute significantly to biomass and productivity, even when at lower abundance than the picocyanobacteria. They are consumed by grazers, forming a link to higher trophic levels, which has a variety of implications for the fate of their fixed carbon.
5. Marine heterotrophic nano- and picoflagellates may often belong to MAST clades (marine stramenopiles). The functional roles of these, and other uncultured groups such as the novel alveolates, remain unclear, but they are likely important members of marine microbial communities, given their widespread distribution.

ACKNOWLEDGMENTS

We are grateful to colleagues who have discussed these topics with us over the years, especially F. Azam, B.J. Binder, R. Massana, N. Simon, and D. Vaulot. In addition, F. Rodriguez helped with pigment tables and HPLC discussion, F. Unrein provided valuable insights on mixotrophy, two undergraduates in A.Z.W.'s laboratory, J. Hilton and A. Engman, provided constructive comments on the manuscript, as did A. Altevogt, M.L. Cuvelier, and P.B. Worden. Two anonymous reviewers also provided helpful comments and corrections. We appreciate very much photographic and data contributions (credited in legends). A.Z.W. was supported by a Gordon and Betty Moore Foundation Young Investigator Award and F.N. by a Marie Curie Intra-European Fellowship ESUMAST (MEIF-CT-2005-025000).

REFERENCES

Agawin, N. S. R., C. M. Duarte, and S. Agusti. 2000. Nutrient and temperature control of the contribution of picoplankton to phytoplankton biomass and production. *Limnol. Oceanogr.* 45: 591–600.

Anderson, R. A., R. R. Bidigare, M. D. Keller, and M. Latasa. 1999. A comparison of HPLC pigment signatures and electron microscopic observations for oligotrophic waters of the North Atlantic and Pacific Oceans. *Deep-Sea Res.* II 43: 517–537.

André, J. M., C. Navarette, J. Blanchot, and M. H. Radenac. 1999. Picophytoplankton dynamics in the equatorial Pacific: Growth and grazing rates from cytometric counts. *J. Geophys. Res. Oceans* 104: 3369–3380.

Aris, R. 1989. *Vectors, Tensors, and the Basic Equations of Fluid Mechanics.* Dover.

Attoui, H., F. M. Jaafar, M. Belhouchet, P. De Micco, X. De Lamballerie, and C. P. Brussaard. 2006. *Micromonas pusilla* reovirus: a new member of the family Reoviridae assigned to a novel proposed genus (Mimoreovirus). *J. Gen. Virol.* 87: 1375–1383.

Azam, F., and A. Worden. 2004. Microbes, molecules, and marine ecosystems. *Science* 303: 1622–1624.

Baldauf, S. L. 2003. The deep roots of eukaryotes. *Science* 300: 1703–1706.

Bass, D., and T. Cavalier-Smith. 2004. Phylum-specific environmental DNA analysis reveals remarkably high global biodiversity of Cercozoa (Protozoa). *Int. J. Syste. Evol. Microbiol.* 54: 2393–2404.

Berg, G., P. Glibert, M. Lomas, and M. Burford. 1997. Organic nitrogen uptake and growth by the chrysophyte *Aureococcus anophagefferens* during a brown tide event. *Mar. Biol.* 129: 377–387.

Berg, G., D. Repeta, and J. Laroche. 2002. Dissolved organic nitrogen hydrolysis rates in axenic cultures of *Aureococcus anophagefferens* (Pelagophyceae): Comparison with heterotrophic bacteria. *Appl. Environ. Microbiol.* 68: 401–404.

Berg, G., D. Repeta, and J. Laroche. 2003. The role of the picoeukaryote *Aureococcus anophagefferens* in cycling of marine high-molecular weight dissolved organic nitrogen. *Limnol. Oceanogr.* 48: 1825–1830.

REFERENCES

Berney, C., J. Fahrni, and J. Pawlowski. 2004. How many novel eukaryotic "kingdoms"? Pitfalls and limitations of environmental DNA surveys. *BMC Biol.* 2: 13.

Bidigare, R. R., and M. E. Ondrusek. 1996. Spatial and temporal variability of phytoplankton distributions in the central equatorial Pacific Ocean. *Deep-Sea Res.* II 43: 809–833.

Binder, B. J., S. W. Chisholm, R. J. Olson, S. L. Frankel, and A. Z. Worden. 1996. Dynamics of pico-phytoplankton, ultra-phytoplankton, and bacteria in the central equatorial Pacific. *Deep-Sea Res.* II 43: 907–931.

Blanchot, J., and M. Rodier. 1996. Picophytoplankton abundance and biomass in the western tropical Pacific Ocean during the 1992 El Nino year: Results from flow cytometry. *Deep-Sea Res.* I 43: 877–895.

Blanchot, J., J. M. Andre, C. Navarette, J. Neveux, and M. H. Radenac. 2001. Picophytoplankton in the equatorial Pacific: Vertical distributions in the warm pool and in high nutrient low chlorophyll conditions. *Deep-Sea Res.* I 48: 297–314.

Booth, B. C., and H. J. Marchant. 1987. Parmales, a new order of marine Chrysophytes, with descriptions of three genera and seven new species. *J. Phycol.* 23: 245–260.

Bravo-Sierra, E., and D. U. Hernandez-Becerril. 2003. Parmales (Chrysophyceae) from the gulf of Tehuantepec, Mexico, including the description of a new species, *Tetraparma insecta* sp. nov., and a proposal to the taxonomy of the group. *J. Phycol.* 39: 577–583.

Brown, S. L., M. R. Landry, R. T. Barber, L. Campbell, D. L. Garrison, and M. M. Gowing. 1999. Picophytoplankton dynamics and production in the Arabian Sea during the 1995 Southwest Monsoon. *Deep-Sea Res.* II 46: 1745–1768.

Brown, S. L., M. R. Landry, S. Christensen, et al. 2002. Microbial community dynamics and taxon specific phytoplankton production in the Arabian Sea during the 1995 monsoon seasons. *Deep-Sea Res.* II 49: 2345–2376.

Brussaard, C. 2004. Viral control of phytoplankton populations—a review. *J. Eukaryot. Microbiol.* 51: 125–138.

Butcher, R. W. 1952. Contribution to our knowledge of the smaller marine algae. *J. Mar. Biol. Assoc. UK* 31: 175–191.

Cachon, M., and B. Caram. 1979. Symbiotic green-alga, *Pedinomonas symbiotica* sp-nov (Prasinophyceae), in the radiolarian *Thalassolampe margarodes*. *Phycologia* 18: 177–184.

Campbell, L., and D. Vaulot. 1993. Photosynthetic picoplankton community structure in the subtropical North Pacific Ocean near Hawaii (Station ALOHA). *Deep-Sea Res.* 40: 2043–2060.

Cavalier-Smith, T. 2004. Only six kingdoms of life. *Proc. Roy. Soc. Lond. Ser. B, Biol. Sci.* 271: 1251–1262.

Chang, F. H. 1983. Winter phytoplankton and microzooplankton populations off the coast of Westland, New Zealand, 1979. *NZ J. Marine Freshwater Res.* 17: 279–304.

Chisholm, S. W., E. V. Armbrust, and R. J. Olson. 1986. The individual cell in phytoplankton ecology: Cell cycles and flow cytometry. *Can. Bull. Fish. Aquat. Sci.* 214: 343–369.

Cochlan, W. P., and P. J. Harrison. 1991. Uptake of nitrate ammonium, and urea by nitrogen-starved cultures of *Micromonas pusilla* (Prasinophyceae): transient responses. *J. Phycol.* 27: 673–679.

Cottrell, M. T., and D. L. Kirchman. 2000. Community composition of marine bacterioplankton determined by 16S rRNA gene clone libraries and fluorescence in situ hybridization. *Appl. Environ. Microbiol.* 66: 5116–5122.

Cottrell, M. T., and C. A. Suttle. 1995. Dynamics of a lytic virus infecting the photosynthetic marine picoflagellate *Micromonas pusilla*. *Limnol. Oceanogr.* 40: 730–739.

Countway, P. D., and D. A. Caron. 2006. Abundance and distribution of *Ostreococcus sp.* in the San Pedro Channel, California, as revealed by quantitative PCR. *Appl. Environ. Microbiol.* 72: 2496–2506.

Courties, C., A. Vaquer, M. Troussellier, et al. 1994. Smallest eukaryotic organism. *Nature* 370: 255.

Crawford, D. W. 1992. Metabolic cost of motility in planktonic protist: Theoretical considerations on size scaling and swimming speed. *Microb. Ecol.* 24: 1–10.

Cuvelier, M. L., A. Ortiz, E. Kim, et al. 2008. Widespread distribution of a unique marine protistan lineage. *Environ. Microbiol.* doi: 10.1111/j.1462-2920.2008.01580.x.

Delwiche, C. F. 1999. Tracing the thread of plastid diversity through the tapestry of life. *Amer. Nat.* 154: S164–S177.

DeLong, E. F., and D. M. Karl. 2005. Genomic perspectives in microbial oceanography. *Nature* 437: 336–342.

Derelle, E., C. Ferraz, S. Rombauts, et al. 2006. Genome analysis of the smallest free-living eukaryote *Ostreococcus tauri* unveils many unique features. *Proc. Natl. Acad. Sci. USA* 103: 11647–11652.

Diez, B., C. Pedros-Alio, and R. Massana. 2001. Study of genetic diversity of eukaryotic picoplankton in different oceanic regions by small-subunit rRNA gene cloning and sequencing. *Appl. Environ. Microbiol.* 67: 2932–2941.

DiTullio, G. R., M. E. Geesey, D. R. Jones, K. L. Daly, L. Campbell, and W. O. Smith. 2003. Phytoplankton assemblage structure and primary productivity along 170°W in the South Pacific Ocean. *Mar. Ecol. Prog. Ser.* 255: 55–80.

Durand, M. D. 2001. Phytoplankton populations dynamics at the Bermuda Atlantic Time-series station in the Sargasso Sea. *Deep-Sea Res.* II 48: 1983–2003.

Durand, M., R. Green, H. Sosik, and R. Olson. 2002. Diel variations in optical properties of *Micromonas pusilla* (Prasinophyceae). *J. Phycol.* 38: 1132–1142.

Eikrem, W., and J. Throndsen. 1990. The ultrastructure of *Bathycoccus* gen. nov. and *Bathycoccus prasinos* sp. nov., a non-motile picoplanktonic alga (Chlorophyta, Prasinophyceae) from the Mediterranean and Atlantic. *Phycologia* 29: 344–350.

Evans, C., S. D. Archer, S. Jaquet, and W. H. Wilson. 2003. Direct estimates of the contribution of viral lysis and microzooplankton grazing to the decline of a *Micromonas* spp. population. *Aquat. Microb. Ecol.* 30: 207–219.

Farinas, B., C. Mary, C. L. D. Manes, Y. Bhaud, G. Peaucellier, and H. Moreau. 2006. Natural synchronisation for the study of cell division in the green unicellular alga *Ostreococcus tauri*. *Plant Mol. Biol.* 60: 277–292.

Felsenstien, J. 2005. PHYLIP (Phylogeny Inference Package) version 3.6. Distributed by the author. Department of Genome Sciences, University of Washington Seattle.

Finlay, B. J. 2004. Protist taxonomy: An ecological perspective. *Philos. Trans. Roy. Soc. Lond. Ser. B, Biol. Sci.* 359: 599–610.

Fuhrman, J. A. 2000. Impact of viruses on bacterial processes. In D. L. Kirchman (ed.), *Microbial Ecology of the Oceans*, 1st edn. Wiley-Liss, pp. 327–350.

Fuhrman, J. A., J. F. Griffith, and M. S. Schwalbach. 2002. Prokaryotic and viral diversity patterns in marine plankton. *Ecol. Res.* 17: 183–194.

Fujiwara, S., M. Tsuzuki, M. Kawachi, N. Minaka, and I. Inouye. 2001. Molecular phylogeny of the Haptophyta based on the *rbc*L gene and sequence variation in the spacer region of the rubisco operon. *J. Phycol.* 37: 121–129.

Fuller, N. J., C. Campbell, D. J. Allen, et al. 2006a. Analysis of photosynthetic picoeukaryote diversity at open ocean sites in the Arabian Sea using a PCR biased towards marine algal plastids. *Aquat. Microb. Ecol.* 43: 79–93.

Fuller, N. J., G. A. Tarran, D. G. Cummings, et al. 2006b. Molecular analysis of photosynthetic picoeukaryote community structure along and Arabian Sea transect. *Limnol. Oceanogr.* 51: 2515–2526.

Gelin, F., I. Boogers, A. A. M. Noordeloos, J. S. Sinninghe Damsté, P. G. Hatcher, and J. W. De Leeuw. 1996. Novel, resistant microalgal polyethers: An important sink of organic carbon in the marine environment? *Geochim. Cosmochim. Acta* 60: 1275–1280.

Giovannoni, S. J., T. B. Britschgi, C. L. Moyer, and K. G. Field. 1990. Genetic diversity in Sargasso Sea bacterioplankton. *Nature* 345: 60–63.

Goericke, R. 1998. Response of phytoplankton community structure and taxon-specific growth rates to seasonally varying physical forcing in the Sargasso Sea off Bermuda. *Limnol. Oceanogr.* 43: 921–935.

Goffredi, S. K., W. J. Jones, C. A. Scholin, et al. 2006. Molecular detection of marine larvae. *Mar. Biotechnol.* 8: 149–160.

Gray, M. W., B. F. Lang, and G. Burger. 2004. Mitochondria of protists. *Annu. Rev. Genet.* 38: 477–524.

Groisillier, A., R. Massana, K. Valentin, D. Vaulotl, and L. Guilloul. 2006. Genetic diversity and habitats of two enigmatic marine alveolate lineages. *Aquat. Microb. Ecol.* 42: 277–291.

Guillou, L., M. Chretiennot-Dinet, S. Boulben, S. Moon-Van Der Staay, and D. Vaulot. 1999a. *Symbiomonas scintillans* gen. et sp nov and *Picophagus flagellatus* gen. et sp nov (Heterokonta): Two new heterotrophic flagellates of picoplanktonic size. *Protist* 150: 383–398.

Guillou, L., M. Chretiennot-Dinet, L. Medlin, H. Claustre, S. Loiseaux-De Goer, and D. Vaulot. 1999b. *Bolidomonas*: A new genus with two species belonging to a new algal class, the Bolidophyceae (Heterokonta). *J. Phycol.* 35: 368–381.

Guillou, L., S. Jacquet, M. Chretiennot-Dinet, and D. Vaulot. 2001. Grazing impact of two small heterotrophic flagellates on *Prochlorococcus* and *Synechococcus*. *Aquat. Microb. Ecol.* 26: 201–207.

Guillou, L., W. Eikrem, M. J. Chretiennot-Dinet, et al. 2004. Diversity of picoplanktonic prasinophytes assessed by direct nuclear SSU rDNA sequencing of environmental samples and novel isolates retrieved from oceanic and coastal marine ecosystems. *Protist* 155: 193–214.

Henley, W., J. L. Hironaka, L. Guillou, M. A. Buchheim, J. A. Buchheim, M. W. Fawley, and K. P. Fawley. 2004. Phylogenetic analysis of the '*Nannochloris*-like' algae and diagnoses of *Picochlorum oklahomensis* gen. et sp nov (Trebouxiophyceae, Chlorophyta). *Phycologia* 43: 641–652.

Hillis, D., C. Moritz, and B. Mable (eds.). 1996. *Molecular Systematics*, 2nd edn. Sinauer.

Iglesias-Rodriguez, M. D., N. A. Nimer, and M. J. Merrett. 1998. Carbon dioxide-concentrating mechanism and the development of extracellular carbonic anhydrase in the marine picoeukaryote *Micromonas pusilla*. *New Phytol.* 140: 685–690.

Jezbera, J., K. Horňák, and K. Šimek. 2005. Food selection by bacterivorous protists: Insight from the analysis of the food vacuole content by means of fluorescence in situ hybridization. *FEMS Microb. Ecol.* 52: 351–363.

Jezbera, J., K. Horňák, and K. Šimek. 2006. Prey selectivity of bacterivorous protists in different size fractions of reservoir water amended with nutrients. *Environ. Microbiol.* 8: 1330–1339.

Jordan, R. W., and A. H. L. Chamberlain. 1997. Biodiversity among haptophyte algae. *Biodiversity Conservation* 6: 131–152.

Karlson, B., D. Potter, M. Kuylenstierna, and R. A. Andersen. 1996. Ultrastructure, pigment composition, and 18S rRNA gene sequence for *Nannochloropsis granulate* sp. nov. (Monodopsidaceae, Eustigmatophyceae), a marine ultraplankter isolated from the Skagerrak, northeast Atlantic Ocean. *Phycologia* 35: 253–260.

Kawachi, M., M. Atsumi, H. Ikemoto, and S. Miyachi. 2002a. *Pinguiochrysis pyriformis* gen. et sp. nov. (Pinguiophyceae), a new picoplanktonic alga isolated from the Pacific Ocean. *Phycol. Res.* 50: 49–56.

Kawachi, M., I. Inouye, D. Honda, C. J. O'Kelly, J. C. Bailey, R. R. Bidigare, and R. A. Andersen. 2002b. The Pinguiophyceae *classis nova*, a new class of photosynthetic stramenopiles whose members produce large amount of omega-3 fatty acids. *Phycol. Res.* 50: 31–47.

Klaveness, D., K. Shalchian-Tabrizi, H. A. Thomsen, W. Eikrem, and K. S. Jakobsen. 2005. *Telonema antarcticum sp nov.*, a common marine phagotrophic flagellate. *Inter. J. Syst. Evol. Microbiol.* 55: 2595–2604.

Knight-Jones, E. W., and P. R. Walne. 1951. *Chromulina pusilla* Butcher; a dominant member of the ultraplankton. *Nature* 167: 445–446.

Landry, M. R., J. Kirshtein, and J. Constantinou. 1996. Abundance and distributions of picoplankton populations in the central equatorial Pacific from 12°N to 12°S, 140°W. *Deep-Sea Res.* II 43: 871–890.

Latasa, M., M. R. Landry, L. Schluter, and R. R. Bidigare. 1997. Pigment-specific growth and grazing rates of phytoplankton in the central equatorial Pacific. *Limnol. Oceanogr.* 42: 289–298.

Lewis, L. A., and R. M. McCourt. 2004. Green algae and the origin of land plants. *Am. J. Botany* 91: 1535–1556.

Li, W. K. W. 1994. Primary production of prochlorophytes, cyanobacteria, and eukaryotic ultraplankton: Measurements from flow cytometric sorting. *Limnol. Oceanogr.* 39: 169–175.

Li, W. K. W., P. M. Dickie, B. D. Irwin, and A. M. Wood. 1992. Biomass of bacteria, cyanobacteria, prochlorophytes and photosynthetic eukaryotes in the Sargasso Sea. *Deep-Sea Res.* 39: 501–519.

Lomas, M., P. M. Glibert, D. A. Clougherty, et al. 2001. Elevated organic nutrient ratios associated with brown tide algal blooms of *Aureococcus anophagefferens* (Pelagophyceae). *J. Plankton Res.* 23: 1339–1344.

Lopez-Garcia, P., D. Moreira, and F. Rodriguez-Valera. 2001a. Diversity of free-living prokaryotes from a deep-sea site at the Antarctic Polar Front. *FEMS Microbiol. Ecol.* 36: 193–202.

Lopez-Garcia, P., F. Rodriguez-Valera, C. Pedrós-Alió, and D. Moreira. 2001b. Unexpected diversity of small eukaryotes in deep-sea Antarctic plankton. *Nature* 409: 603–607.

REFERENCES

Lopez-Garcia, P., F. Rodriguez-Valera, and D. Moreira. 2002. Toward the monophyly of Haeckel's Radiolaria: 18S rRNA environmental data support the sisterhood of Polycystinea and Acantharea. *Mol. Biol. Evol.* 19: 118–121.

Lovejoy, C., R. Massana, and C. Pedrós-Alió. 2006. Diversity and distribution of marine microbial eukaryotes in the Arctic Ocean and adjacent seas. *Appl. Environ. Microbiol.* 72: 3085–3095.

Mackey, D., J. Blanchot, H. Higgins, and J. Neveux. 2002. Phytoplankton abundances and community structure in the equatorial Pacific. *Deep-Sea Res.* I 49: 2561–2582.

Mackey, M. D., D. J. Mackey, H. W. Higgins, and S. W. Wright. 1996. CHEMTAX—a program for estimating class abundances from chemical markers: Application to HPLC measurements of phytoplankton. *Mar. Ecol. Prog. Ser.* 144: 265–283.

Marañón, E., P. M. Holligan, R. Barciela, et al. 2001. Patterns of phytoplankton size structure and productivity in contrasting open-ocean environments. *Mar. Ecol. Prog. Ser.* 216: 43–56.

Marie, D., N. Simon, and D. Vaulot. 2005. Phytoplankton cell counting by flow cytometry. In R. A. Anderson (ed.), *Algal Culturing Techniques.* Elsevier Academic Press, pp. 253–268.

Marie, D., F. Zhu, V. Balague, J. Ras, and D. Vaulot. 2006. Eukaryotic picoplankton communities of the Mediterranean Sea in summer assessed by molecular approaches (DGGE, TTGE, QPCR). *FEMS Microbiol. Ecol.* 55: 403–415.

Massana, R., L. Guillou, B. Diez, and C. Pedrós-Alió. 2002. Unveiling the organisms behind novel eukaryotic ribosomal DNA sequences from the ocean. *Appl. Environ. Microbiol.* 68: 4554–4558.

Massana, R., V. Balague, L. Guillou, and C. Pedrós-Alió. 2004a. Picoeukaryotic diversity in an oligotrophic coastal site studied by molecular and culturing approaches. *FEMS Microbiol. Ecol.* 50: 231–243.

Massana, R., J. Castresana, V. Balague, et al. 2004b. Phylogenetic and ecological analysis of novel marine stramenopiles. *Appl. Environ. Microbiol.* 70: 3528–3534.

Massana, R., R. Terrado, I. Forn, C. Lovejoy, and C. Pedrós-Alió. 2006. Distribution and abundance of uncultured heterotrophic flagellates in the world oceans. *Environ. Microbiol.* 8: 1515–1522.

Massana, R., B. Karniol, T. Pommier, I. Bodaker, and O. Béjà. 2008. Metagenomic retrieval of a ribosomal DNA repeat array from an uncultured marine alveolate. *Environ. Microbiol.* doi: 10.1111/j.1462-2920.2007.01549.x.

Matsumoto, K., K. Furuya, and T. Kawano. 2004. Association of picophytoplankton distribution with ENSO events in the equatorial Pacific between 145°E and 160°W. *Deep-Sea Res.* I 51: 1851–1871.

Medina-Sanchez, J. M., M. Felip, and E. O. Casamayor. 2005. Catalyzed reported deposition-fluorescence in situ hybridization protocol to evaluate phagotrophy in mixotrophic protists. *Appl. Environ. Microbiol.* 71: 7321–7326.

Medlin, L. K., K. Metfies, H. Mehl, K. Wiltshire, and K. Valentin. 2006. Picoeukaryotic plankton diversity at the Helgoland time series site as assessed by three molecular methods. *Microb. Ecol.* 52: 53–71.

Monger, B. C., and M. Landry. 1993. Flow cytometric analysis of marine bacteria with Hoechst 33342. *Appl. Environ. Microbiol.* 59: 905–911.

Moon-Van Der Staay, S., G. Van Der Staay, L. Guillou, D. Vaulot, H. Claustre, and L. Medlin. 2000. Abundance and diversity of prymnesiophytes in the picoplankton community from the equatorial Pacific Ocean inferred from 18S rDNA sequences. *Limnol. Oceanogr.* 45: 98–109.

Moon-Van Der Staay, S. Y., R. Dewachter, and D. Vaulot. 2001. Oceanic 18S rDNA sequences from picoplankton reveal unsuspected eukaryotic diversity. *Nature* 409: 607–610.

Nikolaev, S. I., C. Berney, J. F. Fahrni, et al. 2004. The twilight of Heliozoa and rise of Rhizaria, an emerging supergroup of amoeboid eukaryotes. *Proc. Natl. Acad. Sci. USA* 101: 8066–8071.

Not, F., M. Latasa, D. Marie, T. Cariou, D. Vaulot, and N. Simon. 2004. A single species, *Micromonas pusill* (Prasinophyceae), dominates the eukaryotic picoplankton in the western English Channel. *Appl. Environ. Microbiol.* 70: 4064–4072.

Not, F., R. Massana, M. Latasa. 2005. Late summer community composition and abundance of photosynthetic picoeukaryotes in Norwegian and Barents Seas. *Limnol. Oceanogr.* 50: 1677–1686.

Not, F., R. Gausling, F. Azam, J. F. Heidelberg, and A. Z. Worden. 2007. Vertical distribution of picoeukaryotic diversity in the open ocean. *Environ. Microbiol.* 9: 1233–1252.

Not*, F., K. Valentin*, K. Romari, et al. 2007. The picobiliphytes: A new algal group in the marine picoplankton and a new piece of the puzzle for endosymbiosis. *Science* 315: 253–255. *These authors contributed equally to this work.

Novarino, G. 2003. A companion to the identification of cryptomonad flagellates (Cryptophyceae = Cryptomonadea). *Hydrobiologia* 502: 225–270.

Olson, R. J., E. R. Zettler, S. W. Chisholm, and J. A. Dusenberry. 1991. Advances in oceanography through flow cytometry. In S. Demers (ed.), *Particle Analysis in Oceanography*. Springer-Verlag, pp. 351–399.

Olson, R. J., A. Shalapyonok, and H. M. Sosik. 2003. An automated submersible flow cytometer for analyzing pico- and nanophytoplankton: FlowCytobot. *Deep-Sea Res.* I: 301–315.

Overney, C. C., and J. A. Fuhrman. 1999. Combined microautoradiography 16S rRNA probe technique for determination of radioisotope uptake by specific microbial cell types in situ. *Appl. Environ. Microbiol.* 65: 1746–1752.

Partensky, F., J. Blanchot, and D. Vaulot. 1999. Differential distribution and ecology of *Prochlorococcus* and *Synechococcus* in oceanic waters: A review. In L. Charpy and A. W. D. Larkum (eds.), *Marine Cyanobacteria*. Bulletin de L'Institut Oceanographique, Monaco, Special Paper 19.

Paul, J. H., A. Alfreider, and B. Wawrik. 2000. Micro and macrodiversity in *rbc*L sequences in ambient phytoplankton populations from the Southeastern Gulf of Mexico. *Mar. Ecol. Prog. Ser.* 198: 9–17.

Pérez, V., E. Fernandez, E. Maranon, et al. 2005. Latitudinal distribution of microbial plankton abundance, production and respiration in the Equatorial Atlantic in autumn 2000. *Deep-Sea Res.* I 52: 861–880.

Philippe, H., P. Lopez, H. Brinkmann, et al. 2000. Early-branching or fast-evolving eukaryotes? An answer based on slowly evolving positions. *Proc. Roy. Soc. Lond. Ser. B, Biol. Sci.* 267: 1213–1221.

Posada, D., and K. A. Crandall. 1998. Modeltest: Testing the model of DNA substitution. *Bioinformatics* 14: 817–818.

Raven, J. A. 1986. Physiological consequences of extremely small size for autotrophic organisms in the sea. *Can. Bull. Fish. Aquat. Sci.* 214: 1–70.

Raven, J. A. 1998. Small is beautiful: The picophytoplankton. *Funct. Ecol.* 12: 503–513.

Raven, J. A., Z. V. Finkel, and A. J. Irwin. 2005. Picophytoplankton: Bottom-up and top-down controls of ecology and evolution. *Vie Milieu* 55: 209–215.

Richardson, T. L., and G. A. Jackson. 2007. Small phytoplankton and carbon export from the surface ocean. *Science* 315: 838–840.

Rodriguez, F., E. Derelle, L. Guillou, F. Le Gall, D. Vaulot, and H. Moreau. 2005. Ecotype diversity in the marine picoeukaryote *Ostreococcus* (Chlorophyta, Prasinophyceae). *Environ. Microbiol.* 7: 853–859.

Romari, K., and D. Vaulot. 2004. Composition and temporal variability of picoeukaryote communities at a coastal site of the English Channel from 18S rDNA sequences. *Limnol. Oceanogr.* 49: 784–798.

Rose, J. M., D. A. Caron, M. E. Sieracki, and N. Poulton. 2004. Counting heterotrophic nanoplanktonic protists in cultures and aquatic communities by flow cytometry. *Aquat. Microb. Ecol.* 34: 263–277.

Rossello-Mora, R., and R. Amann. 2001. The species concept for prokaryotes. *FEMS Microbiol. Rev.* 25: 39–67.

Salemi, M., and A. Vandamme (eds.). 2003. *The Phylogenetic Handbook: A Practical Approach to DNA and Protein Phylogeny.* Cambridge University Press.

Sanders, R. W., D. A. Caron, J. M. Davidson, M. R. Dennett, and D. M. Moran. 2001. Nutrient acquisition and population growth of a mixotrophic alga in a axenic and bacterized cultures. *Microb. Ecol.* 42: 513–523.

Schmidt, H., K. Strimmer, M. Vingron, and A. Von Haeseler. 2002. TREE-PUZZLE: Maximum likelihood phylogenetic analysis using quartets and parallel computing. *Bioinformatics* 18: 502–504.

Scholin, C. A., G. J. Doucette, and A. D. Cembella. 2006. Prospects for developing automated systems for in situ detection of harmful algae and their toxins. In M. Babin, C. S. Roesler, and J. J. Cullen (eds.), *Real-Time Coastal Observing Systems for Ecosystem Dynamics and Harmful Algal Bloom.* UNESCO.

Shalapyonok, A., R. J. Olson, and L. S. Shalapyonok. 2001. Arabian Sea phytoplankton during South West and Northeast Monsoons 1995: Composition, size structure and biomass from individual cell properties measured by flow cytometry. *Deep-Sea Res.* II 48: 1231–1261.

Shalchian-Tabrizi, K., W. Eikrem, D. Klaveness, et al. 2006. Telonemia, a new protist phylum with affinity to chromist lineages. *Proc. Roy. Soc. Ser. B, Biol. Sci.* 273: 1833–1842.

Shapiro, H. M. 2003. *Practical Flow Cytometry*, 4 edn. Wiley-Liss.

Sieburth, J. M., V. Smetacek, and J. Lenz. 1978. Pelagic ecosystem structure: Heterotrophic compartments of plankton and their relationship to plankton size fractions. *Limnol. Oceanogr.* 33: 1225–1227.

Sieracki, M. E., C. J. Gobler, T. L. Cucci, E. C. Thier, I. C. Gilg, and M. D. Keller. 2004. Pico- and nanoplankton dynamics during bloom initiation of *Aureococcus* in a Long Island, NY bay. *Harmful Algae* 3: 459–470.

Simon, N., R. G. Barlow, D. Marie, F. Partensky, and D. Vaulot. 1994. Characterization of oceanic photosynthetic picoeukaryotes by flow cytometry. *J. Phycol.* 30: 922–935.

Six*, C., A. Z. Worden*, F. Rodriguez, H. Moreau, and F. Partensky. 2005. New insights into the nature and phylogeny of prasinophyte antenna proteins: *Ostreococcus tauri*, a case study. *Mol. Biol. Evol.* 22: 2217–2230. *These authors contributed equally to this work.

Skovgaard, A., R. Massana, V. Balague, and E. Saiz. 2005. Phylogenetic position of the copepod-infesting parasite *Syndinium turbo* (Dinoflagellata, Syndinea). *Protist* 156: 413–423.

Slapeta, J., P. Lopez-Garcia, and D. Moreira. 2006. Global dispersal and ancient cryptic species in the smallest marine eukaryotes. *Mol. Biol. Evol.* 23: 23–29.

Stoeck, T., B. Hayward, G. T. Taylor, R. Varela, and S. S. Epstein. 2006. A multiple PCR-primer approach to access the microeukaryotic diversity in environmental samples. *Protist* 157: 31–43.

Sweeney, B. M. 1976. *Pedinomonas noctilucae* (Prasinophyceae), the flagellate symbiotic in *Noctiluca* (Dinophyceae) in southeast Asia. *J. Phycol.* 12: 460–464.

Talling, J. F. 1984. Past and contemporary trends and attitudes in work on primary production. *J. Plank. Res.* 6: 203–217.

Taylor, G. T., C. J. Gobler, and S. A. Sanudo-Wilhelmy. 2006. Speciation and concentrations of dissolved nitrogen as determinants of brown tide *Aureococcus anophagefferens* bloom initiation. *Mar. Ecol. Prog. Ser.* 312: 67–83.

Throndsen, J. 1973. Motility in some marine nanoplankton flagellates. *Nor. J. Zool.* 21: 193–200.

Vaillancourt, R. D., J. Marra, M. P. Seki, M. L. Parsons, and R. R. Bidigare. 2003. Impact of a cyclonic eddy on phytoplankton community structure and photosynthetic competency in the subtropical North Pacific Ocean. *Deep-Sea Res.* I 50: 829–847.

Van de Peer, Y., G. Van der Auwera, and R. De Wachter. 1996. The evolution of stramenopiles and alveolates as derived by "substitution rate calibration" of small ribosomal subunit RNA. *J. Mol. Evol.* 42: 201–210.

Vaulot, D., K. Romari, and F. Not. 2002. Are autotrophs less diverse than heterotrophs in marine picoplankton? *Trends Microbiol.* 10: 266–267.

Vaulot, D., F. Le Gall, D. Marie, L. Guillou, and F. Partensky. 2004. The Roscoff Culture Collection (RCC): A collection dedicated to marine picoplankton. *Nova Hedwigia*. 79: 49–70.

Venter, J., K. Remington, J. F. Heidelberg, et al. 2004. Environmental genome shotgun sequencing of the Sargasso Sea. *Science* 304: 66–74.

Ward, B. B. 2005. Molecular approaches to marine microbial ecology and the marine nitrogen cycle. *Annu. Rev. Earth Planet. Sci.* 33: 301–333.

Ward, N. 2006. New directions and interactions in metagenomics research. *FEMS Microbiol. Ecol.* 55: 331–338.

Weithoff, G. 2003. The concepts of "plant functional types" and "functional diversity" in lake phytoplankton—a new understanding of phytoplankton ecology? *Freshwater Biol.* 48: 1669–1675.

Wintzingerode, F., U. B. Gobel, and E. Stackebrandt. 1997. Determination of microbial diversity in environmental samples: Pitfalls of PCR based rRNA analysis. *FEMS Microbiol. Rev.* 21: 213–229.

Worden, A. Z. 2006. Picoeukaryote diversity in coastal waters of the Pacific Ocean. *Aquat. Microb. Ecol.* 43: 165–175.

Worden, A. Z., and B. J. Binder. 2003. Application of dilution experiments for measuring growth and mortality rates among *Prochlorococcus* and *Synechococcus* populations in oligotrophic environments. *Aquat. Microb. Ecol.* 30: 159–174.

Worden, A. Z., J. K. Nolan, and B. Palenik. 2004. Assessing the dynamics and ecology of marine picophytoplankton: The importance of the eukaryotic component. *Limnol. Oceanogr.* 49: 168–179.

Worden, A. Z., M. L. Cuvelier, and D. H. Bartlett. 2006. In-depth analyses of marine microbial community genomics. *Trends Microbiol.* 14: 331–336.

Yuan, J., M.-Y. Chen, P. Shao, H. Zhou, Y.-Q. Chen, and L.-H. Qu. 2004. Genetic diversity of small eukaryotes from the coastal waters of Nansha Islands in China. *FEMS Microbiol. Lett.* 240: 163–170.

Yuasa, T., O. Takahashi, J. K. Dolven, et al. 2006. Phylogenetic position of the small solitary phaeodarians (Radiolaria) based on 18S rDNA sequences by single cell PCR analysis. *Mar. Micropaleo.* 59: 104–114.

Zapata, M., B. Edvardsen, F. Rodríguez, M. Maestro, and J. L. Garrido. 2001. Chlorophyll c2 monogalactosyldiacylglyceride ester (chl c2-MGDG). A novel marker pigment for *Chrysochromulina* species (Haptophyta). *Mar. Ecol. Prog. Ser.* 219: 85–98.

Zapata, M., S. W. Jeffrey, S. W. Wright, F. Rodriguez, J. L. Garrido, and L. Clementson. 2004. Photosynthetic pigments in 37 species (65 strains) of Haptophyta: implications for oceanography and chemotaxonomy. *Mar. Ecol. Prog. Ser.* 270: 83–102.

Zapata, M. 2005. Recent advances in pigment analysis as applied to picophytoplankton. *Vie Milieu* 55: 233–248.

Zhu, F., R. Massana, F. Not, D. Marie, and D. Vaulot. 2005. Mapping of picoeukaryotes in marine ecosystems with quantitative PCR of the 18S rRNA gene. *FEMS Microbiol. Ecol.* 52: 79–92.

Zingone, A., D. Sarno, and G. Forlani. 1999. Seasonal dynamics in the abundance of *Micromonas pusilla* (Prasinophyceae) and its viruses in the Gulf of Naples (Mediterranean Sea). *J. Plank. Res.* 21: 2143–2159.

Zubkov, M. V., M. A. Sleigh, G. A. Tarran, P. H. Burkill, and R. J. G. Leakey. 1998. Picoplankton community structure on an Atlantic transect from 50°N to 50°S. *Deep-Sea Res.* 45: 1339–1355.

ORGANIC MATTER–BACTERIA INTERACTIONS IN SEAWATER

TOSHI NAGATA

Ocean Research Institute, The University of Tokyo, Tokyo, Japan

INTRODUCTION

Organic matter in seawater is a complex mixture of organic compounds with diverse chemical compositions, physical structures, and reactivity. For heterotrophic bacteria in the oceans, organic matter not only serves as a source of carbon and nutrients, but also provides attachment surfaces, depending on its physical dimension. The complexity in composition and structure of organic matter, along with variable supply regimes, is probably one of the major factors that help to maintain a high diversity of prokaryote communities in the oceans. Organic matter–bacteria interactions also exert a large influence on the major properties and patterns of ecosystems, including primary production, food web organization, and biogeochemical fluxes. We now know that bacterial consumption of dissolved organic matter (DOM), and subsequent channeling of bacterial production to protists and viruses, represents a major trophic pathway (microbial loop; Azam 1998) that supports higher trophic levels and drives nutrient regeneration in the upper oceans. Although bacterial use of DOM is highly efficient, a small fraction of DOM resists rapid degradation by bacteria and contributes to the sequestration of enormous amounts of carbon and nutrients in the sea (Hedges 2002). Furthermore, bacterial processing of particulate organic matter (POM) affects greatly the vertical flux of organic carbon from the surface to the ocean interior, which impacts on global cycling of carbon and climate

---

Microbial Ecology of the Oceans, Second Edition. Edited by David L. Kirchman
Copyright © 2008 John Wiley & Sons, Inc.

(Simon et al. 2002). Thus, clarifying the mechanistic basis and controls of organic matter–bacteria systems is the key to better understanding ecosystem processes and biogeochemical cycling in the oceans.

Organic matter–bacteria interactions have been central elements in bacterial biogeochemistry and ecology, covering a broad range of fundamental issues. Here, we focus on interactions between nonliving organic matter (DOM and detritus) and heterotrophic bacteria under aerobic conditions in the oceanic water column. First, a brief overview on the biogeochemistry of organic matter in the ocean will provide a framework of the time scale and functions of different classes of organic matter. The following sections will address how bacteria consume (and transform) DOM and POM in seawater, with an emphasis on possible mechanisms by which biogeochemical fluxes are affected. Finally, we discuss the link between bacterial diversity and the cycling of organic matter.

ORGANIC MATTER INVENTORY AND FLUXES

POM and DOM are defined operationally. The organic matter that is retained on filters (typical nominal retention of 0.7 μm) is POM, whereas that which passes through the filters is DOM; the latter may include small particles and colloids. POM consists of living (plankton) and nonliving (detritus) components. The approximate mass (carbon) ratio of DOM : detritus : plankton in surface water is 200 : 10 : 1, although the mass of detrital POM is not well known (Verity et al. 2000). Clearly, DOM is the dominant form of organic carbon in the oceans. Globally, the stock of carbon in oceanic DOM is 700 Gt, which is similar to the amount of carbon in atmospheric CO_2 (750 Gt). This large size of dissolved organic carbon (DOC) pool implies that even only a 1 percent net annual change of DOC in the ocean is sufficient to generate a CO_2 flux that is comparable to that is caused by fossil fuel combustion per annum (Hedges 2002).

> **Truly Dissolved?**
>
> In chemistry, the term "dissolved" is used if a solid substance (e.g., sodium chloride) is mixed with a solvent (e.g., water) to yield a homogeneous mixture (i.e., solution). However, in the literature of oceanography, "dissolved" organic matter is defined operationally on the basis of filtration. Some DOM components (e.g., free glucose) are "truly dissolved," whereas others are "colloidal," "gel-like," or even "biotic"; some bacteria and most viruses pass through the filters used for examining DOM. Colloids and viruses can be effectively separated by using 1–30 kDa molecular cut-off membranes (ultrafiltration). However, it is still unclear if the DOM that passes through the ultrafiltration membrane (low molecular-weight DOM) is "truly dissolved," given that the chemical identity of its constituents is not well known.

ORGANIC MATTER INVENTORY AND FLUXES

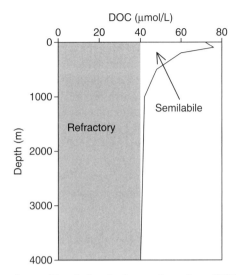

Figure 7.1 A composite profile of dissolved organic carbon (DOC) in the open ocean derived from data from the equatorial, subtropical, and temperate zones. Note that DOC profiles differ depending on oceanic regions. For example, in the Southern Ocean, the vertical gradient of DOC is not as pronounced as indicated in this graph, because of vertical mixing (Ogawa et al. 1999). Refractory DOC is distributed evenly throughout the water column because its turnover time exceeds the time scale of ocean circulation. Semilabile DOC accumulates in the upper ocean and is available for export. The labile DOC pool is not visible in this graph because the concentration is low (<1 μmol/L). Modified from Ogawa and Tanoue (2003).

Although there is a broad continuum of the turnover times of different constituents of DOM (from minutes to millennium), it is useful to distinguish three components based on degradability: labile, semilabile, and refractory DOM (Fig. 7.1 and Table 7.1). This is a conceptual categorization based on oceanic distributions of DOC. The refractory pool dominates deep-water DOC, which ^{14}C-dating indicates

TABLE 7.1 Characteristics of Labile, Semilabile, and Refractory DOC in the Oceans[a]

Pool	Concentration (μmol/L)	Turnover Time	Function	Chemical Identity
Labile	<1	<hours–days	Fueling bacterial production	Includes dissolved free amino acids, free sugars, and labile protein
Semilabile	10–30	Months–years	Export	Unknown
Refractory	40	>1000 years	Storage	Unknown

[a]Carlson and Ducklow (1995); Cherrier et al (1999); Carlson (2002).

has an average age on the order of 4000 years (Bauer 2002). Because this average age exceeds the time scale of deep-water circulation, refractory DOC tends to distribute homogeneously in oceanic basins. A fraction of refractory DOC in surface water is destructed by ultraviolet (UV) irradiation, leading to the formation of DOC that is available for bacterial consumption (Moran and Zepp 2000). The depletion of ^{14}C in DNA extracted from marine bacteria indicates that bacterial production is partly fueled by the carbon that was fixed at distant times in the past (Cherrier et al. 1999).

Semilabile DOM accounts for about a quarter to a half of the surface DOC pool (Carlson 2002). This component turns over on time scales of months to years, because it resists rapid microbial degradation. An important feature of semilabile DOM is that it is "exportable"; the consumption of semilabile DOM can occur at a location distant from where it is produced. The extent of DOC export varies depending on physical and biogeochemical settings of individual regions (Carlson et al. 1994;

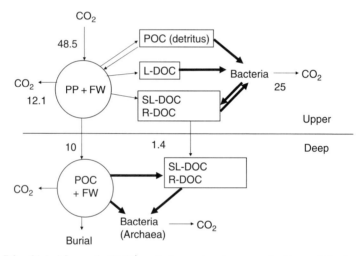

Figure 7.2 Global fluxes (in Gt C/year) of oceanic primary production and its channeling to "dead organic matter"–bacteria pathways. Dissolved organic carbon (DOC) and sinking particulate organic carbon (POC) are produced by several mechanisms that are related to the growth and death of primary producers (PP) and other food web members (FW) (Nagata 2000; Simon et al. 2002). DOC is classified into labile (L-DOC), semilabile (SL-DOC), and refractory (R-DOC) pools, although fluxes of individual components are poorly constrained. On average, bacteria consume about half of primary production in the ocean (Ducklow 2000). In this diagram, it is assumed that all the carbon consumed by bacteria (25 Gt C/year) is eventually respired; microbial loop processes (e.g., flagellate bacterivory) involved in the mineralization are not depicted for simplicity. Respiration by PP + FW (12.1) was estimated as a residual; that is, primary production (48.5: Field et al. 1998) − bacterial respiration (25)−POC export (10: Treguer et al. 2003)−DOC export (1.4: Hansell 2002). These estimates of carbon fluxes have large uncertainties, but they are presented in this diagram to provide a rough idea about the magnitude of the bacterially mediated carbon fluxes addressed in this chapter (bold arrows). Archaea may contribute significantly to carbon fluxes, especially in deep waters.

Hansell et al. 2002). Hansell (2002) has estimated that the global export of DOC is 1.4 Gt C/year.

Labile DOM is typically only a small percentage of the DOM standing stock, accounting for less than 1 percent of the DOC pool. Nonetheless, the consumption by bacteria of labile DOM can be the major flux of energy, carbon and nutrients (e.g., nitrogen and phosphorus) in pelagic ecosystems. Indeed, bacterial carbon consumption of DOM accounts for about half of daily primary production in the euphotic zone (Ducklow 2000; see also Chapter 9). This huge flux of carbon, associated with active regeneration of N and P, is most likely maintained by high turnover of the labile DOM pool, which is intensively replenished by a variety of trophic and nontrophic processes, including phytoplankton excretion, viral lysis, and grazing (Nagata 2000).

POM is available for consumption by phagotrophs in the euphotic zone, thus providing the energy base for grazer food chains. Biogeochemical oceanographers have paid much attention to POM, because it potentially sinks; the sinking of POM is a major component of the "biological pump" that contributes to the sequestration of carbon in deep waters. Globally, the export of organic carbon via sinking of POC below the euphotic zone is about 10 Gt C/year, accounting for 20 percent of primary production in the ocean (Treguer et al. 2003). At a depth of 2000 m, the export flux of organic carbon decreases to about 1 percent of primary production. Microbial processes modulate chemical and physical structures of sinking POM (see below), affecting the length scale of the carbon delivery, that is, how deep organic carbon can reach before it is mineralized.

Figure 7.2 shows global estimates of oceanic primary production and fluxes of "dead organic matter" (DOM and detrital POM), which is produced by several mechanisms that are related to algal physiology and complex trophic interactions (Nagata 2000; Simon et al. 2002). Interactions among POM, DOM, and bacteria are integral parts of oceanic carbon fluxes.

DOM–BACTERIA INTERACTIONS

Labile Low-Molecular Weight (LMW) DOM

In open oceans, free-living bacteria face challenges to exploit extremely dilute, labile DOM in ambient waters. The transport of a monomeric substrate across a membrane could occur passively by diffusion when the ambient concentration of the substrate is high. However, in dilute, marine environments, the substrate needs to be transported actively (i.e., with an expenditure of adenosine triphosphate (ATP)), aided by actions of specific proteins (permeases) tailored for such transport (Davidson 2002). Uptake rates (V, with dimension $M\ L^{-3}\ T^{-1}$) of a given substrate can be related to the substrate concentration (S with dimension $M\ L^{-3}$) by the Michaelis–Menten hyperbolic function (here M, L, and T refer to the dimensions of mass, length, and time):

$$V = V_{max} \frac{S_n}{K_m + S_n}$$

where V_{max} is the maximal uptake rate and K_m is the substrate concentration at which the uptake rate is $\frac{1}{2} V_{max}$. When the substrate concentration in ambient water (S_n) is unknown, kinetic parameters can be derived by conducting radiotracer experiments that examine the relationship between the concentration of added substrate (A) and the fraction of radioisotope tracer taken up (f) during an incubation period t:

$$\frac{t}{f} = \frac{A}{V_{max}} + \frac{K_m + S_n}{V_{max}}$$

The linear regression of t/f against A yields estimates of V_{max} (reverse of slope), $K_m + S_n$ (negative X-intercept), and turnover rate of the substrate (Y-intercept), a model known as the Wright–Hobbie plot (Wright and Hobbie 1966). Especially when $K_m < S_n$, $K_m + S_n$ provides a maximum estimate of the substrate concentration in ambient water.

The introduction of sensitive analytical techniques using high-performance liquid chromatography (HPLC, explained in Chapter 6) for seawater samples (see e.g., Mopper et al. 1992) facilitated the determination of low ambient concentrations of the major classes of monomeric compounds, such as dissolved free amino acids (DFAA) and dissolved free neutral sugars. With such data, bacterial uptake (V with dimension $M\ L^{-3}\ T^{-1}$) can be estimated by multiplying concentration (S, with dimension $M\ L^{-3}$) by the turnover rate constant (k with dimension T^{-1}), the latter being determined as the turnover of radiotracer added at a level much lower than the ambient concentration (Kirchman 2003):

$$V = kS$$

The reciprocal of k is the turnover time of a given compound. Both approaches need corrections for respiration (Suttle et al. 1991) and isotope dilution (Fuhrman 1987). Fuhrman and Ferguson (1986) and Suttle et al. (1991) found that kinetic parameters of DFAA uptake obtained by using the Wright–Hobbie model were generally consistent with the concentrations of DFAA determined by HPLC.

"Modern" Microbial Ecology Began with Radiotracers

The introduction of radiotracer techniques in 1960s and the improvement of epifluorescence microscopy in 1970s provided powerful tools to explore microbial dynamics and regulations of organic matter fluxes in the oceans, which resulted in the establishment of the microbial loop concept in 1980s. We are now equipped with culture-independent (1990s) genomic and metagenomic approaches (2000s), which have begun to unravel oceanic biogeochemistry at a molecular (gene) level (Chapter 3 and 4). In the next decade, microbial ecologists are faced with the challenge of integrating information on oceanic processes and genomic data about microbes in the sea.

Table 7.2 summarizes concentrations and turnover time of DFAA and free sugars in marine waters. Both DFAA and free sugars are very dilute in seawater, with concentrations on the order of nanomolar (10^{-9} mol/L), accounting for only a minor fraction (<1 percent) of bulk DOM. Bacterial permeases are highly efficient in coping with this extremely diluted substrate condition, as indicated by low $K_m + S_n$ values. Because of efficient uptake by bacteria, the turnover times of DFAA and free sugars are typically very short (on the order of minutes to hours), although turnover times can be longer (>100 hours), depending on the environment and substrate. Compound-specific turnover times of LMW DOM can be affected by the availability of nutrients because of constraints posed by carbon and nutrient stoichiometry. For example, in phosphorus-limited surface waters, the availability of inorganic phosphorus can limit bacterial consumption of labile LMW DOM such as glucose, leading to the temporal accumulation of labile DOM (Thingstad et al. 1997). However, there is little evidence in support of the hypothesis that nutrient elements such as nitrogen and iron limit bacterial uptake of labile DOM in the oceans. Kirchman et al. (2000) found that DOM (glucose) rather than iron primarily limited bacterial growth in an iron-limited region.

DFAA and free sugars support a significant, albeit variable (10 to >100 percent) fraction of bacterial production in marine environments (Table 7.2). It is notable that the uptake of a single monomeric compound such as glucose can account for a substantial fraction (>30 percent) of bacterial carbon requirement, depending on the environment (e.g., the equatorial Pacific: Rich et al. 1996). The major mechanisms by which DFAA and free sugars are supplied to meet bacterial demand include the release from phytoplankton, protists, and zooplankton (Nagata 2000), although monomers could be also supplied during the hydrolysis of HMW DOM (Skoog et al. 1999; and see below). In short, DFAA and free sugars represent a relatively small but highly labile pool of DOM, which can support a large fraction of bacterial production in the oceans.

The kinetics of LWM DOM uptake reflect the collective nature of the bulk bacterial community, which consists of numerous species of bacteria (see Chapter 3). Variability in kinetic parameters (particularly K_m) among groups can result in a major discrepancy between observed and predicted values in the model analysis. In fact, earlier studies noted that the data occasionally deviate substantially from the prediction based on the Wright–Hobbie model, particularly in oligotrophic oceans (Hobbie et al. 1972). This discrepancy led to the hypothesis that bacterial populations in oligotrophic waters are more diverse (at least metabolically) than those in eutrophic ones (Williams 1973). Other studies found biphasic (or multiphasic) kinetics of monomer uptake (i.e., curvilinearlity in the Wright–Hobbie plot), suggesting that some bacterial groups in a community adapt to high-substrate conditions (high K_m and high V_{max}), while others adapt to low-substrate conditions (low K_m and low V_{max}) (Riemann and Azam 2002). Recent studies have provided direct evidence in support of the notion that concentration-dependent patterns of amino acid uptake vary among bacterial subpopulations belonging to different phylogenetic groups (Alonso and Pernthaler 2006). An implication of these observations is that DOM concentration in seawater is highly heterogeneous in space and time (Azam 1998).

TABLE 7.2 Concentrations of Dissolved Free Amino Acids (DFAA) and Free Sugars in Marine Systems, Kinetic Parameters (Turnover Time and $K_m + S_n$) and Percentage Contributions of Uptake of These Compounds to Total Bacterial Production (%BP)

Location	Compounds	Conc (nmol/L)	Turnover Time (h)	$K_m + S_n$ (nmol/L)	% BP	Reference
		Open Ocean (Upper)/Coastal				
Sargasso Sea	DFAA (individual)	1.9 ± 3.0	31 ± 36 (7–144)	1.8 ± 1.5	23 ± 52	Suttle et al. (1991)
Sargasso Sea	DFAA (mixture)	1–10	3–4		1–4	Keil and Kirchman (1999)
Gulf of Mexico	DFAA (mixture)		0.4 ± 0.3 (0.1–0.8)	3.6 ± 2.5		Ferguson and Sunda (1984)
Arctic Ocean	DFAA	200	52–101			Rich et al. (1997)
Atlantic coast	DFAA (individual or mixture)	2–33	0.4–1.2	1–38	70–>100	Fuhrman and Ferguson (1986)
Gulf of Mexico	Glucose	2–15	2–5	4–9	1–30	Skoog et al. (1999)
Equatorial Pacific	Glucose	15–38	1.5–2.0		14–51	Rich et al. (1996)
Arctic Ocean	Glucose	42–90[a]	43–140		10–>100	Rich et al. (1997)
Pacific coast, USA	N-Acetyl-glucosamine		38–312	5		Riemann and Azam (2002)
		Estuary				
Chesapeake Bay, USA	4 DFAA	43 ± 26	1.4 ± 0.9 (0.8–2.7)		32 ± 22	Fuhrman (1990)
Kiel Fjord, Germany	Leu		3.5 ± 3.3 (0.2–12)			Hoppe et al. (1988)

Note that concentrations of monomeric pools are low (nmol/L level), but turnover time is high (days or even minutes in coastal waters), resulting in significant contributions to total bacterial production. Data are from the upper 100 m.
[a] Total dissolved neutral monosaccharides.

Extracellular Hydrolytic Enzymes

Bacteria are osmotrophs that obtain nutrients only via the transport of DOM across membranes. Polymeric materials must initially be hydrolyzed by extracellular enzymes, because only small substrate (approximately <600 Da) can pass the outer membrane through porins, which are protein channels in Gram-negative bacteria (Arnosti 2003). Hydrolytic activities of extracellular enzymes are an integral part of the utilization by bacteria of high-molecular-weight (HMW) DOM and POM in marine systems:

$$\text{HMW DOM and POM} \xrightarrow[\text{Extracellular enzymes}]{\text{Hydrolysis}} \text{LMW DOM}_{[out]} \xrightarrow[\text{Permeases}]{\text{Uptake}} \text{LMW DOM}_{[in]}$$

where [out] and [in] indicate the LMW DOM compounds (<600 Da) outside and inside of bacterial cells, respectively. LMW DOM$_{[out]}$ may diffuse away from the cell and mix with the bulk DOM pool (with no gain for the producer of the extracellular enzyme), unless powerful concentrating mechanisms involving permeases effectively recover LMW DOM$_{[out]}$ at a pace comparable to hydrolysis. For the use of HMW DOM, hydrolysis and uptake appear to be generally coupled (Hollibaugh and Azam 1983; Nagata et al. 1998), although Keil and Kirchman (1992) found that about 30 percent of hydrolytic products from algal protein were released as LMW DOM. As we will see later, a substantial uncoupling (excessive hydrolysis relative to uptake) has been documented during the degradation of POM (Smith et al. 1992).

Extracellular enzyme activities have been determined by the use of fluorescent analogs (Hoppe et al. 2002). These analogs include methylumbelliferyl (MUF)-β-1–4-glucopyranoside (for β-glucosidase, BGase) and 4-methylcoumainylamide (MCA)–leucine (for leucyl aminopeptidase, LAPase), although several other analogs for different types of enzymes are available. Upon hydrolysis, fluorophores (MUF or MCA) are liberated. Hydrolytic activities are determined from the increase of fluorescence over time. The activities are related to the substrate concentration by the Michaelis–Menten equation. Because the substrate concentration in ambient water is usually unknown, the half-saturation constant (K_m) derived from the kinetic analysis represents the "apparent" K_m. Kinetic curves display biphasic (or multiphasic) features on occasions (Tholosan et al. 1999), reflecting the diversity of enzymes. In marine systems, the K_m for LAPase and BGase typically varies in the range of 1–50 μmol/L, which tends to increase with increasing productivity (Rath et al. 1993; Hoppe et al. 2002). The maximum velocity (V_{max}) can be determined by adding the substrate at a level that is sufficiently higher than K_m; this saturation level is typically 50–500 μmol/L. In the literature of aquatic microbial ecology, the "activity" of extracellular enzymes often refers to V_{max} which is used as an indicator of the abundance of enzymes in marine waters. Extracellular enzymes in bulk water samples potentially include (1) enzymes associated with cells (ectoenzymes: Chrost 1990), (2) enzymes detached from cells to be associated with detritus particles (detrital enzymes), and (3) enzymes released to ambient water (dissolved enzymes). Generally, ectoenzymes appear to account for the major fraction of bulk hydrolytic activities, but other forms may prevail depending on environments and the type of enzymes (Arnosti 2003).

Field measurements show temporal and spatial variations in relative activities of different types of extracellular enzymes (enzyme profile) over a broad range (Martinez et al. 1996). Variations in enzyme profile can be caused (1) by changes in the expression of enzymes by the same species and (2) by shifts in community compositions. It is not clear which mechanisms are more important in marine systems. At a cellular (species) level, the synthesis of enzymes in general is known to be regulated by several mechanisms, although some enzymes are expressed constitutively (Arnosti 2003). Given the high cost (energy, carbon, and nitrogen) of enzyme synthesis, regulation of the synthesis of extracellular enzymes is tight. On the other hand, enzyme profiles vary greatly among different bacterial strains isolated from marine environments, suggesting that changes in enzyme profile in the field reflect changes in community composition (Martinez et al. 1996).

> **Keep on Working**
>
> Hydrolytic enzymes can be released to ambient waters by bacteria or as a result of viral lysis and grazing. These detrital enzymes, either freely dissolved or associated with nonliving particles, can still catalyze cleavage of chemical bonds. Although the role of these enzymes in transformations of organic matter has been well appreciated in soil and freshwater systems, less attention has been paid regarding similar processes in marine environments (Koike and Nagata 1998). Recently, Ziervogel et al. (2007) found that a polysaccharide hydrolase associated with mineral surface is active and has an extended lifetime in seawater. It remains to be seen if enzyme–organic matter–mineral complexes play a significant role in marine systems as "nonliving organic reactors."

Regardless of the mechanisms, variation in activities of extracellular enzymes can provide insights into the nutritional mode of bacteria and the biochemical composition of polymeric substrates in marine systems. Some features have been noted regarding the patterns in distributions of extracellular enzyme activities:

1. Cell-specific activities of LAPase and BGase are high and low in oligotrophic and eutrophic waters, respectively. They also tend to become high in deep waters (Rath et al. 1993; Hoppe et al. 2002). This pattern presumably reflects increasing importance for bacteria of polymeric substrates relative to monomeric ones from eutrophic through oligotrophic to deep water environments.

2. Activity ratios of extracellular enzymes display systematic variations over large geographic extent. Christian and Karl (1995) found that LPase : BGase ratios were low (0.3) in surface waters of the equatorial Pacific. In contrast, the corresponding ratios were high (>200) in the subtropical Pacific and the Antarctic. The authors suggested that the DOM flux was mainly based on polysaccharide in the equatorial region, whereas a protein-based cycling dominated in other

regions examined. In mesopelagic waters, Fukuda et al. (2000) found that LAPase : BGase ratios systematically increased along a east–west gradient of the subarctic Pacific, suggesting a basin-scale transition in biochemical compositions of sinking POM.

Each broad type (e.g., protease or glucosidase) includes a diverse range of enzymes that can be distinguished by physical (e.g., charge), functional (e.g., type of reaction catalyzed), and genetic (primary sequence of amino acids) characteristics. Recent studies have begun to provide insights into the functional and phylogenetic diversity of extracellular hydrolytic enzymes. Arrieta and Herndl (2002) used capillary electrophoresis zymography to detect up to eight different BGases in a single sample and 11 during an extensive study of a phytoplankton bloom (Fig. 7.3*a*). A close link was found between the succession of BGase diversity and bacterial species richness. The authors suggested that the regulation of the BGase activity and diversity was driven by shifts in the bacterial community structure rather than by induction. Obayashi and Suzuki (2005) used 16 types of fluorogenic substrates to examine diversity of proteolytic enzymes in coastal waters. Seasonal measurements revealed that the activities of trypsin-type endopeptidases (cleaves peptide bonds within a peptide) were consistently higher than those of exopeptidases (cleaves terminal peptide bonds) (Fig. 7.3*b*). This finding is consistent with the hypothesis that the downsizing of polymers by endohydrolases is an important step in polymer degradation (Nunn et al. 2003). Arnosti et al. (2005) examined hydrolysis of six structurally different polysaccharides at eight geographically distant locations. One of the polysaccharides, fucoidan, was not hydrolyzed at any station, while laminarin was hydrolyzed everywhere at variable rates among the stations. These results suggest that the susceptibility to hydrolysis of polysaccharides varies depending on the structure of polysaccharides and metabolic capabilities of microbial community. Cottrell et al. (2000) designed PCR primers for a selected subset of bacterial chitinase genes. Primers were used to amplify DNA in seawater and to construct libraries of chitinase genes. They found that the phylogeny of genes of this class of chitinase does not necessarily correspond to the phylogeny of 16S rRNA genes, presumably because of lateral gene transfer. The libraries were dominated by clones having nucleotide sequences identical to those of chitinase genes of cultured *Alphaproteobacteria*, suggesting that *Alphaproteobacteria* are a principal holder of this class of chitinase gene. Other studies on the biogeography of chitinases have found that sequences of putative chitinase genes vary depending on habitats, suggesting that environments select for specific chitinase genes with different functional features (LeCleir et al. 2004; LeCleir and Hollibaugh 2006).

Polymeric DOM—Protein as a Model

Extracellular enzyme activities examined with simple fluorescent analogs provide information on the cleavage of specific bonds. This approach is useful, but cannot capture the complexity of polymer degradation. Chemically identified components of HMW DOM include proteins, polysaccharides, and lipids, which are major

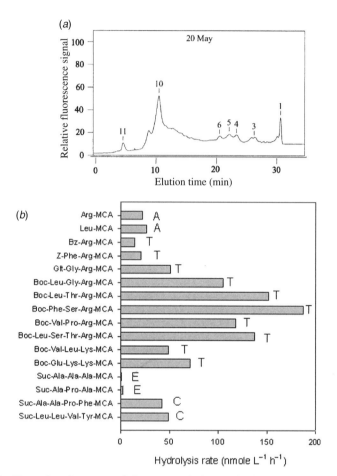

Figure 7.3 Examples of an extracellular enzymatic fingerprint in marine waters. (*a*) Profiles of beta-glucosidase activity in the coastal North Sea. Isozymes of β-glucosidase (βGase) bound to the outside of bacterial cells were extracted and separated by capillary electrophoresis. Peak numbers correspond to different isozymes; there are seven types of βGase in this sample. A substantial variation in enzyme profiles has been observed during a phytoplankton bloom, being related to a shift in bacterial community composition during the same period. (*b*) Profiles of extracellular proteolytic enzymes in the coastal Seto Inland Sea, Japan. Sixteen types of fluorogenic substrates were used to detect different types of proteases: A, substrates for aminopeptidase; T, substrates for trypsin; E, substrates for elastase; C, substrates for chymotrypsin. Relative activities of different types of enzymes may vary spatially and temporally, but Obayashi and Suzuki (2005) found in the coastal Seto Inland Sea that the general pattern varied little over the season, suggesting the importance of trypsin-like endoproteases throughout the year. Part (*a*) from Arrieta and Herndl (2002). Part (*b*) from Obayashi and Suzuki (2005).

cellular constituents with complex structures and diverse compositions. This section focuses on degradation of proteins. Protein can be an important model to study the major features of the complexity in polymer–bacteria interactions in the sea, because protein accounts for a large fraction (60 percent) of algal biomass and because protein can be an important source of carbon and nitrogen for bacteria. Our goal is to examine to what extent dissolved proteins are utilized by bacteria. This section also discusses plausible mechanisms by which degradability of protein decreases to form a semilabile or refractory pool of HMW DOM in seawater.

Studies have suggested that model proteins (e.g., ribulose 1,5-bisphosphate carboxylase) are easily consumed and degraded by bacteria in seawater, with a turnover time that can be comparable to that of DFAA (on the order of hours to days: Hollibaugh and Azam 1983; Keil and Kirchman 1993). These observations are consistent with the finding that a wide variety of proteolytic enzymes are abundant in marine systems (Hoppe et al. 2002) and that dissolved combined amino acids (DCAA) are an important nutritional source for bacteria (Kroer et al. 1994). However, Keil and Kirchman (1994) found that "fresh" protein added to filtered seawater became less easily degraded by bacteria in a short period. They suggested that the added protein adsorbed to colloids and that the adsorbed protein is less easily degraded than free protein. This hypothesis was supported by a study conducted by Nagata and Kirchman (1996), who found that protein adsorbed to colloidal particles is much less easily degradable than the same protein freely dissolved. Borch and Kirchman (1999) and Nagata et al. (1998) found that protein adsorbed to or associated with phospholipid vesicles (liposomes) is protected from bacterial attack. These results could be explained by (1) conformational change of protein hiding the sites that are susceptible to bacterial binding and proteolytic attack and (2) physical protection of protein by a layer of different kinds of polymers. Importantly, the degradation of adsorbed protein differed among different bacterial strains, suggesting that some bacteria are capable of utilizing adsorbed protein more efficiently than others (Nagata and Kirchman 1997).

Protein degradation could be further slowed down by geochemical modification such as glycosylation (Maillard reaction), which involves covalent binding of sugars with primary amino groups of protein. On the basis of the kinetic data obtained by using "fresh" and "glycosylated" proteins, Keil and Kirchman (1993) suggested that DCAA in the Delaware estuary and coastal waters consist of at least three kinetically distinct pools: (1) "fresh" protein that is assimilated almost as fast as DFAA; (2) "modified" protein that is utilized much more slowly than fresh protein; and (3) a kinetically undefined pool. Interestingly, the surface bacterial community was supported more by "fresh" than "modified" protein, whereas the reverse trend was true for deep-water communities (Keil and Kirchman 1999). This result suggests that deep-water bacteria (and probably archaea) are better adapted to utilize "modified" protein than are surface communities. Although there is little evidence from the ^{15}N-NMR spectra that glycosylated protein accounts for a large fraction of HMW DON (McCarthy et al. 1997), the above results still provide insights into interactions of bacteria with "modified" proteins such as glycoprotein (enzymatically modified protein), which has been reported to be abundant in dissolved protein in seawater (Yamada and Tanoue, 2003). Recent studies have suggested fundamental roles of noncovalent (hydrophobic interactions and encapsulation: Nguyen and Harvey 2001, 2003; Zang et al. 2001) and covalent associations (Hsu and Hatcher 2005) of proteins with other organic matter in the long-term preservation of proteins.

Refractory DOM

Most of oceanic DOM (70 percent) has a molecular weight <1000 Da (LMW) and cannot be chemically identified at the molecular level (Benner 2002). The bulk LMW DOM pool is largely refractory, turning over on millennial time scales (Bauer 2002). HMW DOM (>1000 Da) accounts for about 30 percent of DOC and is more rapidly degraded than LMW DOM, as revealed by bioassay experiments (Amon and Benner 1994) and radiocarbon dating (Guo et al. 1994). On the basis of these observations, Amon and Benner (1996) proposed that lability of DOM decreases with decreasing molecular size: the size–reactivity continuum hypothesis. In support of this hypothesis, studies have found that less labile, chemically unidentifiable LMW DOM is produced during early diagenesis of organic matter derived from bacteria (Nagata et al. 2003) and phytoplankton (Hama et al. 2004). Yamashita and Tanoue (2004) found that even chemically identifiable fractions of DOM, such as dissolved combined amino acids, display a compositional signature indicating a successive decrease in degradability (as evaluated by an index of diagenesis) with decreasing molecular size. The above view seems to contradict the other view suggesting that the "limiting step" of organic matter degradation is the hydrolysis of HMW DOM rather than the uptake of hydrolyzates (LMW DOM) and, consequently, that HMW DOM turns over more slowly than LMW DOM. This paradox can be resolved by hypothesizing that there are two distinctive LMW DOM components and different processes occurring at different time scales (Fig. 7.4). On a short time scale, LMW DOM (such as DFAA and free sugars)

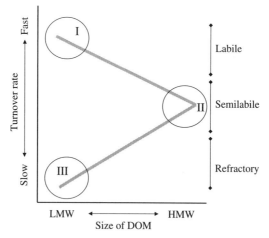

Figure 7.4 Hypothetical diagram explaining the "trend reversal" in the size–lability relationship of dissolved organic matter (DOM) depending on the time scale. LMW, low-molecular-weight; HMW, high-molecular-weight. "Pool II" turns over more slowly than "Pool I," probably because of molecular shielding (protection due to complexation), which impedes hydrolysis (cf. Keil and Kirchman 1994; Nagata and Kirchman 1997). "Pool III" turns over more slowly than "Pool II," probably because of unusual chemical structure (cf. Amon and Benner 1996). The size–reactivity continuum hypothesis suggests that the transition from "Pool II" to "Pool III" occurs during diagenesis (Amon and Benner 1996).

turns over more rapidly than HMW DOM (especially complexed or modified organic polymers, as discussed above), because of the impedance at the step of hydrolysis. However, LMW DOM components produced during diagenetic processes appear to be less accessible for bacteria (i.e., permeases cannot recognize the molecules as substrate), perhaps because of their unusual structures originated from geochemical and enzymatic modifications or because of associations with other compounds, including DOM and minerals.

Recent evidence has suggested that bacteria play a substantial role in the formation of semilabile and refractory DOM in marine systems. The notion that "necromass" (dead cells) of bacteria might be a significant source of DOM is not necessarily new in oceanography. In fact, early studies noted that there are D-amino acids, a biomarker of bacterial cell walls, in marine DOM (Lee and Bada 1977) and that bacteria can transform labile forms of free sugars and amino acids into less degradable materials (see, e.g., Brophy and Carlson 1989). In the mid 1990s, observations made by Tanoue and colleagues (Tanoue 1995; Tanoue et al. 1995; Suzuki et al. 1997) renewed interest among oceanographers, and spurred studies into the role of bacteria as a major source of DOM; these authors found that porins, bacterial channel proteins resides in outer membrane (see above), are highly abundant in DOM in surface and deep oceanic waters. On such porin molecule, which Tanoue et al. (1995) identified on the basis of amino acid sequencing, is that of *Pseudomonas aeruginosa*, a Gram-negative bacterium known as an opportunistic pathogen, inhabiting various environments, including open oceans (Khan et al. 2007). During the past decade, several studies have reported that possible remnants of bacterial cell wall (peptidoglycan) and outer membrane (lipopolysaccharide) components are recovered in significant quantities from oceanic waters; these compounds include D-amino acids (McCarthy et al. 1998; Dittmar et al. 2001; Kaiser and Benner 2008), muramic acids (Benner and Kaiser 2003), and some classes of hydroxyl fatty acids (Wakeham et al. 2003). Remnants of bacterial organic matter have been recovered in sediments (Pedersen et al. 2001; Grutters et al. 2002) in much excess of bacterial biomass (Lomstein et al. 2006; Niggemann and Schubert 2006), suggesting that selective preservation of organic matter derived from bacteria occurs both in water columns and sediments. Experiments have shown that bacterial membrane proteins are more slowly degraded than soluble protein (Nagata et al. 1998) and that peptidoglycan is less easily degradable than protein and other cellular constituents (Nagata et al. 2003, Jørgensen et al. 2003, Veuger et al. 2006), consistent with the assertion that bacterial cell wall and membrane components resist degradation because of steric protection by liposome-like structures (Nagata and Kirchman 1997; Mannino and Harvey 1999; Wakeham et al. 2003) or because of inherent structural complexity (McCarthy et al. 1998; Jørgensen et al. 2003).

Bacterial cellular components are released to the ambient water as DOM and colloids during protist grazing (Koike et al. 1990; Nagata and Kirchman 1996) and viral lysis (Shibata et al. 1997; Middelboe and Jørgensen 2006). In this regard, bacteria–protist and bacteria–virus food chains can be an effective mechanism to produce semilabile and refractory DOM in the oceans (Nagata and Kirchman 1999; Nagata 2000). Other studies, however, have presented a different view by suggesting that bacteria alone can produce DOM (Ogawa et al. 2001; Gruber et al. 2006).

> **Membrane and Cell Wall Components of Prokaryotes**
>
> Gram-negative cells are surrounded by a cell wall (peptidoglycan) and outer membrane (lipopolysaccharide). Peptidoglycan contains D-amino acids. Porins are channel proteins that reside in the outer membrane, which act as pores to deliver specific molecules. Gram-positive bacteria lack an outer membrane, but have a thick cell wall (peptidoglycan) containing teichoic and lipoteichoic acids. In contrast to the membranes of bacteria and eukaryotes, which have ester lipids, archaea have membranes consisting of ether lipids. The composition of archaeal cell wall (pseudopeptidoglycan) is different from that of a bacterial cell wall, although it also contains D-amino acids. Most prokaryotes in the ocean are Gram-negative bacteria, although archaea are abundant in the deep ocean and sometimes in surface waters of polar seas (see Chapter 3).

Kawasaki and Benner (2006) argued that, during exponential growth of bacteria, a fraction of peptidoglycan component is released to ambient water. Studies have suggested that some bacteria, including *P. aeruginosa*, actively release outer membrane vesicles of liposome-like structures containing membrane proteins (Kuehn and Kesty 2005). Thus, there might be a substantial difference in physical and chemical forms of bacteria-derived DOM depending on the production mechanism, namely, the release during bacterial death (protist grazing and viral lysis: Nagata and Kirchman 1999) versus the release from "healthy" bacteria (Kawasaki and Benner 2006). The relative importance of these two pathways probably varies depending on the environment. It remains to be seen if bacterial and archaeal community makeup and microbial food web structures control compositions and turnover of oceanic DOM (hence carbon sequestration in the oceans) through the release of specific cellular components with characteristic structures.

Chemical identity and physical structures of DOM in deep oceanic waters are still poorly understood. Nearly all the nitrogen in HMW DOM in oceanic water columns is chemically bound in amide functional groups (McCarthy et al. 1997), but studies have recovered only a small fraction of nitrogen as chemically identifiable forms such as amino acids and *N*-acetylaminopolysaccharides (i.e., components of protein, peptidoglycan, and chitin), suggesting that amides that resist both chemical (acid) hydrolysis and biological degradation dominate the DOM pool in deep waters (Aluwihare et al. 2005). Recently, Hertkorn et al. (2006) suggested that a major component of deep-water DOM is carboxyl-rich alicyclic molecules (CRAM), which share come structural characteristics found in terpenoids. The authors suggested that CRAM might contribute to the formation of marine gels (Chin et al. 1998; Verdugo et al. 2004) (see below). Mechanisms by which CRAM are produced, preserved, and degraded in seawater have yet to be elucidated.

POM–BACTERIA INTERACTIONS

POM Continuum

Several classes of marine detrital particles with different sizes and properties have been identified (Fig. 7.5). These particles include nanometer-sized particles determined by electron microscopy (Wells and Goldberg 1994) and submicrometer-sized particles determined by resistive pulse particle counters (Koike et al. 1990). Larger classes of particles include transparent exopolymers stained by Alcian Blue (TEP: Alldredge et al. 1986) and proteineous particles stained by Coomassie Brilliant Blue (Long and Azam 1996). Studies have suggested that physical transitions of organic matter across boundaries of POM and DOM occur in seawater (Verdugo et al. 2004). Chin et al. (1998) demonstrated that DOM can spontaneously assemble rapidly (in minutes to hours) into POM (polymer gels) in seawater. Gels are a unique form of molecular organization in which the polymer chains are interconnected by crosslinks aided by cation bridging. Verdugo et al. (2004) suggested that the marine gel phase spans a large size spectrum from colloids to particles of several hundreds of micrometers. Large particles are often called organic aggregates since they are composed of smaller particles. Macroscopic forms of the aggregates ($>500\ \mu$m) are known as marine snow, an important vehicle for the vertical transport of organic matter from the surface to deep waters (Simon et al. 2002).

Some classes of POM and aggregates are known to harbor dense bacterial communities, with abundances occasionally reaching levels one to two orders of magnitude greater than in ambient waters (Alldredge et al. 1986; Herndl 1988). Such kinds of aggregates represent a complex habitat with concentrated substrate, in contrast with dilute, ambient waters. In the Adriatic Sea, Müller-Niklas et al. (1994) found that the concentration of DFAA in amorphous aggregates reached up to 80 μmol/L, a value at least two orders of magnitude higher than the concentration in ambient waters. Development of bacterial colonies, often associated with extracellular polymers, may result in the establishment of bacterial consortia (biofilm) exhibiting various interactions including cooperative (e.g., quorum sensing: Gram et al. 2002) and mutualistic and antagonistic relationships (Long and Azam 2001). In addition, it has been hypothesized that the concentration gradient of substrate that surrounds the surface of POM ("phycosphere": Bell and Mitchell 1972; or "detrisphere": Biddanda and Pomeroy 1988) attracts motile, chemotactic bacteria, leading to patches of bacterial clusters (Grossart et al. 2001).

POM Fluxes

Although aggregates can be "hot spots" of bacterial activities, bacteria attached to POM generally account for only a small fraction (<10 percent) of total bacterial abundance and production in pelagic environments; the vast majority of bacteria in pelagic waters are free-living cells (Table 7.3). However, the contribution of attached bacteria to total bacterial production can be high (>50 percent) in productive coastal

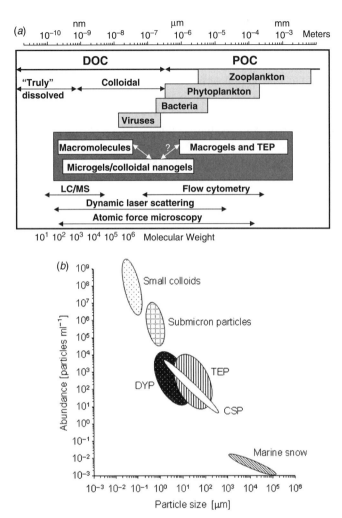

Figure 7.5 Size continuum of organic matter in seawater. (*a*) This diagram emphasizes spontaneous size transitions of "gels," a class of organic polymers that exhibit distinctive physical behavior. Recent studies have suggested that gels and gel-like particles are abundant in seawater (Chin et al. 1998; Verdugo et al. 2004). Also indicated in the diagram are some new technologies that have been applied recently for the analysis of marine particles. (*b*) Approximate ranges of the abundance and size of various marine detrital particles, including small colloidal particles (Wells and Goldberg 1994), submicrometer particles (Koike et al. 1990), DAPI-positive yellow particles (DYP: Mostajir et al. 1995), transparent exopolymers stained by Alcian Blue (TEP: Alldredge et al. 1993), and proteineous particles stained by Coomassie Brilliant Blue (CSP: Long and Azam 1996). In a broad size range, size spectra of oceanic particles generally fit a power function model that assumes large influences of physical particle interactions such as coagulation of small particles and break-up of large aggregates (Wells and Goldberg 1994), but biotic production and consumption can also largely affect size distributions of particles. Part (*a*) from Verdugo et al. (2004). Part (*b*) From Yamasaki et al. (1998).

and estuarine environments (Crump et al. 1998), as well as during phytoplankton blooms (Riemann et al. 2000). In mesocosm experiments, Riemann et al. (2000) found that the major mode of organic matter–bacteria interactions changed during the development and decay of phytoplankton. The transition was characterized by the dominance of free-living bacteria during a prebloom period, followed by the prevalence of attached bacteria during a postbloom period, being associated with changes in bacterial community compositions.

Large aggregates (marine snow) are abundant during phytoplankton blooms. They sink to the deeper layer and play an important role in the vertical transport of materials. The rate at which sinking POM is solubilized and mineralized affects the length scale of carbon delivery in the water column, and thus has a major impact on ocean carbon fluxes. In addition to physical fragmentation of aggregates to colloids and DOM (Karl et al. 1988; Goldthwait et al. 2005), bacterial hydrolysis and degradation of POM play critical roles in the regulation of these fluxes. In fact, attached bacteria exhibit high activities of a range of extracellular enzymes (Simon et al. 2002; Hoppe et al. 2002), consistent with the hypothesis that attached bacteria are largely responsible for the hydrolysis of polymeric components of POM. Smith et al. (1992) found that hydrolysis and uptake are only loosely coupled during the degradation of marine snow. The authors observed a large release of DOM from marine snow collected in the California Bight. The release accounted for 50–98 percent of total degradation (the sum of DOM release and bacterial carbon demand estimated from bacterial production), implying that hydrolysis rate of polymers far exceeded uptake by attached bacteria. Kirchman and White (1999) reached a similar conclusion in their study of the degradation of ^{14}C-labeled chitin particles. They found that hydrolysis exceeded mineralization, leading to the substantial release of ^{14}C-DOM to the ambient water. The release of DOM during degradation of POM could be regarded as enzymatic "sloppy feeding" by attached bacteria, presumably due to active secretion of endohydrolases (which cleaves internal bonds of polymers) to exploit polymeric resources from larger space. Smith et al. (1995) proposed that enzymatic hydrolysis plays key roles in the transformation of polymeric organic matter (mucus) during phytoplankton blooms (Fig. 7.6a).

Attached Bacteria in Deep Waters

Microscopic observations indicate that most prokaryotes are free-living in oceanic water columns (Table 7.3). However, a recent metagenomic analysis conducted at the subtropical Pacific Station ALOHA suggested that a surface-attached lifestyle is important in deep water microbial communities; the study found that genes encoding pilus, polysaccharide, and antibiotic synthesis is more enriched in deeper water samples (DeLong et al. 2006). One hypothesis to explain this discrepancy is that the detachment of cells associated with sinking particles is a major mechanism that supplies free-living prokaryotic cells in deeper waters, although this hypothesis cannot explain taxonomic compositions of deep-water prokaryotic communities, which are distinctive from surface and aggregate-associated communities (Karner et al. 2001).

TABLE 7.3 Contribution of Attached Bacteria to Total Bacterial Abundance and Production[a]

Location	% Attached		Comments	References
	Abundance	Production[a]		
	Surface Open or Coastal Ocean			
North Atlantic Ocean and coast	Very few			Wiebe and Pomeroy (1972)
North Atlantic Ocean	0.1–4.4	2.3–26 (TdR)	>3 mm aggregates collected by SCUBA	Alldredge et al. (1986)
California coast	0.9–3	1.3–13.4 (TdR)	>3 mm aggregates collected by SCUBA	Alldredge et al. (1986)
Northwest Mediterranean	1–3	3–12 (Leu)	>10 μm aggregates (collected by programmed detritus sampler or by SCUBA)	Turley and Stutt (2000)
Atlantic shelf (U.S.A.)	10–20	10–20 (TdR)	Size fractionation (3 μm-pore size)	Ducklow and Kirchman (1983)
	Mesopelagic			
North Pacific subtropical gyre	<5			Cho and Azam (1988)

Northwest Mediterranean (160–380 m)	1	12–48 (Leu)	>10 μm (collected by programmed detritus sampler)	Turley and Stutt (2000)
Subtropical Atlantic (130–550 m)		0.01–0.39 (TdR)	Aggregates collected from the submersible	Alldredge and Youngbluth (1985)
Estuary or Mesocosm				
Columbia River estuary, U.S.A.		13–53 (TdR)	>20 μm particles collected by net	Crump and Baross (1996)
Columbia River estuary, U.S.A.	45	90 (TdR)	Filtration with 3 μm filter	Crump et al. (1998)
Chesapeake Bay, U.S.A. (mid and lower Bay)	<10	5–10 (TdR)	Filtration with 3 μm filter	Griffith et al. (1994)
Mesocosm with Pacific coastal water (prebloom)	7–27	0–35 (Leu)	Filtration with 1 μm filter	Riemann et al. (2000)
Mesocosm with Pacific coastal water (postbloom)	20–58	43–82 (Leu)	Filtration with 1 μm filter	Riemann et al. (2000)

[a]The method for measuring production is given in parenthesis: TdR, thymidine; Leu, leucine.

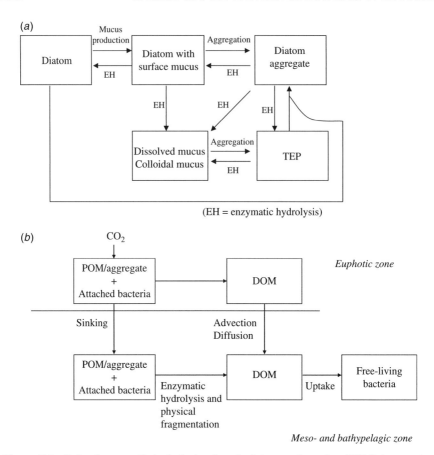

Figure 7.6 Role of enzymatic hydrolysis of particulate organic matter (POM) (aggregates, transparent exopolymers (TEP), marine snow) in marine systems. (*a*) During a diatom bloom, formation of large aggregates (marine snow) is facilitated by sticky polymers (mucus) excreted by algae. Attached bacteria actively produce hydrolytic enzymes to destroy these polymers. Important consequences of this bacterial action are that (1) diatom aggregation is impeded (the sinking particulate organic carbon (POC) flux is reduced) and (2) dissolved organic matter (DOM) is released (free-living bacteria are fed). (*b*) Hydrolysis of POM occurs continuously during the transit of sinking POM in meso- and bathypelagic layers. The DOM released from sinking POM supports production of free-living bacteria. Because of this mechanism, production of free-living bacteria in deep water is largely coupled with sinking POM fluxes (Cho and Azam 1988; Nagata et al. 2000). Part (*a*) from Smith et al. (1995).

The DOM released from POM by physical processes and enzymatic solubilization supports the production of free-living bacteria in deep waters. This hypothesis was supported by the observations that free-living bacterial carbon consumption accounts for a substantial fraction (50–100 percent) of sinking POC fluxes in mesopelogic

(Cho and Azam 1988; Simon et al. 1992) and bathypelagic oceans (Turley and Mackie 1994; Nagata et al. 2000). The depth-dependent distributions of bacterial production are generally consistent with the vertical pattern of sinking POM fluxes, at least at a large scale (Nagata et al. 2000), suggesting that the production of free-living bacteria is largely coupled with sinking POM fluxes. An alternative carbon source for deep-water bacteria is the DOM that is transported to deeper layers due to vertical mixing and advection, which may become an important source, depending on depth and oceanic regions (Carlson et al. 1994). Collectively, the conversion of POM to DOM (by the hydrolysis due to attached bacteria) and subsequent uptake of DOM by free-living bacteria represent an integral part of a large-scale carbon cycling in the oceans (Fig. 7.6b).

Although so far we have regarded attached bacteria as a solubilizer of POM, other perspectives emphasize the role of bacteria as a producer or a source of aggregates. Bacteria are known to form aggregates by excreting polymers to escape from grazing by protists (Hahn and Höfle 2001). Grossart et al. (2006) suggested that bacterial colonization can facilitate the formation of aggregates by diatoms. Fukuda and Koike (2000) found that aggregation of colloids and bacteria are facilitated by feeding currents created by bacterivorous flagellates.

POM–Mineral Interactions

Inorganic matter, such as opal, calcium carbonate, and detrital aluminosilicates, accounts for a substantial fraction (>80 percent) of the total mass of sinking particle mixtures (Hedges et al. 2001). Recently, oceanographers have begun to pay attention to interactions between bacteria and organic–mineral complexes in marine aggregates, because minerals act as "ballast" to enhance the sinking speed of particles and because organic matter affects dissolution of minerals while minerals affect degradation of organic matter (Milliman et al. 1999; Armstrong et al. 2002). Bidle and Azam (1999, 2001) provided compelling evidence that extracellular proteases produced by attached bacteria effectively hydrolyze certain classes of proteins (e.g., frustulins) that provide structural integrity to diatom frustules and protect them against dissolution. The authors have suggested that silicon regeneration in the oceans is largely controlled by bacterial colonization on particles and hydrolytic attack on organic matrices of frustules. Hedges et al. (2001) examined changes in the organic composition of sinking POM during the transit through the oceanic water column. Surprisingly, their measurements from solid-state ^{13}C-NMR spectroscopy showed that minimal changes occur in bulk organic composition, despite extensive (>98 percent) biodegradation; amino-acid-like material predominates throughout the water column. The authors hypothesized that organic matter might be protected from degradation by the inorganic matrix of sinking particles, although other explanations (e.g., selective preservation of bacteria-derived organic matter) were not totally precluded. Indeed, previous studies have shown that bacterial degradation of protein is affected by interactions of protein with inert surfaces (Samuelsson and Kirchman 1990). However, we still have much to learn how organic–mineral interactions affect bacterial consumption of sinking POM.

> **Mineral Protection of Organic Matter**
>
> Studies on organic matter preservation in sediments and soils have led to hypotheses to explain how minerals protect organic matter from enzymatic degradation. One hypothesis is that organic matter resides in small pores (mesopores) or the network of the pores of minerals, being protected from degradation owing to physical exclusion of enzymes (Mayer et al. 2004). Whether or not this model explains organic matter preservation in seawater has yet to be tested.

BACTERIAL COMMUNITY STRUCTURE AND UTILIZATION OF ORGANIC MATTER

Recent studies have begun to explore relationships between bacterial community structure (Chapter 3) and organic matter fluxes in the oceans (Foreman and Covert 2003; Kirchman 2003). There is no a priori reason to expect a tight link between 16S rRNA genes (phylogenetic lineage) and phenotypic traits involved in organic matter cycling (e.g., extracellular enzyme activity), because functional genes (particularly non-essential genes) may be easily transferred across different lineages (lateral gene transfer), as suggested for chitinase genes (Cottrell et al. 2000). In addition, even a single population of bacteria can exhibit differential gene expression in order to exploit different types of organic matter (phenotypic plasticity: Baty et al. 2000), further obscuring the link between taxonomic identity and a particular mode of organic matter utilization. Despite this complexity, recent studies have shown that distribution patterns and some critical ecological traits (e.g., growth) appear to be discernible, if not completely distinctive, among different groups, even at a broad phylogenetic level (Kirchman et al. 2005; Yokokawa et al. 2004; Yokokawa and Nagata 2005), suggesting that different phylogenetic groups of bacteria might express different phenotypic traits to uniquely exploit different organic matter resources. If so, it is reasonable to hypothesize that ocean fluxes of various classes of organic matter represent one of the major forces that affect bacterial community composition and perhaps biodiversity, although there are other potential factors such as temperature, hydrostatic pressure, UV irradiation, nutrient limitation, grazing, and viral infection. Exploring mechanisms by which bacterial community composition is affected is not only imperative in the context of community ecology, but also could contribute to better modeling of ecosystem processes and biogeochemical fluxes in the oceans.

Some researchers have hypothesized that bacterial community composition changes in response to the addition of organic matter. In support of this notion, Pinhassi et al. (1999) found that the community composition changed in mesocosms amended with protein. The increase in bacterial abundance was accompanied by increasing proteolytic activity, which was interpreted as an adaptive trait of responsive species, although there is room for other interpretations. Other investigators have examined whether different groups of bacteria (or prokaryotes) use different kinds of DOM. In order to test this hypothesis, these studies have used an approach that combines microautoradiography with fluorescence in situ hybridization (FISH) to determine the phylogenetic affiliation of individual cells that take up defined DOM components (Cottrell and Kirchman 2000; Elifantz et al. 2005; Teira et al. 2006). In the lower part of Delaware Bay, Cottrell and Kirchman (2000) found that the capacity to assimilate proteins (polymers) and DFAA (monomers) differed among bacterial major groups: *Cytophaga*-like bacteria and *Alphaproteobacteria* dominated in the use of proteins and DFAA, respectively. An important implication of this finding is that the fluxes of HMW DOM and LMW DOM must be more correctly modeled by considering two bacterial components rather than one homogeneous pool. Malmstrom et al. (2004a) supported the above assertion by suggesting that an abundant and ubiquitous alphaproteobacterial clade, SAR11 (Morris et al. 2002), dominated the assimilation of LMW DOM, including amino acids and sugars. Also, *Alphaproteobacteria* appear to dominate the assimilation of dissolved dimethylsulfoniopropionate (DMSP), a precursor of a climatically-active compound, dimethylsulfide (DMS) (Malmstrom et al. 2004b). At a broader level, Teira et al. (2006) found differences in the DOM use in deep-water prokaryote communities; that is, *Archaea* were more active than *Bacteria* in assimilating D-aspartic acid, while these two groups were almost equally active in the assimilation of L-asparatic acid. Although we still do not know if the group-specific use of DOM occurs generally and if it occurs at finer levels of phylogenetic categories, the above data suggest that different phylogenetic groups of bacteria tend to exploit different organic resources and play different roles in the regulation of the turnover of organic matter in marine systems. This resource segregation among different groups implies that trade offs exist between the use of one type of organic matter and that of another (Miki and Yamamura 2005).

FUTURE CHALLENGES

Bacterial degradation and transformation of POM and DOM is a major component of biogeochemical cycling in the oceans. Modelers of marine ecosystems and global carbon cycling are keen to better describe the rates at which different classes of organic matter are remineralized (Christian and Anderson 2002). The predictive power of biogeochemical models would be compromised if we lacked a mechanistic understanding of the key processes that control the major fluxes involved in organic matter transformation and remineralization. During the past three decades, our knowledge of organic matter–bacteria interactions in marine systems has improved significantly, one manifestation being the establishment of the microbial loop concept (Azam 1998). However, we still have much to learn before we can fully explain large variability in turnover rates of different components of organic matter. Controls appear to be intrinsically complex and variable, but we need sound generalizations if microbial processes are to be embedded into the models properly. The roles that semilabile and refractory DOM might play in oceanic ecosystems deserve further consideration and more extensive modeling. Chemical characterization of DOM will continue to be a major challenge (Hedges 2002), but we also need more information about the "tertiary structure" (three-dimensional arrangement) of organic matter, about which a novel concept is now being developed (Verdugo et al. 2004). Even though bulk POM has been chemically characterized reasonably well, especially when compared with DOM, knowledge on particle dynamics (i.e., aggregation and fragmentation of heterogeneous particles accompanied with complex microbial processes) is largely limited, suggesting the need to improve fundamental techniques for exploring in situ interactions between diverse ranges of particles and microbes. We lack a standard technique to separate even living and nonliving particles!

Applications of novel molecular approaches have begun to clarify the dynamic nature of bacterial responses to variations in the organic matter field at both population and community levels, but roles played by individual groups in organic matter fluxes remain to be clarified. Arguably, expanding knowledge on distributions of functional gene repertoires in water columns helps broaden our perspectives of processes occurring in the oceans (DeLong et al. 2006). However, it will be a tremendous challenge for future studies to fill a gap between studies cataloging genes and those to identifying controls of organic matter fluxes in the oceans.

SUMMARY

1. A huge amount of carbon (700 Gt) resides in oceanic DOM. Even only a 1 percent net change in DOM would generate a carbon flux that is comparable to the global annual emission of CO_2 due to fossil fuel combustion.
2. Much of the oceanic DOM is refractory or semilabile, and is only slowly mineralized by microbes. Covalent and noncovalent associations of DOM, combined with microbial modifications, are among the plausible mechanisms by which DOM resists rapid microbial degradation.
3. Labile DOM (e.g., free amino acids and sugars) is dilute in seawater, but bacterial uptake of these compounds is highly efficient and rapid, representing a major flux of carbon and nutrients in pelagic ecosystems.
4. Since only small substrates (<600 Da) can pass through the outer membrane of bacteria, hydrolytic activities of extracellular enzymes are essential in the utilization by bacteria of high-molecular-weight DOM and POM.
5. There is a continuum of oceanic POM, ranging from submicrometer colloids to marine snow. Bacterial processing of POM largely affects distributions and sinking fluxes of POM (biological pump).
6. Novel molecular biology techniques have begun to unravel how organic matter interacts with marine bacteria at community, population, and single-cell levels.

REFERENCES

Alldredge, A. L., and M. J. Youngbluth. 1985. The significance of macroscopic aggregates (marine snow) as sites for heterotrophic bacterial production in the mesopelagic zone of the subtropical Atlantic. *Deep-Sea Res.* 32: 1445–1456.

REFERENCES

Alldredge, A. L., J. J. Cole, and D. A. Caron. 1986. Production of heterotrophic bacteria inhabiting macroscopic organic aggregates (marine snow) from surface waters. *Limnol. Oceanogr.* 31: 68–78.

Alldredge, A. L., U. Passow, and B. E. Logan. 1993. The abundance and significance of a class of large, transparent organic particles in the ocean. *Deep-Sea Res. I* 40: 1131–1140.

Alonso, C., and J. Pernthaler. 2006. Concentration-dependent patterns of leucine incorporation by coastal picoplankton. *Appl. Environ. Microbiol.* 72: 2141–2147.

Aluwihare, L. I., D. J. Repeta, S. Pantoja, and C. G. Johnson. 2005. Two chemically distinct pools of organic nitrogen accumulate in the ocean. *Science* 308: 1007–1010.

Amon, R., M. W., and R. Benner. 1994. Rapid cycling of high-molecular-weight dissolved organic matter in the ocean. *Nature* 369: 549–552.

Amon, R. M. W., and R. Benner. 1996. Bacterial utilization of different size classes of dissolved organic matter. *Limnol. Oceanogr.* 41: 41–51.

Armstrong, R. A., C. Lee, F. I. Hedges, S. Honjo, and S. G. Wakeham. 2002. A new, mechanistic model for organic carbon fluxes in the ocean based on the quantitative association of POC with ballast minerals. *Deep-Sea Res. 2* 49: 219–236.

Arnosti, C. 2003. Microbial extracellular enzymes and their role in dissolved organic matter cycling. In S. E. G. Findlay and R. L. Sinsabaugh (eds.), *Aquatic Ecosystems: Interactivity of Dissolved Organic Matter.* Academic Press, pp. 315–342.

Arnosti, C., S. Durkin, and W. H. Jeffrey. 2005. Patterns of extracellular enzyme activities among pelagic marine microbial communities: Implications for cycling of dissolved organic carbon. *Aquat. Microb. Ecol.* 38: 135–145.

Arrieta, J. M., and G. J. Herndl. 2002. Changes in bacterial β-glucosidase diversity during a coastal phytoplankton bloom. *Limnol. Oceanogr.* 47: 594–599.

Azam, F. 1998. Microbial control of oceanic carbon flux: The plot thickens. *Science* 280: 694–696.

Baty, A. M., C. C. Eastburn, S. Techkarnjanaruk, A. E. Goodman, and G. G. Geesey. 2000. Spatial and temporal variations in chitinolytic gene expression and bacterial biomass production during chitin degradation. *Appl. Environ. Microbiol.* 66: 3574–3585.

Bauer, J. E. 2002. Carbon isotopic composition of DOM. In D. A. Hansell, and C. A. Carlson (eds.), *Biogeochemistry of Marine Dissolved Matter.* Academic Press, pp. 405–453.

Bell, W., and R. Mitchell. 1972. Chemotactic and growth responses of marine bacteria to algal extracellular products. *Biol. Bull.* 143: 265–277.

Benner, R. 2002. Chemical composition and reactivity. In D. A. Hansell, and C. A. Carlson (eds.), *Biochemistry of Marine Dissolved Organic Matter.* Academic Press, pp. 59–90.

Benner, R., and K. Kaiser. 2003. Abundance of amino sugars and peptidoglycan in marine particulate and dissolved organic matter. *Limnol. Oceanogr.* 48: 118–128.

Biddanda, B., and L. Pomeroy. 1988. Microbial aggregation and degradation of phytoplankton-derived detritus in seawater 1. Microbial succession. *Mar. Ecol. Prog. Ser.* 42: 79–88.

Bidle, K. D., and F. Azam. 1999. Accelerated dissolution of diatom silica by marine bacterial assemblages. *Nature* 397: 508–512.

Bidle, K. D., and F. Azam. 2001. Bacterial control of silicon regeneration from diatom detritus: Significance of bacterial ectohydrolases and species identity. *Limnol. Oceanogr.* 46: 1606–1623.

Borch, N. H., and D. L. Kirchman 1999. Protection of protein from bacterial degradation by submicron particles. *Aquat. Microb. Ecol.* 16: 265–272.

Brophy, J. E., and D. J. Carlson. 1989. Production of biologically refractory dissolved organic carbon by natural seawater microbial populations. *Deep-Sea Res.* 36: 497–507.

Carlson, C. A. 2002. Production and removal processes. In D. A. Hansell, and C. A. Carlson (eds.), *Biogeochemistry of Marine Dissolved Organic Matter*. Academic Press, pp. 91–150.

Carlson, C. A., and H. W. Ducklow. 1995. Dissolved organic carbon in the upper ocean of the central equatorial Pacific Ocean. *Deep-Sea Res.* 42: 639–656.

Carlson, C. A., H. W. Ducklow, and A. F. Michaels. 1994. Annual flux of dissolved organic carbon from the euphotic zone in the northwestern Sargasso Sea. *Nature* 371: 405–408.

Cherrier, J., J. E. Bauer, E. R. M. Druffel, R. B. Coffin, and J. P. Chanton. 1999. Radiocarbon in marine bacteria: Evidence for the ages of assimilated carbon. *Limnol. Oceanogr.* 44: 730–736.

Chin, W.-C., M. V. Orellana, and P. Verdugo. 1998. Spontaneous assembly of marine dissolved organic matter into polymer gels. *Nature* 391: 568–572.

Cho, B. C., and F. Azam. 1988. Major role of bacteria in biogeochemical fluxes in the ocean's interior. *Nature* 332: 441–443.

Christian, J. R., and T. R. Anderson. 2002. Modeling DOM biogeochemistry. In D. A. Hansell, and C. A. Carlson (eds.), *Biogeochemistry of Marine Dissolved Organic Matter*. Academic Press, pp. 717–755.

Christian, J., and D. Karl. 1995. Bacterial ectoenzymes in marine waters: Activity ratios and temperature responses in three oceanographic provinces. *Limnol. Oceanogr.* 40: 1042–1049.

Chrost, R. 1990. Microbial ectoenzymes in aquatic environments. In J. Overbeck, and R. Chrost (eds.), *Aquatic Microbial Ecology: Biochemical and Molecular Approaches*. Springer-Valley, pp. 47–78.

Cottrell, M. T., and D. L. Kirchman. 2000. Natural assemblages of marine proteobacteria and members of the *Cytophaga–Flavobacter* cluster consuming low- and high-molecular-weight dissolved organic matter. *Appl. Environ. Microbiol.* 66: 1692–1697.

Cottrell, M. T., D. N. Wood, L. Yu, and D. L. Kirchman. 2000. Selected chitinase genes in cultured and uncultured marine bacteria in the alpha- and gamma-subclasses of the proteobacteria. *Appl. Environ. Microbiol.* 66: 1195–1201.

Crump, B. C., and J. A. Baross. 1996. Particle-attached bacteria and heterotrophic plankton associated with the Columbia River estuarine turbidity maxima. *Mar. Ecol. Prog. Ser.* 138: 265–273.

Crump, B. C., J. A. Baross, and C. A. Simenstad. 1998. Dominance of particle-attached bacteria in the Columbia River estuary, USA. *Aquat. Microb. Ecol.* 14: 7–18.

Davidson, A. L. 2002. Mechanism of coupling of transport to hydrolysis in bacterial ATP-binding cassette transporters. *J. Bacteriol.* 184: 1225–1233.

DeLong, E. F., C. M. Preston, T. Mincer, et al. 2006. Community genomics among stratified microbial assemblages in the ocean's interior. *Science* 311: 496–503.

Dittmar, T., H. P. Fitznar, and G. Kattner. 2001. Origin and biogeochemical cycling of organic nitrogen in the eastern Arctic Ocean as evident from D- and L-amino acids. *Geochim. Cosmochim. Acta* 65: 4103–4114.

Ducklow, H. W. 2000. Bacterial production and biomass in the oceans. In D. L. Kirchman (ed.), *Microbial Ecology of the Oceans*, 1st edn. Wiley-Liss, pp. 85–120.

Ducklow, H. W., and D. L. Kirchman. 1983. Bacterial dynamics and distribution during a spring diatom bloom in the Hudson River plume, USA. *J. Plankton Res.* 5: 333–335.

Elifantz, H., R. R. Malmstrom, M. T. Cottrell, and D. L. Kirchman. 2005. Assimilation of polysaccharides and glucose by major bacterial groups in the Delaware Estuary. *Appl. Environ. Microbiol.* 71: 7799–7805.

Ferguson, R. L., and W. G. Sunda. 1984. Utilization of amino acids by planktonic marine bacteria: Importance of clean technique and low substrate additions. *Limnol. Oceanogr.* 29: 258–274.

Field, C. G., M. J. Behrenfeld, J. T. Randerson, and P. Falkowski. 1998. Primary production of the biosphere: Integrating terrestrial and oceanic components. *Science* 281: 237–240.

Foreman, C. M., and J. S. Covert. 2003. Linkages between dissolved organic matter composition and bacterial community structure. In S. E. G. Findlay, and R. L. Sinsabaugh (eds.), *Aquatic Ecosystems: Interactivity of Dissolved Organic Matter*. Academic Press, pp. 343–362.

Fuhrman, J. 1987. Close coupling between release and uptake of dissolved free amino acids in seawater studied by an isotope dilution approach. *Mar. Ecol. Prog. Ser.* 37: 45–52.

Fuhrman, J. 1990. Dissolved free amino acid cycling in an estuarine outflow plume. *Mar. Ecol. Prog. Ser.* 66: 197–203.

Fuhrman, J. A., and R. L. Ferguson. 1986. Nanomolar concentrations and rapid turnover of dissolved free amino acids in seawater: Agreement between chemical and microbiological measurements. *Mar. Ecol. Prog. Ser.* 33: 237–242.

Fukuda, H., and I. Koike. 2000. Feeding currents of particle-attached nanoflagellates—a novel mechanism for aggregation of submicron particles. *Mar. Ecol. Prog. Ser.* 202: 101–112.

Fukuda, R., Y. Sohrin, N. Saotome, H. Fukuda, T. Nagata, and I. Koike. 2000. East-west gradient in ectoenzyme activities in the subarctic Pacific: Possible regulation by zinc. *Limnol. Oceanogr.* 45: 930–939.

Goldthwait, S. A., C. A. Carlson, G. K. Henderson, and A. L. Alldredge. 2005. Effects of physical fragmentation on remineralization of marine snow. *Mar. Ecol. Prog. Ser.* 305: 59–65.

Gram, L., H.-P. Grossart, A. Schlingloff, and T. Kiorboe. 2002. Possible quorum sensing in marine snow bacteria: Production of acylated homoserine lactones by *Roseobacter* strains isolated from marine snow. *Appl. Environ. Microbiol.* 68: 4111–4116.

Griffith, P., F.-K. Shiah, K. Gloersen, H. W. Ducklow, and M. Fletcher. 1994. Activity and distribution of attached bacteria in Chesapeake Bay. *Mar. Ecol. Prog. Ser.* 108: 1–10.

Grossart, H.-P., L. Riemann, and F. Azam. 2001. Bacterial motility in the sea and its ecological implications. *Aquat. Microb. Ecol.* 25: 247–258.

Grossart, H.-P., T. Kiorboe, K. W. Tang, M. Allgaier, E. M. Yam, and H. Ploug. 2006. Interactions between marine snow and heterotrophic bacteria: Aggregate formation and microbial dynamics. *Aquat. Microb. Ecol.* 42: 19–26.

Gruber, D. F., J.-P. Simjouw, S. P. Seitzinger, and G. L. Taghon. 2006. Dynamics and characterization of refractory dissolved organic matter produced by a pure bacterial culture in an experimental predator-prey system. *Appl. Environ. Microbiol.* 72: 4184–4191.

Grutters, M., W. Van Raaphorst, E. Epping, W. Helder, J. W. de Leeuw, D. P. Glavin, and J. Bada. 2002. Preservation of amino acids from in situ-produced bacterial cell wall peptidoglycan in northeastern Atlantic continental margin sediments. *Limnol. Oceanogr.* 47: 1521–1524.

Guo, L., C. H. J. Coleman, and P. H. Santschi. 1994. The distribution of colloidal and dissolved organic carbon in the Gulf of Mexico. *Mar. Chem.* 45: 105–119.

Hahn, M. W., and M. G. Höfle. 2001. Grazing of protozoa and its effect on populations of aquatic bacteria. *FEMS Microbiol Ecol.* 35: 113–121.

Hama, T., K. Yanagi, and J. Hama. 2004. Decrease in molecular weight of photosynthetic products of marine phytoplankton during early diagenesis. *Limnol. Oceanogr.* 49: 471–481.

Hansell, D. A. 2002. DOC in the global ocean carbon cycle. In D. A. Hansell, and C. A. Carlson (eds.), *Biogeochemistry of Marine Dissolved Organic Matter.* Academic Press, pp. 685–715.

Hansell, D. A., C. A. Carlson, and Y. Suzuki. 2002. Dissolved organic carbon export with North Pacific Intermediate Water formation. *Global Biogeochem. Cycles* 16: 77–84.

Hedges, J. I. 2002. Why dissolved organics matter. In D. A. Hansell, and C. A. Carlson (eds.), *Biochemistry of Marine Dissolved Organic Matter.* Academic Press, pp. 1–33.

Hedges, J. I., J. A. Baldock, Y. Gelinas, C. Lee, M. Peterson, and S. G. Wakeham. 2001. Evidence for non-selective preservation of organic matter in sinking marine particles. *Nature* 409: 801–804.

Herndl, G. 1988. Ecology of amorphous aggregations (marine snow) in the northern Adriatic Sea: 2. Microbial density and activity in marine snow and its implication to overall pelagic processes. *Mar. Ecol. Prog. Ser.* 48: 265–275.

Hertkorn, N., R. Benner, M. Frommberger, P. Schmitt-Kopplin, M. Witt, K. Kaiser, A. Kettrup, and J. I. Hedges. 2006. Characterization of a major refractory component of marine dissolved organic matter. *Geochim. Cosmochim. Acta* 70: 2990–3010.

Hobbie, J. E., O. Holm-Hansen, T. T. Packard, L. R. Pomeroy, R. W. Sheldon, J. P. Thomas, and W. J. Wiebe. 1972. A study of the distribution and activity of microorganisms in ocean water. *Limnol. Oceanogr.* 17: 544–555.

Hollibaugh, J. T., and F. Azam. 1983. Microbial degradation of dissolved proteins in seawater. *Limnol. Oceanogr.* 28: 1104–1116.

Hoppe, H.-G., S.-J. Kim, and K. Gocke. 1988. Microbial decomposition in aquatic environments: Combined process of extracellular enzyme activity and substrate uptake. *Appl. Environ. Microbiol.* 54: 784–790.

Hoppe, H.-G., C. Arnosti, and G. F. Herndl. 2002. Ecological significance of bacterial enzymes in the marine environment. In R. G. Burns, and R. P. Dick (eds.), *Enzymes in the Environment.* Marcel Dekker, pp. 73–107.

Hsu, P.-H., and P. G. Hatcher. 2005. New evidence for covalent coupling of peptides to humic acids based on 2D NMR spectroscopy: A means for preservation. *Geochim. Cosmochim. Acta* 69: 4521–4533.

Jørgensen, N. O. G., R. Stepanaukas, A. G. U. Pedersen, M. Hansen, and O. Nybroe. 2003. Occurrence and degradation of peptidoglycan in aquatic environments. *FEMS Microbiol. Ecol.* 46: 269–280.

Kaiser, K., and R. Benner. 2008. Major bacterial contribution to the ocean reservoir of detrital organic carbon and nitrogen. *Limnol. Oceanogr.* 53: 99–112.

Karl, D. M., G. A. Knauer and J. H. Martin. 1988. Downward flux of particulate organic matter in the ocean: A particle decomposition paradox. *Nature* 332: 438–441.

Karner, M. B., D. F. DeLong and D. M. Karl. 2001. Archaeal dominance in the mesopelagic zone of the Pacific Ocean. *Nature* 409: 507–510.

Kawasaki, N., and R. Benner. 2006. Bacterial release of dissolved organic matter during cell growth and decline: Molecular origin and composition. *Limnol. Oceanogr.* 51: 2170–2180.

Keil, R. G., and D. L. Kirchman. 1992. Bacterial hydrolysis of protein and methylated protein and its implications for studies of protein degradation in aquatic systems. *Appl. Environ. Microbiol.* 58: 1374–1375.

Keil, R. G., and D. L. Kirchman. 1993. Dissolved combined amino acids: Chemical form and utilization by marine bacteria. *Limnol. Oceanogr.* 38: 1256–1270.

Keil, R. G., and D. L. Kirchman. 1994. Abiotic transformation of labile protein to refractory protein in sea water. *Mar. Chem.* 45: 187–196.

Keil, R. G., and D. L. Kirchman. 1999. Utilization of dissolved protein and amino acids in the northern Sargasso Sea. *Aquat. Microb. Ecol.* 18: 293–300.

Khan, N. H., Y. Ishii, N. Kimata, T. Nishio, H. Esaki, M. Nishimura, and K. Kogure. 2007. Isolation of *Pseudomonas aeruginosa* from open ocean and comparison with freshwater, clinical and animal isolates. *Microb. Ecol.* 173–186.

Kirchman, D. L. 2003. The contribution of monomers and other low-molecular weight compounds to the flux of dissolved organic material in aquatic ecosystems. In S. E. G. Findlay, and R. L. Sinsabaugh (eds.), *Aquatic Ecosystems: Interactivity of Dissolved Organic Matter*. Academic Press, pp. 217–241.

Kirchman, D. L., and J. White. 1999. Hydrolysis and mineralization of chitin in the Delaware Estuary. *Aquat. Microb. Ecol.* 18: 187–196.

Kirchman, D. L., B. Meon, M. T. Cottrell, and D. A. Hutchins. 2000. Carbon versus iron limitation of bacterial growth in the California upwelling regime. *Limnol. Oceanogr.* 45: 1681–1688.

Kirchman, D. L., A. I. Dittel, R. R. Malmstrom, and M. T. Cottrell. 2005. Biogeography of major bacterial groups in the Delaware Estuary. *Limnol. Oceanogr.* 50: 1697–1706.

Koike, I., and T. Nagata 1998. High potential activity of extracellular alkaline phosphatase in deep waters of the central Pacific. *Deep-Sea Res.* II 44: 2283–2294.

Koike, I., S. Hara, K. Terauchi, and K. Kogure. 1990. The role of submicrometer particles in the ocean. *Nature* 345: 242–244.

Kroer, N., N. Jørgensen, and R. Coffin. 1994. Utilization of dissolved nitrogen by heterotrophic bacterioplankton—A comparison of 3 ecosystems. *Appl. Environ. Microbiol.* 60: 4116–4123.

Kuehn, M. J., and N. C. Kesty. 2005. Bacterial outer membrane vesicles and the host–pathogen interaction. *Genes Dev.* 19: 2645–2655.

LeCleir, G. R., and J. T. Hollibaugh. 2006. Chitinolytic bacteria from alkaline hypersaline Mono Lake, California, USA. *Aquat. Microb. Ecol.* 42: 255–264.

LeCleir, G. R., A. Buchan, and J. T. Hollibaugh. 2004. Chitinase gene sequences retrieved from diverse aquatic habitats reveal environment-specific distributions. *Appl. Environ. Microbiol.* 70: 6977–6983.

Lee, C., and J. L. Bada. 1977. Dissolved amino acids in the equatorial Pacific, the Sargasso Sea, and Biscayne Bay. *Limnol. Oceanogr.* 22: 502–510.

Lomstein, B. A., B. B. Jørgensen, C. J. Schubert, and J. Niggemann. 2006. Amino acid biogeo- and stereochemistry in coastal Chilean sediments. *Geochim. Cosmochim. Acta* 70: 2970–2989.

Long, R. A., and F. Azam. 1996. Abundant protein-containing particles in the sea. *Aquat. Microb. Ecol.* 10: 213–221.

Long, R. A., and F. Azam. 2001. Antagonistic interactions among marine pelagic bacteria. *Appl. Environ. Microbiol.* 67: 4975–4983.

McCarthy, M., T. Pratum, J. Hedges, and R. Benner. 1997. Chemical composition of dissolved organic nitrogen in the ocean. *Nature* 390: 150–154.

McCarthy, M., D. J. Hedges, I, and R. Benner. 1998. Major bacterial contribution to marine dissolved organic nitrogen. *Science* 281: 231–234.

Malmstrom, R. R., R. P. Kiene, M. T. Cottrell, and D. L. Kirchman. 2004a. Contribution of SAR11 bacteria to dissolved dimethylsulfoniopropionate and amino acid uptake in the North Atlantic Ocean. *Appl. Environ. Microbiol.* 70: 4129–4135.

Malmstrom, R. R., R. P. Kiene, and D. L. Kirchman. 2004b. Identification and enumeration of bacteria assimilating dimethylsulfoniopropionate (DMSP) in the North Atlantic and Gulf of Mexico. *Limnol. Oceanogr.* 49: 597–606.

Mannino, A., and H. R. Harvey. 1999. Lipid composition in particulate and dissolved organic matter in the Delaware Estuary: Sources and diagenetic patterns. *Geochim. Cosmochim. Acta* 63: 2219–2235.

Martinez, J., D. C. Smith, G. F. Steward, and F. Azam. 1996. Variability in ectohydrolytic enzyme activities of pelagic marine bacteria and its significance for substrate processing in the sea. *Aquat. Microb. Ecol.* 10: 223–230.

Mayer, L. M., L. L. Shick, K. R. Hardy, R. Wagai, and J. McCarthy. 2004. Organic matter in small mesopores in sediments and soils. *Geochim. Cosmochim. Acta* 68: 3863–3872.

Middelboe, M., and N. O. G. Jørgensen. 2006. Viral lysis of bacteria: An important source of dissolved amino acids and cell wall compounds. *J. Mar. Biol. Assoc. UK* 86: 605–612.

Miki, T., and N. Yamamura. 2005. Theoretical model of interactions between particle-associated and free-living bacteria to predict functional composition and succession in bacterial communities. *Aquat. Microb. Ecol.* 39: 35–46.

Milliman, J. D., P. J. Troy, W. M. Balch, A. K. Adams, Y. H. Li, and F. T. Mackenzie. 1999. Biologically mediated dissolution of calcium carbonate above the chemical lysocline? *Deep-Sea Res.* I 46: 1653–1669.

Mopper, K., C. A. Schultz, L. Chevolot, C. Germain, R. Revuelta, and R. Dawson. 1992. Determination of sugars in unconcentrated seawater and other natural waters by liquid chromatography and pulsed amperometric detection. *Environ. Sci. Technol.* 26: 133–138.

Moran, M. A., and R. G. Zepp. 2000. UV radiation effects on microbes and microbial processes. In D. L. Kirchman (ed.), *Microbial Ecology of the Oceans*, 1st edn. Wiley-Liss, pp. 201–228.

Morris, R. M., M. S. Rappé, S. A. Connon, K. L. Vergin, W. A. Siebold, C. A. Carlson, and S. J. Giovannoni. 2002. SAR11 clade dominates ocean surface bacterioplankton communities. *Nature* 420: 806–810.

Mostajir, B., J. R. Dolan, and F. Rassoulzadegan. 1995. Seasonal variations of pico-and nano-detrital particles (DAPI yellow particles, DYP) in the Ligurian Sea (NW Mediterranean). *Aquat. Microb. Ecol.* 9: 267–277.

Müller-Niklas, G., G. Schuster, E. Kaltenbock, and G. J. Herndl. 1994. Organic content and bacterial metabolism in amorphous aggregations in the northern Adriatic Sea. *Limnol. Oceanogr.* 39: 58–68.

Nagata, T. 2000. Production mechanisms of dissolved organic matter. In D. L. Kirchman (ed.), *Microbial Ecology of the Oceans*, 1st edn. Wiley-Liss, pp. 121–152.

Nagata, T., and D. L. Kirchman. 1996. Bacterial degradation of protein adsorbed to model submicron particles in seawater. *Mar. Ecol. Prog. Ser.* 132: 241–248.

Nagata, T., and D. L. Kirchman. 1997. Roles of submicron particles and colloids in microbial food webs and biogeochemical cycles within marine environments. *Adv. Microb. Ecol.* 15: 81–103.

Nagata, T., and D. L. Kirchman. 1999. Bacterial mortality: A pathway for the formation of refractory DOM? In M. Brylinsky, C. Bell, and P. Johnson-Green (eds.), *New Frontiers in Microbial Ecology: Proceedings of the 8th International Symposium on Microbial Ecology*. Atlantic Canada Society for Microbial Ecology, pp. 153–158.

Nagata, T., R. Fukuda, I. Koike, K. Kogure, and D. L. Kirchman. 1998. Degradation by bacteria of membrane and soluble protein in seawater. *Aquat. Microb. Ecol.* 14: 29–37.

Nagata, T., H. Fukuda, R. Fukuda, and I. Koike. 2000. Bacterioplankton distribution and production in deep Pacific waters: Large-scale geographic variations and possible coupling with sinking particle fluxes. *Limnol. Oceanogr.* 45: 426–435.

Nagata, T., B. Meon, and D. L. Kirchman. 2003. Microbial degradation of peptidoglycan in seawater. *Limnol. Oceanogr.* 48: 745–754.

Nguyen, R. T., and R. H. Harvey. 2001. Preservation of protein in marine systems: Hydrophobic and other noncovalent associations as major stabilizing forces. *Geochim. Cosmochim. Acta* 65: 1467–1480.

Nguyen, R. T., and R. H. Harvey. 2003. Preservation via macromolecular associations during *Botryococcus braunii* decay: Proteins in the Pula kerogen. *Org. Geochem.* 34: 1391–1403.

Niggemann, J., and C. J. Schubert. 2006. Sources and fate of amino sugars in coastal Peruvian sediments. *Geochim. Cosmochim. Acta* 70: 2229–2237.

Nunn, B. L., A. Norbeck, and R. G. Keil. 2003. Hydrolysis patterns and the production of peptide intermediates during protein degradation in marine systems. *Mar. Chem.* 83: 59–73.

Obayashi, Y., and S. Suzuki. 2005. Proteolytic enzymes in coastal surface seawater: Significant activity of endopeptidases and exopetidases. *Limnol. Oceanogr.* 50: 722–726.

Ogawa, H., and E. Tanoue. 2003. Dissolved organic matter in oceanic waters. *J. Oceanogr.* 59: 129–147.

Ogawa, H., R. Fukuda, and I. Koike. 1999. Vertical distribution of dissolved organic carbon and nitrogen in the Southern Ocean. *Deep-Sea Res. I* 46: 1809–1826.

Ogawa, H., Y. Amagai, I. Koike, K. Kaiser, and R. Benner. 2001. Production of refractory dissolved organic matter by bacteria. *Science* 292: 917–920.

Pedersen, A. G. U., T. R. Thomsen, B. A. Lomstein, and N. O. G. Jørgensen. 2001. Bacterial influence on amino acid enantiomerization in a coastal marine sediment. *Limnol. Oceanogr.* 46: 1358–1369.

Pinhassi, J., F. Azam, J. Hemphala, R. A. Long, J. Martinez, U. L. Zweifel, and Å. Hagström. 1999. Coupling between bacterioplankton species composition, population dynamics, and organic matter degradation. *Aquat. Microb. Ecol.* 17: 13–26.

Rath, J., C. Schiller, and G. J. Herndl. 1993. Ectoenzymatic activity and bacterial dynamics along a trophic gradient in the Caribbean Sea. *Mar. Ecol. Prog. Ser.* 102: 89–96.

Rich, J. H., H. W. Ducklow, and D. L. Kirchman. 1996. Concentrations and uptake of neutral monosaccharides along 140°W in the equatorial Pacific: Contribution of glucose to heterotrophic bacterial activity and the DOM flux. *Limnol. Oceanogr.* 41: 595–604.

Rich, J., M. Gosselin, E. Sherr, B. Sherr, and D. L. Kirchman. 1997. High bacterial production, uptake and concentrations of dissolved organic matter in the Central Arctic Ocean. *Deep-Sea Res.* II 44: 1645–1663.

Riemann, L., and F. Azam. 2002. Widespread N-acetyl-D-glucosamine uptake among pelagic marine bacteria and its ecological implications. *Appl. Environ. Microbiol.* 68: 5554–5562.

Riemann, L., G. F. Steward, and F. Azam. 2000. Dynamics of bacterial community composition and activity during a mesocosm diatom bloom. *Appl. Environ. Microbiol.* 66: 578–587.

Samuelsson, M.-O., and D. L. Kirchman. 1990. Degradation of absorbed protein by attached bacteria in relationship to surface hydrophobicity. *Appl. Environ. Microbiol.* 56: 3643–3648.

Shibata, A., K. Kogure, I. Koike, and K. Ohwada. 1997. Formation of submicron colloidal particles from marine bacteria by viral infection. *Mar. Ecol. Prog. Ser.* 155: 303–307.

Simon, M., N. A. Welschmeyer, and D. L. Kirchman. 1992. Bacterial production and the sinking flux of particulate organic matter in the subarctic Pacific. *Deep-Sea Res.* 39: 1997–2008.

Simon, M., H.-P. Grossart, B. Schweitzer, and H. Ploug. 2002. Microbial ecology of organic aggregates in aquatic ecosystems. *Aquat. Microb. Ecol.* 28: 175–211.

Skoog, A., B. Biddanda, and R. Benner. 1999. Bacterial utilization of dissolved glucose in the upper water column of the Gulf of Mexico. *Limnol. Oceanogr.* 44: 1625–1633.

Smith, D. C., M. Simon, A. L. Alldredge, and F. Azam. 1992. Intense hydrolytic enzyme activity on marine aggregates and implications for rapid particle dissolution. *Nature* 359: 139–142.

Smith, D. C., G. F. Steward, R. A. Long, and F. Azam. 1995. Bacterial mediation of carbon fluxes during a diatom bloom in a mesocosm. *Deep-Sea Res.* II 42: 75–97.

Suttle, C. A., A. M. Chan, and J. A. Fuhrman. 1991. Dissolved free amino acids in the Sargasso Sea: Uptake and respiration rates, turnover times, and concentrations. *Mar. Ecol. Prog. Ser.* 70: 189–199.

Suzuki, S., K. Kogure, and E. Tanoue. 1997. Immunochemical detection of dissolved proteins and their source bacteria in marine environments. *Mar. Ecol. Prog. Ser.* 158: 1–9.

Tanoue, E. 1995. Detection of dissolved protein molecules in oceanic waters. *Mar. Chem.* 51: 239–252.

Tanoue, E., S. Nishiyama, M. Kamo, and A. Tsugita. 1995. Bacterial membranes: Possible source of a major dissolved protein in seawater. *Geochim. Cosmochim. Acta* 59: 2643–2648.

Teira, E., H. van Aken, C. Veth, and G. J. Herndl. 2006. Archaeal uptake of enantiomeric amino acids in the meso- and bathypelagic waters of the North Atlantic. *Limnol. Oceanogr.* 51: 60–69.

Thingstad, T. F., Å. Hagström, and F. Rassoulzadegan. 1997. Accumulation of degradable DOC in surface waters: Is it caused by a malfunctioning microbial loop? *Limnol. Oceanogr.* 42: 398–404.

Tholosan, O., F. Lamy, J. Garcin, T. Polychronaki, and A. Bianchi. 1999. Biphasic extracellular proteolytic enzyme activity in benthic water and sediment in the Northwestern Mediterranean Sea. *Appl. Environ. Microbiol.* 65: 1619–1626.

Treguer, P., L. Legendre, R. T. Rivkin, O. Ragueneau, and N. Dittert. 2003. Water column biogeochemistry below the euphotic zone. In J. R. M. Fasham (ed.), *Ocean Biogeochemistry: A Synthesis of the Joint Global Ocean Flux Study (JGOFS)*. Springer-Verlag, pp. 145–156.

Turley, C. M., and P. J. Mackie. 1994. Biogeochemical significance of attached and free-living bacteria and the flux of particles in the NE Atlantic Ocean. *Mar. Ecol. Prog. Ser.* 115: 191–203.

Turley, C. M., and E. D. Stutt. 2000. Depth-related cell-specific bacterial leucine incorporation rates on particles and its biogeochemical significance in the Northwest Mediterranean. *Limnol. Oceanogr.* 45: 419–425.

Verdugo, P., A. L. Alldredge, F. Azam, D. L. Kirchman, U. Passow, and P. H. Santschi. 2004. The oceanic gel phase: A bridge in the DOM–POM continuum. *Mar. Chem.* 92: 67–85.

Verity, P. G., S. C. Williams, and Y. Hong. 2000. Formation, degradation, and mass:volume ratios of detritus derived from decaying phytoplankton. *Mar. Ecol. Prog. Ser.* 207: 53–68.

Veuger, B., D. V. Oevelen, H. T. S. Boschker, and J. J. Middelburg. 2006. Fate of peptidoglycan in an intertidal sediment: An in situ ^{13}C-labeling study. *Limnol. Oceanogr.* 51: 1572–1580.

Wakeham, S. G., T. K. Pease, and R. Benner. 2003. Hydroxy fatty acids in marine dissolved organic matter as indicators of bacterial membrane material. *Org. Geochem.* 34: 857–868.

Wells, M. L., and E. D. Goldberg. 1994. The distribution of colloids in the North Atlantic and Southern Oceans. *Limnol. Oceanogr.* 39: 286–302.

Wiebe, W. J., and L. R. Pomeroy. 1972. Microorganisms and their association with aggregates and detritus in the sea: A microscopic study. *Mem. Ist. Ital. Idrobiol.* 29: 325–352.

Williams, P. J. leB 1973. The validity of the application of simple kinetic analysis to heterogeneous microbial populations. *Limnol. Oceanogr.* 18: 159–165.

Wright, R. T., and J. E. Hobbie. 1966. Use of glucose and acetate by bacteria and algae in aquatic ecosystems. *Ecology* 47: 447–464.

Yamada, N., and E. Tanoue. 2003. Detection and partial characterization of dissolved glycoproteins in oceanic waters. *Limnol. Oceanogr.* 48: 1037–1048.

Yamasaki, A., H. Fukuda, R. Fukuda, T. Miyajima, T. Nagata, H. Ogawa, and I. Koike. 1998. Submicrometer particles in northwest Pacific coastal environments: Abundance, size distribution, and biological origins. *Limnol. Oceanogr.* 43: 536–542.

Yamashita, Y., and E. Tanoue. 2004. Chemical characteristics of amino acid-containing dissolved organic matter in seawater. *Org. Geochem.* 35: 679–692.

Yokokawa, T., and T. Nagata. 2005. Growth and grazing mortality rates of phylogenetic groups of bacterioplankton in coastal marine environments. *Appl. Environ. Microbiol.* 71: 6799–6807.

Yokokawa, T., T. Nagata, M. T. Cottrell, and D. L. Kirchman. 2004. Growth rate of the major phylogenetic bacterial groups in the Delaware estuary. *Limnol. Oceanogr.* 49: 1620–1629.

Zang, X., R. T. Nguyen, R. H. Harvey, H. Knicker, and P. G. Hatcher. 2001. Preservation of proteinaceous material during the degradation of the green alga *Botryococcus braunii*: A solid state 2D ^{15}N ^{13}C NMR spectroscopy study. *Geochim. Cosmochim. Acta* 65: 3299–3305.

Ziervogel, K., E. Karlsson, and C. Arnosti. 2007. Surface associations of enzymes and of organic matter: Consequences for hydrolytic activity and organic matter remineralization in marine systems. *Mar. Chem.* 104: 241–252.

PHYSIOLOGICAL STRUCTURE AND SINGLE-CELL ACTIVITY IN MARINE BACTERIOPLANKTON

PAUL A. DEL GIORGIO

Département des Sciences Biologiques, Université du Québec à Montréal, CP 8888, succ. Centre Ville, Montréal, Québec, Canada H3C 3P8

JOSEP M. GASOL

Institut de Ciències del Mar, CMIMA (CSIC), Passeig Marítim de la Barceloneta 37–49, 08003 Barcelona, Catalunya, Spain

INTRODUCTION

Generations of microbiologists have pondered over the issue of bacterial activity and whether most bacteria in the environment are "active," "inactive," or "dormant" (Stevenson 1978). This is a question with clear practical implications, in terms of food, sanitary, and clinical microbiology, as well as profound ecological implications. The issue started to be debated in earnest within the aquatic microbial community roughly at the time researchers were developing new methods to estimate the total abundance of aquatic bacteria, since some of the first stains used were thought to respond differently depending on the RNA content of single bacterial cells (see, e.g., Francisco et al. 1973). It was not until bacteria were recognized as being not only very abundant, but also active, that the notion of microbes as important partners in the marine food web gained widespread recognition (Pomeroy 1974). The early studies using radiotracer incorporation showed surprisingly high rates of bacterial

---

Microbial Ecology of the Oceans, Second Edition. Edited by David L. Kirchman
Copyright © 2008 John Wiley & Sons, Inc.

biomass production and of DNA replication, so clearly there had to be active bacterial cells in the oceans. The question then arose as to how this activity was distributed among the 10^5-10^6 cells that are on average present in 1 mL of ocean water (Hoppe 1976; Stevenson 1978). Did all these cells have roughly the same level of metabolic activity, or was the bulk of bacterioplankton activity concentrated in a few key players? Marine microbial ecology has gone a long way since those early studies, but the question is as valid and relevant today as it was back then.

That complex aquatic bacterial assemblages harbor a wide range of single-cell metabolic activities as well as of physiological states is neither surprising nor much contested nowadays. Yet, in spite of the simplicity of this basic premise, the actual description and quantification of this metabolic and physiological heterogeneity have proven elusive, and its underlying mechanisms, as well as its ecological and evolutionary consequences, have been very difficult to address in practice (Mason et al. 1986; Koch 1997; Kell et al. 1998; Smith and del Giorgio 2003). Under the term "bacterial single-cell activity," various authors throughout the years have lumped together a number of very different cellular processes, including cell division rate, single and complex substrate uptake, respiratory activity, protein synthesis, intracellular enzyme activity, membrane potential and polarity, membrane integrity, ultrastructural characteristics, nucleic acid and other macromolecular contents, motility, and the list continues. While all these different aspects of cell structure, metabolism, and physiology are surely connected, the links that exist between them are complex, with feedbacks, compensatory mechanisms, and nonlinear responses, so that information on any particular cell function does not necessarily allow inferences about the others (Smith and del Giorgio 2003). In this chapter, we refer to the ensemble of these single-cell characteristics and processes as the "physiological structure" of bacterial assemblages. As we hope to make evident in this chapter, this structure can be neither described nor understood on the basis of any one single aspect of cell function.

In spite of the very incomplete descriptions of the physiological structure of bacterioplankton that we currently have, some patterns are emerging: the evidence converges to suggest that while many bacterial cells appear to be intact, many of these do not seem to have detectable levels of activity, at least with the methods that are currently available. In addition, in all assemblages studied to date, there seems to be a non-negligible proportion of cells that are either damaged or dead. The presence of significant numbers of these cells in almost all assemblages studied begs the question: Why are there so many apparently inactive, damaged, and dead cells in all aquatic systems? What are the factors that induce cell inactivation and damage? How can these cells persist in the assemblage, and what is their role in community functioning and the maintenance of genetic diversity? How and when does individual cell activity relate to phylogenetic composition? Recent technical and conceptual developments are providing the tools to at least begin to address some of these questions.

In this chapter, we review the current state of knowledge and the technical and conceptual progress that has been made in the past two decades in the study of the single-cell characteristics and of the physiological structure of marine bacterioplankton. There are five main issues concerning the physiological structure of marine bacterial assemblages that we address in this chapter: (1) the approaches used to probe various aspects of bacterial activity and physiology at the single cell level; (2) the connections that exist between these various cell functions; (3) the environmental and biological factors that influence the different aspects of bacterial physiology and metabolism; (4) the empirical findings concerning the distribution of single-cell bacterial activity and the physiological structure of marine bacterioplankton assemblages; and (5) the biogeochemical, ecological, and evolutionary significance of the resulting patterns of cellular activity. It is the goal of this chapter to provide a conceptual framework to help interpret and integrate the data that have been generated in the last few years concerning the physiological structure of bacterioplankton assemblages.

DISTRIBUTION OF PHYSIOLOGICAL STATES IN BACTERIOPLANKTON ASSEMBLAGES

The Concept of "Physiological Structure" of Bacterioplankton Assemblages

The physiological structure of bacterioplankton is the distribution of cells in different physiological categories within the assemblage. The physiological structure is akin to the "phylogenetic structure" of an assemblage, except that the classification of organisms based on "phylogenetic units" is replaced by one based on "physiological state units." The challenge in microbial ecology has precisely been how to define these units. Metabolic activity and cell physiology are generally not discreet variables—they are rather expressed as a continuum that is impossible to capture by any single method. Cells must thus be grouped into artificial categories that are essentially operational, depending directly on the characteristics and thresholds of the methods used (see below).

The physiological structure of bacterioplankton, that is, the distribution within the assemblage of cells in different physiologic categories, is regulated by two major groups of factors. On the one hand, there are the environmental and phylogenetic factors that influence the level of activity and physiological response of individual bacterial cells. It is unlikely that cells from different bacterial strains and taxa respond identically to environmental stimuli, and since in any bacterioplankton assemblage there are many coexisting bacterial phylotypes, for any combination of environmental factors (e.g., salinity, and rate of supply and nature of the organic substrates and nutrients) there must necessarily be a diversity of metabolic and physiological responses.

The second major aspect of the regulation of the physiological structure of bacterioplankton is imposed by factors that regulate not the activity but the loss and persistence of these different categories of cells in the assemblage. For example, there is now evidence that protists may selectively graze on the more active cells (del Giorgio et al. 1996), and that viruses may preferentially infect cells that are growing and have higher levels of metabolism (Weinbauer 2004), such that these cells at the high end of the activity spectrum may be subjected to higher removal rates than the average cell in the assemblage. Thus, the proportion of these cells in a given assemblage is not a simple function of the rates of cell division and activation, but rather depends as well on the loss rates. Likewise, because of their size, and also the relative recalcitrance of their cellular components (see, e.g., McCarthy et al. 1998), dead or inactive bacteria may remain suspended in plankton for long periods and thus accumulate in the assemblage even if the rates at which they are generated are low.

There are no doubt interactions and feedbacks between these two levels of control of physiological structure. For example, viral infection and the resulting bacterial lysis transfers cells from the active to the injured or dead pool, but, in doing so, releases organic carbon and nutrients that may induce the activation of dormant cells, or enhances the growth rates of cells that are already active. It is thus the net balance between levels of regulation (i.e., rates of activation, inactivation, or death) and selective removal, persistence, or loss that determines the actual proportion of cells in different physiological categories found in aquatic bacterioplankton assemblages. We focus on the description of the physiological structure, as well as on these different aspects of regulation in the following sections.

The distribution of cells into different physiological categories is termed the **"physiological structure"** of bacterioplankton:

- Within a bacterial assemblage, there is a continuum of activity: from dead to highly active cells.
- The categories used to describe the physiological structure are operational, and depend on the methods used.
- The physiological structure is related, albeit in complex ways, to the size structure of the community, as well as to the phylogenetic structure.
- The physiological structure is dynamic; that is, the proportions of cells in various physiological states may vary on short time scales and small spatial scales.

Starvation, Dormancy, and Viability in Marine Bacterioplankton

One issue at the center of the concept of the physiological structure concerns the different adaptive strategies developed by aquatic bacteria to cope with a dilute and fluctuating environment, and the cellular physiological states associated with these strategies. In particular, and as we will see below, the evidence to date suggests that, in all aquatic ecosystems, there is a fraction, often large, of bacterial cells that do not have measurable metabolic activity but that nevertheless retain cellular integrity. The marine microbial literature abounds in references to "dormant," "latent," "starved," "quiescent," and "inactive" bacteria to describe these cells. These terms are often used quite loosely to denote absence or low levels of bacterial activity, but dormancy, starvation response, and slow growth are not synonymous, because they are associated with different life strategies and ecological consequences.

Starvation, that is, the presence of organic substrates below the threshold of the uptake capacities or of the minimum cell requirements, is the most obvious cause for the inactivation of bacterial cells in aquatic ecosystems, and is the one that has been traditionally invoked to explain differences in cell activity within and among assemblages (Kjelleberg et al. 1993; Morita 1997). The physiological, molecular, and genetic basis of the starvation response has been studied mostly in laboratory cultures of *Vibrio* spp. and *Escherichia coli* (Morita 1997; Kjelleberg et al. 1993; Koch 1997), and a handful of marine isolates (Morita 1997), including the marine bacterium *Sphingomonas* sp. (Fegatella and Cavicchioli 2000). All cultured strains explored to date appear to have some systematic response to starvation, regulated by several genes, which involves a complex series of morphological as well as molecular and physiological changes: reduction in size and changes in cell shape, changes in macromolecular composition, as well as the synthesis of specific proteins, changes in nucleic acid structure, and cell miniturisation (Nyström et al. 1992; Ihssen and Egli 2005). Some of the changes are geared to maximizing substrate uptake; others are to enable the cell to quickly resume synthesis and growth if conditions become favorable; and yet others are to protect the cell against cell damage or degradation in the absence of sufficient energy and carbon supply; there is also protection against possible sudden carbon surfeit that could have negative consequences to the cells (Koch 1997). In addition, the starvation-related changes may confer protection against other stresses, such as ultraviolet (UV) or temperature shock, viral infection, and even protistan grazing and digestion (Nyström et al. 1992), and this cross-protection does not necessarily exist for cells that have low levels of metabolism and are simply growing slowly.

Dormancy, on the other hand, is a response to stress, not just to carbon starvation, but also to osmotic, temperature, water, and other types of environmental stresses, also involving structural and biochemical changes that function to keep the cell in a state of suspended animation (Koch 1997). In this regard, dormancy could be viewed as the final and most extreme stage of the starvation response (Kjelleberg et al. 1993). Bacteria in the dormant state do not undergo cell division, and function at very low metabolic rates or have complete metabolic arrest (Kjelleberg et al. 1993; Barer and Harwood 1999). When environmental conditions change and become favorable, dormant bacteria can (in theory) resume cell division. Dormancy involves adaptive mechanisms, but not morphological differentiation such as sporulation, and cells may remain in a dormant state for extended periods of time without significant loss of cellular integrity (Roszak and Colwell 1987; Barer and Harwood 1999). As with the starvation-related process, dormant cells may have increased resistance to a range of environmental stresses (Kell et al. 1998).

Aspects of cell dormancy are related to the concept of cell culturability. In traditional microbiology, the capacity of cells to multiply in artificial media was used as the main criteria for bacterial viability. This definition had to be revised when it become clear that there were cells that lost their ability to grow in culture but nevertheless remained alive in a state that was later termed "viable but not culturable" (VBNC) (Roszak and Colwell 1987). A VBNC cell is one that persists intact in the environment, but that does not divide in culture. The cell must retain cellular integrity, an intact membrane, RNA and DNA, and potential for protein synthesis;

a VBNC cell may also show measurable levels of cellular activity in terms of substrate uptake, respiration, or protein turnover (Gribbon and Barer 1995; Kell et al. 1998). Gram-negative bacteria, such as *Vibrio cholerae*, *V. vulnificus*, and *E. coli*, have been reported to enter this VBNC state, from which they are able to return to a culturable state under certain conditions (Colwell 2000; Oliver 2000). There has been debate as to whether the VBNC state actually exists in nature or if it is the result of experimental conditions (Kell et al. 1998; Oliver 2000). The importance of this concept from a sanitary and clinical viewpoint is the realization that culturability cannot be used as the ultimate indicator of cellular viability or survival. Kell et al. (1998) point out that dormancy refers to cells that have negligible activity but that are ultimately culturable, whereas VBNC cells are metabolically active but nonculturable, but this criterion can only be applied to strains that are easily culturable in the laboratory. In this regard, aquatic microbial ecologists recognized long ago that although most marine bacteria cannot be cultured, they should not be considered dead or non-viable (see, e.g., Staley and Konopka 1985; Amann et al. 1990).

Finally, many marine bacterial isolates are naturally slow-growing under ambient conditions (Schut et al. 1997; Eguchi et al. 2001; Rappé et al. 2002), even under optimum conditions. Slow growth can therefore be both a stage in the response to extreme starvation and an adaptive strategy of marine bacteria. In either case, growth (even slow growth) requires the operation and maintenance of a large number of cellular processes, including transport and biosynthesis systems, and there are thermodynamic constraints on how slowly these systems can effectively operate (Chesbro et al. 1990). For example, at very low substrate and energy fluxes, there is a danger associated with incomplete synthesis of macromolecules, or the degradation of newly synthesized compounds, with negative effects on cell growth and survival (Koch 1997). In pure culture, responses to extreme carbon starvation are heterogeneous, with some cells taking up substrate and continuing slow synthesis and growth, and others not, and instead entering dormancy, suggesting that there is a choice of energy expenditure even in situations of extreme lack of substrates or unfavorable environmental conditions (Koch 1997).

What are then the costs and benefits associated with entering dormancy, as opposed to the strategy of maintaining low growth rates and dividing until thermodynamically and physically impossible? The latter strategy implies cell injury and death if unfavorable conditions persist; the former implies decreased cell death over extended exposure to unfavorable conditions but increased expenditures in terms of the dormancy response. This question has recently been explored through modeling. Bär et al. (2002) modeled the outcome of two cyanobacterial populations that are capable or incapable of becoming dormant in response to lack of water. They concluded that under a constant supply of water (even if the supply is extremely low), the dormant strategy does not provide any benefit, and it can, instead, decrease the chances of survival of the population. Konopka (2004) also modeled two bacterial populations, one entering dormancy and the other not making this transition, and found that the cells capable of entering a dormant state would be favored only in situations where the periodicity of substrate inputs is long compared with the minimum doubling time of the bacterium. These results point to the fact that intermittency and microscale patchiness of resource supply probably play as large a role as the actual rate of supply in terms of shaping bacterial responses in aquatic ecosystems.

The starvation response and the processes leading to cell dormancy profoundly alter the physical and biochemical properties of cells relative to slow-growing cells, and this in turn influences how the cell interacts with both its physical and biological environments. Whether bacterioplankton cells are intrinsically slow-growing, in a starvation/survival mode, or dormant sensu stricto is not just a methodological or philosophical issue but rather one that has significant ecological consequences. In practice, however, it is at present difficult, if not impossible, to differentiate in natural communities intrinsically slow-growing cells from those that are undergoing a starvation response. In addition, it is also difficult to differentiate truly dormant cells from those that have low levels of metabolic activity. In the rest of this chapter, we will refer to "inactive" cells to encompass these different physiological categories, acknowledging that this represents a major oversimplification that needs to be corrected in the future.

Dormancy, starvation-survival, latency, slow growth, and **inactivity** are often used interchangeably to denote low levels of cellular activity in marine bacteria, but these terms are not synonyms and refer to different states:

- Under conditions of extreme substrate and energy deprivation, marine bacteria may undergo a "**starvation**" response.
- The **starvation** response is regulated by specific genes, and involves cell miniaturization and profound changes in macromolecular composition, with the synthesis of specialized protective proteins.
- Prolonged starvation may lead to cell "**dormancy**," which is a state of complete metabolic arrest that allows long-term survival under unfavorable conditions. Cells in a dormant state are more resistant to environmental stresses than active cells.
- There are costs and benefits associated with entering dormancy as opposed to maintaining a slow level of metabolic activity and growth as a response to low substrate availability.
- Resource patchiness and temporal variability play a major role in shaping the survival strategies of marine bacteria, whether it be slow growth, starvation response, or dormancy.
- Microbiologists have further described the **viable but nonculturable (VBNC)** state, based on the inability of strains to grow in artificial media.
- The VBNC state is not analogous to cell dormancy, because in the former the cells may have significant metabolic activity, whereas in the latter cells have no measurable metabolic activity but retain the capacity to reproduce on plates. Since most marine bacteria cannot be grown in plates even if actively growing in the ocean, this concept has little applicability to the ecology of naturally occurring marine bacteria.

DESCRIBING THE PHYSIOLOGICAL STRUCTURE OF BACTERIOPLANKTON

A number of cellular characteristics are indicative of the physiological state of the cell and of its level of metabolic activity (Fig. 8.1). In this section, we describe the most ecologically relevant indicators of cell physiology, and the analytical tools that are used to assess these indicators. We then propose a series of operational categories based on these measurements that can be used to describe the physiological structure of bacterioplankton.

Single-Cell Properties and Methodological Approaches

There are increasing numbers of techniques available to probe the activity and characteristics of single bacterioplankton cells. We refer the reader to excellent reviews for technical details (Joux and Lebaron 2000; Brehm-Stecher and Johnson 2004). Here we focus on the adaptation and application of these techniques to the study of bacterioplankton communities. A list of techniques and probes that have been used to assess bacterial single cell characteristics in bacterioplankton can be found in Table 8.1. These techniques target different categories of cell function: (1) morphological integrity; (2) DNA quantification and status; (3) respiration activity; (4) the status of the membrane; (5) the internal enzymatic capacity; (6) the uptake of organic substances; (7) the synthesis of DNA; (8) cell growth; (9) the amount of rRNA (see Fig. 8.1).

Morphological Integrity Marine bacteria that are intact and functional are generally surrounded by a capsular layer, while damaged cells do not have this capsule (Heissenberger et al. 1996). The capsule may play a number of roles in the functioning of the cell: It may prevent viral adhesion, or act as a grazing deterrent (Storedegger and Herndl 2002). Bacteria growing in culture become more hydrophobic, and the capsule may be involved in the increased capacity to attach to

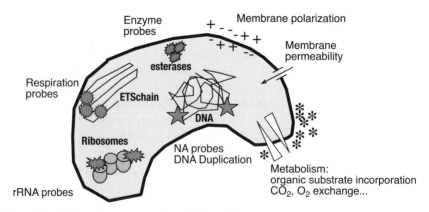

Figure 8.1 Schematic representation of the cellular components targeted by the different activity probes listed and categorized in Table 8.1.

TABLE 8.1 Activity Probes Used for Monitoring Bacterial Viability[a]

Probe(s)[b]	Mode of Action	Method[c]	Key References
Respiration			
INT	Reduction by dehydrogenases	OM	Zimmerman et al. (1978)
CTC, XTT	Indicator of respiratory-chain activity	EF, FC, Spect	del Giorgio et al. (1997)
Membrane Polarization and Membrane Potential (including dye-exclusion probes)			
EthBr, Eth-D2	Excluded by living cells	EF/FC	Schumann et al. (2003)
PI	Excluded by living cells	EF/FC	Williams et al. (1998)
Rh123	Accumulated in live cells	FC	Davey et al. (1999)
Calcafluor white, Tinopal CBS-X,...	Excluded by living cells	FC	Mason et al. (1995)
Carbocyanines (cationic) (JC-1, $DiOC_6(3)$, $DiOC_2(3)$)	Excluded by living cells	FC	Mason et al. (1995)
Oxonols (anionic) ($DiBaC_4(3)$, Oxonol V, VI)	Excluded by living cells	FC	del Giorgio and Bouvier (2002)
TOPRO-1, TOPRO-3 TOPRO-3	Excluded by living cells	EF/FC	Maranger et al. (2002)
Sytox Green	Excluded by living cells	EF/FC	Lebaron et al. (1998)
Intracellular Enzymes			
FDG	β-galactosidase activity	FC	Nir et al. (1990)
FDA and its derivatives: 6CFDA, calcein blue AM, SFDA, BCECF_AM; Chemchrome B, V6	Cleaved by intracellular enzymes	EF/SPC/FC	Quinn (1984) and Catala et al. (1999)

(*Continued*)

251

TABLE 8.1 *Continued*

Probe(s)[b]	Mode of Action	Method[c]	Key References
CellTracker Green CMFDA	Esterases	EF	Schumann et al. (2003)
CMAC-Leu	Aminopeptidases	EF	Schumann et al. (2003)
Nucleic Acid State			
AO	Different color when linked to DNA or other things	FC	Darzynkiewicz and Kapuscinski (1990)
DAPI (or SG1) destaining	Eliminates staining of non-DNA	EF	Zweifel and Hagström (1995)
Syto13, SybrGreen, TOTO TOPRO	Separates at least two populations based on DNA/RNA content	FC	Robertson and Button (1989), Li et al. (1995)
Double/Multiple Staining Protocols			
Double "live/dead" stains BacLight, NADS,....	Live/dead	EF/FC	Joux et al. (1997a), Grégori et al. (2001)
FDA/PI; BCECF_AM/PI	Activity/dead	FC	Yamaguchi and Nasu (1997)
EUB FISH probes/PI/DAPi	Active/dead/inactive/recently dead/live but inactive	EF	Williams et al. (1998), Howard-Jones et al. (2001)
Cell Division/Growth			
DVC	Live cells elongate when in presence of antibiotics	EF/FC	Kogure et al. (1979), Joux and Lebaron (1997)
Microcolony formation	Capability of producing a small colony	OM/EF	Jannasch and Jones (1959), Simu et al. (2005)
BrdU immunodetection	Incorporated into DNA instead of TdR	EF	Urbach et al. (1999), Pernthaler et al. (2002)

Miscellaneous

c-SNARF-1 AM	Intracellular pH	FC	Leyval et al. (1997)
16S rRNA probes	Detection of ribosomes	EF/FC	Amann et al. (1990), Bouvier and del Giorgio (2003)
AAs, glucose, thymidine MAR	Organic substrate incorporation	EF	Hoppe (1976), Tabor and Neihof (1984), Pedrós-Alió and Newell (1989), Karner and Fuhrman (1997), Cottrell and Kirchman (2000)

[a] Updated from Table 3 in Gasol and del Giorgio (2000). An expanded version of this table can be found in the web appendix (www.icm.csic.es/bio/personal/gasol/chapterKirchmanbook).

[b] TNT, 2-(p-iodophenyl)-3-p-(nitrophenyl)-5-phenyltetrazolium chloride; AO, acridine orange; DAPI, 4′,6-diamidino-2-phenylindole; EthBr, ethidium bromide; SG1, SybrGreen 1; CTC, 5-cyano-2,3-ditolyltetrazoliumchloride; XTT, sodium 3′-[1-[(phenylamino)-carbonyl]-3,4-tetrazolium-bis(4-methoxy-6-nitro)benzenesulfonic acid hydrate; Eth-D2, ethidium homodimer-2; PI, propidium iodide; Rh123, rhodamine123; DiBAC$_4$(3), bis(1,2-dibutylbarbituric acid) trimethine oxonol; DIOC, diethyloxacarbocyanine; DiOC$_6$(3), 3,3′-dihexyloxacarbocyanine iodide; 6CFDA, 6-carboxyfluorescein diacetate; FDG, fluorescein–galactopyranose; FDA, fluorescein diacetate; CMAC-Leu, 7-amino-4-chloromethylcoumarin, L-leucine amide, hydrochloride; CMFDA, 5-chloromethylfluorescein diacetate; BCECF-AM, 2′,7′-bis(2-carboxyethyl)-5-(and-6)-carboxyfluorescein acetoxy methyl ester; MAR, microautoradiography; DVC, direct viable count; BrdU, 5-bromo-2′-deoxyuridine; NADS, nucleic acid double-staining protocol.

[c] EF, epifluorescence microscopy; FC, flow cytometry; SPC, solid-phase cytometry; OM, optical microscopy; Spect, spectrophotometry.

surfaces (van Loodsrecht et al. 1987). Production of a capsule might be an indication of nutrient limitation (van Loodsrecht et al. 1987) and could also be a mechanism to enhance nutrient and substrate acquisition. Damaged and dead cells usually lack a capsule, show shrunken or partially deteriorated membranes, and may lack intracellular structures such as cytoplasm and ribosomes (Heissenberger et al. 1996). The capsules can be observed by transmission electron microscopy (TEM), or by a simpler microscopic method using Congo Red or Maneval's stain of cells transferred from filters onto a gelatin-covered slide (Storedegger and Herndl 2001).

Cellular Composition   The ionic composition of the bacterial cells can give hints of relative activity (Fagerbakke et al. 1999), and also the macromolecular composition of bacterial cells is not fixed but rather varies as a function of cell size, activity, physiological state, and even phylogenetic composition (Kjelleberg et al. 1993). There is a wide plasticity in composition within a given bacterial organism, which is evident, for example, during the starvation process (Roszak and Colwell 1987; Kjelleberg et al. 1993).

Some of these properties can be examined by flow cytometery. The amount of intracellular nucleic acids is determined from the fluorescence of stains that attach preferentially to DNA, RNA or both. While DAPI, and to a lesser extent, acridine orange, are the standard fluorochromes for determining bacterial abundance in plankton samples using epifluorescence, a new generation of blue-excitable stains, including Syto13, PicoGreen, TOTO, SybrGreen, and SybrGold, are increasingly being used, particularly in combination with flow cytometry. The latter technique allows the simultaneous determination of bacterial abundance and of the amount of fluorescence emitted, and thus the relative nucleic acid content, of individual cells (del Giorgio et al. 1996; Gasol and del Giorgio 2000).

One of the most interesting findings is that bacterioplankton cells tend to cluster into distinct fractions based on differences in the individual cell fluorescence (related to the nucleic acid content) and in the side and forward light scatter signal (see, e.g., Li et al. 1995). There are at least two major fractions: cells with high nucleic acid content (HNA cells) and cells with low nucleic acid content (LNA cells) (Robertson and Button 1989; Li et al. 1995; Gasol et al. 1999; Gasol and del Giorgio 2000; Lebaron et al. 2002). These cytometric populations are almost always present regardless of the sample and of the different protocols, stains, and type of cytometer used, suggesting that these fractions are not methodological artifacts (Bouvier et al. 2007). The general trend that emerges is that the HNA cells appear to be not only larger (Troussellier et al. 1999) but also more active, with higher growth rates than the LNA cells, and that changes in total bacterial abundance are often associated with changes in this fraction (Gasol and del Giorgio 2000; Lebaron et al. 2002). Several studies have supported the idea that HNA bacteria are the high-growth component of bacterial assemblages (Gasol et al. 1999; Troussellier et al. 1999; Gasol and del Giorgio 2000; Lebaron et al. 2001; Seymour et al. 2004), because they increase in abundance in mesocosm or dilution experiments, while LNA bacteria do not (Gasol et al. 1999; Zubkov et al. 2004).

However, other studies have shown that LNA also develop in marine dilution experiments (Jochem et al. 2004; Longnecker et al. 2006). Cell sorting experiments have yielded mixed results, some showing higher cell-specific activities of the HNA fraction (Servais et al. 1999; Lebaron et al. 2001, 2002), others showing substantial activity in the LNA fraction as well (Zubkov et al. 2004; Longnecker et al. 2005). The evidence to date suggests that the simple dichotomy of HNA = active and LNA = inactive is an oversimplification (Bouvier et al. 2007), and that LNA cannot routinely be considered to be inactive or dead.

Nucleoid Zweifel and Hagström (1995) proposed a de-staining procedure that stains with DAPI only those cells with a compacted nucleoid (NucC, nucleoid-containing cells). The standard DAPI procedure, they claimed, stains the DNA and also other cellular components. An isopropanol rinse following DAPI staining removes the nonspecific staining and leaves only the stain associated with DNA, which appears as a yellowish spot. Choi et al. (1996) proposed some modifications to the protocol of Zweifel and Hagström, yielding higher counts of nucleoid-containing cells.

Electron Transport Activity Probes The activity of the respiratory enzymes can be assayed using tetrazolium salts as indicators of bacterial respiration; the electrons produced from the electron transport chain in actively respiring bacteria reduce these salts from a soluble form to a dense, insoluble, cell-localized precipitate. These compounds compete with oxygen as electron acceptors. Depending on the type of salt used, the precipitate is colored or fluorescent. The first application of the method was with the probe 2-(p-iodophenyl)-3 (phenyl)-5-phenyltetrazolium chloride (INT) (Zimmerman et al. 1978). Rodriguez et al. (1992) introduced a modification of the method, based on the fluorogenic tetrazolium dye 5-cyano-2,3-ditolyltetrazolium chloride (CTC), which is reduced to a red-fluorescent formazan molecule that can be detected both with epifluorescence microscopy and with flow cytometry (del Giorgio et al. 1997; Sieracki et al. 1999; Sherr et al. 1999a). The techniques were initially thought to be applicable only to aerobic bacteria, but in anaerobic conditions, particularly during fermentation, the methods seems to work well (Smith and McFeters 1997), although nonbiological reduction may occur in sediments and other reducing environments (Smith and McFeters 1997).

Some authors have criticized the CTC method, on the basis of the potential cellular toxicity of the CTC, or because of the low numbers of CTC-positive cells often recorded in field studies (Ullrich et al. 1999; Servais et al. 2001). Ullrich et al. (1996) and Servais et al. (2001) both showed inhibitory effects of CTC on bacterial respiration and metabolic activity. On the other hand, Epstein and Rossel (1995) reported that CTC-stained and nonstained benthic bacteria grew equally well and that the ciliate *Cyclidium* sp. can survive on a diet of CTC-stained bacteria. Based on flow-cytometric sorting of red-fluorescent particles (assumed to be CTC-labeled bacteria) after incorporation of a radioactive tracer, Servais et al. (2001) concluded that the CTC dye is not suitable for the detection and enumeration of active bacteria, since the apparent contribution of these cells to total leucine uptake was relatively small, and similar results were later reported by Longnecker et al. (2005). However, recent flow-cytometric observations suggest that part of the discrepancy might be due to the fact that very active cells might explode as the result of the CTC-formazan granule formation, and free the granule to the medium. Cell sorting based only on red fluorescence of the granules would miss the most active cells, as those would have been destroyed before sorting (Gasol and Arístegui 2007). Overall, it would appear that the CTC method targets the cells with the highest respiration rates (Sherr et al. 2001; Smith and del Giorgio 2003; Sieracki et al. 1999).

Membrane Integrity and Functionality The bacterial cell membrane plays a key role in the functioning of the cell, and the state of the membrane thus provides much information on the general physiological condition of the cell (Brehm-Stecher and Johnson 2004). Several aspects of membrane function are of ecological interest, such as membrane integrity, energization and polarity, and the existence of pH and ionic gradients. Membrane potential and membrane integrity are two aspects of cellular function that should be a priori strongly linked, but studies have shown that this expectation is not always borne out, because the loss of membrane integrity is not always accompanied by loss of potential, and vice versa (Novo et al. 2000; del Giorgio and Bouvier 2002).

There are a number of different approaches to probe the state of the cell membrane (Table 8.1). Cell injury and damage results in increased membrane permeability that can be assessed using exclusion stains that bind to nucleic acids, such as propidium iodide (PI), TOPRO, or SYTOX Green (Veldhuis et al. 2001; Roth et al. 1997; Maranger et al. 2002). Because of their molecular weight and structure, these dyes can only enter cells with damaged or "leaky" membranes, and cannot enter cells with intact membranes (Haugland 2005). Cell injury and death is accompanied by loss of membrane potential and polarization. The latter can be assessed with the use of oxonols, negatively charged molecules that are excluded by cells with polarized membranes (Jepras et al. 1995). For example, $DiBAC_4(3)$ is an anionic membrane potential-sensitive dye that enters cells with depolarized plasma membranes and then binds to lipid-containing intracellular material. It has been used to differentiate live and dead cells (Mason et al. 1995) and to assess the physiological changes of bacteria along salinity gradients (del Giorgio and Bouvier 2002).

Intracellular Enzymatic Activity Various fluorescein esters have already been used in bacterial viability assays, including fluorescein diacetate (FDA: Chrzanowski et al. 1984; Diaper et al. 1992), carboxyfluorescein diacetate (CFDA: Dive et al. 1988; Miskin et al. 1998), and Chemchrome B (Clarke and Pinder 1998) and CMFDA (Schumann et al. 2003). These esters are uncharged and nonfluorescent, and are passively transported into cells. Internal enzymes (esterases) then hydrolyze them to fluorescein derivatives, which are charged and fluorescent. In viable cells, the membrane is impermeable to the charged molecules, which, therefore, can accumulate intracellularly and be detected by fluorescence. Organisms with leaky membranes do not retain the fluorescein. Thus, these assays target two aspects of cell function: intracellular enzymatic activity and membrane permeability. The method has some problems: many bacteria are unable to transport FDA, the fluorescence emission tends to be weak, and intracellular esterase activity is not always directly coupled to respiration (Diaper et al. 1992).

Uptake of Organic Substances Microautoradiography (MAR) is one of the earliest single-cell methods; the first reports of its use in aquatic microbial ecology were published in 1959 (Saunders 1959), and Brock (1967) used it to quantify the growth rate of a conspicuous freshwater bacterium. It was used in marine samples as early as 1976 (Hollibaugh 1976; Hoppe 1976). In this technique, microbial assemblages are incubated with a radiolabeled substrate and cells are then placed in contact with an autoradiographic emulsion, and subsequent exposure of the emulsion to the radioactive emissions produces silver grain deposits around the cells that are radioactive. Amino acids, acetate, glucose, along with leucine and thymidine were used in the initial studies, but recent studies have expanded to other substrates, such as chitin, N-acetylglucosamine, and even $^{14}CO_2$. Grain density may provide further information on the level of substrate uptake by the individual cells (Cottrell and Kirchman 2003; Sintes and Herndl 2006). In 1999, two independent studies showed that it was possible to combine this technique with fluorescence in situ hybridization (FISH), providing the first insights into the in situ single-cell activity of specific bacterial groups. A number of acronyms have been proposed for variations of the same approach, including MAR-FISH (microautoradiography–fluorescence in situ hybridization: Lee et al. 1999), STAR-FISH (substrate tracking autoradiography–fluorescence in situ hybridization: Ouverney and Fuhrman 1999), and Micro-FISH (microautoradiography–fluorescence in situ hybridization: Cottrell and Kirchman 2000).

DNA Synthesis and Replication An alternative to using radioactive compounds for measuring bacterial DNA synthesis is to use 5-bromo-2′-deoxyuridine (BrdU), which is a thymidine analog. Cells that are synthesizing DNA, and thus in the process of cell division, incorporate BrdU into DNA, and can be detected by immunofluorescence using anti-bromodeoxyuridine monoclonal antibodies or isolated by immunochemical capture using antibody-coated paramagnetic beads (Urbach et al. 1999; Borneman 1999; Hamasaki et al. 2007). This protocol has been also combined with CARD-FISH to allow the identification of individual growing cells (Pernthaler et al. 2002).

Direct Viable Count (DVC) Method The direct viable count method was first described by Kogure et al. (1979) and was based on incubating samples with the antibiotic nalidixic acid, which inhibits cell division. Active cells continue to grow but not to divide, elongating in the process, and these elongated cells can be detected microscopically (or by flow cytometry). The method has been used extensively to assess activity and viability of bacteria both in culture and in the environment (Roszak and Colwell 1987; Kogure et al. 1987; Yokomaku et al. 2000). The treatment generating elongated bacteria now includes the addition of glycine (Yokomaku et al. 2000) and an antibiotic cocktail (Joux and Lebaron 1997).

> There are currently a variety of techniques that target key categories of bacterial function and physiology at the single-cell level:
>
> - **Morphological integrity**, including the presence of capsules and intracellular contents (determined using transmission electron microscopy)
> - **Macromolecular composition**, including the cell-specific DNA and RNA contents using flow cytometry and fluorescence in situ hybridization (FISH)
> - **Respiratory activity**, determined using redox-sensitive dyes such as CTC, combined with microscopy or flow cytometry
> - **Status of the cell membrane**, using exclusion dyes such as PI or TOPRO, and potential-sensitive dyes such as oxonols, combined with flow cytometry
> - **Internal enzymatic capacity**, determining internal esterase or galactosidase activities
> - **Substrate uptake**, using radiolabeled organic molecules
> - **Synthesis and replication of DNA**, using microautoradiography or the incorporation of nucleotide analogs such as BrdU and immunofluorescence
> - **Cellular growth**, by inhibiting cell division and following the resulting cell elongation using microscopy or flow cytometry
> - **Mixed approaches**, which combine two or more of the above in a single protocol

Combined Approaches There are increasing attempts to combine approaches, and a number of recent combinations have been proposed. Molecular probes introduced the BacLight Live/Dead kit, which is a mixture of a DNA stain (Syto9) that permeates into cells and an exclusion stain (PI). Both are added at the same time and interact so that live cells are stained green and dead cells are stained red. These cells can be counted by epifluorescence microscopy or flow cytometry. The latter allows the separation of cells into categories: green, green plus orange–red, and orange–red cells, which correspond to live, damaged, and dead cells, respectively (Grégori et al. 2001). Recently, Manini and Danovaro (2006) have suggested combining EthD-2 and PI in sediments. EthD-2 has a much higher (i.e., double) molecular weight, and is thus a good candidate to overcome problems encountered with the use of propidium iodide (PI).

Other combination approaches have also been proposed. Williams et al. (1998) combined staining with PI (as a membrane integrity marker), FISH, and DAPI; this protocol differentiates cells that are dead (positive signal due to propidium iodide (PI+), and a negative FISH signal (FISH−), cells that are dead but had been active until recently (PI+, FISH+), cells that are alive and active (PI−, FISH+), and cells that are inactive but not dead (PI−, FISH−). It is, in theory, possible to discriminate dormant bacteria from the difference in the number of NucC cells, and the number of "live" cells determined by a probe that does not permeate membranes (see, e.g., Gasol et al. 1999; Luna et al. 2002). Manini and Danovaro (2006) proposed the destaining of SybrGreen I in a way similar to the destaining of DAPI. They combined this procedure with staining by PI or EthD-2, which allows dead cells (without a visible nucleoid region) to be distinguished from red-stained bacteria with a still-visible nucleoid (NucC) region. These cells (PI+, NucC+) are considered to be either cells that died recently or cells competent for transformation (having large "holes" in the membrane to allow DNA entry). Pirker et al. (2005) combined MAR and PI, to simultaneously assess the metabolic activity and viability of individual bacterioplankton cells in the North Sea.

Operational Categories of Single-Cell Activity

There has been much debate concerning the effectiveness of some of the methods described above, particularly in the context of natural bacterial communities characterized by low levels of activity and also by relatively mild sources of stress, at least compared to laboratory conditions (Ullrich et al. 1996; Karner and Fuhrman 1997; Kell et al. 1998; Colwell 2000; Smith and del Giorgio 2003). The current methods only allow us to place bacteria in very broad physiological categories, that is, active or inactive in substrate uptake, or having and intact or injured membranes. The categories themselves are often not well defined, because all the methods have thresholds of detection, and the thresholds themselves vary with the different protocols used and sample conditions. The physiological categories obtained are thus operational and in no way absolute (Smith and del Giorgio 2003). In addition, there is likely a large range of activity or physiological states within each of these categories. For example, within the cells that are scored positive for substrate uptake, there is probably a large range in the actual rates of substrate uptake (see, e.g., Sintes and Herndl 2006). Likewise, there is often a large range in the level of red fluorescence of cells that are scored as positive to CTC reduction within any given sample (Sherr et al. 1999a; Posch et al. 1997).

There are several approaches that link individual cell activity to phylogenetic composition, including the following:

- **Flow sorting of specific cell fractions** (radiolabeled with substrates or stained with specific dyes) followed by molecular analysis of the DNA of the sorted fractions
- **Autoradiography** (radiolabeled substrate uptake) combined with FISH to determine the identity of substrate-active cells (Micro-FISH, MARFISH)
- Detection of DNA-synthesizing cells through the **incorporation of BrdU followed by immunocytochemical** detection of the target cells with further DNA analysis, or FISH identification

Universe of DAPI+ particles

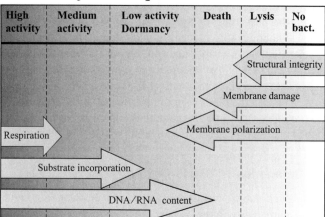

Figure 8.2 Conceptualization of a continuum of physiological states that is targeted at different levels by different methods. The DAPI-positive particles can in theory be classified into the upper-line categories, but can also be classified depending on each method sensitivity and target cellular process.

Other problems arise from the fact that there is not always a tight correlation between categories. For example, a dormant cell may be intact and viable but show no signs of activity whatsoever, whereas a cell that is badly damaged may still show measurable signs of remnant metabolism. In this regard, Smith and del Giorgio (2003) suggested that the fact that current methods target different cell functions and have different intrinsic thresholds, and are thus often not comparable, should be viewed as an advantage rather than as an obstacle, because when taken collectively they provide complementary information on the physiological structure of bacterioplankton. Figure 8.2 offers a simplified view of how the different categories of methods cover different areas of the physiological spectrum of bacterioplankton assemblages. These approaches should be viewed not as independent but rather as providing a nested structure, which allows one to differentiate at least five basic categories of cells: (1) intact, active and growing; (2) intact, active but without apparent growth; (3) intact but no apparent sign of activity; (4) injured, with or without signs of activity; (5) dead. These categories roughly agree with those proposed by other authors (Barer 1997; Kell et al. 1998; Weichart 2000; Nebe-von Caron et al. 2000). Dormant cells would fall somewhere within categories 2 and 3, whereas extremely slow growing cells might fall in categories 2, 3, and 4. We use this basic classification to describe the existing data in the sections below.

Regulation of Physiological Structure of Marine Bacterioplankton

In this section, we briefly review some of the main factors that influence the level of metabolic activity and the physiological state of marine bacterioplankton cells, as well as the factors that influence the persistence and loss of the different physiological fractions of the community. Chapter 10 explores the factors that influence growth at the community level. Here, we focus on the factors that influence the physiological state and the metabolic activity of single bacterial cells.

Factors Influencing Physiological State of Bacterial Cells in Marine Ecosystems

Slow growth and dormancy are certainly not the only physiological states of interest from an ecological point of view. Cells are subject to a wide range of environmental characteristics that either enhance metabolic activity or are deleterious and generate stress, and that may ultimately cause cell death. The response of cells to stress depends greatly on what cellular function is investigated and might vary specifically (Joux et al. 1997a; Agogué et al. 2005). For example, exposure to heat or UV radiation had very different impacts on various cellular functions of *E. coli* (Fiksdal and Tryland 1999). López-Amorós et al. (1995a, b) found that stress and mortality caused by starvation did not result in an increase in membrane permeability, and was thus difficult to detect using traditional exclusion stains. Likewise, the capacity to detect bacteria using FISH is extremely sensitive to osmotic stress and to carbon starvation, but not to cold and other types of physical and chemical stress (Tolker-Nielsen et al. 1997; Oda et al. 2000). There is thus no single or simple response to stress, and loss of a certain function does not necessarily imply loss or degradation of all other functions, except in the most extreme cases.

The availability of energy, mainly in the form of organic carbon (or as certain inorganic compounds or light for some bacteria), is perhaps one of the key factors influencing bacterial singe-cell activity in aquatic ecosystems (Kjelleberg et al. 1993; Morita 1997; Azam 1998; and see Chapter 10). As discussed above, the response of cultured bacteria to extreme carbon limitation has been extensively studied, but the question remains: What is starvation from the viewpoint of a marine bacterium, and is extreme starvation the most common state for marine bacteria? Although much of the ocean is extremely dilute and oligotrophic, this environment is still not analogous to the conditions of complete nutrient deprivation created in laboratory cultures used to study the starvation response. (Fegatella and Cavicchioli 2000). Substrate concentrations in marine waters are indeed extremely low, due in part to low supply rates, and but also as the result of high turnover of these substrates, precisely due to extremely efficient bacterial consumption. It has often been assumed that bacteria in oligotrophic marine systems oscillate between scenarios of feast or famine (Azam 1998) in terms of substrate availability, but the reality for most oceanic bacteria lies probably somewhere in between. In this regard, Ferenci (2001) recently introduced the notion of "hungry" bacteria: those that experience suboptimal resource conditions that are somewhere between feast and extreme famine. Hungry bacteria develop a whole set of adaptations that are in fact distinct from those that characterize the more extreme starvation response.

We do not know to what extent marine bacteria are "hungry" as opposed to "starved," but it is clear that there is no absolute threshold of carbon limitation/starvation that applies to all bacterial types, so a given level of organic substrate availability will trigger the starvation response in some bacteria and not in others (Ferenci 2001). There is increasing evidence that phylotypes within broad phylogenetic groupings may share functional and metabolic traits such as patterns of substrate utilization or intrinsic levels of cellular activity (Cottrell and Kirchman 2000; Zubkov et al. 2001; Yokokawa and Nagata 2005).

Exposure to high substrate concentrations can cause cell death in cells that are dormant and in starvation-survival mode (Koch 1997). In fact, typical oligotrophic marine bacteria, such as *Sphingomonas* spp. (Schut et al. 1997) and *Pelagibacter* (Rappé et al. 2002), are unable to grow in rich media, most likely due to an inability to deal with high respiration rates and the associated oxidative stress. In marine environments bacterial cells need to cope not only with low ambient concentrations of organic substrates and nutrients, but also, and perhaps more importantly, with highly intermittent supply at very different spatial and temporal scales. Some of these fluctuations may exceed the protective mechanisms of the cells, with deleterious or lethal consequences to the microorganisms (Koch 1997).

Cells lacking key inorganic nutrients might still consume and respire organic matter, but biomass accumulation and cell division may be severely impeded under nutrient starvation. This should be reflected in the distribution of bacterial cells with different physiological states, as cells would be actively respiring but unable to synthesize nucleic acids. It has been noted that while carbon starvation often leads to cell miniaturization (Morita 1997), acute nutrient limitation may lead to an increase in cell mass (Kjelleberg et al. 1993). Elemental composition may shift according to the physiological state of cells. For example, bacteria typically have more nitrogen and phosphorus per unit carbon than algae, but unicellular cyanobacteria often have higher carbon in relation to heterotrophic microorganisms, and this has been interpreted as storage of carbon under nutrient deficiency (Heldal et al. 2003). Carbon availability associated with inorganic nutrient limitation can translate into the accumulation of excess reserve polymers, greatly increasing cell size. In the "Winnie-the-Pooh" hypothesis (Thingstad et al. 2005), increase in cell size might be seen as a strategy to "dominate" the system and simultaneously optimize uptake and minimize predation.

There are a number of direct as well as indirect effects of light, and particularly UV, on bacterial physiology. Direct effects include damage to DNA, RNA, and cell membranes (Herndl et al. 1993; Fiksdal and Tryland 1999). Indirect effects include UV-mediated generation of toxic radicals and hydrogen (Xenopoulos and Bird 1997); increased substrate availability through photochemical degradation of refractory dissolved organic carbon (DOC) into more bioavailable forms (Kieber et al. 1989; Moran and Zepp 1997); reduction in viral abundance and infectivity (Suttle and Chen 1992; Noble and Fuhrman 1997; Wilhelm et al. 1998), perhaps countered by induction of the lytic cycle in lysogenic bacteria (Freifelder 1987; Maranger et al. 2002); and declines in the rate of flagellate grazing on bacteria and thus influence on the loss rates of bacteria (Sommaruga et al. 1996). Not all bacterioplankton groups seem to be equally susceptible to UV damage (Joux et al. 1999; Arrieta et al. 2000; Agogué et al. 2005; Alonso-Sáez et al. 2006), and UV appears to have a greater effect on the more active cells (Rae and Vincent 1998; Maranger et al. 2002; Alonso-Sáez et al. 2006). Cell-specific activity seems to be more susceptible to UV damage than membrane structure and potential (Alonso-Sáez et al. 2006).

Membranes, nucleic acids, and certain enzymes are sites prone to damage by heat (Teixeira et al. 1997; Fiksdal and Tryland 1999), but a typical marine bacterium will be rarely exposed to very high temperatures. Cold shock has been shown to generate sublethal injuries and force cells into a dormant state (see, e.g., Kjelleberg et al. 1987; Roszak and Colwell 1987; Munro et al. 1989). Whatever the direct mechanism involved at low temperatures, the general pattern is that more cells enter dormancy, and there are fewer cells capable of forming colonies or reducing CTC (see, e.g., Smith et al. 1994). Temperature and organic carbon availability interact, such that the inhibitory effects of temperature are modulated in the presence of higher substrate concentrations (see, e.g., Wiebe et al. 1992; Pomeroy and Wiebe 2001).

Osmotic shock has also been shown to generate sublethal injuries and force cells into dormancy. The capacity to detect bacteria by FISH, for example, has been shown to be extremely sensitive to osmotic stress (Tolker-Nielsen et al. 1997; Oda et al. 2000). del Giorgio and Bouvier (2002) found in the 0–10 salinity range of the Choptank estuary a dramatic decline in hybridization with FISH probes, a decrease in bacterial production and growth efficiency, and an increase in cells with depolarized and injured membranes. Painchaud et al. (1995) had already identified salinity as one of the main factors related to cell mortality in the St. Lawrence estuary. Changes in the structure and functioning of the microbial food web along gradients of salinity (Schultz and Ducklow 2000; del Giorgio and Bouvier 2002; Bouvier and del Giorgio 2002; Casamayor et al. 2002; Gasol et al. 2004; Kirchman et al. 2005) indicate that this parameter is a variable structuring both the composition and physiology of bacterioplankton.

Finally, there are a wide range of chemical substances that may influence bacterial activity and physiology in marine waters. These include antibiotic and toxic compounds released by bacteria (Long and Azam 2001) and by other planktonic organisms (Adolph et al. 2004), allelopathic substances (Gross 2003), or substances associated with quorum sensing (Miller and Bassler 2001). Contaminants may also interfere with marine bacterial activity (Price et al. 1986). Other factors may influence bacterial physiology, such as barometric pressure, low oxygen, or even localized anoxia (e.g. in particles or guts), and turbulence (e.g. Malits et al. 2004), but, as far as we know, their effects have not been well explored to date.

Factors Influencing Loss and Persistence of Physiological Fractions

Grazing Predation is a major loss mechanism of bacterial cells in all marine systems (see Chapter 11). There is now strong evidence that protistan grazing can be highly selective in terms of the physiological state of the cells. In general, cells that are more active tend to be selectively cropped by protistan grazers (del Giorgio et al. 1996; Šimek et al. 1997; Pernthaler et al. 1997; Tadonléké et al. 2005). The underlying basis of this selectivity is partly cell size, since there is in general a positive relationship between cell size and activity in bacterioplankton

(Gasol et al. 1995; and see below), but is not restricted to it. Grazing preferentially removes active cells, but may produce injured and dead cells, through incomplete digestion in protistan vacuoles and passage through the gut of metazoans. Incomplete feeding also results in cell fragments (Nagata and Kirchman 2000; Nagata 2000a), or incompletely digested egesta (picopellets, Nagata 2000b) that can potentially be counted as normal bacteria by standard protocols.

Cell Size Small size may facilitate survival by being a refuge from predation (Jürgens and Güde 1994). Several empirical studies have shown a preference of larger bacterial prey by cultured and natural assemblages of flagellates (Andersson et al. 1986; González et al. 1990; Jürgens and Güde 1994) and ciliates (Fenchel 1980; Epstein and Shiaris 1992; Šimek et al. 1994). Size, thus, can have a large impact on the fate of bacterial production because it determines susceptibility to grazers and viruses (see also Chapter 11).

Viral Lysis Viral infection has been proposed as the main mechanism generating empty or "ghost" bacterial cells that retain external structure but lack internal material (Heissenberger et al. 1996). There is now increasing evidence that viral infection is selective, towards cells that are either more sensitive, or with higher growth and metabolic rates (Waterbury and Valois 1993; Middelboe et al. 2001; Bouvier and del Giorgio 2007), and, like grazing, viral lysis preferentially removes cells from the most active categories and moves them to the dead or detrital categories. About 35 percent of all bacteria may be infected at a given time, and lysogeny may influence the growth rate and the metabolism of the host cells (Weinbauer 2004). There are feedbacks between processes, so that environmental stress may trigger the lytic cycle in lysogenic bacteria (Weinbauer 2004).

Cell Motility This is another factor that plays a role in both cellular activity and loss. Motility is widespread in marine bacteria (Blackburn et al. 1998; Grossart et al. 2001). It allows a bacterium to track patches of organic matter (Grossart et al. 2001), and, in an environment with a complex microscale structure (sensu Azam 1998), motility should increase the chance of growth and survival (Blackburn et al. 1997). On the other hand, motility increases the probability of encounter with a phage (Bratbak et al. 1992) or with a protist (González et al. 1993; Monger et al. 1999), although very fast bacteria may also be able to avoid predation (Matz and Jürgens 2001).

Cell Degradation The persistence of dead and injured cells in the water column has not been well studied (Mason et al. 1986). Due to their size, these cells remain suspended, and, due to the relative recalcitrance of their membrane components (see, e.g., McCarthy et al. 1998), dead and injured cells, as well as cell fragments, may persist for a long time (Mason et al. 1986). Remains of bacterial cell membranes and cellular fragments have been proposed as the main source of dissolved organic nitrogen in the ocean (McCarthy et al. 1998). The potential interaction between abiotic factors (UV, temperature, water chemistry) and grazing in determining the loss of dead and injured cells has not been well explored to date.

> **Regulation of the physiological structure of bacterioplankton communities** has three main components:
>
> - *Environmental factors* that influence the individual level of metabolic activity and cell integrity and damage, such as substrate and nutrient availability, UV, and temperature
> - *Physical and biological factors* that influence the persistence and loss of the various physiological fractions, such as selective grazing and viral infection, and selective degradation
> - *Intrinsic phylogenetic characteristics* that modulate the response of different bacterial strains to the above factors

DISTRIBUTION OF SINGLE-CELL CHARACTERISTICS IN MARINE BACTERIOPLANKTON ASSEMBLAGES

In this section, we discuss the data generated over the past three decades concerning the distribution of single-cell characteristics in marine bacterioplankton assemblages. We have assembled a comprehensive dataset based on published and unpublished observations of bacterioplankton single-cell activity and physiological state from studies that have applied some of the techniques listed in Table 8.1.

Distribution of Single-Cell Activity and Physiological States in Marine Bacterioplankton

For each category of assay, we present the average proportion of cells scored as positive relative to the total for all the studies that have applied the technique, together with the range of values obtained in the different studies as well as the resulting standard error and distribution quartiles (Fig. 8.3). Most of the published studies have applied a single technique, although there are a handful of marine studies that have combined several approaches. The latter have been summarized in Table 8.2 and are discussed in the following section.

Figure 8.3 shows a clear ordination of methods based on the average proportion of cells scored as positive when applied to marine bacterioplankton assemblages. Although almost all techniques yield a wide range of results, most results for any given technique cluster in a specific range (as evidenced from the standard error around the mean). At one end of the spectrum, methods that target activity tend to yield relative low proportions of reactive cells. For example, researchers find on average less than 20 percent of CTC-positive, esterase-positive, and DVC-positive cells in marine samples. At the other extreme are methods that target membrane integrity or macromolecular composition, such as BacLight, FISH (RNA) or high-DNA cells, which fall consistently in the range between 40 and 60 percent of the total cells. Other approaches, such as MAR (substrate uptake) and ultrastructural analysis

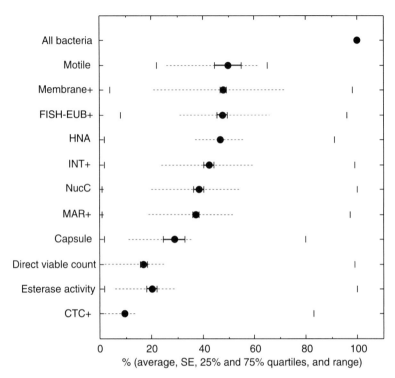

Figure 8.3 The percentage of reactive cells determined by different methods for marine plankton communities (database available at www.icm.csic.es/bio/personal/gasol/chapterKirchmanbook). The average (filled-in circles), standard error (SE: solid lines), range between the 25 and 75 percent quartiles (dotted lines), and ranges of values (spaces between the two vertical lines) are presented. "Membrane +" indicates all membrane polarity and potential probes, including double-staining protocols.

(presence of capsule), generally fall within the range of 20–40 percent. These results would suggest that, on average, marine bacterioplankton assemblages have a relatively high proportion (>50 percent) of cells that appear to have intact membranes and ultrastructural integrity, and detectable levels of DNA and RNA, but that only a fraction of these apparently intact and healthy cells are active in substrate uptake, and that an even smaller fraction have high levels of metabolic activity, expressed as cell growth, cell division, or detectable respiratory activity. Moreover, these results suggest that, on average, a nontrivial proportion (>20 percent) of marine bacterioplankton cells may be either injured or dead.

The pattern in Figure 8.3 further suggests that certain aspects of cell physiology may be more coupled than others. For example, DNA and RNA contents and membrane integrity result in similar proportions of positive cells, whereas respiratory, enzymatic, and substrate uptake activity appear generally clustered in the lower end of the spectrum. The fact that a relatively large fraction of intact cells appears to have extremely low or nondetectable metabolic rates agrees well with recent estimates

TABLE 8.2 Overview of Studies Comparing Different Activity Protocols

	Site			Method (and %)[a]			
Marine Studies							
Hoppe (1976)	Kiel Bay	AO>	MAR+>	cfu			
		100	41	1			
Meyer-Reil (1978)	Kiel Bay	AO>	INT+>	cfu			
		100	31	2			
Kogure et al. (1979)	Pacific Ocean	AO>	DVC>	cfu			
		100	7	0.0001			
Tabor and Neihof (1984)	Chesapeake Bay	AO>	INT+≈	MAR+>	DVC		
		100	46	43	32		
Newell et al. (1986)	Chesapeake Bay	TC>	MAR+≈	INT+>	DVC		
		100	61	56	35		
Choi et al. (1996)	Oregon coast	DAPI>	NucC>	Live+>	CTC+		
		100	25	14	7.5		
			44	21	19		
Karner and Fuhrman (1997)	California coastal	DAPI>	FISH>	MAR+>	NucC>	CTC+	
		100	56	49	29	1	
Joux and Lebaron (1997)	Mediterranean coast	DAPI>	DVC>	CTC+>	CFU		
		100	12	3	0.8		
Gasol et al. (1999)	Mediterranean coast	DAPI>	Live+	NucC>	HNA		
		100	70	68	60		
Catala et al. (1999)	Mediterranean coast	DAPI>	Esterase+>	DVC~	CTC>	cfu	
		100	8	6	5	1	

(*Continued*)

TABLE 8.2 *Continued*

Site	Method (and %)[a]					
Bernard et al. (2000b): Mediterranean coast	SG> 100	CTC+≈ 2	cfu 1.5			
Grossart et al. (2001): California coast	DAPI> 100	Motile> 60	NucC> 33	CTC+ 9		
Sherr et al. (2002): Atlantic shelf	DAPI> 100	NucC> 51	CTC+ 6			
Storedegger and Herndl (2002): North Sea	DAPI> 100	Capsule> 35	CTC 19			
Schumann et al. (2003): Estuarine	DAPI> 100	PI-> 74	Esterase+ 12	CTC+ 10		
Baltic Sea	DAPI> 100	67	8	15		
Schumann et al. (2003): Global	DAPI> 100	EthD2-> 87	Sytox-~ 71	PI-> 70		
Yakimov et al. (2004): Ross Sea	DAPI> 100	FISH+> 87	CTC+ 11			
	100	18	2			
Davidson et al. (2004): Tasmanian coastal	DAPI> 100	PI-> 30	MPN> 23	CFDA+≈ 13	CTC+≈ 12	
Zampino et al. (2004): Mediterranean coast	DAPI> 100	Live+≈ 4	DVC≈ 4	CTC+≈ 2	CTC+ 13	
Pearce et al. (2007): Antarctic coast	DAPI> 100	Esterase+> 37	Live+ 14		cfu 1	Est.+> 10	leu AMP 6

Other Aquatic Studies

Reference	Environment						
Maki and Remsen (1981)	Lakes	AO> 100	INT+≈ 10	DVC 8			
Quinn (1984)	Lakes	AO> 100	INT+≈ 26	DVC> 22			
Porter et al. (1995)	Lake Windermere	AO> 100	CFDA~ 6	CTC> 4	FDA+> 5	cfu 0.4	
Porter et al. (1995)	River Mersey	AO> 100	Chem B> 80	DVC> 71	DVC> 0.6	cfu 0.2	
Yamaguchi and Nasu (1997)	Rivers	DAPI> 100	Esterase+> 63	CTC+ 25	cfu 56		
Kawai et al. (1999)	Purified water	DAPI> 100	6CFDA> 20	DVC≈ 5	CTC+> 5	μcol> 1	cfu 0.001
Yokomaku et al. (2000)	River water	DAPI> 100	Esterase+~ 34	DVC> 17-45	CTC+> 20	cfu 12	
Berman et al. (2001)	Lake Kinneret	DAPI> 100	Live+≈ 9	NucC~ 8	CTC+ 5		
Vollertsen et al. (2001)	Wastewater	AO> 100	DAPI> 75	Live+> 60	MAR+> 40	CTC+ 20	
Luna et al. (2002)	Marine sediments	AO> 100	Live+> 28	NucC> 4	DVC 2		
Schumann et al. (2003)	Freshwater	DAPI> 100	PI-> 65	Esterase+ 9	CTC+ 11		
Freese et al. (2006)	River	DAPI> 100	FISH+≈ 29	Membr+> 27	Esterase+> 5	CTC+ 2	

[a]The methods are ordered according to the percentage of the reference value (first column). A qualitative appreciation is given by symbols =, ≈ or > (indicating equal, similar or higher value). Acronyms are as in Table 8.1, with the addition of cfu, colony-forming units; MAR, microautoradiography; SG, SybrGreen; AB, antibody-positive; leu-AMP, leucine aminopeptidase; B-Gal: β-galactosidase; μ-col, microcolony formation.

of the proportion of marine bacterioplankton cells that are dividing, based on the incorporation of BrdU, which are in the range of less then 1 to 10 percent in coastal assemblages (Urbach et al. 1999; Borneman 1999; Pernthaler et al. 2002).

Simultaneous Determination of Several Aspects of Single-Cell Activity and Physiology

The average distribution of single-cell characteristics presented in Figure 8.3 is derived mostly from studies that have applied individual techniques in different systems, and thus the pattern could be driven to some extent by biases in terms of the origin and characteristics of the samples. To what extent does the average distribution of single-cell characteristics presented in Figure 8.3 represent the real sequence of physiological states that could be found within a single sample if multiple techniques were applied simultaneously? In Table 8.2, we have summarized the marine studies that combined and compared several approaches in the same samples. For reference, we have included in this table some additional studies that simultaneously assessed multiple aspects of single-cell activity in other aquatic ecosystems.

The data summarized in Table 8.2 show a pattern of distribution of single-cell characteristics within individual samples that is similar to the average cross-sample pattern shown in Figure 8.3. Most studies that combined several approaches found relatively high proportions of cells that appear intact in terms of membrane integrity, ultrastructure, and macromolecular component, but a much smaller fraction that appears active in terms of respiration, intracellular enzyme activity, or substrate uptake. Some of these studies found some coherence between the different aspects of single-cell physiology. For example, Karner and Fuhrman (1997) found that NucC values were similar to those from autoradiography and FISH, although these two aspects of cell function were not significantly correlated. Likewise, Gasol et al. (1999) found that the number of HNA bacteria measured by flow cytometry was similar to the number of NucC cells, and the rates of change of both groups were similar. Davidson et al. (2004) and Schumann et al. (2003) found similar number of CTC-positive and CFDA-positive or esterase-active cells, suggesting that these aspects of cell function might be more tightly coupled.

Some of these studies have also revealed apparent discrepancies. For example, Karner and Fuhrman (1997) reported extremely low numbers of CTC-positive cells, in spite of relatively high proportions of MAR and FISH-positive cells. Likewise, Grossart et al. (2001) found that between 22 and 64 percent of bacteria were motile in coastal marine sample. Larger cells were more motile than smaller cells, and there was a correlation between colony-forming units (CFU) and percentage of cells that were motile, but this percentage was about twofold higher than the percentage of NucC-containing cells, and about sixfold higher than the proportion of CTC-positive cells. Since motile bacteria must be respiring, these differences must be due to the different thresholds of detection of each method. Interestingly, Grossart et al. (2001) also found that motile and CTC-positive percentages tended to converge (91 percent for CTC-positive and 85 percent for motile cells) upon the addition of organic substrates, suggesting greater uncoupling of these processes under growth-limiting conditions.

Some of these apparent discrepancies may be methodological, but others may be a reflection of the complexity of the coupling between various aspects of cellular function. A number of laboratory and regrowth studies have revealed complex patterns of response to starvation and stress at the single-cell level. For example, Joux et al. (1997b) described the response of a marine bacterium to prolonged starvation, in which total cell counts and DNA content remained relatively stable during the first 2 months of starvation, whereas there was a clear sequence of changes in other aspects of cell functioning, starting with loss of culturability, followed by a decline in respiratory and enzymatic capacity, and ending in a loss of membrane integrity. Others have shown similar responses following exposure to a variety of stresses in pure bacterial cultures (Fiksdal and Tryland 1999; Caro et al. 1999; Suller and Lloyd 1999; Maalej et al. 2004). These studies collectively confirm that there is a hierarchy of physiological states in plankton assemblages, such that, in practical terms, scoring a cell as "live" does not imply activity, and, conversely, that scoring a cell as "dead" or "injured" does not imply complete inactivity.

Patterns in Distribution of Single-Cell Activity and Physiology Along Marine Gradients

Figure 8.4 shows the relationship between the abundance of seven specific bacterial fractions (MAR-positive, CTC-positive, HNA, membrane-intact (membrane-positive), membrane-impaired, DVC, NucC and esterase-positive) and total bacterial abundance. These relationships include all the individual data points that were used to construct Figure 8.3. The relationships extend over three orders of magnitude in total bacterial abundance (10^5–10^7 cells/mL), thus covering much of the range in bacterial abundance in aquatic waters. Figure 8.4 suggests different patterns in the relative abundance of these different fractions: the log-slopes of the HNA and membrane-positive relationships are not significantly different from unity, suggesting that although there is a large degree of scatter in each of these relationships, the relative contribution of these fractions does not change systematically along gradients of increasing bacterial abundance and thus of total system productivity. In contrast, the log-slopes of the CTC-positive, DVC, NucC and MAR-positive relationships are significantly greater than unity, suggesting that the relative importance of these fractions tends to increase with increasing overall abundance and probably system productivity. The proportion of esterase-positive cells tends to decrease with total bacterial abundance, but the database is not large and seems to be biased towards nutrient-rich aquatic ecosystems.

There appears to be a relative constancy in the proportion of intact versus damaged cells, and of cells with high nucleic acid contents and nucleoid, and a correlation between these fractions. Further, the relative constancy of these fractions suggests that the processes that result in bacterial cell death and damage, as well as in distribution of DNA, are not only present along the entire marine gradient, but are exerting a similar effect throughout. In contrast, the apparent increase in metabolically active

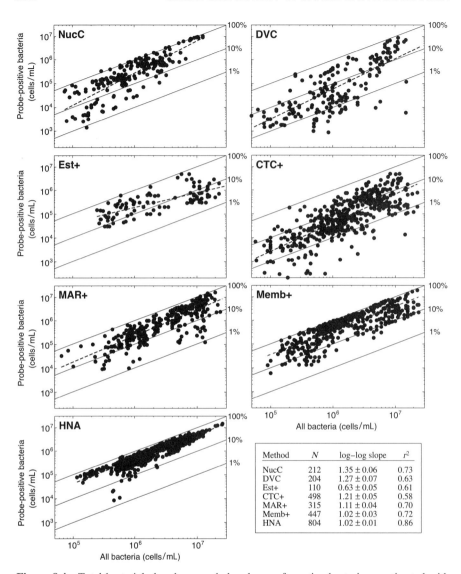

Figure 8.4 Total bacterial abundance and abundance of reactive bacteria as estimated with the most frequently used activity probes or methodologies. The dashed lines are the regression lines. Est+ = esterase+; Memb+ = Membrane+.

bacteria suggests that, along this same gradient, some processes, probably linked to substrate availability, enhance single-cell activity. For example, in any given system, the proportion of CTC-positive bacteria tends to be higher for particle-associated than for free-living bacteria (Sherr et al. 1999a, 2002; Yager et al. 2001). Further, these patterns agree with previous studies reporting that both the number of

CTC-positive cells and the percentage of CTC-positive cells increase with system productivity (del Giorgio and Scarborough 1995; Lovejoy et al. 1996; Søndergaard and Danielsen 2001), and that the abundance of CTC-positive cells is correlated with bacterial production (del Giorgio et al. 1997; Lovejoy et al. 1996; Sherr et al. 1999a) and cell growth (Choi et al. 1996, 1999; Sherr et al. 1999b). The apparent increase in the proportion of highly active cells could also potentially be due to a covariation with other physical factors, such as temperature, which may vary independently from nutrient and substrate availability. However, there are reports of extremely high numbers of substrate-reactive and percentage of CTC-positive cells in very cold but nutrient-rich waters (Yager et al. 2001; Schumman et al. 2003; Freese et al. 2006), so temperature itself does not appear to the main reason for the low apparent proportion of highly active cells in low-productivity systems. Arístegui et al. (2005) found 8 percent of CTC-positive prokaryotes in mesopelagic samples, a fraction similar to that in much warmer surface waters. In addition, the formazan granules of CTC-positive cells in the mesopelagic zone were 30 percent larger than those in cells at the surface, suggesting that the per cell respiration rate was higher in the mesopelagic zone. Likewise, Herndl et al. (2005) reported relatively high proportions of active *Archaea* in deep oceanic waters.

To summarize, our analysis suggests that the relative proportion of intact and live cells does not vary significantly across marine gradients, but that the proportion of active cells (in terms of respiration, enzyme activity, or substrate uptake) tends to increase along the same gradients. These observations suggest (1) that the proportion of inactive or dormant cells declines systematically along a gradient of increasing bacterial abundance and thus of system productivity, (2) that cell death and injury may be regulated by factors that are relatively independent from those that regulate cell activation and inactivation, and (3) that there may be changes in the selective loss processes of specific physiological fractions along these gradients, and in particular, a relatively higher removal of the more active cells in the least productive systems.

Distribution of Activity and Growth Among Bacterial Size Classes

Standard allometric relationships suggest that the smallest bacterial cells should have the highest specific metabolic and growth rates, and early results supported this pattern (Fuhrman 1981), but later studies have shown evidence that the smaller cells tend to have lower specific metabolic rates (Billen et al. 1990; Bjørnsen et al. 1989). For example, Gasol et al. (1995) showed that most CTC-positive cells in a coastal marine assemblage were in the medium to large size range. Their calculations of specific growth rates for each of seven size classes between 0.008 and 0.30 μm^3 showed that growth rates increased from about 0.3/d to 1.4/d and biomass turnover times to decrease with increasing cell size, from about 2.1 days for small bacteria down to 0.5 days for larger bacteria. Others have also observed a relationship between cell size and the capacity to reduce CTC (Søndergaard and Danielsen

2001), and Bernard et al. (2000b) and Lebaron et al. (2002) confirmed this pattern by measuring tritiated leucine incorporation rates by bacteria sorted according to size using flow cytometry. Cottrell and Kirchman (2004) also showed that larger bacteria were more likely to be taking up thymidine than smaller bacteria.

Distribution of Activity Across and Within Major Phylogenetic Groups

An increasing number of studies have attempted to determine the phylogenetic affiliation of cells in different physiological states. Several of these have used cell sorting in order to isolate cells or clone genes, or have used FISH or DNA fingerprinting to analyze specific fractions, such as HNA and LNA cells, or CTC-positive cells. Results have been somewhat contradictory. Some indicate that these fractions are not phylogenetically distinct (Bernard et al. 2000a; Servais et al. 2003; Fuchs et al. 2005; Longnecker et al. 2005), while others suggest that the composition might differ between fractions (Bernard et al. 2000b). Mary et al. (2006) suggested that most LNA cells in open ocean environments belong to the SAR11 clade, and that these cells do not to appear in the HNA fraction. Other studies have combined BrdU incorporation with FISH to assess the composition of substrate-active or dividing cells. For example, Pernthaler et al. (2002) found that only 3 percent of total cells incorporated BrdU, indicating a very small fraction of dividing cells, at least in early autumn in the North Sea. Use of group-specific probes allowed the identification of the dividing cells as belonging to three main populations: SAR86, *Roseobacter*, and *Alteromonas*. Interestingly, the SAR86 group of the *Gammaproteobacteria* accounted for over 50 percent of BrdU-positive bacteria. Hamasaki et al. (2007) combined BrdU uptake and immunocapture followed by DGGE in coastal marine samples, and found that the composition of the BrdU-incorporating cells was substantially different from that of the total community, suggesting that numerically dominant phylotypes may not necessarily be those that are highly active and growing.

An increasing number of studies have used MAR-FISH to explore the phylogenetic affiliation of cells that are active in substrate uptake. We have collected some of the recently published results using this approach (Fig. 8.5), selecting the substrates most likely being universally incorporated, and analyzing their use by the most abundant groups of marine bacteria. These results show (1) a wide range in the proportion of cells active in substrate uptake within each of the major bacterial groups, and (2) major differences in the proportion of cells apparently active in substrate uptake within a given group, depending on the substrate utilized. Certain substrates, such as thymidine, generally lead to relatively low fraction of active cells in most groups, but most other substrates that have been tested, such as amino acids and glucose, result in an extremely wide range of values, both for the same group in different samples and for different phylogenetic groups within the same sample. A portion of this variability is no doubt methodological and related to the sensitivity of the FISH and MAR protocol used (Bouvier and del Giorgio 2003),

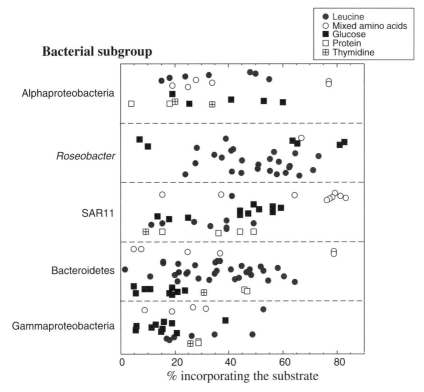

Figure 8.5 Some published MAR-FISH data showing the percentage of cells of the major bacterial groups (*Alpha-* and *Gammaproteobacteria*, and *Bacteroidetes*, and the alphaproteobacterial subgroups SAR11 and *Roseobacter*) active in assimilating different substrates.

but often the same protocol will yield a wide range in the apparent proportion of cells capable of substrate uptake. Studies that have simultaneously assessed the uptake of several substrates have shown that within the major bacterial groups the proportion of cells that take up each substrate varies greatly (Alonso-Sáez and Gasol 2007), so the relatively low percentage of substrate-active bacteria often reported may, in fact, reflect metabolic specialization rather than low metabolic activity. Nevertheless, most reported values of MAR-FISH are below 60 percent, regardless of the substrate used, suggesting that at any given time a relatively large fraction of all major bacterial phylogenic groups may not be active in the uptake of a specific substrate. This in turn suggests that within major phylogenetic groups there may be at any given time a range of cellular metabolic states, and that activation, inactivation, dormancy, and/or injury or death operate at lower phylogenetic levels than those targeted by most MAR-FISH protocols.

What is the Link Between Single-Cell Activity and Phylogenetic Affiliation?

- MAR-FISH analyses show that in most cases there is a mixture of cells that are active and inactive in substrate uptake within any given bacterial group (see, e.g., Fig. 8.5), suggesting that the level of single-cell activity is not intrinsic but rather that members of the same group may express very different levels of activity depending on their microenvironment and of their immediate history.
- This scenario would further suggest that resource microheterogeneity may play a key role in determining the distribution of activity within bacterial assemblages.
- Alternatively, the heterogeneity of single-cell activity detected within broad phylogenetic groups may indicate that within these groups there is a wide range of genetic diversity that is expressed as a wide range in metabolic responses of different cells to the same set of environmental conditions.
- This establishes two extreme scenarios: (1) the physiological structure is entirely due to environmental heterogeneity, microscale patchiness, and temporal variability; (2) physiological heterogeneity is due entirely to genetic and phenotypic diversity. Where along this gradient lie natural bacterioplankton assemblages is still a matter of study.

Dynamics of Single-Cell Activity and Physiological States

The dynamics of cell activation, inactivation, injury, and death are not well understood for marine bacterioplankton assemblages. Cells that are apparently inactive and are thus not scored positive in activity assays can quickly activate following changes in resources and physical conditions. For example, Choi et al. (1996) found in a starved laboratory culture resupplied with nutrients that 100 percent of cells without visible nucleoid developed yellow-fluorescing nucleoids before cell abundance began to increase. Likewise, Choi et al. (1999) reported that substrate additions to a natural marine assemblage resulted in rapid increases in the proportion of CTC-positive cells, from low ambient values (<10 percent), to over 70 percent under enriched conditions, without any significant change in cell abundance. These studies suggest rapid activation upon a change in the resource environment of bacterial cells that are either dormant or have extremely low levels of activity. There is also evidence in situ of rapid changes in the distribution of activity within the bacterioplankton assemblages. There were large and rapid increases in the proportion of CTC-positive bacteria as a response to algal blooms in arctic waters, even at low temperatures (Yager et al. 2001). Gasol et al. (1998) and Hagström et al. (2001) observed diel cycles of NucC-containing cell abundance in several oceanic stations, with increases during the night. The relatively large (14–86 percent) shifts in NucC numbers recorded during the diel cycles suggests that at least some non-NucC bacteria were capable of turning on and becoming NucC-positive on short time scales.

The higher proportion of NucC cells found at night could be an indication of membrane changes and/or DNA synthesis and thus of cell division; the lower proportion of NucC cells during the day could in turn be the result of strong selective grazing or of cell inactivation.

The dynamics in the physiological structure of bacterioplankton assemblages result from a combination of changes in the physiological states of individual cells and of loss processes that selectively influence the different fractions, such as protistan grazing, but these processes are difficult to discern in ambient waters. The dynamics of the physiological structure may be better understood through the manipulation of the resource conditions or the grazing pressure, or both. Figure 8.6 shows an example of an experiment with a coastal marine bacterioplankton assemblage in 100 L tanks. Bacteria initially grew fast, and for some days became decoupled from their protistan grazers. Initially, protistan density was extremely low, and bacterial abundance quickly increased over the first few hours or days, but total abundance dropped steeply as heterotrophic nanoflagellate abundance increased (Fig. 8.6a). At the start of the incubation, the proportions of "live" and CTC-positive cells were low (<50 and <20 percent, respectively), similar to in situ values, but both increased sharply, exceeding 60 percent. The relative abundance of both live and CTC-positive cells dropped sharply to low levels similar to initial

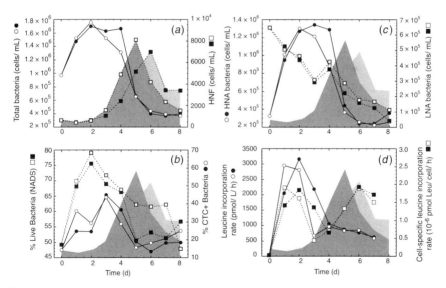

Figure 8.6 Example of the change in bacterial abundance, leucine incorporation rate, flagellate abundance, and some cell-specific indicators of bacterial metabolic status in two replicate (closed and open symbols) laboratory microcosms with Mediterranean coastal waters. (a) Total bacterial and flagellate abundance. (b) Percentage of CTC-positive and membrane-positive bacteria ("live", determined with the NADS protocol). (c) HNA and LNA bacteria. (d) Total leucine incorporation rate and cell-specific leucine incorporation rate.

values, concomitant with the increase in protistan density (Fig. 8.6b). HNA cells followed a similar pattern in density as the total number of cells, whereas LNA cells showed a completely different dynamics during the incubation, apparently independent of protistan grazing (Fig. 8.6c). The highest proportions of LNA cells occurred at the onset of the incubation, and at the peak of protistan grazing, which corresponded to the lowest proportions of live, CTC-positive, and HNA cells. Total leucine incorporation followed well the patterns in total and HNA abundance, but cell-specific leucine followed closely the pattern in the percentage of CTC-positive cells (Fig. 8.6d).

These experiments suggest the following: (1) in the absence of grazing, there is evidence for both cell activation and multiplication; (2) under these circumstances, dormant and inactive cells that were originally in the assemblage are diluted, leading to high proportions of active cells; (3) protistan grazing removes all categories of cells, but preferentially those that are active, alive, and with HNA contents, thus impacting not only the total abundance but the proportion of these various fractions; (4) some bacterial fractions, such as LNA cells, are not selectively removed, and can thus persist in the assemblage, even under high grazing pressure; (5) the variations in cell-specific metabolic activity at the community level result not only from changes in the overall level of activity of all the cells, but also from changes in the relative distribution of highly active versus less active cells; (6) even dilute marine waters can harbor bacterial assemblages with relatively high proportions of active cells. It is difficult to extrapolate from these experiments to natural conditions, but the observations of generally low proportions of active and dividing cells, high proportions of LNA and dead/injured cells, and relatively high protistan abundance in ambient marine waters would suggest that microbial communities in situ resemble more a regrowth assemblage after 4–5 days of incubation (Fig. 8.6) than either the initial or final conditions found in these experiments.

That even extremely oligotrophic waters may support high proportions of highly active cells was also reported by del Giorgio et al. (1996) in experiments carried out in coastal Mediterranean waters. In these experiments, the ambient bacterial assemblages were incubated in situ in dialysis bags with and without protistan grazers, and both the total abundance and the abundance of CTC-positive cells were followed. Similar to what was observed above, the results from these dialysis bag experiments also revealed large increases in the percentage of CTC-positive cells in the absence of grazing, and a strong negative relationship between the percentage of CTC-positive cells and protistan density in the various treatments (del Giorgio et al. 1996). Perhaps more interestingly, these experiments allowed us to explore in more detail the processes underlying the changes in total bacterial abundance, growth rate and production.

If only total abundance is considered (as is usually the case in most studies), the results suggested a relatively modest net growth rate and grazing rate for the entire assemblage (0.08/d and -0.4/d) (Fig. 8.7). If, alternatively, one considers not just total abundance but also the dynamics of CTC-positive cells, the scenario that emerged was much more complex: most of the growth of the assemblage appeared

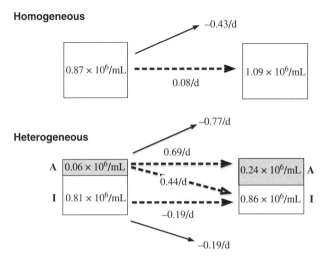

Figure 8.7 Comparison of an approach that considers the populations homogeneous with one that considers them heterogeneously formed by an active (A) and an inactive (I) pool. The values in the boxes are bacterial abundances, and the values with the arrows are rates of change (growth), of loss to predators (arrows pointing away from the boxes), or of transformation from the A to the I pool. The model used in the original publication (del Giorgio et al. 1996) assumed that no cells could move from the I to the A pool.

due to the multiplication of a very small, highly active fraction that was present at the onset of the experimental incubations; this fraction had much higher growth rates ($0.7/d$) than those estimated for the bulk assemblage, and also had much higher predation loss rates ($-0.8/d$) (Fig. 8.7). In addition, modeling also indicated that there was not only selective removal of the highly active cells during the course of the experiment, but also inactivation of at least a portion of these newly formed cells (at a rate of $0.4/d$), which then fed the pool of inactive bacteria and allowed this pool to remain relatively constant in numbers (Fig. 8.7). Although the two scenarios shown in Figure 8.7 share the same starting conditions and result in the same final total bacterial abundance, the one that takes into consideration the internal heterogeneity within the bacterial assemblages, in spite of its simplicity, provides a more realistic depiction of the complex interactions, mechanisms, and controls that underlie the changes at the community level, than the one that considers the assemblage as being homogenous.

ECOLOGICAL IMPLICATIONS OF PATTERNS IN BACTERIOPLANKTON SINGLE-CELL ACTIVITY

The evidence from studies that have targeted single-cell activity suggests that bacterioplankton assemblages should be considered a dynamic mosaic of activity, and that there may be as much variability in cell-specific metabolism within any single bacterioplankton assemblage as there is in the average activity between assemblages along environmental gradients and between ecosystems (ultraoligotrophic gyres to highly productive coastal and estuarine regimes). In brief, the evidence suggests that (1) there are dead and injured cells in all assemblages assayed so far, (2) there are significant numbers of cells that appear to be dormant, (3) the abundance as well as the proportion of active cells increases with system productivity, (4) the metabolic and growth rates of the active fraction also tend to increase, and (5) the loss processes affect these different fractions very differently. We explore possible ecological consequences of these patterns in the sections below.

Community Versus Individual Cell Growth and Metabolic Rates

The concepts of specific production, growth rate, and biomass turnover are generally applied in aquatic microbial ecology at the population and more often at the community level. Total metabolism is scaled to total bacterial abundance, and thus the resulting growth and turnover rates represent an average for the various phylogenetic and physiological groups that comprise the assemblage. These bulk growth and specific production rates have been the focus of research for the past three decades, and much has been learnt in terms of the magnitude and regulation of bacterial growth (reviewed in Chapters 9 and 10). Bacterial abundance is by far one of the least variable parameters in oceanography, and variations in bacterial abundance in general explain very little of the patterns in bulk metabolism and growth. On the other hand, there is an overall range of at least three to four orders of magnitude in the average specific growth rates (estimated as total bacterial production divided by total abundance), and variation in specific respiration (bacterial respiration divided by abundance) is somewhat smaller; the ranges for both are even greater if deep ocean assemblages are considered.

The evidence presented in the previous sections would suggest that at least a portion of this range in the average specific activity across communities is due to changes in the proportion of different physiological fractions. The key question is to what extent this large range in cell-specific activity is due to shifts in the specific activity of the entire assemblage or to shifts in the relative proportion of cells in different physiological states. For example, we could hypothesize a bacterioplankton assemblage composed of cells that are mostly dormant and in metabolic arrest, except for a small fraction of cells that upon encounter of local patches of higher substrate concentration quickly resumes growth and divide. Bulk approaches would rightly conclude that biomass doubling time of this assemblage is extremely long, and that growth rates for the entire assemblage are therefore very low, and yet these estimates would mask the fact that the division rate of the cells that are actually dividing may be on the order of a few days or even hours.

The average growth rates (μ) for an assemblage are the net result of the contributions of the various physiological fractions. In the simplest of scenarios, such as

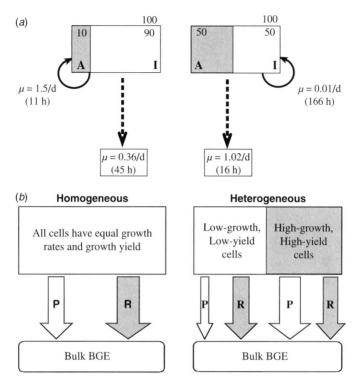

Figure 8.8 (a) Simulation of the bulk growth rate of a bacterial assemblage formed by different proportions of very active (A, growing with a growth rate 1.5/d) and not very active (I, with growth rate 0.01/d) bacterial cells. (b) The bacterial growth efficiency (BGE) of a bacterial assemblage that is homogeneous or is composed of two subpopulations: one growing with a high yield and another with a lower yield.

the one presented in Figure 8.8a, cells are distributed into two physiological categories, one fast-growing (in our example $\mu = 1.5/d$), and the other slow-growing (i.e., $\mu = 0.01/d$). The growth rates measured for the entire assemblage could thus vary over two orders of magnitude, from 0.01/d (all slow-growing cells) to 1.5/d (all fast-growing cells), and this large variation in community growth rates could be generated with no changes in cell density, and no changes in the actual levels of activity within each fraction. This scenario could be extended to the more realistic situation of multiple coexisting physiological categories, all varying not in their level of activity but rather in their contribution to total biomass and activity. Under this alternative scenario, bacterial assemblages have cells along roughly the same range of specific production (i.e., from dormant to highly active), and what differs between assemblages are not the actual levels of activity but rather the proportion of components with different activity.

Community Versus Population or Single-Cell Based Growth Estimates

- Scaling total bacterial activity (i.e., biomass production or cell division) to the entire community or to the active (and growing) portion of the community has profound consequences on the resulting patterns of bacterial growth.
- Low community-level growth rates may result from low overall growth rate within the assemblage, or from the coexistence of a small number of highly active cells within a large background of extremely slow-growing and dormant bacteria.
- Current information suggests that both scenarios coexist in marine bacterioplankton communities: the dominant groups often have intrinsically low growth rates, but the real growth rates of the active fractions are probably greatly underestimated by dilution with large numbers of inactive bacteria.
- Field and experimental studies have shown that the dynamics observed at the community level in total production and growth and biomass turnover rates are not the result of just changes in overall abundance and growth rate, but rather are underlain by complex shifts in the distribution of activity and physiological states within the assemblage.

Linking the Distribution of Single-Cell Parameters and the Bulk Assemblage Response

There are several ways by which bacterioplankton assemblages can shift their overall metabolism and growth rates: (1) there can simply be more or fewer cells doing the same thing; (2) the cell number may not change substantially, but all the cells in the assemblage may shift up or down their metabolic rate; (3) the number may not change substantially, but the proportions of cells in different physiological states may vary. Although this is not explicitly stated, the underlying assumption in many aquatic microbial studies is that variations in total community metabolism or growth are the result of some combination of scenarios 1 and 2, and scenario 3 has seldom been considered as an alternative, yet the evidence presented in the previous sections would suggest that this may in fact be one of the key underlying mechanisms in the patterns of bacterial metabolism in marine systems.

There have been relatively few studies that have attempted to link the patterns in the distribution of single cell activity with the patterns of bulk community metabolism. Several authors have shown that the rates of bacterial production and respiration relate more closely to the abundance of CTC-positive cells and HNA cells than to total counts (Lovejoy et al. 1996; del Giorgio et al. 1997; Smith 1998; Sherr et al. 1999a). Both the electron transport system measure of respiration and the CTC count of respiring bacteria were found to be positively correlated in at least one field study (Berman et al. 2001), and estuarine bacterial respiration was also strongly correlated to the number of CTC-positive cells (Smith 1998). In other studies, total red CTC fluorescence correlated to CO_2 production in mixed bacterial assemblages in aerobic bioreactors (Cook and Garland 1997), and there was

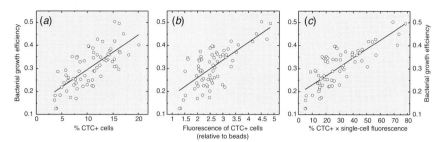

Figure 8.9 Relationship between percentage of CTC-positive cells and bacterial growth efficiency (BGE) of the bulk assemblage (*a*), average fluorescence of the CTC-positive cells (*b*), and a combination of both parameters (*c*). The data are from various estuaries and lakes in Quebec and eastern United States. Equations are: (*a*) BGE = 0.14 + 0.015 CTC%, r^2 = 0.53, N = 88; (*b*) BGE = 0.10 + 0.077F, r^2 = 0.49, N = 88; (*c*) BGE = 0.19 + 0.00389 %CTC × F, r^2 = 0.64, N = 88 where F = fluorescence.

coherence between the patterns in CTC-positive cell abundance and CO_2 production in *E. coli* cultures under varying conditions (Créach et al. 2003).

In theory, not only growth rate, but also other key processes, such as bacterial growth efficiency (BGE), should be linked to the distribution of growth and activity within the community. For example, Figure 8.8*b* shows two scenarios: one community where all cells have similar growth rates and growth efficiencies and thus contribute equally to production and respiration, and another community that is composed of groups with different growth rates and growth efficiencies. In this case, the community growth rate and BGE are a function of the proportions of each physiological group. This latter scenario was explored by del Giorgio and Newell (in preparation), who hypothesized that highly active, CTC-positive bacteria have both a high growth rate and higher growth efficiency. This hypothesis was tested with coastal and salt marsh bacterioplankton communities with a wide range in both bulk bacterial growth rate and growth efficiency. In this study, BGE was positively correlated with the proportion of CTC-positive cells (Fig. 8.9*a*), suggesting that these cells may have higher BGE than less active, CTC-negative cells, and that community BGE was the direct result of the relative contributions of these groups. Furthermore, these authors found that BGE was also positively related to the fluorescence of the CTC cells (orange or red: Fig. 8.9*b*), which is itself an index of single-cell metabolic activity. These results further suggest that even within the CTC-positive cells there is a range of single-cell activity and that this may also be linked to their individual growth rate and efficiency. The best predictor of BGE in these communities was in fact a combination of both the percentage of CTC-positive cells and their average fluorescence (Fig. 8.9*c*), exemplifying the interaction between the proportion of cells in different physiological states, and the level of activity within these categories, and their roles in shaping metabolism at the community level.

Ecological Role of Different Physiological Fractions

There are some conceptual as well as practical differences between the ecology of microbes and that of other organisms. Marine and terrestrial ecologists working on eukaryotes and metazoans do not include dead or injured organisms in population and community budgets, growth rate estimations, or genetic models. In animal populations, predators generally do not attack and consume dead prey, and it is often the slowest and smallest that are more vulnerable. Things are quite different within microbial assemblages: dead and dying cells persist in the environment, and current techniques target not only the active fractions but also the inactive and even the injured and dead cell pools (cf. Pedrós-Alió 2006), so the latter are included in budgets and calculations. Poor health and small size actually represent refuges in the microbial world. Many of the cells we enumerate in marine samples are there not by virtue of their growth but because they have avoided predation and thus reflect inefficient removal mechanisms. On the other hand, the cells that do grow are probably maintained at low levels through selective grazing and viral infection (see, e.g., Bouvier and del Giorgio 2007), with episodic "blooms" (see, e.g., Beardsley et al. 2003), and yet these cells are probably responsible for much of the carbon processing even if present at very low densities.

The ecological notion of different bacterial pools, some very abundant but turning over very slowly, others much less abundant but turning over very rapidly, is in many ways analogous to the biogeochemical notion of coexistence of different DOC pools (see Chapter 7): a large refractory pool that turns over slowly and thus persists in the ecosystem, and a small, highly labile pool that turns over rapidly and never builds up. In the same way that each of these DOC pools plays a role in the global carbon dynamics, so do each of the bacterial fractions in the functioning of bacterioplankton assemblages. It has been hypothesized that the more recalcitrant pools of DOC play a stabilizing role in ecosystem function, by fueling a slow but continuous background metabolism (del Giorgio and Williams 2005). Do the dormant and slow-growing bacterial fractions similarly play a role in stabilizing the functioning of microbial food webs?

CONCLUDING REMARKS

The physiological structure probably plays a central role in the maintenance of bacterial function and diversity in marine ecosystems, and as such it is not accidental but rather is an adaptive characteristic of bacterioplankton. In the oligotrophic, extremely heterogeneous and fluctuating marine environment, cell persistence is almost as important as cell growth. Marine microbial ecology has traditionally focused on the magnitude and regulation of bacterial production, growth, and loss at the community level, and much less on the issue of persistence and its related processes at the single-cell level. The dynamics observed at the community level in total bacterial abundance, production, and average growth and biomass turnover rates are underlain by complex shifts in the distribution of activity and physiological states within the assemblage. The physiological structure is a reflection not just of environmental conditions, but also of key biological interactions within microbial food webs, and the physiological states assumed by bacteria allow growth when possible, and persistence when necessary, thus playing a key role in both shaping community metabolism, and maintaining microbial genetic diversity.

SUMMARY

1. Marine bacterioplankton assemblage have relative large proportions of cells with extremely low metabolic rates, and which are dormant, injured or dead.
2. The low proportion of measurably active bacteria observed in many marine assemblages is not simply the result of oligotrophy and resource limitation. Selective removal processes, such as protistan grazing and viral infection, influence the physiological structure of bacterioplankton.
3. The range in physiological states reflects both the large physiological and metabolic versatility that characterizes individual marine bacterial strains and the high genetic and phenotypic diversity of bacterioplankton communities.
4. The proportion of highly active cells increases, and that of inactive or dormant cells declines systematically, along gradients of increasing system productivity.
5. The selective loss processes of specific physiological fractions change along these gradients, and, in particular, there is a relatively higher removal of the more active cells in the least productive systems. Cell injury and death may be regulated by factors that are relatively independent of those that regulate cell activation and inactivation.
6. Single-cell variations in physiological state may in fact be the predominant strategy both to cope with scarce and fluctuating resources and to escape from predation, viral infection, and other types of negative biological interactions. Low activity and eventually dormancy are part of long-term survival and protection strategies, whereas high metabolic activity increases the potential for growth, but also vulnerability to predation and death.

ACKNOWLEDGMENTS

We thank Thierry Bouvier, Erik Smith, and Laura Alonso-Sáez for providing us with the data of their compilations, Jerome Comte for ideas for Figure 8.1, and K. Jürgens and an anonymous reviewer for useful comments on the manuscript. Thomas Lefort and Lisa Fateaux helped collect the papers reviewed. We also thank Dave Kirchman for providing this opportunity to review the topic of bacterial single-cell activity in the oceans. This work was supported by the National Science and Engineering Research Council of Canada (PdG), and projects MODIVUS (CTM2005-04795/MAR) and EU NoE MARBEF (JMG).

REFERENCES

Adolph, S., S. Bach, M. Blondel, A. Cueff, M. Moreau, G. Pohnert, S. A. Poulet, T. Wichard, and A. Zuccaro. 2004. Cytotoxicity of diatom-derived oxylipins in organisms belonging to different phyla. *J. Exp. Biol.* 207: 2935–2946.

Agogué, H., F. Joux, I. Obernosterer, and P. Lebaron. 2005. Resistance of marine bacterioneuston to solar radiation. *Appl. Environ. Microbiol.* 71: 5282–5289.

Alonso-Sáez, L., and J. M. Gasol. 2007. Seasonal variation in the contribution of different bacterial groups to the uptake of low molecular weight-compounds in NW Mediterranean coastal waters. *Appl. Environ. Microbiol.* 73: 3528–3535.

Alonso-Sáez, L., J. M. Gasol, T. Lefort, J. Höfer, and R. Sommaruga. 2006. Impact of sunlight radiation on bacterial activity and differential sensitivity of natural bacterioplankton groups in NW Mediterranean coastal waters. *Appl. Environ. Microbiol.* 72: 5806–5813.

Amann, R. J., B. J. Binder, R. J. Olson, S. W. Chisholm, R. Devereux, and D. A. Stahl. 1990. Combination of 16S rRNA-targeted oligonucleotide probes with flow cytometry for analyzing mixed microbial populations. *Appl. Environ. Microbiol.* 56: 1919–1925.

Andersson, A., U. Larsson, and Å. Hagström. 1986. Size-selective grazing by a microflagellate on pelagic bacteria. *Mar. Ecol. Prog. Ser.* 33: 51–57.

Arístegui, J., C. M. Duarte, J. M. Gasol, and L. Alonso-Sáez. 2005. Active mesopelagic prokaryotes support high respiration in the subtropical northeast Atlantic Ocean. *Geophys. Res. Lett.* 32: L03608–L03612.

Arrieta, J. M., M. G. Weinbauer, and G. J. Herndl. 2000. Interspecific variability in sensitivity to UV radiation and subsequent recovery in selected isolates of marine bacteria. *Appl. Environ. Microbiol.* 66: 1468–1473.

Azam, F. 1998. Microbial control of oceanic carbon flux: The plot thickens. *Science* 280: 694–696.

Bär, M., J. von Hardenberg, E. Meron, and A. Provenzale. 2002. Modelling the survival of bacteria in drylands: The advantage of being dormant. *Proc. Roy. Soc. Lond. Ser. B, Biol. Sci.* 269: 937–942.

Barer, M. R. 1997. Viable but non-culturable and dormant bacteria: Time to resolve an oxymoron and a misnomer? *J. Med. Microbiol.* 46: 629–631.

Barer, M. R., and C. R. Harwood. 1999. Bacterial viability and culturability. *Adv. Microb. Ecol.* 41: 93–137.

Beardsley, C., J. Pernthaler, W. Wosniok, and R. Amann. 2003. Are readily culturable bacteria in coastal North Sea waters suppressed by selective grazing mortality? *Appl. Environ. Microbiol.* 69: 2624–2630.

Berman, T., B. Kaplan, S. Chava, Y. Viner, B. F. Sherr, and E. B. Sherr. 2001. Metabolically active bacteria in Lake Kinneret. *Aquat. Microb. Ecol.* 23: 213–224.

Bernard, L., C. Courties, P. Servais, M. Troussellier, M. Petit, and P. Lebaron. 2000a. Relationships among bacterial cell size, productivity, and genetic diversity in aquatic environments using cell sorting and flow cytometry. *Microb. Ecol.* 40: 148–158.

Bernard, L., H. Schäfer, F. Joux, C. Courties, G. Muyzer, and P. Lebaron. 2000b. Genetic diversity of total, active and culturable marine bacteria in coastal seawater. *Aquat. Microb. Ecol.* 23: 1–11.

Billen, G., P. Servais, and S. Becquevort. 1990. Dynamics of bacterioplankton in oligotrophic and eutrophic aquatic environments: Bottom-up or top-down control? *Hydrobiologia* 207: 37–42.

Bjørnsen, P. K., B. Riemann, J. Pock-Steen, T. G. Nielsen, and S. J. Horsted. 1989. Regulation of bacterioplankton production and cell volume in a eutrophic estuary. *Appl. Environ. Microbiol.* 55: 1512–1518.

Blackburn, N., F. Azam, and Å. Hagström. 1997. Spatially explicit simulation of a microbial food web. *Limnol. Oceanogr.* 42: 613–622.

Blackburn, N., T. Fenchel, and J. Mitchell. 1998. Microscale nutrient patches in planktonic habitats shown by chemotactic bacteria. *Science* 282: 2254–2256.

Borneman, J. 1999. Culture-independent identification of microorganisms that respond to specified stimuli. *Appl. Environ. Microbiol.* 65: 3398–3400.

Bouvier, T., and P. A. del Giorgio. 2002. Compositional changes in free-living bacterial communities along a salinity gradient in two temperate estuaries. *Limnol. Oceanogr.* 47: 453–470.

Bouvier, T., and P. A. del Giorgio. 2003. Factors influencing the detection of bacterial cells using fluorescence in situ hybridization (FISH): A quantitative review of published reports. *FEMS Microbiol. Ecol.* 44: 3–15.

Bouvier, T., and P. A. del Giorgio. 2007. Key role of selective viral-induced mortality in determining marine bacterial community composition. *Environ. Microbiol.* 9: 287–297.

Bouvier, T., P. A. del Giorgio, and J. M. Gasol. 2007. A comparative study of the cytometric characteristics of high and low nucleic-acid bacterioplankton cells from different aquatic ecosystems. *Environ. Microbiol.* 9: 2050–2066.

Bratbak, G., M. Heldal, T. F. Thingstad, B. Riemann, and O. H. Haslund. 1992. Incorporation of viruses into the budget of microbial C-transfer. A first approach. *Mar. Ecol. Prog. Ser.* 83: 273–280.

Brehm-Stecher, B. F., and E. A. Johnson. 2004. Single-cell microbiology: Tools, technologies, and applications. *Microbiol. Molec. Biol. Rev.* 68: 538–559.

Brock, T. D. 1967. Mode of filamentous growth of *Leucothrix mucor* in pure culture and in nature, as studied by tritiated thymidine autoradiography. *J. Bacteriol.* 93: 985–990.

Caro, A., P. Got, and B. Baleux. 1999. Physiological changes of *Salmonella typhimurium* cells under osmotic and starvation conditions by image analysis. *FEMS Microbiol. Lett.* 179: 265–273.

Casamayor, E. O., R. Massana, S. Benlloch, et al. 2002. Changes in archaeal, bacterial and eukaryal assemblages along a salinity gradient by comparison of genetic fingerprinting methods in a multipond solar saltern. *Environ. Microbiol.* 4: 338–348.

Catala, P., N. Parthuisot, L. Bernard, J. Baudart, K. Lemarchand, and P. Lebaron. 1999. Effectiveness of CSE to counterstain particles and dead bacterial cells with permeabilised membranes: Application to viability assessment in waters. *FEMS Microbiol. Lett.* 178: 219–226.

Chesbro, W. M., M. Arbige, and R. Eifert. 1990. When nutrient limitation places bacteria in the domains of slow growth: Metabolic, morphologic and cell cycle behavior. *FEMS Microbiol. Ecol.* 74: 103–120.

Choi, J. W., E. B. Sherr, and B. F. Sherr. 1996. Relation between presence-absence of a visible nucleoid and metabolic activity in bacterioplankton cells. *Limnol. Oceanogr.* 41: 1161–1168.

Choi, J. W., B. F. Sherr, and E. B. Sherr. 1999. Dead or alive? A large fraction of ETS-inactive marine bacterioplankton cells, as assessed by reduction of CTC, can become ETS-active with incubation and substrate addition. *Aquat. Microb. Ecol.* 18: 105–115.

Chrzanowski, T. H., R. D. Crotty, J. G. Hubbard, and R. P. Welch. 1984. Applicability of the fluorescein diacetate method of detecting active bacteria in freshwater. *Microb. Ecol.* 10: 179–185.

Clarke, R. G., and A. C. Pinder. 1998. Improved detection of bacteria by flow cytometry using the combination of antibody and viability markers. *J. Appl. Microb.* 84: 577–584.

Colwell, R. R. 2000. Bacterial death revisited. In R. R. Colwell and D. J. Grimes (eds.), *Nonculturable Microorganisms in the Environment.* ASM Press, pp. 325–343.

Cook, K. L., and J. L. Garland. 1997. The relationship between electron transport activity as measured by CTC reduction and CO_2 production in mixed microbial communities. *Microb. Ecol.* 34: 237–247.

Cottrell, M. T., and D. L. Kirchman. 2000. Natural assemblages of marine proteobacteria and members of *Cytophaga-Flavobacter* cluster consuming low- and high-molecular-weight dissolved organic matter. *Appl. Environ. Microbiol.* 66: 1692–1697.

Cottrell, M. T., and D. L. Kirchman. 2003. Contribution of major bacterial groups to bacterial biomass production (thymidine and leucine incorporation) in the Delaware estuary. *Limnol. Oceanogr.* 48: 168–178.

Cottrell, M. T., and D. L. Kirchman. 2004. Single-cell analysis of bacterial growth, cell size, and community structure in the Delaware estuary. *Aquat. Microb. Ecol.* 34: 139–149.

Créach, V., A.-C. Baudoux, G. Bertru, and B. Le Rouzic. 2003. Direct estimate of active bacteria: CTC use and limitations. *J. Microbiol. Methods* 52: 19–28.

Darzynkiewicz, Z., and J. Kapuscinski. 1990. Acridine orange: A versatile probe of nucleic acids and other cell constituents. In M. R. Melamed, T. Lindmo, and M. L. Mendelsohn (eds.), *Flow Cytometry and Cell Sorting*, 2nd edn. Wiley-Liss, pp. 291–314.

Davey, H. M., D. H. Weichart, D. B. Kell, and A. S. Kaprelyants. 1999. Estimation of microbial viability using flow cytometry. *Curr. Prot. Cytometry* 11.3.1–11.3.20.

Davidson, A. T., P. G. Thompson, K. Westwood, and R. van den Enden. 2004. Estimation of bacterioplankton activity in Tasmanian coastal waters and between Tasmania and Antarctica using stains. *Aquat. Microb. Ecol.* 37: 33–45.

del Giorgio, P. A., and T. C. Bouvier. 2002. Linking the physiologic and phylogenetic successions in free-living bacterial communities along an estuarine gradient. *Limnol. Oceanogr.* 47: 471–486.

del Giorgio, P. A., and G. Scarborough. 1995. Increase in the proportion of metabolically active bacteria along gradients of enrichment in freshwater and marine plankton: Implications for estimates of bacterial growth and production rates. *J. Plankton Res.* 17: 1905–1924.

del Giorgio, P. A., and P. J. LeB. Williams. 2005. The global significance of respiration in aquatic ecosystems: From single cells to the biosphere. In P. A. del Giorgio and P. J. LeB. Williams (eds.), *Respiration in Aquatic Ecosystems*. Oxford University Press, pp. 267–273.

del Giorgio, P. A., J. M. Gasol, D. Vaqué, P. Mura, S. Agustí, and C. M. Duarte. 1996. Bacterioplankton community structure: Protists control net production and the proportion of active bacteria in a coastal marine community. *Limnol. Oceanogr.* 41: 1169–1179.

del Giorgio, P. A., Y. T. Prairie, and D. F. Bird. 1997. Coupling between rates of bacterial production and the abundance of metabolically active bacteria in lakes, enumerated using CTC reduction and flow cytometry. *Microb. Ecol.* 34: 144–154.

Diaper, J. P., K. Tither, and C. Edwards. 1992. Rapid assessment of bacterial viability by flow cytometry. *Appl. Microbiol. Biotechnol.* 38: 268–272.

Dive, C., H. Cox, J. V. Watson, and P. Workman. 1988. Polar fluorescein derivatives as improved substrate probes for flow cytoenzymological assay of cellular esterases. *Mol. Cell Probes* 2: 131–145.

Eguchi, M., M. Ostrowski, F. Fegatella, J. Bowman, D. Nichols, T. Nishino, and R. Cavicchioli. 2001. *Sphingomonas alaskensis* strain AFO1, an abundant oligotrophic ultramicrobacterium from the North Pacific. *Appl. Environ. Microbiol.* 67: 4945–4954.

Epstein, S. S., and J. Rossel. 1995. Methodology of in situ grazing experiments: Evaluation of a new vital dye for preparation of fluorescently labeled bacteria. *Mar. Ecol. Prog. Ser.* 128: 143–150.

Epstein, S. S., and M. P. Shiaris. 1992. Size-selective grazing of coastal bacterioplankton by natural assemblages of pigmented flagellates, colorless flagellates, and ciliates. *Microb. Ecol.* 23: 211–225.

Fagerbakke, K. M., S. Norland, and M. Heldal. 1999. The inorganic ion content of native aquatic bacteria. *Can. J. Microbiol.* 45: 304–311.

Fegatella, F., and R. Cavicchioli. 2000. Physiological responses to starvation in the marine oligotrophic ultramicrobacterium *Sphingomonas* sp. strain RB2256. *Appl. Environ. Microbiol.* 66: 2037–2044.

Fenchel, T. 1980. Suspension feeding in ciliated protozoa: Feeding rates and their ecological significance. *Microb. Ecol.* 6: 13–25.

Ferenci, T. 2001. Hungry bacteria—definition and properties of a nutritional state. *Environ. Microbiol.* 3: 605–611.

Fiksdal, L., and I. Tryland. 1999. Effect of U.V. light irradiation, starvation and heat on *Escherichia coli* β-D-galactosidase activity and other potential viability parameters. *J. Appl. Microbiol.* 87: 62–71.

Francisco, D. E., R. A. Mah, and A. C. Rabin. 1973. Acridine orange-epifluorescence technique for counting bacteria in natural waters. *Trans. Am. Microsc. Soc.* 93: 416–421.

Freese, H. M., U. Karsten, and R. Schumann. 2006. Bacterial abundance, activity, and viability in the eutrophic river Warnow, Northeast Germany. *Microb. Ecol.* 51: 117–127.

Freifelder, D. 1987. *Microbial Genetics*. Jones and Bartlett.

Fuchs, B., D. Woebken, M. V. Zubkov, P. Burkill, and R. Amann. 2005. Molecular identification of picoplankton populations in contrasting waters of the Arabian Sea. *Aquat. Microb. Ecol.* 39: 145–157.

Fuhrman, J. A. 1981. Influence of methods on the apparent size distribution of bacterioplankton cells: Epifluorescence microscopy compared to scanning electron microscopy. *Mar. Ecol. Prog. Ser.* 5: 103–106.

Gasol, J. M., and J. Arístegui. 2007. Cytometric evidence reconciling the toxicity and usefulness of CTC as a marker of bacterial activity. *Aquat. Microb. Ecol.* 46: 71–83.

Gasol, J. M., and P. A. del Giorgio. 2000. Using flow cytometry for counting natural planktonic bacteria and understanding the structure of planktonic bacterial communities. *Scientia Marina* 64: 197–224.

Gasol, J. M., P. A. del Giorgio, R. Massana, and C. M. Duarte. 1995. Active versus inactive bacteria: Size-dependence in a coastal marine plankton community. *Mar. Ecol. Prog. Ser.* 128: 91–97.

Gasol, J. M., M. D. Doval, J. Pinhassi, J. I. Calderón-Paz, N. Guixa-Boixareu, D. Vaqué, and C. Pedrós-Alió. 1998. Diel variations in bacterial heterotrophic production in the Nortwestern Mediterranean Sea. *Mar. Ecol. Prog. Ser.* 164: 125–133.

Gasol, J. M., U. L. Zweifel, F. Peters, J. A. Fuhrman, and Å. Hagström. 1999. Significance of size and nucleic acid content heterogeneity as measured by flow cytometry in natural planktonic bacteria. *Appl. Environ. Microbiol.* 65: 4475–4483.

Gasol, J. M., I. Joint, K. Garde, et al. 2004. Control of heterotrophic prokaryotic abundance and growth rate in hypersaline planktonic environments. *Aquat. Microb. Ecol.* 34: 193–206.

González, J. M., E. B. Sherr, and B. F. Sherr. 1990. Size-selective grazing on bacteria by natural assemblages of estuarine flagellates and ciliates. *Appl. Environ. Microbiol.* 56: 583–589.

González, J. M., E. B. Sherr, and B. F. Sherr. 1993. Differential feeding by marine flagellates on growing versus starving, and on motile versus non-motile, bacterial prey. *Mar. Ecol. Prog. Ser.* 102: 257–267.

Grégori, G., S. Citterio, A. Ghiani, M. Labra, S. Sgorbati, S. Brown, and M. Denis. 2001. Resolution of viable and membrane-compromised bacteria in freshwater and marine waters based on analytical flow cytometry and nucleic acid double staining. *Appl. Environ. Microbiol.* 67: 4662–4670.

Gribbon, L. T., and M. R. Barer. 1995. Oxidative metabolism in nonculturable *Heliobacter pylori* and *Vibrio vulnificus* cells studied by substrate-enhanced tetrazolium reduction and digital image processing. *Appl. Environ. Microbiol.* 61: 3379–3384.

Gross, E. M. 2003. Allelopathy of aquatic autotrophs. *Crit. Rev. Plant Sci.* 22: 313–339.

Grossart, H.-P., L. Riemann, and F. Azam. 2001. Bacterial motility in the sea and its ecological implications. *Aquat. Microb. Ecol.* 25: 247–258.

Hagström, Å., J, Pinhassi, and U. L. Zweifel. 2001. Marine bacterioplankton show bursts of rapid growth induced by substrate shifts. *Aquat. Microb. Ecol.* 24: 109–115.

Hamasaki, K., A. Taniguchi, Y. Tada, R. A. Long, and F. Azam. 2007. Actively growing bacteria in the Inland Sea of Japan, identified by combined bromodeoxyuridine immunocapture and denaturing gradient gel electrophoresis. *Appl. Environ. Microbiol.* 73: 2787–2798.

Haugland, R. P. 2005. *A Guide to Fluorescent Probes and Labeling Technologies*, 10th edn. Invitrogen Corp.

Heissenberger, A., G. G. Leppard, and G. J. Herndl. 1996. Relationship between the intracellular integrity and the morphology of the capsular envelope in attached and free-living marine bacteria. *Appl. Environ. Microbiol.* 62: 4521–4528.

Heldal, M., D. J. Scanlan, S. Norland, F. Thingstad, and N. H. Mann. 2003. Elemental composition of single cells of various strains of marine *Prochlorococcus* and *Synechococcus* using X-ray microanalysis. *Limnol. Oceanogr.* 48: 1732–1743.

Herndl, G. J., G. Müller-Niklas, and J. Frick. 1993. Major role of ultraviolet-B in controlling bacterioplankton growth in the surface layer of the ocean. *Nature* 361: 717–719.

Herndl, G. J., T. Reinthaler, E. Teira, H. van Aken, C. Veth, A. Pernthaler, and J. Pernthaler. 2005. Contribution of *Archaea* to total prokaryotic production in the deep Atlantic Ocean. *Appl. Environ. Microbiol.* 71: 2303–2309.

Hollibaugh, J. T. 1976. The biological degradation of arginine and glutamic acid in seawater in relation to the growth of phytoplankton. *Mar. Biol.* 36: 303–312.

Hoppe, H.-G. 1976. Determination and properties of actively metabolizing heterotrophic bacteria in the sea, investigated by means of micro-autoradiography. *Mar. Biol.* 36: 291–302.

Howard-Jones, M. H., M. E. Frischer, and P. G. Verity. 2001. Determining the physiological status of individual bacterial cells. In J. Paul (ed.), *Methods in Microbiology*. Academic Press, pp. 175–205.

Ihssen, J., and T. Egli. 2005. Global physiological analysis of carbon- and energy-limited growing *Escherichia coli* confirms a high degree of catabolic flexibility and preparedness for mixed substrate utilization. *Environ. Microbiol.* 7: 1568–1581.

Jannasch, H. W., and G. E. Jones. 1959. Bacterial populations in sea water as determined by different methods of enumeration. *Limnol. Oceanogr.* 4: 128–139.

Jepras, R. I., J. Carter, S. C. Pearson, F. E. Paul, and M. J. Wilkinson. 1995. Development of a robust flow cytometric assay for determining numbers of viable bacteria. *Appl. Environ. Microbiol.* 61: 2696–2701.

Jochem, F. J., P. J. Lavrentyev, and M. R. First. 2004. Growth and grazing rates of bacteria groups with different apparent DNA content in the Gulf of Mexico. *Mar. Biol.* 145: 1213–1225.

Joux, F., and P. Lebaron. 1997. Ecological implications of an improved direct viable count method for aquatic bacteria. *Appl. Environ. Microbiol.* 63: 3643–3647.

Joux, F., and P. Lebaron. 2000. Use of fluorescent probes to assess physiological functions of bacteria at single-cell level. *Microbes Infect.* 2: 1523–1535.

Joux, F., P. Lebaron, and M. Troussellier. 1997a. Succession of cellular states in a *Salmonella typhimurium* population during starvation in artificial seawater microcosms. *FEMS Microbiol. Ecol.* 22: 65–76.

Joux, F., P. Lebaron, and M. Troussellier. 1997b. Changes in cellular states of the marine bacterium *Deleya aquamarina*. *Appl. Environ. Microbiol.* 63: 2686–2694.

Joux, F., W. H. Jeffrey, P. Lebaron, and D. L. Mitchell. 1999. Marine bacterial isolates display diverse responses to UV-B radiation. *Appl. Environ. Microbiol.* 65: 3820–3827.

Jürgens, K., and H. Güde. 1994. The potential importance of grazing-resistant bacteria in planktonic systems. *Mar. Ecol. Prog. Ser.* 112: 169–188.

Karner, M., and J. A. Fuhrman. 1997. Determination of active marine bacterioplankton: a comparison of universal 16S rRNA probes, autoradiography, and nucleoid staining. *Appl. Environ. Microbiol.* 63: 1208–1213.

Kawai, M., N. Yamaguchi, and M. Masu. 1999. Rapid enumeration of physiologically active bacteria in purified water used in the pharmaceutical manufacturing process. *J. Appl. Microbiol.* 86: 496–504.

Kell, D. B., A. S. Kaprelyants, D. H. Weichart, C. R. Harwood, and M. R. Barer. 1998. Viability and activity in readily culturable bacteria: a review and discussion of the practical issues. *Antonie van Leeuwenhoek* 73: 169–187.

Kieber, D. J., J. McDaniel, and K. Mopper. 1989. Photochemical source of biological substrates in sea water: Implications for carbon cycling. *Nature* 341: 637–639.

Kirchman, D. L., A. I. Dittel, R. R. Malmstrom, and M. T. Cottrell. 2005. Biogeography of major bacterial groups in the Delaware estuary. *Limnol. Oceanogr.* 50: 1697–1706.

Kjelleberg, S., M. Hermansson, P. Mården, and G. W. Jones. 1987. The transient phase between growth and nongrowth of heterotrophic bacteria, with emphasis on the marine environment. *Annu. Rev. Microbiol.* 41: 25–49.

Kjelleberg, S., N. Albertson, K. Flärdh, L. Holmquist, A. Jouper-Jaan, R. Marouga, J. Ostling, B. Svenblad, and D. Weichart. 1993. How do non-differentiating bacteria adapt to starvation? *Antonie van Leeuwenhoek* 63: 333–341.

Koch, A. L. 1997. Microbial physiology and ecology of slow growth. *Microbiol. Mol. Biol. Rev.* 61: 1092–2172.

Kogure, K, U. Simidu, and N. Taga. 1979. A tentative direct microscopic method for counting living marine bacteria. *Can. J. Microbiol.* 25: 415–420.

Kogure, K., U. Simidu, N. Taga, and R. R. Colwell. 1987. Correlation of direct viable count with heterotrophic activity for marine bacteria. *Appl. Environ. Microbiol.* 53: 2332–2337.

Konopka, A. 2004. Theoretical analysis of the starvation response under subtrate pulses. *Microb. Ecol.* 38: 321–329.

Lebaron, P., P. Catala, and N. Parthuisot. 1998. Effectiveness of SYTOX Green Stain for bacterial viability assessment. *Appl. Environ. Microbiol.* 64: 2697–2700.

Lebaron, P., P. Servais, H. Agogué, C. Courties, and F. Joux. 2001. Does the high nucleic acid content of individual bacterial cells allow us to discriminate between active cells and inactive cells in aquatic systems? *Appl. Environ. Microbiol.* 67: 1775–1782.

Lebaron, P., P. Servais, A.-C. Baudoux, M. Bourrain, C. Courties, and N. Parthuisot. 2002. Variations of bacterial-specific activity with cell size and nucleic acid content assessed by flow cytometry. *Aquat. Microb. Ecol.* 28: 131–140.

Lee, N., P. H. Nielsen, K. H. Andreasen, S. Juretschko, J. L. Nielsen, K.-H. Schleifer, and M. Wagner. 1999. Combination of fluorescent in situ hybridization and microautoradiography—a new tool for structure-function analyses in microbial ecology. *Appl. Environ. Microbiol.* 65: 1289–1297.

Leyval, D., F. Debay, J.-M. Engasser, and J.-L. Goergen. 1997. Flow cytometry for the intracellular pH measurement of glutamate producing *Corynebacterium glutamicum*. *J. Microbiol. Meth.* 29: 121–127.

Li, W. K. W., J. F. Jellett, and P. M. Dickie. 1995. DNA distributions in planktonic bacteria stained with TOTO or TO-PRO. *Limnol. Oceanogr.* 40: 1485–1495.

Long, R. A., and F. Azam. 2001. Antagonistic interactions among marine pelagic bacteria. *Appl. Environ. Microbiol.* 67: 4975–4983.

Longnecker, K., B. F. Sherr, and E. B. Sherr. 2005. Activity and phylogenetic diversity of bacterial cells with High and Low nucleic acid content and electron transport system activity in an upwelling ecosystem. *Appl. Environ. Microbiol.* 71: 7737–7749.

Longnecker, K., D. S. Homen, E. B. Sherr, and B. F. Sherr. 2006. Similar community structure of biosynthetically active prokaryotes across a range of ecosystem trophic states. *Aquat. Microb. Ecol.* 42: 265–276.

López-Amorós, R., J. Comas, and J. Vives-Rego. 1995a. Flow cytometric assessment of *Escherichia coli* and *Salmonella typhimurium* starvation-survival in seawater using rhodamine 123, propidium iodine, and oxonol. *Appl. Environ. Microbiol.* 61: 2521–2526.

López-Amorós, R., D. J. Mason, and D. Lloyd. 1995b. Use of two oxonols and a fluorescent tetrazolium dye to monitor starvation of *Escherichia coli* in seawater by flow cytometry. *J. Microbiol. Meth.* 22: 165–176.

Lovejoy, C., L. Legendre, B. Klein, J.-É. Tremblay, R. G. Ingram, and J.-C. Therriault. 1996. Bacterial activity during early winter mixing (Gulf of St. Lawrence, Canada). *Aquat. Microb. Ecol.* 10: 1–13.

Luna, G. M., E. Manini, and R. Danovaro. 2002. Large fraction of dead and inactive bacteria in coastal marine sediments: Comparison of protocols for determination and ecological significance. *Appl. Environ. Microbiol.* 68: 3509–3513.

Maalej, A., M. Denis, and S. Dukan. 2004. Temperature and growth-phase effects on *Aeromonas hydrophila* survival in natural seawater microcosms: Role of protein synthesis and nucleic acid content on viable but temporarily nonculturable response. *Microbiology* 150: 181–187.

McCarthy, M. D., J. I. Hedges, and R. Benner. 1998. Major bacterial contribution to marine dissolved organic nitrogen. *Science* 281: 231–234.

REFERENCES

Maki, J. S., and C. C. Remsen. 1981. Comparison of two direct-count methods for determining metabolizing bacteria in freshwater. *Appl. Environ. Microbiol.* 41: 1132–1138.

Malits, A., F. Peters, M. Bayer-Giraldi, C. Marrasé, A. Zoppini, Ò. Guadayol, and M. Alcaraz. 2004. Effects of small-scale turbulence on bacteria: A matter of size. *Microb. Ecol.* 48: 287–299.

Manini, E., and R. Danovaro. 2006. Synoptic determination of living/dead and active/dormant bacterial fractions in marine sediments. *FEMS Microbiol. Ecol.* 55: 416–423.

Maranger, R., P. A. del Giorgio, and D. F. Bird. 2002. Accumulation of damaged bacteria and viruses in lake water exposed to solar radiation. *Aquat. Microb. Ecol.* 28: 213–227.

Mary, I., J. L. Heywood, B. M. Fuchs, R. Amann, G. A. Tarran, P. H. Burkill, and M. V. Zubkov. 2006. SAR11 dominance among metabolically active low nucleic acid bacterioplankton in surface waters along an Atlantic meridional transect. *Aquat. Microb. Ecol.* 45: 107–113.

Mason, C. A., G. Hamer, and J. D. Bryers. 1986. The death and lysis of microorganisms in environmental processes. *FEMS Microbiol. Lett.* 39: 373–384.

Mason, D. J., R. López-Amorós, R. Allman, J. M. Stark, and D. Lloyd. 1995. The ability of membrane potential dyes and calcafluor white to distinguish between viable and non-viable bacteria. *J. Appl. Bacteriol.* 78: 309–315.

Matz, C., and K. Jürgens. 2001. Effects of hydrophobic and electrostatic cell surface properties of bacteria on feeding rates of heterotrophic nanoflagellates. *Appl. Environ. Microbiol.* 67: 814–820.

Meyer-Reil, L.-A. 1978. Autoradiography and epifluorescence microscopy combined for the determination of number and spectrum of actively metabolizing bacteria in natural waters. *Appl. Environ. Microbiol.* 36: 506–512.

Middelboe, M, Å. Hagström, N. Blackburn, B. Sinn, U. Fischer, N. H. Borch, J. Pinhassi, K. Simu, and M. G. Lorenz. 2001. Effects of bacteriophages on the population dynamics of four strains of pelagic marine bacteria. *Microb. Ecol.* 42: 395–406.

Miller, M. B., and B. L. Bassler. 2001. Quorum sensing in bacteria. *Annu. Rev. Microbiol.* 55: 165–199.

Miskin, I., G. Rhodes, K. Lawlor, J. R. Saunders, and R. W. Pickup. 1998. Bacteria in post-glacial freshwater sediments. *Microbiology* 144: 2427–2439.

Monger, B. C., M. R. Landry, and S. L. Brown. 1999. Feeding selection of heterotrophic marine nanoflagellates based on the surface hydrophobicity of their picoplankton prey. *Limnol. Oceanogr.* 44: 1917–1927.

Moran, M. A., and R. G. Zepp. 1997. Role of photoreactions in the formation of biologically labile compounds from dissolved organic matter. *Limnol. Oceanogr.* 42: 1307–1316.

Morita, R. Y. 1997. *Bacteria in Oligotrophic Environments*. Chapman and Hall.

Munro, P. M., M. J. Gauthier, V. A. Breittmayer, and J. Bongiovanni. 1989. Influence of osmoregulation processes on starvation survival of *Escherichia coli* in seawater. *Appl. Environ. Microbiol.* 55: 2017–2024.

Nagata, T. 2000a. "Picopellets" produced by phagotrophic nanoflagellates: Role in material cycling within marine environments. In N. Handa, E. Tanoue, and T. Hama (eds.), *Dynamics and Characterization of Marine Organic Matter*. Terrapub/Kluwer, pp. 241–256.

Nagata, T. 2000b. Production mechanisms of dissolved organic matter. In D. L. Kirchman (ed.), *Microbial Ecology of the Oceans*, 1st edn. Wiley-Liss, pp. 121–152.

Nagata, T., and D. L. Kirchman. 2000. Bacterial mortality: A pathway for the formation of refractory DOM? In C. R. Bell, M. Brylinsky, and P. Johnson-Green (eds.), *Microbial Biosystems: New Frontiers—Proceedings of the 8th International Symposium on Microbial Ecology.* Atlantic Canada Society for Microbial Ecology, pp. 153–158.

Nebe-von Caron, G., P. J. Stephens, C. J. Hewitt, J. R. Powell, and R. A. Badley. 2000. Analysis of bacterial function by multi-colour fluorescence flow cytometry and single cell sorting. *J Microbiol. Meth.* 42: 97–114.

Newell, S. Y., R. D. Fallon, and P. S. Tabor. 1986. Direct microscopy of natural assemblages. In J. S. Poindexter, and E. R. Leadbetter (eds.), *Bacteria in Nature*, Vol. 1. Plenum Press, pp. 1–48.

Nir, R., Y. Yisraeli, R. Lamed, and E. Sahar. 1990. Flow cytometry sorting of viable bacteria and yeasts according to β-galactosidase activity. *Appl. Environ. Microbiol.* 56: 3861–3866.

Noble, R. T., and J. A. Fuhrman. 1997. Virus decay and its causes in coastal waters. *Appl. Environ. Microbiol.* 63: 77–83.

Novo, D., N. G. Perlmutter, R. H. Hunt, and H. M. Shapiro. 1999. Accurate flow cytometric membrane potential measurement in bacteria using diethyloxacarbocyanine and a radiometric technique. *Cytometry* 35: 55–63.

Nyström, T., R. M. Olsson, and S. Kjelleberg. 1992. Survival, stress resistance and alterations in protein expression in the marine *Vibrio* sp. S14 during starvation for different individual nutrients. *Appl. Environ. Microbiol.* 58: 55–65.

Oda, Y., S.-J. Slagman, W. G. Meijer, L. J. Forney, and J. C. Gottschal. 2000. Influence of growth rate and starvation on fluorescent in situ hybridization of *Rhodopseudomonas palustris*. *FEMS Microbiol. Ecol.* 32: 205–213.

Oliver, J. D. 2000. The viable but nonculturable state and cellular resuscitation. In C. R. Bell, M. Brylinsky, and P. Johnson-Green (eds.), *Microbial Biosystems: New Frontiers— Proceedings of the 8th International Symposium on Microbial Ecology.* Atlantic Canada Society for Microbial Ecology, pp. 723–730.

Ouverney, C. C., and J. A. Fuhrman. 1999. Combined microautoradiography-16S rRNA probe technique for determination of radioisotope uptake by specific microbial cell types in situ. *Appl. Environ. Microbiol.* 65: 1746–1552.

Painchaud, J., J.-C. Therriault, and L. Legendre. 1995. Assessment of salinity-related mortality of freshwater bacteria in the Saint Lawrence estuary. *Appl. Environ. Microbiol.* 61: 205–208.

Pearce, I., A. T. Davidson, E. M. Bell, and S. Wright. 2007. Seasonal changes in the concentration and metabolic activity of bacteria and viruses at an Antarctic coastal site. *Aquat. Microb. Ecol.* 47: 11–23.

Pedrós-Alió, C. 2006. Marine microbial diversity: Can it be determined? *Trends Microbiol.* 14: 257–262.

Pedrós-Alió, C., and S. Y. Newell. 1989. Microautoradiographic study of thymidine uptake in brackish waters around Sapelo Island, Georgia, USA. *Mar. Ecol. Prog. Ser.* 55: 83–94.

Pernthaler, J., T. Posch, K. Šimek, J. Vrba, R. Amann, and R. Psenner. 1997. Contrasting bacterial strategies to coexist with a flagellate predator in an experimental microbial assemblage. *Appl. Environ. Microbiol.* 63: 596–601.

Pernthaler, A., J. Pernthaler, M. Schattenhofer, and R. Amann. 2002. Identification of DNA-synthesizing bacterial cells in coastal North Sea plankton. *Appl. Environ. Microbiol.* 68: 5728–5736.

Pirker, H., C. Pausz, K. E. Storedegger, and G. J. Herndl. 2005. Simultaneous measurement of metabolic activity and membrane integrity in marine bacterioplankton determined by confocal laser-scanning microscopy. *Aquat. Microb. Ecol.* 39: 225–233.

Pomeroy, L. R. 1974. The ocean's food web, a changing paradigm. *BioScience* 24: 499–504.

Pomeroy, L. R., and W. J. Wiebe. 2001. Temperature and substrates as interactive limiting factors for marine heterotrophic bacteria. *Aquat. Microb. Ecol.* 23: 187–204.

Porter, J., J. Diaper, C. Edwards, and R. Pickup. 1995. Direct measurements of natural planktonic bacterial community viability by flow cytometry. *Appl. Environ. Microbiol.* 61: 2783–2786.

Posch, T., J. Pernthaler, A. Alfreider, and R. Psenner. 1997. Cell-specific respiratory activity of aquatic bacteria studied with the tetrazolium reduction method, cyto-clear slides, and image analysis. *Appl. Environ. Microbiol.* 63: 863–873.

Price, N. M., P. J. Harrison, M. R. Landry, F. Azam, and K. J. F. Hall. 1986. Toxic effects of latex and Tygon tubing on marine phytoplankton, zooplankton and bacteria. *Mar. Ecol. Prog. Ser.* 34: 41–49.

Quinn, J. P. 1984. The modification and evaluation of some cytochemical techniques for the enumeration of metabolically active heterotrophic bacteria in the aquatic environment. *J. Appl. Bacteriol.* 57: 51–57.

Rae, R., and W. F. Vincent. 1998. Effects of temperature and ultraviolet radiation on microbial foodweb structure: potential responses to global change. *Freshwater Biol.* 40: 747–758.

Rappé, M. S., S. A. Connon, K. L. Vergin, and S. J. Giovanonni. 2002. Cultivation of the ubiquitous SAR11 marine bacterioplankton clade. *Nature* 418: 630–633.

Robertson, B. R., and D. K. Button. 1989. Characterizing aquatic bacteria according to population, cell size, and apparent DNA content by flow cytometry. *Cytometry* 10: 70–76.

Rodríguez, G. G., D. Phipps, K. Ishiguro, and H. F. Ridgway. 1992. Use of a fluorescent redox probe for direct visualization of actively respiring bacteria. *Appl. Environ. Microbiol.* 58: 1801–1808.

Roszak, D. B., and R. R. Colwell. 1987. Survival strategies of bacteria in the natural environment. *Microbiol. Rev.* 51: 365–379.

Roth, B. L., M. Poot, S. T. Yue, and P. J. Millard. 1997. Bacterial viability and antibiotic susceptibility testing with SYTOX green nucleic acid stain. *Appl. Environ. Microbiol.* 63: 2421–2431.

Saunders, G. W. Jr. 1959. The application of radioactive tracers to the study of lake metabolism. PhD Thesis, University of Michigan.

Schultz, G. E., and H. Ducklow. 2000. Changes in bacterioplankton metabolic capabilities along a salinity gradient in the York River estuary, Virginia, USA. *Aquat. Microb. Ecol.* 22: 163–174.

Schumann, R., U. Schiewer, U. Karsten, and T. Rieling. 2003. Viability of bacteria from different aquatic habitats. II. Cellular fluorescent markers for membrane integrity and metabolic activity. *Aquat. Microb. Ecol.* 32: 137–150.

Schut, F., R. A. Prins, and J. C. Gottschal. 1997. Oligotrophy and pelagic marine bacteria: Facts and fiction. *Aquat. Microb. Ecol.* 12: 177–202.

Servais, P., C. Courties, P. Lebaron, and M. Troussellier. 1999. Coupled bacterial activity measurements with cell sorting by flow cytometry. *Microb. Ecol.* 38: 180–189.

Servais, P., E. O. Casamayor, C. Courties, P. Catala, N. Parthuisot, and P. Lebaron. 2003. Activity and diversity of bacterial cells with high and low nucleic acid content. *Aquat. Microb. Ecol.* 33: 41–51.

Servais, P., H. Agogué, C. Courties, F. Joux, and P. Lebaron. 2001. Are the actively respiring cells (CTC+) those responsible for bacterial production in aquatic environments? *FEMS Microbiol. Ecol.* 35: 171–179.

Seymour, J. R., J. G. Mitchell, and L. Seuront. 2004. Microscale heterogeneity in the activity of coastal bacterioplankton communities. *Aquat. Microb. Ecol.* 35: 1–16.

Sherr, B. F., P. A. del Giorgio, and E. B. Sherr. 1999a. Estimating abundance and single-cell characteristics of respiring bacteria via the redox dye CTC. *Aquat. Microb. Ecol.* 18: 117–131.

Sherr, E. B., B. F. Sherr, and C. T. Sigmon. 1999b. Activity of marine bacteria under incubated and in situ conditions. *Aquat. Microb. Ecol.* 20: 213–223.

Sherr, E. B., B. F. Sherr, and T. J. Cowles. 2001. Mesoscale variability in bacterial activity in the Northeast Pacific Ocean off Oregon, USA. *Aquat. Microb. Ecol.* 25: 21–30.

Sherr, E. B., B. F. Sherr, and P. G. Verity. 2002. Distribution and relation of total bacteria, active bacteria, bacterivory, and volume of organic detritus in Atlantic continental shelf waters off Cape Hatteras NC, USA. *Deep-Sea Res.* 49: 4571–4585.

Sieracki, M. E., T. L. Cucci, and J. Nicinski. 1999. Flow cytometric analysis of 5-cyano-2,3-ditolyl tetrazolium chloride activity of marine bacterioplankton in dilution cultures. *Appl. Environ. Microbiol.* 65: 2409–2417.

Šimek, K., J. Vrba, and P. Hartman. 1994. Size-selective feeding by *Cyclidium* sp. on bacterioplankton and various sizes of cultured bacteria. *FEMS Microbiol. Ecol.* 14: 157–168.

Šimek, K., J. Vrba, J. Pernthaler, T. Posch, P. Hartman, J. Nedoma, and R. Psenner. 1997. Morphological and compositional shifts in an experimental bacterial community influenced by protists with contrasting feeding modes. *Appl. Environ. Microbiol.* 63: 587–595.

Simu, K., K. Holmfeldt, U. L. Zweifel, and Å. Hagström. 2005. Culturability and coexistence of colony-forming and single-cell marine bacterioplankton. *Appl. Environ. Microbiol.* 71: 4793–4800.

Sintes, E., and G. J. Herndl. 2006. Quantifying substrate uptake by individual cells of marine bacterioplankton by catalyzed reporter deposition fluorescence in situ hybridization combined with microautoradiography. *Appl. Environ Microbiol.* 72: 7022–7028.

Smith, E. M. 1998. Coherence of microbial respiration rate and cell-specific bacterial activity in a coastal planktonic community. *Aquat. Microb. Ecol.* 16: 27–35.

Smith, E. M., and P. A. del Giorgio. 2003. Low fractions of active bacteria in natural aquatic communities? *Aquat. Microb. Ecol.* 31: 203–208.

Smith, J. J., and G. A. McFeters. 1997. Mechanisms of INT (2-(4-iodophenyl)-3-(4-nitrophenyl)-5-phenyl tetrazolium chloride) and CTC (5-cyano-2,3-ditolyl tetrazolium chloride) reduction in *Escherichia coli* K-12. *J. Microbiol. Meth.* 29: 161–175.

Smith, J. J., J. P. Howington, and G. A. McFeters. 1994. Survival, physiological response, and recovery of enteric bacteria exposed to a polar marine environment. *Appl. Environ. Microbiol.* 60: 2977–2984.

Sommaruga, R., A. Oberleiter, and R. Psenner. 1996. Effect of UV radiation on the bacterivory of a heterotrophic nanoflagellate. *Appl. Environ. Microbiol.* 62: 4395–4400.

Søndergaard, M., and M. Danielsen. 2001. Active bacteria (CTC+) in temperate lakes: Temporal and cross-system variations. *J. Plankton Res.* 23: 1195–1206.

Staley, J. T., and A. Konopka. 1985. Measurement of in situ activities of nonphotosynthetic microorganisms in aquatic and terrestrial habitats. *Annu. Rev. Microbiol.* 39: 321–346.

Stevenson, L. N. 1978. A case for bacterial dormancy in aquatic systems. *Microb. Ecol.* 4: 127–133.

Storedegger, K. E., and G. J. Herndl. 2001. Visualization of the exopolysaccharide bacterial capsule and its distribution in oceanic environments. *Aquat. Microb. Ecol.* 26: 195–199.

Storedegger, K. E., and G. J. Herndl. 2002. Distribution of capsulated bacterioplankton in the North Atlantic and North Sea. *Microb. Ecol.* 44: 154–163.

Suller, M. T. E., and D. Lloyd. 1999. Fluorescence monitoring of antibiotic-induced bacterial damage using flow cytometry. *Cytometry* 35: 235–241.

Suttle, C. A., and F. Chen. 1992. Mechanisms and rates of decay of marine viruses in seawater. *Appl. Environ. Microbiol.* 58: 3721–3729.

Tabor, P. S., and R. A. Neihof. 1984. Direct determination of activities for microorganisms of Chesapeake Bay populations. *Appl. Environ. Microbiol.* 48: 1012–1019.

Tandoléké, R. D., D. Planas, and M. Lucotte. 2005. Microbial food webs in boreal humic lakes and reservoirs: Ciliates as a major factor related to the dynamics of the most active bacteria. *Microb. Ecol.* 49: 325–341.

Teixeira, P., H. Castro, C. Mohácsi-Farkas, and R. Kirby. 1997. Identification of sites of injury in *Lactobacillus bulgaricus* during heat stress. *J. Appl. Microbiol.* 83: 219–226.

Thingstad, T. F., L. Øvreås, J. K. Egge, T. Løvdal, and M. Heldal. 2005. Use of non-limiting substrates to increase size; a generic strategy to simultaneously optimize uptake and minimize predation in pelagic osmotrophs? *Ecol. Lett.* 8: 675–682.

Tolker-Nielsen, T., M. H. Larn, H. Kyed, and S. Molin. 1997. Effects of stress treatments on the detection of *Salmonella typhimurium* by in situ hybridization. *Int. J. Food Microbiol.* 34: 251–258.

Troussellier, M., C. Courties, P. Lebaron, and P. Servais. 1999. Flow cytometric discrimination of bacterial populations in seawater based on SYTO 13 staining of nucleic acids. *FEMS Microbiol. Ecol.* 29: 319–330.

Ullrich, S., B. Karrasch, H.-G. Hoppe, K. Jeskulke, and M. Mehrens. 1996. Toxic effects on bacterial metabolism of the redox dye 5-cyano-2,3-ditolyl tetrazolium chloride *Appl. Environ. Microbiol.* 62: 4587–4593.

Ullrich, S., B. Karrasch, and H.-G. Hoppe. 1999. Is the CTC dye technique an adequate approach for estimating active bacterial cells? *Aquat. Microb. Ecol.* 17: 207–209.

Urbach, E., K. L. Vergin, and S. J. Giovannoni. 1999. Immunochemical detection and isolation of DNA from metabolically active bacteria. *Appl. Environ. Microbiol.* 65: 1207–1213.

Van Loosdrecht, M. C. M., J. Lyklema, W. Norde, G. Schraa, and A. J. B. Zehnder. 1987. The role of bacterial cell wall hydrophobicity in adhesion. *Appl. Environ. Microbiol.* 53: 1893–1897.

Veldhuis, M. J. W., G. W. Kraay, and K. R. Timmermans. 2001. Cell death in phytoplankton: Correlation between changes in membrane permeability, photosynthetic activity, pigmentation and growth. *Eur. J. Phycol.* 36: 167–177.

Vollertsen, J., A. Jahn, J. L. Nielsen, T. Hvitved-Jacobsen, and P. H. Nielsen. 2001. Comparison of methods for determination of microbial biomass in wastewater. *Water Res.* 35: 1649–1658.

Waterbury, J. B., and F. W. Valois. 1993. Resistance to co-occurring phages enables marine *Synechococcus* communities to coexist with cyanophages abundant in seawater. *Appl. Environ. Microbiol.* 59: 3393–3399.

Weichart, D. H. 2000. Stability and survival of VBNC cells—conceptual and practical implications. In C. R. Bell, M. Brylinsky, and P. Johnson-Green (eds.), *Microbial Biosystems: New Frontiers—Proceedings of the 8th International Symposium on Microbial Ecology.* Atlantic Canada Society for Microbial Ecology, pp. 731–736.

Weinbauer, M. G. 2004. Ecology of prokaryotic viruses. *FEMS Microbiol. Rev.* 28: 127–181.

Wiebe, W. J., W. M. Sheldon Jr., and L. R. Pomeroy. 1992. Bacterial growth in the cold: Evidence for an enhanced substrate requirement. *Appl. Environ. Microbiol.* 58: 359–364.

Wilhelm, S. W., M. G. Weinbauer, C. A. Suttle, R. J. Pledger, and D. L. Mitchell. 1998. Measurements of DNA damage and photoreactivation imply that most viruses in marine surface waters are infective. *Aquat. Microb. Ecol.* 14: 215–222.

Williams, S. C., Y. Hong, D. C. A. Danavall, M. H. Howard-Jones, D. Gibson, M. E. Frischer, and P. G. Verity. 1998. Distinguishing between living and nonliving bacteria: Evaluation of the vital stain propidium iodine and its combined use with molecular probes in aquatic samples. *J. Microbiol. Meth.* 32: 225–236.

Xenopoulos, M. A., and D. F. Bird. 1997. Effect of acute exposure to hydrogen peroxide on the production of phytoplankton and bacterioplankton in a mesohumic lake. *Photochem. Photobiol.* 66: 471–478.

Yager, P. L., T. L. Connelly, B. Mortazavi, K. E. Wommack, N. Bano, J. E. Bauer, S. Opsahl, and J. T. Hollibaugh. 2001. Dynamic bacterial and viral response to an algal bloom at subzero temperatures. *Limnol. Oceanogr.* 46: 790–801.

Yakimov, M. M., G. Gentile, V. Bruni, S. Cappello, G. D'Auria, P. N. Golyshin, and L. Giuliano. 2004. Crude oil-induced structural shift of coastal bacterial communities of rod bay (Terra Nova Bay, Ross Sea, Antarctica) and characterization of cultured cold-adapted hydrocarbonoclastic bacteria. *FEMS Microbiol. Ecol.* 49: 419–432.

Yamaguchi, N., and M. Nasu. 1997. Flow cytometric analysis of bacterial respiratory and enzymatic activity in the natural aquatic environment. *J. Appl. Microbiol.* 83: 43–52.

Yokokawa, T., and T. Nagata. 2005. Growth and grazing mortality rates of phylogenetic groups of bacterioplankton in coastal marine environments. *Appl. Environ. Microbiol.* 71: 6799–6807.

Yokomaku, D., N. Yamaguchi, and M. Nasu. 2000. Improved direct viable count procedure for quantitative estimation of bacterial viability in freshwater environments. *Appl. Environ. Microbiol.* 66: 5544–5548.

Zampino, D., R. Zaccone, and R. La Ferla. 2004. Determination of living and active bacterioplankton: A comparison of methods. *Chem. Ecol.* 20: 411–422.

Zimmermann, R., R. Iturriaga, and J. Becker-Birck. 1978. Simultaneous determination of the total number of aquatic bacteria and the number thereof involved in respiration. *Appl. Environ. Microbiol.* 36: 926–935.

Zubkov, M. V., B. M. Fuchs, P. H. Burkill, and R. Aman. 2001. Comparison of cellular and biomass specific activities of dominant bacterioplankton groups in stratified waters of the Celtic Sea. *Appl. Environ. Microbiol.* 67: 5210–5218.

Zubkov, M. V., J. I. Allen, and B. M. Fuchs. 2004. Coexistence of dominant groups in marine bacterioplankton community—a combination of experimental and modelling approaches. *J. Mar. Biol. Assoc. UK* 84: 519–529.

Zweifel, U. L., and Å. Hagström. 1995. Total counts of marine bacteria include a large fraction of non-nucleoid-containing bacteria (ghosts). *Appl. Environ. Microbiol.* 61: 2180–2185.

HETEROTROPHIC BACTERIAL RESPIRATION

CAROL ROBINSON[*]

Plymouth Marine Laboratory, Prospect Place, The Hoe, Plymouth, PL1 3DH, U.K.

INTRODUCTION

Marine heterotrophic bacteria are ubiquitous and abundant, occurring from the sea surface microlayer to the deepest ocean depths, at densities measured in millions of cells per milliliter of seawater. They are an essential component of the marine biota, contributing significantly to biogeochemical cycles of the major elements and representing a considerable proportion of microbial genetic diversity.

Respiration is the physiological process in which reduced organic substrates are oxidized to release energy, and entails electron flow through membrane-associated transport systems from a donor to an electron acceptor. In aerobic respiring bacteria, the terminal electron acceptor is molecular oxygen, and the oxidation process results in the formation of carbon dioxide. In anaerobic bacteria, electron acceptors include nitrate, manganese, iron, and sulfate (Chapter 1). Several recent reviews describe the ecophysiology of aerobic, suboxic, and anaerobic respiration (Williams and del Giorgio 2005; King 2005; Codispoti et al. 2005). This chapter will focus on aerobic bacterial respiration. I will use "bacteria" to mean chemoheterotrophic bacteria, although photoheterotrophic microbes may also use dissolved organic material (DOM) (Chapter 5). In the meso- and bathypelagic regions of the oceans, *Bacteria*

---

[*]Present address: School of Environmental Sciences, University of East Anglia, Norwich, NR4 7TJ, UK.

Microbial Ecology of the Oceans, Second Edition. Edited by David L. Kirchman
Copyright © 2008 John Wiley & Sons, Inc.

coexist with *Archaea* (Karner et al. 2001; Herndl et al. 2005). The routine method of determining bacterial production, from the incorporation of leucine, overestimates "bacterial" production in these regions, as both *Bacteria* and *Archaea* incorporate leucine. It is not known to what extent *Archaea* contribute to the respiration of the deep-water prokaryotic community. Archaea are not abundant in the surface layer of most oceanic regimes examined here, and so for simplicity I will use the term "bacteria" throughout (Church et al. 2003; Alderkamp et al. 2006; and see Chapter 3).

The DOM required for bacterial respiration and production is provided from the activity of the contemporary microbial community, from the activity of a microbial community that occurred prior to the current one in time and/or distant from the current one in space, from atmospheric deposition, or from allochthonous supply from rivers and coastal regions. Processes within the microbial community that produce DOM include phytoplankton excretion, protozoan and metazoan egestion, excretion and "sloppy feeding," bacterial exo-enzymatic transformation of particulate organic material and virally induced cell lysis (Nagata 2000). Bacterial activity is both a source of carbon dioxide to the water column, that is, a "sink" or "drain" of organic carbon from the microbial foodweb, and a food source for microbial grazers, that is, a "link" in the microbial foodweb (Pomeroy 1974; Ducklow et al. 1986, 1987). Understanding the flow of carbon through the bacterial portion of the marine microbial foodweb requires quantification of both production and respiration. This allows the derivation of bacterial carbon demand (BCD) and bacterial growth efficiency (BGE) (see the text box).

Bacterial carbon demand (BCD) is the total amount of carbon required for both bacterial respiration (BR) and bacterial production (BP): BCD = BP + BR.

Bacterial growth efficiency (BGE) is the proportion of the bacterial carbon demand that is used for bacterial production: BGE = BP/BCD.

Primary production (PP) is the production of organic carbon from dissolved inorganic carbon, principally through the process of photosynthesis.

Community respiration (CR) is the respiration of the entire algal, bacterial, archaeal, and zooplankton community present in a water sample. Operationally the sample is usually collected in a bottle of approximately 100 mL and is not prefiltered, so the sampled plankton community is nominally in the less than 200 μm size range.

However, until recently, the international research focus has been on bacterial and primary production rather than on bacterial or community respiration. This is probably due to the relatively simple and sensitive, though interpretatively complex, methods of determining bacterial and primary production. However, this has created a significant gap in our ability to budget the ocean carbon cycle, and has led to continued calls for bacterial and community respiration measurements to become routine in marine biogeochemical studies (Williams 1984, 2000; Jahnke and Craven 1995; del Giorgio and Cole 1998; Williams and del Giorgio 2005).

For this chapter, I have collated published and some unpublished direct measurements of bacterial and community respiration in a freely accessible database at http://web.pml.ac.uk/amt/data/respiration.xls (Robinson and Williams 2005).

This database will continue to be updated as data become available. This chapter will describe methods for measuring and predicting bacterial respiration, synthesize our current knowledge on the spatiotemporal variability of bacterial respiration and the influence of environmental and ecological factors on this variability, and estimate the range of the contribution of bacterial respiration to community respiration. I will compare approaches for predicting bacterial respiration, and then compare predictions of bacterial respiration with direct measurements. The suggestion that in a major proportion of the world's oceans bacterial respiration exceeds primary production (del Giorgio et al. 1997; Hoppe et al. 2002) stimulated an active debate (Duarte and Agustí 1998; Duarte et al. 1999, 2001; Williams 1998; Williams and Bowers 1999; Serret et al. 2001, 2002, 2006; del Giorgio and Duarte 2002; Hansell et al. 2004; Williams et al. 2004; López-Urrutia et al. 2006). I will analyze a recent Atlantic Ocean transect of concurrent measurements of primary production, bacterial production, and temperature to compare the predictions of bacterial respiration with primary production. Finally, in a changing environment, it is prudent to speculate how bacterial respiration and production may change with predicted increases in temperature, dissolved carbon dioxide, and potentially inorganic nutrient and organic substrate supply.

MEASUREMENT OF BACTERIAL RESPIRATION AND PRODUCTION

Routine Measurement Techniques for Bacterial Respiration and Their Limitations

Direct measurement of bacterial respiration requires the separation of bacteria from the rest of the plankton community. Bacterial respiration is then measured as oxygen consumption (or occasionally pCO_2 or dissolved inorganic carbon (DIC) production) in the "bacterial" portion of the water sample in the dark. Separation is usually achieved by "gentle" filtration through a 0.8 µm filter (Williams 1981). However, there appears to be little standardization of the methodology (see text box), and the recommended controls on changes to bacterial abundance, community structure, and organic and inorganic nutrient concentrations due to filtration (Gasol and Morán 1999) are not routinely reported.

> ### Size Fractionation
>
> While some research groups use a custom-built, reverse-flow, gravity-fed filtration system incorporating 142-mm polycarbonate filters (e.g., Smith and Kemp 2001; Robinson et al. 2002a; González et al. 2003), others use 47-mm polycarbonate or glass fiber filters (e.g., Lemée et al. 2002; Navarro et al. 2004; Reinthaler and Herndl 2005), filter cartridges (Obernosterer et al. 2003) or cross-flow filtration units (Daneri et al. 1994; Blight et al. 1995) under low vacuum (<100-mm Hg). The pore size of the filters used is usually 0.8-µm, but ranges from 0.6 to 3-µm (Table 9.1).

TABLE 9.1 Microbial Activity During Incubation of a Size-Fractionated Water Sample

Location	Fraction (μm)	Bacterial Respiration (mmol $O_2/m^3/d$)	Comments	Reference
Coastal mesocosm	3.0	5	Linear consumption[a]	Williams (1981)
Canadian Arctic	1.0	5	No effect of incubation	Harrison (1986)
Chesapeake Bay	3.0	1	Linear consumption[a]	Sampou and Kemp (1994)
Gulf of Mexico	1.0	1–10	Linear consumption[a]	Biddanda et al. (1994)
Welsh coast	0.8	5	Linear consumption[a]	Blight et al. (1995)
Ross Sea	0.8	3–8[b]	Bacterial growth not different from in situ	Carlson et al. (1999)
Chesapeake Bay	3.0	2–36	Linear consumption[a]	Smith and Kemp (2001)
North Sea	0.8	4	Linear consumption[a]	Robinson et al. (2002a)
Delaware Estuary	0.8	0.08–20	Linear consumption[a]	Preen and Kirchman (2004)
New Caledonia lagoon	0.6	5–72	Linear consumption in 9 out of 27 experiments	Briand et al. (2004)
North Atlantic	0.6	0.06–0.3	Linear consumption[a]	Reinthaler et al. (2006a)

[a]Oxygen consumption was reported to be linear during these incubations.
[b]Calculated from production of dissolved inorganic carbon (DIC) and a respiratory quotient (RQ) of 1.

Bacterial abundance in some 0.8-μm filtrates can range from 60 to 90 percent of that in the unfiltered sample (Lemée et al. 2002; Reinthaler and Herndl 2005). The concurrent determination of bacterial abundance and respiration in the filtrate allows the calculation of cell specific bacterial respiration (=bacterial respiration/bacterial abundance), which could then be used to calculate the bacterial respiration of the unfiltered sample. However, since filtration efficiencies are not routinely reported, this "correction" is also not routine. Carlson et al. (1999) and Robinson et al. (2002a) describe how to minimize (<10 percent) any increase in dissolved organic carbon (DOC) in the filtrate caused by cell lysis, and González et al. (2003) found no significant difference in DOC concentration, nitrate concentration, or bacterial abundance between filtered and unfiltered samples from the Bay of Biscay when less than 20 L sample volume was filtered.

Separation of heterotrophic bacteria from autotrophic picoplankton using size fractionation alone is problematic. In ocean regions where autotrophic picoplankton are abundant, use of a 0.8 μm size fraction may lead to a significant overestimate of heterotrophic bacterial respiration (Geider 1997). Chlorophyll a-containing cells should be enumerated in the filtrate in order to constrain the error on the estimate of bacterial respiration. However this is not routinely reported. It is possible to minimize bacterial activity (and so potentially determine the respiration of the autotrophic picoplankton) using single antibiotics or mixtures of antibiotics (see, e.g., Fahnenstiel et al. 1991). However, bacterial resistance, phytoplankton sensitivity, and even bacterial use of the antibiotic as a nitrogen source (Attrassi et al. 1993; Prabaharan et al. 1994; Huan et al. 2000) have so far precluded their routine use in this context.

Isolation of the bacterial fraction necessarily breaks the trophic linkages between bacteria and their predators and organic and inorganic nutrient suppliers. In bacterial populations tightly controlled by bacterivores, separation can potentially enable increased bacterial growth, and so overestimate respiration. Prolonged incubation (>24 h) of the bacterial fraction could result in nutrient limitation, leading to an unrepresentative measurement of respiration. However, when time course experiments of size fractionated water samples are conducted, most studies (about 70 percent; Table 9.1) report that oxygen consumption is linear during the incubation, indicating that these potential problems are relatively unimportant. Paradoxically, despite linear decreases in oxygen (i.e., a constant rate of bacterial respiration), several authors report concurrent increases in bacterial production (Pomeroy et al. 1994; Blight et al. 1995; Arístegui et al. 2005b; Alonso-Sáez et al. 2007). Comparison of oxygen consumption in unfiltered and filtered water samples will also give an indication of whether oxygen consumption in the filtrate is a gross overestimate of bacterial respiration. In the dataset collated for this chapter, bacterial respiration was greater than 80 percent of community respiration in only 20 percent of cases.

Since carbon is used as the currency for ecosystem models and budgets, respiration measured by oxygen consumption must be converted to respiration in units of carbon dioxide produced using the appropriate respiratory quotient (RQ).

The **respiratory quotient (RQ)** is the molar ratio of CO_2 produced to O_2 consumed during respiration. Depending on the organic substrate, the RQ can range from 0.67 (fatty acids) to 1.33 (glycolic acid), with a value of 0.89 derived from synthesis of typical algal cell material (40 percent protein, 40 percent carbohydrate, 15 percent lipid, and 5 percent nucleic acid) (Williams and del Giorgio 2005). Hence, no single RQ is likely to be applicable for all ecosystems at all times. Few direct measurements of the RQ of natural bacterial populations exist (see, e.g., Robinson et al. 2002a). For comparative purposes, a value of 1.0 will be used in this chapter.

Routine Measurement Techniques for Bacterial Production and Their Limitations

Bacterial production is routinely determined from the incorporation of tritiated thymidine ([^{3}H]TdR) (Fuhrman and Azam 1980) and/or leucine ([^{3}H]Leu or [^{14}C]Leu) (Kirchman et al. 1985; Simon and Azam 1989) into bacterial cells during a short (approximately 1–3 h) incubation in the dark at in situ temperature. For a comprehensive review of methodologies, see Ducklow (2000). The limitation of the approach lies in converting thymidine or leucine incorporation rates into rates of bacterial carbon production (Kirchman and Ducklow 1993). The required conversion factors can either be derived from the total DNA and protein content, the ratio of thymidine to DNA or leucine to protein in a "typical" cultured bacterial cell, and the intracellular isotope dilution of the incorporated labeled compound (often called the theoretical conversion factor) or derived empirically for each environment studied. Empirical conversion factors are calculated by comparing the incorporation of the radiolabeled compound with the increase in bacterial abundance or biomass over time. This involves isolating the bacterial fraction of the natural water sample either by filtration (0.6–1.0 μm filter) or dilution (with 0.2 μm filtered water), or both, and monitoring bacterial abundance and [^{3}H]TdR, [^{3}H]Leu, or [^{14}C]Leu incorporation over time, to derive the number of cells produced per mole of thymidine or leucine incorporated. Production is then calculated by multiplying this by the carbon content of a "typical" bacterial cell.

Commonly used empirical and theoretical conversion factors range over at least a factor of 3: 0.7–3.1 kg C/mol leucine (Simon and Azam 1989; Kirchman 1993; González et al. 2003; Pérez et al. 2005), 0.67–3.5 × 10^{18} cells/mol thymidine (Fuhrman and Azam 1980; Riemann et al. 1987; Ducklow et al. 1993; Sherry et al. 2002) and 7–30 fg C/cell (Ducklow et al. 1993; Fukuda et al. 1998; Carlson et al. 1999; Zubkov et al. 2001; Sherry et al. 2002). The thymidine and leucine methods measure different, though related, physiological processes: DNA and protein synthesis, respectively. Leucine : thymidine incorporation ratios can vary by more than an order of magnitude, depending on bacterial physiological state and in situ temperature, as well as on the choice of conversion factor (Kirchman 1992; Li et al. 1993; Pomroy and Joint 1999; Ducklow 2000). This range of conversion factors and leucine : thymidine incorporation ratios complicates comparisons of bacterial production between ecosystems (Ducklow and Carlson 1992) and over long periods of time (approximately 20 years; this chapter).

Perhaps less ambiguously, bacterial production can also be derived from the increase in bacterial abundance (determined by epifluorescence microscopy or flow cytometry: Zubkov et al. 1999) or bacterial particulate organic carbon (POC, determined by high temperature combustion: Carlson et al. 1999) during incubations of the "bacterial" fraction of a water sample.

MAGNITUDE AND VARIABILITY OF BACTERIAL RESPIRATION

In their comprehensive review of bacterial growth efficiencies, del Giorgio and Cole (2000) compiled a database of bacterial respiration measurements in open ocean ($n = 62$), coastal ($n = 123$) and estuarine ($n = 54$) environments from 26 studies published between 1980 and 1999. The mean bacterial respiration was 6.26 mmol C/m^3/d (3.13 ± 0.55 µg C/L/h) in open ocean regions, 20.48 mmol C/m^3/d (10.24 ± 1.10 µg C/L/h) in coastal areas and 17.44 mmol C/m^3/d (8.72 ± 1.51 µg C/L/h) in estuaries.

Units for Respiration Rates

Bacterial processes occur on the microscale, and estimates of bacterial production are derived from 1–3 h incubations of 1–2 mL samples. However, due to the sensitivity of the techniques used to measure bacterial respiration (Robinson and Williams 2005; Reinthaler et al. 2006b), incubations lasting up to 24 h are required. Published values of bacterial processes are given in a variety of units, including g O$_2$/m^3/d, µg C/L/h, and mmol C/m^3/d. For comparative purposes, all data in this chapter (except respiration per bacterial cell) are given in mmol C/m^3/d.

For this chapter, I have added published and unpublished measurements of bacterial respiration to the del Giorgio and Cole (2000) database, but have restricted the analysis to open ocean and coastal environments (see Table 9.2; $n = 420$). Studies were not included if community respiration was measured rather than bacterial respiration, or if the filtrate was amended with organic or inorganic nutrients or stripped of DIC (e.g., Bjørnsen and Kuparinen 1991; Cherrier et al. 1996). One study on the Georgia shelf (Griffith et al. 1990) was not included in the analysis as rates of bacterial respiration were about 20-fold higher than in any of the other coastal studies. Nearly all (97 percent) of the respiration data are derived from oxygen consumption, while the rest are from DIC production (2 percent) or the difference between DOC consumption and POC production in a size-fractionated water sample (1 percent). In 92 percent of cases the filters had a pore size less than 1 µm, and in 8 percent of cases a 3 µm filter was used. Where available, the bacterial respiration data are supported by concurrent measurements of community respiration, bacterial production, bacterial abundance, bacterial biomass, DOC, chlorophyll a (a measure of phytoplankton biomass), temperature, salinity, and inorganic nutrients (ammonia, phosphate, nitrate, and nitrite). The medians (average ± standard deviation) bacterial respiration for open ocean and coastal regions, respectively, are 0.5 (1.3 ± 2.3, $n = 105$) and 2.7 (7.1 ± 12.2, $n = 315$) mmol C/m^3/d, a factor of 12 and 8 lower than in the del Giorgio and Cole (2000) dataset, respectively.

As with the global database of community respiration measurements (Robinson and Williams 2005), the striking thing about the bacterial respiration database is just how few direct measurements there are: an estimated two to three orders of magnitude less than the global number of primary or bacterial production measurements (Williams and del Giorgio 2005). The bacterial respiration data are biased with respect to sample depth, latitude, and season; 81 percent of the data were collected in the surface 10 m, 87 percent were collected between 30°N and 60°N, and 80 percent were collected between April and October. Despite this quantitative limitation, the data do show trends in temporal and spatial variability.

TABLE 9.2 Bacterial Respiration, Contribution to Community Respiration (%CR) and Bacterial Growth Efficiency (BGE)

Location	Bacterial Respiration (mmol O_2/m^3/d)	%CR	BGE	Method	Reference
Open Ocean					
North Atlantic	12.2 ± 2.1	ND	0.05 ± 0.01	DOC and bacterial POC	Kirchman et al. (1991)
South Atlantic	1.1 ± 0.9	44 ± 24	ND	Oxygen	Boyd et al. (1995)
Sargasso Sea	0.45 ± 0.01	ND	0.07 ± 0.03	DIC and [^{3}H]TdR	Hansell et al. (1995)
South Atlantic	0.7 ± 0.9	42 ± 35	ND	Oxygen	Blight (1996)
Sargasso Sea	1.3–2.0	ND	0.07–0.19	DOC and bacterial POC	Carlson and Ducklow (1996)
Ross Sea, Antarctica	0.44 ± 0.12	ND	0.21 ± 0.12	DIC and bacterial POC	Carlson et al. (1999)
Subarctic NE Pacific	1.2 ± 2.5	51 ± 30	0.24 ± 0.18	Oxygen, [^{3}H]TdR, and [^{14}C]Leu	Sherry et al. (1999)
Subtropical Atlantic	0.6 ± 0.4	ND	ND	Oxygen	Obernosterer et al. (2001)
NW Mediterranean	1.0 ± 0.9	76 ± 67	0.11 ± 0.12	Oxygen and [^{3}H]Leu	Lemée et al. (2002)
Sargasso Sea	0.4 ± 0.1	23 ± 5	ND	Oxygen	Obernosterer et al. (2003)
Bay of Biscay	1.8 ± 1.8	74 ± 14	0.01–0.08	Oxygen and [^{3}H]Leu	González et al. (2003)
North and South Atlantic	1.3 ± 1.0	74 ± 17	ND	Oxygen	Robinson, C. (unpublished)
Coastal and Shelf Seas					
Coastal Arctic	2.0 ± 1.8	41 ± 27	ND	Oxygen	Harrison (1986)
Georgia shelf[a]	75 ± 39	76 ± 26	0.04 ± 0.05	Oxygen and [^{3}H]TdR	Griffith et al. (1990)

Location				Method	Reference
Danish fjord	11 ± 4	40–50	0.46 ± 0.06	Oxygen and [³H]TdR	Sand-Jensen et al. (1990)
Louisiana shelf	4.6–22.2	ND	0.24 ± 0.14	Oxygen, [³H]TdR, and [¹⁴C]Leu	Chin-Leo and Benner (1992)
Santa Rosa Sound	37 ± 37	ND	0.15 ± 0.11	Oxygen and bacterial POC	Coffin et al. (1993)
Knebel and Aarhus Bay, Denmark	11.9 ± 8.8	ND	0.21 ± 0.12	Oxygen, [³H]TdR, and [¹⁴C]Leu	Daneri et al. (1994)
Louisiana shelf	6.5 ± 1.1	49	0.38 ± 0.13	Oxygen, [³H]TdR, and [³H]Leu	Biddanda et al. (1994)
Menai Straits, U.K.	4.8 ± 4.1	48 ± 22	0.15 ± 0.11	Oxygen and [³H]TdR	Blight et al. (1995)
Chesapeake Bay	10.2 ± 8.3	ND	ND	Oxygen	Smith (1998)
Coastal Antarctic	5.6	53	ND	Oxygen	Robinson et al. (1999)
Spanish Ria	4.8 ± 5.8	ND	ND	Oxygen	Moncoiffé et al. (2000)
Funka Bay, Japan	3.8 ± 2.6	ND	0.07 ± 0.06	Oxygen and bacterial POC	Lee et al. (2002)
NW Mediterranean	3.0 ± 1.5	52 ± 26	ND	Oxygen	Navarro et al. (2004)
New Caledonia lagoon	23.3 ± 15.9	ND	0.06 ± 0.06	Oxygen and bacterial POC	Briand et al. (2004)
Southern North Sea	1.5 ± 1.6	ND	0.2 ± 0.1	Oxygen and [¹⁴C]Leu	Reinthaler and Herndl (2005)
Western English Channel	2.5 ± 2.1	43 ± 30	ND	Oxygen	Robinson, C. (unpublished)

Data are expressed as mean ± SD when available, otherwise as the range. ND, not determined.
[a]Not included in data analysis.

Temporal Variability

Bacterial activity can vary over a diel cycle due to variation in phytoplankton extracellular release of DOC and competition with phytoplankton for inorganic nitrogen and phosphorus (Gasol et al. 1998; Kuipers et al. 2000). However, as far as I am aware, there are no published diel studies of bacterial respiration in coastal or open ocean regions.

Several studies have determined the seasonal variability of bacterial respiration and its relationship with environmental factors such as light and temperature and with ecological factors such as chlorophyll a, primary production, community respiration, and bacterial abundance and production. The seasonal variability in surface water bacterial respiration, as a fraction of community respiration, for four northern hemisphere studies (Blight et al. 1995; Lemée et al. 2002; Navarro et al. 2004; Reinthaler and Herndl 2005) is summarized in Figure 9.1. The seasonal cycles of bacterial respiration in the mesotrophic Menai Straits, U.K. and the southern North Sea show one maximum in June (approximately 8 mmol $O_2/m^3/d$) and August (approximately 3.5 mmol $O_2/m^3/d$), respectively (Blight et al. 1995; Reinthaler and Herndl 2005; see Fig. 9.1a,d), while the seasonal pattern of bacterial respiration in the oligotrophic Mediterranean Sea is more complex. In the Bay of Palma, Mallorca, bacterial respiration was highest (approximately 4 mmol $O_2/m^3/d$) in February and June 2002; at the DYFAMED station offshore from Nice, bacterial respiration peaked (approximately 2 mmol $O_2/m^3/d$) in both February and August 1999 (Navarro et al. 2004; Lemée et al. 2002; see Fig. 9.1c,b).

The seasonal range in bacterial respiration is approximately 4 mmol $O_2/m^3/d$ in the southern North Sea and Mediterranean Sea and approximately 8 mmol $O_2/m^3/d$ in the Menai Straits. Sherry et al. (1999) measured a similar range in bacterial respiration (approximately 8 mmol $C/m^3/d$) between September and February or

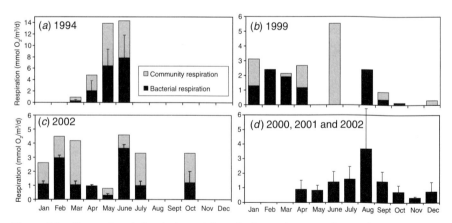

Figure 9.1 Seasonal variation in bacterial respiration and community respiration. Data are averages per month, error bars indicate 1 standard deviation of the mean and $n = 4-28$ for the different months. Redrawn from (a) Blight et al. (1995), (b) Lemée et al. (2002), (c) Navarro et al. (2004), and (d) Reinthaler and Herndl (2005).

May/June on the slope at 50°N in the subarctic northeast Pacific. A smaller (about 10-fold) range in bacterial respiration was observed at an open ocean station at the same latitude (Sherry et al. 1999).

As bacterial respiration is not routinely measured during international time-series studies, such as the Hawaii Ocean Time-series (HOT) and Bermuda Atlantic Time-series Study (BATS), we have no information on interannual variability of bacterial respiration.

Spatial Variability

Since bacterial activity is influenced by plankton community structure, substrate, and nutrient supply and environmental factors such as temperature, pressure, and light, bacterial respiration may vary spatially both vertically (with depth from the surface microlayer to mesopelagic depths) and horizontally (across gradients of light, temperature, and trophic structure and activity) at scales from less than 1 to more than 1000 km (Longhurst 1998; Martin et al. 2005).

The sea-surface microlayer (SML) is the boundary (10–250 μm) between the ocean and the atmosphere, and its chemical composition and microbiological activity can influence ocean–atmosphere gas exchange (Obernosterer et al. 2005; Calleja et al. 2005; Joux et al. 2006). The SML receives intense solar radiation, especially in the UVB wavelength range (300–320 nm), which may determine the microbial community structure and activity in this unique environment. In order to test whether bacterial activity in the SML is independent of bacterial activity in the underlying water (ULW), Reinthaler (2005) measured bacterial respiration and production, DOC, inorganic nutrients, and dissolved amino acids in the SML and ULW of the subtropical Atlantic Gyre and the western Mediterranean Sea. Amino acids and dissolved organic nitrogen and phosphorus (DON and DOP) were significantly enriched in the SML as compared with the ULW. Bacterial respiration was elevated by a factor of 15 in the SML above the 0.8 ± 0.8 mmol $O_2/m^3/d$ in the ULW, while bacterial production was 0.1- to 0.7-fold lower in the SML than in the ULW. The bacterial growth efficiency in the SML was therefore extremely low at 0.2–2 percent. Nutrient limitation caused by differential photochemical conversion of DOC, DON, and DOP to more labile constituents may be one reason for the elevated bacterial respiration in the SML (Reinthaler 2005).

Relatively little is known about bacterial respiration below 200 m in the water column due to the scarcity of direct measurements, despite this realm comprising about 75 percent of the volume of the global ocean. Among many complications in studying the deep ocean, the abundance of *Archaea* tends to increase with depth below 200 m (Karner et al. 2001). Therefore, measurements of "bacterial" abundance, respiration, and production below 200 m will include *Archaea*. For simplicity, however, I will continue to use "bacteria" in the following discussion.

Deep-water bacterial respiration can be deduced from bacterial production measurements and an assumed BGE, estimates of organic carbon supply from the epipelagic (0–200 m) zone, the distribution of dissolved oxygen and tracers such as ^{228}Ra, ^{3}H, ^{3}He, or CFCs that determine the apparent age of the water mass, or bacterial electron transport system activity (ETS). Arístegui et al. (2003, 2005a) used calculations of community respiration in the deep ocean to derive estimates of 0.01–0.02 mmol $C/m^3/d$ and approximately 0.003 mmol $C/m^3/d$ for mesopelagic (100–1000 m) and bathypelagic (1000 m–ocean floor) community respiration, respectively. However, recent direct measurements of community respiration (Arístegui et al. 2005b) and bacterial respiration (0.6 μm filtrate; Reinthaler et al. 2006a) in the 100–3000 m depth range of the North Atlantic Ocean (0.06–0.2 mmol $C/m^3/d$) are up to an order of magnitude higher than these previous estimates. The reason(s) for this discrepancy may be that (1) the earlier estimates of mesopelagic respiration based on converting ETS activity to in situ respiration are underestimates, as they use an inappropriate empirical relationship derived from senescent bacterial cultures; (2) the earlier estimates of respiration derived from the assumption that the supply of organic carbon to the mesopelagic zone is equivalent to the maximum amount of respiration are underestimates, due to underestimating the input of organic carbon (e.g., lateral supply of organic carbon from upwelling regions is very difficult to quantify); or (3) the more recent measured rates of respiration are overestimates due to stimulation caused by decompression of the samples prior to incubation (Reinthaler et al. 2006a). Bacterial respiration in the mesopelagic zone is a major component of the carbon budget for the biosphere, and, as such, significant international effort should be expended to improve knowledge of its magnitude and variability through an increase in the number of direct measurements and improved estimates of organic carbon supply (Arístegui et al. 2005a; IMBER 2005).

Bacterial respiration has been shown to vary spatially across gradients in phytoplankton production, and this is also consistent with the approximately fivefold higher median bacterial respiration in coastal compared with oceanic regions (2.7 and 0.5 mmol $O_2/m^3/d$, respectively). Reinthaler et al. (2006b) found that euphotic zone bacterial respiration decreased (from 1.22 to 0.20 mmol $O_2/m^3/d$) and the contribution of bacterial respiration to community respiration increased from 36 to 76 percent along a transect from the Mauritanian upwelling into the oligotrophic subtropical Atlantic. Sherry et al. (1999) also found a significant decrease in bacterial respiration (from 9.2 to 0.4 mmol $C/m^3/d$) but a small decrease in the percentage of community respiration due to bacteria (from 60 to 40 percent) along a transect from slope to open ocean waters in the northeast Pacific in September. Bacterial respiration and percentage of community respiration increased (from 0.3 to 0.8 mmol $C/m^3/d$ and from 25 to 65 percent) along the same transect in May–June, highlighting the interaction of seasonal and spatial variability.

Thirteen, predominantly coastal and shelf, studies also measured chlorophyll a, allowing an assessment of the relationship between phytoplankton biomass and bacterial activity at these locations. Bacterial respiration was not significantly related to chlorophyll a ($r^2 = 0.02$, $n = 244$), indicating that DOM from sources other than concurrent phytoplankton production contributes to the substrate supply for bacterial respiration in these regions. Community respiration is often weakly related to chlorophyll a in coastal, oceanic and estuarine waters, with about 30 percent of the variability in community respiration explained by the variability in chlorophyll a (Robinson et al. 2002a,b; Robinson and Williams 2005; Hopkinson and Smith 2005).

RELATIONSHIP BETWEEN BACTERIAL RESPIRATION AND ENVIRONMENTAL AND ECOLOGICAL FACTORS

Due to the strong temperature effect on cellular metabolism, one would expect a significant relationship between bacterial respiration and in situ temperature. As Figure 9.2 shows, variability in temperature explains about 30 percent of the variability in bacterial respiration. Similar relationships were derived between community respiration and temperature in estuarine, coastal, and open ocean waters (Hopkinson and Smith 2005; Robinson and Williams 2005). Temperature- bacterial respiration relationships within individual studies vary, likely due to limitation by other factors such as substrate supply (Pomeroy and Wiebe 2001; López-Urrutia and Morán 2007). Lemée et al. (2002) found no significant relationship between bacterial respiration and temperature during a seasonal study in the NW Mediterranean Sea, while temperature explained 23 percent of the variation in bacterial respiration during a spatial and temporal survey of the southern North Sea (Reinthaler and Herndl 2005).

Few studies have linked bacterial respiration with inorganic nutrient supply or limitation. Lemée et al. (2002) found that bacterial respiration was not related to in situ concentrations of nitrate or phosphate in the NW Mediterranean Sea, whereas bacterial production was. Obernosterer et al. (2003) showed that bacterial respiration was phosphate-limited at 50-m in the Sargasso Sea; addition of 0.1 mmol PO_4/m^3 (final concentration) increased bacterial respiration by 2.5-fold to 3.5 mmol O_2/m^3 during a 5-day incubation.

The effect of water pressure on bacterial respiration will remain ambiguous until respiration can be measured at in situ pressure. Decompression of deep water samples prior to measuring production and respiration may stimulate (Jannasch and Wirsen 1982) or inhibit (Tamburini et al. 2003) bacterial activity. Therefore, the applicability of bacterial respiration measurements derived from decompressed samples is currently unclear (Reinthaler et al. 2006a).

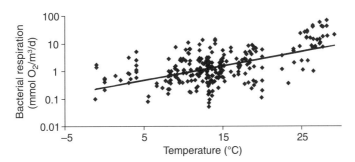

Figure 9.2 Relationship between bacterial respiration (BR) and temperature (T). All data except temperature were log-transformed and lines are the ordinary least-squares regressions: $\log_{10} BR = 0.05\ T - 0.59$, $n = 286$, $p < 0.001$, $r^2 = 0.32$.

Microbial community structure impacts bacterial respiration through production of organic substrates by phytoplankton excretion, zooplankton "sloppy feeding," and viral infection. Unfortunately, bacterial respiration has not been measured directly at the same time as these processes. Several studies measured DOC concentrations concurrently with bacterial respiration, although only a small and variable fraction of bulk DOC can be used by bacteria on short time scales (see Chapter 7). González et al. (2003) found bacterial respiration to be significantly related to DOC in the Bay of Biscay during summer, Reinthaler (2005) found a weak relationship between bacterial respiration and DOC in a dataset for the North Sea, Mediterranean Sea, and South Atlantic Ocean, and Lemée et al. (2002) found no significant relationship during a seasonal study of the NW Mediterranean Sea, whereas bacterial production and DOC were highly correlated. For the dataset compiled for this chapter, variability in DOC explained less than 20 percent of the variability in bacterial respiration (Fig. 9.3a).

Up to 25 percent of photosynthetically fixed carbon in the ocean may be transformed into DOC by phytoplankton cell lysis due to viral infection (Wilhelm and Suttle 1999; and see Chapter 12). However, the impact on bacterial respiration of this "viral shunt" of particulate organic carbon (POC) into the dissolved pool has

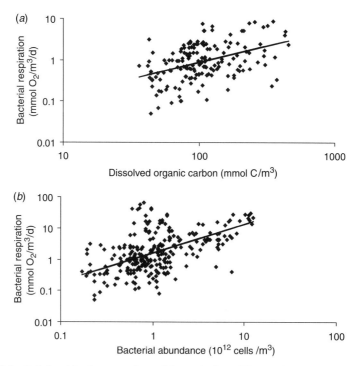

Figure 9.3 Relationship between bacterial respiration (BR) and (a) dissolved organic carbon (DOC) and (b) bacterial abundance (BA). Lines are ordinary least-squares regressions of the log-transformed data: $BR = 0.02\ DOC^{0.80}$, $n = 167$, $p < 0.001$, $r^2 = 0.18$; $BR = 1.62\ BA^{0.81}$, $n = 260$, $p < 0.001$, $r^2 = 0.27$.

not yet been quantified. Fuhrman and Suttle (1993) suggested that bacterial respiration can increase by 33 percent due to lysis of virally infected bacteria. Two studies assessed the effect of the addition of a viral concentrate to natural microbial assemblages on community and bacterial respiration (Eissler and Quiñones 2003; Eissler et al. 2003). Unfortunately, the results were inconclusive, as in four experiments bacterial respiration increased after addition of natural viral particles and in three experiments bacterial respiration decreased. These experiments are complicated to interpret, as the viral infection cycle may cause bacterial respiration to increase just before lysis, and after lysis a balance will exist between decreased bacterial respiration due to cell mortality and increased bacterial respiration of the remaining healthy cells due to the stimulatory effect of increased organic substrate.

Biodiversity can influence rates of microbial processes, yet few studies have assessed the impact of changes in marine bacterioplankton community composition on respiration. Reinthaler et al. (2005) used terminal restriction fragment length polymorphism (T-RFLP) to determine bacterioplankton community composition alongside measurements of bacterial respiration in the southern North Sea over a seasonal cycle. Major changes in bacterioplankton richness were apparent from April to December associated with significant changes in cell-specific bacterial production, with smaller changes in richness occurring between July and December. However, bacterial respiration was independent of shifts in the bacterial community composition. These authors also found a negative relationship between bacterioplankton richness and bacterial carbon demand and cell-specific bacterial respiration and production, suggestive of a highly active species-poor bacterial assemblage in spring, when higher bacterial respiration was observed, and a species-rich bacterial assemblage with reduced productivity and respiration later in the year.

If cell-specific respiration is relatively constant, then bacterial respiration should be related to bacterial abundance. However, due to a combination of the difficulty in accurately determining abundance from microscopy and the physiological flexibility of bacterial metabolism, abundance explains less than 30 percent of the variability in bacterial respiration (Fig. 9.3b). Individual studies have highlighted the approximately 10-fold variability in cell-specific bacterial respiration. González et al. (2003) found the mean cell-specific bacterial respiration before a summer upwelling event in the Bay of Biscay (3.0 ± 0.9 fmol O_2/cell/h) to be eightfold greater than that after the upwelling event (0.4 ± 0.07 fmol O_2/cell/h). Reinthaler et al. (2005) report a range of 0.02–0.15 fmol O_2/cell/h for bacterioplankton assemblages in the North Sea, and Blight et al. (1995) measured a range of 0.02–0.3 fmol O_2/cell/h during a seasonal cycle in the Menai Straits. Bacterial energetics are discussed in further detail in Chapter 8.

The coupling between bacterial respiration and production is extensively reviewed in del Giorgio and Cole (1998, 2000). Figure 9.4 shows the relationship between the concurrent determinations of bacterial respiration and production collated for this chapter. Bacterial respiration and production range over 3.5 orders of magnitude, with the production range being an order of magnitude lower than that of respiration (bacterial production 0.005–10 mmol C/m^3/d, respiration 0.05–100 mmol O_2/m^3/d). As del Giorgio and Cole (2000) point out, the large amount of variance around the BR : BP relationship means that any given value of

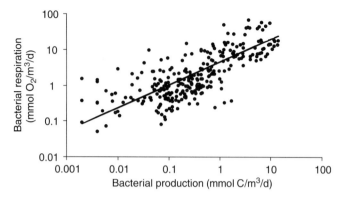

Figure 9.4 Relationship between bacterial respiration (BR) and bacterial production (BP). The line is the ordinary least-squares regression of the log-transformed data: BR = 3.69 BP$^{0.58}$, $n = 286$, $p < 0.001$, $r^2 = 0.56$.

production can have an associated bacterial respiration ranging over two to three orders of magnitude. This variability is likely a combination of the wide range of conversion factors used to estimate bacterial production (see the supplementary Table 2 at http://webpml.ac.uk/amt/data/respiration.xls), the choice of experimental approach in concurrently measuring bacterial production and respiration (see the text box), and the influence of environmental factors on bacterial growth efficiencies. Until we can reduce or understand the causes of this variability, it remains very difficult to predict bacterial respiration from bacterial production.

Time Scales for Measuring Respiration and Production

While bacterial respiration is derived from typically 24-hour (or longer) incubations of filtered water samples, bacterial production can be measured over much shorter time scales (30 minutes to 2 hours) on unfiltered water samples. Hence, for any one bacterial respiration measurement, bacterial production could be (a) measured in the unfiltered water sample, (b) measured in the filtered water sample at the start of the bacterial respiration incubation, (c) measured in the filtered water sample at the end of the bacterial respiration incubation or (d) derived from an average of bacterial production measured in the filtered water sample throughout the 24-hour bacterial respiration incubation.

The concurrent measurement of bacterial respiration and production allows the determination of the bacterial growth efficiency (BGE). Table 9.2 summarizes the range in calculated BGE determined during 17 oceanic and coastal studies. The means for the individual studies range over an order of magnitude from about 0.04 to 0.46, with usually a twofold range within any one field program. The median BGE values for open ocean and coastal regions are 0.08 (average 0.14 ± 0.14 SD, $n = 69$) and 0.16 (0.19 ± 0.16, $n = 240$), respectively, which are not substantially different from the estimates of del Giorgio and Cole (2000), who calculated median BGE values of 0.09 (0.15 ± 0.12, $n = 62$) and 0.25 (0.27 ± 0.18, $n = 123$) for open ocean and coastal areas, respectively.

BGE is influenced at a cellular level by the availability of organic and inorganic substrates and the energetic costs associated with growth in a particular environment, and so tends to increase with increasing ecosystem productivity and supply of nutrients (del Giorgio and Cole 2000). Several studies published since the review of del Giorgio and Cole (2000) corroborate this. In a seasonal study at the DYFAMED station in the Mediterranean Sea, Lemée et al. (2002) found BGE to range between less than 0.01 and more than 0.40 and to be positively correlated with DOC and chlorophyll *a*. González et al. (2003) found the BGE to be significantly lower before (0.027 ± 0.013) than after (0.08 ± 0.006) a summer upwelling in the Bay of Biscay. Reinthaler and Herndl (2005), in a seasonal study in the southern North Sea, found that BGE increased from 0.06 ± 0.03 in winter to 0.25 ± 0.09 in the spring and summer. Alonso-Sáez et al. (2007) found a higher BGE (average 0.30) in the upwelling area off the North West African coast than in the open-ocean waters of the North Atlantic subtropical gyre (average 0.09). The proliferation of BGE measurements since 2000 has also enabled an assessment of the relative influence of production and respiration on BGE in regions of differing trophic status (i.e., eutrophic, mesotrophic, and oligotrophic), based on inorganic nutrient load. Reinthaler (2005) measured concurrent bacterial production and respiration in the southern North Sea, subtropical Atlantic and western Mediterranean Sea using the same methods and conversion factors (mean BGE = 0.19 ± 0.9, $n = 144$), and found that BGE in eutrophic surface waters is more influenced by production, whereas in oligotrophic surface waters, most of the variability in BGE is explained by respiration. This concurs with the finding that under warm, oligotrophic conditions, there is strong resource limitation of bacterial production but not of bacterial respiration (López-Urrutia and Morán 2007).

BACTERIAL RESPIRATION AS A PROPORTION OF COMMUNITY RESPIRATION

Robinson and Williams (2005) compiled bacterial respiration measurements from five open ocean and coastal studies published before 1999 ($n = 62$) to show that bacterial respiration was on average 56 percent of community respiration. Table 9.2 summarizes the percentage contribution of bacterial to community respiration (percent CR) determined during 15 studies. Apart from data collected in the Sargasso Sea, where the mean percent CR was 23 ± 5 ($n = 3$), (Obernosterer et al. 2003), the mean percent CR for each individual study is either in the 40–50 percent range or the 70–80 percent range. However, within each study, there is substantial variation around this mean, with evidence that the relative contribution of bacterial respiration to community respiration varies with season and trophic status. Sherry et al. (1999) report that bacteria account for 20 percent of community respiration in spring,

60 percent in summer, and 100 percent in winter along a transect in the NE Pacific. Similarly, Lemée et al. (2002), in a seasonal study in the NW Mediterranean Sea, found that bacteria accounted for 78 percent of community respiration between January and July and 33 percent between August and December. Figure 9.1 shows the change in proportion of community respiration attributable to bacteria during three studies completed over an annual cycle (Blight et al. 1995; Lemée et al. 2002; Navarro et al. 2004). The low bacterial contribution to community respiration in the Sargasso Sea (Obernosterer et al. 2003) is surprising given that the heterotrophic bacterial abundance in the 0.6 μm filtrate accounted for 78 percent of the abundance in unfiltered samples. However, a similar mismatch was observed occasionally in the study of Lemée et al. (2002). Until a standard method incorporating routine checks for filtration efficiency is adopted for the measurement of bacterial respiration, it is difficult to be sure that this variability is real and not an artifact of the filtration procedure.

Regression analysis of concurrent bacterial and community respiration measurements generates a slope equivalent to bacterial respiration being 45 percent of community respiration (Fig. 9.5). Robinson and Williams (2005) showed that the bacterial contribution to community respiration derived from a range of approaches, including biomass calculations, food web models, and direct measurements, converged to a value of 42 percent. This is consistent with a recent review of microzooplankton grazing, which suggests that microzooplankton respiration can account for a similar proportion of community respiration as that ascribed to bacterial respiration, and may vary by a factor of two depending on the number of grazing links before protists are consumed by metazoans (Calbet and Landry 2004).

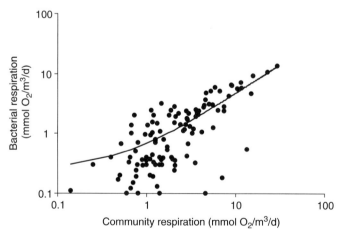

Figure 9.5 Relationship between bacterial respiration (BR) and community respiration (CR). The line is the ordinary least-squares regression of the log-transformed data: $BR = 0.45\, CR^{0.93}$, $n = 119$, $p < 0.001$, $r^2 = 0.54$.

PREDICTING BACTERIAL RESPIRATION

The relationships shown in Figures 9.2–9.5 suggest that bacterial respiration can be predicted to a greater or lesser extent from temperature, DOC, community respiration, and bacterial abundance and production. The most common methods of predicting bacterial respiration are either from measured bacterial production and an estimate of BGE for the particular ecosystem under study (Robinson et al. 1999; Robinson and Williams 1999, Søndergaard et al. 2000) or from a published relationship (del Giorgio and Cole 2000) between bacterial production and respiration (Robinson et al. 2002a,b). Unfortunately, BGE varies greatly and there is significant variability in the relationship between bacterial respiration and production (see Fig. 9.4). del Giorgio and Cole (1998, 2000) compiled an extensive dataset of direct concurrent measurements of bacterial production and respiration (in µg C/L/h) made in marine, freshwater, estuarine, and riverine systems, and derived equations to estimate bacterial respiration from the ordinary least-squares (OLS) regressions of the log-transformed data. Converting their data into mmol $C/m^3/d$ and recalculating the OLS regressions gives the equations in the text box. The relationship between bacterial respiration and production, calculated from the coastal and open ocean dataset compiled for this chapter, is also given.

Equations for Calculating Bacterial Respiration (BR) and Bacterial Growth Efficiency (BGE)

All rates are in mmol $C/m^3/d$ and log-transformed, except temperature (T).

$BR = 5.57\ BP^{0.41}$, $r^2 = 0.46$, $n = 328$ (del Giorgio and Cole 1998)
$BR = 7.09\ BP^{0.36}$, $r^2 = 0.14$, $n = 195$ (del Giorgio and Cole 2000)
$BGE = 0.374 - 0.0104\ T$; $r^2 = 0.54$, $n = 107$ (Rivkin and Legendre 2001)
$BR = 3.69\ BP^{0.58}$, $r^2 = 0.56$, $n = 286$ (this chapter)

Rivkin and Legendre (2001) found a significant inverse relationship between temperature (T) and BGE that would allow bacterial respiration to be computed from production and temperature ($BR = [BP/(0.374 - 0.0104\ T)] - BP$). This relationship suggests that a larger fraction of ingested DOC is respired by bacteria at low than at high latitudes. Using a dataset of measured community respiration, Rivkin and Legendre (2001) showed that bacterial respiration computed from their bacterial production and temperature relationship was a valid proxy for community respiration over a wide range of observed temperatures (-1.4 to $29\ °C$) and bacterial production (0.2–415 mg $C/m^3/d$). The fact that bacterial production often correlates with temperature and chlorophyll a (White et al. 1991) also leads to the possibility that primary production, community respiration, and bacterial production and respiration could be estimated from remotely sensed temperature and chlorophyll a values.

Unfortunately, the scatter in the relationship between BGE and temperature (Fig. 1 in Rivkin and Legendre 2001) means that for any particular temperature,

BGE can range from 0.1 to 0.8. For the same bacterial production (e.g., 0.3 mmol C/m^3/d), bacterial respiration would range from 0.8 to 2.7 mmol C/m^3/d. This makes it difficult to apply the Rivkin and Legendre (2001) model to regional studies with small temperature fluctuations (Robinson et al. 2002a,b; Reinthaler and Herndl 2005).

Figure 9.6 shows the differences between the models of del Giorgio and Cole (1998, 2000) and Rivkin and Legendre (2001) when predicting bacterial production over a range of temperatures from 1 to 25 °C and a range of bacterial production consistent with that measured in coastal and open ocean systems (Fig. 9.4). The greatest discrepancy between the models occurs at temperatures above 20 °C and bacterial production above 10 mmol C/m^3/d and at temperatures below 10 °C and bacterial production below 5 mmol C/m^3/d.

In order to investigate the effects of temperature and resource availability on bacterial production and respiration, and so on BGE, López-Urrutia and Morán (2007) compiled a dataset of bacterial respiration and production ($n = 205$) and derived equations to estimate these parameters from temperature, chlorophyll a, and bacterial abundance. The best prediction of bacterial respiration is described by BR $= 3.21 \times 10^{11} \times e^{-0.589/kT}$ where BR is in fgC/cell/d, T is the absolute temperature in Kelvin, and k is Boltzmann's constant (8.62×10^{-5} eV/K). Future datasets of concurrent temperature, bacterial abundance, and bacterial respiration are required to test the applicability of this model.

Allometric equations based on the relationship between metabolic activity (e.g., respiration), biomass (size), and temperature have been successfully used to derive community respiration from the size structure of bacterioplankton, phytoplankton, and micro- and meso-zooplankton (López-Urrutia et al. 2006). Robinson et al. (1999, 2002a) used the equation of Caron et al. (1990) relating oxygen consumption and mean cell volume to determine heterotrophic nanoflagellate, heterotrophic

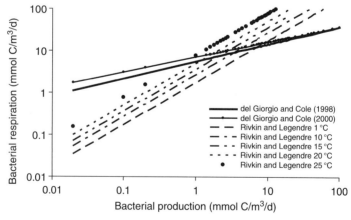

Figure 9.6 Bacterial respiration (BR) derived from the equations of del Giorgio and Cole (1998, 2000: solid lines) and Rivkin and Legendre (2001: dashed and dotted lines) for a range of temperatures from 1 to 25 °C and a range of bacterial production from 0.02 to 20 mmol C/m^3/d.

dinoflagellate, and ciliate respiration. The ability to determine bacterial cell size with shipboard flow cytometers either in discrete mode or connected to a continuous surface seawater supply would enable the determination of bacterial respiration at the high spatiotemporal resolutions normally only associated with hydrographic and chemical properties. However, the relatively small range in bacterial cell size (0.3–0.8 μm), the large uncertainty in estimating bacterial cell size from flow cytometry, and the large range in bacterial respiration (3.5 orders of magnitude) may preclude this approach.

Since temperature, DOC, primary production, bacterial production and respiration, and bacterivore and viral biomass and activity all influence or are influenced by bacterial respiration, an ecosystem approach to estimating bacterial respiration is sometimes possible. Inverse analysis searches for the value of the metabolic rate that has not been measured (in this case bacterial respiration) that best fits the constraints of the metabolic rates and biomass (such as chlorophyll a, primary production, and bacterial production and abundance) measured in a particular field study. The efficacy of the approach is obviously weakened if several parameters (rather than just one) have not been measured. Ducklow et al. (2000) estimated bacterial respiration in the Ross Sea by assuming that bacterial respiration was less than the measured community respiration and choosing values for BGE. A BGE of 0.35–0.45 and a leucine conversion factor of 1.5 kg C/mol were required if bacterial respiration were to amount to 10 percent of community respiration. Anderson and Ducklow (2001) used a steady-state microbial loop model to investigate the relationships between primary production, bacterial respiration and production, BGE, and phytoplankton excretion of DOC at three open-ocean sites. Model solutions supported the conclusion that oceanic systems are characterized by a BGE of approximately 0.15 (as in del Giorgio and Cole 2000) and low rates of bacterial production and phytoplankton excretion, and so bacterial respiration is a major carbon sink. In a similar modeling analysis, this time of the oligotrophic eastern Mediterranean Sea, Anderson and Turley (2003) found that the observed bacterial production to primary production ratio (BP : PP) of 0.22 and a BGE of 0.15 could only be maintained if DOC was supplied to the microbial loop from sources other than concurrent primary production. Since a terrestrial source of DOC is unlikely in the Cretan Sea, then bacterial production was overestimated, or primary production was underestimated, or BGE was in the range 0.16–0.20 and phytoplankton exudation was high (30–40 percent). Inverse analysis, therefore, can constrain estimates of primary production, bacterial production and respiration, and BGE.

COMPARISON BETWEEN MEASUREMENTS AND PREDICTIONS OF BACTERIAL RESPIRATION

Figure 9.7 compares measured values of bacterial respiration with bacterial respiration predicted from the del Giorgio and Cole (2000) and Rivkin and Legendre (2001) equations, using data known not to be included in the formation of either model. This subset of the database has a median temperature of 13.2 °C (mean

13.6 ± 4.2 SD °C), a median bacterial respiration of 1.0 mmol C/m^3/d (2.0 ± 2.5 mmol C/m^3/d), and a median bacterial production of 0.2 mmol C/m^3/d (0.4 ± 0.8 mmol C/m^3/d). Within these ranges of temperature and bacterial respiration and production, the del Giorgio and Cole (2000) model tends to overestimate respiration, while the Rivkin and Legendre (2001) model tends to underestimate it. For any given measured value of bacterial respiration, the predicted value covers a fivefold range. This variability is particularly noticeable at high rates, and one study was not included in Figure 9.7 for this reason. Briand et al. (2004) measured bacterial respiration and production ($n = 24$) in a tropical lagoon in the South Pacific Ocean, where temperature ranged from 25.6 to 28.2 °C, bacterial production ranged from 0.2 to 11.6 mmol C/m^3/d, and bacterial respiration ranged from 7 to 67 mmol C/m^3/d (average 20.9 ± 15.4 SD). In this instance, bacterial respiration predicted from the del Giorgio and Cole (2000) model ranged from 4 to 17 mmol C/m^3/d (average 10.5 ± 3.3) and bacterial respiration derived from the

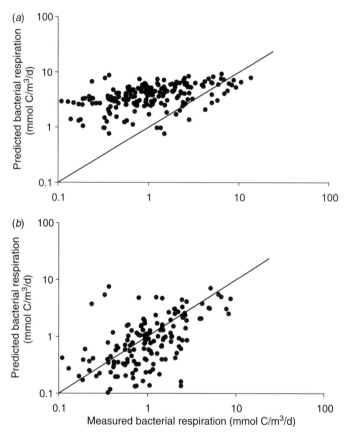

Figure 9.7 Relationship between measured bacterial respiration and bacterial respiration predicted from the equations of (*a*) del Giorgio and Cole (2000) and (*b*) Rivkin and Legendre (2001). Lines are 1 : 1.

MAGNITUDE OF BACTERIAL RESPIRATION

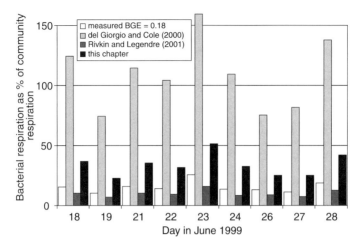

Figure 9.8 Bacterial respiration as a percentage of community respiration during a coccolithophore bloom in the North Sea in June 1999 (Robinson et al. 2002a) predicted using the measured BGE of 0.18 and the equations of del Giorgio and Cole (2000), Rivkin and Legendre (2001), and this chapter.

Rivkin and Legendre (2001) model ranged from 2 to 132 mmol $C/m^3/d$ (average 35.7 ± 30.5).

These differences in magnitude of predicted bacterial respiration can have a significant impact on the estimate of community respiration attributable to bacteria. During a study of the dynamics of an *Emiliania huxleyi* bloom in the northern North Sea, Robinson et al. (2002a) found that the difference between the proportion of community respiration attributable to bacteria derived from the del Giorgio and Cole (1998) model and that derived from the Rivkin and Legendre (2001) equation was an average 51 percent. Recalculating the bacterial respiration as a percentage of community respiration using the latest del Giorgio and Cole equation (2000) gives an unrealistically high estimate (median 107 percent), while using the model of Rivkin and Legendre (2001) gives an unrealistically low estimate (median 9 percent). A similarly low median estimate of 14 percent is calculated from the bacterial growth efficiency (0.18) directly measured during this *E. huxleyi* bloom. However, the predictive equation derived in this chapter gives a more realistic estimate of the proportion of community respiration attributable to bacteria during this North Sea study (median 34 percent) (Fig. 9.8).

MAGNITUDE OF BACTERIAL RESPIRATION IN RELATION TO PRIMARY PRODUCTION

As Anderson and Ducklow (2001) point out, the ideal measure of bacteria as a sink for organic carbon is bacterial respiration as a fraction of total primary production (see the text box). This fraction cannot exceed unity if the system is in steady state and is "isolated" in the sense that it does not receive carbon from a source "external" to the contemporary microbial community. Determination of this fraction will therefore give an indication of how much phytoplankton-derived carbon is respired by bacteria to carbon dioxide and whether bacterial respiration is wholly supported by concurrent phytoplankton production. Although counterintuitive, the more frequently derived ratios BCD : PP and BP : PP can exceed unity without requiring a source of DOC in addition to that from concurrent phytoplankton production (Strayer 1988).

Primary Production

Total primary production is the sum of particulate (PP_{POC}) and dissolved (PP_{DOC}) primary production. PP_{POC} is the production remaining in the particulate phase, while PP_{DOC} is the portion of production released as DOM. Dissolved primary production can be derived from the equation of Teira et al. (2001):

$$\log_{10} PP_{DOC} = 0.75 \pm 0.15 \times \log_{10} PP_{POC} - 0.82 \pm 0.09$$
$$(n = 54, r^2 = 0.33, p < 0.0001).$$

Gross production (GP) is the amount of production retained in the particulate phase, plus that released into the dissolved organic phase plus that respired to carbon dioxide.

Unfortunately, few, if any, datasets of directly measured concurrent bacterial respiration and total primary production exist. Therefore, in order to assess the relationship between bacterial respiration and total primary production in the open ocean and to compare the predictive ability of the equations of del Giorgio and Cole (2000), Rivkin and Legendre (2001), and this chapter, I used a dataset of concurrent measurements of euphotic zone temperature, bacterial, particulate primary and gross production rates, and community respiration collected along a north–south transect in the Atlantic Ocean (Serret et al. 2002; Pérez et al. 2005). Total primary production was calculated from the equation of Teira et al. (2001). Data were pooled into five latitudinal bands including 40°N–27°N and 10°S–20°S, which correspond to the North Atlantic Subtropical Gyre–East (NAST-E) and South Atlantic Subtropical Gyre (SATL) provinces of Longhurst (1998), respectively (Table 9.3).

Bacterial respiration predicted from the del Giorgio and Cole (2000) relationship exceeded measured community respiration (Serret et al. 2002, 2006) in all latitudinal bands by a factor of 116–514 percent, while the bacterial contribution derived from the equations of Rivkin and Legendre (2001) accounted for a surprisingly small amount of community respiration (7–18 percent) (Table 9.3). Bacterial respiration derived from the predictive equation calculated in this chapter was equivalent to a more realistic 24–86 percent of the measured community respiration. Similarly, bacterial respiration derived from the del Giorgio and Cole (2000) equation always exceeded the concurrent estimate of total primary production (by a factor of 137–810 percent), while bacterial respiration derived from the Rivkin and Legendre (2001) model was always less than about 30 percent of the concurrent estimate of

TABLE 9.3 Measured and Predicted Parameters of Plankton Activity Along a North–South Transect of the Atlantic Ocean

Latitudinal Band	PP_{total}	GP	CR	BP	del Giorgio and Cole (2000)	Rivkin and Legendre (2001)	This Chapter
		(mmol C/m²/d)				Bacterial Respiration	
47°N–40°N ($n = 2$)	22 ± 5	ND	ND	0.7 ± 0.1	83 ± 5	3 ± 0.2	16 ± 1
40°N–27°N (NAST-E) ($n = 4$)	17 ± 4	39 ± 15	55 ± 39	0.7 ± 0.2	105 ± 21	4 ± 1.4	19 ± 5
20°N–10°S ($n = 7$)	67 ± 29	153 ± 43	79 ± 21	0.8 ± 0.2	92 ± 16	6 ± 0.8	19 ± 3
10°S–20°S (SATL) ($n = 5$)	19 ± 5	58 ± 19	39 ± 24	0.9 ± 0.3	154 ± 4	6 ± 2	27 ± 3
(20°S–30°S) ($n = 3$)	30 ± 8	61 ± 18	22 ± 6	0.7 ± 0.1	113 ± 20	4 ± 0.3	19 ± 1

Bacterial production (BP), gross production (GP), and community respiration (CR) are from Serret et al. (2002), Pérez et al. (2005), and the Atlantic Meridional Transect database held at the British Oceanographic Data Centre (www.bodc.ac.uk). Bacterial respiration is predicted from the relationships of del Giorgio and Cole (2000), Rivkin and Legendre (2001), and this chapter, and total primary production (PP_{total}) is derived from the equation of Teira et al. (2001). Data are mean ± SD. ND, not determined. 40°N–27°N and 10°S–20°S are equivalent to the North Atlantic Subtropical Gyre–East (NAST-E) and South Atlantic Subtropical Gyre (SATL) provinces of Longhurst (1998).

total primary production. Bacterial respiration predicted from the relationship with bacterial production computed in this chapter was 28–73 percent less than concurrent total primary production in the latitudinal bands 47°N–40°N, 20°N–10°S, and 20°S–30°S, equivalent to total primary production in the NAST and higher (by a factor of 142 percent) than total primary production in the SATL. Direct measurements along this transect suggest that community respiration was lower (by 36–67 percent) than gross production in all latitudinal bands except the NAST-E (Serret et al. 2002, 2006). Of the three derivations for the relationship between total primary production and bacterial respiration, therefore, the balance using bacterial respiration predicted from this chapter's equation lies closest to the measured balance between gross production and community respiration (Table 9.3). The difference between the ratio of bacterial respiration to total primary production and the ratio of community respiration to gross production within the SATL may be due to the surprisingly low total primary production measured there. In the SATL, total primary production amounted to about 30 percent of gross production, whereas it was equivalent to about 50 percent of gross production in the other latitudinal bands. Therefore, based on the ratios of bacterial respiration to total primary production derived from data analysis within this chapter, there is no strong evidence that bacterial respiration is not wholly supported by organic carbon produced within the contemporary microbial community along the north–south Atlantic Ocean transect.

BACTERIAL RESPIRATION IN A CHANGING ENVIRONMENT

Some of the predicted changes in ocean physics and chemistry due to global change include higher sea surface temperature and CO_2 concentrations, lower pH, and enhanced upper ocean stratification (Solomon et al. 2007). These changes may have direct (e.g., temperature) or indirect (e.g., changes in DOM production by phytoplankton due to increased seawater acidity) consequences on bacterial production. However, model simulations, experimental manipulations, and direct observation of long-term shifts in plankton distribution and activity have focused on the impact on phyto- and zooplankton rather than on bacterioplankton. Where bacterial activity has been included in an assessment of the impact of a changing environment, the emphasis has been on bacterial production rather than on respiration (see, e.g., Grossart et al. 2006).

The significant relationship with temperature (Fig. 9.2) suggests that bacterial respiration will increase with increasing water temperature. López-Urrutia and Morán (2007) show that bacterial respiration depends on temperature, and that this temperature dependence is similar to that of bacterial production. Thus, BGE would not change with increasing temperature, in contrast to the prediction of Rivkin and Legendre (2001).

López-Urrutia et al. (2006), using metabolic theory and plankton community size structure, predict that community respiration will increase relatively more than gross production with increasing seawater temperature. These authors found that the threshold value of gross production at which gross production exceeds community respiration increases with increasing temperature, ranging from less than 1 mmol $O_2/m^3/d$ in cold environments to more than 4 mmol $O_2/m^3/d$ at temperatures above 15 °C. This increase in net heterotrophy equates, by the end of the century, to a reduction of 4 Gt C/year (21 percent) in the amount of CO_2 captured by the marine biota. Such an effect is not currently incorporated into coupled atmosphere–ocean climate models, yet would significantly modify predicted global change.

Since a significant component of the organic carbon substrate for bacterial respiration is derived from phytoplankton activity, any shift in the magnitude of primary production or the relative proportion of dissolved and particulate production due to climate change is likely to impact bacterial respiration. Model simulations under warmer and more stratified conditions suggest that reduced nutrient availability would prevail in low and mid latitudes and shallower mixed layers would occur at high latitudes, together leading to an approximately 9 percent decrease in global primary production (Bopp et al. 2001). However, regional changes in phytoplankton community structure, such as a shift to diatoms with increased atmospheric iron deposition or a shift to calcifying organisms in a shallower mixed layer, may have a greater impact on the provision of bacterial substrates than a change in global primary production per se (Boyd and Doney 2002).

The absorption by the world's oceans of one-third of the anthropogenic carbon emissions each year is expected to cause surface water CO_2 concentrations to increase almost threefold and pH to decrease by up to 0.35 by the end of this century compared with pre-industrial values (Solomon et al. 2007). Such changes may have a profound effect on phytoplankton community structure, due to impaired biogenic calcification by coccolithophores (Riebesell et al. 2000). Little is known about the consequences of ocean acidification on bacterioplankton community structure or activity. In a mesocosm study, Grossart et al. (2006) found that bacterial production and hydrolytic ectoenzyme activity were highest at levels of pCO_2 equivalent to 700 ppm. This increase in bacterial activity was mainly due to the activity of particle-attached bacteria. The abundances of free and attached bacteria were highest in the high-pCO_2 mesocosm during the decline of the algal bloom, indicating that increased pCO_2 may have a larger impact when DOM is plentiful. Engel et al. (2004) found significant differences in the quality and quantity of phytoplankton-derived DOM between mesocosms maintained at past-, present-, and future-relevant concentrations of pCO_2. The relationship between pCO_2 and bacterial activity is, therefore, most likely an indirect one caused by pCO_2-induced changes in DOM produced by phytoplankton. Unfortunately, bacterial respiration was not directly measured during this study.

Atmospheric deposition of volatile organic compounds and photochemical breakdown of DOC to low-molecular-weight compounds such as acetate, formate, pyruvate, and methanol are also potentially significant sources of bacterial substrate that may change with alterations in climate (wind patterns, precipitation, and irradiance) (Duarte et al. 2006; Epp et al. 2007).

The complex and dynamic interactions between the various predicted physicochemical changes due to increasing emissions of CO_2 and marine bacterial activity are not well understood, and so our ability to predict the magnitude and sign of the response of bacterial respiration to global change is uncertain. Reducing the uncertainty in predicting the impact of global change on the functioning of the marine microbial foodweb will require a concerted international effort, and should be a high-priority research objective.

SUMMARY

1. The database of bacterial respiration measurements has improved recently, with more than 250 observations published since 2000. The median rates of bacterial respiration for open oceans and coastal regions are 0.5 and 2.7 mmol $C/m^3/d$, respectively. This represents a reduction in the mean bacterial respiration of 80 and 30 percent respectively, since a similar analysis was undertaken in 2000, likely due to an increase in data from oceanic and coastal sites with low productivity. However, this is still a woefully low number of direct measurements of bacterial respiration from which to describe a major, if not the dominant, route of carbon flow in the marine environment.
2. The percentage contribution of bacteria to community respiration varies with season and phytoplankton activity, with the mean for a particular study lying in either the 40–50 or the 70–80 percent range. Regression analysis of 14 studies suggests that bacterial respiration accounts on average for 45 percent of community respiration.
3. The median BGEs for open oceans and coastal regions are 0.08 and 0.16, respectively. Whereas BGE in eutrophic waters is more influenced by bacterial production, recent evidence suggests that in oligotrophic waters most of the variability in BGE is explained by bacterial respiration.
4. Published relationships between bacterial respiration and bacterial production and between BGE and temperature appear to be able to predict bacterial respiration to no better than a factor of five. The empirical relationship between bacterial respiration and bacterial production derived in this chapter predicts values of bacterial respiration midway between those derived from either previous model, likely due to improved coverage of the latest dataset. However, direct measurement at a range of time, space, and trophic scales remains the best approach to understanding how environment and community structure influence the variability of bacterial respiration.
5. For one open-ocean latitudinal study, estimates of the ratio of total primary production to bacterial respiration derived from two published predictive relationships differed substantially from the measured ratio between gross production and community respiration. Bacterial respiration predicted from the equation derived in this chapter provides ratios of total primary production to bacterial respiration that are closest to the measured ratios of gross production to community respiration. The data suggest that bacterial respiration could be wholly supported by concurrent phytoplankton production.
6. In a warmer and more acidic marine environment, caused by sustained anthropogenic emissions of carbon dioxide, indirect evidence suggests that bacterial respiration will increase due to the direct effect of increasing temperature and

the indirect effect of increasing availability of labile DOM caused by shifts in phytoplankton community structure and increased production of DOM by phytoplankton. However, considerable uncertainty exists in our ability to predict the complex interactions likely to occur between climate and the activity of the marine microbial biota.

ACKNOWLEDGMENTS

I thank Josep Gasol and Thomas Reinthaler for constructive comments that improved this chapter. Many thanks are due to Thomas Reinthaler, Rodolphe Lemée, Gwen Moncoiffé, Nuria Navarro, Natalia González, Pablo Serret, Emilio Fernández, Begona Castro, and Stephen Blight for access to their original data, to Paul del Giorgio for access to the database compiled for del Giorgio and Cole (2000), to Claudia Castellani and Gwen Moncoiffé for compiling the Atlantic Meridional Transect (AMT) data held at the British Oceanographic Data Centre (BODC), and to Reg Uncles and Gavin Tilstone for mathematical advice. This work was supported by the U.K. Natural Environment Research Council through the Plymouth Marine Laboratory Core Strategic Research Programme.

REFERENCES

Alderkamp, A. C., E. Sintes, and G. J. Herndl. 2006. Abundance and activity of major groups of prokaryotic plankton in the coastal North Sea during spring and summer. *Aquat. Microb. Ecol.* 45: 237–246.

Alonso-Sáez, L., J. M. Gasol, J. Arístegui, J. C. Vilas, D. Vagué, C. M. Duarte, and S. Agustí. 2007. Large-scale variability in surface bacterial carbon demand and growth efficiency in the subtropical northeast Atlantic Ocean. *Limnol. Oceanogr.* 52: 533–546.

Anderson, T. R., and H. W. Ducklow. 2001. Microbial loop carbon cycling in ocean environments studied using a simple steady-state model. *Aquat. Microb. Ecol.* 26: 37–49.

Anderson, T. R., and C. M. Turley. 2003. Low bacterial growth efficiency in the oligotrophic eastern Mediterranean Sea: A modeling analysis. *J. Plankton Res.* 25: 1011–1019.

Arístegui, J., S. Agustí, and C. M. Duarte. 2003. Respiration in the dark ocean. *Geophys. Res. Lett.* 30: 1041, doi: 10.1029/2002GL016227.

Arístegui, J., S. Agustí, J. J. Middelburg, and C. M. Duarte. 2005a. Respiration in the mesopelagic and bathypelagic zones of the oceans. In P. A. del Giorgio and P. J. leB. Williams (eds.), *Respiration in Aquatic Ecosystems*. Oxford University Press, pp. 181–205.

Arístegui, J., C. M. Duarte, J. M. Gasol, and L. Alonso-Sáez. 2005b. Active mesopelagic prokaryotes support high respiration in the subtropical northeast Atlantic Ocean. *Geophys. Res. Lett.* 32: L03608, doi: 10.1029/2004GL021863.

Attrassi, B., M. Saghi, and G. Flatau. 1993. Multiple antibiotic resistance of bacteria in Atlantic coast (Morocco). *Environ. Technol.* 14: 1179–1186.

Biddanda, B., S. Opsahl, and R. Benner. 1994. Plankton respiration and carbon flux through bacterioplankton on the Louisiana shelf. *Limnol. Oceanogr.* 39: 1259–1275.

Bjørnsen, P. K., and J. Kuparinen. 1991. Determination of bacterioplankton biomass, net production and growth efficiency in the Southern Ocean. *Mar. Ecol. Prog. Ser.* 71: 185–194.

Blight, S. P. 1996. Microbial metabolism and temperature. Comparative studies in the Southern Ocean and a temperate coastal ecosystem. PhD Thesis. University of Wales, Bangor, UK.

Blight, S. P., T. L. Bentley, D. Lefèvre, C. Robinson, R. Rodrigues, J. Rowlands, and P. J. leB. Williams. 1995. Phasing of autotrophic and heterotrophic plankton metabolism in a temperate coastal ecosystem *Mar. Ecol. Prog. Ser.* 128: 61–75.

Bopp, L., P. Monfray, O. Aumont, J.-L. Dufresne, H. Le Treut, G. Madec, L. Terray, and J. C. Orr. 2001. Potential impact of climate change on marine export production. *Global Biogeochem. Cycles* 15: 81–99.

Boyd, P. W., and S. C. Doney. 2002. Modelling regional responses by marine pelagic ecosystems to global climate change. *Geophys. Res. Lett.* 29: 1806–1809.

Boyd, P. W., C. Robinson, G. Savidge, and P. J. leB. Williams. 1995. Water column and sea-ice primary production during Austral spring in the Bellingshausen Sea. *Deep-Sea Res.* II 42: 1177–1200.

Briand, E., O. Pringault, S. Jacquet, and J. P. Torréton. 2004. The use of oxygen microprobes to measure bacterial respiration for determining bacterioplankton growth efficiency. *Limnol. Oceanogr. Meth.* 2: 406–416.

Calbet, A., and M. R. Landry. 2004. Phytoplankton growth, microzooplankton grazing, and carbon cycling in marine systems. *Limnol. Oceanogr.* 49: 51–57.

Calleja, M. L., C. M. Duarte, N. Navarro, and S. Agustí. 2005. Control of air–sea CO_2 disequilibria in the subtropical NE Atlantic by planktonic metabolism under the ocean skin. *Geophys. Res. Lett.* 32: L08606, doi: 10.1029/2004GL022120.

Carlson, C. A., and H. W. Ducklow. 1996. Growth of bacterioplankton and consumption of dissolved organic carbon in the Sargasso Sea. *Aquat. Microb. Ecol.* 10: 69–85.

Carlson, C. A., N. R. Bates, H. W. Ducklow, and D. A. Hansell. 1999. Estimation of bacterial respiration and growth efficiency in the Ross Sea, Antarctica. *Aquat. Microb. Ecol.* 19: 229–244.

Caron, D. A., J. C. Goldman, and T. Fenchel. 1990. Protozoan respiration and metabolism. In G. M. Capriulo (ed.), *Ecology of Marine Protozoa*. Oxford University Press, pp. 307–322.

Cherrier, J., J. E. Bauer, and E. R. M. Druffel. 1996. Utilization and turnover of labile dissolved organic matter by bacterial heterotrophs in eastern North Pacific surface waters. *Mar. Ecol. Prog. Ser.* 139: 267–279.

Chin-Leo, G., and R. Benner. 1992. Enhanced bacterioplankton production and respiration at intermediate salinities in the Mississippi River plume. *Mar. Ecol. Prog. Ser.* 87: 87–103.

Church, M. J., E. F. DeLong, H. W. Ducklow, M. B. Karner, C. M. Preston, and D. M. Karl. 2003. Abundance and distribution of planktonic Archaea and Bacteria in the waters west of the Antarctic Peninsula. *Limnol. Oceanogr.* 48: 1893–1902.

Codispoti, L. A., T. Yoshinari, and A. H. Devol. 2005. Suboxic respiration in the oceanic water column. In P. A. del Giorgio and P. J. leB. Williams (eds.), *Respiration in Aquatic Ecosystems*. Oxford University Press, pp. 225–247.

Coffin, R. B., J. P. Connolly, and P. S. Harris. 1993. Availability of dissolved organic carbon to bacterioplankton examined by oxygen utilization. *Mar. Ecol. Prog. Ser.* 101: 9–22.

Daneri, G., B. Riemann, and P. J. leB. Williams. 1994. In situ bacterial production and growth yield measured by thymidine, leucine and fractionated dark oxygen uptake. *J. Plankton Res.* 16: 105–113.

del Giorgio, P. A., and J. J. Cole. 1998. Bacterial growth efficiency in natural aquatic systems. *Ann. Rev. Ecol. Systematics* 29: 503–541.

del Giorgio, P. A., and J. J. Cole. 2000. Bacterial energetics and growth efficiency. In D. L. Kirchman (ed.), *Microbial Ecology of the Oceans*, 1st edn. Wiley-Liss, pp. 289–325.

del Giorgio, P. A., and C. M. Duarte. 2002. Respiration in the open ocean. *Nature* 420: 379–384.

del Giorgio, P. A., J. J. Cole, and A. Cimbleris. 1997. Respiration rates in bacteria exceed phytoplankton production in unproductive systems. *Nature* 385: 148.

Duarte, C. M., and S. Agustí. 1998. The CO_2 balance of unproductive aquatic ecosystems. *Science* 281: 234–236.

Duarte, C. M., S. Agustí, P. A. del Giorgio, and J. J. Cole. 1999. Regional carbon imbalances in the oceans. Response. *Science* 284: 1735b.

Duarte, C. M., S. Agustí, J. Arístegui, N. González, and R. Anadón. 2001. Evidence for a heterotrophic subtropical northeast Atlantic. *Limnol. Oceanogr.* 46: 425–428.

Duarte, C. M., J. Dachs, M. Llabres, P. Alonso-Laita, J. M. Gasol, A. Tovar-Sanchez, S. Sanudo-Wilhemy, and S. Agustí. 2006. Aerosol inputs enhance new production in the subtropical northeast Atlantic. *J. Geophys. Res. Biogeosci.* 111: G04006.

Ducklow, H. 2000. Bacterial production and biomass in the oceans. In D. L. Kirchman (ed.), *Microbial Ecology of the Oceans*, 1st edn. Wiley-Liss, pp. 85–120.

Ducklow, H. W., and C. A. Carlson. 1992. Oceanic bacterial production. *Adv. Microbial Ecol.* 12: 113–181.

Ducklow, H. W., D. A. Purdie, P. J. leB. Williams, and J. H. Davies. 1986. Bacterioplankton—a sink for carbon in a coastal marine plankton community. *Science* 232: 865–867.

Ducklow, H. W., D. A. Purdie, P. J. leB. Williams, and J. H. Davies. 1987. Bacteria—link or sink—response. *Science* 235: 88–89.

Ducklow, H. W., D. L. Kirchman, H. L. Quinby, C. A. Carlson, and H. G. Dam. 1993. Stocks and dynamics of bacterioplankton carbon during the spring bloom in the eastern North Atlantic Ocean. *Deep-Sea Res.* II 40: 245–263.

Ducklow, H. W., M.-L. Dickson, D. L. Kirchman, G. Steward, J. Orchardo, J. Marra, and F. Azam. 2000. Constraining bacterial production, conversion efficiency and respiration in the Ross Sea, Antarctica, January–February, 1997. *Deep-Sea Res.* II 47: 3227–3247.

Eissler, Y., and R. A. Quiñones. 2003. The effect of viral concentrate addition on the respiration rate of Chaetoceros gracilis cultures and microplankton from a shallow bay (Coliumo, Chile). *J. Plankton Res.* 25: 927–938.

Eissler, Y., E. Sahlsten, and R. A. Quiñones. 2003. Effects of virus infection on respiration rates of marine phytoplankton and microplankton communities. *Mar. Ecol. Prog. Ser.* 262: 71–80.

Engel, A., B. Delille, S. Jacquet, U. Riebesell, E. Rochelle-Newall, A. Terbrüggen, and I. Zondervan. 2004. Transparent exopolymer particles and dissolved organic carbon production by *Emiliania huxleyi* exposed to different CO_2 concentrations: A mesocosm experiment. *Aquat. Microb. Ecol.* 34: 93–104.

Epp, R. G., D. J. Erickson, N. D. Paul, and B. Sulzberger. 2007. Interactive effects of solar UV radiation and climate change on biogeochemical cycling. *Photochem. Photobiol. Sci.* 6: 286–300.

Fahnenstiel, G. L., H. J. Carrick, and R. Iturriaga. 1991. Physiological characteristics and food web dynamics of *Synechococcus* in Lakes Huron and Michigan. *Limnol. Oceanogr.* 36: 219–234.

Fuhrman, J. A., and F. Azam. 1980. Bacterioplankton secondary production estimates for coastal waters of British Columbia, Antarctica and California. *Appl. Environ. Microbiol.* 39: 1085–1095.

Fuhrman, J. A., and C. A. Suttle. 1993. Viruses in marine planktonic systems. *Oceanography* 6: 51–63.

Fukuda, R., H. Ogawa, T. Nagata, and I. Koike. 1998. Direct determination of carbon and nitrogen contents of natural bacterial assemblages in marine environments. *Appl. Environ. Microbiol.* 64: 3352–3358.

Gasol, J. M., and X. A. G. Morán. 1999. Effects of filtration on bacterial activity and picoplankton community structure as assessed by flow cytometry. *Aquat. Microb. Ecol.* 16: 251–264.

Gasol, J. M., M. D. Doval, J. Pinhassi, J. I. Calderón-Paz, N. Guixa-Boixareu, D. Vaqué, and C. Pedrós-Alió. 1998. Diel variations in bacterial heterotrophic activity and growth in the northwestern Mediterranean Sea. *Mar. Ecol. Prog. Ser.* 164: 107–124.

Geider, R. J. 1997. Photosynthesis or planktonic respiration? *Nature* 388: 132.

González, N., R. Anadón, and L. Viesca. 2003. Carbon flux through the microbial community in a temperate sea during summer: role of bacterial metabolism. *Aquat. Microb. Ecol.* 33: 117–126.

Griffith, P. C., D. J. Douglas, and S. C. Wainwright. 1990. Metabolic activity of size fractionated microbial plankton in estuarine, nearshore, and continental shelf waters of Georgia. *Mar. Ecol. Prog. Ser.* 59: 263–270.

Grossart, H.-P., M. Allgaier, U. Passow, and U. Riebesell. 2006. Testing the effect of CO_2 concentration on the dynamics of marine heterotrophic bacterioplankton. *Limnol. Oceanogr.* 51: 1–11.

Hansell, D. A., N. R. Bates, and K. Gundersen. 1995. Mineralization of dissolved organic carbon in the Sargasso Sea. *Mar. Chem.* 51: 201–212.

Hansell, D. A., H. W. Ducklow, A. M. Macdonald, and M. O. Baringer. 2004. Metabolic poise in the North Atlantic Ocean diagnosed from organic matter transports. *Limnol. Oceanogr.* 49: 1084–1094.

Harrison, W. G. 1986. Respiration and its size dependence in microplankton populations from surface waters of the Canadian Arctic. *Polar Biol.* 6: 145–152.

Herndl, G. J., T. Reinthaler, E. Teira, H. van Aken, C. Veth, A. Pernthaler, and J. Pernthaler. 2005. Contribution of Archaea to total prokaryotic production in the deep Atlantic Ocean. *Appl. Environ. Microbiol.* 71: 2303–2309.

Hopkinson Jr., C. S., and E. M. Smith. 2005. Estuarine respiration: an overview of benthic, pelagic, and whole system respiration. In P. A. del Giorgio and P. J. leB. Williams (eds.), *Respiration in Aquatic Ecosystems*. Oxford University Press, pp. 122–146.

Hoppe, H.-G., K. Gocke, R. Koppe, and C. Begler. 2002. Bacterial growth and primary production along a north-south transect of the Atlantic Ocean. *Nature* 416: 168–171.

Huan, J., X. X. Tang, X. Z. Gong, J. X. Dai, and Z. H. Chen. 2000. Selective marker of marine microalgal genetic engineering. *Acta Bota. Sin.* 42: 841–844.

IMBER. 2005. Science Plan and Implementation Strategy. IGBP Report No. 52, IGBP Secretariat, Stockholm.

Jahnke, R. A., and D. B. Craven. 1995. Quantifying the role of heterotrophic bacteria in the carbon cycle: The need for respiration rate measurements. *Limnol. Oceanogr.* 40: 436–441.

Jannasch, H. J., and C. O. Wirsen. 1982. Microbial activities in undecompressed microbial populations from deep seawater samples. *Appl. Environ. Microbiol.* 43: 1116–1124.

Joux, F., H. Agogue, I. Obernosterer, C. Dupuy, T. Reinthaler, G. J. Herndl, and P. Lebaron. 2006. Microbial community structure in the sea surface microlayer at two contrasting coastal sites in the northwestern Mediterranean Sea. *Aquat. Microb. Ecol.* 42: 91–104.

Karner, M. B., E. F. Delong, and D. M. Karl. 2001. Archaeal dominance in the mesopelagic zone of the Pacific Ocean. *Nature* 409: 507–510.

King, G. M. 2005. Ecophysiology of microbial respiration. In P. A. del Giorgio and P. J. leB. Williams (eds.), *Respiration in Aquatic Ecosystems*. Oxford University Press, pp. 18–35.

Kirchman, D. L. 1992. Incorporation of thymidine and leucine in the subarctic Pacific: Application to estimating bacterial production. *Mar. Ecol. Prog. Ser.* 82: 301–309.

Kirchman, D. L. 1993. Leucine incorporation as a measure of biomass production by heterotrophic bacteria. In P. F. Kemp, B. F. Sherr, E. B. Sherr, and J. J. Cole (eds.), *Handbook of Methods in Aquatic Microbial Ecology*. Lewis Publishers, pp. 509–512.

Kirchman, D. L., and H. W. Ducklow. 1993. Estimating conversion factors for the thymidine and leucine methods for measuring bacterial production. In P. Kemp, B. Sherr, E. Sherr, and J. J. Cole (eds.), *Handbook of Methods in Microbial Ecology*. Lewis Publishers, pp. 513–517.

Kirchman, D. L., E. K'Nees, and R. Hodson. 1985. Leucine incorporation and its potential as a measure of protein synthesis by bacteria in natural aquatic systems. *Appl. Environ. Microbiol.* 49: 599–607.

Kirchman, D. L., Y. Suzuki, C. Garside, and H. W. Ducklow. 1991. High turnover rates of dissolved organic carbon during a spring phytoplankton bloom. *Nature* 352: 612–614.

Kuipers, B., G. J. van Noort, J. Vosjan, and G. J. Herndl. 2000. Diel periodicity of bacterioplankton in the euphotic zone of the subtropical Atlantic Ocean. *Mar. Ecol. Prog. Ser.* 201: 13–25.

Lee, C. W., I. Kudo, T. Yokokawa, M. Yanada, and Y. Maita. 2002. Dynamics of bacterial respiration and related growth efficiency, dissolved nutrients and dissolved oxygen concentration in a subarctic coastal embayment. *Mar. Freshwater Res.* 53: 1–7.

Lemée, R., E. Rochelle-Newall, F. Van Wembeke, M.-D. Pizay, P. Rinaldi, and J.-P. Gattuso 2002. Seasonal variation of bacterial production, respiration and growth efficiency in the open NW Mediterranean Sea. *Aquat. Microb. Ecol.* 29: 227–237.

Li, W. K. W., P. M. Dickie, W. G. Harrison, and B. D. Irwin. 1993. Biomass and production of bacteria and phytoplankton during the spring bloom in the western North Atlantic Ocean. *Deep-Sea Res.* II 40: 307–327.

Longhurst, A. 1998. *Ecological Geography of the Sea*. Academic Press.

López-Urrutia, Á., and X. A. G. Morán. 2007. Resource limitation of bacterial production distorts the temperature dependence of oceanic carbon cycling. *Ecology* 88: 817–822.

López-Urrutia, Á., E. San Martin, R. P. Harris, and X. Irigoien. 2006. Scaling the metabolic balance of the oceans. *Proc. Natl. Acad. Sci. USA* 103: 8739–8744.

Martin, A. P., M. V. Zubkov, P. H. Burkill, and R. J. Holland. 2005. Extreme spatial variability in marine picoplankton and its consequences for interpreting Eulerian time-series. *Biol. Lett.* 1: 366–369.

Moncoiffé, G., X. A. Alvarez-Salgado, F. G. Figueiras, and G. Savidge. 2000. Seasonal and short-time-scale dynamics of microplankton community production and respiration in an inshore upwelling system. *Mar. Ecol. Prog. Ser.* 196: 111–126.

Nagata, T. 2000. Production mechanisms of dissolved organic matter. In D. L. Kirchman (ed.), *Microbial Ecology of the Oceans*, 1st edn. Wiley-Liss, pp. 121–152.

Navarro, N., S. Agustí, and C. M. Duarte. 2004. Plankton metabolism and dissolved organic carbon use in the Bay of Palma, NW Mediterranean Sea. *Aquat. Microb. Ecol.* 37: 47–54.

Obernosterer, I., P. Ruardij, and G. J. Herndl. 2001. Spatial and diurnal dynamics of dissolved organic matter (DOM) fluorescence and H_2O_2 and the photochemical oxygen demand of surface water DOM across the subtropical Atlantic Ocean. *Limnol. Oceanogr.* 46: 632–643.

Obernosterer, I., N. Kawasaki, and R. Benner. 2003. P-limitation of respiration in the Sargasso Sea and uncoupling of bacteria from P-regeneration in size-fractionation experiments. *Aquat. Microb. Ecol.* 32: 229–237.

Obernosterer, I., P. Catala, T. Reinthaler, G. J. Herndl, and P. Lebaron. 2005. Enhanced heterotrophic activity in the surface microlayer of the Mediterranean Sea. *Aquat. Microb. Ecol.* 39: 293–302.

Pérez, V., E. Fernández, E. Marañón, et al. 2005. Latitudinal distribution of microbial plankton abundance, production, and respiration in the Equatorial Atlantic in autumn 2000. *Deep-Sea Res.* I 52: 861–880.

Pomeroy, L. R. 1974. Oceans food web, a changing paradigm. *BioScience* 24: 499–504.

Pomeroy, L. R., and W. J. Wiebe. 2001. Temperature and substrates as interactive limiting factors for marine heterotrophic bacteria. *Aquat. Microb. Ecol.* 23: 187–204.

Pomeroy, L. R., J. E. Sheldon, and W. M. Sheldon. 1994. Changes in bacterial numbers and leucine assimilation during estimations of microbial respiratory rates in seawater by the precision Winkler method. *Appl. Environ. Microbiol.* 60: 328–332.

Pomroy, A., and I. Joint. 1999. Bacterioplankton activity in the surface waters of the Arabian Sea during and after the 1994 SW monsoon. *Deep-Sea Res.* II 46: 767–794.

Prabaharan, D., M. Sumathi, and G. Subramanian. 1994. Ability to use ampicillin as a nitrogen source by the marine cyanobacterium *Phormidium valderianum* BDU-30501. *Curr. Microbiol.* 28: 315–320.

Preen, K., and D. L. Kirchman. 2004. Microbial respiration and production in the Delaware Estuary. *Aquat. Microb. Ecol.* 37: 109–119.

Reinthaler, T. 2005. Prokaryotic respiration and production in the open ocean. PhD Thesis, NIOZ.

Reinthaler, T., and G. L. Herndl. 2005. Seasonal dynamics of bacterial growth efficiencies in relation to phytoplankton in the southern North Sea. *Aquat. Microb. Ecol.* 39: 7–16.

Reinthaler, T., C. Winter, and G. J. Herndl. 2005. Relationship between bacterioplankton richness, respiration and production in the southern North Sea. *Appl. Environ. Microbiol.* 71: 2260–2266.

Reinthaler, T., H. van Aken, C. Veth, J. Arístegui, C. Robinson, P. J. leB. Williams, P. Lebaron, and G. J. Herndl. 2006a. Prokaryotic respiration and production in the meso-and bathypelagic realm of the eastern and western North Atlantic basin. *Limnol. Oceanogr.* 51: 1262–1273.

Reinthaler, T., K. Bakker, R. Manuels, J. van Ooijen, and G. J. Herndl. 2006b. Fully automated spectrophotometric approach to determine oxygen concentrations in seawater via continuous flow analysis. *Limnol. Oceanogr. Meth.* 4: 358–366.

Riebesell, U., I. Zondervan, B. Rost, P. D. Tortell, R. E. Zeebe, and F. M. M. Morel. 2000. Reduced calcification in marine plankton in response to increased atmospheric CO_2. *Nature* 407: 634–637.

Riemann, B., P. K. Bjørnsen, S. Newell, and R. Fallon. 1987. Calculation of cell production of coastal marine bacteria based on measured incorporation of [^{3}H]thymidine. *Limnol. Oceanogr.* 32: 471–476.

Rivkin, R. B., and L. Legendre. 2001. Biogenic carbon cycling in the upper ocean: Effects of microbial respiration. *Science* 291: 2398–2400.

Robinson, C., and P. J. leB. Williams. 1999. Plankton net community production and dark respiration in the Arabian Sea during September 1994. *Deep-Sea Res*. II 46: 745–765.

Robinson, C., and P. J. leB. Williams. 2005. Respiration and its measurement in surface marine waters. In P. A. del Giorgio and P. J. leB. Williams (eds.), *Respiration in Aquatic Ecosystems*, Oxford University Press.

Robinson, C., S. D. Archer, and P. J. leB. Williams. 1999. Microbial dynamics in coastal waters of East Antarctica: Plankton production and respiration. *Mar. Ecol. Prog. Ser.* 180: 23–36.

Robinson, C., C. E. Widdicombe, M. V. Zubkov, G. A. Tarran, A. E. J. Miller, and A. P. Rees. 2002a. Plankton community respiration during a coccolithophore bloom. *Deep-Sea Res*. II 49: 2929–2950.

Robinson, C., P. Serret, G. Tilstone, E. Teira, M. V. Zubkov, A. P. Rees, and E. M. S. Woodward. 2002b. Plankton respiration in the Eastern Atlantic. *Deep-Sea Res*. I 49: 787–813.

Sampou, P., and W. M. Kemp. 1994. Factors regulating plankton community respiration in Chesapeake Bay. *Mar. Ecol. Prog. Ser.* 110: 249–258.

Sand-Jensen, K., L. M. Jensen, S. Marcher, and M. Hansen. 1990. Pelagic metabolism in eutrophic coastal waters during a late summer period. *Mar. Ecol. Prog. Ser.* 65: 63–72.

Serret, P., C. Robinson, E. Fernández, E. Teira, and G. Tilstone. 2001. Latitudinal variation of the balance between plankton photosynthesis and respiration in the eastern Atlantic Ocean. *Limnol. Oceanogr.* 46: 1642–1652.

Serret, P., E. Fernández, and C. Robinson. 2002. Biogeographic differences in the net ecosystem metabolism of the open ocean. *Ecology* 83: 3225–3234.

Serret, P., E. Fernández, C. Robinson, E. M. S. Woodward, and V. Pérez. 2006. Local productivity does not control the balance between plankton photosynthesis and respiration in the eastern Atlantic Ocean. *Deep-Sea Res*. II 53: 1611–1628.

Sherry, N. D., P. W. Boyd, K. Sugimoto, and P. J. Harrison. 1999. Seasonal and spatial patterns of heterotrophic bacterial production, respiration and biomass in the subarctic NE Pacific. *Deep-Sea Res*. II 46: 2557–2578.

Sherry, N. D., B. Imanian, K. Sugimoto, P. W. Boyd, and P. J. Harrison. 2002. Seasonal and interannual trends in heterotrophic bacterial processes between 1995 and 1999 in the subarctic NE Pacific. *Deep-Sea Res*. II 49: 5775–5791.

Simon, M., and F. Azam. 1989. Protein content and protein synthesis rates of planktonic marine bacteria. *Mar. Ecol. Prog. Ser.* 51: 201–213.

Smith, E. M. 1998. Coherence of microbial respiration rate and cell-specific bacterial activity in a coastal planktonic community. *Aquat. Microb. Ecol.* 16: 27–35.

Smith, E. M., and W. M. Kemp. 2001. Size structure and the production/respiration balance in a coastal plankton community. *Limnol. Oceanogr.* 46: 473–485.

Solomon, S., D. Qin, M. Manning, Z. Chen, M. Marquis, K. B. Averyt, M. Tignor, and H. L. Miller (eds.). 2007. *Climate Change 2007: The Physical Science Basis. Contribution of Working Group I to the Fourth Assessment Report of the Intergovernmental Panel on Climate Change.* Cambridge University Press.

Søndergaard, M., P. J. leB. Williams, G. Cauvet, B. Riemann, C. Robinson, S. Terzic, E. M. S. Woodward, and J. Worm. 2000. Net accumulation and flux of dissolved organic carbon and dissolved organic nitrogen in marine plankton communities. *Limnol. Oceanogr.* 45: 1097–1111.

Strayer, D. 1988. On the limits to secondary production. *Limnol. Oceanogr.* 33: 1217–1220.

Tamburini, C., J. Garcin, and A. Bianchi. 2003. Role of deep-sea bacteria in organic matter mineralization and adaptation to hydrostatic pressure conditions in the NW Mediterranean Sea. *Aquat. Microb. Ecol.* 32: 209–218.

Teira, E., M. J. Pazó, P. Serret, and E. Fernández. 2001. Dissolved organic carbon (DOC) production by microbial populations in the Atlantic Ocean. *Limnol. Oceanogr.* 46: 1370–1377.

White, P. A., J. Kalff, J. B. Rasmussen, and J. M. Gasol. 1991. The effect of temperature and algal biomass on bacterial production and specific growth rate in freshwater and marine habitats. *Microb. Ecol.* 21: 99–118.

Wilhelm, S. W., and C. A. Suttle. 1999. Viruses and nutrient cycles in the sea. *BioScience* 49: 781–788.

Williams, P. J. leB. 1981. Microbial contribution to overall marine plankton metabolism: Direct measurements of respiration. *Oceanol. Acta* 4: 359–364.

Williams, P. J. leB. 1984. A review of measurements of respiration rates of marine plankton populations. In J. A. Hobbie and P. J. leB. Williams (eds.), *Heterotrophic Activity in the Sea*. Plenum Press, pp. 357–389.

Williams, P. J. leB. 1998. The balance of plankton respiration and photosynthesis in the open oceans. *Nature* 394: 55–57.

Williams, P. J. leB. 2000. Net production, gross production and respiration: What are the interconnections and what controls what? In R. B. Hanson, H. W. Ducklow, and J. G. Field (eds.), *The Changing Ocean Carbon Cycle: A Midterm Synthesis of the Joint Global Ocean Flux Study*. Cambridge University Press, pp. 37–60.

Williams, P. J. leB., and D. Bowers. 1999. Regional carbon imbalances in the oceans. *Science* 284: 1735.

Williams, P. J. leB., and P. del Giorgio. 2005. Respiration in aquatic ecosystems: History and background. In P. A. del Giorgio and P. J. leB. Williams (eds.), *Respiration in Aquatic Ecosystems*, Oxford University Press, pp. 1–17.

Williams, P. J. leB., P. J. Morris, and D. M. Karl. 2004. Net community production and metabolic balance at the oligotrophic ocean site, station ALOHA. *Deep-Sea Res.* I 51: 1563–1578.

Zubkov, M. V., B. M. Fuchs, H. Eilers, P. H. Burkill, and R. Amann. 1999. Determination of total protein content of bacterial cells by SYPRO staining and flow cytometry. *Appl. Environ. Microbiol.* 65: 3251–3257.

Zubkov, M. V., M. A. Sleigh, and P. H. Burkill. 2001. Heterotrophic bacterial turnover along the 20°W meridian between 59°N and 37°N in July 1996. *Deep-Sea Res.* II 48: 987–1001.

10

RESOURCE CONTROL OF BACTERIAL DYNAMICS IN THE SEA

MATTHEW J. CHURCH

Department of Oceanography, University of Hawaii, Honolulu, HI 96822, U.S.A.

INTRODUCTION

Understanding the types of processes that limit the growth and accumulation of aquatic microbes is central to our knowledge of ocean ecosystem dynamics. However, such studies are often challenged by the physical, biological, and chemical complexity of the marine environment. Numerous factors can regulate microorganism population dynamics, often simultaneously. Such controls include biotic factors such as predation, disease, and competition; and abiotic controls such as temperature, pH, nutrient availability, and sunlight. In the marine environment, these factors regulate microbial biomass and growth over spatial and temporal scales ranging from the cellular to the ecosystem level (Cullen 1991).

When describing those factors that control plankton population dynamics, it is important to determine whether the limiting resource controls the standing biomass, or whether a limiting resource restricts an organism's growth rate. In his treatise titled *Organic Chemistry in Its Applications to Agriculture and Physiology*, Liebig (1840) described conditions controlling agricultural crop yields, postulating that yields were limited by the resource in shortest supply relative to the crop's requirement for the resource. This concept, referred to as the law of the minimum

---

Microbial Ecology of the Oceans, Second Edition. Edited by David L. Kirchman
Copyright © 2008 John Wiley & Sons, Inc.

(see the text box), has been widely used to characterize situations where a specific resource limits the abundance or standing stock of planktonic populations (de Baar 1994). In addition to controlling standing biomass, resources can also restrict rates of population metabolism or growth (Blackman 1905). In ocean ecosystems, the scant availability of certain nutrient resources can limit both standing biomass and depress microorganism growth rates. However, determining the types of resources that control microorganism standing stocks or metabolism in situ is seldom straightforward: versatility in bacterial physiology, low microorganism abundances, complex nutrient substrate bioavailability, and heterogeneity (both spatially and temporally) in resource availability make characterizing the specific factors limiting plankton dynamics a daunting task. In addition, in many oceanic ecosystems, multiple resources appear to colimit plankton growth and biomass, further complicating identification of specific resources limiting plankton assemblages (Terry et al. 1985; Egli 1991; Arrigo 2005; Saito et al. 2008).

Terms for Two Types of Limitations

Law of the Minimum Characterization of biomass limitation by a single limiting resource; the law states that the resource in shortest supply relative to its demand limits biomass yield.

Blackman Limitation Resource control of microorganism physiology or specific growth rate; the Monod growth model characterizes the Blackman-type growth limitation.

This chapter will discuss some of the processes controlling the growth, population size, and diversity of free-living heterotrophic bacteria living in the well-lit regions of the open ocean. Heterotrophic bacteria play interlinked roles as both consumers and producers of organic matter (inseparably linking them to higher trophic levels), and recyclers of mineral nutrients. For the purpose of this review, I restrict my discussion to resources that control "heterotrophic" bacterial dynamics, which I define as those prokaryotic members of the picoplankton (<2 μm in diameter) that assimilate organic matter as a principle source of energy and carbon.

GROWTH IN THE SEA

With the exception of hydrothermal systems, sunlight is the ultimate source of energy to marine ecosystems. The harvesting of energy via biologically mediated phototrophic processes fuels all subsequent metabolic reactions in planktonic food webs. Aside from the narrow vertical section of the upper ocean that receives radiant energy, the vast majority of the ocean's interior is deprived of this energy source. In this respect, we need look no further in our search to define the specific resources that control microbial dynamics in the sea: in most of the ocean, energy limits bacterioplankton growth and metabolism.

It is important to emphasize a central distinction between the flow of energy and the movement of elements through the marine environment: energy follows a unidirectional path, while bioelements cycle (Fig. 10.1). Energy enters aquatic systems in the form of solar radiation and undergoes biological conversion to chemical energy, eventually being dissipated to the system as heat. In this sense, ecosystems serve as complex systems for dissipating energy (Reynolds 2001). In contrast to this unidirectional energy flow, bioelements are cycled through microbial assimilation and regeneration. All aspects of the hierarchical organization of ocean ecosystems, from genes to cells to populations to communities, shape the pathways that energy and material travel through the environment. The time required for a complete turn of the loop depicted in Figure 10.1 is determined by the complexity of the food web and the amount of energy harvested by cellular metabolism. As a result, energy flow and elemental cycling are inseparable; the efficiency of energy harvesting depends in part on the availability of nutrients, and the turnover of the nutrient pools depends on the flow of energy.

In the ocean, energy input and material cycles are often separated in space. The primary source of energy to plankton food webs is sunlight; however, the pull of gravity results in a downward flux of photosynthetically fixed organic matter and hence delivery of this biologically fixed energy to the deep sea. The high-energy region where light impinges on the surface ocean tends to be nutrient-poor, while the low-energy region of the ocean remains relatively enriched in mineral nutrients.

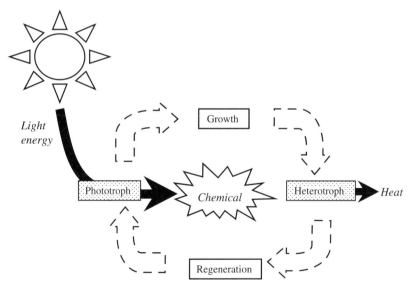

Figure 10.1 Depiction of the interconnected unidirectional flow of energy and cyclic flow of matter through ocean ecosystems. The dark solid lines represent energy flow, while dashed lines are elemental cycling. Radiant energy is harvested by biology, stored as ATP, and used to drive cell production and growth. Organic matter is produced via assimilation of mineral nutrients that are recycled through food web processes.

The thermocline forms a density barrier, restricting replenishment of nutrients into the energy-rich euphotic zone (Fig. 10.2). Thus, although the majority of the ocean may be limited by energy, in the well-lit upper ocean, organic matter production is often limited by the availability of nutrients.

Heterotrophic bacteria obtain energy via the oxidation of dissolved organic matter (DOM). DOM is a bioactive reservoir of reduced carbon substrates; many of these substrates also contain essential nutrients that support bacterial growth, including nitrogen, phosphorus, sulfur, and iron. DOM comprises an enormous potential source of energy and reduced carbon for bacterial growth; however, the vast majority of DOM is biochemically resistant, presumably reflecting its poor nutritional and energetic content (Williams 2000). Concentrations of DOM are typically elevated in the productive upper ocean, decreasing approximately twofold to the bottom of the sea (deep-sea DOC concentrations typically range from 40–45 μmol/L).

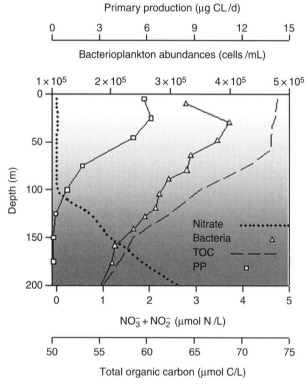

Figure 10.2 Vertical profiles of inorganic nutrients ($NO_3^- + NO_2^-$), primary production (PP), bacterioplankton abundance, and total organic carbon (TOC) in the upper ocean at Station ALOHA in the central North Pacific. Rates of PP and concentrations of TOC and bacterioplankton abundances are greatest in the high-energy region of the upper ocean. In contrast, the dimly lit region of the lower euphotic zone supports lower productivity and decreasing concentrations of bacterioplankton and TOC.

Bacterial abundances are usually greatest in the upper ocean where organic matter fluxes are high (Fig. 10.2). In pelagic ecosystems, DOM production is constrained by photosynthesis, and a substantial fraction (generally 10–20 percent) of photosynthetically fixed energy channels directly into the DOM pool via photosynthetic exudation or excretion (Baines and Pace 1991; Nagata 2000). In addition, trophodynamic processes introduce additional sources of DOM through sloppy feeding or viral lysis (Nagata 2000; Carlson 2002). Depending on the ecosystem, 50–100 percent of the photosynthetically fixed organic matter eventually channels through the DOM pool to fuel bacterial growth.

As obligate osmotrophs (see the text box), bacteria are confined to utilize relatively small solutes, roughly less than 500 Da. DOM consists of a broad spectrum of reduced carbon compounds, most of which remain chemically uncharacterized; however, among the identifiable compounds, polymers such as polysaccharides and protein comprise a large (20–30 percent) fraction of the bulk pool (Pakulski and Benner 1994; Hedges 2002). The structural complexity and size of these biopolymers precludes direct bacterial transport, requiring mobilization of hydrolytic enzymes to cleave the polymeric substrates into monomers that are readily transported into the cell.

Osmotroph An organism that obtains nutrients and energy via passive or active transport of low-molecular-weight (<500 Da) substrates across the cell membrane.

Growth and Nutrient Uptake Kinetics

The diverse physiological capabilities found among bacterioplankton suggest that competition for resources in the oceans is intense. Bacteria obtain most of their nutritional and energetic requirements through active transport of inorganic and organic substrates. Transport systems are functionally characterized based on the mode of transport and the energy source that fuels the enzymatic activity (Saier 2003). The general classes of transporters include channel proteins; primary, secondary, and group translocaters; and ligand shuttles. Channel proteins are energy-independent, passive systems that control diffusion of solutes into the cell; such systems are generally rare among bacteria (Paulsen et al. 2000). Primary transport systems harness energy from light, redox potential, or chemical energy and hydrolyze adenosine triphosphate (ATP) during solute uptake; these systems include ATP-binding cassette (ABC) transporters. Secondary carriers are a diverse group of transporters that rely on chemiosmotic energy established by ion pumping across cell membranes; these systems catalyze uni-, anti-, and symport of solutes (Paulsen et al. 2000; Konings 2006). Group translocation or phosphotransferase systems couple substrate phosphorylation with transport across the cytoplasmic membrane. Hydrophilic substrates such as mineral salts and amino acids are frequently transported by primary or secondary carriers, while group translocation is used for transport of certain sugars. Ligand shuttles bind substrate outside the cell, and both the substrate and carrier protein are assimilated; such shuttles are common for acquisition of trace metals such as iron (Button 1985).

In the nutrient-limited open ocean, bacterial growth depends on two processes: the rate that cells transport and assimilate growth-limiting nutrients and the rate that cells metabolize intracellular substrates (Button 1985). Under steady state, nutrient-limited bacterial growth depends on the transport rate of the growth-limiting nutrient. Since nutrient transport is catalyzed by enzymatic reactions, transport can be described using a Michaelis–Menten kinetic model:

$$V = \frac{-dS_{out}}{dt} = \frac{V_{max}[S_{out}]}{K_t + [S_{out}]}$$

where V is the rate of nutrient transport, V_{max} is the maximum rate of nutrient transport, S_{out} is the concentration of limiting nutrient in the environment, and K_t is the substrate concentration where $V = \frac{1}{2}V_{max}$. The kinetic constants V_{max} and K_t vary depending on the availability of growth-limiting substrate in the environment.

If the nutrient (or energy source) limiting bacterial growth is known, growth can be characterized as a function of extracellular nutrient concentration, as described by the Monod growth model:

$$\mu = \frac{\mu_{max}[S_{out}]}{K_\mu + [S_{out}]}$$

where μ is the specific growth rate of the population, μ_{max} is the maximum growth rate of the population under nutrient sufficient conditions, S_{out} is the concentration of the limiting substrate outside the cell, and K_μ is the concentration of substrate where $\mu = \frac{1}{2}\mu_{max}$. At steady state, any reduction in μ below μ_{max} is defined as a function of substrate availability. The Monod model is an example of the Blackman-type limitation, where substrate concentration regulates μ. Note that this hyperbolic function also applies to energy-limited cell growth, where production depends on acquisition of energy. At low concentrations of a limiting resource ($S_{out} < K_\mu$), μ increases rapidly and bacterial growth scales linearly with the availability of the resource. With increasing availability of the resource ($S_{out} > K_\mu$), μ "saturates" (reaches a maximum) and becomes independent of the concentration of the resource. In this case, growth becomes limited by other resources or processes.

Bacteria respond to fluctuating nutrient availability by regulating the number of active transporters used to acquire substrates. Under conditions of low substrate availability, a greater fraction of the total cellular energy can be devoted to substrate uptake, thereby increasing an organism's affinity for that substrate. Both organic and inorganic substrate concentrations in much of the world's oceans are very low (nanomolar to micromolar ranges), suggesting that bacteria growing in low nutrient seawater would favor allocation of cellular energy to acquiring resources (via synthesis of transport proteins) at the expense of growth. In contrast, bacteria growing under more eutrophic conditions or in close proximity to substrate sources might maximize growth while decreasing their affinity for specific substrates. The observed variability in marine bacterial cellular stoichiometry, specifically N:P ratios, may mirror such adaptive responses to nutrient availability. In low-nutrient environments, greater allocation of resources towards synthesis of N-rich transport proteins would tend to increase cellular N:P ratios. In contrast, under nutrient-replete conditions where growth is favored, increased production of P-rich nucleic acids would lower cellular N:P ratios (Klausmeier et al. 2004; Arrigo 2005).

Marine bacteria are able to utilize very low concentrations of inorganic and organic solutes. Button (1998) concluded that the dominant groups of pelagic marine bacteria possess multiple high-affinity transport systems that enable them to simultaneously transport several types of nutrient substrates. Although the kinetic parameter K_t is often used to describe substrate affinity, at low substrate concentrations K_t become sensitive to variations in V_{max}. Because of this, Button (1991) redefined substrate affinity from the biomass normalized initial slope ($\alpha°$) of the kinetic response to substrate concentration. Low values of $\alpha°$ reflect lower substrate affinity (i.e., maximal growth is achieved at elevated substrate concentrations), while high values of $\alpha°$ indicate an increase in substrate affinity (maximal growth occurs at low nutrient concentrations).

The allocation of cellular energy towards biosynthesis and transport may also control competition and population succession amongst the bacterioplankton. Figure 10.3 depicts several hypothetical kinetic responses of two populations of bacteria to variable concentrations of a single growth limiting resource. In (Fig. 10.3a) the two populations have identical maximal growth rates, but population A has a greater affinity ($\alpha_A > \alpha_B$ and $K_{tA} < K_{tB}$) for the substrate than population B; at low substrate concentrations, population A would out compete B for the limiting resource. In (Fig. 10.3b), populations A and B have similar substrate affinities ($\alpha_A° = \alpha_B°$), but differing K_t and μ_{max} values ($K_{tA} < K_{tB}$, $\mu_{maxB} > \mu_{maxA}$). Under this scenario, population A would dominate at low substrate concentrations, but population B would increase as resources become more available. In (Fig. 10.3c), both populations differ in μ_{max} and substrate affinities. At low resource availability, population A will dominate, with population B out-competing A along a gradient in increasing resource availability. Competitive exclusion would suggest that, in equilibrium, two species competing for the same resource pools cannot stably coexist (Hardin 1960); however, assemblages of marine bacteria typically contain diverse coexisting populations. This apparent exception to the competitive exclusion principle has been coined the "paradox of the plankton" (Hutchinson 1961). The spatial and temporal complexity of the seascape appears to drive diversification through some combination of bottom-up and top-down processes. The dynamic interaction of multiple growth limiting resources and selective mortality drive physiological specialization and permit stable coexistence of numerous bacterial populations.

The Principle of Competitive Exclusion This is the ecological theory that two species competing for the same resources cannot stably coexist; competition for identical resource pools leads to the exclusive dominance of one species.

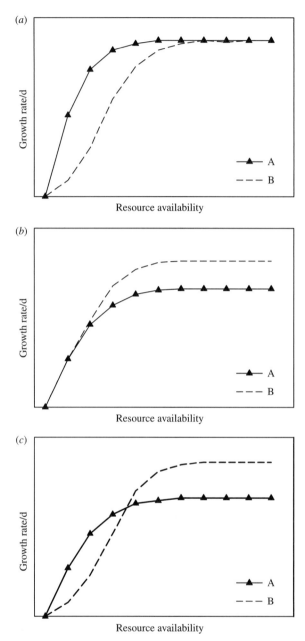

Figure 10.3 Hypothetical kinetic responses of two populations (A and B) competing for limiting resources. In (*a*), populations A and B have identical maximal growth rates but differing substrate affinities. In (*b*), both populations have equivalent substrate affinity ($\alpha°$), but population B has a greater maximal growth rate. In (*c*), population A demonstrates increased substrate affinity but a lower growth rate than population B. See the text for additional description. Modified from Lalli and Parsons (1993).

APPROACHES TO UNDERSTANDING RESOURCE CONTROL OF GROWTH

Comparative Approaches

Following the realization that bacteria play important roles in marine food webs (Pomeroy 1974; Azam et al. 1983), microbial ecologists began an impressive campaign to quantify the contributions of bacteria to ocean ecosystem biomass and productivity. These efforts have provided large datasets useful for evaluating controls on bacterioplankton dynamics. Development of conceptual and mechanistic models (Fasham et al. 1990; Anderson and Ducklow 2001) and comparative ecosystem analyses (Gasol and Duarte 2000; Li et al. 2004) utilizing these data have provided predictive approaches for assessing the relative importance of top down and bottom up processes (see the text box) in controlling bacterioplankton dynamics.

Comparative ecosystem analyses are used to define relationships between two ecosystem properties or processes through assessment of a statistical relationship. The resulting relationship can be used to define a predictive range of variability between two ecosystem properties (Gasol and Duarte 2000). There are a number of excellent syntheses that identify the comparative ecology of bacteria in the oceans (Ducklow and Carlson 1992; Gasol and Duarte 2000; Li et al. 2004), and the reader is referred to these studies for more comprehensive treatment of the subject. A few important conclusions from these studies will be highlighted here.

Two Types of Mechanisms Controlling Populations

Bottom-up Control of microorganism growth or population size by resource availability, including nutrient or energy sources.

Top-down Control of microorganism biomass by mortality, including factors such as predation or viral lysis.

One of the more robust trends identified from comparative analyses is that across diverse aquatic ecosystems, bacterial abundance generally demonstrates a positive relationship to chlorophyll (Bird and Kalff 1984; Cole et al. 1988; Ducklow and Carlson 1992; Gasol and Duarte 2000). The nature of this relationship, generally taken as the power slope of the regression of bacterial abundance versus chlorophyll concentration, has been used to identify two general trends: (1) the upper limit of bacterioplankton biomass is controlled by the availability of phytoplankton derived resources; (2) bacterial biomass becomes proportionally more important in ecosystems where photosynthetic biomass is low.

Li et al. (2004) compared the relationship between abundance and chlorophyll across diverse marine ecosystems (13,973 paired measurements). In their study, the power slope of the abundance–chlorophyll regression averaged 0.46. The proportionally lower bacterial abundance relative to chlorophyll suggests that bacterial biomass may be under stricter top down control than phytoplankton biomass. Ducklow (1999)

TABLE 10.1 Ranges of Surface-Water Temperature, Nutrients, Chlorophyll, and Bacterioplankton Biomass and Production in Several Oceanic Ecosystems

Region	Temperature (°C)	NO_3^- (μmol/L)	PO_4^{3-} (μmol/L)	Fe (nmol/L)	Chl a (μg Chl/L)	BB (μg C/L)	BP (μg C/L/d)
Oligotrophic							
HOT[a]	23–27	0.001–0.019	0.006–0.19	0.2–0.7	0.02–0.2	2.3–13	0.04–1.2
BATS[b]	19–29	<0.05–0.65	<0.03–0.11	0.09–2.0	0.02–0.4	1.8–20	0.1–6.2
Mediterranean Sea[c]	13–25	0.001–2.0	<0.002–0.23	<0.2–4.8	0.09–0.4	2.2–10	0.1–5.2
Temperate							
NABE[d]	7–15	<1.0–9.0	0.1–1.0	0.07–0.10	0.4–3.1	5–40	3.6–9
HNLC							
Subarctic North Pacific[e]	5–18	6–17	0.2–1.5	0.04–0.16	0.1–1.8	7–25	0.5–2.0
Equatorial Pacific[f]	24–30	3–11	0.2–0.9	<0.03–0.05	0.1–0.6	14–18	0.6–2.4

High-latitude

Ross Sea[g]	−1.8–0.01	8–31	0.4–2.0	<0.05–0.25	0.03–4.4	0.4–11	0.02–1.4
Arctic Ocean[h]	−1.8–0.2	<0.2–15	0.8–2.0	0.9–4.5	0.07–4.3	2.6–13	0.3–15

NO_3^-, nitrate; PO_4^{3-}, soluble reactive phosphate; Fe, dissolved iron; (Chl *a*), chlorophyll *a*; BB, bacterial biomass; BP, bacterial production.

[a]Temperature, NO_3^-, PO_4^{3-}, Chl *a*, and BB from Hawaii Ocean Time-series (HOT) program public data (http://hahana.soest.hawaii.edu/hot/hot-dogs/); Fe from Boyle et al. (2005); BB calculated from assuming 15 fg C per cell; BP calculated from Church et al. (2006), assuming 1.5 kg C/mol leucine incorporated.

[b]Temperature, Chl *a*, BB, and BP from Bermuda Atlantic Time-series Study (BATS) public data (http://bats.bios.edu/); BB calculated assuming 15 fg C per cell, BP calculated from [^{3}H]thymidine incorporation rates assuming 2.0×10^{18} cells/mol thymidine incorporated and 15 fg C per cell; NO_3^-, PO_4^{3-} from Cavendar-Bares et al. (2001); Fe from Sedwick et al. (2005).

[c]Temperature, NO_3^-, PO_4^{3-}, Chl *a*, BB, and BP from Sala et al. (2002), Thingstad et al. (2005), and Zohary and Roberts (1998); Fe from Sarthou and Jeandel (2001) and Bonnet and Guieu (2006).

[d]Temperature, PO_4^{3-}, and Chl *a* from publicly available U.S.J.G.O.F.S. North Atlantic Bloom Experiment data (http://usjgofs.whoi.edu/jg/dir/jgofs/nabe/atlantisII/); NO_3^- from Garside and Garside (1993); Fe from Martin et al. (1993); BB and BP from Ducklow et al. (1993) and Li et al. (1993).

[e]Temperature, NO_3^-, PO_4^{3-}, and Chl *a* from offshore stations along Line P data (http://www-sci.pac.dfo-mpo.gc.ca/osap/data/linep/linepselectdata_e.htm); Fe from Martin et al. (1988); BB and BP from Kirchman et al. (1993).

[f]Temperature, NO_3^-, PO_4^{3-}; Fe, and Chl *a* from publicly available U.S.J.G.O.F.S. Equatorial Pacific process cruise data (http://usjgofs.whoi.edu/jg/dir/jgofs/eqpac/); BB and BP from Ducklow et al. (1995) and Kirchman et al. (1995).

[g]Temperature, NO_3^-, PO_4^{3-}, Chl *a*, and Fe from publicly available U.S.J.G.O.F.S. Ross Sea process cruise data (http://usjgofs.whoi.edu/jg/dir/jgofs/southern/); BB and BP from Ducklow et al. (2001).

[h]Temperature, NO_3^-, PO_4^{3-}, Chl *a*, BB, and BP from Cota et al. (1996), Sherr et al. (2003), and Sherr and Sherr (2003); Fe from Measures (1999).

reached a similar conclusion in his comparison of bacterial and phytoplankton biomass across several ocean ecosystems. Table 10.1 provides ranges of bacterial biomass and production in the surface waters (<10 m) of several ocean ecosystems. The observed ranges in these data suggest that surface water bacterial biomass generally varies by less than an order of magnitude, with greater variability being observed in more seasonally forced systems and lower variability in seasonally stable systems.

Another striking finding emerging from these comparative studies is that bacterial biomass becomes an increasingly important component of total plankton biomass in those environments where chlorophyll is low (Gasol et al. 1997; Cotner and Biddanda 2002). Such analyses indicate bacterial biomass may rival or exceed photosynthetic biomass in unproductive aquatic systems where chlorophyll is less than 0.05–1 μg/L. It is important to note that these conclusions are highly dependent on the choice of conversion factors used to estimate bacterial (carbon per cell) and phytoplankton (C:Chl) biomass. In addition, many estimates of bacterial abundance are derived by epifluorescent microscopy, and thus likely include unicellular cyanobacteria such as *Prochlorococcus* in estimates of heterotrophic biomass. This is particularly problematic in the tropical and subtropical gyres, which also are generally systems demonstrating low chlorophyll concentrations.

The general trend of increasing bacterial biomass relative to total plankton biomass can be seen in Figure 10.4. In the examples shown, bacterial biomass becomes increasingly important as ecosystem productivity declines. During the highly productive North Atlantic Bloom (Fig. 10.4c), bacterial biomass was a lower fraction of total plankton biomass than has been observed in other ecosystems such as the high-nutrient-low-chlorophyll (HNLC) Equatorial Pacific or the subtropical North Pacific.

Correlative analyses have also been used to assess the dependence of volumetric (units of per liter or per m^3) and areal (depth-integrated, units of per m^2) rates of bacterial production on primary production. The strength of these relationships is generally assumed to reflect the degree of coupling between bacterial growth and primary production. Over large space and time scales, variations in the magnitude of primary production appear mirrored in bacterial production, with euphotic zone bacterial production equivalent to 10–30 percent of primary production (Cole et al. 1988; Ducklow 1999). In a review on the contributions of bacterial biomass and production to the upper ocean, Ducklow (1999) examined the relationships between bacteria and photoautotrophic plankton in highly seasonal (Ross Sea), temperate (North Atlantic and subarctic North Pacific), and seasonally stable (central gyres and Equatorial Pacific) ecosystems. Across these diverse environments, bacterial biomass varied between 217 and 1500 mg C/m^2, with the lowest bacterial biomass being found in the seasonally hyperproductive Ross Sea and peak biomass observed in the central North Pacific. In contrast, photoautotrophic biomass in these same ecosystems demonstrated much greater variability (ranging between 447 and 11,450 mg C/m^2). The magnitudes of primary production and bacterial production were about the same (465–1548 mg $C/m^2/d$ and 55–285 mg $C/m^2/d$, respectively).

APPROACHES TO UNDERSTANDING RESOURCE CONTROL OF GROWTH

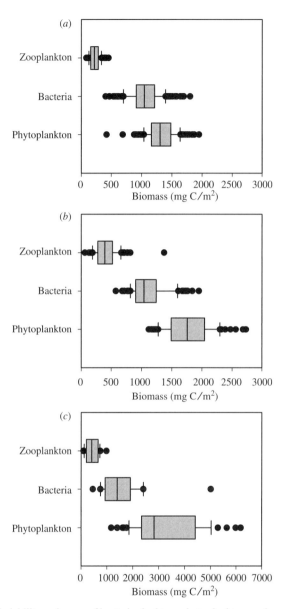

Figure 10.4 Variability and range of bacterioplankton, phytoplankton, and zooplankton euphotic zone biomass in three ocean ecosystems: (*a*) Hawaii Ocean Time-series (HOT); (*b*) equatorial Pacific; (*c*) and North Atlantic Bloom Experiment (NABE). The upper and lower boundaries of each box represent the 75th and 25th percentiles of the data; the line inside each box represents the median value; capped bars show the 90th and 10th percentiles of the data; filled circles are data lying outside the 90th and 10th percentiles. Phytoplankton biomass calculated based on euphotic zone chlorophyll *a* inventories assuming a carbon : chlorophyll ratio of 50 : 1. Bacterial biomass is calculated from cellular abundances assuming 15 fg C per cell. All data were obtained from the U.S. J.G.O.F.S. data server (http://usjgofs.whoi.edu/jg/dir/jgofs/).

> **Bacterial production** is the rate of synthesis of bacterial biomass.
>
> **Bacterial carbon demand** is the total amount of carbon required to support bacterial growth; it includes carbon consumed in respiration and synthesis of new biomass.

We currently have no direct method of determining the flux of DOM that supports bacterial growth; however, there are proxies useful for constraining this flux (see Chapter 8). The total flux of carbon required to support bacterial growth, termed bacterial carbon demand (BCD: see the text box), includes the carbon used for biosynthesis (production, BP) and respiration. If respiration and production can be estimated, we have a reasonable measure of the BCD (Ducklow 2000). Bacterial growth efficiency (BGE) is defined as the yield of bacterial biomass relative to the total amount of material utilized for growth, or mathematically:

$$BGE = BP/BCD.$$

Measurements of BGE in the oceans suggest that on average about 15 percent of the total carbon consumed by bacteria fuels biomass production. Assuming a BGE= 0.15 for all the ecosystems examined by Ducklow (1999), BCD was equivalent to about 29–175 percent of contemporaneous primary production (Fig. 10.5).

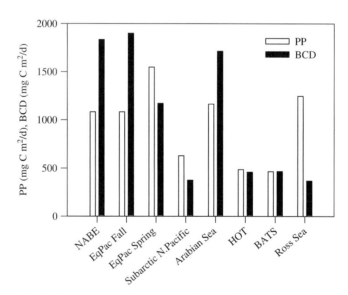

Figure 10.5 Euphotic zone rates of primary production (PP) and bacterial carbon demand (BCD) for the ocean ecosystems evaluated by Ducklow (1999). BCD was calculated assuming 15 percent bacterial growth efficiency for all ecosystems. NABE, North Atlantic Bloom Experiment; EqPac, Equatorial Pacific; HOT, Hawaii Ocean Time-series; BATS, Bermuda Atlantic Time-series Study.

Inclusion of respiratory losses demonstrates that bacterial growth consumes a large fraction of the photosynthetically fixed carbon. It is important to note that these estimates of primary production do not include direct estimates of DOM production, which can vary from 10 to 50 percent of the measured rate of particulate primary production (Carlson 2002). In this regard, the comparison provides a lower limit for the actual rate of primary production.

This exercise highlights a recurring and perplexing observation derived from comparative analyses; in aquatic ecosystems where primary production is low (<70 μg C/L/d or $<$about 1 g C/m^2/d), BCD often appears to exceed measured primary production (del Giorgio et al. 1997). Euphotic zone rates of primary production throughout much of the open ocean fall below these rates, implying that bacterial growth (including production and respiration) in some of the largest ecosystems on Earth consumes more carbon than is locally produced. Various experimental studies support this hypothesis (Duarte and Agustí 1998; del Giorgio and Duarte 2002; Williams et al. 2004); if shown to be true, this would have profound implications toward our understanding of energy flow and nutrient cycling through food webs of the open sea.

Experimental Approaches for Defining Limitation of Bacterial Growth

The most common way to evaluate controls on bacterioplankton dynamics is by direct experiments. Experiments can be useful for evaluating controls on microbial physiology and biomass across scales ranging from individual genes to ecosystems. The most common experimental approach to assessing resource limitation of bacterial growth is to measure changes in one or more proxies of microbial biomass (most often cell abundance), production, or metabolism (e.g., respiration) over some time period following the addition of potentially growth limiting resources.

Figure 10.6 depicts three hypothetical responses to experimental treatments aimed at evaluating the relative influence of top-down and bottom-up processes in controlling bacterial population size. In the first example (Fig. 10.6a), the addition of nutrients stimulates bacterial growth, while removal of predators has no influence on bacterial abundance (relative to the unamended control). Eventually, bacterial abundance stabilizes as the population size consumes the available resources. Such a scenario exemplifies a Liebig-like limitation of bacterial growth (Cullen 1991). In the second example (Fig. 10.6b), the addition of nutrients has no appreciable influence on bacterial growth, while removal of predators relieves top-down control on population size. In the final example (Fig. 10.6c), bacterial growth increases upon addition of nutrients, but the resulting size of the population is controlled by predation. Thus, when predators are removed and nutrients are added, the population size exceeds that of the nutrient-only treatment. In natural systems, the upper limit of bacterial population size is defined by the availability of resources, but the actual or realized size of the population is controlled by mortality through predation or disease (Carpenter et al. 1985; McQueen et al. 1986; Ducklow and Carlson 1992).

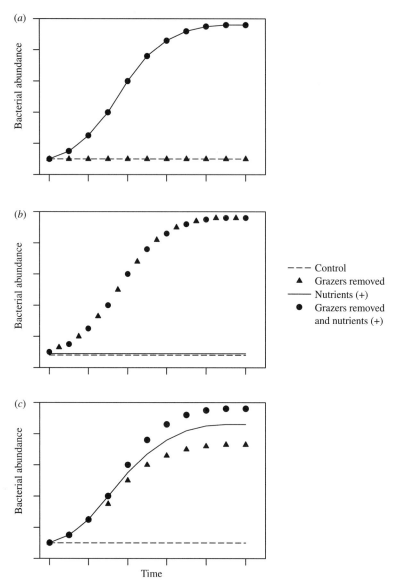

Figure 10.6 Hypothetical responses in bacterioplankton population size to nutrient-dose and grazer-removal experiments. In (*a*), bacterial abundance increases in the two treatments receiving nutrients, but does not respond to grazer removal; this example demonstrates bottom-up control of bacterioplankton growth. In (*b*) bacterial abundance increases when predators are excluded, demonstrating top-down control of population size. In (*c*), bacterial abundance increases by addition of nutrients and exclusion of predators; however, the combined influence of adding nutrients and removing predators permits largest expansion of population size; this is an example of bottom-up growth limitation with top-down control of population size. Modified from Cullen (1991).

LIMITATION BY DISSOLVED ORGANIC MATTER

DOM constitutes one of the largest sources of nutrients and energy available to marine bacteria. The range of reduced carbon compounds available to support bacterial growth includes everything from simple one-carbon-containing substrates such as methane to complex multi-carbon-containing solutes such as aromatic hydrocarbons. The bulk DOM pool contains a suite of bioelements, most notably carbon, nitrogen, and phosphorous, and various trace elements. In the open ocean, planktonic organisms are the dominant source for DOM, being produced as a byproduct of cellular metabolism or as a consequence of cell death and decomposition. In some cases, DOM production is facilitated by bacterial enzymatic hydrolysis (Smith et al. 1992) or via direct solubilization of particulate material, the latter process being governed by physical processes, and the chemical and thermodynamic equilibrium between the particle and the surrounding media. Each pathway is likely to produce different types of DOM—in some cases substantially modified from the initial cellular material, while in other cases, the macromolecular constituents of the cells would be largely intact.

Despite its importance to fueling bacterial nutrition and energy, we have only limited knowledge of the processes that control the cycling of DOM and even less information on the specific types of DOM compounds in seawater. The vast majority (70–95 percent) of the oceanic DOM pool remains chemically uncharacterized (Benner 2002), providing little information about its potential nutritional or energetic value for supporting bacterial growth. The complexity of naturally occurring substrates combined with dilute concentrations makes evaluating the turnover and reactivity of individual substrates difficult. Our inability to quantify and chemically determine the composition of oceanic DOM is one of the single largest barriers to advancing our understanding of how microbes intersect with biogeochemical cycles. It is likely no coincidence that past efforts to cultivate marine bacteria have met with poor successes, given that we understand so little about the types of substrates fuelling bacterial metabolism in nature.

Concentrations of the bulk pools of dissolved organic carbon, nitrogen, and phosphorus (DOC, DON, and DOP, respectively) tend to be elevated in the upper ocean and decline with increasing distance (vertically) from their photosynthetic production source. DOC concentrations typically range between 60 and 90 μmol/L (Fig. 10.7) and decline to about 40 μmol/L in the deep sea (Hansell 2002). DON and DOP exhibit similar depth-dependent patterns, with upper ocean concentrations typically 5–6 μmol/L and 0.1–0.5 μmol/L, respectively (Fig. 10.7). Concentrations of DON and DOP in the deep sea are more difficult to measure, but reported deep-ocean DON concentrations typically range between 2 and 4 μmol/L, while DOP ranges from 0.02 to 0.2 μmol/L (Bronk 2002; Karl and Björkman 2002). Much of the focus of DOM research has been on DOC and on evaluating microbial control of ocean carbon pools; however, DON and DOP play crucial roles in bacterial nutrition and as sources of energy.

DOM is often characterized based on its reactivity and apparent biological availability, both of which likely mirror the energy and nutritional value of the

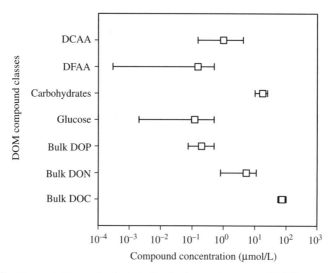

Figure 10.7 Concentrations of selected dissolved organic matter (DOM) compounds from various marine ecosystems. Data for dissolved organic carbon (DOC), nitrogen (DON), and phosphorus (DOP) are from the U.S. J.G.O.F.S. data server, with individual compound class data from Bronk (2002), Benner (2002), Skoog et al. (2002), Rich et al. (1996, 1997). Symbols represent mean values and error bars are ranges of data. DCAA, dissolved combined amino acids; DFAA, dissolved free amino acids.

substrates. Three broadly defined pools of DOM have been characterized based on their turnover times (see the text box): labile, semilabile, and refractory. The least abundant pool of DOM includes the labile substrates, with utilization time scales ranging from minutes to days. Concentrations of labile DOM are very low (<1 μmol/L), comprising less than 1 percent of the total organic carbon in the upper ocean. These low concentrations are a consequence of the rapid utilization of labile DOM, which appear to meet the majority of bacterial growth demands in the sea. In general, labile DOM consists of compounds that can be readily assimilated or salvaged for biosynthesis, or those compounds that are readily oxidized for energy, including simple sugars and amino acids. Interestingly, although the majority of the DOM pool consists of low-molecular-weight (<1000 Da) solutes, most of these solutes appear to resist microbial degradation over long time scales (thousands of years: Benner et al. 1992).

The Three-Class Model for DOM in the Oceans

Labile DOM   Substrates that are rapidly consumed (minutes to days) to support bacterial growth; concentrations of labile substrates are generally very low (nanomolar) due to rapid bacterial consumption.

Semilabile DOM   Substrates that are consumed over time scales ranging from weeks to months.

Refractory DOM   Substrates that persist for years to millennia.

Approximately half of the DOM inventory in the surface ocean is classified as semilabile, demonstrating turnover times that range from weeks to months. Concentrations of semilabile DOC are often 20–30 μmol/L, so, over seasonal or annual time scales, semilabile DOM must support a large fraction of bacterial growth (Carlson et al. 1994; Repeta and Aluwihare 2006). The largest pool of DOM compounds are the refractory substrates that turnover on time scales approaching or exceeding ocean circulation. Concentrations of refractory DOC average about 40 μmol/L and demonstrate little variation with depth. The slow turnover times suggest that refractory DOM supports little (if any) of the nutrient or energy demands of bacterioplankton.

Bacterial Growth on Bulk DOM Pools

Most studies on limitation of bacterial growth by DOM have focused on compound-specific bioassay experiments. However, there are examples of studies identifying bacterial degradation and utilization of bulk DOM pools, including studies in high-latitude, high-nutrient systems and low-latitude, low-nutrient systems. In general, upper-ocean concentrations of DOM tend to be greater in the low-nutrient physically stable ocean gyres, with high-nutrient well-mixed systems generally displaying lower concentrations of DOM (Hansell 2002). This trend derives in part from the different partitioning of particulate and DOM pools among ecosystems, and reflects the physical mixing regime in these different environments. In a comparative study of bacterial growth and DOM degradation in the Sargasso Sea and the Ross Sea, Carlson et al. (1998) found that nearly all (89 percent) of the total organic matter produced during the summer growing season in the Ross Sea was retained as particulate carbon, and what little carbon was partitioned to DOM was rapidly consumed by bacteria. In contrast, in the Sargasso Sea, accumulated semilabile DOM accounts for the majority (86 percent) of the organic matter inventory, but a large fraction (about 50 percent) of this DOM resists microbial degradation over monthly or even annual time scales. It remains unclear why the DOM produced in high-nutrient ecosystems should be more biologically reactive than DOM produced in low-nutrient environments. One possible explanation is that the types of substrates available to support bacterial growth have greater nutritional value (lower $C:N:P$ ratios) than those substrates available in low-nutrient systems.

A number of both experimental and modeling approaches have sought to understand the processes that hamper bacterial consumption of DOM. Nutrient-amended bioassay experiments suggest that bacterial growth may frequently be limited by one or more inorganic nutrients, thereby restricting bacterial utilization of semilabile DOM. Thingstad et al. (1997) proposed that the interaction of inorganic nutrient limitation and tight predation control of bacterial biomass results in accumulation of utilizable DOM in nutrient poor environments. In regions of the North Atlantic and Mediterranean Sea, nitrate (NO_3^-):phosphate (PO_4^{3-}) ratios are found in excess of about 40:1, and DON:DOP ratios range from about 30:1 to 200:1 (Cavender-Bares et al. 2001; Krom et al. 2005; Thingstad et al. 2005). Experimental studies in these regions suggest that bacterial growth may be restricted by the availability of PO_4^{3-} (Cotner et al. 1997; Rivkin and Anderson 1997; Thingstad et al. 1998; Zohary and Robarts 1998; Caron et al. 2000), supporting the hypothesis that semilabile DOM accumulates as a result of inorganic nutrient limitation of bacterial growth.

However, in many of the same ocean ecosystems where PO_4^{3-} has been hypothesized to limit bacterial growth, several studies have found no clear evidence that semilabile DOM accumulates as a result of inorganic nutrient limitation of bacterial growth. Instead, these studies generally suggest that semilabile DOM provides a poor growth substrate and therefore resists bacterial degradation. In a series of experiments designed to test the influence of inorganic and organic nutrients on bacterial utilization of the seasonally accumulated semilabile DOM in the Sargasso Sea, Carlson et al. (1996, 2002, 2004) concluded that bacterial growth in the Sargasso Sea is most closely controlled by the flux of labile DOM. These studies provided no evidence that inorganic nutrients, specifically PO_4^{3-}, restricted bacterial growth or, more importantly, controlled the consumption of the seasonally accumulated DOM. They argue that the poor energetic and nutritional quality of the seasonally produced semilabile substrates prohibits rapid bacterial consumption. Moreover, the consortia of microbes in the mesopelagic more rapidly degraded the semilabile DOM than groups of microbes found in the upper ocean, suggesting that utilization of the semilabile DOM pool depends in part on the structure of the microbial community (Carlson et al. 2004). Perhaps, the solution to these diverging views of the processes controlling bacterial consumption of DOM rests with the temporal and spatial heterogeneity of ocean ecosystems. The intermittent nature of mesoscale processes may result in high-frequency fluctuations in the types of resources limiting bacterial growth in the sea.

Limitation by Specific DOM Compounds

Among the chemically characterized constituents of oceanic DOM, the most common compound classes include carbohydrates (poly- and monosaccharides), proteins, and lipids (Fig. 10.7). The combined inventories of these three general compound classes comprise 20–40 percent of the bulk DOM pool (Benner et al. 1992; Pakulski and Benner 1994). Among the identifiable components of the bulk DOM pool that support bacterial growth, simple sugars, amino acids, carbohydrates, and proteins have received the most attention. In the open ocean, concentrations of monosaccharides are generally low (ranging from 0.002 to 0.8 μmol C/L), with glucose typically being the most abundant of the simple sugars. In many marine environments, glucose turns over rapidly (hours to days); however, in other ecosystems, glucose turns over more slowly (on the order of hundreds of days) suggesting that glucose plays a variable role in supporting BCD.

Carbohydrates are the most abundant identifiable component of DOM, with polysaccharides forming the dominant fraction of total carbohydrates (Benner 2002). Concentrations of dissolved polysaccharides (methodologically defined as dissolved combined neutral sugars) in seawater are generally at least 10-fold greater than individual sugar monomers (Fig. 10.7), suggesting that these substrates cycle more slowly than simple sugars; nonetheless, these biopolymers fuel a substantial fraction of bacterial metabolism in specific oceanic systems.

Among the nitrogen-containing reduced carbon substrates, dissolved free amino acids (DFAA) and protein (methodologically defined as dissolved combined amino acids, DCAA) appear to support the greatest fraction of bacterial growth. Information emerging from several of the recently completed bacterial genome sequences indicates a substantial cellular allocation to amino acid and peptide transport, suggesting these substrates play an important role in meeting bacterial nitrogen demands (Table 10.2). Nearly half of all the transporters encoded by the alphaproteobacterium *Pelagibacter ubique* are high-affinity ABC transporters for assimilating low-molecular-weight reduced nitrogenous compounds such as urea, amino acids, spermidine, and putrescine (Giovannoni et al. 2005b).

Oceanic concentrations of DFAA tend to be very low, ranging from about 0.001 to 0.1 μmol/L (Benner 2002; Bronk 2002), and, similar to glucose, rapid bacterial uptake of DFAA often results in fast turnover of these substrates (Fuhrman and Ferguson 1986; Fuhrman 1987; Suttle et al. 1991; Rich et al. 1997). The most prevalent individual DFAA compounds found in seawater include glycine, arginine, alanine, phenylalanine, and serine (Benner 2002). Despite rapid cycling, the total contribution of DFAA to total bacterial nitrogen demand appears to be less than 40 percent (Kirchman 2000), with the remaining nitrogen requirements being met via assimilation of DCAA, ammonium (NH_4^+), and NO_3^-.

Numerous studies have examined how concentrations and fluxes of DFAA and glucose control bacterial growth. Most often, nutrient-dose bioassay experiments are used to evaluate the bacterial response to additions of these substrates. Typically, such experiments amend seawater samples with one or more organic or inorganic substrates and examine the response of bacterial production and abundance over some incubation period (usually several days). In numerous oceanic ecosystems, additions of labile DOM to whole seawater or predator-reduced seawater treatments have been shown to stimulate bacterial growth (Table 10.3). Often, stimulation of bacterial growth by organic matter appears to be independent of inorganic nutrient concentrations; for example, additions of glucose have been shown to stimulate bacterial production in both high-nutrient low-chlorophyll (HNLC) regions of the Equatorial Pacific and subarctic North Pacific (Kirchman 1990; Kirchman and Rich 1997) and in the nutrient starved waters of the Sargasso Sea and the Mediterranean Sea (Carlson and Ducklow 1996; Zweifel 1999; Pinhassi et al. 2006).

In a series of nutrient-dose experiments in the subarctic North Pacific, Kirchman (1990) found that additions of 0.1–0.5 μmol/L DFAA stimulated bacterial production by 30–300 percent relative to unamended controls. In these experiments, additions of DFAA stimulated bacterial growth to a greater extent than additions of glucose (0.5 μmol/L), glucose plus NH_4^+ (0.5 μmol/L each), or protein. Such results are consistent with bacterial assimilation of DFAA via salvage pathways, thereby conserving cellular energy (Kirchman 1990). Similar results have been obtained in nutrient-amendment experiments in the Sargasso Sea (Carlson and Ducklow 1996), the Southern Ocean (Church et al. 2000), and the Northeast Pacific (Cherrier et al. 1996; Cherrier and Bauer 2004).

TABLE 10.2 Nitrogen Sources Available to Various Groups of Marine Bacteria Deduced from Genome Sequences

Isolated Organism	Phylogenetic Affiliation	Region Isolated	Putative Nitrogen Substrates
Pelagibacter ubique	*Alphaproteobacteria* SAR11	Oregon coast	NH_4^+, urea, amino acids, nucleic acids, spermidine, putrescine
Silicibacter pomeroyi	*Alphaproteobacteria*	Southeastern U.S.A. coastal seawater	NH_4^+, urea, spermidine, putrescine, taurine
Erythrobacter sp. NAP1	*Alphaproteobacteria*	North Atlantic	NH_4^+, NO_2^-, amino acids, oligo/di/tripeptides
Marinomonas sp. MED121	*Gammaproteobacteria*	Mediterranean Sea	NH_4^+, NO_2^-, NO_3^-, amino acids, urea/short-chain amides
Marine *Actinobacteria* (PHSC20C1)	*Actinobacteria*	Antarctic coastal water	NH_4^+, NO_3^-, amino acids, oligo/dipeptides, nucleotides, spermidine/putrescine,
Croceibacter atlanticus (HTCC2559)	*Bacteroidetes/Chlorobium*	Sargasso Sea	NH_4^+, oligo/dipeptides, nucleosides, amino acids, spermidine/putrescine

TABLE 10.3 Results from Selected Nutrient-Dose Experiments in Various Ocean Ecosystems Demonstrating the Influence of DOM, Inorganics, Iron, and Temperature on Bacterioplankton Growth

Region	Organic	Effect?	Inorganic	Effect?	Fe	Effect?	Temperature Y/N	Notes	Study
Oligotrophic									
Eastern North Pacific	Glu, DFAA, PE	Y Y Y	NH_4^+	N	*		*	Largest response in DFAA, PE, or glucose + NH_4^+	Cherrier et al. (1996)
Sargasso Sea (BATS)	Glu, DFAA, PE	Y Y Y	NH_4^+, PO_4^{3-}	N, N	*		*	Cell abundance and production increased with DOM addition	Carlson and Ducklow (1996)
Caribbean, Sargasso Sea, Gulf Stream	Glu	Y/N	NH_4^+, PO_4^{3-}	N, Y/N	*		*	Increases in cell abundance and growth by DOM and PO_4^{3-}; responses variable in space	Rivkin and Anderson (1997)
Mediterranean Sea	*		PO_4^{3-}	Y/N	Fe + EDTA, Fe + EDTA + PO_4^{3-}	Y/N Y	*	Increases in cell abundance and growth; response variable in space	Zohary and Robarts (1998)

(*Continued*)

TABLE 10.3 Continued

Region	Organic	Effect?	Inorganic	Effect?	Fe	Effect?	Temperature Y/N	Notes	Study
Temperate									
Subarctic North Pacific	Glu, DFAA	Y, Y	NH_4^+, PO_4^{3-}	N, N	*		*	Large increases in DFAA treatments with weak response to Glu treatment	Kirchman (1990)
California Current	Glu	Y	*		Fe, Glu + Fe	N, Y	*	Limited by labile DOM with Fe stress	Kirchman et al. (2000)
Subantarctic	Glu, DFAA	Y, Y	$NH_4^+ +$ PO_4^{3-}	N	Fe, Glu + Fe, DFAA + Fe	N, Y, Y	*	Limited by labile DOM with Fe stress	Church et al. (2000)
Low-latitude									
Equatorial Pacific	Glu, DFAA	Y, Y	NH_4^+	N	*		Y	Largest responses increases in growth in Glu, Glu + NH_4^+; larger response at elevated temperatures	Kirchman and Rich (1997)

358

High-latitude

Region	Substrate	Response	Inorganic nutrients	Response	Temperature	Comment	Reference
Arctic Ocean	Glu, DFAA, BSA, PE	Y, Y, N, Y	$NH_4^+ + PO_4^{3-}$	N	*	Larger bacterial response to DOM additions at elevated temperatures; variable responses to temperature and DOM	Kirchman et al. (2005)

Modified and updated from Williams (2000).
Whether an effect was observed is indicated by Y (yes) or N (no); * indicates that the treatment was not examined. Glu, glucose; DFAA, dissolved free amino acids; BSA, bovine serum albumin; PE, plankton extract.

In other examples, bacterial growth appears limited by the availability of easily catabolized sources of energy. In the Northeast Pacific Cherrier and Bauer (2004) observed rapid bacterial utilization of DFAA, protein, and monosaccharides; however, bacterial consumption of DFAA was coupled to production of NH_4^+, suggesting the DFAA substrates were catabolized for energy rather than assimilated via salvage pathways. These authors conclude that the quality of available DOM substrates limited bacterial growth and bacterial DOM utilization (Cherrier and Bauer 2004). Kirchman and Rich (1997) found that additions of glucose $+ NH_4^+$ (1 µmol/L each) increased rates of bacterial production by fourfold relative to the unamended controls; in contrast, additions of DFAA stimulated production to a lesser degree (about twofold), suggesting that growth may have been limited by the availability of easily catabolized substrates.

Although the overall conclusion that heterotrophic bacterial growth rates are frequently controlled by the flux of labile DOM appears generally consistent across varied ocean ecosystems, there are subtle but important differences in the nature of the bacterial responses to DOM amendments in these ecosystems (Table 10.3). There are many examples where DFAA and glucose supply a major fraction of the carbon required to support bacterial production. For example, Rich et al. (1997) found that despite high concentrations of both DFAA and glucose, glucose met upwards of 100 percent of the bacterial production in the Arctic Ocean. In the Equatorial Pacific, Rich et al. (1996) measured variable rates of glucose uptake (ranging 3 from 148 nmol/L/d), and concluded that glucose alone could support up to half (15–47 percent) of production and a large fraction of the respiratory demands of the cells. In the Gulf of Aqaba in the Red Sea, Grossart and Simon (2002) estimated that in some locations bacterial use of DFAA could support up to 100 percent of production, while in other areas glucose uptake accounted for closer to half of bacterial production.

However, there are also several notable exceptions where glucose and DFAA support only a small fraction of bacterial growth. In the Sargasso Sea, Keil and Kirchman (1999) found that glucose and DFAA accounted for less than 20 percent of the total bacterial carbon and nitrogen demands. Similarly, in the subarctic North Pacific, Keil and Kirchman (1991) estimated that bacterial utilization of DFAA accounted for about 24 percent of the total cellular nitrogen demands. In the Gulf of Mexico, Skoog et al. (1999) measured very low glucose uptake rates (<3 nmol glucose) and concluded that glucose supported less than 10 percent of bacterial production. Similarly, on a transect between $33°N$ and $45°N$ in the North Pacific Ocean, Skoog et al. (2002) found that glucose supported a variable but generally small fraction (0.7–22 percent) of the bacterial carbon demands. In the Ross Sea, Kirchman et al. (2001) determined that glucose met less than 10 percent of the carbon demand by bacterial production. These studies indicate that although a large fraction of bacterial energy and carbon and nitrogen demands can be supported by consumption of monomer substrates such as amino acids and glucose, other substrate pools must also be important in fueling bacterial production; polysaccharides and proteins appear likely candidates.

Phosphorus plays a key role as a component of genetic material (DNA and RNA), cell membranes (lipids), and nucleotides. Many of these cellular constituents find their way into the DOM pool, delivered via exudation, grazing, and viral lysis of plankton cells. In low-nutrient ocean ecosystems, DOP generally accounts for 70–80 percent of the total phosphorus inventory in the upper ocean. A number of studies have examined bacterial utilization of the bulk DOP pool and bacterial turnover of specific DOP compounds. In general, DOP appears to play a prominent role in supporting microbial growth in the sea. For example, in a recent study of microbial phosphorus utilization in the central North Pacific, Björkman and Karl (2003) estimated that, on average, 50 percent of the total microbial phosphorus demand in the upper ocean derived from utilization of DOP, with the remaining fraction from assimilation of PO_4^{3-}.

Like other constituents of the DOM pool, the bulk DOP pool remains poorly characterized. The limited amount of information on specific DOP compound classes complicates determining the bioavailability of DOP substrates that support bacterial growth. Some of the major identifiable DOP compound classes include phosphate esters (C–O–P), phosphonates (C–P), and polyphosphates (Karl and Björkman 2002). Dissolved nucleotides form an important component of the DOP, with ATP probably the most widely studied of the nucleotides. Concentrations of dissolved ATP are generally elevated (typically in the 10–100 picomolar range) in the upper ocean, where production and biomass are greatest. Dissolved ATP turns over very rapidly (hours to days), and the vast majority channels into anabolic pathways fueling biosynthesis (Azam and Hodson 1977; Björkman and Karl 2005). Björkman and Karl (1994) and Björkman et al. (2000) examined the bioavailability of PO_4^{3-} and several DOP-containing compounds in open ocean and nearshore waters of the central North Pacific. These authors found that PO_4^{3-} was consistently the preferred source of phosphorus for microbial growth, but among the DOP compounds examined, the nucleotides (ATP and guanosine triphosphate) and ribulose-1,5-bisphosphate were more bioavailable than several phosphate monoesters.

LIMITATION BY INORGANIC NUTRIENTS

In the sunlit portions of the open oceans, restricted physical supply of nutrients together with rapid biological nutrient assimilation combine to keep nutrient concentrations low. Concentrations of inorganic nutrients in the euphotic zone of open ocean ecosystems typically range from subnanomolar to micromolar (Table 10.1). The physical stability of the low-latitude central gyres restricts nutrient input, while enhanced entrainment of nutrients by the wind-driven mixing in higher-latitude ecosystems generally results in elevated concentrations of inorganic nutrients. Trace-element (e.g., iron and zinc) availability also plays an important role in controlling plankton dynamics in various ocean ecosystems (Cullen 1991; Bruland and Lohan 2004).

Nitrogen

Inorganic nitrogen exists in a variety of forms ranging across eight redox states (Capone 2000). Thus, not only is nitrogen required as an essential nutrient for growth, but several nitrogen substrates (NH_4^+, NO_3^-, and DON) also serve as important bacterial energy sources. In addition, under suboxic or anoxic conditions, several oxidized forms of nitrogen (NO_3^- and NO_2^-) can also serve as electron acceptors during bacterial oxidization of organic matter (denitrification) or ammonia (anammox) (Capone 2000; Kuypers et al. 2003), as discussed in Chapter 14. The availability of nitrogen-containing substrates in the oceans varies considerably in both space and time. Dissolved dinitrogen (N_2) is never limiting in the open ocean (concentrations of about 1 mmol/L), with concentrations largely varying as a function of gas solubility (Capone 2000). However, only selected groups of microorganisms (diazotrophs) have the capability of utilizing N_2 as a nutrient source (see Chapter 13). The relatively large energy demands associated with N_2 fixation appear to have largely relegated this process to phototrophic microorganisms found in the well-lit regions of the upper ocean. However, several *Gammaproteobacteria* and *Alphaproteobacteria*, as well as several presumed anaerobic bacteria, appear to have the capability of fixing N_2 in the open ocean (Zehr et al. 1998; Braun et al. 1999; Church et al. 2005), although the potential importance of heterotrophic N_2 fixation remains unknown.

Unlike dissolved N_2, NO_3^- demonstrates a classic nutrient-like distribution, with concentrations being low in the euphotic zone (<0.01–10 μmol/L) and high in the deep sea (30–40 μmol/L). In tropical and subtropical ecosystems (excluding the near-equatorial waters) NO_3^- is often found at exceedingly low concentrations (<10 nmol/L; Table 10.1), suggesting that NO_3^- could frequently limit plankton growth. Bacterial utilization of NO_3^- as a nutrient source is energetically demanding, requiring intracellular reduction to NH_4^+ prior to assimilation (Table 10.4; Vallino et al. 1996). Photoautotrophic plankton have historically been viewed as the primary consumers of NO_3^- in the sea. However, a number of studies indicate that, under some conditions, heterotrophic bacteria compete effectively with phytoplankton for NO_3^-. In the subarctic North Pacific, Kirchman and Wheeler (1998) found that bacterial uptake accounted for about 30 percent of the total planktonic uptake of NO_3^-, and bacterial assimilation of NO_3^- supported a substantial fraction of bacterial production. During the spring bloom in the North Atlantic, bacterial NO_3^- uptake accounted for 4–14 percent of total bacterial nitrogen assimilation, and comprised 2–8 percent of bacterial nitrogen demands (Kirchman et al. 1994). Analyses of

TABLE 10.4 Summary of Enzymes and Energetic Expenses Required for Transformation of Selected Nitrogen Substrates during Assimilatory Nitrogen Reduction

Substrate	Enzyme	Reaction	Electrons	ATP
N_2	Nitrogenase	$N_2 \rightarrow NH_4^+$	8	16
NO_3^-	Nitrate reductase	$NO_3^- \rightarrow NO_2^-$	2	0
NO_2^-	Nitrite reductase	$NO_2^- \rightarrow NH_4^+$	6	0
NH_4^+	Glutamine synthetase, glutamate synthetase	$NH_4^+ \rightarrow$ glutamate	2	1

Adapted from García-Fernández et al. (2004).

nasA genes (encoding a structural component of assimilatory nitrate reductase) revealed distinct groups of *Gammaproteobacteria* capable of assimilating NO_3^- (Allen et al. 2001). Interestingly, *Gammaproteobacteria* similar to *Vibrio* spp. and *Pseudoalteromonas* spp. are often among the first bacterial groups to respond to labile DOM input (Pinhassi et al. 1999; Eilers et al. 2000; Carlson et al. 2004), suggesting a possible linkage between labile DOM substrate availability and NO_3^- utilization by bacterioplankton.

Nitrite (NO_2^-) is an important intermediate in biological nitrogen cycling, formed during dissimilatory nitrate reduction, ammonium oxidation, and assimilatory nitrate reduction. NO_2^- concentrations are typically very low (<0.1–60 nmol/L) in the open ocean (Dore and Karl 1996). The recently completed genome sequences of several marine microorganisms have revived interest in NO_2^- as a potential limiting nutrient for plankton growth. In particular, isolates of *Prochlorococcus* spp. are unable to grow on NO_3^-, but several of the low-light isolates encode nitrite reductases, indicating that NO_2^- might supply part of the cellular nitrogen demands for these cyanobacteria (Moore et al. 2002; Rocap et al. 2003). In addition, recent cultivation-dependent and -independent techniques have revealed that the abundant groups of marine *Crenarchaea* play an important role in ocean nitrification, highlighting an additional potential source of NO_2^- (Konneke et al. 2005; Wuchter et al. 2006).

While these studies emphasize the essential role of microbes in NO_3^- and NO_2^- cycling, few studies have evaluated direct limitation of heterotrophic bacterial growth by these substrates. From a thermodynamical perspective, the high energetic costs associated with growth on NO_3^- and NO_2^- indicate that, given a choice of nitrogen-containing substrates, microorganisms should grow more efficiently on reduced nitrogen substrates such as DFAA and NH_4^+ (Table 10.4). Perhaps not surprisingly, many more studies have examined potential NH_4^+ limitation of bacterial growth. At concentrations found in marine ecosystems, assimilation of nitrogen into biosynthetic pathways occurs primarily via the glutamine synthetase/glutamate synthetase (GS/GOGAT) system, where NH_3 is combined with glutamate to form glutamine. The GS pathway requires four atoms of carbon for each atom of nitrogen assimilated; thus, for those ecosystems where NH_4^+ is a primary resource, quantifying NH_4^+ assimilation also provides information on carbon demands.

Ammonium concentrations are generally less than 0.05 μmol/L in oligotrophic waters and somewhat higher (<0.5 μmol/L) in nearshore ecosystems. Bacterial assimilation of NH_4^+ appears to be an important component of ocean nitrogen cycles (Wheeler and Kirchman 1986; Kirchman et al. 1989; Kirchman and Wheeler 1998). Marine bacteria appear to have very efficient NH_4^+ uptake systems; moreover, the proportion of total NH_4^+ uptake attributed to heterotrophic bacteria tends to increase with decreasing NH_4^+ concentrations (Suttle et al. 1990), perhaps as a consequence of the increased relative proportion of heterotrophic bacteria to ecosystem biomass in low productivity systems. Regardless, it appears bacterioplankton are efficient competitors for NH_4^+ at the low concentrations found in most oceanic systems.

Numerous studies have evaluated the potential for NH_4^+ to limit bacterial growth. In general, these studies support the idea that NH_4^+ assimilation is stoichiometrically tightly coupled to carbon uptake and possibly regulated by the bioavailability of energy sources. In both high nutrient–low chlorophyll and oligotrophic oceanic ecosystems, addition of NH_4^+ has been shown to have little or no appreciable influence on bacterial growth (Kirchman 1990; Carlson and Ducklow 1996; Cherrier et al. 1996; Kirchman and Rich 1997; Rivkin and Anderson 1997; Church et al. 2000). However, in all these studies, the addition of labile organic matter (glucose) combined with NH_4^+ generally stimulated bacterial growth.

Phosphorus

As previously described, phosphorus is an essential nutrient that serves a vital role in cellular energy storage and transformations, membrane structure, and genetic exchange. Orthophosphate concentrations in the upper ocean range from nanomolar to micromolar and vary like other nutrients with depth. In the physically stable central gyres and marginal seas, where vertical nutrient exchange is low, euphotic zone concentrations of PO_4^{3-} are very low, often less than 30 nmol/L (Karl and Tien 1992; Björkman et al. 2000; Wu et al. 2000; Van Den Broeck et al. 2004). In these systems, PO_4^{3-} turnover times are highly variable, ranging from rapid (<2 hours) to hundreds of days. Many bacteria possess high-affinity PO_4^{3-} transport proteins that appear to provide a competitive advantage, and in many systems a large portion (averaging 60 percent) of total PO_4^{3-} uptake appears to be controlled by bacteria (Björkman and Karl 1994; Dolan et al. 1995; Zohary and Robarts 1998). The observation that bacteria compete with phytoplankton for PO_4^{3-} would suggest that in some systems bacterial phosphorus uptake may indirectly limit primary production (Joint and Morris 1982; Joint et al. 2002). However, the dependence of bacterial production on photosynthetic byproducts suggests that such competition would not be a successful evolutionary strategy.

A number of studies have demonstrated direct PO_4^{3-} limitation of bacterial growth in oligotrophic ecosystems. The low concentrations and rapid PO_4^{3-} cycling observed in various oligotrophic ecosystems provide compelling examples that PO_4^{3-} availability may play an important role in controlling bacterial growth and DOC consumption. In both the Mediterranean and Sargasso Sea, euphotic zone concentrations of PO_4^{3-} can be less than 2 nmol/L (Cavendar-Bares et al. 2002), and various nutrient-dose bioassay experiments conducted in these ecosystems suggest that PO_4^{3-} availability can directly limit bacterial growth (Cotner et al. 1997; Rivkin and Anderson 1997; Caron et al. 2000; Obernosterer et al. 2003) and consumption of DOM (Zweifel 1999). In one example, Rivkin and Anderson (1997) found that additions of PO_4^{3-} to surface waters of the Sargasso Sea stimulated bacterial growth by up to sixfold relative to unamended controls. However, as previously described, a number of nutrient-dose experiments have not observed consistent stimulation of bacterial growth or DOM consumption by the addition of PO_4^{3-} (Carlson and Ducklow 1996; Carlson et al. 2002; Sala et al. 2002). To date, the role of PO_4^{3-} as a growth-limiting nutrient remains unclear, perhaps due in part to spatial and temporal variability in its input to ocean ecosystems.

Trace Nutrients

Throughout much of the world's oceans, bioessential trace metals, including manganese, iron, zinc, cobalt, and nickel, are found at concentrations ranging from picomolar to nanomolar (Brand et al. 1983; Bruland et al. 1991; Bruland and Lohan 2004). These low concentrations restrict phytoplankton growth in large regions of the open sea. In particular, a considerable amount of research has explored the connectivity between iron availability and ocean production. Iron is the fourth most abundant element in the Earth's crust; however, in remote areas of the open ocean, iron delivery is largely restricted to either atmospheric input of terrestrially derived material or mixing of iron from the deep sea. As a result, in many regions of the world's oceans, iron concentrations are extremely low (subnanomolar). In such regions, phytoplankton biomass and rates of primary production are low despite relatively high concentrations of inorganic nutrients. These regions have been termed high nutrient–low chlorophyll (HNLC) ecosystems, and they include the Southern Ocean, the subarctic North Pacific, and the Equatorial Pacific (Martin and Fitzwater 1988; Coale et al. 1996; de Baar et al. 2005). Nutrient bioassays and mesoscale in situ enrichment experiments have demonstrated that the HNLC condition derives, at least in part, from low concentrations of iron relative to that required by phytoplankton.

To date, most research on iron limitation of HNLC systems has focused on phytosynthetic plankton, with relatively little attention being paid to heterotrophic bacteria. However, bacterioplankton are effective competitors for iron at low concentrations, with bacterial uptake often accounting for over 50 percent of the total planktonic iron uptake (Maldonado and Price 1999). Heterotrophic bacteria have a larger iron requirement than most phytoplankton; Fe:C ratios of heterotrophic bacteria appear to range between about 6 and 8 μmol Fe/mol C, while phytoplankton are closer to 3–4 μmol Fe/mol C (Tortell et al. 1996). Iron is an essential component of the electron-transport chain, with upwards of 94 percent of the total cellular iron supporting respiratory activity (Tortell et al. 1999). As a result, reduced iron availability could decrease BGE and result in increased catabolic demands and deceased biosynthetic production by the cells. Consistent with this hypothesis, Tortell et al. (1996) observed that iron-limited bacterial cultures had lower BGE than iron-sufficient cells. Moreover, these experiments suggested that iron-induced depression of BGE could increase BCD and result in simultaneous limitation of bacterial growth by both iron and DOM (Tortell et al. 1999).

In various experiments designed to examine the response of planktonic assemblages to iron enrichment, bacterioplankton generally respond to increased iron concentrations, but it is often difficult to decipher whether this is a direct response to iron or a secondary response to the enhanced flux of DOM following alleviation of iron limitation of primary producers. To date, relatively few studies have directly examined limitation of bacterial growth by iron in the oceans. In one example, iron was shown to directly limit bacteria growth in the high-nutrient waters off Antarctica (Pakulski et al. 1996). In other studies, bacterial growth in HNLC regions appears limited by labile DOM more than by iron directly. In a study in the Southern Ocean, Church et al. (2000) observed no bacterial response to treatments containing iron alone; however, additions of glucose plus iron stimulated bacterial growth to a greater extent than additions of glucose alone. Similar results were observed in iron-limited waters off California; in this study, Kirchman et al. (2000) concluded that bacterial growth was suppressed by DOM availability, but when DOM limitation was relieved, growth rapidly became iron-limited.

Over the past decade, mesoscale iron-fertilization experiments have conclusively demonstrated the important role of iron in controlling planktonic growth in HNLC regions. In situ iron-enrichment experiments have been conducted in different regions of the Southern Ocean, the subarctic Pacific, and the Equatorial Pacific. A number of these studies have considered the bacterial response to mesoscale iron enrichments. During IronEx II in the Equatorial Pacific, Cochlan (2001) examined time-dependent changes in bacterial production and abundance in the iron-enriched patch relative to the unamended surrounding waters. Despite a nearly 17-fold increase in chlorophyll during the experiment, bacterial abundance increased a modest 2-fold, despite nearly 5-fold changes in growth rates. Such results imply that bacterial growth was closely coupled to mortality (predation or viral lysis) during this experiment. A similar disproportionate increase in phytoplankton biomass relative to bacteria was observed during the Southern Ocean fertilization experiments (Hall and Safi 2001; Arrieta et al. 2004; Oliver et al. 2004). In these experiments, bacterial biomass demonstrated modest increases (about 2-fold or less), while phytoplankton biomass increased 6–27 times following the addition of iron. Such studies highlight the integral role of bacteria in pelagic food webs; close coupling between bacterial growth and predation suggests that although bacterial growth is often limited by bottom-up forcing, mortality keeps close pace with fluctuations in growth.

In addition to iron, bacteria have an obligate requirement for several other metals, most of which are utilized for their redox potential by specific enzymes. Recent progress on the biochemistry of metalloenzymes has demonstrated that many eukaryotic marine phytoplankton are able to substitute specific metals in the active sites of certain enzymes. Various species of marine diatoms, coccolithophores, and prymnesiophytes can substitute cobalt or cadmium for zinc in the active site of the enzyme carbonic anhydrase (Morel et al. 1994; Sunda and Huntsman 1995; Saito et al. 2008). The extent to which such metal substitutions may occur among marine bacteria remains largely unknown. In the ocean environment, many of the bioessential metals (manganese is an exception) are found as organic complexes, complicating interpretation of the concentration of the metal that is truly bioavailable. The extent to which heterotrophic bacteria may selectively utilize such complexes to meet their cellular metal demands remains unclear.

TEMPERATURE–DOM INTERACTIONS

Temperature has been shown to interact with various resource controls on bacterial growth. Physiological changes to temperature are characterized by a dimensionless index called Q_{10}, which describes the change in a metabolic rate following a 10 °C increase in temperature. The limited data available suggest that Q_{10} values for oceanic bacteria range between 1 and 3 (Robinson and Williams 1993; Kirchman et al. 1995, 2005; Kirchman and Rich 1997). Based on a series of temperature shift experiments, Wiebe et al. (1992, 1993) hypothesized that at low temperatures, bacteria require higher concentrations of substrates to overcome suppression of transport enzymes and to maintain growth. The kinetic parameters describing bacterial substrate transport are temperature-dependent (Nedwell and Rutter 1994; Reay et al. 1999), so increasing substrate affinity (increases in $\alpha°$ or decreases in K_t) could overcome the thermodynamic barrier imposed by low temperatures.

Several studies have used cross-ecosystem analyses to investigate how temperature influences bacterial metabolism in the ocean. A number of these studies suggest that bacterial production comprises a smaller fraction of total ecosystem metabolism in cold, high-latitude oceans compared with more temperate or tropical ecosystems. Pomeroy et al. (1991) suggest that this observation derives from a stricter dependence of bacterial metabolism on temperature compared with phytoplankton. Based on dynamics of bacterial production and respiration in the cold waters off Newfoundland, Pomeroy and Deibel (1986) hypothesized that the low temperatures found in high-latitude marine ecosystems suppress bacterial activity and result in a larger fraction of primary production funneling toward higher trophic levels rather than being respired by microbial food webs. This hypothesis was partly supported by laboratory experiments using cold-adapted marine bacteria; Wiebe et al. (1992) found that at cold temperatures (-1.5 to $0\,°C$), psychrophilic bacterial growth required higher concentrations of labile organic substrates.

However, unlike Pomeroy and Deibel (1986), Li and Dickie (1987) measured bacterial and primary production over an annual cycle in coastal waters off Nova Scotia where temperatures ranged between -1 and $20\,°C$. These authors concluded that temperature did not differentially influence bacterial activity relative to primary production. Similarly, comparative analyses of bacterial growth in high-latitude and temperate marine ecosystems failed to support the hypothesis that temperature regulates bacterial growth and the subsequent passage of carbon through microbial food webs (Rivkin et al. 1996). The influence of temperature on BGE is also not always clear; Rivkin and Legendre (2001) found that temperature could account for approximately half of the variability in measured BGE among different ecosystems, with BGE being greatest in high-latitude, cold ocean waters and lowest in tropical ecosystems. Rivkin and Legendre (2002) found a weak dependence of bacterial growth on temperature, but a stronger dependence of respiration on temperature.

Field-based experiments have also sought to evaluate the interactive influences of temperature and DOM. A series of temperature shift experiments in the Equatorial Pacific failed to demonstrate an enhanced substrate requirement when bacteria were grown at cooler temperatures (Kirchman and Rich 1997). These authors concluded that bacteria respond more rapidly and to a greater extent to elevated substrate concentrations in warmer waters than in 5 °C cooler water. Similar experiments in the perennially cold water of the Arctic Ocean provided additional insight into the interactive influences of temperature and substrate (Kirchman et al. 2005). In these experiments, increases in bacterial production by the addition of DOM were the same or less at low temperatures relative to warmer temperatures. The issue of temperature regulation of bacterial growth and subsequent control on DOM availability remains uncertain.

LIGHT

Until recently, the role of sunlight in controlling heterotrophic bacterial growth was largely overlooked, aside from selected studies focused on the microbiological and biogeochemical influences of ultraviolet radiation (Moran and Zepp 2000; Mopper and Kieber 2002). Historically, oxidation of DOM was assumed to be the principal pathway for heterotrophic bacteria to acquire energy; however, several recent discoveries focus attention on sunlight as a potential source of energy for various groups of heterotrophic microorganisms (see Chapter 5). The discovery that numerous and diverse marine bacteria encode a functional rhodopsin demonstrated a previously unrecognized photophysiological mechanism of planktonic energy harvesting (Béjà et al. 2000, 2001). Following closely on this discovery, Kolber et al. (2000, 2001) quantified the productivity and abundance of aerobic, anoxygenic photosynthetic (AAnP) bacteria in the central Pacific Ocean. Despite previous success in cultivating AAnP bacteria (Shiba et al. 1991), these microbes had been largely overlooked in studies of bacterioplankton dynamics. Finally, the realization that the most abundant photosynthetic organisms on Earth are able to assimilate selected DOM substrates (Palenik et al. 2003; Rocap et al. 2003; Zubkov et al. 2003, 2004) requires reassessment of the types of microorganisms that contribute to heterotrophic activity and DOM utilization.

These recent discoveries suggest that sunlight could play an important role in regulating heterotrophic activity and bacterial growth in the oceans. To date, however, the ecological and biogeochemical significance of these recently discovered phototrophic microbes remains largely unknown. Experiments off the coast of Southern California designed to evaluate the influences of light on bacterioplankton community structure indicated that photoautotrophic plankton were most affected by changes in light availability (Schwalbach et al. 2005). In a laboratory study using the rhodopsin-bearing alphaproteobacterium *Pelagibacter ubique*, Giovannoni et al. (2005a) observed constitutive expression of rhodopsin independent of the light environment. Moreover, these authors found no difference in the growth rates or cell yields of *P. ubique* grown in the light or dark. However, substrate consumption was not measured, so it is not possible to discern whether sunlight altered the growth efficiency of *P. ubique*. Enhanced growth efficiency would reduce the total carbon demand, but might not alter the cell growth rates or yields.

In other examples, light energy has been shown to have a positive influence on the growth and production of heterotrophic bacteria. Time-series measurements at Station ALOHA in the subtropical Pacific revealed consistent photostimulation of [^{3}H]leucine incorporation (Church et al. 2006). The response of [^{3}H]leucine incorporation to irradiance was quantified using a photosynthetic-irradiance model, demonstrating that the kinetic response of bacterial production to irradiance was similar to light-driven photoautotrophic fixation of carbon dioxide (Church et al. 2004).

Laboratory experiments with rhodopsin-bearing isolates of the *Bacteriodetes* phylum (*Donkdonia* sp. MED134) demonstrate enhancement of growth on exposure to light (Gómez-Consarnau et al. 2007). Similarly, laboratory studies using *Escherichia coli* that had been transformed with an expression vector containing a proteorhodopsin gene demonstrated that the energy gained from light harvesting was sufficient to replace the energy gained by purely heterotrophic respiration (Walter et al. 2007). These exciting discoveries open the door to numerous new research directions on the role of light energy in fueling ocean metabolism.

RESOURCE CONTROL OF SPECIFIC BACTERIAL POPULATIONS IN THE SEA

One of the overriding goals of microbial ecology is to evaluate how microbial diversity regulates ecosystem functioning, including understanding how species diversity modifies population sizes and productivity. Application of molecular tools to microbial oceanography has moved us closer to realizing this goal. Of particular interest for this review, several types of methods are now routinely used to link identification of microorganisms with specific biogeochemical processes. Fluorescence in situ hybridization (FISH) has been used extensively to enumerate the abundances of phylogenetically distinct groups of marine microbes (DeLong et al. 1989; Amann et al. 1990; Glöckner et al. 1999; Karner et al. 2001; Morris et al. 2002). More recently, this approach has been coupled with microautoradiography (Cottrell and Kirchman 2003), flow-cytometric sorting (Zubkov et al. 2003), or stable isotope probing (Orphan et al. 2001) to examine patterns of substrate utilization by specific groups of microbes. Among the successes of these techniques are studies that have identified phylogenetic differences in microbial assimilation of organic and inorganic substrates such as bicarbonate, methane, chitin, dimethylsulfide propionate (DMSP), nucleotides, and simple monomers such as glucose and DFAA.

Single-cell identification approaches have shed insight into the question of whether specialists (bacteria with high specificity substrate transport systems) or generalists (bacteria able to utilize a diverse suite of substrate classes) dominate in different aquatic ecosystems. Using a combined microautoradiography and FISH (MICRO-FISH) technique, Cottrell and Kirchman (2000) examined ^{3}H-DOM uptake by *Alphaproteobacteria*, *Betaproteobacteria*, and *Gammaproteobacteria*, as well as members of the *Cytophaga–Flavobacteria* subdivision of *Bacteriodetes* in the Delaware Bay. These authors found that *Alphaproteobacteria* dominated uptake of DFAA and *N*-acetylglucosamine, while members of the *Cytophaga–Flavobacteria* group were the dominant consumers of chitin, and protein. Similarly, Malmstrom et al. (2004, 2005) found that the members of the SAR11 clade of *Alphaproteobacteria* were responsible for approximately half of the total DFAA uptake and glucose assimilation in the North Atlantic. Such results may explain the prevalence of *Alphaproteobacteria* in marine ecosystems, where fluxes of labile DOM (specifically glucose and DFAA) often support a substantial fraction of bacterial metabolism.

Moreover, these results suggest that different groups of bacteria are adapted to utilize specific DOM substrate types; thus, the types of resources available could form a prominent control on competition and species succession.

Organic sulfur may also play an important role in controlling the activities of various *Alphaproteobacteria*. Several clades of *Alphaproteobacteria* are often numerically dominant members of oceanic and near-shore bacterioplankton assemblages (González and Moran 1997; Morris et al. 2002). By combining genetic interrogations with substrate assimilation techniques, several studies have identified possible control of the ocean's reduced sulfur cycle by members of *Alphaproteobacteria*, *Roseobacter* and SAR11. While considerable attention has focused on bacterial control of the climate-sensitive gas dimethylsulfide (DMS), recent studies suggest alphaproteobacterial assimilation of DMSP may also form an important mechanism to introduce sulfur into marine food webs. Upwards of 40 percent of the marine bacterioplankton in the upper ocean are capable of demethylating DMSP to methyl mercaptopropionate, thereby retaining sulfur in the marine food web rather than producing DMS (Howard et al. 2006).

Several other studies have identified patterns of bacterioplankton abundance consistent with specialized utilization of specific substrates. In a series of enrichment experiments in the North Sea, Eilers et al. (2000) observed consistent stimulation of various subgroups of *Gammaproteobacteria* (including *Vibrio* spp. and *Alteromonas/Colwellia*) when seawater was amended with mixtures of simple sugar and amino acid monomers. In contrast, other members of the *Gammaproteobacteria* (SAR86 and *Oceanospirillum*), *Bacteriodetes*, and the alphaproteobacterial group, *Roseobacter*, either did not respond or demonstrated only weak response to substrate additions (Eilers et al. 2000). Additions of protein to coastal California seawater triggered a large response by members of the *Cytophaga–Flavobacteria* (Pinhassi et al. 1999). Seasonal dynamics of bacterial phyla in the Baltic Sea suggest that *Alphaproteobacteria* are better adapted to low-nutrient conditions, while members of the *Cytophaga–Flavobacteria* are adapted to higher productivity (Pinhassi and Hagström 2000). In a time-series assessment of bacterial diversity in the oligotrophic Sargasso Sea, Morris et al. (2005) identified prominent temporal patterns of bacterial succession in the upper ocean, finding that members of the *Alphaproteobacteria* and *Gammaproteobacteria* such as SAR11, SAR86, and OCS116 were dominant during periods of high stratification when nutrient concentrations were low. In contrast, the contributions of members of marine *Actinobacteria* and selected *Alphaproteobacteria* (SAR11 and OCS116) were more prevalent at the base of the mixed layer following wintertime convective overturn of the upper ocean.

Taken together, these patterns suggest that different groups of microorganisms play specialized roles in their utilization of substrates in the sea. Thus, understanding substrate and resource flux and availability also requires knowledge of the specific groups of microorganisms mediating turnover of these resource pools. The black box approach of describing marine bacteria as an equivalent functional group appears antiquated and possibly misdirected; we now need approaches that merge organisms with ecological function and biogeochemical activities.

SUMMARY

1. In low-nutrient systems, bacteria appear to allocate a greater fraction of cellular resources toward substrate acquisition at the expense of growth. This strategy may have strong influence on cellular stoichiometry.
2. The availability and flux of readily assimilated DOM often limits bacterial growth in the sea.
3. Bacterial growth and productivity vary by approximately an order of magnitude, while biomass appears more stable. The apparent stability in biomass in the face of variable growth suggests tight top-down control of bacterioplankton population sizes.
4. Many bacteria possess high-affinity substrate transport systems allowing them to grow on very low substrate concentrations.
5. Sunlight may play an important and largely uncharacterized role in controlling bacterial growth and production in the sea.
6. The spatial and temporal complexity of marine ecosystems results in fluctuating availability of growth-limiting resources. Such dynamics likely control oscillations in the dominant bacterial phyla.

ACKNOWLEDGMENTS

I am thankful for the helpful suggestions and comments of Karin Björkman, Carol Robinson, and Toshi Nagata. I also acknowledge the guidance, patience, and careful editing of David Kirchman. Finally, I am grateful for the numerous far-ranging discussions with David Karl that fueled many of the ideas presented here. I acknowledge support during the writing of this chapter from the National Science Foundation (OCE 04-25363 and EF 04-24599), and the Gordon and Betty Moore Foundation (via an award to David Karl).

REFERENCES

Allen, A. E., M. G. Booth, M. E. Frischer, P. G. Verity, J. P. Zehr, and S. Zani. 2001. Diversity and detection of nitrate assimilation genes in marine bacteria. *Appl. Environ. Microbiol.* 67: 5343–5348.

Amann, R. I., L. Krumholz, and D. A. Stahl. 1990. Fluorescent oligonucleotide probing of whole cells for determinative, phylogenetic, and environmental studies in microbiology. *J. Bacteriol.* 172: 762–770.

Anderson, T. R., and H. W. Ducklow. 2001. Microbial loop carbon cycling in ocean environments studied using a simple steady-state model. *Aquat. Microb. Ecol.* 26: 37–49.

Arrieta, J. M., M. G. Weinbauer, C. Lute, and G. J. Herndl. 2004. Response of bacterioplankton to iron fertilization in the Southern Ocean. *Limnol. Oceanogr.* 49: 799–808.

Arrigo, K. R. 2005. Marine microorganisms and global nutrient cycles. *Nature* 438: 122–122.

Azam, F., and R. Hodson. 1977. Dissolved ATP in the sea and its utilization by marine bacteria. *Nature* 267: 696–698.

Azam, F., T. Fenchel, J. G. Field, J. S. Gray, L. A. Meyer-Reil, and F. Thingstad. 1983. The ecological role of water column microbes in the sea. *Mar. Ecol. Prog. Ser.* 10: 257–263.

Baines, S. B., and M. L. Pace. 1991. The production of dissolved organic matter by phytoplankton and its importance to bacteria—patterns across marine and fresh water systems. *Limnol. Oceanogr.* 36: 1078–1090.

Béjà, O., L. Aravind, E. V. Koonin, M. T. Suzuki, A. Hadd, L. P. Nguyen, S. Jovanovich, C. M. Gates, R. A. Feldman, J. L. Spudich, E. N. Spudich, and E. F. DeLong. 2000. Bacterial rhodopsin: Evidence for a new type of phototrophy in the sea. *Science* 289: 1902–1906.

Béjà, O., E. N. Spudich, J. L. Spudich, M. Leclerc, and E. F. DeLong. 2001. Proteorhodopsin phototrophy in the ocean. *Nature* 411: 786–789.

Benner, R. 2002. Chemical composition and reactivity. In D. A. Hansell and C. A. Carlson (eds.), *Biogeochemistry of Marine Dissolved Organic Matter.* Academic Press, pp. 59–90.

Benner, R., J. D. Pakulski, M. McCarthy, J. I. Hedges, and P. G. Hatcher. 1992. Bulk chemical characteristics of dissolved organic matter in the ocean. *Science* 255: 1561–1564.

Bird, D. F., and J. Kalff. 1984. Empirical relationships between bacterial abundance and chlorophyll concentration in fresh and marine waters. *Can. J. Fish. Aquat. Sci.* 41: 1015–1023.

Björkman, K., and D. M. Karl. 1994. Bioavailability of inorganic and organic phosphorus compounds to natural assemblages of microorganisms in Hawaiian coastal waters. *Mar. Ecol. Prog. Ser.* 111: 265–273.

Björkman, K. M., and D. M. Karl. 2003. Bioavailability of dissolved organic phosphorus in the euphotic zone at station ALOHA, North Pacific Subtropical Gyre. *Limnol. Oceanogr.* 48: 1049–1057.

Björkman, K. M., and D. M. Karl. 2005. Presence of dissolved nucleotides in the North Pacific Subtropical Gyre and their role in cycling of dissolved organic phosphorus. *Aquat. Microb. Ecol.* 39: 193–203.

Björkman, K., A. L. Thomson-Bulldis, and D. M. Karl. 2000. Phosphorus dynamics in the North Pacific subtropical gyre. *Aquat. Microb. Ecol.* 22: 185–198.

Blackman, F. F. 1905. Optima and limiting factors. *Ann. Bot.* 19: 281–295.

Bonnet, S., and C. Guieu. 2006. Atmospheric forcing on the annual iron cycle in the western Mediterranean Sea: A 1-year survey. *J. Geophys. Res.* 111: C09010, doi:10.1029/2005JC003213.

Boyle, E. A., B. A. Bergquist, R. A. Kayser, and N. Mahowald. 2005. Iron, manganese, and lead at Hawaii Ocean Time-series station ALOHA: Temporal variability and an intermediate water hydrothermal plume. *Geochim. Cosmochim. Acta* 69: 933–952.

Brand, L. E., W. G. Sunda, and R. R. L. Guillard. 1983. Limitation of marine phytoplankton reproductive rates by zinc, manganese, and iron. *Limnol. Oceangr.* 28: 1182–1195.

Braun, S. T., L. M. Proctor, S. Zani, M. T. Mellon, and J. P. Zehr. 1999. Molecular evidence for zooplankton-associated nitrogen-fixing anaerobes based on amplification of the *nifH* gene. *FEMS Microbiol. Ecol.* 28: 273–279.

REFERENCES

Bronk, D. A. 2002. Dynamics of DON. In D. A. Hansell and C. A. Carlson (eds.), *Biogeochemistry of Marine Dissolved Organic Matter*. Academic Press, pp. 153–249.

Bruland, K. W., and M. C. Lohan. 2004. Controls of trace metals in seawater. In H. D. Holland and K. K. Turekian (eds.), *Treatise on Geochemistry*, Vol. 6. Elsevier, pp. 23–47.

Bruland, K. W., J. R. Donat, and D. A. Hutchins. 1991. Interactive influences of bioactive trace metals on biological production in oceanic waters. *Limnol. Oceangr.* 36: 1555–1577.

Button, D. K. 1985. Kinetics of nutrient-limited transport and microbial growth. *Microbiol. Mol. Biol. Rev.* 49: 270–297.

Button, D. K. 1991. Biochemical basis for whole cell uptake kinetics—specific affinity, oligotrophic capacity, and the meaning of the Michaelis constant. *Appl. Environ. Microbiol.* 57: 2033–2038.

Button, D. K. 1998. Nutrient uptake by microorganisms according to kinetic parameters from theory as related to cytoarchitecture. *Microbiol. Mol. Biol. Rev.* 62: 636–645.

Capone, D. G. 2000. The marine microbial nitrogen cycle. In D. L. Kirchman (ed.), *Microbial Ecology of the Oceans*, 1st edn. Wiley-Liss, pp. 455–493.

Carlson, C. A. 2002. Production and removal processes. In D. A. Hansell and C. A. Carlson (eds.), *Biogeochemistry of Marine Dissolved Organic Matter*. Academic Press, pp. 91–151.

Carlson, C. A., and H. W. Ducklow. 1996. Growth of bacterioplankton and consumption of dissolved organic carbon in the Sargasso Sea. *Aquat. Microb. Ecol.* 10: 69–85.

Carlson, C. A., H. W. Ducklow, and A. F. Michaels. 1994. Annual flux of dissolved organic carbon from the euphotic zone in the northwestern Sargasso Sea. *Nature* 371: 405–408.

Carlson, C. A., H. W. Ducklow, D. A. Hansell, and W. O. Smith. Jr. 1998. Organic carbon partitioning during spring phytoplankton blooms in the Ross Sea polynya and the Sargasso Sea. *Limnol. Oceanogr.* 43: 375–386.

Carlson, C. A., S. J. Giovannoni, D. A. Hansell, S. J. Goldberg, R. Parsons, M. P. Otero, K. Vergin, and B. R. Wheeler. 2002. Effect of nutrient amendments on bacterioplankton production, community structure, and DOC utilization in the northwestern Sargasso Sea. *Aquat. Microb. Ecol.* 30: 19–36.

Carlson, C. A., S. J. Giovannoni, D. A. Hansell, S. J. Goldberg, R. Parsons, and K. Vergin. 2004. Interactions among dissolved organic carbon, microbial processes, and community structure in the mesopelagic zone of the northwestern Sargasso Sea. *Limnol. Oceanogr.* 49: 1073–1083.

Caron, D. A., E. L. Lim, R. W. Sanders, M. R. Dennett, and U. G. Berninger. 2000. Responses of bacterioplankton and phytoplankton to organic carbon and inorganic nutrient additions in contrasting oceanic ecosystems. *Aquat. Microb. Ecol.* 22: 175–184.

Carpenter, S. R., J. F. Kitchell, and J. R. Hodgson. 1985. Cascading trophic interactions and lake productivity. *BioScience* 35: 634–639.

Cavender-Bares, K. K., D. M. Karl, and S. W. Chisholm. 2001. Nutrient gradients in the western North Atlantic Ocean: Relationship to microbial community structure and comparison to patterns in the Pacific Ocean. *Deep-Sea Res.* II 48: 2373–2395.

Cherrier, J., and J. E. Bauer. 2004. Bacterial utilization of transient plankton-derived dissolved organic carbon and nitrogen inputs in surface ocean waters. *Aquat. Microb. Ecol.* 35: 229–241.

Cherrier, J., J. E. Bauer, and E. R. M. Druffel. 1996. Utilization and turnover of labile dissolved organic matter by bacterial heterotrophs in eastern north Pacific surface waters. *Mar. Ecol. Prog. Ser.* 139: 267–279.

Church, M. J., D. A. Hutchins, and H. W. Ducklow. 2000. Limitation of bacterial growth by dissolved organic matter and iron in the Southern Ocean. *Appl. Environ. Microbiol.* 66: 455–466.

Church, M. J., H. W. Ducklow, and D. M. Karl. 2004. Light dependence of ^{3}H-leucine incorporation in the oligotrophic North Pacific Ocean. *Appl. Environ. Microbiol.* 70: 4079–4087.

Church, M. J., C. M. Short, B. D. Jenkins, D. M. Karl, and J. P. Zehr. 2005. Temporal patterns of nitrogenase gene (*nifH*) expression in the oligotrophic North Pacific Ocean. *Appl. Environ. Microbiol.* 71: 5362–5370.

Church, M. J., H. W. Ducklow, R. M. Letelier, and D. M. Karl. 2006. Temporal and vertical dynamics in picoplankton photoheterotrophic production in the subtropical North Pacific Ocean. *Aquat. Microb. Ecol.* 45: 41–53.

Coale, K. H., K. S. Johnson, S. E. Fitzwater, et al. 1996. A massive phytoplankton bloom induced by an ecosystem-scale iron fertilization experiment in the equatorial Pacific Ocean. *Nature* 383: 495–501.

Cochlan, W. P. 2001. The heterotrophic bacterial response during a mesoscale iron enrichment experiment (IronEx II) in the eastern equatorial Pacific Ocean. *Limnol. Oceanogr.* 46: 428–435.

Cole, J. J., S. Findlay, and M. L. Pace. 1988. Bacterial production in fresh and saltwater ecosystems—a cross-system overview. *Mar. Ecol. Prog. Ser.* 43: 1–10.

Cota, G. F., L. R. Pomeroy, W. G. Harrison, E. P. Jones, F. Peters, W. M. Sheldon Jr., and T. R. Weingartner. 1996. Nutrients, primary production and microbial heterotrophs in the southeastern Chukchi Sea: Arctic summer nutrient depletion and heterotrophy. *Mar. Ecol. Prog. Ser.* 135: 247–258.

Cotner, J. B., and B. A. Biddanda. 2002. Small players, large role: Microbial influence on biogeochemical processes in pelagic aquatic ecosystems. *Ecosystems* 5: 105–121.

Cotner, J. B., J. W. Ammerman, E. R. Peele, and E. Bentzen. 1997. Phosphorus-limited bacterioplankton growth in the Sargasso Sea. *Aquat. Microb. Ecol.* 13: 141–149.

Cottrell, M. T., and D. L. Kirchman. 2000. Natural assemblages of marine proteobacteria and members of the *Cytophaga–Flavobacter* cluster consuming low- and high-molecular-weight dissolved organic matter. *Appl. Environ. Microbiol.* 66: 1692–1697.

Cottrell, M. T., and D. L. Kirchman. 2003. Contribution of major bacterial groups to bacterial biomass production (thymidine and leucine incorporation) in the Delaware estuary. *Limnol. Oceanogr.* 48: 168–178.

Cullen, J. J. 1991. Hypotheses to explain high-nutrient conditions in the open sea. *Limnol. Oceanogr.* 36: 1578–1599.

de Baar, H. J. W. 1994. von Liebig's law of the minimum and plankton ecology. *Prog. Oceanogr.* 33: 347–386.

de Baar, H. J. W., P. W. Boyd, K. H. Coale, et al. 2005. Synthesis of iron fertilization experiments: From the iron age in the age of enlightenment. *J. Geophys. Res.* 110: C09S16, doi: 10.1029/2004JC002601.

del Giorgio, P. A., and C. M. Duarte. 2002. Respiration in the open ocean. *Nature* 420: 379–384.

del Giorgio, P. A., J. J. Cole, and A. Cimbleris. 1997. Respiration rates in bacteria exceed phytoplankton production in unproductive aquatic systems. *Nature* 385: 148–151.

DeLong, E. F., G. S. Wickham, and N. R. Pace. 1989. Phylogenetic stains—ribosomal RNA based probes for the identification of single cells. *Science* 243: 1360–1363.

Dolan, J. R., T. F. Thingstad, and F. Rassoulzadegan. 1995. Phosphate transfer between microbial size fractions in Villefranche Bay (NW Mediterranean Sea), France in autumn 1992. *Ophelia* 41: 71–85.

Dore, J. E., and D. M. Karl. 1996. Nitrite distributions and dynamics at Station ALOHA. *Deep-Sea Res.* II 43: 385–402.

Duarte, C. M., and S. Agustí. 1998. The CO_2 balance of unproductive aquatic ecosystems. *Science* 281: 234–236.

Ducklow, H. W. 1999. The bacterial component of the oceanic euphotic zone. *FEMS Microbiol. Ecol.* 30: 1–10.

Ducklow, H. W. 2000. Bacterial production and biomass in the ocean. In D. L. Kirchman (ed.), *Microbial Ecology of the Oceans*, 1st edn. Wiley-Liss, pp. 85–120.

Ducklow, H. W., and C. A. Carlson. 1992. Oceanic bacterial production. *Adv. Microb. Ecol.* 12: 113–181.

Ducklow, H. W., D. L. Kirchman, H. L. Quinby, C. A. Carlson, and H. G. Dam. 1993. Stocks and dynamics of bacterioplankton carbon during the spring bloom in the eastern North Atlantic Ocean. *Deep-Sea Res.* II 40: 245–263.

Ducklow, H. W., H. L. Quinby, and C. A. Carlson. 1995. Bacterioplankton dynamics in the Equatorial Pacific during the 1992 El Nino. *Deep-Sea Res.* II 42: 621–638.

Ducklow, H., C. Carlson, M. Church, D. Kirchman, D. Smith, and G. Steward. 2001. The seasonal development of the bacterioplankton bloom in the Ross Sea, Antarctica, 1994–1997. *Deep-Sea Res.* II 48: 4199–4221.

Egli, T. 1991. On multiple-nutrient-limited growth of microorganisms, with special reference to dual limitation by carbon and nitrogen substrates. *Antonie van Leeuwenhoek* 60: 225–234.

Eilers, H., J. Pernthaler, and R. Amann. 2000. Succession of pelagic marine bacteria during enrichment: A close look at cultivation-induced shifts. *Appl. Environ. Microbiol.* 66: 4634–4640.

Fasham, M. J. R., H. W. Ducklow, and S. M. McKelvie. 1990. A nitrogen-based model of plankton dynamics in the oceanic mixed layer. *J. Mar. Res.* 48: 591–639.

Fuhrman, J. 1987. Close coupling between release and uptake of dissolved free amino acids in seawater studied by an isotope dilution approach. *Mar. Ecol. Prog. Ser.* 37: 45–52.

Fuhrman, J. A., and R. L. Ferguson. 1986. Nanomolar concentrations and rapid turnover of dissolved free amino acids in seawater: Agreement between chemical and microbiological measurements. *Mar. Ecol. Prog. Ser.* 33: 237–242.

García-Fernández, J. M., N. T. de Marsac, and J. Diez. 2004. Streamlined regulation and gene loss as adaptive mechanisms in *Prochlorococcus* for optimized nitrogen utilization in oligotrophic environments. *Microbiol. Mol. Biol. Rev.* 68: 630–638.

Garside, C., and J. Garside. 1993. The "f-ratio" on 20°W during the North Atlantic Bloom Experiment. *Deep-Sea Res.* II 40: 75–90.

Gasol, J. M., and C. M. Duarte. 2000. Comparative analyses in aquatic microbial ecology: How far do they go? *FEMS Microbiol. Ecol.* 31: 99–106.

Gasol, J. M., P. A. del Giorgio, and C. M. Duarte. 1997. Biomass distribution in marine planktonic communities. *Limnol. Oceanogr.* 42: 1353–1363.

Giovannoni, S. J., L. Bibbs, J. C. Cho, et al. (2005a). Proteorhodopsin in the ubiquitous marine bacterium SAR11. *Nature* 438: 82–85.

Giovannoni, S. J., H. J. Tripp, S. Givan, et al. (2005b). Genome streamlining in a cosmopolitan oceanic bacterium. *Science* 309: 1242–1245.

Glöckner, F. O., B. M. Fuchs, and R. Amann. 1999. Bacterioplankton compositions of lakes and oceans: A first comparison based on fluorescence in situ hybridization. *Appl. Environ. Microbiol.* 65: 3721–3726.

Gómez-Consarnau, L., J. M. González, M. Coll-Llado, P. Gourdon, T. Pascher, R. Neutze, C. Pedrós-Alió, and J. Pinhassi. 2007. Light stimulates growth of proteorhodopsin-containing marine Flavobacteria. *Nature* 445: 210–213.

González, J. M., and M. A. Moran. 1997. Numerical dominance of a group of marine bacteria in the alpha-subclass of the class Proteobacteria in coastal seawater. *Appl. Environ. Microbiol.* 63: 4237–4242.

Grossart, H. P., and M. Simon. 2002. Bacterioplankton dynamics in the Gulf of Aqaba and the Northern Red Sea in early spring. *Mar. Ecol. Prog. Ser.* 239: 263–276.

Hall, J. A., and K. Safi. 2001. The impact of in situ iron fertilisation on the microbial food web in the Southern Ocean. *Deep-Sea Res.* II 48: 2591–2613.

Hansell, D. A. 2002. DOC in the global ocean carbon cycle. In D. A. Hansell and C. A. Carlson (eds.), *Biogeochemistry of Marine Dissolved Organic Matter*. Academic Press, pp. 685–715.

Hardin, G. 1960. The competitive exclusion principle. *Science* 131: 1292–1298.

Hedges, J. I. 2002. Why dissolved organics matter? In D. A. Hansell and C. A. Carlson (eds.), *Biogeochemistry of Marine Dissolved Organic Matter*. Academic Press, pp. 1–33.

Howard, E. C., J. R. Henriksen, A. Buchan, et al. 2006. Bacterial taxa that limit sulfur flux from the ocean. *Science* 314: 649–652.

Hutchinson, G. E. 1961. The paradox of the plankton. *Amer. Nat.* 95: 137–145.

Joint, I. R., and R. J. Morris. 1982. The role of bacteria in the turnover of organic matter in the sea. *Oceanogr. Mar. Biol. Annu. Rev.* 20: 65–118.

Joint, I., P. Henriksen, G. A. Fonnes, D. Bourne, T. F. Thingstad, and B. Riemann. 2002. Competition for inorganic nutrients between phytoplankton and bacterioplankton in nutrient manipulated mesocosms. *Aquat. Microb. Ecol.* 29: 145–159.

Karl, D. M., and K. M. Björkman. 2002. Dynamics of DOP. In D. A. Hansell and C. A. Carlson (eds.), *Biogeochemistry of Marine Dissolved Organic Matter*. Academic Press, pp. 249–366.

Karl, D. M., and G. Tien. 1992. MAGIC—a sensitive and precise method for measuring dissolved phosphorus in aquatic environments. *Limnol. Oceanogr.* 37: 105–116.

Karner, M. B., E. F. DeLong, and D. M. Karl. 2001. Archaeal dominance in the mesopelagic zone of the Pacific Ocean. *Nature* 409: 507–510.

Keil, R. G., and D. L. Kirchman. 1991. Contribution of dissolved free amino acids and ammonium to the nitrogen requirements of heterotrophic bacterioplankton. *Mar. Ecol. Prog. Ser.* 73: 1–10.

Keil, R. G., and D. L. Kirchman. 1999. Utilization of dissolved protein and amino acids in the northern Sargasso Sea. *Aquat. Microb. Ecol.* 18: 293–300.

Kirchman, D. L. 1990. Limitation of bacterial growth by dissolved organic matter in the subarctic Pacific. *Mar. Ecol. Prog. Ser.* 62: 47–54.

Kirchman, D. L. 2000. Uptake and regeneration of inorganic nutrients by marine heterotrophic bacteria. In D. L. Kirchman (ed.), *Microbial Ecology of the Oceans*, 1st edn. Wiley-Liss, pp. 261–288.

Kirchman, D. L., and J. H. Rich. 1997. Regulation of bacterial growth rates by dissolved organic carbon and temperature in the Equatorial Pacific Ocean. *Microb. Ecol.* 33: 11–20.

Kirchman, D. L., and P. A. Wheeler. 1998. Uptake of ammonium and nitrate by heterotrophic bacteria and phytoplankton in the sub-Arctic Pacific. *Deep-Sea Res.* I 45: 347–365.

Kirchman, D. L., R. G. Keil, and P. A. Wheeler. 1989. The effect of amino acids on ammonium utilization and regeneration by heterotrophic bacteria in the subarctic Pacific. *Deep-Sea Res.* I 36: 1763–1776.

Kirchman, D. L., R. G. Keil, M. Simon, and N. A. Welschmeyer. 1993. Biomass and production of heterotrophic bacterioplankton in the oceanic subarctic Pacific. *Deep-Sea Res.* I 40: 967–988.

Kirchman, D. L., H. W. Ducklow, J. J. McCarthy, and C. Garside. 1994. Biomass and nitrogen uptake by heterotrophic bacteria during the spring phytoplankton bloom in the North Atlantic Ocean. *Deep-Sea Res.* I 41: 879–895.

Kirchman, D. L., J. H. Rich, and R. T. Barber. 1995. Biomass and biomass production of heterotrophic bacteria along 140°W in the Equatorial Pacific: Effect of temperature on the microbial loop. *Deep-Sea Res.* II 42: 603–619.

Kirchman, D. L., B. Meon, M. T. Cottrell, D. A. Hutchins, D. Weeks, and K. W. Bruland. 2000. Carbon versus iron limitation of bacterial growth in the California upwelling regime. *Limnol. Oceanogr.* 45: 1681–1688.

Kirchman, D. L., B. Meon, H. W. Ducklow, C. A. Carlson, D. A. Hansell, and G. F. Steward. 2001. Glucose fluxes and concentrations of dissolved combined neutral sugars (polysaccharides) in the Ross Sea and Polar Front Zone, Antarctica. *Deep-Sea Res.* II 48: 4179–4197.

Kirchman, D. L., R. R. Malmstrom, and M. T. Cottrell. 2005. Control of bacterial growth by temperature and organic matter in the Western Arctic. *Deep-Sea Res.* II 52: 3386–3395.

Klausmeier, C. A., E. Litchman, T. Daufresne, and S. A. Levin. 2004. Optimal nitrogen-to-phosphorus stoichiometry of phytoplankton. *Nature* 429: 171–174.

Kolber, Z. S., C. L. Van Dover, R. A. Niederman, and P. G. Falkowski. 2000. Bacterial photosynthesis in surface waters of the open ocean. *Nature* 407: 177–179.

Kolber, Z. S., F. G. Plumley, A. S. Lang, et al. 2001. Contribution of aerobic photoheterotrophic bacteria to the carbon cycle in the ocean. *Science* 292: 2492–2495.

Konings, W. N. 2006. Microbial transport: Adaptations to natural environments. *Antonie van Leeuwenhoek* 90: 325–342.

Konneke, M., A. E. Bernhard, J. R. de la Torre, C. B. Walker, J. B. Waterbury, and D. A. Stahl. 2005. Isolation of an autotrophic ammonia-oxidizing marine archaeon. *Nature* 437: 543–546.

Krom, M. D., E. M. S. Woodward, B. Herut, et al. 2005. Nutrient cycling in the south east Levantine basin of the eastern Mediterranean: Results from a phosphorus starved system. *Deep-Sea Res.* II 52: 2879–2896.

Kuypers, M. M. M., A. O. Sliekers, G. Lavik, M. Schmid, B. B. Jorgensen, J. G. Kuenen, J. S. S. Damste, M. Strous, and M. S. M. Jetten. 2003. Anaerobic ammonium oxidation by anammox bacteria in the Black Sea. *Nature* 422: 608–611.

Lalli, C., and T. Parsons. 1993. *Biological Oceanography: An Introduction*. Open University.

Li, W. K. W., and P. M. Dickie. 1987. Temperature characteristics of photosynthetic and heterotrophic activities—seasonal variations in temperate microbial plankton. *Appl. Environ. Microbiol.* 53: 2282–2295.

Li, W. K. W., P. M. Dickie, W. G. Harrison, and B. D. Irwin. 1993. Biomass and production of bacteria and phytoplankton during the spring bloom in the western North Atlantic Ocean. *Deep-Sea Res.* II 40: 307–327.

Li, W. K. W., E. J. H. Head, and W. G. Harrison. 2004. Macroecological limits of heterotrophic bacterial abundance in the ocean. *Deep-Sea Res.* I 51: 1529–1540.

Liebig, J. von. 1840. *Chemistry in its Application to Agriculture and Physiology*. Taylor & Walton.

McQueen, D. J., J. R. Post, and E. L. Mill. 1986. Trophic relationships in freshwater pelagic ecosystems. *Can. J. Fish. Aquat. Sci.* 43: 1571–1581.

Maldonado, M. T., and N. M. Price. 1999. Utilization of iron bound to strong organic ligands by plankton communities in the subarctic Pacific Ocean. *Deep-Sea Res.* II 46: 2447–2473.

Malmstrom, R. R., R. P. Kiene, and D. L. Kirchman. 2004. Identification and enumeration of bacteria assimilating dimethylsulfoniopropionate (DMSP) in the North Atlantic and Gulf of Mexico. *Limnol. Oceanogr.* 49: 597–606.

Malmstrom, R. R., M. T. Cottrell, H. Elifantz, and D. L. Kirchman. 2005. Biomass production and assimilation of dissolved organic matter by SAR11 bacteria in the northwest Atlantic Ocean. *Appl. Environ. Microbiol.* 71: 2979–2986.

Martin, J. H., and S. E. Fitzwater. 1988. Iron deficiency limits phytoplankton growth in the northeast Pacific subarctic. *Nature* 331: 341–343.

Martin, J. H., S. E. Fitzwater, R. M. Gordon, C. N. Hunter, and S. J. Tanner. 1993. Iron, primary production and carbon nitrogen flux studies during the JGOFS North Atlantic Bloom Experiment. *Deep-Sea Res.* II 40: 115–134.

Measures, C. I. 1999. The role of entrained sediments in sea ice in the distribution of aluminium and iron in the surface waters of the Arctic Ocean. *Mar. Chem.* 68: 59–70.

Moore, L. R., A. F. Post, G. Rocap, and S. W. Chisholm. 2002. Utilization of different nitrogen sources by the marine cyanobacteria *Prochlorococcus* and *Synechococcus*. *Limnol. Oceanogr.* 47: 989–996.

Mopper, K., and D. J. Kieber. 2002. Photochemistry and the cycling of carbon, sulfur, nitrogen, and phosphorus. In D. A. Hansell and C. A. Carlson (eds.), *Biogeochemistry of Marine Dissolved Organic Matter*. Academic Press, pp. 455–508.

Moran, M. A., and R. G. Zepp. 2000. UV radiation effects on microbes and microbial processes. In D. L. Kirchman (ed.), *Microbial Ecology of the Oceans*, 1st edn. Wiley-Liss, pp. 201–227.

Morel, F. M. M., J. R. Reinfelder, S. B. Roberts, C. P. Chamberlain, J. G. Lee, and D. Yee. 1994. Zinc and carbon co-limitation of marine phytoplankton. *Nature* 369: 740–742.

Morris, R. M., M. S. Rappé, S. A. Connon, K. L. Vergin, W. A. Siebold, C. A. Carlson, and S. J. Giovannoni. 2002. SAR11 clade dominates ocean surface bacterioplankton communities. *Nature* 420: 806–810.

REFERENCES

Morris, R. M., K. L. Vergin, J. C. Cho, M. S. Rappé, C. A. Carlson, and S. J. Giovannoni. 2005. Temporal and spatial response of bacterioplankton lineages to annual convective overturn at the Bermuda Atlantic Time-series Study site. *Limnol. Oceanogr.* 50: 1687–1696.

Nagata, T. 2000. Production mechanisms of dissolved organic matter. In D. L. Kirchman (ed.), *Microbial Ecology of the Oceans*, 1st edn. Wiley-Liss, pp. 121–152.

Nedwell, D. B., and M. Rutter. 1994. Influence of temperature on growth rate and competition between two psychrotolerant Antarctic bacteria—low temperature diminishes affinity for substrate uptake. *Appl. Environ. Microbiol.* 60: 1984–1992.

Obernosterer, I., N. Kawasaki, and R. Benner. 2003. P-limitation of respiration in the Sargasso Sea and uncoupling of bacteria from P-regeneration in size-fractionation experiments. *Aquat. Microb. Ecol.* 32: 229–237.

Oliver, J. L., R. T. Barber, W. O. Smith, Jr, and H. W. Ducklow. 2004. The heterotrophic bacterial response during the Southern Ocean Iron Experiment (SOFeX). *Limnol. Oceanogr.* 49: 2129–2140.

Orphan, V. J., C. H. House, K. U. Hinrichs, K. D. McKeegan, and E. F. DeLong. 2001. Methane-consuming archaea revealed by directly coupled isotopic and phylogenetic analysis. *Science* 293: 484–487.

Pakulski, J. D., and R. Benner. 1994. Abundance and distribution of carbohydrates in the ocean. *Limnol. Oceanogr.* 39: 930–940.

Pakulski, J. D., R. B. Coffin, C. A. Kelley, S. L. Holder, R. Downer, P. Aas, M. M. Lyons, and W. H. Jeffrey. 1996. Iron stimulation of Antarctic bacteria. *Nature* 383: 133–134.

Palenik, B., B. Brahamsha, F. W. Larimer, et al. 2003. The genome of a motile marine *Synechococcus*. *Nature* 424: 1037–1042.

Paulsen, I. T., L. Nguyen, M. K. Sliwinski, R. Rabus, and M. H. Saier. 2000. Microbial genome analyses: Comparative transport capabilities in eighteen prokaryotes. *J. Mol. Biol.* 301: 75–100.

Pinhassi, J., and Å. Hagström. 2000. Seasonal succession in marine bacterioplankton. *Aquat. Microb. Ecol.* 21: 245–256.

Pinhassi, J., F. Azam, J. Hemphaelae, R. A. Long, J. Martinez, U. L. Zweifel, and Å. Hagström. 1999. Coupling between bacterioplankton species composition, population dynamics, and organic matter degradation. *Aquat. Microb. Ecol.* 17: 13–26.

Pinhassi, J., L. Gómez-Consarnau, L. Alonso-Saez, M. M. Sala, M. Vidal, C. Pedrós-Alió, and J. M. Gasol. 2006. Seasonal changes in bacterioplankton nutrient limitation and their effects on bacterial community composition in the NW Mediterranean Sea. *Aquat. Microb. Ecol.* 44: 241–252.

Pomeroy, L. R. 1974. The ocean's food web, a changing paradigm. *BioScience* 24: 499–504.

Pomeroy, L. R., and D. Deibel. 1986. Temperature regulation of bacterial activity during the spring bloom in Newfoundland coastal waters. *Science* 233: 359–361.

Pomeroy, L. R., W. J. Wiebe, D. Deibel, R. J. Thompson, G. T. Rowe, and J. D. Pakulski. 1991. Bacterial responses to temperature and substrate concentration during the Newfoundland spring bloom. *Mar. Ecol. Prog. Ser.* 75: 143–159.

Reay, D. S., D. B. Nedwell, J. Priddle, and J. C. Ellis-Evans. 1999. Temperature dependence of inorganic nitrogen uptake: Reduced affinity for nitrate at suboptimal temperatures in both algae and bacteria. *Appl. Environ. Microbiol.* 65: 2577–2584.

Repeta, D. J., and L. I. Aluwihare. 2006. Radiocarbon analysis of neutral sugars in high-molecular-weight dissolved organic carbon: Implications for organic carbon cycling. *Limnol. Oceanogr.* 51: 1045–1053.

Reynolds, C. S. 2001. Emergence in pelagic communities. *Sci. Mar.* 65: 5–30.

Rich, J. H., H. W. Ducklow, and D. L. Kirchman. 1996. Concentrations and uptake of neutral monosaccharides along 140°W in the Equatorial Pacific: Contribution of glucose to heterotrophic bacterial activity and the DOM flux. *Limnol. Oceanogr.* 41: 595–604.

Rich, J., M. Gosselin, E. Sherr, B. Sherr, and D. L. Kirchman. 1997. High bacterial production, uptake and concentrations of dissolved organic matter in the Central Arctic Ocean. *Deep-Sea Res.* II 44: 1645–1663.

Rivkin, R. B., and M. R. Anderson. 1997. Inorganic nutrient limitation of oceanic bacterioplankton. *Limnol. Oceanogr.* 42: 730–740.

Rivkin, R. B., and L. Legendre. 2001. Biogenic carbon cycling in the upper ocean: Effects of microbial respiration. *Science* 291: 2398–2400.

Rivkin, R. B., and L. Legendre. 2002. Roles of food web and heterotrophic microbial processes in upper ocean biogeochemistry: Global patterns and processes. *Ecol. Res.* 17: 151–159.

Rivkin, R. B., M. R. Anderson, and C. Lajzerowicz. 1996. Microbial processes in cold oceans. 1. Relationship between temperature and bacterial growth rate. *Aquat. Microb. Ecol.* 10: 243–254.

Robinson, C., and P. J. L. Williams. 1993. Temperature and Antarctic plankton community respiration. *J. Plankton Res.* 15: 1035–1051.

Rocap, G., F. W. Larimer, J. Lamerdin, et al. 2003. Genome divergence in two *Prochlorococcus* ecotypes reflects oceanic niche differentiation. *Nature* 424: 1042–1047.

Saier, M. H. 2003. Tracing pathways of transport protein evolution. *Mol. Microbiol.* 48: 1145–1156.

Saito, M. A., T. J. Goepfert, and J. T. Ritt. 2008. Some thoughts on the concept of colimitation: three definitions and the importance of bioavailability. *Limnol. Oceangr.* 53: 276–290.

Sala, M. M., F. Peters, J. M. Gasol, C. Pedrós-Alió, C. Marrase, and D. Vaque. 2002. Seasonal and spatial variations in the nutrient limitation of bacterioplankton growth in the northwestern Mediterranean. *Aquat. Microb. Ecol.* 27: 47–56.

Sarthou, G., and C. Jeandel. 2001. Seasonal variations of iron concentrations in the Ligurian Sea and iron budget in the Western Mediterranean Sea. *Mar. Chem.* 74: 115–129.

Schwalbach, M. S., M. Brown, and J. A. Fuhrman. 2005. Impact of light on marine bacterioplankton community structure. *Aquat. Microb. Ecol.* 39: 235–245.

Sedwick, P., T. M. Church, A. R. Bowie, et al. 2005. Iron in the Sargasso Sea (Bermuda Atlantic Time-series Study region) during summer: eolian imprint, spatiotemporal variability, and ecological implications. *Global Biogeochem. Cycles* 19: GB4006, doi: 4010.1029/2004GB002445.

Sherr, B. F., and E. B. Sherr. 2003. Community respiration/production and bacterial activity in the upper water column of the central Arctic Ocean. *Deep-Sea Res.* I 50: 529–542.

Sherr, E. B., B. F. Sherr, P. A. Wheeler, and K. Thompson. 2003. Temporal and spatial variation in stocks of autotrophic and heterotrophic microbes in the upper water column of the central Arctic Ocean. *Deep-Sea Res.* I 50: 557–571.

Shiba, T., Y. Shioi, K. Takamiya, D. C. Sutton, and C. R. Wilkinson. 1991. Distribution and physiology of aerobic bacteria containing bacteriochlorophyll alpha on the east and west coasts of Australia. *Appl. Environ. Microbiol.* 57: 295–300.

Skoog, A., B. Biddanda, and R. Benner. 1999. Bacterial utilization of dissolved glucose in the upper water column of the Gulf of Mexico. *Limnol. Oceanogr.* 44: 1625–1633.

Skoog, A., K. Whitehead, F. Sperling, and K. Junge. 2002. Microbial glucose uptake and growth along a horizontal nutrient gradient in the North Pacific. *Limnol. Oceanogr.* 47: 1676–1683.

Smith, D. C., M. Simon, A. L. Alldredge, and F. Azam. 1992. Intense hydrolytic enzyme activity on marine aggregates and implications for rapid particle dissolution. *Nature* 359: 139–142.

Sunda, W., and S. A. Huntsman. 1995. Cobalt and zinc interreplacement in marine phytoplankton: Biological and geochemical implications. *Limnol. Oceangr.* 40: 1404–1417.

Suttle, C. A., J. A. Fuhrman, and D. G. Capone. 1990. Rapid ammonium cycling and concentration-dependent partitioning of ammonium and phosphate: Implications for carbon transfer in planktonic communities. *Limnol. Oceanogr.* 35: 424–433.

Suttle, C. A., A. M. Chan, and J. A. Fuhrman. 1991. Dissolved free amino acids in the Sargasso Sea: Uptake and respiration rates, turnover times, and concentrations. *Mar. Ecol. Prog. Ser.* 70: 189–199.

Terry, K. L., J. Hirata, and E. A. Laws. 1985. Light-limited, nitrogen-limited, and phosphorus-limited growth of *Phaeodactylum tricornutum* Bohlin Strain TFX-1—Chemical composition, carbon partitioning, and the diel periodicity of physiological processes. *J. Exp. Mar. Biol. Ecol.* 86: 85–100.

Thingstad, T. F., Å. Hagström, and F. Rassoulzadegan. 1997. Accumulation of degradable DOC in surface waters: Is it caused by a malfunctioning microbial loop? *Limnol. Oceanogr.* 42: 398–404.

Thingstad, T. F., U. Li Zweifel, and F. Rassoulzadegan. 1998. P limitation of heterotrophic bacteria and phytoplankton in the northwest Mediterranean. *Limnol. Oceanogr.* 43: 88–94.

Thingstad, T. F., M. D. Krom, R. F. C. Mantoura, et al. 2005. Nature of phosphorus limitation in the ultraoligotrophic eastern Mediterranean. *Science* 309: 1068–1071.

Tortell, P. D., M. T. Maldonado, and N. M. Price. 1996. The role of heterotrophic bacteria in iron-limited ocean ecosystems. *Nature* 383: 330–332.

Tortell, P. D., M. T. Maldonado, J. Granger, and N. M. Price. 1999. Marine bacteria and biogeochemical cycling of iron in the oceans. *FEMS Microbiol. Ecol.* 29: 1–11.

Vallino, J. J., C. S. Hopkinson, and J. E. Hobbie. 1996. Modeling bacterial utilization of dissolved organic matter: Optimization replaces Monod growth kinetics. *Limnol. Oceanogr.* 41: 1591–1609.

Van Den Broeck, N., T. Moutin, M. Rodier, and A. Le Bouteiller. 2004. Seasonal variations of phosphate availability in the SW Pacific Ocean near New Caledonia. *Mar. Ecol. Prog. Ser.* 268: 1–12.

Walter, J. M., D. Greenfield, C. Bustamante, and J. Liphardt. 2007. Light-powering *Escherichia coli* with proteorhodopsin. *Proc. Natl. Acad. Sci. USA* 104: 2408–2412.

Wheeler, P. A., and D. L. Kirchman. 1986. Utilization of inorganic and organic nitrogen by bacteria in marine systems. *Limnol. Oceanogr.* 31: 998–1009.

Wiebe, W. J., W. M. Sheldon, and L. R. Pomeroy. 1992. Bacterial growth in the cold—Evidence for an enhanced substrate requirement. *Appl. Environ. Microbiol.* 58: 359–364.

Wiebe, W. J., W. M. Sheldon, and L. R. Pomeroy. 1993. Evidence for an enhanced substrate requirement by marine mesophilic bacterial isolates at minimal growth temperatures. *Microb. Ecol.* 25: 151–159.

Williams, P. J. leB. 2000. Heterotrophic bacteria and the dynamics of dissolved organic material. In D. L. Kirchman (ed.), *Microbial Ecology of the Oceans*, 1st edn. Wiley-Liss, pp. 153–200.

Williams, P. J. leB., P. J. Morris, and D. M. Karl. 2004. Net community production and metabolic balance at the oligotrophic ocean site, station ALOHA. *Deep-Sea Res.* I 51: 1563–1578.

Wu, J. F., W. Sunda, E. A. Boyle, and D. M. Karl. 2000. Phosphate depletion in the western North Atlantic Ocean. *Science* 289: 759–762.

Wuchter, C., B. Abbas, M. J. L. Coolen, et al. 2006. Archaeal nitrification in the ocean. *Proc. Natl. Acad. Sci. USA* 103: 12317–12322.

Zehr, J. P., M. T. Mellon, and S. Zani. 1998. New nitrogen-fixing microorganisms detected in oligotrophic oceans by amplification of nitrogenase (*nifH*) genes. *Appl. Environ. Microbiol.* 64: 3444–3450.

Zohary, T., and R. D. Robarts. 1998. Experimental study of microbial P limitation in the eastern Mediterranean. *Limnol. Oceanogr.* 43: 387–395.

Zubkov, M. V., B. M. Fuchs, G. A. Tarran, P. H. Burkill, and R. Amann. 2003. High rate of uptake of organic nitrogen compounds by *Prochlorococcus* cyanobacteria as a key to their dominance in oligotrophic oceanic waters. *Appl. Environ. Microbiol.* 69: 1299–1304.

Zubkov, M. V., G. A. Tarran, and B. M. Fuchs. 2004. Depth related amino acid uptake by *Prochlorococcus* cyanobacteria in the Southern Atlantic tropical gyre. *FEMS Microbiol. Ecol.* 50: 153–161.

Zweifel, U. L. 1999. Factors controlling accumulation of labile dissolved organic carbon in the Gulf of Riga. *Est. Coast. Shelf Sci.* 48: 357–370.

11

PROTISTAN GRAZING ON MARINE BACTERIOPLANKTON

KLAUS JÜRGENS

Leibniz Institute for Baltic Sea Research, 18119 Rostock, Germany

RAMON MASSANA

Institut de Ciències del Mar, CMIMA (CSIC), Passeig Marítim de la Barceloneta 37–49, 08003 Barcelona, Catalunya, Spain

INTRODUCTION

Bacterial growth rates are often high, yet bacterial numbers are remarkably constant in pelagic systems, implying that bacterial mortality rates have to be in the same range as bacterial production. Physical transport processes cannot be considered as important removal processes, because suspended bacteria do not sink, and there is no evidence for substantial physiological death of planktonic bacteria. This necessitates biological removal agents. Bacterioplankton constitutes a potentially important food resource that, in terms of total biomass, is in the same range as phytoplankton at least in oligotrophic systems (Simon et al. 1992). However, bacteria cannot be efficiently gathered by marine mesozooplankters (mostly copepods) and microzooplankters (mostly ciliates and dinoflagellates). The missing bacterial grazers were found when the smallest planktonic unpigmented protists, heterotrophic nanoflagellates, were studied not only in terms of taxonomy of isolated forms, as has been done by dedicated protozoologists for a long time, but by microbial ecologists with an ecological perspective. By studying the bioenergetics, growth, and feeding of different flagellate isolates, and in combination with field data, Fenchel (1982a–d) showed

Microbial Ecology of the Oceans, Second Edition. Edited by David L. Kirchman
Copyright © 2008 John Wiley & Sons, Inc.

that heterotrophic nanoflagellates were indeed adapted for sustained growth on suspended bacteria in the ocean and potentially capable of controlling bacterial numbers. Several field studies in different marine systems corroborated the importance of flagellate bacterivory (Sieburth 1984).

The transfer of organic carbon from the dissolved fraction to higher trophic levels in a linear heterotrophic food chain via bacteria, nanoflagellates, and ciliates was formalized as the "microbial loop" concept (Azam et al. 1983). In the following years, different methods were developed to obtain estimates of protist grazing on planktonic bacteria. They confirmed the view that this process was mainly performed by nanoplanktonic cells, mostly heterotrophic flagellates in the size range 2–5 μm. This was true not only for marine and freshwater pelagic habitats but also for groundwater, soils, and sediments, although these systems have not been studied as intensively. It was shown that bacterial production and bacterial grazing losses, although both measured with a considerable degree of uncertainty, were roughly in balance (McManus and Fuhrman 1988).

The picture outlined in the microbial loop concept is still correct, although it has to be augmented by many additional facets. In many marine systems, primary production is largely based on picoplanktonic cells (both cyanobacteria and picoeukaryotes, as discussed in Chapters 1 and 6). Therefore, phagotrophic protists were discovered to be also the main herbivores, and these trophic interactions take place in a complex microbial food web (Sherr and Sherr 1994). With bacteriophages, another biotic controlling agent has been discovered that can be quantitatively nearly as important as protistan grazing (see Chapter 12) but has different implications for bacterial communities and food webs. In addition, there are situations, particularly in coastal and estuarine systems, where bacterivory might be dominated by ciliates (Sherr and Sherr 1987) or even metazoans, for example when appendicularians (e.g., *Oikopleura*) are abundant (Deibel and Lee 1992).

There is an ongoing debate as to whether bacterial numbers and biomass are mainly controlled by substrate supply and nutrients (bottom-up: see Chapter 10) or by predation (top-down), and how these modes of control interact (Thingstad 2000) and shift over time (Psenner and Sommaruga 1992). Different lines of support for these hypotheses come from experimental studies and empirical models analyzing published data sets. Most studies use the so-called "black box" approach, which considers bacterial communities as homogeneous assemblages of cells. A different view may develop taking into account the large heterogeneity of bacterial communities, in terms of bacterial phylogenetic composition (see Chapters 3–5 and 7) and physiological states (Chapter 8), but also with respect to phenotypic properties that affect their vulnerability towards grazers. These new insights represented a change in research focus, from quantitative studies concentrating on carbon flow budgets to studies focusing on determining the specific populations of bacteria and the underlying mechanisms of predator–prey interactions, including predator foraging and prey selection, prey vulnerability, and grazing resistance. Most of these studies were done in freshwater systems, in which the whole spectrum of approaches, from defined laboratory experiments with cultivated

INTRODUCTION

model organisms to chemostat, mesocosm, and field studies with complex communities, revealed a rather comprehensive picture on grazing impacts and feedback mechanisms (Jürgens and Matz 2002). Studies in marine systems have lagged behind, but presumably many of the results derived from freshwater plankton are valid also for marine communities. Nevertheless, there are also important differences between freshwater and marine systems, both in terms of abiotic constraints (salinity, trophic level, allochthonous carbon influence, etc.) and in the phylogenetic composition of the microbial assemblages. The newly emerged picture is that grazing on bacteria is a strong selection pressure in planktonic systems, thereby shaping the phenotypic and genetic composition of prokaryotic communities in the world's oceans (Pernthaler 2005). One can even speculate that the dominance and global success of some prokaryotic groups may result from this selective force and adaptations to it.

The black box was not only illuminated on the bacterial side but recently also on the side of the bacterivorous protists (Moon-van der Staay et al. 2001). The molecular tools that had been very useful for examining the phylogenetic composition of bacterial communities can also be applied in the study of protist diversity. This has resulted in some surprising new insights into the composition and distribution of phototrophic and heterotrophic marine protists. As for prokaryotes, many of the newly identified and globally distributed taxa are yet uncultured (Massana et al. 2006b; see also Chapter 6). A fundamental question remains whether the results on bacteria–protist interactions and on the autecology of bacterivores obtained with some easy to culture nanoflagellate species are representative for the species really dominating in the ocean.

Today, marine protists are studied with three fundamentally different approaches, performed by scientists from different schools. Plankton ecologists focus on quantitative data on abundance, biomass, growth, and grazing rates of protists. To achieve a carbon budget and pinpoint the major carbon and nutrient pathways, the resolution is necessarily low and marine protists are lumped into one, or a few, functional groups. Classical protozoologists have been studying the ultrastructure and ecophysiology of many protistan species for more than two centuries. Their approach focuses on the phylogenetic and functional diversity, mostly depending on enrichments and isolations. Even when the composition of field samples is directly inspected, quantitative data on the different groups is generally lacking. A third approach, which was adopted from studies of aquatic bacteria, uses molecular techniques to investigate the composition of marine protists. Unfortunately, there are still only a few attempts to combine the different approaches for analyzing the structure and function of marine protistan assemblages.

This chapter gives an overview of the phylogenetic and functional groups of bacterivorous protists in marine systems, and on their interactions with prokaryotic communities. The term "bacterivory" is used here for the consumption of prokaryotic picoplankton in general, including heterotrophic and autotrophic bacteria as well as archaea. The fact that different groups of protists such as ciliates and dinoflagellates are often also the major herbivores of marine phytoplankton is well established

(Calbet and Landry 2004; Sherr and Sherr 1994, 2002) and will not be explored further here (see also Chapter 1). We will first present the main bacterivorous protist groups and then the feeding behavior and bioenergetics of heterotrophic flagellates, the main bacterial grazers in the sea. We will then summarize the quantitative impact of protistan grazing on bacterial communities and the approaches by which it has been measured. Protistan predation acts as a strong selection force on planktonic bacteria, and we will discuss recent findings on how this influences the phylogenetic and phenotypic bacterial composition. We will end the chapter by presenting the novel possibilities opened by the use of molecular tools to elucidate the most important bacterivores in the ocean.

Some Definitions of Organisms

- **Eukaryotes** Organisms that contain cells where the genetic material is organized in a membrane-bound nucleus. In contrast, Bacteria and Archaea are prokaryotes.
- **Protists** Unicellular eukaryotes, including some algae, protozoa, and fungi. They account for most of the eukaryotic diversity, and are found in the six phylogenetic super groups of eukaryotes (Fig. 11.1). See Chapter 6.
- **Protozoa** Unpigmented, phagotrophic protists, generally motile.
- **Heterotrophic Flagellates** Protozoa with one or few flagella used for feeding and motility. In marine plankton, most cells are between 2 and 20 μm in size, so they are called heterotrophic nanoflagellates (HNF).

NEW INSIGHTS INTO PHYLOGENETIC ORGANIZATION

During the second half of the 20th century, molecular developments provided a systematic way to relate all living organisms through DNA sequence comparisons, initially using the small-subunit ribosomal RNA gene (SSU rDNA): 16S rDNA in prokaryotes and 18S rDNA in eukaryotes (Woese 1987) (see also Chapters 3 and 6). The first eukaryotic phylogenetic trees had a top crown of lineages and several basal lineages, often amitochondrial (without mitochondria), which were considered primitive eukaryotes (Schlegel 1994). Recent phylogenies using concatenated protein genes show that virtually all eukaryotes can be assigned to one of six supergroups (Adl et al. 2005; Baldauf 2003), which form a crown radiation without clear ranking among them and with an uncertain root (Fig. 11.1). Few morphological or ultrastructural characters unite the diverse lineages within each supergroup, but the phylogenetic signatures are robust. Previous basal lineages are now placed within these supergroups, as exemplified by the amitochondrial microsporidia, now known as fast-evolving fungi. A consequence of the molecular framework is that many incertae sedis protists (Patterson and Zölffel 1991) are finding their phylogenetic position in the eukaryotic tree, some recent examples being the Telonemia (Schalchian-Tabrizi et al. 2006) and the katablepharids (Okamoto and

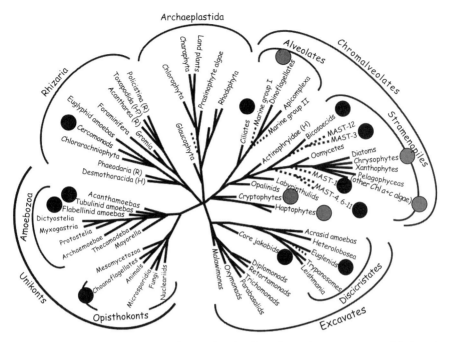

Figure 11.1 Schematic phylogenetic tree showing the eukaryotic supergroups and their main lineages. Dots indicate lineages that contain marine bacterivorous protists (dark if the group is exclusively heterotrophic and gray if it contains also phototrophs). Modified from Baldauf (2003).

Inouye 2005). In addition, sequences of the 18S rRNA gene of cultured representatives are crucial in placing protists within this phylogenetic context (Cavalier-Smith and Chao 2003a, b; Scheckenbach et al. 2005).

Phagotrophy, which is the capacity to ingest living organisms, is assumed to have played a central role in the origin and evolution of the eukaryotic cell (Margulis 1981), and it is therefore not surprising that it remains widely dispersed throughout the eukaryotic world (Raven 1997; Sherr and Sherr 2002). Only the supergroup Archaeplastida does not seem to have any phagotrophic member, although there are two reports of the prasinophytes *Micromonas* and *Pyramimonas* ingesting bacteria (Bell and Laybourn-Parry 2003; González et al. 1993a). In marine systems, the collection of protists that have bacteria in their diet is astonishingly diverse (Fig. 11.1). They range from forms that base their nutrition exclusively on bacteria to forms that ingest bacteria to complement their diet to varying degrees. Based on general morphology and on the motility and feeding mechanisms, bacterivorous protists can be separated into amoeboids, flagellates, and ciliates. Amoeboid protists have flexible shapes and move and feed with temporary projections, called pseudopods. Flagellates have an oval or spherical shape and one or more flagella that are used for propulsion and to create feeding currents.

TABLE 11.1 Common Genera of Marine Heterotrophic Bacterivorous Protists

Overall Form	Supergroup	Taxonomic Group	Common Genera
Ciliates	Alveolates	Ciliates	*Colpidium, Cyclidium, Strombidium*
Amoeboids	Amoebozoa	Flabellinid amoebas	*Flabellula*
		Tubulinid amoebas	*Platyamoeba, Vannella*
	Rhizaria	Cercomonads	*Cercomonas, Massisteria, Metromonas*
Flagellates	Opisthokonts	Choanoflagellates	*Diaphanoeca, Monosiga,*
	Stramenopiles	Bicosoecids	*Caecitellus, Cafeteria, Pseudobodo*
		Chrysomonads	*Paraphysomonas, Spumella*
		Pedinellids	*Ciliophrys, Pteridomonas*
	Discicristates	Euglenids/Bodonids	*Bodo, Neobodo, Rhynchomonas*
		Core jakobids	*Jakoba*
	Alveolates	Dinoflagellates	*Gymnodinium, Katodinium*
	Chromalveolates	Cryptophytes	*Goniomonas*
	Incertae sedis	Apusomonads	*Amastigomonas, Ancyromonas, Apusomonas*

Ciliates can have very elaborated morphologies and possess many cilia (ultrastructurally analogous to flagella). Examples of common marine genera within each group are given in Table 11.1.

Ciliates are ecologically important protists in marine systems (Pierce and Turner 1992), and all of them belong to the same phylogenetic group within the alveolates. Besides the numerous cilia covering the cell body, they are generally characterized by having two nuclei: a diploid micronucleus involved in reproduction and a polyploid macronucleus involved in cell regulation. These microbes come in a large range of sizes (from < 10 μm to > 200 μm) and feed mainly on smaller organisms, from bacteria to algae and other protists (Jonsson 1986), although many are mixotrophs and retain functional chloroplasts of ingested prey (Stoecker et al. 1987). One species is an obligate phototroph (Crawford and Lindholm 1997). The most abundant marine species are the spirotrichs, mostly naked forms (Sherr and Sherr 2000), although loricated forms such as the tintinnids have attracted much attention, in part due to their easier identification.

Amoeboid protists are found within the supergroups unikonts (Amoebozoa) and Rhizaria. Marine Amoebozoa include naked lobose amoeba, with sizes between 10 and 20 μm, which live in suspended aggregates and can be important grazers of surface-attached bacteria (Rogerson et al. 2003). Because of the difficulty in observing and quantifying these cells, very little ecological information exists regarding their abundance, diversity and activity. Marine Rhizaria include diverse lineages in constant phylogenetic rearrangements, such as the Cercozoa (Cavalier-Smith and Chao 2003a), Foraminifera (Archibald et al. 2003), Radiolaria (Polet et al. 2004) and Heliozoa (Nikolaev et al. 2004). The first lineage unites disparate groups such as the cercomonads, euglyphid amoebas and chlorarachniophytes, whereas the later two are polyphyletic (labelled with (R) and (H) in Fig. 11.1). Rhizaria contain a substantial organismal diversity, with naked and flagellated cells (cercomonads), cells with calcareous tests (foraminifers), cells with internal mineral skeletons (radiolarians), and cells with axopods radiating from the cell surface (heliozoans). Among all these varied groups, cercomonads are clearly bacterivorous and can be relatively important in aquatic systems (Arndt et al. 2000; Patterson and Lee 2000), whereas the other cells are often larger, less abundant, likely algivorous, and with unknown impacts on marine food webs (Caron and Swanberg 1990).

Flagellates are the most important bacterivorous forms in aquatic environments, and are very diverse (Arndt et al. 2000; Boenigk and Arndt 2002; Patterson and Larsen 1991). The main marine groups, choanoflagellates, stramenopiles, bodonids, and apusomonads, all belong to different eukaryotic supergroups, unikonts, chromalveolates, excavates, and incertae sedis, respectively (Fig. 11.1), so they have a completely different evolutionary history. Choanoflagellates, related to animals and fungi, are easily recognized by their single flagellum surrounded by a collar of pseudopodial filaments (Thomsen and Buck 1991). They are relatively small (3–10 μm) and have a spherical to ovoid shape. Some are naked, and others (acanthoecids) possess a siliceous lorica surrounding the cell on which a detailed taxonomy has been based and that is used for identification and quantification of these flagellates in marine samples (Leakey et al. 2002). All choanoflagellates are heterotrophs, mostly bacterivorous.

Stramenopile flagellates have two flagella of unequal size, one of them covered by tripartite hairs (Preisig et al. 1991). The stramenopile group (synonymous to heterokonts, Patterson 1989) comprises many different lineages, and includes some of the most prominent marine protists such as diatoms, in which the gametes have the typical flagella. Bacterivorous stramenopile flagellates belong to exclusively heterotrophic lineages (bicosoecids) or to lineages with heterotrophic, mixotrophic, and phototrophic taxa (chrysomonads and pedinellids). They have a spherical to oval shape and are often small (2–10 μm) and difficult to identify, although some have conspicuous features like the cup-shaped lorica of *Bicosoeca* or the siliceous scales of *Paraphysomonas*. Chrysomonads and bicosoecids are assumed to account for a large fraction of heterotrophic flagellates in marine systems (Brandt and Sleigh 2000), but most stramenopile sequences detected in molecular surveys are unrelated to these two groups (Massana et al. 2006b).

Bodonids are heterotrophic flagellates that are distinguished by the presence of the kinetoplast, a DNA-containing granule located within a single mitochondrion (Zhukov 1991). The cells are generally oval, between 4 and 10 μm in size, with one of the two flagella often used to glide along or attach to surfaces. Many bodonids seem specialized to feed on bacteria attached to surfaces or in aggregates. Jakobid flagellates are phylogenetically related to bodonids, and some have been retrieved from the marine pelagial. The apusomonads include biflagellated gliding bacterivores that seem to be common in the sea (Scheckenbach et al. 2005) but have an undefined phylogenetic placement (Cavalier-Smith and Chao 2003b).

Besides these bacterivorous flagellates, there are larger marine heterotrophic flagellates that are mostly algivorous but can complement their diet with bacteria. Some examples within the chromalveolates are the katablepharids (Okamoto and Inouye 2005), the Telonemia (Schalchian-Tabrizi et al. 2006), and the dinoflagellates (Larsen and Sournia 1991). Dinoflagellates are particularly important and common in marine systems, and play different ecological roles (see Chapter 1). They have two flagella, often a sort of armor called theca composed of cellulosic plates, and a large nucleus with permanently condensed chromosomes. They cover a huge size range, from small naked cells slightly larger than 5 μm to large and conspicuous cells of up to 2 mm. Dinoflagellates include important primary producers, voracious predators on different prey items (mostly algae and other protists, but also bacteria), and species that cause economic impacts, such as those responsible for harmful algal blooms (Jeong 1999).

Finally, many marine phototrophic flagellates also have the capacity to ingest bacteria, thus being mixotrophs that derive their nutrition from photosynthesis and prey ingestion (Jones 2000; Sanders 1991). Mixotrophic flagellates are common within the chrysomonads, pedinellids, dinoflagellates, and cryptomonads, lineages that also contain purely heterotrophic forms. In these lineages, mixotrophy seems to be the ancient character, and chloroplast loss in the heterotrophic forms the derived character (Raven 1997). Another very important marine algal group, the haptophytes, consists so far only of pigmented organisms that often are able to ingest bacteria (Green 1991).

FUNCTIONAL SIZE CLASSES OF PROTISTS

Another way to classify aquatic organisms, besides by morphology, nutritional mode, and phylogenetic affiliation, is by size. Cell size has many implications for the physiology and bioenergetics of protists, such as growth and grazing rates, as well as for the flow of organic carbon (paths and efficiency) through microbial food webs (Legendre and Lefevre 1995; Verity and Smetacek 1996). As discussed in Chapter 2, there is a well-established division of aquatic microorganisms into logarithmic size classes, proposed three decades ago (Sieburth et al. 1978): microplankton (cells of 20–200 μm), nanoplankton (2–20 μm), and picoplankton (0.2–2 μm). In marine systems, microplankton includes ciliates, large algae, and small metazoans; the nanoplankton are a diverse collection of flagellates together with small ciliates and amoebas; and picoplankton comprises heterotrophic and phototrophic bacteria and very small protists (Sherr and Sherr 2000). Examples of bacterivorous organisms can be found in the three size classes, but not all size classes are equally important as bacterial grazers.

The predator : prey size ratio varies among different groups (Hansen et al. 1994) and limits the maximal size difference between bacteria and their grazers. The marine environment presents additional constraints, imposed by the typical small size of bacteria (around 0.05 $μm^3$) and low abundance (around $0.5–1.0 \times 10^6$ cells/mL). In such conditions, physical and hydrodynamic considerations theoretically restrict bacterivory to small grazers, typically within the nanoplankton. Whereas nanociliates appear to be important only in relatively enriched conditions (Sherr and Sherr 1987) and amoebas in particulate flocs (Arndt 1993; Rogerson et al. 2003), it appears that most bacterivorous protists in the marine pelagic zone are flagellates (Fenchel 1986a; Laybourn-Parry and Parry 2000). This functional group, generally in the size range 2–5 μm, is commonly called the heterotrophic nanoflagellates (HNF). The predominance of HNF as marine bacterivores has been confirmed by manipulations with size-fractionated natural assemblages (Calbet et al. 2001; Wikner and Hagström 1988) and by direct observation of protists with ingested fluorescent bacteria (Sherr and Sherr 1991; Unrein et al. 2007). HNF abundances typically increase with the trophic state of the system, and are on average 1000 times less abundant than bacteria (Sanders et al. 1992). Table 11.2 gives HNF counts from a monthly sampling at the Blanes Bay Microbial Observatory (Mediterranean Sea), and cruises in the Norwegian Sea, in the Indian Ocean, and around the Antarctic peninsula. The values found in the four systems are strikingly similar; on average, HNF represent 20 percent of the protistan cells, numbering 850 cells/mL, and most of them (76 percent) are very small (≤ 3 μm).

The heterotrophic microplankton, mostly ciliates and large dinoflagellates, are considered the main herbivores in the open sea (Sherr and Sherr 2002). Some ciliates (e.g., spirotrichs) are omnivorous and can exert a significant grazing impact on bacteria (particularly large bacteria and picocyanobacteria) in some coastal and estuarine systems (Sherr and Sherr 1987). Some large ciliates have evolved to live strictly on bacteria by developing a filter membrane able to collect submicrometer particles. However, the low porosity of these filters imposes low clearance rates, so these ciliates can only develop in sites with high bacterial abundance, such as eutrophic lakes, sediments, and decomposing organic material (Fenchel 1980). At the lower size range, heterotrophic flagellates smaller than 2 μm, and thus belonging to the picoplankton, have been isolated by providing bacteria as food (Guillou et al. 1999) and observed by direct inspection of natural assemblages. However, these minute predators do not seem to be the dominant grazers in the marine pelagic zone.

NATURAL ASSEMBLAGES OF MARINE HETEROTROPHIC NANOFLAGELLATES

It is assumed that the dominant heterotrophic nanoflagellates in marine samples are choanoflagellates and stramenopiles, both bicosoecids and chrysomonads, whereas other taxa are less common (Arndt et al. 2000; Fenchel 1982a; Patterson et al. 1993; Vørs et al. 1995). This general view is based on inspecting natural cells by different methods (Fig. 11.2), each having its limitations. Electron microscopy allows ultrastructural studies and taxonomic identification, but does not provide quantitative data on natural samples (Vørs et al. 1995). Live observations allow a certain level of species identification due to characteristic motility (Arndt et al. 2000, 2003), but are intrinsically difficult and rarely done. Epifluorescence microscopy is the only method that provides quantitative data on heterotrophic flagellate abundance (Fenchel 1982c), but provides little morphological information, and only two groups can be

TABLE 11.2 Abundance of Heterotrophic (HF) and Phototrophic (PF) Flagellates in the Photic Zone of Four Marine Areas[a]

	Dates	Samples	HF (cells/mL)			PF (cells/mL)			HF/(HF + PF)
			Mean	Range	≤3 μm	Mean	Range	≤3 μm	
Blanes Bay	2001–2006	56	1040	150–2040	83%	5600	700–37400	83%	24%
Barents Sea	August 2002	30	830	450–1620	63%	7300	1500–17800	87%	13%
Indian Ocean	June 2003	22	500	260–780	84%	2600	600–13700	91%	23%
Antarctica	December 2002	21	1010	200–2180	75%	3800	2500–5400	75%	20%
Average	—	—	850	260–1650	76%	4800	1300–18,600	84%	20%

[a]R. Massana, unpublished data.

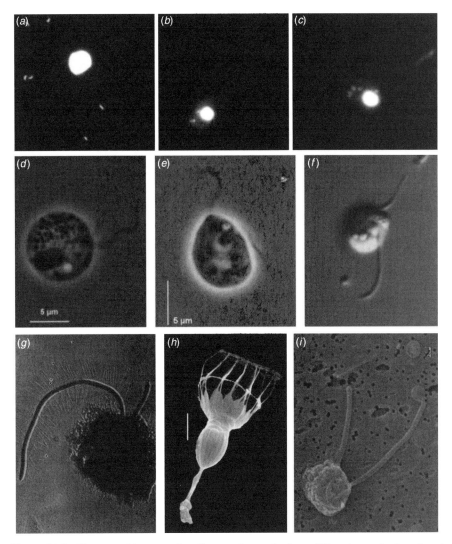

Figure 11.2 Heterotrophic nanoflagellates observed by different methods. (*a–c*) Epifluorescence microscopy: (*a*) *Spumella*; (*b*) choanoflagellate; (*c*) unidentified cell. (*d–f*) Phase contrast microscopy: (*d*) *Spumella*; (*e*) *Cafeteria*; (*f*) *Ancyromonas*. (*g–i*) Electron microscopy: (*g*) *Spumella*; (*h*) *Diplotheca*; (*i*) *Cafeteria*. Photographs are from the authors except (*b*) (F. Unrein), (*d–f*) (micro*scope web page http:/srtarcentral.mbl.edu/microscop), and (*h*) (F. Nitsche and H. Arndt, University of Cologne).

clearly distinguished: the choanoflagellates by the conspicuous collar and the single flagella, and the dinoflagellates by the granulated appearance of the nucleus. Often, cells having two flagella of unequal length are considered stramenopiles (Arndt et al. 2000), but even this is not very informative, given the huge phylogenetic and functional diversity of this group (Cavalier-Smith and Chao 2006).

Nevertheless, some studies have reported the abundance of several HNF groups in the euphotic zone of different marine areas by epifluorescence microscopy. In the Southern Ocean across the Polar Front, choanoflagellates accounted for 20 percent of cell numbers on average, dinoflagellates contributed to 17 percent, whereas the rest remained unidentified (Leakey et al. 2002). In the Equatorial Pacific, the averaged values for these three groups were 7, 4, and 89 percent, respectively (Vørs et al. 1995). In both studies, dinoflagellates dominated the community biomass, reflecting their relatively larger size. In Southampton Water, choanoflagellates and dinoflagellates accounted for 4 and 10 percent of abundance, whereas stramenopiles (45 percent of chrysomonads and 7 percent of bicosoecids) and bodonids (6 percent) were also identified, leaving 28 percent of cells unidentified (Brandt and Sleigh 2000). In the dataset from different marine systems presented in Table 11.2, most heterotrophic flagellates were very small, had few morphological features, and could not be identified. From the information available, it thus seems obvious that a large fraction of heterotrophic flagellates in the oceans remains unidentified and that direct inspection does not reveal the dominant taxa. Molecular approaches are starting to provide new data on this issue (see the final section of this chapter).

FUNCTIONAL ECOLOGY OF BACTERIVOROUS FLAGELLATES

Living in a Dilute Environment

In most studies, nanoflagellate grazing accounted for most of the protistan bacterivory in planktonic systems. This implies that HNF have evolved adaptations to efficiently capture bacterial cells from a relatively dilute environment and experiencing low Reynolds numbers, where viscous forces dominate over inertial ones (Fenchel 1986a; Lighthill 1976). Fenchel (1982a) proposed three different mechanisms of concentrating and capturing suspended food particles: direct interception feeding (also named raptorial), filter feeding, and diffusion feeding (Fig. 11.3). Filtration and direct interception, the two most common modes in HNF, are achieved by undulating flagella that produce a water current transporting particles towards the cell. In filter-feeding flagellates (choanoflagellates and some pedinellids), bacteria are collected by sieving through a filter, whose porosity is determined by the spacing between the pseudopodial tentacles. Low porosity, as occurs in choanoflagellates, allows very small particles to be captured, but implies high hydrodynamic pressure across the filter and reduced clearance rates (Fenchel 1986b). In direct interception feeders (e.g., most stramenopiles) particles are carried along flow lines in the feeding current and intercepted at the flagellate surface. As bacterial cells are handled and processed individually, this feeding mode can potentially involve recognition and prey selection mechanisms and bacterial escape responses. The predator : prey size ratio is critical for interception feeders, as their efficiency in capturing bacteria declines rapidly with increasing flagellate size (Fenchel 1984). In diffusion feeders,

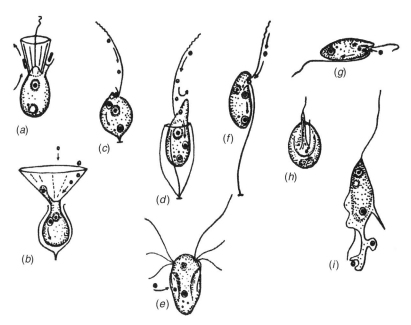

Figure 11.3 Feeding types in bacterivorous nanoflagellates. (*a*) Filter-feeding: choanoflagellate *Monosiga*. (*b*) Sedimentation: choanoflagellate *Choanoeca perplexa*. (*c*–*f*) Interception feeding: (*c*) chrysomonad *Spumella*; (*d*) bicosoecid *Bicosoeca*; (*e*) diplomonad; (*f*) bodonid *Bodo saltans*. (*g*–*i*) Raptorial feeding by a pharynx (*g*: bodonid) or by pseudopod-like structures (*h*: apusomonad *Apusomonas*; *i*: cercomonad *Cercomonas*). Modified from Boenigk and Arndt (2002).

such as the pedinellid *Ciliophrys*, motile prey collide with the sticky pseudopodial tentacles of the motionless predator. An additional feeding mode is used by some HNF taxa (e.g., bodonids) that prey on bacteria attached to biofilms or aggregates, often aided by a pseudopod-like structure (pharynx) to detach bacteria. Within each of these feeding mechanisms, further variability in terms of feeding behavior and selection strategies can be observed among different species (Boenigk and Arndt 2002).

The behavioral and morphological adaptations of HNF to concentrate suspended food particles allow maximal clearance rates in the range of 10^5–10^6 times their body volume per hour (Fenchel 1982a), with the lower value probably being more realistic (Boenigk and Arndt 2002). These specific clearance rates are much larger than those of suspension-feeding metazoans, and nanoprotists are indeed the only grazers that can survive exclusively on a bacterial diet in the marine pelagic zone. For a typical HNF 3 μm in size, these values would translate to clearance rates of 1.4–14 nL/h. At typical values of bacterial abundance (around 10^6 cells/mL) and bacterial cell volume (0.1 μm^3), and a flagellate growth efficiency of 20 percent (Straile 1997), this would enable growth rates in the range of 0.05–0.5 divisions per day.

In oligotrophic systems, with lower bacterial concentrations and smaller cell sizes, growth rates would be even slower. The implications of these rough calculations are that marine flagellates can thrive on suspended bacteria, but also that food limitation might be frequent, so other strategies and adaptations to maximize food uptake or to survive starvation might become important.

An adaptation that allows higher HNF clearance rates is loose attachment to seston particles. It has been calculated that the hydrodynamic flow field of attached suspension feeders makes possible higher feeding flow rates and thus higher clearance rates (Lighthill 1976). This has also been confirmed experimentally for HNF (Christensen-Dalsgaard and Fenchel 2003) and tintinnid ciliates (Jonsson et al. 2004). Cultured flagellates tend to attach to surfaces (Boenigk and Arndt 2000; Fenchel 1986a), and a high potential for colonization of marine snow particles by HNF has been demonstrated (Kiørboe et al. 2003). This suggests that a certain proportion of planktonic HNF might be associated with particles, either to exploit locally enhanced bacterial concentrations in these microniches or to use them as drift anchors to remove bacterial cells from the surrounding water with higher clearance rates. However, it is at present not possible to resolve the small-scale microbial distribution in the particle continuum and possibly gel-like structure of marine plankton (Azam 1998) and to clearly distinguish between freely suspended and loosely attached HNF.

Another adaptation of HNF in marine systems is to use a wide range of food items, since all types of particles within the prey size range can potentially be eaten via phagocytosis. At the larger end, the prey can be algae and other flagellates (Arndt et al. 2000), and at the lower end viruses (González and Suttle 1993), colloid molecules (Sherr 1988), and detritus particles (Scherwass et al. 2005). An interesting implication of the fact that dissolved organic carbon (DOC) can spontaneously form particulate structures (Kerner et al. 2003) is that nanoflagellates might circumvent bacteria and have direct access to part of the DOC pool. Besides heterotrophic bacteria, most HNF in the euphotic zone could potentially supplement their diet with photosynthetic prokaryotes and algae smaller than 3 µm, such as *Ostreococcus* and *Micromonas*. All of these sources probably relax the food limitation of HNF. Although bacteria are generally the major food source, probably most HNF are in fact omnivorous feeders (Arndt et al. 2000).

As already mentioned, many protistan groups include mixotrophic species that contain chloroplasts and are capable of phagocytosis. The contribution of mixotrophic flagellates to bacterivory is large in many marine systems (Arenovski et al. 1995; Sanders et al. 2000; Unrein et al. 2007). Mixotrophic flagellates are mostly haptophytes and dinoflagellates in marine systems and chrysophytes in freshwater systems (Sanders 1991). The degree of phototrophic and phagotrophic nutrition varies greatly among different mixotrophs (Caron 2000; Jones 2000). Some are primarily heterotrophs, using phototrophy only when prey concentrations are limiting (Caron et al. 1990), whereas others are primarily phototrophs, using phagotrophy to gain carbon under light-limited conditions (Jones 2000), or when inorganic nutrients (Nygaard and Tobiesen 1993) or trace elements (Maranger et al. 1998) become limiting.

Judging the Ecological Relevance of Isolated HNF

Once in culture, growth and feeding parameters can provide hints about whether the isolate is a potentially important bacterivore in the pelagic environment. Some examples are:

- **Growth on Bacteria** The numerical response describes the dependence of HNF growth rate (μ) on bacterial concentration (B). After inoculation of HNF on different concentrations of (non-growing) bacteria in batch culture, μ is determined for the exponential increase of HNF. Plots of μ versus B can be fitted to a hyperbolic function of the form of Michaelis–Menten kinetics, from which the maximum specific growth rate (μ_{max}) and the half-saturation constant (K_m: the prey concentration where $\mu = 50$ percent of μ_{max}) are derived. Minimal bacterial concentration after exponential growth of HNF indicates threshold levels below which feeding becomes inefficient.
- **Bacterial Consumption** The functional response describes the dependence of HNF ingestion rates on bacterial concentration. It can be fitted to the same hyperbolic function as above, and yields the maximum specific ingestion rate.
- **Prey Size Spectrum** Size-dependent feeding efficiency can be determined with differently sized labeled bacteria or artificial particles (fluorescent microspheres), and indicates whether the HNF are able to harvest small planktonic bacteria.
- **Resistance to Starvation** Rate of decline and persistence of HNF after exhaustion of bacterial prey.

Using Culture Experiments to Infer the Ecological Role of HNF

At present, the only reliable way to obtain information on the physiological properties and ecological adaptations of HNF is to use isolated strains. The first detailed studies in this respect were done by Fenchel (1982b) with isolates from different taxonomic groups growing in batch cultures. These experiments confirmed the capability of HNF to prey efficiently on suspended bacteria, and were the basis for the appreciation of their ecological role in microbial food webs. Meanwhile, a number of laboratory studies with isolated marine and freshwater HNF taxa have been performed to investigate feeding and growth rates, nutrient regeneration and excretion, temperature responses, and other autecological aspects (see the examples in Table 11.3). These investigations served principally two purposes: First, they revealed information on the autecology and performance of the different taxonomic protists. Second, they were used as model systems, with the goal of drawing general conclusions with respect to the impact of HNF (as a functional group) on natural bacterial communities and their general ecological functions.

From HNF growth on bacteria in batch culture experiments, essential parameters such as growth rate, bacterial grazing, clearance rate, and yield can be derived

TABLE 11.3 Published Autecological Laboratory Studies with
Isolated Marine HNF Taxa

HNF Genera	Goals of Study	References
Chrysomonads		
Paraphysomonas spp.	Growth, grazing, herbivory, nutrient cycling, excretion, temperature range, turbulence	(1–13)
Oikomonas sp.	Growth, grazing	(14)
Spumella sp.	Grazing, motility	(15)
Bicosoecids		
Cafeteria sp.	Growth, impact on bacteria	(16,17)
Pseudobodo sp.	Growth, grazing	(1,14)
Caecitellus parvulus	Grazing	(18)
Bodonids		
Bodo sp.	Growth, grazing, motility, nutrient cycling	(4,8,14,15,19)
Rhynchomonas sp.	Growth, grazing	(14)
Choanoflagellates		
Codosiga sp.	Growth	(19)
Diaphanoeca sp.	Grazing	(20,21)
Monosiga sp.	Growth, grazing, impact on bacteria	(1,17)
Stephanoeca sp.	Growth, nutrient cycling	(8,14,22)
Pedinellids		
Pteridonomas danica	Growth, grazing, excretion, temperature range	(5,10,18,20,23)
Ciliophrys sp.	Growth	(5,19)
Actinomonas mirabilis	Growth, grazing, temperature range	(1,5)
Jakobids		
Jakoba libera	Growth, impact on bacteria, nutrient cycling	(8,16,17,19)

References: (1) Fenchel (1982b); (2) Caron et al. (1985); (3) Goldman et al. (1985); (4) Sibbald and Albright (1988); (5) Tobiesen (1990); (6) Nagata and Kirchman (1991); (7) Choi and Peters (1992); (8) Eccleston-Parry and Leadbeater (1995); (9) Peters and Gross (1994); (10) Ishigaki and Sleigh (2001); (11) Delaney (2003); (12) Selph et al. (2003); (13) Choi (1994); (14) Davis and Sieburth (1984); (15) Kiørboe et al. (2004); (16) Mohapatra and Fukami (2004a); (17) Vázquez-Domínguez et al. (2005); (18) Zubkov and Sleigh (1999); (19) Eccleston-Parry and Leadbeater (1994); (20) Fenchel (1986b); (21) Andersen (1989); (22) Geider and Leadbeater (1988); (23) Pelegrí et al. (1999).

(Fenchel 1982b). The results generally supported the notion that HNF should be able to control bacterial numbers in marine plankton. Growth and feeding parameters have been determined meanwhile for a range of different taxa (Table 11.3), and a compilation of data published up to 1994 can be found in Eccleston-Parry and Leadbeater (1994). The numerical response, that is, the dependence of HNF growth rate on bacterial concentration, yields the maximum growth rate (μ_{max}) and the half-saturation constant (K_m), which is the bacterial concentration at which the growth rate is 50 percent of μ_{max} and is the concentration to which the flagellates are adapted. Maximum growth rates typically corresponded to doubling times in the range of 3–7 hours (Eccleston-Parry and Leadbeater 1994), but can vary greatly with the bacterial prey (Mohapatra and Fukami 2004a). Growth rates were lowest for some choanoflagellates and *Ciliophrys*. The reported K_m values are generally in the range $1-10 \times 10^6$ bacteria/mL (Eccleston-Parry and Leadbeater 1994; Mohapatra and Fukami 2004a), which is well above the bacterial concentrations found in most of the ocean. On the other hand, low threshold levels of bacterial concentrations, $2-10 \times 10^4$/mL, have been determined (Eccleston-Parry and Leadbeater 1994).

Although these investigations were extremely helpful in understanding the functional role of HNF, it is questionable whether the results can really be transferred to the natural environment, as there are some limitations inherent in these studies. First, most of our knowledge on feeding, growth, and excretion by HNF was derived from studies with a few model organisms such as *Paraphysomonas imperforata*, *Bodo designis*, and *Pteridomonas danica* (see Table 11.3). High growth rates and rapid responses towards high bacterial concentrations are typical of most of these species, and suggest that they are adapted to exploit patches of enhanced bacterial concentrations (e.g., in marine snow). Only a few HNF taxa examined so far seem to exhibit growth characteristics and survival patterns that resemble a k-strategy (e.g., the freshwater *Poterioochromonas malhamensis* in Boenigk et al. 2006). Meanwhile, we know that the HNF taxa listed in Table 11.3, and most of those available in public culture collections (Cowling 1991), play quantitatively only a marginal role in the oceans (Massana et al. 2006b) and are probably selectively enriched during the isolation process (Lim et al. 1999). Second, the bacteria generally offered as prey are not the strains dominating in the ocean, have cell volumes that can be 10–50 times larger than typical bacterioplankton, and are fed to HNF at unrealistic abundances. HNF probably grow at much lower rates, and are often food-limited in oligotrophic marine systems.

Culturing Heterotrophic Nanoflagellates

Stock cultures of HNF are generally maintained on sterilized seawater or artificial media supplemented with an organic carbon source to promote bacterial growth (e.g., yeast extract or cereal grains). For experimental work, two cultivation techniques are used:

- **Batch Culture** HNF are inoculated on a bacterial suspension, and follow different growth phases (exponential, stationary, decline) concomitant with bacterial decrease. Several autecological parameters can be derived from the development of bacteria and HNF numbers (e.g., HNF growth and consumption rate, and cell yield).
- **Continuous Culture or Chemostat** These are based in a continuous supply of fresh media that enables continuous growth of HNF. In single-stage chemostats supplied with bacterial growth medium, bacteria and HNF oscillate, and the permanent strong predation pressure on bacteria can be used to analyze bacterial resistance mechanisms. Two-stage chemostats, with bacteria in the first stage and HNF in the second, can be used to obtain steady-state conditions at defined growth rates (determined by the dilution rate) of HNF.

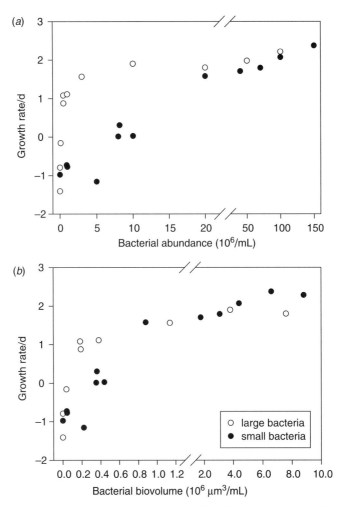

Figure 11.4 Growth rates of the heterotrophic nanoflagellate *Spumella* sp. on large and small bacteria, plotted as bacterial abundance (*a*) or as bacterial biovolume (*b*). Data are compiled from Boenigk et al. (2006) and Pfandl and Boenigk (2006). The large bacterial strain was *Listonella pelagia* CB5 (mean volume 0.38 μm^3) and the small bacterial strain was *Polynucleobacter* sp. (mean volume 0.044 μm^3).

One approach to assess more realistically the in situ conditions for feeding and growth studies would be to use natural bacterioplankton as the food source in these experiments. The use of continuous culture systems (chemostats) in which both bacterial prey and HNF predator can be maintained at low growth rates, and where bacteria remain small, is also an option (Caron 1990), particularly when natural mixed assemblages are used (Massana and Jürgens 2003; Selph et al. 2003). Another valuable approach would be to use as prey bacterial strains that permanently form small cell sizes and that are widespread in bacterioplankton

(Hahn 2003; Hahn et al. 2003; Rappé et al. 2002). This has been done in laboratory experiments with freshwater HNF and "ultramicrobacteria" belonging to the *Polynucleobacter* genus (Boenigk et al. 2006). This approach could be used to examine to what extent HNF growth rates are affected when the same bacterial biomass (assuming similar prey food quality) is distributed among many small cells compared with larger cells (Boenigk et al. 2006). Obviously, the growth rate of HNF is greatly reduced when the same bacterial abundance is provided as small cells compared with large bacteria (Fig. 11.4a). However, when considering bacterial biovolume, HNF growth rates on small bacteria were also significantly reduced at lower (and relevant) food concentrations (Fig. 11.4b). This suggests energetic constrains for HNF in concentrating small-sized prey at low bacterial concentrations. Similar studies with marine isolates of ultramicrobacteria and HNF are still lacking.

IMPACT OF PROTISTAN BACTERIVORY ON MARINE BACTERIOPLANKTON

Search for the Perfect Method to Quantify Protistan Bacterivory

For an understanding of the cycling of organic matter in the sea, it is essential to have information on rates of bacterivory. The first grazing estimates of HNF were obtained from comparing the increase in flagellate numbers with the decrease in bacterial numbers in laboratory cultures (Fenchel 1982b). Range of techniques have been developed for in situ measurements of protistan bacterivory (see reviews by Landry 1994; Strom 2000). The techniques fall into three categories: (1) tracer techniques that follow labeled bacteria or prey surrogates into the predators; (2) community manipulation techniques that uncouple predator and prey; (3) direct inferences from characteristics of natural populations. The first two categories, which comprise the most commonly used techniques, have also been termed class I and II methods (Strom 2000). All techniques have their drawbacks, and the most appropriate will depend on the system to be studied and the particular questions.

The most widespread tracer technique is based on the use of fluorescently labeled, heat-killed bacteria (FLB) (Sherr et al. 1987). FLB are added to a water sample in short-term experiments (<1 hour) and, after fixation and staining, ingested FLB are counted inside the protistan food vacuoles. Ideally, FLB are prepared with bacteria from the same studied habitat and are applied at "tracer" concentrations (5–10 percent of bacteria). The major problems associated with this technique are that selection against fluorochrome-labeled bacteria can occur (Fu et al. 2003; Landry et al. 1991), fixatives can cause egestion of vacuole content (Sanders et al. 1989), and there can be statistical problems in obtaining reliable counts of ingested FLB, particularly at low grazer densities (McManus and Okubo 1991). Although the FLB technique is tedious and time-consuming, it is the best technique to provide specific ingestion rates for certain size or taxonomic classes of protists. An alternative in oligotrophic systems with low grazer concentrations is to follow the disappearance of added FLB in incubations of 24–48 hours (Marrasé et al. 1992), which can also be assessed by flow cytometry (Vázquez-Domínguez et al. 1999). This approach yields a community grazing rate. An additional step, which avoids prelabeling of bacteria, is to use live bacterial strains and immunofluorescence (Christoffersen et al. 1997) or fluorescence in situ hybridization (FISH) to detect ingested specific bacteria inside the food vacuoles, or to use bacterial strains that express fluorescent proteins (Fu et al. 2003).

The main other tracer technique is based on radiolabeled bacteria (Nygaard and Hessen 1990). Ideally, a labelled compound (e.g., [^{3}H] thymidine) is added to a water sample, and is first incorporated by bacteria and later by the bacterivores ingesting labelled bacteria. By size-fractionation at the beginning and the end of the incubation, the transfer of the label to the grazer size fraction can be quantified and converted to community grazing rates. The method is quite sensitive, but requires a clear separation of bacteria and grazers (e.g., by 1 μm pore-size filters) and short incubation times to minimize bacterial digestion. An interesting modification is the dual-labeling technique (Zubkov and Sleigh 1995), in which bacteria are labeled simultaneously by two compounds (e.g., [^{3}H]thymidine and [^{14}C]leucine) that have a different fate after ingestion and correspond either to excretion (^{3}H) or to assimilation (^{14}C) of bacterial macromolecules.

Community manipulation techniques are based on measuring bacterial growth in incubations in which grazing is suppressd by dilution, size-fractionation, or metabolic inhibitors. The dilution technique, in which natural assemblages are progressively diluted with filtered seawater and the response in prey populations monitored during 12–24 hour incubations (Landry and Hassett 1982), has become very important for measuring microzooplankton herbivory in the ocean (Dolan et al. 2000). It has also been used to assess bacterivory (Tremaine and Mills 1987) and grazing on picocyanobacteria (Worden and Binder 2003). It has been modified to assess in parallel bacterial mortality due to protistan grazing and viral lysis (Evans et al. 2003). A potential problem is that dilution and containment often change the physiology and growth of bacteria compared with in situ conditions. Another approach is to add either bacterial or eukaryotic antibiotics to natural assemblages to stop bacterial growth or to inhibit grazers, respectively (Taylor and Pace 1987). This approach has the problem that either their effect is not complete (at low concentrations) or the antibiotics impact nontarget organisms when applied at high concentrations (Tremaine and Mills 1987). Size-fractionation experiments in which most of the bacterivores are removed (e.g., by a 1 μm pore-size filter), allowing bacterial growth in the absence of predators, are easy to perform, and were among the first approaches that demonstrated top-down control on bacterioplankton (Wright and Coffin 1984). However, a clear size separation between bacteria and their consumers is not always possible, and several artefacts are common in oligotrophic systems, such as organic matter enrichment during the filtration step.

The third category, inference of grazing rates from direct inspection of natural bacterivores, would be the ideal one, as no manipulation of the sample is needed. One example is the measurement of the activity of digestive enzymes necessary for the breakdown of bacterial peptidoglycan, such as different lysozymes

(González et al. 1993a; Vrba et al. 1993). The major challenges here are to obtain reliable standardization to convert enzyme activity to grazing rates and interference with phytoplankton enzyme activity (Sherr and Sherr 1999). A valuable, although time-consuming approach, is microscopic analysis of protistan food vacuoles for prey items that are either autofluorescent or can be specifically stained, for instance using specific FISH probes (Jezbera et al. 2005). By estimating experimentally the digestion rates of these prey populations, grazing rates on specific prey by different protistan taxa can be obtained, as shown for *Synechococcus* (Dolan and Šimek 1998).

Rates of Protistan Bacterivory in the Sea

A review of the published data on bacterivory in planktonic systems revealed some interesting patterns (Vaqué et al. 1994). The observed ingestion rates using the FLB approach were generally in the range of 2–20 bacteria/HNF/h (9.7 on average). These values are significantly lower than most estimates derived from cultured HNF (Eccleston-Parry and Leadbeater 1994; Fenchel 1982b), and give a good picture of the bacterial grazing losses that can be expected at a given HNF abundance. Community grazing rates ranged between 1.8×10^4 and 5.8×10^4 bacteria/mL/h with tracer methods (mostly FLB uptake) and $4.7-6.5 \times 10^4$ bacteria/mL/h with community manipulation methods. The lower values obtained with tracer methods (25 percent lower on average) seem reasonable, since these methods assess only specific protistan bacterivory, whereas the community manipulaton methods provide estimates of community bacterivory. An empirical model, using published datasets, predicts community grazing rate (GT, in bacteria/mL/h) from the independent variables HNF abundance (HNF, in cells/mL), temperature (T, in °C) and bacterial abundance (BAC, in cells/mL) (Vaqué et al. 1994):

$$\log GT = -3.21 + 0.99 \log HNF + 0.028T + 0.55 \log BAC$$

This equation can be used to obtain a rough estimate of HNF bacterivory. In a recent study on seasonal protistan bacterivory, the values obtained with three methods—FLB uptake, FLB disappearance, and this empirical model—were astonishingly similar (Unrein et al. 2007).

Depending on the method, data on the contribution of different protistan groups can be obtained in addition to total grazing losses. For example, the FLB technique can provide information on the size of the grazers as well as on the contribution of mixotrophic protists to bacterivory. One example is the seasonal study performed by Unrein et al. (2007) at an oligotrophic coastal site in the Mediterranean Sea (Fig. 11.5). By using FLB uptake data, it was shown that flagellates smaller than 5 µm accounted for about 80 percent of total bacterivory. Moreover, mixotrophic flagellates had lower specific ingestion rates than HNF, but, due to their higher abundance, they were important grazers thoughout the season and accounted for 35–65 percent (mean 50 percent) of the measured total bacterivory (Fig. 11.5).

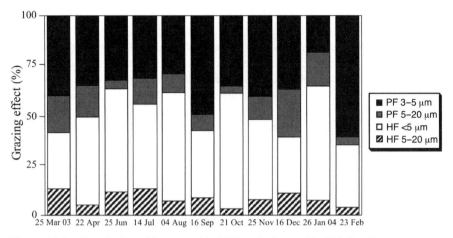

Figure 11.5 Contribution of heterotrophic (HF) and mixotrophic (PF) flagellates, each separated into two size classes, to total bacterivory in Blanes Bay during a seasonal cycle. Modified from Unrein et al. (2007).

A further step would be to use FLB uptake to quantify bacterial consumption by different nanoflagellate groups, as far as the distinction is possible by epifluorescence microscopy. This has been done to some extent in freshwater systems, and has revealed dominant types of grazers as well as seasonal changes in their grazing rates (Cleven and Weisse 2001; Domaizon et al. 2003). Similar studies for marine systems are lacking so far.

Balance of Bacterial Production and Protistan Grazing

In the first compilation of published data on bacterial production and grazing mortality, a fairly good correspondence between bacterial growth and grazing loss rates was found (Sanders et al. 1992). Values for a wide range of aquatic systems aggregated around the 1:1 line, suggesting a steady-state like situation in which all bacterial production is consumed by bacterivores. However, for single ecosystems, there was considerable deviation from the 1:1 line, and it was the averaging across systems with orders of magnitude difference in microbial abundance and production that produced strong relationships between bacterial and HNF parameters (Sanders et al. 1992). Additionally, measurements of both bacterial production and bacterial grazing vary greatly, depend on a range of conversion factors and assumptions, and therefore are probably uncertain by a factor of two or more (Sherr et al. 1989).

Including more recent datasets has revealed a clear trend for deviations from the 1:1 line, both for the most oligotrophic (Gasol et al. 2002) and for the most productive (Strom 2000) marine regimes. In oligotrophic systems, estimates of grazing losses were generally higher than bacterial production rates, and it has been speculated that this is because of methodological biases (Gasol et al. 2002). In more eutrophic systems, on the other hand, grazing losses were often significantly lower than bacterial production. It has been proposed that here viral lysis becomes important. For some coastal marine systems, viral infection contributed as much to bacterial mortality as protistan bacterivory (Fuhrman and Noble 1995), whereas viral infection seems to be less important in open-ocean oligotrophic sites (Guixa-Boixareu et al. 1999). Other environments in which viral mortality seems to become particularly important are those where abiotic conditions are unfavorable for protistan growth, such as anoxic and hypersaline water bodies (Pedrós-Alió et al. 2000; Weinbauer and Höfle 1998). Nevertheless, comparative analysis across a range of systems has revealed that bacterivory is generally the dominating bacterial loss factor for planktonic bacteria (Pedrós-Alió et al. 2000).

Bottom-Up Versus Top-Down Control of Bacteria and Bacterivorous Protists

There has been some discussion whether planktonic bacteria are controlled by supply of resources such as organic carbon or inorganic nutrients (bottom-up control, discussed in Chapter 10) or predation by bacterivores (top-down control). The relative importance of competition and predation for population regulation is an old topic in community ecology (Hairston et al. 1960). There are arguments for both types of controlling mechanisms (Thingstad 2000), and there is no satisfactory accepted theory of the regulation of bacterial stocks and production in pelagic systems. The low variability across temporal and spatial scales of bacterial numbers and the close coupling between bacteria and bacterivores in experiments were used initially as evidence to argue for efficient top-down control by protistan grazers (Ducklow 1983). On the other hand, the positive correlations between phytoplankton (as a major source of dissolved organic matter, DOM) and bacteria in cross-system comparisons are typical resource-consumer relationships and thus a sign for bottom-up control. Also, the relationships between bacterial production and bacterial abundance were used to derive conclusions on the prevailing control mode in a given system (Billen et al. 1990; Gasol et al. 2002). When grazing is not very important, more bacterial production is converted to bacterial biomass, resulting in steeper slopes of this relationship, whereas under high grazing pressure, an increase in bacterial production is not reflected in an increase in bacterial biomass. Therefore, the slopes of this relationship can indicate the relative importance of top-down and bottom-up mechanisms (Gasol et al. 2002). Measured slopes differ between datasets from different marine systems and different seasons, indicating shifting modes of control. There is a tendency for steeper slopes in coastal and estuarine systems, indicating stronger bottom-up control, whereas open-ocean data have lower slopes, indicating top-down control (Gasol et al. 2002). This finding is in contrast to the intuition that substrate control of bacteria should be more prevalent in oligotrophic systems. Further, there is also experimental evidence for a low bacterial response after grazer removal in oligotrophic systems, which contradicts a tight bacteria–HNF coupling (Jürgens et al. 2000).

Another empirical model to derive information on the relative importance of top-down and bottom-up controls is based on the observed relationship between bacterial and HNF abundance (Gasol 1994). It originated from a compilation of published datasets (Gasol and Vaqué 1993) that served to compute a theoretical maximal HNF abundance at given bacterial abundances. The distance between the maximal and the realized HNF abundance (D) was termed the "degree of uncoupling" between bacteria and their predators. The datasets analyzed in Gasol et al. (2002) indicate that D increases with increasing bacterial abundance. This would confirm the notion stated before that in eutrophic systems the importance of grazing is reduced and possibly viral lysis becomes more important. Besides indicating a release of grazing pressure of bacteria from HNF, the D-value also indicates the strength of top-down control on HNF. This is highly probable, as HNF are within the prey-size spectrum of most metazoan and larger protozoan grazers (Jürgens et al. 1996). Effects of higher trophic levels on bacteria–protist relationships became obvious in size-fractionated incubations in which top-down control on HNF was relieved by removing larger predators (Calbet and Landry 1999; Wikner and Hagström 1988). There is even evidence that mesozooplankton predation impacts can cascade down a food chain consisting of copepods, ciliates, and HNF until the level of planktonic bacteria (Zöllner et al. in preparation).

Besides system productivity, heterogeneity in prey edibility has also been recognized as an important factor in the balance of top-down and bottom-up controls (Abrams and Walters 1996; Bohannan and Lenski 1999). If a part of the prey population has a refuge from predation, the biomass of this population would increase with productivity, whereas the edible prey would behave as mentioned earlier (its densities would remain constant with productivity and would be converted to a higher predator biomass). Therefore the existence of inedible prey can increase the importance of bottom-up control also in productive systems. Experimentally, such a pattern has been shown with vulnerable and resistant bacteria and bacteriophages (Bohannan and Lenski 1999). Comparable mechanisms are imaginable for the bacteria–flagellate interaction in the plankton, although this is difficult to assess in field samples. The abundance of predation-resistant bacteria could increase with the productivity of the system and be mainly bottom-up-controlled, whereas edible bacteria would be top-down-controlled. Assuming a trade-off between competitive ability (e.g., in substrate capture) and defense mechanisms, resistant bacteria would require higher substrate input and appear above a certain threshold productivity level. Thus, for understanding the in situ regulatory mechanisms and the interplay of bottom-up and top-down control, it is important to resolve the heterogeneity of the bacterial communities with respect to their edibility by bacterivores.

Ecological Functions of Bacterial Grazers

The carbon flow into bacterioplankton, as estimated from bacterial respiration and bacterial production, is without doubt a major pathway in marine systems (Chapter 9) and supports a considerable biomass of heterotrophic bacteria. However, in steady-state models of carbon cycling through the pelagic microbial food web, the predicted carbon fluxes via protistan grazing and viral lysis constitute only a low fraction of the total carbon budget (Anderson and Ducklow 2001). The underlying reason is the relatively low growth efficiency (10–15 percent) of marine bacteria (del Giorgio and Cole 2000), which implies that most organic carbon used by bacteria is respired and not converted into biomass. Although growth efficiencies of heterotrophic bacterivores are probably slightly higher (Straile 1997), we cannot expect to find a substantial carbon flow via the microbial loop when only heterotrophic bacteria form the food resource. An exception would be systems that are fueled mainly by the input of allochtonous organic carbon, such as some dystrophic lakes (Hessen 1992).

A function of bacterivorous protists more important than transferring bacterial carbon to higher trophic levels is to remineralize bacterially bound nutrients (nitrogen and phosphorus) mainly in form of ammonium and orthophosphate (Caron and Goldman 1990). Excretion of nitrogen and phosphorus by phagotrophic protists is thought to be one of the major mechanisms that regenerates inorganic nutrients from the particulate phase. The actual nutrient excretion rates increase with decreasing protistan growth efficiency and depend on the relative stoichiometry of bacteria and protists (Kirchman 2000), with bacteria being relatively enriched in nitrogen and phosphorus. Protistan grazers are also important for the regeneration of micronutrients such as bioavailable iron (Barbeau et al. 2001). Besides inorganic nutrient excretion, protists release a range of organic molecules as excretory products, such as urea and purines (Caron and Goldman 1990), dissolved amino acids (Nagata and Kirchman 1991), refractory organic material (Nagata and Kirchman 1992), dissolved DNA (Alonso et al. 2000), hydrolytic enzymes (Mohapatra and Fukami 2004b), and surface-active organic matter (Kujawinski et al. 2002). On average, protists egest 10–30 percent of the ingested organic matter, some in colloidal forms (picopellets) that likely represent undigested remains of bacteria, such as cell walls (Nagata 2000). Therefore, HNF and other phagotrophic protists contribute significantly to the pool of dissolved organic matter in the sea and mediate the chemical environment of planktonic bacteria.

Besides heterotrophic bacteria, picocyanobacteria (mainly the genera *Prochlorococcus* and *Synechococcus*) and picoeukaryotic autotrophs (e.g., *Ostreococcus*) are within the prey size range of nanoflagellated protists. Thus, HNF can simultaneously become important herbivores in marine plankton. Grazing rates on picocyanobacteria have been estimated by the dilution technique (Worden and Binder 2003) and by inspection of protistan vacuoles. The latter approach revealed that clearance rates of HNF grazing on *Synechococcus* and on bacteria are similar. Christaki et al. (2001) estimated that HNF consume daily 1–45 percent of *Synechococcus* standing stock in the western Mediterranean Sea. However, some HNF strains do not feed on *Prochlorococcus* or *Synechococcus* (Guillou et al. 2001), and there is evidence that some strains of these cyanobacteria are of poor food quality and support only slow growth of protists (Christaki et al. 2002; Guillou et al. 2001). The physiological and genetic variability among the genera *Synechococcus* and *Prochlorococcus* in the ocean is large (Fuller et al. 2006), implying also variable phenotypic properties. The structure and composition of the cell surface of *Synechococcus* might play not only a role in motility and physiological functions (Brahamsha 1996) but also in the recognition and selection by protistan grazers (Postius and Ernst 1999) and defensive responses.

Another important ecological role of bacterivorous protists is the stimulation of bacterial activity (Posch et al. 1999) and of bacterial decomposition of organic material (Fenchel and Harrison 1976; Ribblett et al. 2005; Sherr et al. 1982). The underlying mechanisms of this process are still not fully understood, but high bacterial turnover and the excretion of nutrients and other growth-stimulating substances by the grazers might be involved. Grazer impacts on specific bacterial functions and biogeochemical transformations are possible, but have seldom been studied in detail (Lavrentyev et al. 1997). This is obviously an important field, in which further investigations are required.

GRAZING AS A SHAPING FORCE OF BACTERIAL ASSEMBLAGES

Many new and detailed insights into mechanisms of bacteria–protist interactions have been gained in the last few years. These involve mechanisms of foraging, food spectrum, and prey selection by protistan bacterivores (Boenigk and Arndt 2002), and grazing resistance properties of bacterial prey (Hahn and Höfle 2001; Jürgens and Güde 1994; Jürgens and Matz 2002). If protistan grazing is a major mortality factor for planktonic bacteria, it becomes a strong selective force. The underlying mechanisms in this predator–prey interaction might be key factors for many of the observed phenotypic properties of bacterial assemblages (Fig. 11.6) as well as for their taxonomic composition.

Bacterial Cell Size Determines Vulnerability Towards Grazers

Pelagic systems offer few refuges from predation, and communities from bacteria to fish are subjected to size-structured predator–prey interactions (Verity and Smetacek 1996). Size-selective grazing by bacterivorous flagellates was also the first qualitative grazing factor found to be relevant for planktonic bacterial assemblages (Chrzanowski and Šimek 1990; González et al. 1990). Size-selectivity occurs in nearly all bacterivores, and has several underlying mechanisms that do not require active predator choice but can be explained by the functional morphology of the feeding mechanisms. In filter-feeding taxa (e.g., choanoflagellates), the porosity of the filter structures determines the particle size that can be retained (Fenchel 1986a). However, size-selective food uptake also occurs in interception-feeding flagellates that encounter and engulf single bacterial cells. It is explained by simple geometric models in which the encounter rate between prey and predator increases with the size of the prey particle until a certain limit (Fenchel 1982a). This gives a good approximation of the size dependence of particle ingestion, although the values might also depend on repulsive and attractive forces between prey and predator (Monger and Landry 1990) as well as on the exact prey geometry (Posch et al. 2001).

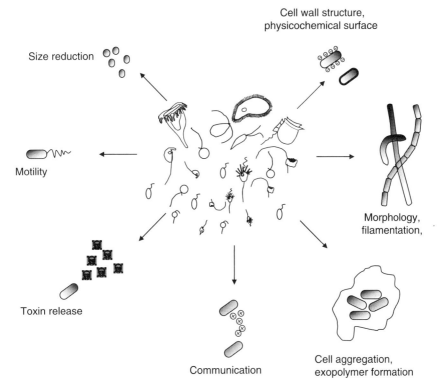

Figure 11.6 Overview of bacterial resistance mechanisms that might provide protection from predation by heterotrophic protists. Modified from Pernthaler (2005).

For cultured protists, particle-size selection can be examined with fluorescent beads, as shown for three species of interception-feeding HNF of slightly different sizes (Fig. 11.7). The obvious size-dependent feeding efficiencies reveal several interesting features: (1) the maximum clearance rate increases approximately with the square of the particle radius until a certain size limit is reached; and (2) even within this narrow prey size range, niche differentiation of bacterivorous predators exists, and each species has a characteristic size-dependent feeding efficiency curve with different minimum, optimum, and maximum ingestible particle sizes (Jürgens and Matz 2002).

However, size-selective grazing is only one underlying reason for the dominance of small bacteria in planktonic systems. The other mechanism is miniaturization of cell size during adaptation to carbon starvation, enabling cells to survive long term at reduced metabolism (dormancy) (Kjelleberg et al. 1993). Recently, some dominant bacteria of aquatic systems that are characterized by small cell size also during active growth have been identified and cultured: in marine systems, the clade of the alpha-proteobacterium *Pelagibacter ubique* (SAR11) (Rappé et al. 2002), and in limnic systems, the betaproteobacterial genus *Polynucleobacter* (Hahn 2003) and the

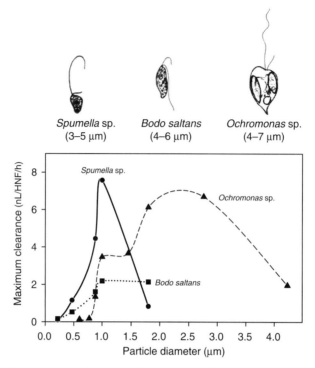

Figure 11.7 Size-selectivity feeding of three species of heterotrophic flagellates studied in short-term uptake experiments with fluorescent beads. Modified from Jürgens and Matz (2002).

gram-positive *Actinobacteria* (Hahn et al. 2003). These bacteria seem to be true oligotrophs, and are characterized by slow growth and low cell yield in culture, and they cannot grow in the presence of high organic matter concentrations.

However, although some bacterial taxa obviously remain small during active growth, many planktonic bacteria have to reach a certain size before cell division, and some measurements by flow cytometry revealed a clear correlation between bacterial size and activity (Bernard et al. 2000). This suggests that size-selective grazing also has major effects on the activity and production of bacterial assemblages; the actively growing portion of the bacterial assemblage is highly susceptible to grazing and therefore preferentially eliminated. This principle has been demonstrated in several experimental studies with marine bacterial assemblages (Gasol et al. 1995; Sherr et al. 1992).

If a large portion of small bacteria in pelagic habitats is either dormant or growing at very low rates, this has another interesting implication for their interactions with bacterivorous grazers. The reduced vulnerability of the smallest bacterial size classes does not enable long-term survival in the presence of grazers, as many flagellates can ingest very small particles, down to the size of viruses and colloids (González and Suttle 1993; Sherr 1988). Therefore, additional defense mechanisms should be expected, such as resistance to digestive enzymes inside protistan food vacuoles, similar to what is known for some pathogenic bacteria (King et al. 1988). This resistance would be similar to the resistance of starvation-adapted cells to various abiotic stress factors, such as ultraviolet light, radicals, and temperature (Nyström et al. 1992).

A refuge from grazing is possible not only at the lower end of size classes but also at the upper end, when cells become too large to be ingested by small protists. Many bacterial taxa can, under appropriate growth conditions, exceed the size limit that can be handled and engulfed by heterotrophic nanoflagellates. The development of grazing-resistant bacterial morphologies, such as filamentous, spiral-shaped, or aggregated cells during high protistan grazing, seems to be a common phenomenon in productive environments such as eutrophic lakes (Güde 1989; Jürgens and Stolpe 1995), but it also occurs in coastal marine (Havskum and Hansen 1997) and brackish (Engström-Ost et al. 2002) waters. In freshwater plankton, where this phenomenon is particularly important, the development of complex bacterial morphologies has been more intensively studied. Filamentous bacteria are found in all major phylogenetic groups (Jürgens et al. 1999; Langenheder and Jürgens 2001; Šimek et al. 1999), and single taxa of resistant morphotypes can become for short periods the dominant component of bacterial biomass (Pernthaler et al. 2004). The development of resistant morphologies can be mediated by phenotypic plasticity, as has been shown for bacterial strains that produce microcolonies (Hahn et al. 2000; Matz et al. 2002b) or filaments (Hahn and Höfle 1999; Hahn et al. 1999). The regulating mechanisms seem to differ between different bacterial strains. In some strains, the development of resistant morphotypes is favored by high growth rates (Hahn et al. 1999), whereas in others it is stimulated by chemicals released during bacterivory (Corno and Jürgens 2006). It has to be kept in mind, however, that these predation-resistant morphologies are only temporary refuges, as aggregates are colonized by HNF adapted to graze on attached bacteria, and larger bacterial morphologies are finally consumed by larger protozoan and metazoan grazers.

Other Antipredator Traits of Prokaryotes

In addition to bacterial cell size, a range of bacterial properties influence the vulnerability at least against some of the different protistan groups acting as bacterivores (Fig. 11.6). For example, the production of chemical feeding deterrents, either lethal toxins or inhibitory substances, seems to be more widespread in bacteria than previously thought, and has been found in bacteria growing in biofilms (Matz et al. 2005) as well as in suspension (Matz et al. 2004b). For bacteria growing in biofilms, the protected growth in an extracellular polysaccharide matrix can be regulated by cell–cell communication (quorum sensing) (Matz et al. 2004a), which might be influenced by predators.

Bacterial motility was known to enhance encounter rates with predators and therefore also ingestion rates (González et al. 1993b). Matz and Jürgens (2005) showed that at high bacterial swimming velocity (approximately >40 μm/s) bacteria cannot be captured by HNF. Bacterial motility, which seems to be common in marine bacterioplankton (Grossart et al. 2001), might therefore not only serve to locate micropatches of enhanced substrate concentrations (Blackburn et al. 1998) and to colonize aggregates (Kiørboe et al. 2002) but also to enhance survival in the presence of protistan grazing.

The physicochemical surface properties of prokaryotes are probably very important for a whole range of bacterial functions, and probably also mediate interactions with grazers. Bacterial hydrophobicity and surface charge are variable in the environment (Stoderegger and Herndl 2005) and have been proposed as factors influencing prey uptake by protistan grazers. However, there are contradictory results with regard to these factors (Matz and Jürgens 2001; Monger et al. 1999), complicated by the fact that they change with the physiological state of the bacteria (Stoderegger and Herndl 2004). The chemical coating of prey particles seems to have a strong influence on the ingestion efficiency of nanoflagellates (Matz et al. 2002a), and there is evidence for biochemical prey recognition (e.g., by glycoproteins or lectins) in interception-feeding protists (Sakaguchi et al. 2001; Wootton et al. 2007). Some components in the outer surface structure (e.g., within *Synechococcus* cells) probably trigger the egestion of cells that are already inside food vacuoles (Boenigk et al. 2001b). It has been also speculated that certain cell wall structures might be resistant to the digestive enzymes within protistan food vacuoles (Jürgens and Matz 2002). However, the widespread occurrence of capsules in marine bacteria (Stoderegger and Herndl 2002) and its interactions with grazers (Matz et al. 2002a) remains to be studied in more detail.

A better understanding of how different bacterial phenotypic traits affect the vulnerability towards bacterivorous protists has been obtained using high-resolution videomicroscopy, by which the fate of single bacterial cells from well-characterized cultures can be followed during the different stages of the feeding process of bacterivorous protists (Boenigk and Arndt 2000). Such an analysis has provided evidence for a complex protistan selection behavior and insights on how bacterial properties interfere at different stages of the process of food acquisition (encounter, capture, and particle handling, ingestion, digestion, and assimilation). A schematic overview on how bacterial resistance mechanisms interfere with the feeding behavior of interception-feeding flagellates is shown in Figure 11.8.

The encounter rate between bacteria and protists is affected by bacterial size and motility and by bacterial exudates operating as attractants or repellents in chemoreception prey location for protists (Blackburn et al. 1998). Once a bacterial cell is encountered by a flagellate predator, bacterial motility and surface charge seem to be effective as initial mechanisms of capture avoidance (Fig. 11.8*a*). High swimming speed provides a mechanism to escape from predators. Prey particles with extremely negative surface charges reduce capture probabilities due to repulsive surface forces (Matz and Jürgens 2001). Bacterial cells captured by a flagellate can still potentially escape, favored by several bacterial features (Fig. 11.8*b*); very large and very small bacteria, as well as complex shape and morphology, cause handling problems. Further, extreme surface charge or "distasteful" biochemical surface compounds of the bacterial prey may result in active rejection of the captured bacteria. Even during the process of food vacuole formation (Fig. 11.8*c*), oversized bacteria might be rejected and highly motile cells may escape from the phagocytosis process.

GRAZING AS A SHAPING FORCE OF BACTERIAL ASSEMBLAGES

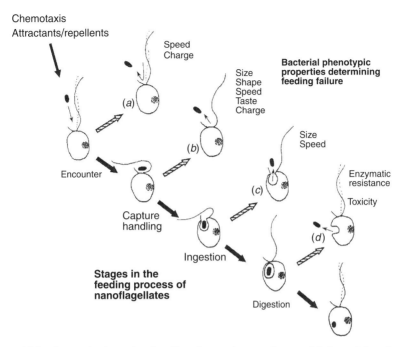

Figure 11.8 Stages in bacteria–flagellate interactions and potential bacterial resistance mechanisms affecting feeding success at each stage. Modified from Jürgens and Matz (2002).

Finally, resistance can also occur in the last stage (Fig. 11.8*d*), and unpalatable prey bacteria can be prematurely egested (Boenigk et al. 2001a) or be resistant to enzymatic digestion. Some bacterial strains produce highly toxic compounds (e.g., the pigment violacein) that result in immediate death of the predators when these compounds are released into the food vacuoles (Matz et al. 2004b).

Bacterial resistance mechanisms are not efficient against all types of predators; there are probably always protistan taxa that are able to cope with bacterial defenses (Wu et al. 2004), and a succession of defense mechanisms in response to changes in the grazer community can be observed (Salcher et al. 2005). Additionally, there are constraints imposed by the actual growth-limiting conditions of the bacteria, and the prevailing grazing resistance mechanisms differ for example under carbon- and phosphorus-limited growth conditions (Matz and Jürgens 2003). A challenging future goal is to understand the adaptive trade-offs that bacteria experience with respect to different predators and different abiotic factors, and how the main selective pressures interact.

Grazing-Mediated Changes in Bacterial Community Composition

Considering that planktonic prokaryotes possess diverse mechanisms to reduce grazing losses by phagotrophic protists and that these properties are not equally distributed among the different phylogenetic lineages of prokaryotes, its is evident that grazing is also a structuring force for bacterial community composition (BCC) that is at least as important as substrate supply. Therefore, it is not surprising that changes in predation pressure on bacteria, either naturally occurring or experimentally enhanced, are generally reflected in changes of BCC. This has been mainly observed in limnic microbial communities, both in model chemostat systems (Hahn and Höfle 1999; Pernthaler et al. 1997) and in food web manipulation experiments (Jürgens et al. 1999; Langenheder and Jürgens 2001; Šimek et al. 1999) and seasonal plankton successions in lakes (Pernthaler et al. 2004). The observed shifts in BCC in response to predation pressure seem to follow consistent patterns, depending on the balance between bacterial production and bacterivory (Šimek et al. 2002).

There have been only a few comparable studies in marine systems where the impact of changing predation regimes on BCC has been studied, but so far it seems that trophic interactions are not as strongly reflected in changes of BCC as in limnic systems (Yokokawa and Nagata 2005; Suzuki 1999). It remains to be investigated whether it is a consistent distinction between freshwater and marine systems and whether it is due to differences in phylogenetic compositions or available resistance mechanisms. Changes in BBC due to size-selective removal of intermediate-sized bacteria is probably a generally valid mechanism, and has been shown to be responsible for the suppression of readily culturable marine *Gammaproteobacteria* (Beardsley et al. 2003). An intriguing question is whether avoidance or reduction of grazing losses is a decisive ecological trait for marine pelagic bacteria and thus an underlying factor for the observed dominance of bacterial clusters such as SAR11 (Morris et al. 2002) or *Roseobacter* (Selje et al. 2004). For limnic planktonic bacteria there is already evidence that the dominant groups of small-sized bacteria (*Polynucleobacter* and *Actinobacteria*) experience significantly lower grazer mortality (Boenigk et al. 2006; Hahn et al. 2003; Jezbera et al. 2006).

MOLECULAR TOOLS FOR PROTISTAN ECOLOGY

Culturing Bias and Molecular Approaches

Given the ecological importance of marine heterotrophic flagellates and their potentially large phylogenetic and functional diversity, it is remarkable that we still know so little about the composition of natural assemblages. Perhaps this is due to the assumption from early studies that small flagellates in culture represented the dominant species in the sea (Fenchel 1982d), and thus their physiological characterization would directly apply to natural cells. However, past and present data are disputing this optimistic view. Comparisons of culture-based estimates and direct counts in a variety of systems revealed that only about 1 percent of heterotrophic flagellates could be cultured (Caron et al. 1989), a striking similarity to the "Great Plate Anomaly" in bacteria (Staley and Konopka 1985), as discussed in Chapter 3. Another similarity to the bacterial world is that cultured flagellates do not seem to be dominant in situ. Lim et al. (1999) studied the abundance of *Paraphysomonas imperforata* in a set of enrichments by using a molecular probe. The incubations were systematically dominated by *P. imperforata*, but this flagellate was rare in the original sample, indicating a strong bias by the enrichment. The standard approach to enrich and culture heterotrophic flagellates is to provide organic matter (e.g., wheat grains) to promote growth of bacteria that become food for the grazers (Cowling 1991). This creates a situation very different from natural conditions (high levels of organic matter, large bacterial cells, high bacterial densities, and the presence of particles) that could explain the culturing bias. There is a need for culturing attempts using more realistic bacterial abundances and sizes, as has been done in freshwater systems with ultramicrobacteria or natural bacteria in moderate concentrations (Boenigk et al. 2005).

Molecular techniques provide a way to circumvent the need for culturing to assess microbial diversity (Caron et al. 2004). They have been very successful in identifying the dominant marine *Bacteria* (Giovannoni et al. 1990) and *Archaea* (Delong 1992), and have recently been applied also to marine protists (Díez et al. 2001; López-García et al. 2001; Moon-van der Staay et al. 2001). These molecular techniques and studies are discussed in Chapters 3 and 6. Given that protists appear in all eukaryotic supergroups (Fig. 11.1), it is not possible to design polymerase chain reaction (PCR) primers covering only protistan lineages. Therefore, universal eukaryotic primers are regularly used, and the prefiltration step becomes essential to remove large protists and metazoans that would be also amplified. Alternatively, group-specific primers can be used to investigate the diversity of particular protistan groups (Table 11.4). Clone libraries give a list of protists in the sample that can further serve to design oligonucleotide probes to be applied by fluorescence in situ hybridization (FISH; see Table 11.4 for protistan probes) to assess the abundance of specific taxa and to obtain some hints on the size and shape of the targeted cells.

One of the first habitats to be investigated was the surface water of coastal and open sea environments (Díez et al. 2001; Moon-van der Staay et al. 2001). The two main findings were the large phylogenetic diversity and the presence of novel lineages, likely representing uncultured protists. In libraries of surface marine picoeukaryotes, 39 percent of the clones clustered with well-known groups of phototrophic protists, 22 percent with well-known groups of heterotrophic protists, 23 percent with novel lineages within the alveolates, and 16 percent with novel lineages within the stramenopiles (Massana et al. 2004). Recent data from other oceanic sites are consistent with these figures (Countway et al. 2005; Worden 2006). Novel alveolates are related to dinoflagellates, and form two lineages with a substantial genetic diversity (Groisillier et al. 2006), and novel stramenopiles (MASTs, for marine stramenopiles: Massana et al. 2004) form independent lineages at the base of the stramenopile radiation (Fig. 11.9a; see also Chapter 6). Besides the oceanic surface, other marine habitats have been investigated such as the bathypelagic zone (López-García et al. 2001), anoxic waters and sediments (Dawson and Pace 2002; Stoeck and Epstein 2003), and hydrothermal vents (Edgcomb et al. 2002). Even though each habitat can have specific sequences, a common trait of all of them is the large diversity and the presence of lineages of unknown affiliation, highlighting the large protistan diversity still to be discovered.

An obvious endeavor is to identify the organisms responsible for the novel sequences, particularly when these are represented by many clones in clone libraries

TABLE 11.4 PCR Primers and FISH Probes Designed for Marine Heterotrophic Flagellates

Primer/Probe Name	Sequence (5' to 3')	Position[a]	Target Group	Taxonomic Group	Reference
PCR Primers					
EukA	AACCTGGTTGATCCTGCCAGT	1–21	All eukaryotes		(1)
EukB	TGATCCTTCTGCAGGTTCACCTAC	1795–1772	All eukaryotes		(1)
Gonio42F	GATTAAGCCATGCATGTCTAAGTGTAAATAAG	41–73	*Goniomonas*	Cryptophyceae	(2)
1256R	GCACCACCACCCAYAGAATCAAGAAAGAWCTTC	1270–1237	Most cercozoans	Cercozoa	(3)
Chryso-R	CCAACAAAATAGACCAAGG	838–818	Most chrysophytes	Chrysophyceae	(4)
kineto14F	CTGCCAGTAGTCATATGCTTGTTTTCAAGGA	14–43	Most kinetoplastids	Kinetoplastea	(5)
kineto2026R	GATCCTTCTGCAGGTTCACCTACAGCT	1794–1768	Most kinetoplastids	Kinetoplastea	(5)
243F	CCAATGCACCCTCTGGGTGGTT	223–244	Clade A	Cercozoa	(6)
1259F	GGTCCRGACAYAGTRAGGATTGACAGATTGAAG	1208–1241	Most cercozoans	Cercozoa	(6)
FISH Probes					
Euk502 m	GCACCAGACTTGCCCTCC	565–548	All eukaryotes		(7)
PV1	TAAAACCATCCTATTATATC	661–642	*Paraphysomonas vestita*	Chrysophyceae	(8)
PV2	TTCCGTATGCCAGTCAGA	198–184	*Paraphysomonas vestita*	Chrysophyceae	(8)
PV3	AGTATAAATATCACAGTCCGA	1695–1675	*Paraphysomonas vestita*	Chrysophyceae	(8)
PV4	ATATAATCTTTTCGATGATGA	719–699	*Paraphysomonas vestita*	Chrysophyceae	(8)
PV5	CCCATCCTATTATATCAGAAA	656–637	*Paraphysomonas vestita*	Chrysophyceae	(8)
Pband 635	TGAGGGATGGACCGGTTGCC	667–648	*Parahysomonas bandaiensis*	Chrysophyceae	(9)

Name	Sequence	Position	Target	Group	Ref
Pband 663	GGCCGCAGAAACCTGGTACACA	695–674	Parahysomonas bandaiensis	Chrysophyceae	(9)
Pband 1683	CCGATCCGCGGTCCGAAA	1690–1673	Parahysomonas bandaiensis	Chrysophyceae	(9)
Pband 706	CCCACACCAGACAACTCAAT	732–713	Parahysomonas bandaiensis	Chrysophyceae	(9)
Pimp 635	TGAGGGGCGGACCGGTCGCC	667–648	Parahysomonas imperforata	Chrysophyceae	(9)
Pimp 663	GGACGCAGAGACCAGGTGCACA	695–674	Parahysomonas imperforata	Chrysophyceae	(9)
Pimp 1683	CCAAGCCGCAGTCCGAGA	1690–1673	Parahysomonas imperforata	Chrysophyceae	(9)
Pbiv 1537	CAAGATTCAGAATTGCAAAAA	1566–1546	P. bandaiensis, imperforata, vestita	Chrysophyceae	(9)
NS4	TACTTCGGTCTGCAAACC	855–837	MAST-4	Stramenopiles	(10)
PED1646	TTACCTTCGATCGGTAGG	1663–1646	Pedinellids/Rhizochromulinales	Dictyochophyceae	(11)
FV18TS-650	TGTCGTGCGAACCCATAG	668–650	Actuariola	Kinetoplastea	(12)
NS1A	ATTACCTCGATCCGCAAA	857–839	MAST-1A	Stramenopiles	(13)
NS1B	AACGCAAGTCTCCCCGCG	729–710	MAST-1B	Stramenopiles	(13)
NS1C	GTGTTCCCTAACCCGAC	739–723	MAST-1C	Stramenopiles	(13)
NS2	CGGGTCCCGAGCACGACA	738–722	MAST-2	Stramenopiles	(13)
MAST-12Nor2	TACAGTGCCAATGGAGAC	863–845	KKTS_D3 (MAST-12)	Stramenopiles	(14)
CAF01	ACAGTGCTGACACCCTGT	1077–1060	Cafeteria roenbergensis	Bicosoecida	(15)
CET01	CAGCTCAATACGGACACC	895–878	Caecitellus parvulus/paraparvulus	Bicosoecida	(15)

References: (1) Medlin et al. (1988); (2) Heyden et al. (2004); (3) Bass and Cavalier-Smith (2004); (4) Berglund et al. (2005); (5) Heyden and Cavalier-Smith (2005); (6) Karpov et al. (2006); (7) Lim et al. (1993); (8) Rice et al. (1997); (9) Caron et al. (1999); (10) Massana et al. (2002); (11) Beardsley et al. (2005); (12) Stoeck et al. (2005); (13) Massana et al. (2006b); (14) Kolodziej and Stoeck (2007); (15) Massana et al. (2007).

[a] positions refer to the 18S rDNA of *S. cerevisiae*.

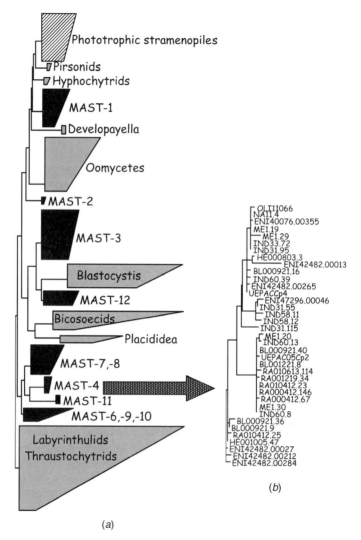

Figure 11.9 (*a*) Phylogenetic tree of stramenopiles, showing MAST lineages (black boxes) among known heterotrophic groups (gray boxes). Phototrophic groups are in the cross-hatched box. (*b*) Phylogenetic tree of MAST-4 sequences. See Chapter 6 for another version of this tree. (*b*) Modified from Massana et al. (2006b).

and are widely distributed. Novel alveolates are now thought to be mainly parasites of a varied array of marine organisms (Dolven et al. 2007; Groisillier et al. 2006), opening an unexpected role of parasitism in pelagic environments. On the other hand, some MAST groups have been identified as free-living bacterivorous heterotrophic flagellates (Massana et al. 2006b) through a combination of FISH with specific probes and other measurements. These MAST cells, of sizes ranging from 2 to 8 µm, appear to be widely distributed in the world oceans (Table 11.5).

TABLE 11.5 Characteristics of Five MAST Groups
(Modified from Massana et al. 2006b)

	MAST-1A	MAST-1B	MAST-1C	MAST-2	MAST-4
Morphology					
Shape	Spherical	Spherical	Spherical	Spherical	Spherical
Mean diameter (μm)	7.7 ± 0.3	4.1 ± 0.1	5.6 ± 0.3	6.2 ± 0.6	2.3 ± 0.1
Number of flagella	0	1	1	1	1
Size of flagellum (μm)	—	8.7 ± 0.6	12.0 ± 1.2	5.7 ± 0.4	3.0 ± 0.1
Trophic Mode					
Growth in the dark	Yes	Yes	Yes	Yes	Yes
Bacterivory	Presumed	Observed	Observed	Presumed	Observed
Chloroplasts	Not seen	Not seen	Not seen	Not seen	Not seen
Distribution and Abundance					
Samples found	17/24	16/24	23/24	11/24	17/24
Maximum (cells/mL)	86	274	201	10	569
Mean (cells/mL)	11	39	31	4	131
Mean % of HNF	1.0	2.8	2.7	0.3	9.2

For instance, MAST-4 averages 130 cells/mL in oceanic regimes examined so far (except the polar ones) and represents 9 percent of HNF counts. MAST-1 cells contributed less to HNF counts (5 percent), but in one Antarctic sample reached 35 percent. It is interesting to note that the genetic variability of MAST-4 sequences (Fig. 11.9b) gives room for significant ecological differentiation within the group, which could explain its wide distribution. Other abundant members of HNF assemblages on a global scale could be MAST lineages that have not yet been probed, or other phylogenetic groups. Molecular surveys detected few clones related to cultured heterotrophic flagellates (a few sequences affilated to *Paraphysomonas*, *Cafeteria*, *Caecitellus*, and *Cryothecomonas*). In an effort to investigate the abundance of some of these flagellates, FISH probes against *Cafeteria* and *Caecitellus* were designed, and these flagellates were below detection (<1 cell/mL) in most samples investigated (del Campo and Massana, in preparation). Thus, both clone libraries and preliminary FISH results converge to suggest that many (if not all) heterotrophic flagellates in culture do not represent the dominant ones in natural assemblages.

Environmental sequencing can also be used to compare the composition of heterotrophic flagellates in marine and freshwater systems, which were assumed to harbor rather similar assemblages (Arndt et al. 2000; Boenigk and Arndt 2002). The data obtained so far suggest that stramenopiles might account for a large fraction of unidentified flagellates in both systems, but these sequences are mostly chrysomonads and bicosoecids in freshwater (Richards et al. 2005) and MASTs in marine systems (Massana et al. 2004). These surveys did not reveal the existence of any phylotype common to both systems. In addition, when particular groups have been carefully studied, such as heterotrophic cryptomonads (Heyden et al. 2004) or bodonids

(Heyden and Cavalier-Smith 2005), the molecular signatures revealed distinct freshwater and marine clusters. Thus, freshwater and marine assemblages of heterotrophic flagellates appear to be clearly different, as is known for most other planktonic organisms as well.

Global Distribution and Diversity of Marine Protists

There is considerable debate about the extent of diversity and biogeographical distribution of microorganisms (Hughes Martiny et al. 2006). It has been argued that, given the huge population abundances, small cell sizes, and apparent absence of dispersal barriers, most protistan species must be cosmopolitan and the number of total species relatively low (Finlay 2002; Patterson and Lee 2000). However, this global-distribution but low-diversity view arises by using the morphospecies concept, and it is doubtful that this concept has enough resolving power to categorize the relevant scales of variation within protists (Mitchell and Meisterfeld 2005). It is known that strains from the same protistan morphospecies can belong to genetically, reproductively, and ecologically different groups, which are named "ecotypes" or "cryptic species" (Rodríguez et al. 2005; Vargas et al. 1999). However, it is often possible to find ultrastructural features that differentiate cryptic species, which then become "pseudocryptic" (Hausmann et al. 2006). The most elegant way to show the limitation of the morphospecies concept is a mating experiment between strains of the same morphospecies. Since only strains genetically very close can interbreed, strains not able to mate would indicate that a given morphospecies contains several biological species, which could be traced with genetic markers (Amato et al. 2007; Coleman 2000). Thus, for investigating the global distribution and diversity of protists, molecular tools for comparing assemblages from distant systems can obviously provide more systematic and objective data for the debate. Nevertheless, the morphospecies concept is still useful, as it provides information on the ecological role and functional adaptations of protistan taxa.

Clone libraries of protistan assemblages from distant pelagic marine sites generally retrieved similar phylogenetic groups (Díez et al. 2001; Moon-van der Staay et al. 2001; Worden 2006). This suggests that the overall structure of these assemblages is comparable on a worldwide scale, with dinoflagellates, novel alveolates, MASTs, and prasinophytes being well represented. In addition, when inspecting the level of 18S rRNA gene sequence similarity, no particular geographic separation appears. For instance, the heterotrophic flagellate MAST-4 has been found in marine samples from all oceans (Massana et al. 2006b), and identical sequences have been retrieved from systems as distant as the North Atlantic, the Mediterranean Sea, and the Indian Ocean (NA11.4, ME1.19, and IND33.72 in Fig. 11.9b). Examples of identical sequences from distant sites also exist within the prasinophytes, dinoflagellates, or novel alveolates (Groisillier et al. 2006; Guillou et al. 2004). Altogether, we find little sign of geographical barriers in the marine environment, and most groups seem to be roughly globally distributed, supporting the cosmopolitan view mentioned above. It is true, however, that by using phylogenetic markers with a higher resolving power, we could perhaps find locally adapted populations, such as population geneticists do for larger organisms. This global-distribution view only applies to marine habitats that are roughly equivalent, since one already expects to find different protists in different habitats (such as polar and template waters). Moreover, habitats that are discontiguous, such as lakes, hydrothermal vents, or hot springs, can be more prone to harbor endemic populations due to geographical barriers (Whitaker et al. 2003).

Global distribution has also been used as an argument for low protistan diversity, since it prevents allopatric speciation, which is considered to be the most important speciation mechanism (although speciation in sympatry seems more common than previously thought: Culotta and Pennisi 2005). Environmental molecular data

---

Biogeography and Diversity Extent in Protists

This debate perhaps started with the sentence, applied to microorganisms, that "everything is everywhere, but, the environment selects" (Baas Becking 1934), and often is unfocused because of poor definitions or misuse of some important terms:

- **Ecological Niche** This is the multidimensional space of resources (i.e., light, nutrients, biotic interactions, etc.) available to (and specifically used by) organisms. According to the above sentence, two identical ecological niches somewhere in the world would be occupied by the same microorganisms. However, it is very difficult to be sure that geographically separated habitats represent the same ecological niche. So, if these habitats present (slightly) different organisms, it is not clear if these differences are due to speciation and restricted dispersal or due to the fact that the niches have some differences that we do not see.
- **Species Concept** There are many definitions of "species." In eukaryotes, the biological concept, based in sexual compatibility, is the most accepted. However, in protists, the most used is the morphospecies concept, based on morphological characters. It is known that the two definitions do not have the same taxonomic depth.
- **Dispersal Range and Biogeographic Barriers** Given the small cell size and huge population size of microorganisms, they seem to disperse globally without geographic barriers. This probably applies to the marine pelagic environment, but not necessarily to other uncontiguous habitats.
- **Speciation Mechanisms** The two extreme mechanisms are allopatry (speciation in complete geographic isolation) and sympatry (in geographic coexistence). Allopatric speciation is assumed to be the main mechanism, but sympatric speciation is increasingly being considered.
- **Extent of Microbial Diversity** The number of known species increases from large to small living beings up to a point (around 2 mm) where it decreases again, with relatively few described microbial species. It has been claimed that allopatric microbial speciation is prevented by global dispersion. However, molecular analysis of environmental communities indicates a huge level of molecular diversity with currently unknown biological and ecological implications, but suggests a huge number of microbial species.

strongly disagree with the view of low diversity, since virtually all marine groups detected show a very large sequence divergence at the 18S rDNA level. To give two examples, Figure 11.9*b* shows many different sequences within the same group, MAST-4, with up to 3 percent sequence divergence. In an even more striking example, Heyden and Cavalier-Smith (2005) reported a huge diversity of bodonids, with global estimates of thousands of different rRNA sequences. The ecological significance of this huge genetic diversity within a given group is at present not understood. It is unlikely that it represents simply the accumulation of neutral mutations, as has been suggested (Fenchel 2005), since small changes in the 18S rDNA might imply millions of years of evolutionary divergence and thus are signs of other important genomic changes. For instance, *Ostreococcus* strains differing 2 percent in their 18S rDNA represent ecotypes adapted to the light regimes at different water depths (Rodríguez et al. 2005). An intriguing open question is to understand how this huge phylogenetic protistan diversity, particularly of heterotrophic groups (Vaulot et al. 2002), translates into functional diversity in the open sea and by which mechanisms it is generated and maintained. One possible underlying reason is that the seemingly homogeneous pelagic environment can be viewed as a continuum of particulate organic matter where a diverse (in size, growth strategies, and surface properties) assemblage of bacteria with different food qualities and different vulnerabilities to bacterivores thrive. This could set up a plethora of different ecological niches for specialized bacterivores, thus enabling high protistan diversity at small spatial scales. In summary, we argue for a global distribution of marine protists and a large genetic diversity with potentially important ecological implications, which, however, still have to be investigated.

Linking Diversity and Function for Uncultured Heterotrophic Flagellates

Cell cultures yield fundamental ecophysiological information, so the first obvious step is to try to culture the as-yet uncultured groups. A promising strategy would be to use culture media simulating natural conditions, typically more oligotrophic than standard media. This strategy was very successful for marine bacteria (Rappé et al. 2002), but to our knowledge has never been tried for marine heterotrophic flagellates. Once these heterotrophic flagellates are in pure culture, morphological and functional parameters can be estimated, such as ultrastructure by electron microscopy, functional and numerical responses, food size spectra, temperature optima, growth efficiency, competition experiments, and survival responses, similarly to the way in which it has been done for the commonly cultured taxa (Table 11.3).

An intermediate approach before culturing is to promote the growth of uncultured heterotrophic flagellates in unamended dark incubations (Massana et al. 2006a). The setup is easy and includes a prefiltration through a 3 μm filter to remove larger predators, and a dark incubation that prevents the growth of phototrophs. In most samples, this simple setup results in a succession of bacteria and heterotrophic flagellates, and among them uncultured groups abundant in the sea, such as the MAST groups, have been observed (Massana et al. 2006a). The growth of these abundant, uncultured groups was probably due to keeping bacteria and flagellates at realistic abundances and sizes in the unamended incubations. The increase of flagellates was moderate (10–100-fold), sufficient to measure growth rates of uncultured groups, and to provide excellent material for activity measurements. This simple approach could help in the first stages in isolating these previously uncultured protists.

An interesting way to link diversity and activity is by comparing genetic libraries from the same sample prepared from environmental rDNA and rRNA (Stoeck et al. 2007). It has been suggested that the rRNA content reflects activity, so organisms identified in the rRNA library would be the most active, while those identified in the rDNA library would be present but not necessarily active. Some groups, such as the dinoflagellates and novel alveolates I, were only found in the rDNA library, so appeared to be inactive, nonindigeneous members of the community (Stoeck et al. 2007). This is relevant, since these groups are commonly found in rDNA clone libraries (Díez et al. 2001; Worden 2006). However, in addition to activity, it is obvious that the copy number of the rDNA operon, which can vary by orders of magnitude among groups (Zhu et al. 2005), might affect the differences between rRNA and rDNA libraries.

Even if some uncultured protists can finally be cultured, direct measurements with natural assemblages will need to be performed with new techniques, since it is not realistic to expect that all natural variability will ever be represented in culture collections and become subject to laboratory experiments. One example of an innovative approach is the combination of FISH and electron microscopy for obtaining ultrastructural details of uncultured protists (Kolodziej and Stoeck 2007; Stoeck et al. 2003). Other approaches are being developed to obtain grazing rates of uncultured heterotrophic flagellates in the complex scenario of natural assemblages. For instance, short-term FLB ingestion experiments can be combined with FISH with probes for uncultured protists. This will reveal the in situ grazing rates of uncultured grazers and will allow calculation of the grazing impact for separate groups. Further, similar short-term incubations using live bacteria (thus retaining the behavior and surface properties) from different phylogenetic groups and/or sizes can be performed to investigate grazing selectivity of predators and grazing resistance of bacteria in natural assemblages. Linking diversity and function for natural microorganisms—bacteria and protists—is a major challenge in microbial ecology, and requires the synergistic efforts of the three schools of taxonomic, ecological, and molecular expertise, in order to achieve a better understanding of the complexity of natural populations and of their activities and interactions.

SUMMARY

1. The capacity to ingest bacteria is widely distributed in the eukaryotic world. Theoretical considerations together with experimental observations support the notion that the most important bacterial grazers in marine systems are heterotrophic nanoflagellates (HNF) of sizes between 2 and 5 μm.

2. The phylogenetic and functional diversity within heterotrophic flagellate assemblages is very large. These assemblages include cells with different evolutionary histories, feeding mechanisms, prey selection, and behavior. The autecology of HNF has been studied for a set of cultured strains that can be easily isolated and that generally grow rapidly on suspended bacteria. As these taxa are not the ones dominating in the environment, the question remains how representative these results are for natural environments.
3. Protistan grazing on bacteria has been shown to be roughly in balance with bacterial production, thus being the major loss factor for bacteria in most marine systems. Other ecological functions of HNF, besides bacterivory, are herbivory on picosized algae, nutrient remineralization, and production of dissolved and colloidal organic matter.
4. Protistan predation acts as a strong selective force for planktonic bacteria, and phenotypic properties that lower grazing vulnerability become important in pelagic systems. Potential resistance mechanisms that act at different stages of the predator–prey interactions are size, motility, biochemical surface properties, and toxicity. Increased protist predation pressure can trigger phenotypic and diversity changes in natural bacterial communities.
5. Molecular tools can be successfully applied to protistan ecology to reveal the identity of natural populations, and their abundance and distribution. One of the most striking results has been the discovery of novel, as-yet uncultured lineages, such as the MAST lineages that account for a significant fraction of HNF assemblages.
6. Protistan species are distributed globally in the marine pelagic environment, with no signs of geographic segregation. Moreover, local and global protistan diversity appears to be very high, and the ecological significance of this large diversity remains largely underexplored.

ACKNOWLEDGMENTS

We thank J. Pernthaler, F. Nitsche, H.Arndt, J. Boenigk, and F. Unrein for providing figures, and D. Kirchman, H. Arndt, J. Gasol, and two reviewers for comments on earlier drafts.

REFERENCES

Abrams, P. A., and C. J. Walters. 1996. Invulnerable prey and the paradox of enrichment. *Ecology* 77: 1125–1133.

Adl, S. M., A. G. B. Simpson, M. A. Farmer, et al. 2005. The new higher level classification of eukaryotes with emphasis on the taxonomy of protists. *J. Eukaryot. Microbiol.* 52: 399–451.

REFERENCES

Alonso, M. C., V. Rodriguez, J. Rodriguez, and J. J. Borrego. 2000. Role of ciliates, flagellates and bacteriophages on the mortality of marine bacteria and on dissolved-DNA concentration in laboratory experimental systems. *J. Exp. Mar. Biol. Ecol.* 244: 239–252.

Amato, A., W. H. C. F. Kooistra, J. H. L. Ghiron, D. G. Mann, T. Pröschold, and M. Montresor. 2007. Reproductive isolation among sympatric cryptic species in marine diatoms. *Protist* 158: 193–207.

Andersen, P. 1989. Functional biology of the choanoflagellate *Diaphanoeca grandis* Ellis. *Mar. Microb. Food Webs* 3: 35–50.

Anderson, T. R., and H. W. Ducklow. 2001. Microbial loop carbon cycling in ocean environments studied using a simple steady-state model. *Aquat. Microb. Ecol.* 26: 37–49.

Archibald, J. M., D. Longet, J. Pawlowski, and P. J. Keeling. 2003. A novel polyubiquitin structure in Cercozoa and Foraminifera: evidence for a new eukaryotic supergroup. *Mol. Biol. Evol.* 20: 62–66.

Arenovski, A. L., E. L. Lim, and D. A. Caron. 1995. Mixotrophic nanoplankton in oligotrophic surface waters of the Sargasso Sea may employ phagotrophy to obtain major nutrients. *J. Plankton Res.* 17: 801–820.

Arndt, H. 1993. A critical review of the importance of rhizopods (naked and testate amoebae) and actinopods (heliozoa) in lake plankton. *Mar. Microb. Food Webs* 7: 3–29.

Arndt, H., D. Dietrich, B. Auer, et al. 2000. Functional diversity of heterotrophic flagellates in aquatic ecosystems. In B. Leadbeater and J. Green (eds.), *The Flagellates*. Taylor & Francis, pp. 240–268.

Arndt, H., K. Hausmann, and M. Wolf. 2003. Deep-sea heterotrophic nanoflagellates of the Eastern Mediterranean Sea: Qualitative and quantitative aspects of their pelagic and benthic occurrence. *Mar. Ecol. Prog. Ser.* 256: 45–56.

Azam, F. 1998. Microbial control of oceanic carbon flux: The plot thickens. *Science* 280: 694–696.

Azam, F., T. Fenchel, J. G. Field, J. S. Gray, L. A. Meyer-Reil, and F. Thingstad. 1983. The ecological role of water-column microbes in the sea. *Mar. Ecol. Prog. Ser.* 10: 257–263.

Baldauf, S. L. 2003. The deep roots of eukaryotes. *Science* 300: 1703–1706.

Barbeau, K., E. B. Kujawinski, and J. W. Moffett. 2001. Remineralization and recycling of iron, thorium and organic carbon by heterotrophic marine protists in culture. *Aquat. Microb. Ecol.* 24: 69–81.

Baas Becking, L. G. M. 1934. *Geobiologie of inleiding tot de milieukunde*. W. P. Van Stockum.

Bass, D., and T. Cavalier-Smith. 2004. Phylum-specific environmental DNA analysis reveals remarkably high global biodiversity of Cercozoa (Protozoa). *Int. J. Syst. Evol. Microbiol.* 54: 2393–2404.

Beardsley, C., J. Pernthaler, W. Wosniok, and R. Amann. 2003. Are readily culturable bacteria in coastal North Sea waters suppressed by selective grazing mortality? *Appl. Environ. Microbiol.* 69: 2624–2630.

Beardsley, C., K. Knittel, R. Amann, and J. Pernthaler. 2005. Quantification and distinction of aplastidic and plastidic marine nanoplankton by fluorescence in situ hybridization. *Aquat. Microb. Ecol.* 41: 163–169.

Bell, E. M., and J. Laybourn-Parry. 2003. Mixotrophy in the antarctic phytoflagellate, *Pyramimonas gelidicola* (Chlorophyta, Prasinophyceae). *J. Phycol.* 39: 644–649.

Berglund, J., K. Jürgens, I. Bruchmüller, M. Wedin, and A. Andersson. 2005. Use of group-specific PCR primers for identification of chrysophytes by denaturing gradient gel electrophoresis. *Aquat. Microb. Ecol.* 39: 171–182.

Bernard, L., C. Courties, P. Servais, M. Troussellier, M. Petit, and P. Lebaron. 2000. Relationships among bacterial cell size, productivity, and genetic diversity in aquatic environments using cell sorting and flow cytometry. *Microb. Ecol.* 40: 148–158.

Billen, G., P. Servais, and S. Becquevort. 1990. Dynamics of bacterioplankton in oligotrophic and eutrophic aquatic: Bottom-up or top-down control? *Hydrobiologia* 207: 37–42.

Blackburn, N., T. Fenchel, and J. Mitchell. 1998. Microscale nutrient patches in planktonic habitats shown by chemotactic bacteria. *Science* 282: 2254–2256.

Boenigk, J., and H. Arndt. 2000. Particle handling during interception feeding by four species of heterotrophic nanoflagellates. *J. Eukaryot. Microbiol.* 47: 350–358.

Boenigk, J., and H. Arndt. 2002. Bacterivory by heterotrophic flagellates: Community structure and feeding strategies. *Antonie van Leeuwenhoek* 81: 465–480.

Boenigk, J., C. Matz, K. Jürgens, and H. Arndt. 2001a. Confusing selective feeding with differential digestion in bacterivorous nanoflagellates. *J. Eukaryot. Microbiol.* 48: 425–432.

Boenigk, J., C. Matz, K. Jürgens, and H. Arndt. 2001b. The influence of preculture conditions and food quality on the ingestion and digestion process of three species of heterotrophic nanoflagellates. *Microb. Ecol.* 42: 168–176.

Boenigk, J., K. Pfandl, P. Stadler, and A. Chatzinotas. 2005. High diversity of the "*Spumella*-like" flagellates: An investigation based on the SSU rRNA gene sequences of isolates from habitats located in six different geographic regions. *Environ. Microbiol.* 7: 685–697.

Boenigk, J., K. Pfandl, and P. J. Hansen. 2006. Exploring strategies for nanoflagellates living in a "wet desert." *Aquat. Microb. Ecol.* 44: 71–83.

Bohannan, B. J. M., and R. E. Lenski. 1999. Effect of prey heterogeneity on the response of a model food chain to resource enrichment. *Am. Nat.* 153: 73–82.

Brahamsha, B. 1996. An abundant cell-surface polypeptide is required for swimming by the non-flagellated marine cyanobacterium *Synechococcus*. *Proc. Natl. Acad. Sci. USA* 93: 6504–6509.

Brandt, S. M., and M. A. Sleigh. 2000. The quantitative occurrence of different taxa of heterotrophic flagellates in Southampton water, U.K. *Est. Coast. Shelf Sci.* 51: 91–102.

Calbet, A., and M. R. Landry. 1999. Mesozooplankton influences on the microbial food web: Direct and indirect trophic interactions in the oligotrophic open ocean. *Limnol. Oceanogr.* 44: 1370–1380.

Calbet, A., and M. R. Landry. 2004. Phytoplankton growth, microzooplankton grazing, and carbon cycling in marine systems. *Limnol. Oceanogr.* 49: 51–57.

Calbet, A., M. R. Landry, and S. Nunnery. 2001. Bacteria-flagellate interactions in the microbial food web of the oligotrophic subtropical North Pacific. *Aquat. Microb. Ecol.* 23: 283–292.

Caron, D. A. 1990. Growth of two species of bacterivorous nanoflagellates in batch and continous culture, and implications for their planktonic existence. *Mar. Mar. Microb. Food Webs* 4: 143–159.

Caron, D. A. 2000. Symbiosis and mixotrophy among pelagic microorganisms. In D. L. Kirchman (ed.), *Microbial Ecology of the Oceans*, 1st edn. Wiley-Liss, pp. 495–523.

REFERENCES

Caron, D. A., and J. C. Goldman. 1990. Protozoan nutrient regeneration. In G. M. Capriulo (ed.), *Ecology of Marine Protozoa*. Oxford University Press, pp. 283–306.

Caron, D. A., and N. R. Swanberg. 1990. The ecology of planktonic sarcodines. *Rev. Aquat. Sci.* 3: 147–180.

Caron, D. A., J. C. Goldman, O. K. Andersen, and M. R. Dennett. 1985. Nutrient cycling in a microflagellate food chain: 2. Population dynamics and carbon cycling. *Mar. Ecol. Prog. Ser.* 24: 243–254.

Caron, D. A., P. G. Davis, and J. M. Sieburth. 1989. Factors responsible for the differences in cultural estimates and direct microscopical counts of populations of bacterivorous nanoflagellates. *Microb. Ecol.* 18: 89–104.

Caron, D. A., K. G. Porter, and R. W. Sanders. 1990. Carbon, nitrogen and phosphorus budgets for the mixotrophic phytoflagellate *Poteriochromonas malhamensis* (Chrysophyceae) during bacterial ingestion. *Limnol. Oceanogr.* 35: 433–443.

Caron, D. A., E. L. Lim, M. R. Dennett, R. J. Gast, C. Kosman, and E. F. Delong. 1999. Molecular phylogenetic analysis of the heterotrophic chrysophyte genus *Paraphysomonas* (Chrysophyceae), and the design of rRNA-targeted oligonucleotide probes. *J. Phycol.* 35: 824–837.

Caron, D. A., P. D. Countway, and M. V. Brown. 2004. The growing contributions of molecular biology and immunology to protistan ecology: Molecular signatures as ecological tools. *J. Eukaryot. Microbiol.* 51: 38–48.

Cavalier-Smith, T., and E. E.-Y. Chao. 2003a. Phylogeny and classification of phylum Cercozoa (Protozoa). *Protist* 154: 341–358.

Cavalier-Smith, T., and E. E.-Y. Chao. 2003b. Phylogeny of Choanozoa, Apusozoa, and other protozoa and early eukaryote megaevolution. *J. Mol. Evol.* 56: 540–563.

Cavalier-Smith, T., and E. E.-Y. Chao. 2006. Phylogeny and megasystematics of phagotrophic heterokonts (Kingdom Chromista). *J. Mol. Evol.* 62: 388–420.

Choi, J. W. 1994. The dynamic nature of protistan ingestion response to prey abundance. *J. Eukaryot. Microbiol.* 41: 137–146.

Choi, J. W., and F. Peters. 1992. Effects of temperature on two psychrophilic ecotypes of a heterotrophic nanoflagellate, *Paraphysomonas imperforata. Appl. Environ. Microbiol.* 58: 593–599.

Christaki, U., A. Giannakourou, F. Van Wambeke, and G. Gregori. 2001. Nanoflagellate predation on auto- and heterotrophic picoplankton in the oligotrophic Mediterranean Sea. *J. Plankton Res.* 23: 1297–1310.

Christaki, U., C. Courties, H. Karayanni, et al. 2002. Dynamic characteristics of *Prochlorococcus* and *Synechococcus* consumption by bacterivorous nanoflagellates. *Microb. Ecol.* 43: 341–352.

Christensen-Dalsgaard, K. K., and T. Fenchel. 2003. Increased filtration efficiency of attached compared to free-swimming flagellates. *Aquat. Microb. Ecol.* 33: 77–86.

Christoffersen, K., O. Nybroe, K. Jürgens, and M. Hansen. 1997. Measurement of bacterivory by heterotrophic nanoflagellates using immunofluorescence labelling of ingested cells. *Aquat. Microb. Ecol.* 13: 127–134.

Chrzanowski, T. H., and K. Šimek. 1990. Prey-size selection by freshwater flagellated protozoa. *Limnol. Oceanogr.* 35: 1429–1436.

Cleven, E. J., and T. Weisse. 2001. Seasonal succession and taxon-specific bacterial grazing rates of heterotrophic nanoflagellates in Lake Constance. *Aquat. Microb. Ecol.* 23: 147–161.

Coleman, A. W. 2000. The significance of a coincidence between evolutionary landmarks found in mating affinity and a DNA sequence. *Protist* 151: 1–9.

Corno, G., and K. Jürgens. 2006. Direct and indirect effects of protist predation on population size structure of a bacterial strain with high phenotypic plasticity. *Appl. Environ. Microbiol.* 72: 78–86.

Countway, P. D., R. J. Gast, P. Savai, and D. A. Caron. 2005. Protistan diversity estimates based on 18S rDNA from seawater incubations in the western North Atlantic. *J. Eukaryot. Microbiol.* 52: 95–106.

Cowling, A. J. 1991. Free-living heterotrophic flagellates: Methods of isolation and maintenance, including sources of strains in culture. In D. Patterson and J. Larsen (eds.), *The Biology of Free-Living Heterotrophic Flagellates*. Oxford University Press, pp. 477–492.

Crawford, D. W., and T. Lindholm. 1997. Some observations on vertical distribution and migration of the phototrophic ciliate *Mesodinium rubrum* (=*Myrionecta rubra*) in a stratified brackish inlet. *Aquat. Microb. Ecol.* 13: 267–274.

Culotta, E., and E. Pennisi. 2005. Evolution in action. Breakthrough of the year. *Science* 310: 1878–1879.

Davis, P. G., and J. M. C. N. Sieburth. 1984. Estuarine and oceanic microflagellate predation of actively growing bacteria: Estimation by frequency of dividing-divided bacteria. *Mar. Ecol. Prog. Ser.* 19: 237–246.

Dawson, S. C., and N. R. Pace. 2002. Novel kingdom-level eukaryotic diversity in anoxic environments. *Proc. Natl. Acad. Sci. USA* 99: 8324–8329.

Deibel, D., and S. H. Lee. 1992. Retention efficiency of submicrometer particles by the pharyngeal filter of the pelagic tunicate *Oikopleura vanhoeffeni*. *Mar. Ecol. Prog. Ser.* 81: 25–30.

del Giorgio, P., and J. J. Cole. 2000. Bacterial energetics and growth efficiency. In D. L. Kirchman (ed.), *Microbial Ecology of the Oceans*, 1st edn. Wiley-Liss, pp. 289–325.

Delaney, M. P. 2003. Effects of temperature and turbulence on the predator-prey interactions between a heterotrophic flagellate and a marine bacterium. *Microb. Ecol.* 45: 218–225.

Delong, E. F. 1992. Archaea in coastal marine environments. *Proc. Natl. Acad. Sci. USA* 89: 5685–5689.

Díez, B., C. Pedrós-Alió, and R. Massana. 2001. Study of genetic diversity of eukaryotic picoplankton in different oceanic regions by small-subunit rRNA gene cloning and sequencing. *Appl. Environ. Microbiol.* 67: 2932–2941.

Dolan, J. R., and K. Šimek. 1998. Ingestion and digestion of an autotrophic picoplankter, *Synechococcus*, by a heterotrophic nanoflagellate, *Bodo saltans*. *Limnol. Oceanogr.* 43: 1740–1746.

Dolan, J. R., C. L. Gallegos, and A. Moigis. 2000. Dilution effects on microzooplankton in dilution grazing experiments. *Mar. Ecol. Prog. Ser.* 200: 127–139.

Dolven, J. K., C. Lindqvist, V. A. Albert, et al. 2007. Molecular diversity of alveolates associated with neritic North Atlantic radiolarians. *Protist* 158: 65–76.

Domaizon, I., S. Viboud, and D. Fontvieille. 2003. Taxon-specific and seasonal variations in flagellates grazing on heterotrophic bacteria in the oligotrophic Lake Annecy—importance of mixotrophy. *FEMS Microbiol. Ecol.* 46: 317–329.

Ducklow, H. W. 1983. Production and fate of bacteria in the oceans. *BioScience* 33: 494–501.

REFERENCES

Eccleston-Parry, J. D., and B. S. C. Leadbeater. 1994. A comparison of the growth kinetics of 6 marine heterotrophic nanoflagellates fed with one bacterial species. *Mar. Ecol. Prog. Ser.* 105: 167–177.

Eccleston-Parry, J. D., and B. S. C. Leadbeater. 1995. Regeneration of phosphorus and nitrogen by four species of heterotrophic nanoflagellates feeding on three nutritional states of a single bacterial strain. *Appl. Environ. Microbiol.* 61: 1033–1038.

Edgcomb, V. P., D. T. Kysela, A. Teske, A. D. V. Gomez, and M. L. Sogin. 2002. Benthic eukaryotic diversity in the Guaymas Basin hydrothermal vent environment. *Proc. Natl. Acad. Sci. USA* 99: 7658–7662.

Engström-Ost, J., M. Koski, K. Schmidt, et al. 2002. Effects of toxic cyanobacteria on plankton assemblage: Community development during decay of *Nodularia spumigena*. *Mar. Ecol. Prog. Ser.* 232: 1–14.

Evans, C., S. D. Archer, S. Jacquet, and W. H. Wilson. 2003. Direct estimates of the contribution of viral lysis and microzooplankton grazing to the decline of a *Micromonas* spp. population. *Aquat. Microb. Ecol.* 30: 207–219.

Fenchel, T. 1980. Relation between particle size selection and clearance in suspension-feeding ciliates. *Limnol. Oceanogr.* 25: 733–738.

Fenchel, T. 1982a. Ecology of heterotrophic microflagellates. I. Some important forms and their functional morphology. *Mar. Ecol. Prog. Ser.* 8: 211–223.

Fenchel, T. 1982b. Ecology of heterotrophic microflagellates. II. Bioenergetics and growth. *Mar. Ecol. Prog. Ser.* 8: 225–231.

Fenchel, T. 1982c. Ecology of heterotrophic microflagellates. III. Adaptations to heterogeneous environments. *Mar. Ecol. Prog. Ser.* 9: 25–33.

Fenchel, T. 1982d. Ecology of heterotrophic microflagellates. IV. Quantitative occurrence and importance as bacterial consumers. *Mar. Ecol. Prog. Ser.* 9: 35–42.

Fenchel, T. 1984. Suspended marine bacteria as a food source. In M. J. Fasham (ed.), *Flows of Energy and Materials in Marine Ecosystems*. Plenum Press, pp. 301–315.

Fenchel, T. 1986a. The ecology of heterotrophic microflagellates. *Adv. Microb. Ecol.* 9: 57–97.

Fenchel, T. 1986b. Protozoan filter feeding. *Prog. Protist* 1: 65–113.

Fenchel, T. 2005. Cosmopolitan microbes and their "cryptic" species. *Aquat. Microb. Ecol.* 41: 49–54.

Fenchel, T., and P. Harrison. 1976. The significance of bacterial grazing and mineral cycling for the decomposition of particulate detritus. In J. M. Anderson and A. Macfadyen (eds.), *The Role of Terrestrial and Aquatic Organisms in Decomposition Processes*. Blackwell, pp. 285–299.

Finlay, B. 2002. Global dispersal of free-living microbial eukaryote species. *Science* 296: 1061–1063.

Fu, Y., C. O'Kelly, M. Sieracki, and D. L. Distel. 2003. Protistan grazing analysis by flow cytometry using prey labeled by in vivo expression of fluorescent proteins. *Appl. Environ. Microbiol.* 69: 6848–6855.

Fuhrman, J. A., and R. T. Noble. 1995. Viruses and protists cause similar bacterial mortality in coastal seawater. *Limnol. Oceanogr.* 40: 1236–1242.

Fuller, N. J., G. A. Tarran, M. Yallop, K. M. Orcutt, and D. J. Scanlan. 2006. Molecular analysis of picocyanobacterial community structure along an Arabian Sea transect reveals distinct spatial separation of lineages. *Limnol. Oceanogr.* 51: 2515–2526.

Gasol, J. M. 1994. A framework for the assessment of top-down vs bottom-up control of heterotrophic nanoflagellate abundance. *Mar. Ecol. Prog. Ser.* 113: 291–300.

Gasol, J. M., and D. Vaqué. 1993. Lack of coupling between heterotrophic nanoflagellates and bacteria: A general phenomenon across aquatic systems? *Limnol. Oceanogr.* 38: 657–665.

Gasol, J. M., P. A. del Giorgio, R. Massana, and C. M. Duarte. 1995. Active versus inactive bacteria: Size-dependence in a coastal marine plankton community. *Mar. Ecol. Prog. Ser.* 128: 91–97.

Gasol, J. M., C. Pedrós-Alió, and D. Vaqué. 2002. Regulation of bacterial assemblages in oligotrophic plankton systems: Results from experimental and empirical approaches. *Antonie van Leeuwenhoek* 81: 435–452.

Geider, R. J., and B. S. C. Leadbeater. 1988. Kinetics and energetics of growth of the marine choanoflagellate *Stephanoeca diplocostata*. *Mar. Ecol. Prog. Ser.* 47: 169–177.

Giovannoni, S. J., T. B. Britschgi, C. L. Moyer, and K. G. Field. 1990. Genetic diversity in Sargasso Sea bacterioplankton. *Nature* 35: 60–63.

Goldman, J. C., D. A. Caron, O. K. Andersen, and M. R. Dennett. 1985. Nutrient cycling in a microflagellate food chain: 1. Nitrogen dynamics. *Mar. Ecol. Prog. Ser.* 24: 231–242.

González, J. M., and C. A. Suttle. 1993. Grazing by marine nanoflagellates on viruses and virus-sized particles—ingestion and digestion. *Mar. Ecol. Prog. Ser.* 94: 1–10.

González, J. M., E. B. Sherr, and B. F. Sherr. 1990. Size-selective grazing on bacteria by natural assemblages of estuarine flagellates and ciliates. *Appl. Environ. Microbiol.* 56: 583–589.

González, J. M., B. F. Sherr, and E. B. Sherr. 1993a. Digestive enzyme activity as a quantitative measure of protistan grazing: The acid lysozyme assay for bacterivory. *Mar. Ecol. Prog. Ser.* 100: 197–206.

González, J. M., E. B. Sherr, and B. F. Sherr. 1993b. Differential feeding by marine flagellates on growing versus starving, and on motile versus nonmotile, bacterial prey. *Mar. Ecol. Prog. Ser.* 102: 257–267.

Green, J. C. 1991. Phagotrophy in prymnesiophyte flagellates. In D. J. Patterson and J. Larsen, (eds.), *The Biology of Free-Living Heterotrophic Flagellates*. Clarendon Press, pp. 401–414.

Groisillier, A., R. Massana, K. Valentin, D. Vaulot, and L. Guillou. 2006. Genetic diversiy and habitats of two enigmatic marine alveolate lineages. *Aquat. Microb. Ecol.* 42: 277–291.

Grossart, H. P., L. Riemann, and F. Azam. 2001. Bacterial motility in the sea and its ecological implications. *Aquat. Microb. Ecol.* 25: 247–258.

Güde, H. 1989. The role of grazing on bacteria in plankton succession. In U. Sommer (ed.), *Plankton Ecology. Succession in Plankton Communities*. Springer-Verlag, pp. 337–364.

Guillou, L., M. J. Chretiennot-Dinet, S. Boulben, S. Y. Moon-van der Staay, and D. Vaulot. 1999. *Symbiomonas scintillans* gen. et sp. nov. and *Picophagus flagellatus* gen. et sp. nov. (Heterokonta): Two new heterotrophic flagellates of picoplanktonic size. *Protist* 150: 383–398.

Guillou, L., S. Jacquet, M. J. Chretiennot-Dinet, and D. Vaulot. 2001. Grazing impact of two small heterotrophic flagellates on *Prochlorococcus* and *Synechococcus*. *Aquat. Microb. Ecol.* 26: 201–207.

Guillou, L., W. Eikrem, M.-J. Chrétiennot-Dinet, et al. 2004. Diversity of picoplanktonic Prasinophyceae assessed by direct SSU rDNA sequencing of environmental samples and novel isolates retrieved from oceanic and coastal marine ecosystems. *Protist* 155: 193–214.

Guixa-Boixareu, N., D. Vaqué, J. M. Gasol, and C. Pedrós-Alió. 1999. Distribution of viruses and their potential effect on bacterioplankton in an oligotrophic marine system. *Aquat. Microb. Ecol.* 19: 205–213.

Hahn, M. W. 2003. Isolation of strains belonging to the cosmopolitan *Polynucleobacter necessarius* cluster from freshwater habitats located in three climatic zones. *Appl. Environ. Microbiol.* 69: 5248–5254.

Hahn, M. W., and M. G. Höfle. 1999. Flagellate predation on a bacterial model community: Interplay of size-selective grazing, specific bacterial cell size, and bacterial community composition. *Appl. Environ. Microbiol.* 65: 4863–4872.

Hahn, M. W., and M. G. Höfle. 2001. Grazing of protozoa and its effect on populations of aquatic bacteria. *FEMS Microbiol. Ecol.* 35: 113–121.

Hahn, M. W., E. R. B. Moore, and M. G. Höfle. 1999. Bacterial filament formation, a defense mechanism against flagellate grazing, is growth rate controlled in bacteria of different phyla. *Appl. Environ. Microbiol.* 65: 25–35.

Hahn, M. W., E. R. B. Moore, and M. G. Höfle. 2000. Role of microcolony formation in the protistan grazing defense of the aquatic bacterium *Pseudomonas* sp. MWH1. *Microb. Ecol.* 39: 175–185.

Hahn, M. W., H. Lunsdorf, Q. Wu, et al. 2003. Isolation of novel ultramicrobacteria classified as Actinobacteria from five freshwater habitats in Europe and Asia. *Appl. Environ. Microbiol.* 69: 1442–1451.

Hairston, N. G., F. E. Smith, and L. B. Slobodkin. 1960. Community structure, population control, and competition. *Am. Nat.* 94: 421–425.

Hansen, B., P. K. Björnsen, and P. J. Hansen. 1994. The size ratio between planktonic predators and their prey. *Limnol. Oceanogr.* 39: 395–403.

Hausmann, K., P. Selchow, F. Scheckenbach, M. Weitere, and H. Arndt. 2006. Cryptic species in a morphospecies complex of heterotrophic flagellates: The case study of *Caecitellus* spp. *Acta Protozool.* 45: 415–431.

Havskum, H., and A. S. Hansen. 1997. Importance of pigmented and colourless nano-sized protists as grazers on nanoplankton in a phosphate-depleted norwegian fjord and in enclosures. *Aquat. Microb. Ecol.* 12: 139–151.

Hessen, D. O. 1992. Dissolved organic carbon in a humic lake: Effects on bacterial production and respiration. *Hydrobiologia* 229: 115–123.

Heyden, S. V. D., and T. Cavalier-Smith. 2005. Culturing and environmental DNA sequencing uncover hidden kinetoplastid biodiversity and a major marine clade within ancestrally freshwater *Neobodo designis*. *Int. J. Syst. Evol. Microbiol.* 55: 2605–2621.

Heyden, S. V. D., E. E. Chao, and T. Cavalier-Smith. 2004. Genetic diversity of goniomonads: An ancient divergence between marine and freshwater species. *Eur. J. Phycol.* 39: 350.

Hughes Martiny, J. B., B. J. M. Bohannan, J. H. Brown, et al. 2006. Microbial biogeography: Putting microorganisms on the map. *Nat. Rev. Microbiol.* 4: 102–112.

Ishigaki, T., and M. A. Sleigh. 2001. Grazing characteristics and growth efficiencies at two different temperatures for three nanoflagellates fed with *Vibrio* bacteria at three different concentrations. *Microb. Ecol.* 41: 264–271.

Jeong, H. J. 1999. The ecological roles of heterotrophic dinoflagellates in marine planktonic community. *J. Eukaryot. Microbiol.* 46: 390–396.

Jezbera, J., K. Hornak, and K. Šimek. 2005. Food selection by bacterivorous protists: Insight from the analysis of the food vacuole content by means of fluorescence in situ hybridization. *FEMS Microbiol. Ecol.* 52: 351–363.

Jezbera, J., K. Hornak, and K. Šimek. 2006. Prey selectivity of bacterivorous protists in different size fractions of reservoir water amended with nutrients. *Environ. Microbiol.* 8: 1330–1339.

Jones, R. I. 2000. Mixotrophy in planktonic protists: An overview. *Freshwater Biol.* 45: 219–226.

Jonsson, P. R. 1986. Particle size selection, feeding rates and growth dynamics of marine planktonic oligotrichous ciliates (Ciliophora: Oligotrichina). *Mar. Ecol. Prog. Ser.* 33: 265–277.

Jonsson, P., M. Johansson, and R. Pierce. 2004. Attachment to suspended particles may improve foraging and reduce predation risk for tintinnid ciliates. *Limnol. Oceanogr.* 49: 1907–1914.

Jürgens, K., and H. Güde. 1994. The potential importance of grazing-resistant bacteria in planktonic systems. *Mar. Ecol. Prog. Ser.* 112: 169–188.

Jürgens, K., and G. Stolpe. 1995. Seasonal dynamics of crustacean zooplankton, heterotrophic nanoflagellates and bacteria in a shallow, eutrophic lake. *Freshwater Biol.* 33: 27–38.

Jürgens, K., and C. Matz. 2002. Predation as a shaping force for the phenotypic and genotypic composition of planktonic bacteria. *Antonie van Leeuwenhoek* 81: 413–434.

Jürgens, K., J. M. Gasol, and D. Vaqué. 2000. Bacteria—flagellate coupling in microcosm experiments in the Central Atlantic Ocean. *J. Exp. Mar. Biol. Ecol.* 245: 127–147.

Jürgens, K., S. A. Wickham, K. O. Rothhaupt, and B. Santer. 1996. Feeding rates of macro- and microzooplankton on heterotrophic nanoflagellates. *Limnol. Oceanogr.* 41: 1833–1839.

Jürgens, K., J. Pernthaler, S. Schalla, and R. Amann. 1999. Morphological and compositional changes in a planktonic bacterial community in response to enhanced protozoan grazing. *Appl. Environ. Microbiol.* 65: 1241–1250.

Karpov, S. A., D. Bass, A. P. Mylnikov, and T. Cavalier-Smith. 2006. Molecular phylogeny of Cercomonadidae and kinetid patterns of *Cercomonas* and *Eocercomonas* gen. nov. (Cercomonadida, Cercozoa). *Protist* 157: 125–158.

Kerner, M., H. Hohenberg, S. Ertl, M. Reckermann, and A. Spitzy. 2003. Self-organization of dissolved organic matter to micelle-like microparticles in river water. *Nature* 422: 150–154.

King, C. H., E. B. Shotts, R. E. Wooley, and K. G. Porter. 1988. Survival of coliforms and bacterial pathogens within protozoa during chlorination. *Appl. Environ. Microbiol.* 54: 3023–3033.

Kiørboe, T., H.-P. Grossart, H. Ploug, and K. Tang. 2002. Mechanisms and rates of bacterial colonization of sinking aggregates. *Appl. Environ. Microbiol.* 68: 3996–4006.

Kiørboe, T., K. Tang, H.-P. Grossart, and H. Ploug. 2003. Dynamics of microbial communities on marine snow aggregates: Colonization, growth, detachment, and grazing mortality of attached bacteria. *Appl. Environ. Microbiol.* 69: 3036–3047.

Kiørboe, T., H. P. Grossart, H. Ploug, K. Tang, and B. Auer. 2004. Particle-associated flagellates: swimming patterns, colonization rates, and grazing on attached bacteria. *Aquat. Microb. Ecol.* 35: 141–152.

REFERENCES

Kirchman, D. 2000. Uptake and regeneration of inorganic nutrients by marine heterotrophic bacteria. In D. Kirchman (ed.), *Microbial Ecology of the Oceans*, 1st edn. Wiley-Liss, pp. 261–288.

Kjelleberg, S., N. Albertson, K. Flärdh, et al. 1993. How do non-differentiating bacteria adapt to starvation? *Antonie van Leeuwenhoek* 63: 333–341.

Kolodziej, K., and T. Stoeck. 2007. Cellular identification of a novel uncultured marine stramenopile (MAST-12 clade) small-subunit rRNA gene sequence from a Norwegian estuary by use of fluorescence in situ hybridization-scanning electron microscopy. *Appl. Environ. Microbiol.* 73: 2718–2726.

Kujawinski, E. B., J. W. Farrington, and J. W. Moffett. 2002. Evidence for grazing-mediated production of dissolved surface-active material by marine protists. *Mar. Chem.* 77: 133–142.

Landry, M. R. 1994. Methods and controls for measuring the grazing impact of planktonic protists. *Mar. Microb. Food Webs* 8: 37–57.

Landry, M. R., and R. P. Hassett. 1982. Estimating the grazing impact of marine microzooplankton. *Mar. Biol.* 67: 283–288.

Landry, M. R., J. M. Lehner-Fournier, J. A. Sundstrom, V. L. Fagerness, and K. E. Selph. 1991. Discrimination between living and heat-killed prey by a marine zooflagellate *Paraphysomonas vestita* Stokes. *J. Exp. Mar. Biol. Ecol.* 146: 139–152.

Langenheder, S., and K. Jürgens. 2001. Regulation of bacterial biomass and community structure by metazoan and protozoan predation. *Limnol. Oceanogr.* 46: 121–134.

Larsen, J., and A. Sournia. 1991. The diversity of heterotrophic dinoflagellates. In D. J. Patterson and J. Larsen (eds.), *The Biology of Free-Living Heterotrophic Flagellates*. Clarendon Press, pp. 313–332.

Lavrentyev, P. J., W. S. Gardner, and J. R. Johnson. 1997. Cascading trophic effects on aquatic nitrification: experimental evidence and potential implications. *Aquat. Microb. Ecol.* 13: 161–175.

Laybourn-Parry, J., and J. Parry. 2000. Flagellates and the microbial loop. In B. S. C. Leadbeater and J. C. Green, (eds.), *The Flagellates: Unity, Diversity and Evolution*. Taylor & Francis, pp. 216–239.

Leakey, R. J. G., B. S. C. Leadbeater, E. Mitchell, S. M. M. McCready, and A. W. A. Murray. 2002. The abundance and biomass of choanoflagellates and other nanoflagellates in waters of contrasting temperature to the north-west of South Georgia in the Southern Ocean. *Eur. J. Protistol.* 38: 333–350.

Legendre, L., and J. Lefevre. 1995. Microbial food webs and the export of biogenic carbon in oceans. *Aquat. Microb. Ecol.* 9: 69–77.

Lighthill, J. 1976. Flagellar hydrodynamics. *SIAM Rev.* 18: 161–230.

Lim, E. L., L. A. Amaral, D. A. Caron, and E. F. Delong. 1993. Application of rRNA-based probes for observing marine nanoplanktonic protists. *Appl. Environ. Microbiol.* 59: 1647–1655.

Lim, E. L., M. R. Dennett, and D. A. Caron. 1999. The ecology of *Paraphysomonas imperforata* based on studies employing oligonucleotide probe identification in coastal water samples and enrichment cultures. *Limnol. Oceanogr.* 44: 37–51.

López-García, P., F. Rodríguez-Valera, C. Pedrós-Alió, and D. Moreira. 2001. Unexpected diversity of small eukaryotes in deep-sea Antarctic plankton. *Nature* 409: 603–607.

McManus, G. B., and J. A. Fuhrman. 1988. Control of marine bacterioplankton populations: Measurement and significance of grazing. *Hydrobiologia* 159: 51–62.

McManus, G., and A. Okubo. 1991. On the use of surrogate food particles to measure protistan ingestion. *Limnol. Oceanogr.* 36: 613–617.

Maranger, R., D. F. Bird, and N. M. Price. 1998. Iron acquisition by photosynthetic marine phytoplankton from ingested bacteria. *Nature* 396: 248–251.

Margulis, L. 1981. *Symbiosis in Cell Evolution: Life and its Environments on the Early Earth.* W. H. Freeman.

Marrasé, C., E. L. Lim, and D. A. Caron. 1992. Seasonal and daily changes in bacterivory in a coastal plankton community. *Mar. Ecol. Prog. Ser.* 82: 281–289.

Massana, R., and K. Jürgens. 2003. Composition and population dynamics of planktonic bacteria and bacterivorous flagellates in seawater chemostat cultures. *Aquat. Microb. Ecol.* 32: 11–22.

Massana, R., L. Guillou, B. Díez, and C. Pedrós-Alió. 2002. Unveiling the organisms behind novel eukaryotic ribosomal DNA sequences from the ocean. *Appl. Environ. Microbiol.* 68: 4554–4558.

Massana, R., J. Castresana, V. Balague, et al. 2004. Phylogenetic and ecological analysis of novel marine stramenopiles. *Appl. Environ. Microbiol.* 70: 3528–3534.

Massana, R., L. Guillou, R. Terrado, I. Forn, and C. Pedrós-Alió. 2006a. Growth of uncultured heterotrophic flagellates in unamended seawater incubations. *Aquat. Microb. Ecol.* 45: 171–180.

Massana, R., R. Terrado, I. Forn, C. Lovejoy, and C. Pedrós-Alió. 2006b. Distribution and abundance of uncultured heterotrophic flagellates in the world oceans. *Environ. Microbiol.* 8: 1515–1522.

Massana, R., J. del Campo, C. Dinter, and R. Sommaruga. 2007. Crash of a population of the marine heterotrophic flagellate *Cafeteria roenbergensis* by viral infection. *Environ. Microbiol.* 9: 2660–2669.

Matz, C., and K. Jürgens. 2001. Effects of hydrophobic and electrostatic cell surface properties of bacteria on feeding rates of heterotrophic nanoflagellates. *Appl. Environ. Microbiol.* 67: 814–820.

Matz, C., and K. Jürgens. 2003. Interaction of nutrient limitation and protozoan grazing determines the phenotypic structure of a bacterial community. *Microb. Ecol.* 45: 384–398.

Matz, C., and K. Jürgens. 2005. High motility reduces grazing mortality of planktonic bacteria. *Appl. Environ. Microbiol.* 71: 921–929.

Matz, C., J. Boenigk, H. Arndt, and K. Jürgens. 2002a. Role of bacterial phenotypic traits in selective feeding of the heterotrophic nanoflagellate *Spumella* sp. *Aquat. Microb. Ecol.* 27: 137–148.

Matz, C., P. Deines, and K. Jürgens. 2002b. Phenotypic variation in *Pseudomonas* sp. CM10 determines microcolony formation and survival under protozoan grazing. *FEMS Microbiol. Ecol.* 39: 57–65.

Matz, C., T. Bergfeld, S. A. Rice, and S. Kjelleberg. 2004a. Microcolonies, quorum sensing and cytotoxicity determine the survival of *Pseudomonas aeruginosa* biofilms exposed to protozoan grazing. *Environ. Microbiol.* 6: 218–226.

Matz, C., P. Deines, J. Boenigk, H. Arndt, L. Eberl, S. Kjelleberg, and K. Jürgens. 2004b. Impact of violacein-producing bacteria on survival and feeding of bacterivorous nanoflagellates. *Appl. Environ. Microbiol.* 70: 1593–1599.

Matz, C., D. McDougald, A. M. Moreno, P. Y. Yung, F. H. Yildiz, and S. Kjelleberg. 2005. Biofilm formation and phenotypic variation enhance predation-driven persistence of *Vibrio cholerae*. *Proc. Natl. Acad. Sci. USA* 102: 16819–16824.

Medlin, L., H. J. Elwood, S. Stickel, and M. L. Sogin. 1988. The characterization of enzymatically amplified eukaryotic 16S-like rRNA-coding regions. *Gene* 71: 491–499.

Mitchell, E. A. D., and R. Meisterfeld. 2005. Taxonomic confusion blurs the debate on cosmopolitanism versus local endemism of free-living protists. *Protist* 156: 263–267.

Mohapatra, B. R., and K. Fukami. 2004a. Comparison of the numerical grazing response of two marine heterotrophic nanoflagellates fed with different bacteria. *J. Sea Res.* 52: 99–107.

Mohapatra, B. R., and K. Fukami. 2004b. Production of aminopeptidase by marine heterotrophic nanoflagellates. *Aquat. Microb. Ecol.* 34: 129–137.

Monger, B. C., and M. R. Landry. 1990. Direct-interception feeding by marine zooflagellates: the importance of surface and hydrodynamic forces. *Mar. Ecol. Prog. Ser.* 65: 123–140.

Monger, B. C., M. R. Landry, and S. L. Brown. 1999. Feeding selection of heterotrophic marine nanoflagellates based on the surface hydrophobicity of their picoplankton prey. *Limnol. Oceanogr.* 44: 1917–1927.

Moon-van der Staay, S. Y., R. De Wachter, and D. Vaulot. 2001. Oceanic 18S rDNA sequences from picoplankton reveal unsuspected eukaryotic diversity. *Nature* 409: 607–610.

Morris, R. M., M. S. Rappé, S. A. Connon, et al. 2002. SAR11 clade dominates ocean surface bacterioplankton communities. *Nature* 420: 806–810.

Nagata, T. 2000. Production mechanisms of dissolved organic matter. In D. Kirchman (ed.), *Microbial Ecology of the Oceans*, 1st edn. Wiley-Liss, pp. 121–152.

Nagata, T., and D. L. Kirchman. 1991. Release of dissolved free and combined amino acids by bacterivorous marine flagellates. *Limnol. Oceanogr.* 36: 433–443.

Nagata, T., and D. L. Kirchman. 1992. Release of macromolecular organic complexes by heterotrophic marine flagellates. *Mar. Ecol. Prog. Ser.* 83: 233–240.

Nikolaev, S. I., C. Berney, J. F. Fahrni, et al. 2004. The twilight of Heliozoa and rise of Rhizaria, an emerging supergroup of amoeboid eukaryotes. *Proc. Natl. Acad. Sci. USA* 101: 8066–8071.

Nygaard, K., and D. O. Hessen. 1990. Use of ^{14}C-protein-labelled bacteria for estimating clearance rates by heterotrophic and mixotrophic flagellates. *Mar. Ecol. Prog. Ser.* 68: 7–14.

Nygaard, K., and A. Tobiesen. 1993. Bacterivory in algae: A survival strategy during nutrient limitation. *Limnol. Oceanogr.* 38: 273–279.

Nyström, T., R. M. Olsson, and S. Kjelleberg. 1992. Survival, stress resistance, and alterations in protein expression in the marine *Vibrio* sp. strain S14 during starvation for different individual nutrients. *Appl. Environ. Microbiol.* 58: 55–65.

Okamoto, N., and I. Inouye. 2005. The Katablepharids are a distant sister group of the Cryptophyta: A proposal for Katablepharidophyta divisio nova/Katablepharida phylum novum based on SSU rDNA and beta-tubulin phylogeny. *Protist* 156: 163–179.

Patterson, D. J. 1989. Stramenopiles: Chromphytes from a protistan perspective. In J. C. Green, B. S. C. Leadbeater and W. L. Diver (eds.), *Chromphyte Algae: Problems and Perspectives*. Clarendon Press, pp. 357–379.

Patterson, D. J., and J. Larsen (eds.) 1991. *The Biology of Free-Living Heterotrophic Flagellates*. Clarendon Press.

Patterson, D. J., and W. J. Lee. 2000. Geographic distribution; diversity of free-living heterotrophic flagellates. In B. S. C Leadbeater and J. C. Green (eds.), *The Flagellates: Unity, Diversity and Evolution.* Taylor & Francis, pp. 269–287.

Patterson, D. J., and M. Zölffel. 1991. Heterotrophic flagellates of uncertain taxonomic position. In D. J. Patterson and J. Larsen (eds.), *The Biology of Free-Living Heterotrophic Flagellates.* Clarendon Press, pp. 427–275.

Patterson, D. J., K. Nygaard, G. Steinberg, and C. M. Turley. 1993. Heterotrophic flagellates and other protists associated with oceanic detritus throughout the water column in the mid North Atlantic. *J. Mar. Biol. Assoc. UK* 73: 67–95.

Pedrós-Alió, C., J. I. Calderón-Paz, and J. M. Gasol. 2000. Comparative analysis shows that bacterivory, not viral lysis, controls the abundance of heterotrophic prokaryotic plankton. *FEMS Microbiol. Ecol.* 32: 157–165.

Pelegri, S. P., U. Christaki, J. Dolan, and F. Rassoulzadegan. 1999. Particulate and dissolved organic carbon production by the heterotrophic nanoflagellate *Pteridomonas danica* (Patterson and Fenchel). *Microb. Ecol.* 37: 276–284.

Pernthaler, J. 2005. Predation on prokaryotes in the water column and its ecological implications. *Nat. Rev. Microbiol.* 3: 537–546.

Pernthaler, J., T. Posch, K. Šimek, J. Vrba, R. Amann, and R. Psenner. 1997. Contrasting bacterial strategies to coexist with a flagellate predator in an experimental microbial assemblage. *Appl. Environ. Microbiol.* 63: 596–601.

Pernthaler, J., E. Zöllner, F. Warnecke, and K. Jürgens. 2004. Bloom of filamentous bacteria in a mesotrophic lake: Identity and potential controlling mechanism. *Appl. Environ. Microbiol.* 70: 6272–6281.

Peters, F., and T. Gross. 1994. Increased grazing rates of microplankton in response to small-scale turbulence. *Mar. Ecol. Prog. Ser.* 115: 299–307.

Pfandl, K., and J. Boenigk. 2006. Stuck in the mud: Suspended sediments as a key issue for survival of chrysomonad flagellates. *Aquat. Microb. Ecol.* 45: 89–99.

Pierce, R. W., and J. T. Turner. 1992. Ecology of planktonic ciliates in marine food webs. *Rev. Aquat. Sci.* 6: 139–181.

Polet, S., C. Berney, J. Fahrni, and J. Pawlowski. 2004. Small-subunit ribosomal RNA gene sequences of Phaeodarea challenge the monophyly of Haeckel's radiolaria. *Protist* 155: 53–63.

Posch, T., K. Šimek, J. Vrba, J. Pernthaler, J. Nedoma, B. Sattler, B. Sonntag, and R. Psenner. 1999. Predator-induced changes of bacterial size-structure and productivity studied on an experimental microbial community. *Aquat. Microb. Ecol.* 18: 235–246.

Posch, T., J. Jezbera, J. Vrba, K. Šimek, et al. 2001. Size selective feeding in *Cyclidium glaucoma* (Ciliophora, Scuticociliatida) and its effects on bacterial community structure: A study from a continuous cultivation system. *Microb. Ecol.* 42: 217–227.

Postius, C., and A. Ernst. 1999. Mechanisms of dominance: Coexistence of picocyanobacterial genotypes in a freshwater ecosystem. *Arch. Microbiol.* 172: 69–75.

Preisig, H. R., N. Vørs, and G. Hallfors. 1991. Diversity of heterotrophic heterokont flagellates. In D. J. Patterson and J. Larsen (eds.), *The Biology of Free-Living Heterotrophic Flagellates.* Clarendon Press, pp. 361–399.

Psenner, R., and R. Sommaruga. 1992. Are rapid changes in bacterial biomass caused by shifts from top-down to bottom-up control? *Limnol. Oceanogr.* 37: 1092–1100.

Rappé, M. S., S. A. Connon, K. L. Vergin, and S. J. Giovannoni. 2002. Cultivation of the ubiquitous SAR11 marine bacterioplankton clade. *Nature* 418: 630–633.

Raven, J. A. 1997. Phagotrophy in phototrophs. *Limnol. Oceanogr.* 42: 198–205.

Ribblett, S. G., M. A. Palmer, and D. W. Coats. 2005. The importance of bacterivorous protists in the decomposition of stream leaf litter. *Freshwater Biol.* 50: 516–526.

Rice, J., C. D. O'Connor, M. A. Sleigh, P. H. Burkill, I. G. Giles, and M. V. Zubkov. 1997. Fluorescent oligonucleotide rDNA probes that specifically bind to a common nanoflagellate, *Paraphysomonas vestita*. *Microbiology* 143: 1717–1727.

Richards, T. A., A. A. Vepritskiy, D. E. Gouliamova, and S. A. Nierzwicki-Bauer. 2005. The molecular diversity of freshwater picoeukaryotes from an oligotrophic lake reveals diverse, distinctive and globally dispersed lineages. *Environ. Microbiol.* 7: 1413–1425.

Rodríguez, F., E. Derelle, L. Guillou, F. L. Gall, D. Vaulot, and H. Moreau. 2005. Ecotype diversity in the marine picoeukaryote *Ostreococcus* (Chlorophyta, Prasinophyceae). *Environ. Microbiol.* 7: 853–859.

Rogerson, A., O. R. Anderson, and C. Vogel. 2003. Are planktonic naked amoebae predominately floc associated or free in the water column? *J. Plankton Res.* 25: 1359–1365.

Sakaguchi, M., H. Murakami, and T. Suzaki. 2001. Involvement of a 40-kDA glycoprotein in food recognition, prey capture, and induction of phagocytosis in the protozoon *Actinophrys sol*. *Protist* 152: 33–41.

Salcher, M. M., J. Pernthaler, R. Psenner, and T. Posch. 2005. Succession of bacterial grazing defense mechanisms against protistan predators in an experimental microbial community. *Aquat. Microb. Ecol.* 38: 215–229.

Sanders, R. W. 1991. Mixotrophic protists in marine and freshwater ecosystems. *J. Protozool.* 38: 76–81.

Sanders, R. W., K. G. Porter, S. J. Bennett, and A. E. Debiase. 1989. Seasonal patterns of bacterivory by flagellates, ciliates, rotifers, and cladocerans in a freshwater planktonic community. *Limnol. Oceanogr.* 34: 673–687.

Sanders, R. W., D. A. Caron, and U. G. Berninger. 1992. Relationships between bacteria and heterotrophic nanoplankton in marine and fresh waters—an inter-ecosystem comparison. *Mar. Ecol. Prog. Ser.* 86: 1–14.

Sanders, R. W., U. G. Berninger, E. L. Lim, P. F. Kemp, and D. A. Caron. 2000. Heterotrophic and mixotrophic nanoplankton predation on picoplankton in the Sargasso Sea and on Georges Bank. *Mar. Ecol. Prog. Ser.* 192: 103–118.

Schalchian-Tabrizi, K., W. Eikrem, D. Klaveness. 2006. Telonemia, a new protist phylum with affinity to chromist lineages. *Proc. Roy. Soc. Lond., Sec. B, Biol. Sci.* 273: 1833–1842.

Scheckenbach, F., C. Wylezich, M. Weitere, K. Hausmann, and H. Arndt. 2005. Molecular identity of strains of heterotrophic flagellates isolated from surface waters and deep-sea sediments of the South Atlantic based on SSU rDNA. *Aquat. Microb. Ecol.* 38: 239–247.

Scherwass, A., Y. Fischer, and H. Arndt. 2005. Detritus as a potential food source for protozoans: utilization of fine particulate plant detritus by a heterotrophic flagellate, *Chilomonas paramecium*, and a ciliate, *Tetrahymena pyriformis*. *Aquat. Ecol.* 39: 439–445.

Schlegel, M. 1994. Molecular phylogeny of eukaryotes. *Trends Ecol. Evol.* 9: 330–335.

Selje, N., M. Simon, and T. Brinkhoff. 2004. A newly discovered *Roseobacter* cluster in temperate and polar oceans. *Nature* 427: 445–448.

Selph, K. E., M. R. Landry, and E. A. Laws. 2003. Heterotrophic nanoflagellate enhancement of bacterial growth through nutrient remineralization in chemostat culture. *Aquat. Microb. Ecol.* 32: 23–37.

Sherr, E. B. 1988. Direct use of high molecular weight polysaccharide by heterotrophic flagellates. *Nature* 335: 348–351.

Sherr, E. B., and B. F. Sherr. 1987. High rates of consumption of bacteria by pelagic ciliates. *Nature* 325: 710–711.

Sherr, B. F., and E. B. Sherr. 1991. Proportional distribution of total numbers, biovolume and bacterivory among size classes of 2–20 µm nonpigmented marine flagellates. *Mar. Microb. Food Webs* 5: 227–237.

Sherr, E. B., and B. F. Sherr. 1994. Bacterivory and herbivory: Key roles of phagotrophic protists in pelagic food webs. *Microb. Ecol.* 28: 223–235.

Sherr, E. B., and B. F. Sherr. 1999. β-Glucosaminidase activity in marine microbes. *FEMS Microbiol. Ecol.* 28: 111–119.

Sherr, E., and B. Sherr. 2000. Marine microbes. An overview. In D. L. Kirchman (ed.), *Microbial Ecology of the Oceans*, 1st edn. Wiley-Liss, pp. 13–46.

Sherr, E. B., and B. F. Sherr. 2002. Significance of predation by protists in aquatic microbial food webs. *Antonie van Leeuwenhoek* 81: 293–308.

Sherr, B. F., E. B. Sherr, and T. Berman. 1982. Decomposition of organic detritus: A selective role for microflagellate protozoa. *Limnol. Oceanogr.* 27: 765–769.

Sherr, B. F., E. B. Sherr, and R. D. Fallon. 1987. Use of monodispersed, fluorescently labeled bacteria to estimate in situ protozoan bacterivory. *Appl. Environ. Microbiol.* 53: 958–965.

Sherr, B. F., E. B. Sherr, and C. Pedrós-Alió. 1989. Simultaneous measurement of bacterioplankton production and protozoan bacterivory in estuarine water. *Mar. Ecol. Prog. Ser.* 54: 209–219.

Sherr, B. F., E. B. Sherr, and J. McDaniel. 1992. Effect of protistan grazing on the frequency of dividing cells in bacterioplankton assemblages. *Appl. Environ. Microbiol.* 58: 2381–2385.

Sibbald, M. J., and L. J. Albright. 1988. Aggregated and free bacteria as food sources for heterotrophic microflagellates. *Appl. Environ. Microbiol.* 54: 613–616.

Sieburth, J. M. 1984. Protozoan bacterivory in pelagic marine waters. In J. E. Hobbie and P. J. l. Williams (eds.), *Heterotrophic Activity in the Sea.* Plenum Press, pp. 405–444.

Sieburth, J. M., V. Smetacek, and J. Lenz. 1978. Pelagic ecosystem structure: Heterotrophic compartments of the plankton and their relationship to plankton size fractions. *Limnol. Oceanogr.* 23: 1256–1263.

Šimek, K., P. Kojecka, J. Nedoma, P. Hartman, J. Vrba, and J. R. Dolan. 1999. Shifts in bacterial community composition associated with different microzooplankton size fractions in a eutrophic reservoir. *Limnol. Oceanogr.* 44: 1634–1644.

Šimek, K., J. Nedoma, J. Pernthaler, T. Posch, and J. R. Dolan. 2002. Altering the balance between bacterial production and protistan bacterivory triggers shifts in freshwater bacterial community composition. *Antonie van Leeuwenhoek* 81: 453–463.

Simon, M., B. C. Cho, and F. Azam. 1992. Significance of bacterial biomass in lakes and the ocean: Comparison to phytoplankton biomass and biogeochemical implications. *Mar. Ecol. Prog. Ser.* 86: 103–110.

Staley, J. T., and A. Konopka. 1985. Measurement of in situ activities of nonphotosynthetic microorganisms in aquatic and terrestrial habitats. *Annu. Rev. Microbiol.* 39: 321–346.

Stoderegger, K. E., and G. J. Herndl. 2002. Distribution of capsulated bacterioplankton in the North Atlantic and North Sea. *Microb. Ecol.* 44: 154–163.

Stoderegger, K. E., and G. J. Herndl. 2004. Dynamics in bacterial cell surface properties assessed by fluorescent stains and confocal laser scanning microscopy. *Aquat. Microb. Ecol.* 36: 29–40.

Stoderegger, K. E., and G. J. Herndl. 2005. Dynamics in bacterial surface properties of a natural bacterial community in the coastal North Sea during a spring phytoplankton bloom. *FEMS Microbiol. Ecol.* 53: 285–294.

Stoeck, T., and S. Epstein. 2003. Novel eukaryotic lineages inferred from small-subunit rRNA analyses of oxygen-depleted marine environments. *Appl. Environ. Microbiol.* 69: 2657–2663.

Stoeck, T., W. H. Fowle, and S. S. Epstein. 2003. Methodology of protistan discovery: From rRNA detection to quality scanning electron microscope images. *Appl. Environ. Microbiol.* 69: 6856–6863.

Stoeck, T., M. V. J. Schwarz, J. Boenigk, M. Schweikert, S. V. D. Heyden, and A. Behnke. 2005. Cellular identity of an 18S rRNA gene sequence clade within the class Kinetoplastea: the novel genus Actuariola gen. nov. (*Neobodonida*) with description of the type species *Actuariola framvarensis* sp. nov. *Int. J. Syst. Evol. Microbiol.* 55: 2623–2635.

Stoeck, T., A. Zuendorf, H.-W. Breiner, and A. Behnke. 2007. A molecular approach to identify active microbes in environmental eukaryote clone libraries. *Microb. Ecol.* 53: 328–339.

Stoecker, D. K., A. E. Michaels, and L. H. Davis. 1987. Large proportion of marine planktonic ciliates found to contain functional chloroplasts. *Nature* 326: 790–792.

Straile, D. 1997. Gross growth efficiencies of protozoan and metazoan zooplankton and their dependence on food concentration, predator-prey weight ratio, and taxonomic group. *Limnol. Oceanogr.* 42: 1375–1385.

Strom, S. L. 2000. Bacterivory: Interactions between bacteria and their grazers. In D. L. Kirchman (ed.), *Microbial Ecology of the Oceans*, 1st edn. Wiley-Liss, pp. 351–386.

Suzuki, M. T. 1999. Effect of protistan bacterivory on coastal bacterioplankton diversity. *Aquat. Microb. Ecol.* 20: 261–272.

Taylor, G. T., and M. L. Pace. 1987. Validity of eukaryote inhibitors for assessing production and grazing mortality of marine bacterioplankton. *Appl. Environ. Microbiol.* 53: 119–128.

Thingstad, T. 2000. Control of bacterial growth in idealized food webs. In D. L. Kirchman (ed.), *Microbial Ecology of the Oceans*, 1st edn. Wiley-Liss, pp. 229–260.

Thomsen, H. A., and K. R. Buck. 1991. Choanoflagellate diversity with particular emphasis on the Acanthoecidae. In: D. J. Patterson and J. Larsen (eds.), *The Biology of Free-Living Heterotrophic Flagellates*. Clarendon Press, pp. 259–284.

Tobiesen, A. 1990. Temperature dependent filtration rates and size selection in some heterotrophic microflagellates and one dinoflagellate. *Ergeb. Limnol.* 34: 293–304.

Tremaine, S. C., and A. L. Mills. 1987. Inadequacy of the eucaryote inhibitor cycloheximide in studies of protozoan grazing on bacteria at the freshwater-sediment interface. *Appl. Environ. Microbiol.* 53: 1969–1972.

Unrein, F., R. Massana, L. Alonso-Sáez, and J. M. Gasol. 2007. Significant year-round effect of small mixotrophic flagellates on bacterioplankton in an oligotrophic coastal system. *Limnol. Oceanogr.* 52: 456–469.

Vaqué, D., J. M. Gasol, and C. Marrasé. 1994. Grazing rates on bacteria—the significance of methodology and ecological factors. *Mar. Ecol. Prog. Ser.* 109: 263–274.

Vargas, C. D., R. Norris, L. Zaninetti, S. W. Gibb, and J. Pawlowski. 1999. Molecular evidence of cryptic speciation in planktonic foraminifers and their relation to oceanic provinces. *Proc. Natl. Acad. Sci. USA* 96: 2864–2868.

Vaulot, D., K. Romari, and F. Not. 2002. Are autotrophs less diverse than heterotrophs in marine picoplankton? *Trends Microbiol.* 10: 266–267.

Vázquez-Domínguez, E., F. Peters, J. M. Gasol, and D. Vaqué. 1999. Measuring the grazing losses of picoplankton: Methodological improvements in the use of fluorescently labeled tracers combined with flow cytometry. *Aquat. Microb. Ecol.* 20: 119–128.

Vázquez-Domínguez, E., E. O. Casamayor, P. Català, and P. Lebaron. 2005. Different marine heterotrophic nanoflagellates affect differentially the composition of enriched bacterial communities. *Microb. Ecol.* 49: 474–485.

Verity, P. G., and V. Smetacek. 1996. Organism life cycles, predation, and the structure of marine pelagic ecosystems. *Mar. Ecol. Prog. Ser.* 130: 277–293.

Vørs, N., K. R. Buck, F. P. Chavez, et al. 1995. Nanoplankton of the equatorial Pacific with emphasis on the heterotrophic protists. *Deep-Sea Res.* II 42: 585–602.

Vrba, J., K. Šimek, J. Nedoma, and P. Hartman. 1993. 4-Methylumbelliferyl-β-N-acetylglucosaminide hydrolysis by a high-affinity enzyme: A putative marker of protozoan bacterivory. *Appl. Environ. Microbiol.* 59: 3091–3101.

Weinbauer, M. G., and M. G. Höfle. 1998. Significance of viral lysis and flagellate grazing as factors controlling bacterioplankton production in a eutrophic lake. *Appl. Environ. Microbiol.* 64: 431–438.

Whitaker, R. J., D. W. Grogan, and J. W. Taylor. 2003. Geographic barriers isolate endemic populations of hyperthermophilic archaea. *Science* 301: 976–978.

Wikner, J., and Å. Hagström. 1988. Evidence for a tightly coupled nanoplanktonic predator-prey link regulating the bacterivores in the marine environment. *Mar. Ecol. Prog. Ser.* 50: 137–145.

Woese, C. R. 1987. Bacterial evolution. *Microbiol. Rev.* 51: 221–271.

Wootton, E. C., M. V. Zubkov, et al. 2007. Biochemical prey recognition by planktonic protozoa. *Environ. Microbiol.* 9: 216–222.

Worden, A. Z. 2006. Picoeukaryote diversity in coastal waters of the Pacific Ocean. *Aquat. Microb. Ecol.* 43: 165–175.

Worden, A. Z., and B. J. Binder. 2003. Application of dilution experiments for measuring growth and mortality rates among *Prochlorococcus* and *Synechococcus* populations in oligotrophic environments. *Aquat. Microb. Ecol.* 30: 159–174.

Wright, R. T., and R. B. Coffin. 1984. Measuring microzooplankton grazing on planktonic marine bacteria by its impact on bacterial production. *Microb. Ecol.* 10: 137–149.

Wu, Q. L., J. Boenigk, and M. W. Hahn. 2004. Successful predation of filamentous bacteria by a nanoflagellate challenges current models of flagellate bacterivory. *Appl. Environ. Microbiol.* 70: 332–339.

Yokokawa, T., and T. Nagata. 2005. Growth and grazing mortality rates of phylogenetic groups of bacterioplankton in coastal marine environments. *Appl. Environ. Microbiol.* 71: 6799–6807.

Zhu, F., R. Massana, F. Not, D. Marie, and D. Vaulot. 2005. Mapping of picoeucaryotes in marine ecosystems with quantitative PCR of the 18S rRNA gene. *FEMS Microbiol. Ecol.* 52: 79–92.

Zhukov, B. F. 1991. The diversity of bodonids. In D. J. Patterson and J. Larsen (eds.), *The Biology of Free-Living Heterotrophic Flagellates*. Clarendon Press, pp. 177–185.

Zubkov, M. V., and M. A. Sleigh. 1995. Ingestion and assimilation by marine protists fed on bacteria labeled with radioactive thymidine and leucine estimated without separating predator and prey. *Microb. Ecol.* 30: 157–170.

Zubkov, M. V., and M. A. Sleigh. 1999. Growth of amoebae and flagellates on bacteria deposited on filters. *Microb. Ecol.* 37: 107–115.

MARINE VIRUSES: COMMUNITY DYNAMICS, DIVERSITY AND IMPACT ON MICROBIAL PROCESSES

MYA BREITBART
College of Marine Science, University of South Florida, St. Petersburg, FL 33701, U.S.A.

MATHIAS MIDDELBOE
Marine Biological Laboratory, University of Copenhagen, DK-3000 Helsingør, Denmark

FOREST ROHWER
Department of Biology, San Diego State University, San Diego, CA 92182, U.S.A.

INTRODUCTION

Viruses are ubiquitous, abundant, and dynamic components of environmental communities and are found in all the world's ecosystems from deep-sea sediments to the atmosphere. There are approximately 10^7 viruses per milliliter of seawater and 10^9 viruses per gram of soil or sediment (Wommack and Colwell 2000). Most environmental viruses infect bacteria and are also called phages. Extrapolation from the predicted number of bacteria on the planet (10^{30}: Whitman et al. 1998) and a rough ratio of ten viruses per bacterial cell (Wommack and Colwell 2000; Weinbauer 2004) suggests that there are at least 10^{31} viruses on the planet. This

---

Microbial Ecology of the Oceans, Second Edition. Edited by David L. Kirchman
Copyright © 2008 John Wiley & Sons, Inc.

makes viruses the most common biological entities on Earth. The world's oceans and underlying sediments account for about half of the total viruses in the world.

Despite their small size (of the order of 100 nm; 10 attograms (ag) = 10^{-17} g each), viruses constitute a large biomass on a global scale. Global marine viral biomass is estimated to be approximately 200 Mt (2×10^{11} kg) of carbon (Hambly and Suttle 2005), which makes viruses, in particular phages, the ocean's second largest biomass, exceeded only by the total biomass of bacteria. By infecting and subsequently killing a large fraction of marine bacteria, viruses have a significant influence on the fate of marine primary production and secondary bacterial production. Virally driven bacterial mortality also influences bacterial community dynamics, diversity, and the cycling of carbon and nutrients. The past 15 years of research have shown marine viruses to be important players in oceanic processes and have fundamentally changed our conceptual understanding of marine food webs and biogeochemical cycling. Recent evidence also suggests that marine viral communities comprise the largest reservoir of genetic diversity in the ocean, a diversity that we are only just beginning to comprehend (Rohwer 2003).

Thus, even though they act at the smallest scale of biological activity, viruses have significant influences on biological processes at all levels in the marine environment, from microscale community dynamics to global nutrient cycling. This chapter summarizes the present knowledge on the roles of viruses in the marine environment, the diversity and dynamics of marine viral communities, and some of the methods commonly used to study marine viruses. It focuses mainly on phages, which dominate marine viral communities. For a review of viruses infecting marine eukaryotes, see Munn (2006).

VIRUSES AND THE MARINE MICROBIAL FOOD WEB

Direct Counts and Viral Numbers

Initially, studies of environmental microbiology required culturing the bacteria. Early culture-based experiments suggested that most environments contained relatively few bacteria, and it was believed that bacteria played minor roles in global biogeochemical cycles. The introduction of direct microscopic counts (Francisco et al. 1973; Zimmermann and Meyer-Reil 1974; Daley and Hobbie 1975; Ferguson and Rublee 1976; Hobbie et al. 1977) changed that view by showing that there were up to a million bacteria per milliliter of seawater, and that culturing had underestimated bacterial abundances by 100–1000-fold. Direct counts and measurements of bacterial activity using radiolabeled compounds (Wright and Hobbie 1966; Williams and Askew 1968; Azam and Hodson 1977; Riemann 1978; Fuhrman and Azam 1982; Karl and Bossard 1985) revealed the importance of the marine microbial food web, which consists of the autotrophic and heterotrophic microbes, as well as their predators, that are found throughout the world's oceans (Fig. 12.1). The microbial food web plays a major role in global carbon and nutrient cycling, and regulates the transfer of energy and nutrients to higher trophic levels (Pomeroy 1974; Azam et al. 1983; Kirchman 1994). In the ocean, at least 50 percent of the

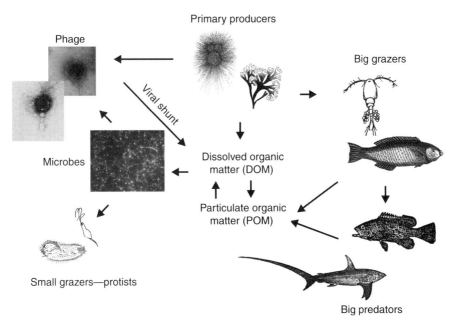

Figure 12.1 The viral shunt and the marine food web. Energy, in the form of fixed carbon, is provided to the marine environment via photosynthesis by the primary producers (cyanobacteria, diatoms, dinoflagellates, kelps, seagrasses, etc.). The fixed carbon, or photosynthate, supports new biomass and respiration of the primary producers. In turn, the primary producers are consumed by grazers (copepods, fish, etc.), who are eaten by bigger predators (sharks, barracuda, etc.). A significant amount of photosynthate is also released as dissolved organic matter (DOM), which supports heterotrophic microbial growth (both bacteria and archaea). While the exact composition of DOM remains one of the major outstanding questions of microbial ecology, it is known to contain both energy (in the form of fixed carbon) and nutrients (such as phosphorus and nitrogen). The viruses and protists kill similar proportions of the microbes, and the lysed cells then join the DOM pool, which feeds more heterotrophic microbes. The viral shunt, where viruses lyse bacteria, creating more DOM as food for bacteria, increases CO_2 respiration of the ecosystem, because every time the organic matter is converted from one pool to another, about 90 percent of the fixed carbon is released as CO_2 (as a very rough estimate, based on basic ecological phenomena). This rapid turnover of CO_2 is one hurdle that needs to be overcome if we want to sequester more CO_2 in the ocean. Similar viral–host dynamics are surely at work in all ecosystems, although in most cases the relative impact of viruses has not been well studied.

carbon fixed by photosynthesis each day ends up supporting bacterial respiration and the production of new bacteria (Azam 1998) (see also Chapter 9).

If there are millions of actively growing bacteria per milliliter, then there must be balancing forces of bacterial mortality. The first recognized bacterial predators in the oceans were nanoflagellates (Fuhrman and McManus 1984; Sherr et al. 1984; Berman et al. 1987; Bjørnsen et al. 1988; Caron and Goldman 1988). Bergh et al. (1989) used transmission electron microscopy (TEM) to show that there were approximately 10 million virus-like particles (VLPs) per milliliter of seawater.

TABLE 12.1 Approximate Abundances of Different Biological Guilds[a] Found in Seawater

Guild	No. in 1 mL of Surface Seawater
Viruses	10^7
Heterotrophic bacteria	10^6
Autotrophic bacteria	10^5
Protozoa	4×10^3
Algae	3×10^3
Zooplankton	<1
Great white sharks	10^{-19}

[a]A guild is a group of organisms that carry out similar functions.

These studies were performed using an ultracentrifuge to pellet viruses from seawater onto TEM grids. TEM counts of viruses are time-consuming, but they provide information about the numbers and types of viruses in a sample. Most of the environmental viruses appear to be phages, which makes sense because bacteria are the most common prey items in the marine environment (Table 12.1). Direct observations of virally infected bacteria verified this assumption and were used to estimate virally induced mortality rates of heterotrophic bacteria and cyanobacteria (Proctor and Fuhrman 1990). Consistent with these observations, viral abundance correlates strongly with bacterial abundance (Wommack and Colwell 2000). In this and other field studies, "bacterial" abundance may include some archaea (see Chapter 3).

Epifluorescent techniques, similar to those used for bacterial counts, were first used for viruses in 1990 (Suttle et al. 1990). These approaches are ideal for rapidly assessing numbers of viruses, but they do not provide information for classifying the viruses. Now the nucleic acid stains such SYBR-Gold/Green and YoPro are routinely used to count viruses (Hennes and Suttle 1995; Noble and Fuhrman 1998). To do this, a small volume (typically 1 mL) of seawater is vacuum-filtered onto a

Small Scale Heterogeneity

On average, there are approximately 10 times as many viruses as prokaryotic cells, resulting in approximately 10^7 viruses per milliliter of seawater. However, the ocean is not a homogeneous mixture. Rather, the ocean can be thought of as a heterogeneous gel, consisting of a continuum between large particles of rich marine snow to vast areas with extremely low carbon and nutrient concentrations (Kiørboe and Jackson 2001). Recent studies have demonstrated significant small-scale spatial heterogeneity in the numbers of viruses and the virus : bacteria ratio in the marine environment (Seymour et al. 2006). Microscale patchiness needs to be considered in future models of marine viral ecology, since these local interactions drive the overall relatively consistent average viral numbers and virus : bacteria ratios in surface seawater.

VIRAL PRODUCTION AND DECAY RATES

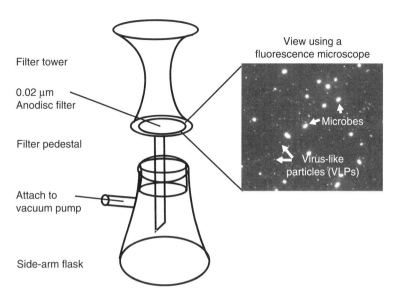

Figure 12.2 Direct count of viruses, bacteria, and other microbes in seawater revolutionized the field of microbial ecology. First, the organisms are concentrated on a two-dimensional surface by gentle filtration. The cells and viruses are then incubated with a nucleic acid stain such as SYBR-Gold. When viewed under the fluorescence microscope, the large spots of light are microbial cells (bacteria and archaea) and the small pinpricks of light are viruses. These groups can be easily distinguished and rapidly counted.

25 mm Anodisc filter with a 0.02 μm pore size. The filter is laid on a drop of the stain for 5–10 minutes, then mounted on a slide and viewed under a fluorescence microscope. The setup and an example of a resulting epifluorescent micrograph are shown in Figure 12.2. The large, bright spots are prokaryotic cells (both bacteria and archaea), while the small, pinpricks of light are viruses. Flow cytometry can also be used for enumerating marine viruses (Marie et al. 1999).

The take-home message from direct counts of viruses from all over the world's oceans is that there are approximately 10^7 viruses per milliliter of surface seawater and 10^9 viruses per gram of sediment. Viral numbers steadily decrease below the thermocline and can be as low as 10^4 viruses per milliliter in the abyss. There are consistently 5–20 times as many viruses as prokaryotic cells (Wommack and Colwell 2000). This ratio may be lower in deeper sediments (Danovaro and Serresi 2000; Danovaro et al. 2001; Danovaro et al. 2002).

VIRAL PRODUCTION AND DECAY RATES

Viral Decay and Rates of Production in Pelagic Systems

The observation that marine viruses are extremely common led directly to questions about their production rates and how fast they killed bacteria. Initial attempts to address this question used the frequency of visibly infected bacteria observed by electron microscopy to estimate the fraction of bacteria that were killed by viruses (Proctor and Fuhrman 1990). The number of visibly infected cells is multiplied by a conversion factor based on the assumption that mature virus particles are only visible in the cells for a brief period (about 10 percent) at the end of the latent period and that the latent period equals bacterial generation time (Proctor and Fuhrman 1990; Weinbauer et al. 2002). This method has been widely used to estimate the impact of viruses on bacterial mortality in pelagic systems (reviewed by Weinbauer 2004).

Viral activity has also been estimated using viral decay rates as a proxy for viral production rates. In these studies, production of new viruses is inhibited by poisoning the bacterial cells, or removing the cells by filtration (Heldal and Bratbak 1991). Viral numbers are then measured over time. In the absence of new production, the decrease in viral numbers over time is due to decay. In a steady-state system, the production of new viruses in natural communities is assumed to equal the decay rate.

To determine how many bacterial cells need to die to maintain the standing stock of viruses, it is also necessary to know how many viruses are produced when one cell lyses. This number is known as the burst size. The average burst size calculated for cultured marine phage is 185 phage particles per cell lysed (Børsheim 1993), which is significantly higher than the average burst size of 25 that has been observed for natural marine communities (Wommack and Colwell 2000), probably reflecting the greater size and biomass of cultured bacteria (Middelboe 2000).

Viral production can also be directly measured using a variety of methods, including incorporation of labeled thymidine, viral reduction by the dilution method, and fluorescently labeled virus (FLV) tracer methods. Among the incubation-based methods, the dilution method is the most widely used protocol for estimating virus production in pelagic systems (Helton et al. 2005).

Phage production using [^{3}H]thymidine incorporation is based on the fact that [^{3}H]thymidine is taken up by bacteria and incorporated into their intracellular nucleotide pool through a salvage pathway that is not used by eukaryotes (Steward et al. 1992a, b). When the bacteria are infected by lytic viruses, the cells continue to take up the isotope during the latent period (Middelboe 2000), and part of the labeled nucleotide is drawn from the host's internal pool and incorporated into viral DNA. When the cells burst, the [^{3}H]thymidine-labeled viruses are released to the environment and virus production can be estimated from the amount of [^{3}H]thymidine incorporated into free virus particles.

The virus dilution method uses virus-free water to dilute the concentration of free viruses to 10–25 percent of natural concentrations (Wilhelm et al. 2002; Winget et al. 2005). Reduction of ambient viruses lowers contact rates, ensuring that all new virus particles produced during the incubation period originate from cells that were infected prior to dilution. The increase in the number of free virus particles over time then reflects the rate of viral production.

For the FLV tracer method, viruses are concentrated from an environmental sample, and fluorescently labeled using a nucleic acid stain such as SYBR-Green I (Noble and Fuhrman 1998). After washing away the excess stain, labeled viruses are added back to natural water samples at tracer levels (<10 percent of ambient virus concentration). After various incubation times, fluorescence microscopy is used to enumerate both labeled viruses and total viruses. Labeled and unlabeled viruses are removed at equal rates, but all new viruses produced are unlabeled. The decrease in the proportion of labeled viruses over time allows for the calculation of viral production and removal rates.

Measurements of Viral Production in Marine Sediments

Viral production in sediments has typically been measured in various types of aerobic slurry incubations, where a sediment sample is diluted with virus-free seawater and the production of viruses is measured as the net increase in viral abundance during a 12–24-hour incubation (Hewson et al. 2003; Mei and Danovaro 2004). Since the perturbation of the sediment in such incubations significantly stimulates benthic microbial activity relative to that seen in undisturbed sediment (Hansen et al. 2000), the slurry approach probably overestimates the in situ benthic viral production. More recently, attempts have been made to reduce this bias by incubating the sediment samples undiluted and homogenized in anaerobic bags (Glud and Middelboe 2004; Middelboe and Glud 2006), which causes less stimulation of bacterial activity than aerobic slurries. Under these conditions, the obtained rates of viral production have generally been lower than those measured in slurries. However, the actual impact of the incubation procedure on the obtained values of benthic viral production is not clear, and there is a need for development of methods for direct measurements of viral production in undisturbed sediment samples.

General Rates of Viral Production

Despite the broad variety of approaches used to estimate viral activity, the general conclusions about the rates of viral production remain the same. Pelagic viral production typically ranges from 2×10^3 to 3×10^6 viruses/mL/h (see Weinbauer and Rassoulzadegan 2004), with a general decrease in production rates from coastal areas to open waters. On average, the total marine viral community turns over every 1–6 days (Wommack and Colwell 2000). The high rates of viral production suggest that viruses are major bacterial predators in the ocean. In planktonic systems, 5–30 percent of heterotrophic bacteria and cyanobacteria are infected by viruses at any time, and viruses cause lysis of a substantial fraction (4–50 percent) of daily bacterial and cyanobacterial production (Fuhrman 1999; Wilhelm and Suttle 1999; Wommack and Colwell 2000).

The production of viruses is dependent on the metabolic state of the host population. A minimum metabolic activity is needed before new viruses are produced from an infected population, and generally there is a positive correlation between the rate of viral lysis and host cell growth rate (Middelboe 2000). Consequently, the influence of viruses on bacterial mortality is expected to be related to trophic conditions and system productivity, which is to some extent supported by field data (Fuhrman 1999; Weinbauer 2004).

In marine sediments, the production of viruses is generally even higher than in the water column (typical ranges of $10-400 \times 10^6$ viruses/mL/h in aerobic sediment (Fischer et al. 2003; Hewson and Fuhrman 2003; Mei and Danovaro 2004) and $1-25 \times 10^6$ viruses/mL/h in deeper anaerobic sediment layers (Glud and Middelboe 2004; Mei and Danovaro 2004; Middelboe and Glud 2006). These results suggest that virally mediated mortality in benthic systems may account for 6–40 percent of bacterial production (Glud and Middelboe 2004; Mei and Danovaro 2004). As in pelagic environments, viral production in sediments correlates positively with measurements of bacterial diagenetic activity such as rates of sulfate reduction and total benthic respiration, indicating that viral activity is regulated by bacterial metabolism (Glud and Middelboe 2004; Middelboe and Glud 2006).

ROLE OF VIRUSES IN BIOGEOCHEMICAL CYCLING

In the nearshore marine environment, bacterial numbers can be controlled by protists (eukaryotic grazers; see Chapter 11) and phages, which account for comparable amounts of bacterial mortality (Proctor and Fuhrman 1988; Wilcox and Fuhrman 1994; Fuhrman and Noble 1995). Figure 12.3 shows an idealized version of the types of experiments used to determine the impacts of predators on bacterial communities. Grazing and viral lysis are predicted to have very different effects on biogeochemical cycling. Grazing by protists can transfer carbon and nutrients to higher trophic levels (Sherr and Sherr 1984; Fenchel 1986), whereas lysis by phages fuels the microbial loop by releasing dissolved organic matter, which is used by heterotrophic bacteria (Fuhrman 1999). Thus, in contrast to protozoan grazing, viral activity converts particulate organic matter into a dissolved form, which is again available for bacterial uptake (viral shunt in Fig 12.1). The released cellular components are rich in nitrogen and phosphorus, and viral activity therefore speeds up the recycling rate of carbon and nutrients within the bacterial community. Compared to a system without viruses, the presence of viruses stimulates bacterial carbon uptake, since a larger fraction of the organic carbon is supplied to the dissolved fraction by viral lysis. As there is a respiratory loss of carbon each time the organic carbon is metabolized by a bacterium, the end result of this organic matter recycling is the production of carbon dioxide and inorganic nutrients, as well as some residual fraction of refractory organic compounds.

Inclusion of viruses in theoretical food-web models suggests that if viral lysis of bacteria is significant and if lysates are consumed by other bacteria, then bacterial growth and respiration would be enhanced at the expense of biomass at higher trophic levels. In fact, the model inferred that viral activity would reduce the export of bacterial carbon by 37 percent but at the same time enhance bacterial carbon consumption by 27 percent, compared with a theoretical food web without viruses. These theoretical considerations have now been verified experimentally by demonstrating that viral lysates are efficiently recycled by bacteria, may cover a substantial fraction of bacterial carbon demand, and significantly reduce the net

ROLE OF VIRUSES IN BIOGEOCHEMICAL CYCLING

(a) Separate protists, bacteria, viruses, and dissolved organic matter (DOM) using filtering

(b) Steady state typical of near-shore

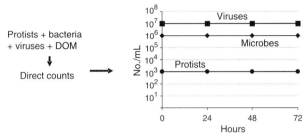

(c) Remove all predators

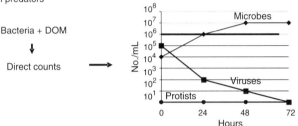

Top-down control: the carrying capacity of the system, in terms of microbial numbers, is higher than observed in the steady state

(d) Addition of viral predators

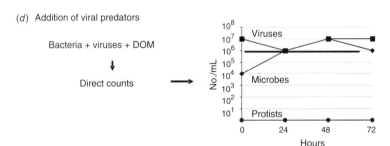

Viruses alone can control microbial populations
Protists alone have the same effect (not shown)

Figure 12.3 Top-down control of bacterial communities by viral and protist predators. Experiments of the type outlined here show that marine bacteria can escape their phage predators if the frequency of infection is reduced by dilution into virus-free seawater. This occurs because the viruses cannot find hosts before they degrade. Importantly, the bacteria grow to a higher concentration than is normally seen under steady-state conditions, indicating that the carrying capacity of the ecosystem is higher than the numbers reached in the presence of predators. The fact that bacteria can grow to higher concentrations in the absence of their predators suggests that the bacterial community as a whole is not resource-limited. Addition of intact viral communities (or protists) will restore the steady-state number of bacteria, indicating that their numbers are maintained by the predators. This may not be true for offshore bacterial communities. This approach has also been used to show that in coastal surface seawater most phage are produced though lytic infections (Wilcox and Fuhrman 1994) and that nonmarine viruses are able to replicate in the marine environment (Sano et al. 2004).

output of bacterial biomass available for grazers (Wilhelm and Suttle 1999; Middelboe and Lyck 2002).

Based on measurements of bacterial and viral production, Wilhelm and Suttle (2000) calculated that viral lysates potentially supplied 4–30 percent of bacterial carbon demand in coastal waters and in the Gulf of Mexico and up to 80–95 percent of the carbon demand in stratified locations in the Strait of Georgia, Canada. These results support the idea that viral lysis may be a key mechanism supplying substrate to the microplankton and stimulates the rates of recycling of dissolved organic matter (DOM) in marine systems.

The immediate consequences of this process are increased bacterial respiration and mineralization of nutrients, reduced transfer of carbon to higher trophic levels and reduced export of particulate material to the seafloor. However, these viral activities are to some extent counteracted by remineralization of nitrogen and phosphorus, which are expected to stimulate overall phytoplankton production, part of which is channeled to higher trophic levels or exported to the sediment. Unfortunately, such feedback mechanisms, as well as the other direct and indirect effects of viruses on microbial processes and interactions, are still not understood in sufficient detail to allow incorporation into quantitative models of marine element cycling. On a global scale, the virally mediated DOM release equals approximately 3–20 Gt dissolved organic carbon per year (Wilhelm and Suttle 1999). Considering that global phytoplankton primary production is approximately 50 Gt C/year, viral lysis of bacteria is a quantitatively important process in the marine carbon cycle.

IMPACT OF VIRUSES ON BACTERIAL DIVERSITY AND COMMUNITY DYNAMICS

Viruses infect bacteria by attachment to specific receptors on the cell surface and injection of viral DNA or RNA into the host. Following infection, the virus may directly enter a lytic cycle (lytic viruses), where the viral genome takes control of the host cell to produce new viruses and subsequently lyses the cell, releasing new viruses to the environment (Fig. 12.4). Alternatively, the viral genetic material can be incorporated into the host genome (temperate viruses), where it may stay in a dormant stage (provirus or prophage) for generations until a lytic cycle may be induced by some external factor. With a third type of viral life cycle, chronic viruses are released from the host cell by extrusion or budding without killing the host (Ackermann and DuBow 1987a, b). These viruses are more like parasites than predators. The impact of chronic viruses in the marine environment is completely unknown.

Another major difference between protozoan and viral modes of bacterial mortality is their impact on the structure of the bacterial community. Omnivorous heterotrophic protists can graze on a wide variety of similar sized microbes, ranging from archaea to picoeukaryotes (see Chapter 11), while phages typically attack a narrower range of hosts, and thus influence bacterial community composition

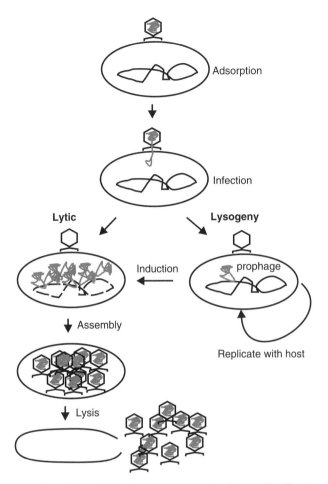

Figure 12.4 Predation of bacteria by viruses occurs on two time scales. The shortest is a lytic infection, where the virus attaches to the outside of the bacterium. This attachment consists of a nonspecific, reversible step, where the virus "rolls" along the outer cell wall of the bacterium. If the bacterium is a species that the virus can infect, the virus will bind to a specific receptor in a second, nonreversible step. The virus then injects its genomic DNA, and the DNA is transcribed and used to make viral proteins. During this time, most bacterial metabolism is shut down and the cell becomes a viral factory. The viral genomic DNA is replicated multiple times and then packaged into new virions. Finally, virally encoded enzymes (e.g., lysozyme) and peptides (e.g., holins) destroy the host cell wall and membrane, respectively. This causes the cell to burst and release the newly assembled virions. In a lysogenic infection, the injected viral DNA is either incorporated into the host chromosome or replicated as an independent element (e.g., as a plasmid). At this point, the viral DNA is called a provirus or prophage, and the microbial host carrying the provirus is called a lysogen. A virus that can establish lysogeny is called temperate. During the proviral stage, some genes are expressed that can dramatically change the behavior of the host in a process called lysogenic conversion. The provirus then replicates with the host until it re-enters the lytic cycle in a process called induction.

through lysis of specific host organisms. The effect of viruses on individual bacterial populations depends on the abundance and activity of the specific host populations and their susceptibility to the co-occurring viruses. The viral infection rate is dependent on host–virus contact rates (Murray and Jackson 1992), suggesting that dominant members of bacterial communities are more likely to be controlled by viral lysis than less significant populations. However, it should be noted that phages infecting rare members of the bacterial community can consistently be isolated from the marine environment, and one study has shown that rare marine bacterial groups may be more susceptible to virally induced mortality (Bouvier and del Giorgio 2007).

How Host-Specific are Marine Viruses?

Many marine phages are often extremely host-specific, incapable of infecting even different strains of the same host species (see, e.g., Moebus and Nattkemper 1981; Moebus 1992). However, some marine phages have been shown to have broad host ranges (Sullivan et al. 2003). Since all of these studies have been conducted using a limited number of cultured phage–host systems, it is completely unknown whether the majority of marine phages have narrow or wide host ranges.

The process by which phages specifically lyse the most abundant bacterial host, creating an open niche and allowing another bacterial genotype to become abundant, has been termed "kill the winner" (Thingstad and Lignell 1997; Thingstad 2000) (Fig. 12.5). This conceptual model predicts that the mechanism of selective virally induced bacterial mortality is a driving force for fluctuations in bacterial communities and contributes to maintaining a high bacterial diversity by selective suppression of numerically dominant populations. Experimental and field data support the hypothesis that viruses influence the dynamics and composition of bacterial and phytoplankton communities, although the effects are variable between locations and experiments (Waterbury and Valois 1993; van Hannen et al. 1999; Wommack et al. 1999a, b; Fuhrman and Schwalbach 2003; Øvreås et al. 2003; Schwalbach et al. 2004; Winter et al. 2004; Hewson and Fuhrman 2006; Hewson et al. 2006; Bouvier and del Giorgio 2007). A number of factors affect how viruses influence bacterial diversity and community dynamics and add further complexity to virus–host interactions than is included in present conceptual models. These include (1) induction of lysogens, (2) development of resistance against viral attack (Middelboe et al. 2001; Thyrhaug et al. 2003), (3) large clonal diversity within bacterial populations (Waterbury and Valois 1993; Tarutani et al. 2000; Riemann and Middelboe 2002b), (4) the input of labile substrate from viral lysates (Middelboe et al. 2003b), (5) possibility of viral loss by protistan grazing (Miki and Yamamura 2005), and (6) possible benefits of infection for the host (e.g., lysogenic conversion).

Does "Kill the Winner" Happen in Natural Marine Communities?

The "kill the winner" hypothesis suggests that viruses lyse the most abundant hosts, allowing for less abundant bacteria to become dominant. Experimental tests of this hypothesis in natural communities through enrichment or depletion of ambient viral communities have produced highly variable results. One major complication with these experiments is the large degree of sample manipulation involved (including containment under laboratory conditions), which can result in changes in bacterial community structure. Although it is clear that viruses influence host community structure, these effects are highly variable and require further investigation.

Resistance mutations may cause alterations in bacterial membrane proteins or lipopolysaccharides, which reduce or prevent phage adsorption. The transformation of receptor molecules may also reduce the cell's competitiveness, because these molecules are involved in bacterial metabolism (Schwartz 1980; Lenski and Levin 1985; Lenski 1988). The trade-off between resistance and competitiveness determines the success of the emerging resistant clones. Theoretically, this

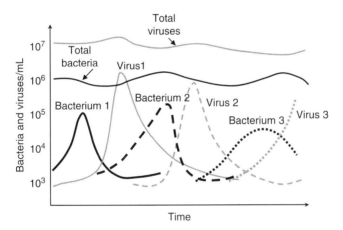

Figure 12.5 Diagram illustrating "kill the winner" dynamics. Although the total concentration of bacteria and viruses remains relatively stable over time, the concentrations of individual bacterial and viral strains change dramatically. As a bacterial species becomes abundant (Bacterium 1), viruses that can infect those bacteria (Virus 1) will specifically lyse that host. This will lead to a decrease in the population size of Bacterium 1 and an increase in the population size of Virus 1. However, as the host population declines, there will no longer be a host to produce more viruses, which will lead to a subsequent decrease in the population size of Virus 1. The virally mediated decline of the host population creates an open niche, allowing another bacterium to become abundant (Bacterium 2). A virus that can infect Bacterium 2 then becomes abundant, and the cycle continues. Selective virally induced bacterial mortality is a driving force for fluctuations in the structure of bacterial communities and contributes to maintaining high bacterial diversity.

mechanism results in dynamic fluctuations of sensitive and resistant bacterial clones, depending on substrate conditions and the composition of the viral community (Riemann and Middelboe 2002b). Consequently, in addition to influencing bacterial population dynamics, viruses probably drive a large clonal variation within populations of bacteria and phytoplankton with different patterns of susceptibility to different co-occurring viruses (see, e.g., Tarutani et al. 2000; Middelboe et al. 2001).

Viruses may also affect bacterial community composition indirectly by mechanisms such as release of new substrates (the cell lysates), altering the dynamics and composition of the pool of DOM available for bacteria, and hence the bacterial community composition (van Hannen et al. 1999; Middelboe et al. 2003b). This substrate transformation mechanism may promote coexistence of diverse bacteria in natural communities.

In addition to their effects on bacterial population dynamics, phages are also important mediators of genetic exchange in the environment, influencing bacterial diversity via lysogenic conversion and via generalized (Paul et al. 1991b; Jiang and Paul 1998a, b; Paul 1999; Clokie et al. 2003) and specialized (Waldor and Mekalanos 1996; Acheson et al. 1998; Faruque et al. 2000) transduction. Using heterotrophic marine isolates and mixed marine bacterial communities, Jiang and Paul (1998a, b) found frequencies between 1.33×10^{-7} and 5.13×10^{-9} transductants per plaque-forming unit. Based on these frequencies, numerical modeling suggests that marine phages transduce 10^{28} base pairs of DNA per year in the world's oceans (Paul et al. 2002). In addition, some phages infecting marine *Synechococcus* strains are capable of encapsidating host DNA in their capsids (at a frequency of about 1 in 10^4 *Synechococcus* phage particles), suggesting that cyanophage are also potential agents of horizontal gene transfer (Clokie et al. 2003). Recent studies have demonstrated the presence of gene transfer agents (GTAs) in marine bacterial genomes. GTAs are virus-like particles that carry random pieces of host genomic DNA, and can transfer this DNA between host cells, but do not encode viral genes (Lang and Beatty 2007). The significance of GTA-mediated gene transfer in the marine environment is currently unknown.

Lysogenic conversion is characterized by changes in the properties of the host cells after establishment of lysogeny. Viral DNA is then integrated into the bacterial genome as a prophage and its lytic action is postponed. As measured by prophage induction experiments in natural communities and cultured isolates, lysogeny is common in the marine environment (Jiang and Paul 1996, 1998a, b; Weinbauer and Suttle 1996, 1999; McDaniel et al. 2002; Ortmann et al. 2002; Williamson et al. 2002; Weinbauer et al. 2003). In addition, prophages are present in most sequenced marine bacterial genomes (Canchaya et al. 2003; Chen et al. 2006). Bacteria that are lysogenized by a given phage are immune to infection by homologous viruses (Ackermann and DuBow 1987a, b), and therefore a large part of the marine bacterial community may be immune to some of the co-occurring viruses. Other effects may include changes in the morphological, immunological, or metabolic properties of the hosts (Weinbauer and Rassoulzadegan 2004). For example, the cholera toxin that enables *Vibrio cholerae* to cause disease is carried on a lysogenic phage (Waldor and Mekalanos 1996).

> **To Lyse or Wait?**
>
> Lysogeny is prevalent in the marine environment; however, the molecular mechanisms and environmental controls on lysogeny are still poorly understood. Lysogeny can provide advantages to the host, such as gaining new metabolic capabilities or immunity to infection by similar phages. By being temperate, a phage may be able to survive during times of low host abundance or growth, waiting for conditions to improve before becoming lytic (see, e.g., Williamson et al. 2002; McDaniel and Paul 2005). Most studies of lysogeny involve artificially inducing phages to become lytic through the use of a DNA-damaging agent such as mitomycin C or ultraviolet light. These methods have been used to demonstrate clear seasonal variations in lysogeny in some marine environments (McDaniel et al. 2002; Laybourn-Parry et al. 2007). Future research will hopefully shed light on the environmental conditions that influence the lysogenic decision in natural marine communities.

In a comprehensive survey of lysogeny in the marine environment, the frequency of lysogenic cells was found to be negatively related to bacterial abundance and production (Weinbauer et al. 2003). This is consistent with seasonal studies from Tampa Bay, Florida, in which lysogeny was primarily detected in winter months, during periods of low host cell density (McDaniel et al. 2002; Williamson et al. 2002). Thus, lysogeny seems to be a survival strategy at low host abundance and activity, whereas high host abundance and activity seems to favor the lytic life cycle (Weinbauer et al. 2003). The environmental factors controlling lysogeny in the marine environment are still largely unknown, although links to nutrients and salinity have been suggested (Wilson et al. 1996; McDaniel and Paul 2005; Williamson and Paul 2006). Even less is known about the molecular mechanisms of lysogeny in marine phages. It is therefore essential to expand our knowledge on these issues in order to improve our understanding of the role of lysogeny in marine communities, and to develop more sophisticated models of virus–host dynamics.

MARINE VIRAL DIVERSITY

Despite their abundance and importance, we are only just beginning to understand the diversity and biogeography of environmental viruses. Both culture-based and culture-independent approaches for studying the diversity of environmental viruses are discussed below.

Methods for Examining Marine Viral Diversity

Culture-Based

1. *Genomic Sequencing*. Complete genome sequencing of cultured phage genomes has provided insight into the ecology and evolution of marine phages (Rohwer et al. 2000; Paul and Sullivan 2005).

2. *Host-Range Studies.* Cultured phages are tested against a wide range of related hosts in order to determine their degree of host specificity (Moebus and Nattkemper 1981; Moebus 1992; Sullivan et al. 2003).

Culture-Independent

1. *Transmission Electron Microscopy (TEM).* TEM allows for the discrimination of some phages based on morphology (Bergh et al. 1989)
2. *Pulsed-Field Gel Electrophoresis (PFGE).* Viral communities can be separated based on genome size, allowing for comparison of overall viral fingerprints between different times or locations (Steward et al. 2000; Wommack et al. 1999a). PFGE can be combined with hybridization or amplification of signature genes to connect gene presence with genome size class (Wommack et al. 1999b; Sandaa and Larsen 2006).
3. *Signature Gene Amplification.* Although no single gene is found in all viruses, there are structural or functional genes that are conserved among restricted groups of viruses. These conserved genes can be used to examine the diversity and biogeographical distribution of certain viral groups (Breitbart and Rohwer 2004; Short and Suttle 2005; Culley et al. 2003). Signature genes can be cloned and sequenced, or analyzed with profiling methods such as denaturing gradient gel electrophoresis (DGGE) (Short and Suttle 1999, 2002).
4. *Metagenomic Sequencing.* Metagenomic (whole-community) sequencing analyzes the collective viral genomes contained within a marine sample. This method provides information about the gene content of the most abundant viral genomes, and allows for prediction of marine viral community structure and diversity (Breitbart et al. 2002, 2004).

Culture-Based Studies of Viral Diversity

To study phage by culturing, it is first necessary to grow the bacterial host in liquid culture and then mix it with a sample containing the phage. The mixture is added to top agar and poured on top of the normal agar plate. As the host grows, the infected cells are lysed, and the phages diffuse through the top agar and attack nearby bacteria, leaving a clear area called a plaque.

Culture-based studies have shown that there is at least one phage that can infect any given marine bacterial host (Moebus and Nattkemper 1981; Moebus 1991, 1992; Suttle and Chan 1993; Kellogg et al. 1995; Sullivan et al. 2003). Furthermore, these phages are often (but not always) specific to certain hosts (Kellogg et al. 1995; Rohwer et al. 2000; Sullivan et al. 2003). There are an estimated 6 million free-living (Curtis et al. 2002) and 3–5 million eukaryote-associated (Novotny et al. 2002) prokaryotic species on the planet. If each of these bacteria is a host for 10 different phage species, then there are over 100 million phage species on the planet (Rohwer 2003).

Once phages have been isolated, they are typically described based on physical properties of the phage particles, such as morphology, nucleic acid type, and resistance to chemical solvents (Murphy et al. 1995; Ackermann 2001). Cultured phages can be further characterized by determination of host range, restriction mapping, hybridization analyses, and genome sequencing. As of this writing, the complete genomes of 284 phages have been reported in the GenBank database (http://www.ncbi.nlm.nih.gov/genomes/VIRUSES/viruses.html), but only 18 of these are marine phages (Lohr et al. 2005; Paul and Sullivan 2005). Most (60–80 percent) of the open reading frames (ORFs; often assumed to be protein-coding regions) from cultured marine phage genomes are unknown, but analyses of cultured marine phage genomes have revealed the presence of several metabolic genes that may explain why certain phages are successful in the marine environment (Sullivan et al. 2005).

Photosynthesis by the cyanobacteria *Synechococcus* and *Prochlorococcus* accounts for approximately one-third of the carbon fixed in the marine environment. Cyanophages infecting *Synechococcus* and *Prochlorococcus* have acquired genes involved in photosynthesis (Mann et al. 2003; Lindell et al. 2004). For example, the *psbA* gene, which encodes the D1 protein, has been found in the genomes of several *Prochlorococcus* and *Synechococcus* phages (Mann et al. 2003, 2005; Lindell et al. 2004; Millard et al. 2004; Sullivan et al. 2005, 2006) and in marine viral metagenomes (Angly et al. 2006). D1 is a rate-limiting photosynthesis protein, and host-derived D1 concentrations dramatically drop during phage infection (Bailey et al. 2004). Expression of the phage-encoded *psbA* gene enables cyanophages to maintain photosynthesis throughout the infection cycle (Lindell et al. 2005; Clokie et al. 2006), presumably providing the bacteria with energy while the phages replicate. In the open ocean, the energy benefit associated with the acquisition of photosynthesis genes by cyanophage was probably a key evolutionary step.

Similarly, phosphate is a major limiting nutrient in parts of the ocean, and many marine phages encode enzymes involved in phosphate metabolism. For example, the phosphate-inducible *phoH* gene has been identified in several cultured marine phage genomes (Rohwer et al. 2000; Miller et al. 2003; Sullivan et al. 2005) and in uncultured marine viral communities (Rohwer et al., unpublished). Genes involved in nucleotide scavenging and metabolism are also found in marine phages. Nucleotide scavenging may be especially important in nutrient-poor environments, such as the ocean (Rohwer et al. 2000). Genes encoding ribonucleotide reductase, which catalyzes a thioredoxin-mediated reduction of diphosphates during nucleotide metabolism, have been identified in the genomes of several cultured marine phages (Rohwer et al. 2000; Chen and Lu 2002; Sullivan et al. 2005), and are abundant in metagenomic libraries of uncultured marine viruses (Angly et al. 2006). Experimental evidence has shown that some cultured marine phages use host-derived nucleotides during phage replication (Wikner et al. 1993), but this has yet to be investigated in natural communities.

The Need for Culture-Independent Methods

The disagreement between the number of bacteria that can be cultured from an environment and the number of bacteria observed by direct microscopic counts has been termed the "great plate-count anomaly" (Staley and Konopka 1985), and describes one of the greatest challenges to the study of environmental phages. Classical studies of phages require that they be grown in culture on a bacterial host. However, it is estimated that over 99 percent of all environmental bacteria cannot be cultured using standard techniques (Fuhrman and Campbell 1998; Rappé and Giovannoni 2003). Polymerase chain reaction (PCR) amplification and sequencing of the 16S ribosomal DNA (16S rDNA) locus directly from environ mental samples has shown that bacteria isolated using standard culturing techniques are rarely the dominant bacteria present in the natural environment (Giovannoni and Rappé 2000; Hugenholtz 2002; and see Chapter 3). Because less than 1 percent of environmental hosts have been cultured, the phages that infect those uncultured bacteria are almost completely undescribed.

New methods have allowed researchers to culture a wide variety of environmental bacteria, including those that are numerically dominant in the environment and bacterial groups without any previously cultured representatives (Connon and Giovannoni 2002; Kaeberlein et al. 2002; Rappé et al. 2002; Zengler et al. 2002; Joseph et al. 2003; Cho and Giovannoni 2004; Page et al. 2004; Stevenson et al. 2004; Selje et al. 2005; Simu et al. 2005). As these novel bacteria are obtained in culture, it will be interesting to isolate and characterize phages capable of infecting them. An alternative approach is to study natural viral communities without the requirement for culturing.

Culture-Independent Studies of Viral Diversity Using Transmission Electron Microscopy

Natural viral communities have been characterized based on physical parameters of virus particles, such as capsid diameter and tail length (Torella and Morita 1979; Børsheim et al. 1990; Hara et al. 1991; Cochlan et al. 1993; Alonso et al. 2001). Transmission electron microscopy (TEM) analyses have demonstrated that the morphological diversity observed in natural environmental viral communities is substantially different from the morphological diversity represented by cultured phage isolates (Ackermann 2001). For example, natural soil virus communities contain more filamentous viruses and viruses with elongated capsids than are known among the cultured phage isolates (Williamson et al. 2005). In addition, the capsids of cultured phage isolates are larger on average than those found in the environment, supporting the idea that cultured isolates are not representative of natural communities (Børsheim 1993).

While viral morphology in the water column is dominated by the tailed icosahedral morphotypes, morphological analysis of the viral community in a coastal benthic system indicated that interstitial viral communities were dominated by long (>1 μm) filamentous forms with helical symmetry (Middelboe et al. 2003a, b). Several types of these filamentous forms were observed, which resembled the morphology of known filamentous viruses, characterized by a chronic infection cycle. Filamentous forms are rarely found in the water column, which suggests that they are adapted to the benthic environment and specific to the interstitial hosts (Middelboe et al. 2003a). In the interstitium, where particle adsorption and the impeded transport will counteract efficient spreading of virus particles from lytic infections (Maranger and Bird 1996), such transportation and spreading of the viruses along with their host cells may be an advantageous strategy. It can be speculated that this is the underlying reason for the dominance of filamentous viruses in the investigated sediment. However, little is known about the life strategies of benthic viruses.

Whole-Genome Profiling of Viral Communities Based on Genome Size

Whole viral communities have been profiled by separating viruses based on genome size using pulsed-field gel electrophoresis (PFGE) (Klieve and Swain 1993; Swain et al. 1996; Wommack et al. 1999a, b; Steward et al. 2000; Steward 2001; Riemann and Middelboe 2002; Jiang et al. 2003; Sandaa et al. 2003; Sandaa and Larsen 2006). PFGE creates an overview of the viral community, where the number of bands seen on the gel indicates the number of dominant genome size classes, and the relative intensity of each band indicates the number of viruses with genomes of that size. Changes in the genome size profile of viral communities can be examined over temporal and spatial gradients, as well as gradients of nutrient concentration, carbon availability, and microbial diversity. In marine samples, between 7 and 35 bands can be observed by PFGE (Wommack et al. 1999a, b; Steward et al. 2000; Larsen et al. 2001; Steward 2001; Riemann and Middelboe 2002a; Øvreås et al. 2003). Although there are changes in the presence or intensity of individual bands between different marine samples, a few viral genome size classes are consistently abundant in seawater (Steward et al. 2000). Steward et al. (2000) showed that the major size classes found in seawater were 31–36 kilobases (kb) and 58–63 kb, with more than 50 percent of the viral genomes in the 28–45 kb size range and more than 90 percent of the viral genomes in the 26–69 kb size range. However, genome sizes as large as 500 kb have been identified in the marine environment (Sandaa and Larsen 2006). A disadvantage of this method is that different viral genotypes with similar genome sizes cannot be distinguished by PFGE. Wommack et al. (1999b) increased the resolution of this approach by separating viral communities using PFGE and then hybridizing these gels with DNA probes specific for an individual virus or group of viruses. This study showed geographically localized episodic changes in the abundance of specific viruses, consistent with the "kill the winner" hypothesis (Thingstad and Lignell 1997; Wommack et al. 1999b).

Studies of Viral Diversity Using Signature Genes

Studies of specific prokaryotic species in the environment were limited by culturing biases until 16S rDNA analyses of uncultured bacteria from environmental samples were introduced (Lane et al. 1985). Using this method, bacterial diversity is examined directly from environmental samples using a variety of molecular methods, including PCR amplification, sequencing, fingerprinting, microarrays, and in situ hybridization (see Chapters 3–6). Similar methods are also available for analyses of archaea (Baker et al. 2003; and see Chapter 3) and eukarya (Zhu et al. 2003; and see Chapter 6). However, it is not possible to study total viral diversity using approaches analogous to 16S rDNA profiling, because no single genetic element is shared by all viruses (Rohwer and Edwards 2002).

Whole-genome comparisons have shown, however, that there are conserved genes (signature genes) shared among members of certain viral taxonomic groups. Signature genes for viral groups, as described by the Phage Proteomic Tree (Rohwer and Edwards 2002), or by other criteria, can therefore be used to study the diversity, distribution, and dynamics of viruses in the environment. Cyanophages, for example, have been studied using capsid-encoding sequences (Fuller et al. 1998; Zhong et al. 2002; Dorigo et al. 2004; Short and Suttle 2005), and the diversity of algal viruses and T7-like podophage has been examined by sequencing DNA polymerase genes (Chen and Suttle 1996; Chen et al. 1996; Breitbart and Rohwer 2004). This approach has led to the identification of at least two novel groups of marine picorna-like viruses (Culley et al. 2003). The genetic diversity of conserved genes can be determined through direct sequencing (Zhong et al. 2002; Breitbart and Rohwer 2004), or profiling and sequencing (Chen and Suttle 1995; Chen et al. 1996; Short and Suttle 1999, 2000, 2002, 2005; Culley et al. 2003; Frederickson et al. 2003; Muhling et al. 2005; Sandaa and Larsen 2006). Signature gene studies suggest that environmental viral diversity is high and that most environmental phages are not closely related to cultured isolates. In addition, signature gene studies have shown that many phages are widely distributed in the environment, even across different biomes (Breitbart and Rohwer 2004; Short and Suttle 2005).

Metagenomic Studies of Viral Diversity

A drawback to the use of conserved genes for diversity analyses is that it does not enable the discovery of completely novel groups of viruses. Recently, metagenomic approaches have been used to study uncultured environmental microbial and viral communities (reviewed by Riesenfeld et al. 2004; Edwards and Rohwer 2005). The term "metagenomics" refers to the functional and sequence-based analysis of the collective genomes contained in an environmental sample (Handelsman et al. 1998). For metagenomic studies, total nucleic acids are isolated from an environmental sample, fragmented, and then cloned into a vector and transformed. Metagenomic analyses can involve sequencing of random clones (Breitbart et al. 2002, 2003, 2004; Tyson et al. 2004; Venter et al. 2004; Breitbart and Rohwer 2005; Cann et al. 2005; Tringe et al. 2005; Culley et al. 2006; Edwards et al. 2006), or selectively sequencing clones containing phylogenetic anchors (Stein et al. 1996; Suzuki et al. 2004) or genes of interest (Béjà et al. 2000, 2001; Ginolhac et al. 2004; Liebeton and Eck 2004; Rhee et al. 2005). The application of metagenomic approaches is discussed in Chapter 4.

One advantage of using viral communities for metagenomic analyses lies in the small genome size of viruses (50 kb on average versus about 2000 kb for a bacterial genome: Steward et al. 2000; Wommack and Colwell 2000), which decreases the amount of sequencing required to adequately sample the community. However, there are numerous challenges associated with sequencing DNA from viral communities. The small genome size of viruses means that the low amount of viral DNA can be overwhelmed and obscured by the often higher concentrations of cellular and free DNA in the environment (DeFlaun et al. 1986, 1987; Paul et al. 1987, 1991b). Cloning viral DNA is also difficult due to modified DNA bases and the presence of lethal genes such as holins and lysozymes (Wang et al. 2000). These limitations have been overcome through concentration of viruses from large-volume samples using tangential flow filtration, purification of viruses from free nucleic acids using density-dependent gradient centrifugation and nuclease treatment, and construction of linker-amplified shotgun libraries (LASLs; Figs. 12.6 and 12.7). LASLs overcome many of the difficulties encountered when creating a shotgun library from viral DNA. The initial fragmentation step disrupts lethal genes, preventing their expression. Addition of PCR amplification increases the total amount of DNA available for cloning and replaces modified DNA bases with unmodified nucleotides.

A new method that is being used for environmental metagenomics is pyrosequencing. The 454 Life Sciences Corporation (Branford, CT) has developed a novel form of rapid, high-throughput sequencing that involves an emulsion-based method to isolate and amplify DNA fragments in vitro combined with picoliter-sized pyrosequencing reactions (Margulies et al. 2005). The pyrosequencing method presents a 25-fold decrease in cost and a 100-fold increase in throughput compared with current Sanger sequencing technology. This method has proven successful with mixtures of cultured phage (F. Rohwer, unpublished data), and environmental microbial and phage communities (Edwards et al. 2006; Angly et al. 2006). Application of pyrosequencing technology to viral metagenomics will allow the inexpensive, rapid accumulation of a large amount of sequence data.

Once metagenomic sequences have been obtained, the identity of the environmental viruses is determined by BLAST comparison against the GenBank nonredundant (nr) database. Significant hits to GenBank entries are classified into groups (i.e., phage, virus, mobile, repeat, bacteria, archaea, and eukarya), based on sequence annotation. Bacterial hits are examined manually for prophages as annotated by Casjens (2003). Significant hits to phages are further classified into phage families according to their position on the Phage Proteomic Tree (Rohwer and Edwards 2002) (if available) or their classification by the International Committee on Taxonomy of Viruses (ICTV) (Murphy et al. 1995). The Phage Proteomic Tree can be used to determine the relationships of phages based on their DNA sequences, and allows for statistical comparisons of viral communities based on phylogeny, as have been performed for microbes (Martin 2002; Breitbart et al. 2004).

The structure and diversity of environmental viral communities can also be predicted from metagenomic sequence data (Breitbart et al. 2002). This analysis requires assembly of the metagenomic sequence fragments and production of a contig spectrum (i.e., distribution of overlapping sequences). Sequences that assemble together into a contiguous sequence (contig) are considered a resampling of the same viral genotype. More contigs would therefore be expected from a sample with low viral diversity than one with high diversity. The contig spectrum can be

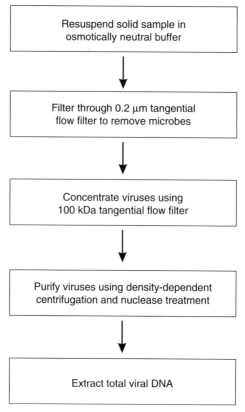

Figure 12.6 Overview of the procedure for purifying viral DNA from an environmental sample. Before metagenomic sequencing, viruses must be purified from cells and free nucleic acids, both of which occur at very high concentrations in seawater. Filtration is used to remove bacteria and eukaryotic cells. Free nucleic acids are removed using nucleases and density-dependent centrifugation in cesium chloride. Density-dependent centrifugation works because free DNA is denser than phage particles, which are about 50 percent protein and 50 percent DNA. Any contaminating cells that passed through the filtration step are also removed by the density-dependent centrifugation, since cells are less dense due to their membranes.

used to mathematically model the structure and diversity of the viral community from which the sequences were obtained (Lander and Waterman 1988; Breitbart et al. 2002, 2004; Angly et al. 2005).

Metagenomic studies of marine viruses have shown that uncultured viral communities are some of the most diverse communities ever observed. There are an estimated 5000 viral genotypes in 200 L of seawater (Breitbart et al. 2002) and possibly a million different viral genotypes in 1 kg of marine sediment (Breitbart et al. 2004). Approximately 75 percent of the metagenomic sequences have no significant similarity to anything that has been sequenced before, suggesting that the vast majority of environmental viral diversity is novel (Edwards and Rohwer 2005;

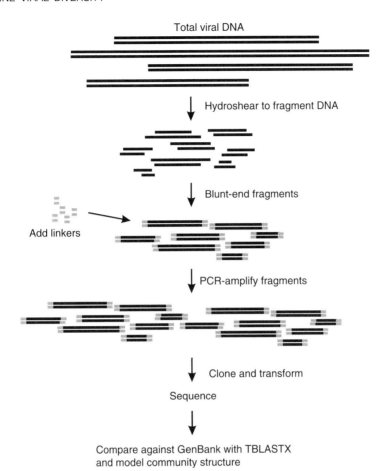

Figure 12.7 Construction of a linker-amplified shotgun library (LASL) from viral DNA. By incorporating fragmentation and amplification steps, this method overcomes many of the obstacles (i.e., low DNA content, modified nucleotides, and lethal genes) associated with cloning viral DNA. The LASL method produces a high-coverage random library containing over a million clones from less than 1 ng of input viral DNA. Briefly, total viral community DNA is randomly sheared into 1–2 kb fragments and the fragments are end-repaired. Short dsDNA linkers are then ligated to the ends of the fragments. Finally, the fragments are PCR-amplified using the high-fidelity Vent DNA polymerase with primers to the linkers. Addition of a random-primed strand-displacement amplification step can be used to create a LASL from single-stranded DNA viruses (Breitbart and Rohwer 2005) and RNA viruses can be recovered by first synthesizing cDNA using random-primed reverse transcriptase (Culley et al. 2006; Zhang et al. 2006). The resulting fragments are cloned, sequenced, and analyzed bioinformatically.

Culley et al. 2006). Some phage groups (T7-like podophage, λ-like siphophage, and T4-like myophage) are abundant in all marine viral communities, while other groups are never observed, which suggests a common phylogenetic origin of marine phage populations (Breitbart et al. 2004).

A VISION FOR THE FUTURE

Viruses are ubiquitous, abundant, diverse and important components in the marine environment. Recent research has yielded insight into the composition and ecological roles of marine viruses, but there are still many unanswered questions. Here we detail three important future research areas.

1. **How Much Does Viral Lysis Actually Contribute to Marine Carbon and Nutrient Cycling, and How Does that Role Vary Between Environments?**

    Viruses potentially have significant impacts on marine carbon and nutrient cycling due to recycling of matter, but there is a gap in our knowledge of the actual quantitative role of viruses on biogeochemical cycling. Future studies need to gather quantitative information on the effect of lysate-mediated recycling of inorganic nutrients for system productivity. It has almost been a mantra in discussions of viral impact on nutrient cycling that viral activity leads to increased respiration and reduced transfer of organic matter to higher trophic levels and the sea floor. However, current models of the influence of viral activity on biogeochemical cycling do not take into account that an increased bacterial mineralization of nutrients such as nitrogen and phosphorus in viral lysates will also stimulate primary production, which in fact may increase production at higher trophic levels rather than reducing it. In the oligotrophic ocean, primary production is often driven by nutrients that are regenerated in the surface layers; and in such systems, viral activity may in fact be a driving force for total system productivity by stimulation of bacterial nutrient mineralization. The speculative nature of the quantitative importance of many aspects of viral activity emphasizes the need for further studies of the biogeochemical roles of viruses in different environments. It is essential to improve our knowledge on these processes to be able to include viruses in global biogeochemical models. This will require the development of solid and standardized methods for estimating rates of viral production and the contribution of lysates to bacterial carbon demand.

2. **What Characteristics Define Marine Viruses?**

    Metagenomic studies have shown that certain groups of viruses are more abundant in the marine environment, and phylogenetic analyses suggest that marine phages share a common origin. Numerous "specialization" genes identified in marine phages indicate that these phages are adapted to life in the oceans. However, most viral sequence is completely unknown, and future studies will need to identify the genes that are common among marine viruses, and have the daunting task of determining the function of these genes. Most studies of marine viruses have focused on double-stranded DNA viruses, but new techniques are revealing abundant and diverse communities of other groups of viruses (RNA and single-stranded DNA) in the oceans.

The ecology of these viral groups is largely unknown. Future studies need to further examine the relationships between marine viruses and those in other environments, and determine if there are geographically restricted populations of marine viruses. Finally, there are abundant groups of marine organisms (e.g., copepods and marine archaea) for which viruses have not been described. This is likely due to a lack of investigation, as well as methodological limitations (e.g., lack of pure cultures). Future studies need to focus efforts on determining the impact of viral infection on these groups.

3 **Are Viruses Controlling Bacterial Community Composition and Driving Microbial Diversification?**

There are relatively few studies examining the temporal and spatial variability in viral community composition. Future studies need to document this variability (e.g., seasonal patterns) and determine the importance of viral diversity for structuring bacterial communities. Studies have revealed mixed effects of viral communities on bacterial community structure—ranging from having no effect to showing significant correlations between viral and bacterial community composition. It is likely that there are many physical, chemical, and biological factors that influence the degree of impact that viruses have on bacterial communities. In addition, it is difficult to differentiate whether viruses are driving changes in bacterial diversity, or the viruses are changing in response to a shift in bacterial community composition. There seems to be a huge diversity of bacterial clones within specific populations (i.e., bacteria with identical 16S rRNA gene sequences), which all show different susceptibilities to the variety of viruses that infect them. We need to examine the importance of viruses as driving forces of the clonal succession and functional diversity in natural populations of marine bacteria. This has implications for our understanding of the role of viruses for bacterial community dynamics, and such complexities need to be better understood so that they can be incorporated into models.

SUMMARY

1. There are approximately 10 million viruses per milliliter of seawater, making viruses the most abundant biological entities in the oceans. Viruses are important, dynamic, and diverse components of the marine environment.
2. Top-down control, through viral infection and grazing, controls the abundance of marine bacteria. Virally mediated bacterial death creates more dissolved organic matter for bacterial consumption, and increases the cycling rates of carbon and nutrients in the marine environment. Through lysis of specific bacterial strains, viruses can influence bacterial community composition, and may drive speciation. Horizontal gene transfer by viruses is an important mechanism of gene exchange in the marine environment.

3. Marine viral communities are extremely diverse, although most viral sequence diversity remains unknown. Certain viral groups are more prevalent in the marine environment, and some of these viruses carry genes involved in adaptation to life in the oceans.

REFERENCES

Acheson, D. W. K., J. Reidl, X. Zhang, G. T. Keusch, J. J. Mekalanos, and M. K. Waldor. 1998. In vivo transduction with Shiga toxin 1-encoding phage. *Infect. Immun.* 66: 4496–4498.

Ackermann, H. W. 2001. Frequency of morphological phage descriptions in the year 2000. *Arch. Virol.* 146: 843–857.

Ackermann, H.-W., and M. S. DuBow. 1987a. *Viruses of Prokaryotes; Volume I: General Properties of Bacteriophages.* CRC Press.

Ackermann, H.-W., and M. S. DuBow. 1987b. *Viruses of Prokaryotes; Volume II: Natural Groups of Bacteriophages.* CRC Press.

Alonso, M., F. Jimenez-Gomez, J. Rodriguez, and J. Borrego. 2001. Distribution of virus-like particles in an oligotrophic marine environment (Alboran Sea, Western Mediterranean). *Microb. Ecol.* 42: 407–415.

Angly, F., B. Rodriguez-Brito, D. Bangor, et al. 2005. PHACCS, an online tool for estimating the structure and diversity of uncultured viral communities using metagenomic information. *BMC Bioinformatics* 6: 41.

Angly, F., B. Felts, M. Breitbart, et al. 2006. The marine viromes of four oceanic regions. *PLoS Biol.* 4: e368. doi: 10.1371/journal.pbio.0040368.

Azam, F. 1998. Microbial control of oceanic carbon flux: The plot thickens. *Science* 280: 694–696.

Azam, F., and R. E. Hodson. 1977. Dissolved ATP in the sea and its utilization by marine bacteria. *Nature* 267: 696–698.

Azam, F., T. Fenchel, J. G. Field, J. S. Gray, L. A. Meyer-Reil, and F. Thingstad. 1983. The ecological role of water-column microbes in the sea. *Mar. Ecol. Prog. Ser.* 10: 257–263.

Bailey, S., M. Clokie, A. Millard, and N. Mann. 2004. Cyanophage infection and photoinhibition in marine cyanobacteria. *Res. Microbiol.* 155: 720–725.

Baker, G., J. Smith, and D. Cowan. 2003. Review and re-analysis of domain-specific 16S primers. *J. Microbiol. Meth.* 55: 541–555.

Béjà, O., L. Aravind, E. V. Koonin, et al. 2000. Bacterial rhodopsin: Evidence for a new type of phototropy in the sea. *Science* 289: 1902–1906.

Béjà, O., E. Spudich, J. Spudich, M. Leclerc, and E. DeLong. 2001. Proteorhodopsin phototrophy in the ocean. *Nature* 411: 786–789.

Bergh, Ø., K. Y. Børsheim, G. Bratbak, and M. Heldal. 1989. High abundance of viruses found in aquatic environments. *Nature* 340: 467–468.

Berman, T., M. Nawrocki, G. Taylor, and D. Karl. 1987. Nutrient flux between bacteria, bacterivorous nanoplanktonic protists and algae. *Mar. Microb. Food Webs* 2: 69–82.

REFERENCES

Bjørnsen, P. K., B. Riemann, S. J. Horsted, T. G. Nielsen, and J. Pock-Sten. 1988. Trophic interactions between heterotrophic nanoflagellates and bacterioplankton in manipulated seawater enclosures. *Limnol. Oceanogr.* 33: 409–420.

Børsheim, K. 1993. Native marine bacteriophages. *FEMS Microbiol. Ecol.* 102: 141–159.

Børsheim, K. Y., G. Bratbak, and M. Heldal. 1990. Enumeration and biomass estimation of planktonic bacteria and viruses by transmission electron microscopy. *Appl. Environ. Microbiol.* 56: 352–356.

Bouvier, T., and del Giorgio, P. A. 2007. Key role of selective viral-induced mortality in determining marine bacterial community composition. *Environ. Microbiol.* 9: 287–297.

Breitbart, M., and F. Rohwer. 2004. Global distribution of nearly identical phage-encoded DNA sequences. *FEMS Microbiol. Lett.* 236: 245–252.

Breitbart, M., and F. Rohwer. 2005. Method for discovering novel DNA viruses in blood using viral particle selection and shotgun sequencing. *Biotechniques* 39: 729–736.

Breitbart, M., P. Salamon, B. Andresen, J. M. Mahaffy, A. M. Segall, D. Mead, F. Azam, and F. Rohwer. 2002. Genomic analysis of uncultured marine viral communities. *Proc. Natl. Acad. Sci. USA* 99: 14250–14255.

Breitbart, M., I. Hewson, B. Felts, J. Mahaffy, J. Nulton, P. Salamon, and F. Rohwer. 2003. Metagenomic analyses of an uncultured viral community from human feces. *J. Bacteriol.* 85: 6220–6223.

Breitbart, M., B. Felts, S. Kelley, J. Mahaffy, J. Nulton, P. Salamon, and F. Rohwer. 2004. Diversity and population structure of a nearshore marine sediment viral community. *Proc. Roy. Soc., Ser. B, Biol. Sci.* 271: 565–574.

Canchaya, C., C. Proux, G. Fournous, A. Bruttin, and H. Brussow. 2003. Prophage genomics. *Microbiol. Mol. Biol. Rev.* 67: 238–276.

Cann, A., S. Fandrich, and S. Heaphy. 2005. Analysis of the virus population present in equine faeces indicates the presence of hundreds of uncharacterized virus genomes. *Virus Genes* 30: 151–156.

Caron, D., and J. Goldman. 1988. Dynamics of protistan carbon and nutrient cycling. *J. Protozool.* 35: 247–249.

Casjens, S. 2003. Prophages and bacterial genomics: What have we learned so far? *Mol. Microbiol.* 49: 277–300.

Chen, F., and J. Lu. 2002. Genomic sequence and evolution of marine cyanophage P60: A new insight on lytic and lysogenic phages. *Appl. Environ. Microbiol.* 68: 2589–2594.

Chen, F., and C. Suttle. 1995. Amplification of DNA polymerase gene fragments from viruses infecting microalgae. *Appl. Environ. Microbiol.* 61: 1274–1278.

Chen, F., and C. A. Suttle. 1996. Evolutionary relationships among large double-stranded DNA viruses that infect microalgae and other organisms as inferred from DNA polymerase genes. *Virology* 219: 170–178.

Chen, F., C. A. Suttle, and S. M. Short. 1996. Genetic diversity in marine algal virus communities as revealed by sequence analysis of DNA polymerase genes. *Appl. Environ. Microbiol.* 62: 2869–2874.

Chen, F., K. Wang, J. Stewart, and R. Belas. 2006. Induction of multiple prophages from a marine bacterium: A genomic approach. *Appl. Environ. Microbiol.* 72: 4995–5001.

Cho, J., and S. Giovannoni. 2004. Cultivation and growth characteristics of a diverse group of oligotrophic marine *Gammaproteobacteria*. *Appl. Environ. Microbiol.* 70: 432–440.

Clokie, M., A. Millard, W. Wilson, and N. Mann. 2003. Encapsidation of host DNA by bacteriophages infecting marine *Synechococcus* strains. *FEMS Microbiol. Ecol.* 46: 349–352.

Clokie, M. R. J., J. Y. Shan, S. Bailey, Y. Jia, H. M. Krisch, S. West, and N. H. Mann. 2006. Transcription of a "photosynthetic" T4-type phage during infection of a marine cyanobacterium. *Environ. Microbiol.* 8: 827–835.

Cochlan, W. P., J. Wikner, G. F. Steward, D. C. Smith, and F. Azam. 1993. Spatial distribution of viruses, bacteria, and chlorophyll a in neritic, oceanic, and estuarine environments. *Mar. Ecol. Prog. Ser.* 92: 77–87.

Connon, S., and S. Giovannoni. 2002. High-throughput methods for culturing microorganisms in very low nutrient media yield diverse new marine isolates. *Appl. Environ. Microbiol.* 68: 3878–3885.

Culley, A., A. Lang, and C. Suttle. 2003. High diversity of unknown picorna-like viruses in the sea. *Nature* 424: 1054–1057.

Culley, A. I., A. S. Lang, and C. A. Suttle. 2006. Metagenomic analysis of coastal RNA virus communities. *Science* 312: 1795–1798.

Curtis, T., W. Sloan, and J. Scannell. 2002. Estimating prokaryotic diversity and its limits. *Proc. Natl. Acad. Sci. USA* 99: 10494–10499.

Daley, R. J., and J. E. Hobbie. 1975. Direct counts of aquatic bacteria by a modified epifluorescence technique. *Limnol. Oceanogr.* 20: 875–882.

Danovaro, R., and M. Serresi. 2000. Viral density and virus-to-bacterium ratio in deep-sea sediments of the Eastern Mediterranean. *Appl. Environ. Microbiol.* 66: 1857–1861.

Danovaro, R., A. Dell'Anno, A. Trucco, M. Serresi, and S. Vanucci. 2001. Determination of virus abundance in marine sediments. *Appl. Environ. Microbiol.* 67: 1384–1387.

Danovaro, R., E. Manini, and A. Dell'Anno. 2002. Higher abundance of bacteria than viruses in deep Mediterranean sediments. *Appl. Environ. Microbiol.* 66: 1857–1861.

DeFlaun, M., J. Paul, and D. Davis. 1986. Simplified method for dissolved DNA determination in aquatic environments. *Appl. Environ. Microbiol.* 52: 654–659.

DeFlaun, M., J. Paul, and W. Jeffrey. 1987. Distribution and molecular weight of dissolved DNA in subtropical estuarine and oceanic environments. *Mar. Ecol. Prog. Ser.* 38: 65–73.

Dorigo, U., S. Jacquet, and J. Humbert. 2004. Cyanophage diversity, inferred from g20 gene analyses, in the largest natural lake in France, Lake Bourget. *Appl. Environ. Microbiol.* 70: 1017–1022.

Edwards, R., and F. Rohwer. 2005. Viral metagenomics. *Nat. Rev. Microbiol.* 3: 504–510.

Edwards, R., B. Rodriguez-Brito, L. Wegley, et al. 2006. Using pyrosequencing to shed light on deep mine microbial ecology under extreme hydrogeological conditions. *BMC Genomics* 7: 57 doi: 10.1186/1471-2164-7-57.

Faruque, S. M., Asadulghani, M. M. Rahman, M. K. Waldor, and D. A. Sack. 2000. Sunlight-induced propagation of the lysogenic phage encoding cholera toxin. *Infect. Immun.* 68: 4795–4801.

Fenchel, T. 1986. The ecology of heterotrophic microflagellates. *Adv. Microb. Ecol.* 9: 57–97.

Ferguson, R. L., and P. Rublee. 1976. Contribution of bacteria to standing crop of coastal plankton. *Limnol. Oceanogr.* 21: 141–145.

Fischer, U., C. Wieltschnig, A. Kirschner, and B. Velimirov. 2003. Does viral-induced lysis contribute significantly to bacterial mortality in the oxygenated sediment layer of shallow oxbow lakes? *Appl. Environ. Microbiol.* 69: 5821–5829.

Francisco, D. E., R. A. Mah, and A. C. Rabin. 1973. Acridine orange–epifluorescence technique for counting bacteria in natural waters. *Trans. Am. Micro. Soc.* 92: 416–421.

Frederickson, C., S. Short, and C. Suttle. 2003. The physical environment affects cyanophage communities in British Columbia inlets. *Microb. Ecol.* 46: 348–357.

Fuhrman, J. 1999. Marine viruses: Biogeochemical and ecological effects. *Nature* 399: 541–548.

Fuhrman, J. A., and F. Azam. 1982. Thymidine incorporation as a measure of heterotrophic bacterioplankton production in marine surface waters: Evaluation and field results. *Mar. Biol.* 66: 109–120.

Fuhrman, J. A., and L. Campbell. 1998. Marine ecology: Microbial microdiversity. *Nature* 393: 410–411.

Fuhrman, J. A., and G. B. McManus. 1984. Do bacteria-sized marine eucaryotes consume significant bacterial production? *Science* 224: 1257–1260.

Fuhrman, J. A., and R. T. Noble. 1995. Viruses and protists cause similar bacterial mortality in coastal seawater. *Limnol. Oceanogr.* 40: 1236–1242.

Fuhrman, J., and M. Schwalbach. 2003. Viral influence on aquatic bacterial communities. *Biol. Bull.* 204: 192–195.

Fuller, N. J., W. H. Wilson, I. R. Joint, and N. H. Mann. 1998. Occurrence of a sequence in marine cyanophages similar to that of T4 g20 and its application to PCR-based detection and quantification techniques. *Appl. Environ. Microbiol.* 64: 2051–2060.

Ginolhac, A., C. Jarrin, B. Gillet, P. Robe, P. Pujic, K. Tuphile, H. Bertrand, T. Vogel, G. Perriere, P. Simonet, and R. Nalin. 2004. Phylogenetic analysis of polyketide synthase I domains from soil metagenomic libraries allows selection of promising clones. *Appl. Environ. Microbiol.* 70: 5522–5527.

Giovannoni, S., and M. Rappé. 2000. Evolution, diversity, and molecular ecology of marine prokaryotes. In D. L. Kirchman (ed.), *Microbial Ecology of the Oceans*, 1st edn. Wiley-Liss, pp. 47–84.

Glud, R., and M. Middelboe. 2004. Virus and bacteria dynamics of a coastal sediment: Implications for benthic carbon cycling. *Limnol. Oceanogr.* 49: 2073–2081.

Hambly, E., and C. Suttle. 2005. The viriosphere, diversity, and genetic exchange within phage communities. *Curr. Opin. Microbiol.* 8: 444–450.

Handelsman, J., M. Rondon, S. Brady, J. Clardy, and R. Goodman. 1998. Molecular biological access to the chemistry of unknown soil microbes: A new frontier for natural products. *Chem. Biol.* 5: R245–249.

Hansen, J. W., B. Thamdrup, and B. B. Jørgensen. 2000. Anoxic incubation of sediment in gas-tight plastic bags: A method for biogeochemical process studies. *Mar. Ecol. Prog. Ser.* 208: 273–282.

Hara, S., K. Terauchi, and I. Koike. 1991. Abundance of viruses in marine waters: Assessment by epifluorescence and transmission electron microscopy. *Appl. Environ. Microbiol.* 57: 2731–2734.

Heldal, M., and G. Bratbak. 1991. Production and decay of viruses in aquatic environments. *Mar. Ecol. Prog. Ser.* 72: 205–212.

Helton, R., M. Cottrell, D. Kirchman, and K. Wommack. 2005. Evaluation of incubation-based methods for estimating virioplankton production in estuaries. *Aquat. Microb. Ecol.* 41: 209–219.

Hennes, K., and C. Suttle. 1995. Direct counts of viruses in natural waters and laboratory cultures by epifluorescence microscopy. *Limnol. Oceanogr.* 40: 1050–1055.

Hewson, I., and J. Fuhrman. 2003. Viriobenthos production and virioplankton sorptive scavenging by suspended sediment particles in coastal and pelagic waters. *Microb. Ecol.* 46: 337–347.

Hewson, I., and J. A. Fuhrman. 2006. Viral impacts upon marine bacterioplankton assemblage structure. *J. Mar. Biol. Assoc. UK* 86: 577–589.

Hewson, I., G. Vargo, and J. Fuhrman. 2003. Bacterial diversity in shallow oligotrophic marine benthos and overlying waters: Effects on virus infection, containment, and nutrient enrichment. *Microb. Ecol.* 46: 322–336.

Hewson, I., D. M. Wingett, K. E. Williamson, J. A. Fuhrman, and K. E. Wommack. 2006. Viral and bacterial assemblage covariance in oligotrophic waters of the West Florida Shelf (Gulf of Mexico). *J. Mar. Biol. Assoc. UK* 86: 591–603.

Hobbie, J., R. Daley, and S. Jasper. 1977. Use of nuclepore filters for counting bacteria by epifluorescence microscopy. *Appl. Environ. Microbiol.* 33: 1225–1228.

Hugenholtz, P. 2002. Exploring prokaryotic diversity in the genomic era. *Genome Biol.* 3: 3.1–3.8.

Jiang, S. C., and J. H. Paul. 1996. Occurrence of lysogenic bacteria in marine microbial communities as determined by prophage induction. *Mar. Ecol. Prog. Ser.* 142: 27–38.

Jiang, S., and J. Paul. 1998a. Significance of lysogeny in the marine environment: Studies with isolates and a model of lysogenic phage production. *Microb. Ecol.* 35: 235–243.

Jiang, S. C., and J. H. Paul. 1998b. Gene transfer by transduction in the marine environment. *Appl. Environ. Microbiol.* 64: 2780–2787.

Jiang, S., W. Fu, W. Chu, and J. Fuhrman. 2003. The vertical distribution and diversity of marine bacteriophage at a station off Southern California. *Microb. Ecol.* 45: 399–410.

Joseph, S., P. Hugenholtz, P. Sangwan, C. Osburne, and P. Janssen. 2003. Laboratory cultivation of widespread and previously uncultured microbes. *Appl. Environ. Microbiol.* 69: 7210–7215.

Kaeberlein, T., K. Lewis, and S. Epstein. 2002. Isolating 'uncultivable' microorganisms in pure culture by a simulated natural environment. *Science* 296: 1127–1129.

Karl, D. M., and P. Bossard. 1985. Measurement and significance of ATP and adenine nucleotide pool turnover in microbial cells and environmental samples. *J. Microbiol. Meth.* 3: 125–139.

Kellogg, C. A., J. B. Rose, S. C. Jiang, J. M. Turmond, and J. H. Paul. 1995. Genetic diversity of related vibriophages isolated from marine environments around Florida and Hawaii, USA. *Mar. Ecol. Prog. Ser.* 120: 89–98.

Kiorboe, T., and G.A., Jackson. 2001. Marine snow, organic solute plumes, and optimal chemosensory behavior of bacteria. *Limnol. Oceanogr.* 46: 1309–1318.

Kirchman, D. L. 1994. The uptake of inorganic nutrients by heterotrophic bacteria. *Microb. Ecol.* 28: 255–271.

Klieve, A. V., and R. A. Swain. 1993. Estimation of ruminal bacteriophage numbers by pulsed-field gel electrophoresis and laser densitometry. *Appl. Environ. Microbiol.* 59: 2299–2303.

Lander, E. S., and M. S. Waterman. 1988. Genomic mapping by fingerprinting random clones: A mathematical analysis. *Genomics* 2: 231–239.

Lane, D., B. Pace, G. Olsen, D. Stahl, M. Sogin, and N. Pace. 1985. Rapid determination of 16S rRNA sequences for phylogenetic analysis. *Proc. Natl. Acad. Sci. USA* 82: 6955–6959.

Lang, A. S., and J. T. Beatty. 2007. Importance of widespread gene transfer agent genes in α-proteobacteria. *Trends Microbiol.* 15: 54–62.

Larsen, A., T. Castberg, R. Sandaa, C. Brussard, J. Egge, M. Heldal, A. Paulino, R. Thyrhaug, E. van Hannen, and G. Bratbak. 2001. Population dynamics and diversity of phytoplankton, bacteria, and viruses in a seawater enclosure. *Mar. Ecol. Prog. Ser.* 221: 47–57.

Laybourn-Parry, J., W. A. Marshall, and N. J. Madan. 2007. Viral dynamics and patterns of lysogeny in saline Antarctic lakes. *Polar Biol.* 30: 351–358.

Lenski, R. E. 1988. Dynamics of interactions between bacteria and virulent bacteriophage. In K. C. Marshall (ed.), *Advances in Microbial Ecology*. Plenum Press, pp. 1–44.

Lenski, R. E., and B. R. Levin. 1985. Constraints on the coevolution of bacteria and virulent phage: A model, some experiments, and predictions for natural communities. *Am. Nat.* 125: 585–602.

Liebeton, K., and J. Eck. 2004. Identification and expression in *E. coli* of novel nitrile hydratases from the metagenome. *Eng. Life Sci.* 4: 557–562.

Lindell, D., J. Jaffe, Z. Johnson, G. Church, and S. Chisholm. 2005. Photosynthesis genes in marine viruses yield proteins during host infection. *Nature* 438: 86–89.

Lindell, D., M. Sullivan, Z. Johnson, A. Tolonen, F. Rohwer, and S. Chisholm. 2004. Photosynthesis genes in *Prochlorococcus* cyanophage. *Proc. Natl. Acad. Sci. USA* 101: 11013–11018.

Lohr, J., F. Chen, and R. Hill. 2005. Genomic analysis of bacteriophage φJL001: Insights into its interaction with a sponge-associated alpha-proteobacterium. *Appl. Environ. Microbiol.* 71: 1598–1609.

McDaniel, L., and J. H. Paul. 2005. Effect of nutrient addition and environmental factors on prophage induction in natural populations of marine *Synechococcus* species. *Appl. Environ. Microbiol.* 71: 842–850.

McDaniel, L., L. Houchin, and J. H. Paul. 2002. Plankton blooms: Lysogeny in natural *Synechococcus* populations. *Nature* 415: 496.

Mann, N., A. Cook, A. Millard, S. Bailey, and M. Clokie. 2003. Marine ecosystems: Bacterial photosynthesis genes in a virus. *Nature* 424: 741.

Mann, N. H., M. R. J. Clokie, A. Millard, A. Cook, W. H. Wilson, P. J. Wheatley, A. Letarov, and H. M. Krisch. 2005. The genome of S-PM2, a "photosynthetic" T4-type bacteriophage that infects marine *Synechococcus* strains. *J. Bacteriol.* 187: 3188–3200.

Maranger, R., and D. F. Bird. 1996. High concentrations of viruses in the sediments of Lake Gilbert, Quebec. *Microb. Ecol.* 31: 141–151.

Margulies, M., M. Egholm, W. Altman, et al. 2005. Genome sequencing in microfabricated high-density picolitre reactors. *Nature* 437: 376–380.

Marie, D., C. P. D. Brussaard, R. Thyrhaug, G. Bratbak, and D. Vaulot. 1999. Enumeration of marine viruses in culture and natural samples by flow cytometry. *Appl. Environ. Microbiol.* 65: 45–52.

Martin, A. 2002. Phylogenetic approaches for describing and comparing the diversity of microbial communities. *Appl. Environ. Microbiol.* 68: 3673–3682.

Mei, M., and R. Danovaro. 2004. Virus production and life strategies in aquatic sediments. *Limnol. Oceanogr.* 49: 459–470.

Middelboe, M. 2000. Microbial growth rate and marine virus-host dynamics. *Microb. Ecol.* 40: 114–124.

Middelboe, M., and R. Glud. 2006. Viral activity along a trophic gradient in continental margin sediments off central Chile. *Mar. Biol. Res.* 2: 41–51.

Middelboe, M., and P. Lyck. 2002. Regeneration of dissolved organic matter by viral lysis in marine microbial communities. *Aquat. Microb. Ecol.* 27: 187–194.

Middelboe, M., Å. Hagström, N. Blackburn, B. Sinn, U. Fischer, N. Borch, J. Pinhassi, K. Simu, and M. Lorenz. 2001. Effects of bacteriophages on the population dynamics of four strains of pelagic marine bacteria. *Microb. Ecol.* 42: 395–406.

Middelboe, M., R. Glud, and K. Finster. 2003a. Distribution of viruses and bacteria in relation to diagenetic activity in an estuarine sediment. *Limnol. Oceanogr.* 48: 1447–1456.

Middelboe, M., L. Riemann, G. F. Steward, V. Hansen, and O. Nybroe. 2003b. Virus-induced transfer of organic carbon between marine bacteria in a model community. *Aquat. Microb. Ecol.* 33: 1–10.

Miki, T., and N. Yamamura. 2005. Intraguild predation reduces bacterial species richness and loosens the viral loop in aquatic systems: "Kill the killer of the winner" hypothesis. *Aquat. Microb. Ecol.* 40: 1–12.

Millard, A., M. R. J. Clokie, D. A. Shub, and N. H. Mann. 2004. Genetic organization of the psbAD region in phages infecting marine *Synechococcus* strains. *Proc. Natl. Acad. Sci. USA* 101: 11007–11012.

Miller, E., J. Heidelberg, J. Eisen, et al. 2003. Complete genome sequence of the broad-host-range vibriophage KVP40: Comparative genomics of a T4-related bacteriophage. *J. Bacteriol.* 185: 5220–5233.

Moebus, K. 1991. Preliminary observations on the concentration of marine bacteriophages in the water around Helgoland. *Helgo. Meeresunters* 45: 411–422.

Moebus, K. 1992. Further investigations on the concentration of marine bacteriophages in the water around Helgoland, with reference to the phage-host systems encountered. *Helgo. Meeresunters* 46: 275–292.

Moebus, K., and H. Nattkemper. 1981. Bacteriophage sensitivity patterns among bacteria isolated from marine waters. *Helgo. Meeresunters* 34: 375–385.

Muhling, M., N. Fuller, A. Millard, et al. 2005. Genetic diversity of marine *Synechococcus* and co-occurring cyanophage communities: Evidence for viral control of phytoplankton. *Environ. Microbiol.* 7: 499–508.

Munn, C. B. 2006. Viruses as pathogens of marine organisms—from bacteria to whales. *J. Mar. Biol. Assoc. UK* 86: 453–467.

Murphy, F. A., C. M. Fauquet, D. H. L. Bishop, S. A. Ghabrial, A. W. Jarvis, G. P. Martelli, M. A. Mayo, and M. D. Summers (eds.). 1995. *Virus Taxonomy: Sixth Report of the International Committee on Taxonomy of Viruses.* Springer-Verlag.

Murray, A. G., and G. S. Jackson. 1992. Viral dynamics: A model of the effects of size, shape, motion and abundance of single-celled planktonic organisms and other particles. *Mar. Ecol. Prog. Ser.* 89: 103–116.

Noble, R. T., and J. A. Fuhrman. 1998. Use of SYBR Green I for rapid epifluorescence counts of marine viruses and bacteria. *Aquat. Microb. Ecol.* 14: 113–118.

Novotny, V., Y. Basset, S. Miller, G. Weiblen, B. Bremer, L. Cizek, and P. Drozd. 2002. Low host specificity of herbivorous insects in a tropical forest. *Nature* 416: 841–844.

Ortmann, A. C., J. E. Lawrence, and C. A. Suttle. 2002. Lysogeny and lytic viral production during a bloom of the cyanobacterium *Synechococcus* spp. *Microb. Ecol.* 43: 225–231.

Øvreås, L., D. Bournde, R. Sandaa, E. Casamayor, S. Benlloch, V. Goddard, G. Smerdon, M. Heldal, and T. Thingstad. 2003. Response of bacterial and viral communities to nutrient manipulations in seawater mesocosms. *Aquat. Microb. Ecol.* 31: 109–121.

Page, K., S. Connon, and S. Giovannoni. 2004. Representative freshwater bacterioplankton isolated from Crater Lake, Oregon. *Appl. Environ. Microbiol.* 70: 6542–6550.

Paul, J. H. 1999. Microbial gene transfer: An ecological perspective. *J Mol. Microbiol. Biotechnol.* 1: 45–50.

Paul, J., and M. Sullivan. 2005. Marine phage genomics: What have we learned? *Curr. Opin. Biotechnol.* 16: 299–307.

Paul, J., W. Jeffrey, and M. DeFlaun. 1987. Dynamics of extracellular DNA in the marine environment. *Appl. Environ. Microbiol.* 53: 170–179.

Paul, J., S. Jiang, and J. Rose. 1991a. Concentration of viruses and dissolved DNA from aquatic environments by vortex flow filtration. *Appl. Environ. Microbiol.* 57: 2197–2204.

Paul, J. H., M. E. Frischer, and J. M. Thurmond. 1991b. Gene transfer in marine water column and sediment microcosms by natural plasmid transformation. *Appl. Environ. Microbiol.* 57: 1509–1515.

Paul, J., M. Sullivan, A. Segall, and F. Rohwer. 2002. Marine phage genomics. *Comp. Biochem. Physiol.* B 133: 463–476.

Pomeroy, L. R. 1974. The ocean's food web, a changing paradigm. *BioScience* 24: 499–504.

Proctor, L. M., and J. A. Fuhrman. 1988. Marine bacteriophages and bacterial mortality. *EOS* 69: 1111–1112.

Proctor, L. M., and J. A. Fuhrman. 1990. Viral mortality of marine bacteria and cyanobacteria. *Nature* 343: 60–62.

Rappé, M., and S. Giovannoni. 2003. The uncultured microbial majority. *Annu. Rev. Microbiol.* 57: 369–394.

Rappé, M., S. Connon, K. Vergin, and S. Giovannoni. 2002. Cultivation of the ubiquitous SAR11 marine bacterioplankton clade. *Nature* 418: 630–633.

Rhee, J., D. Ahn, Y. Kim, and J. Oh. 2005. New thermophilic and thermostable esterase with sequence similarity to the hormone-sensitive lipase family, cloned from a metagenomic library. *Appl. Environ. Microbiol.* 71: 817–825.

Riemann, B. 1978. Differentiation between heterotrophic and photosynthetic plankton by size fractionation, glucose uptake, ATP and chlorophyll content. *Oikos* 31: 358–367.

Riemann, L., and M. Middelboe. 2002a. Stability of bacterial and viral community compositions in Danish coastal waters as depicted by DNA fingerprinting techniques. *Aquat. Microb. Ecol.* 27: 219–232.

Riemann, L., and M. Middelboe. 2002b. Viral lysis of marine bacterioplankton: Implications for organic matter cycling and bacterial clonal composition. *Ophelia* 56: 57–68.

Riesenfeld, C., P. Schloss, and J. Handelsman. 2004. METAGENOMICS: Genomic analysis of microbial communities. *Annu. Rev. Genet.* 38: 525–552.

Rohwer, F. 2003. Global phage diversity. *Cell* 113: 141.

Rohwer, F., and R. Edwards. 2002. The phage proteomic tree: A genome based taxonomy for phage. *J. Bacteriol.* 184: 4529–4535.

Rohwer, F., A. M. Segall, G. Steward, V. Seguritan, M. Breitbart, F. Wolven, and F. Azam. 2000. The complete genomic sequence of the marine phage Roseophage SIO1 shares homology with non-marine phages. *Limnol. Oceanogr.* 42: 408–418.

Sandaa, R. A., and A. Larsen. 2006. Seasonal variations in virus-host populations in Norwegian coastal waters: Focusing on the cyanophage community infecting marine *Synechococcus* spp. *Appl. Environ. Microbiol.* 72: 4610–4618.

Sandaa, R.-A., E. Skjoldal, and G. Bratbak. 2003. Virioplankton community structure along a salinity gradient in a solar saltern. *Extremophiles* 7: 347–351.

Sano, E., S. Carlson, L. Wegley, and F. Rohwer. 2004. Movement of viruses between biomes. *Appl. Environ. Microbiol.* 70: 5842–5846.

Schwalbach, M., I. Hewson, and J. Fuhrman. 2004. Viral effects on bacterial community composition in marine plankton microcosms. *Aquat. Microb. Ecol.* 34: 117–127.

Schwartz, M. 1980. Interactions of phages with their receptor proteins. In L. Randall and L. Philipson (eds.), *Virus Receptors (Receptors and recognition)*. Chapman & Hall, pp. 61–91.

Selje, N., T. Brinkhoff, and M. Simon. 2005. Detection of abundant bacteria in the Weser estuary using culture-dependent and culture-independent approaches. *Aquat. Microb. Ecol.* 39: 17–34.

Seymour, J. R., L. Seuront, M. Doubell, R. L. Waters, and J. G. Mitchell. 2006. Microscale patchiness of virioplankton. *J. Mar. Biol. Assoc. UK* 86: 551–561.

Sherr, B. F., and E. B. Sherr. 1984. Role of heterotrophic protozoa in carbon and energy flow in aquatic ecosystems. In M. J. Klugg and C. A. Reddy (eds.), *Current Perspectives in Microbial Ecology*. American Society for Microbiology, pp. 412–423.

Sherr, B. F., E. B. Sherr, and S. Y. Newell. 1984. Abundance and productivity of heterotrophic nanoplankton in Georgia coastal waters. *J. Plankton Res.* 6: 195–202.

Short, C. M., and C. A. Suttle. 2005. Nearly identical bacteriophage structural gene sequences are widely distributed in marine and freshwater environments. *Appl. Environ. Microbiol.* 71: 480–486.

Short, S. M., and C. A. Suttle. 1999. Use of the polymerase chain reaction and denaturing gradient gel electrophoresis to study diversity in natural virus communities. *Hydrobiologia* 401: 19–32.

Short, S. M., and C. A. Suttle. 2000. Denaturing gradient gel electrophoresis resolves virus sequences amplified with degenerate primers. *Biotechniques* 28: 20–26.

Short, S. M., and C. A. Suttle. 2002. Sequence analysis of marine virus communities reveals that groups of related algal viruses are widely distributed in nature. *Appl. Environ. Microbiol.* 68: 1290–1296.

Simu, K., K. Holmfeldt, U. Zweifel, and Å. Hagström. 2005. Culturability and coexistence of colony-forming and single-cell marine bacterioplankton. *Appl. Environ. Microbiol.* 71: 4793–4800.

Staley, J. T., and A. Konopka. 1985. Measurement of in situ activities of nonphotosynthetic microorganisms in aquatic and terrestrial habitats. *Annu. Rev. Microbiol.* 39: 321–46.

Stein, J. L., T. L. Marsh, K. Y. Wu, H. Shizuya, and E. F. Delong. 1996. Characterization of uncultivated prokaryotes: Isolation and analysis of a 40-kilobase-pair genome fragment from a planktonic marine archaeon. *J. Bacteriol.* 178: 591–599.

Stevenson, B., S. Eichorst, J. Wertz, T. Schmidt, and J. Breznak. 2004. New strategies for cultivation and detection of previously uncultured microbes. *Appl. Environ. Microbiol.* 70: 4748–4755.

Steward, G. 2001. Fingerprinting viral assemblages by pulsed field gel electrophoresis. In J. Paul (ed.), *Methods in Microbiology*. Academic Press, pp. 85–103.

Steward, G. F., J. Wikner, W. P. Cochlan, D. C. Smith, and F. Azam. 1992a. Estimation of virus production in the sea: I. Method development. *Mar. Microb. Food Webs* 6: 57–78.

Steward, G. F., J. Wikner, W. P. Cochlan, D. C. Smith, and F. Azam. 1992b. Estimation of virus production in the sea: II. Field results. *Mar. Microb. Food Webs* 6: 79–90.

Steward, G., J. Montiel, and F. Azam. 2000. Genome size distributions indicate variability and similarities among marine viral assemblages from diverse environments. *Limnol. Oceanogr.* 45: 1697–1706.

Sullivan, M., J. Waterbury, and S. Chisholm. 2003. Cyanophages infecting the oceanic cyanobacterium *Prochlorococcus*. *Nature* 424: 1047–1051.

Sullivan, M., M. Coleman, P. Weigele, F. Rohwer, and S. Chisholm. 2005. Three *Prochlorococcus* cyanophage genomes: Signature features and ecological interpretations. *PLoS Biol.* 3: 790–806.

Sullivan, M., D. Lindell, J. Lee, L. Thompson, J. Bielawski, and S. Chisholm. 2006. Prevalence and evolution of core photosystem II genes in marine cyanobacterial viruses and their hosts. *PLoS Biol.* 4: e234. doi: 10.1371/journal.pbio.0040234.

Suttle, C. A., and A. M. Chan. 1993. Marine cyanophages infecting oceanic and coastal strains of *Synechococcus*—abundance, morphology, cross-infectivity and growth characteristics. *Mar. Ecol. Prog. Ser.* 92: 99–109.

Suttle, C. A., A. M. Chan, and M. T. Cottrell. 1990. Infection of phytoplankton by viruses and reduction of primary productivity. *Nature* 347: 467–469.

Suzuki, M., C. Preston, O. Béjà, J. de la Torre, G. Steward, and E. DeLong. 2004. Phylogenetic screening of ribosomal RNA gene-containing clones in bacterial artificial chromosome (BAC) libraries from different depths in Monterey Bay. *Microb. Ecol.* 48: 473–488.

Swain, R., J. Nolan, and A. Klieve. 1996. Natural variability and diurnal fluctuations within the bacteriophage population of the rumen. *Appl. Environ. Microbiol.* 62: 994–997.

Tarutani, K., K. Nagasaki, and M. Yamaguchi. 2000. Viral impact on total abundance and clonal composition of the harmful bloom-forming phytoplankton *Heterosigma akashiwo*. *Appl. Environ. Microbiol.* 66: 4916–4920.

Thingstad, T. 2000. Elements of a theory for the mechanisms controlling abundance, diversity, and biogeochemical role of lytic bacterial viruses in aquatic systems. *Limnol. Oceanogr.* 45: 1320–1328.

Thingstad, T. F., and R. Lignell. 1997. Theoretical models for the control of bacterial growth rate, abundance, diversity and carbon demand. *Aquat. Microb. Ecol.* 13: 19–27.

Thyrhaug, R., A. Larsen, T. Thingstad, and G. Bratbak. 2003. Stable coexistence in marine algal host-virus systems. *Mar. Ecol. Prog. Ser.* 254: 27–35.

Torella, F., and R. Y. Morita. 1979. Evidence by electron micrographs for a high incidence of bacteriophage particles in the waters of Yaquina Bay, Oregon: Ecological and taxonomical implications. *Appl. Environ. Microbiol.* 37: 774–778.

Tringe, S., C. von Mering, A. Kobayashi, et al. 2005. Comparative metagenomics of microbial communities. *Science* 308: 554–557.

Tyson, G., J. Chapman, P. Hugenholtz, et al. 2004. Community structure and metabolism through reconstruction of microbial genomes from the environment. *Nature* 428: 37–43.

van Hannen, E., G. Zwart, M. van Agterveld, H. Gons, J. Ebert, and H. Laanbroek. 1999. Changes in bacterial and eukaryotic community structure after mass lysis of filamentous cyanobacteria associated with viruses. *Appl. Environ. Microbiol.* 65: 795–801.

Venter, J., K. Remington, J. Heidelberg, et al. 2004. Environmental genome shotgun sequencing of the Sargasso Sea. *Science* 304: 66–74.

Waldor, M. K., and J. J. Mekalanos. 1996. Lysogenic conversion by a filamentous phage encoding cholera toxin. *Science* 272: 1910–1914.

Wang, I., D. Smith, and R. Young. 2000. HOLINS: The protein clocks of bacteriophage infections. *Annu. Rev. Microbiol.* 54: 799–825.

Waterbury, J., and F. Valois. 1993. Resistance to co-occurring phages enables marine *Synechococcus* communities to coexist with cyanophages abundant in seawater. *Appl. Environ. Microbiol.* 59: 3393–3399.

Weinbauer, M. 2004. Ecology of prokaryotic viruses. *FEMS Microbiol. Rev.* 28: 127–181.

Weinbauer, M., and F. Rassoulzadegan. 2004. Are viruses driving microbial diversification and diversity? *Environ. Microbiol.* 6: 1–11.

Weinbauer, M. G., and C. A. Suttle. 1996. Potential significance of lysogeny to bacteriophage production and bacterial mortality in coastal waters of the Gulf of Mexico. *Appl. Environ. Microbiol.* 62: 4375.

Weinbauer, M. G., and C. A. Suttle. 1999. Lysogeny and prophage induction in coastal and offshore bacterial communities. *Aquat. Microb. Ecol.* 21: 99–118.

Weinbauer, M., C. Winter, and M. Höfle. 2002. Reconsidering transmission electron microscopy based estimates of viral infection of bacterio-plankton using conversion factors derived from natural communities. *Aquat. Microb. Ecol.* 27: 103–110.

Weinbauer, M. G., I. Brettar, and M. G. Höfle. 2003. Lysogeny and virus-induced mortality of bacterioplankton in surface, deep, and anoxic marine waters. *Limnol. Oceanogr.* 48: 1457–1465.

Whitman, W., D. Coleman, and W. Wiebe. 1998. Prokaryotes: The unseen majority. *Proc. Natl. Acad. Sci. USA* 95: 6578–6583.

Wikner, J., J. J. Vallino, G. F. Steward, D. C. Smith, and F. Azam. 1993. Nucleic acids from the host bacterium as a major source of nucleotides for three marine bacteriophages. *FEMS Microbiol. Ecol.* 12: 237–248.

Wilcox, R. M., and J. A. Fuhrman. 1994. Bacterial viruses in coastal seawater: Lytic rather than lysogenic production. *Mar. Ecol. Prog. Ser.* 114: 35–45.

Wilhelm, S., and C. Suttle. 1999. Viruses and nutrient cycles in the sea. *BioScience* 49: 781–788.

Wilhelm, S., and C. Suttle. 2000. Viruses as regulators of nutrient cycles in the sea. In C. R. Bell, M. Brylinshri, and P. Johnson-Green (eds.), *Proceedings of the 8th International Symposium of Microbial Ecology.* Atlantic Canada Society for Microbial Ecology, pp. 551–556.

Wilhelm, S., S. Brigden, and C. Suttle. 2002. A dilution technique for the direct measurement of viral production: A comparison in stratified and tidally mixed coastal waters. *Microb. Ecol.* 43: 168–173.

REFERENCES

Williams, P. J. L., and C. Askew. 1968. A method of measuring the mineralization by microorganisms of organic compounds in seawater. *Deep-Sea Res.* 15: 365–375.

Williamson, K., M. Radosevich, and K. Wommack. 2005. Abundance and diversity of viruses in six Delaware soils. *Appl. Environ. Microbiol.* 71: 3119–3125.

Williamson, S., and J. Paul. 2006. Environmental factors that influence the transition from lysogenic to lytic existence in the φHSIC/*Listonella pelagia* marine phage–host system. *Microb. Ecol.* 52: 217–225.

Williamson, S. J., L. A. Houchin, L. McDaniel, and J. H. Paul. 2002. Seasonal variation in lysogeny as depicted by prophage induction in Tampa Bay, Florida. *Appl. Environ. Microbiol.* 68: 4307–4314.

Wilson, W. H., N. G. Carr, and N. H. Mann. 1996. The effect of phosphate status on the kinetics of cyanophage infection in the oceanic cyanobacterium *Synechococcus sp.* WH7803. *J. Phycol.* 32: 506–516.

Winget, D., K. Williamson, R. Helton, and K. Wommack. 2005. Tangential flow diafiltration: an improved technique for estimation of virioplankton production. *Aquat. Microb. Ecol.* 31: 221–232.

Winter, C., A. Smit, G. Herndl, and M. Weinbauer. 2004. Impact of virioplankton on archaea and bacterial community richness as assessed in seawater batch cultures. *Appl. Environ. Microbiol.* 70: 804–813.

Wommack, K., and R. Colwell. 2000. Virioplankton: Viruses in aquatic ecosystems. *Microbiol. Mol. Biol. Rev.* 64: 69–114.

Wommack, K., J. Ravel, R. Hill, J. Chun, and R. Colwell. 1999a. Population dynamics of Chesapeake Bay virioplankton: Total-community analysis by pulsed-field gel electrophoresis. *Appl. Environ. Microbiol.* 65: 231–240.

Wommack, K., J. Ravel, R. T. Hill, and R. R. Colwell. 1999b. Hybridization analysis of Chesapeake Bay virioplankton. *Appl. Environ. Microbiol.* 65: 241–250.

Wright, R. T., and J. E. Hobbie. 1966. Use of glucose and acetate by bacteria and algae in aquatic ecosystems. *Ecology* 47: 447–468.

Zengler, K., G. Toledo, M. Rappé, J. Elkins, E. Mathur, J. Short, and M. Keller. 2002. Cultivating the uncultured. *Proc. Natl. Acad. Sci. USA* 99: 15681–15686.

Zhang, T., M. Breitbart, W. Lee, et al. 2006. RNA viral community in human feces: Prevalence of plant pathogenic viruses. *PLoS Biol.* 4: e3.

Zhong, Y., F. Chen, S. W. Wilhelm, L. Poorvin, and R. E. Hodson. 2002. Phylogenetic diversity of marine cyanophage isolates and natural virus communities as revealed by sequences of viral capsid assembly protein gene g20. *Appl. Environ. Microbiol.* 68: 1576–1584.

Zhu, F., R. Massana, F. Not, D. Marie, and D. Vaulot. 2003. Mapping of picoeucaryotes in marine ecosystems with quantitative PCR of the 18S rRNA gene. *FEMS Microbiol. Ecol.* 52: 79–92.

Zimmermann, R., and L.-A. Meyer-Reil. 1974. A new method for fluorescence staining of bacterial populations on membrane filters. *Kieler Meeresforsch.* 30: 24–26.

13

MOLECULAR ECOLOGICAL ASPECTS OF NITROGEN FIXATION IN THE MARINE ENVIRONMENT

JONATHAN P. ZEHR

Ocean Sciences Department, University of California, Santa Cruz, CA 95064, U.S.A.

HANS W. PAERL

Institute of Marine Sciences, University of North Carolina at Chapel Hill, Morehead City, NC 28557, U.S.A.

INTRODUCTION

Although nitrogen (N) as dinitrogen gas (N_2) is the most abundant element in the atmosphere, N in this form is unavailable to most organisms. N is a major component of many metabolites and structural molecules (amino acids, proteins, and nucleic acids). N_2 fixation is the conversion of the inert atmospheric gas N_2 into biologically available compounds such as ammonium and nitrogen oxides. Most organisms obtain N from combined forms of N, including nitrate, nitrite, urea, ammonium, or organic N, and these compounds are often in low concentrations relative to other elements required for growth. N_2-fixing microorganisms, also called diazotrophs, are able to draw upon the large reservoir of N_2 in the atmosphere, since it readily dissolves in seawater (about 400–575 µmol/L N_2: Sharp 1983).

Microbial Ecology of the Oceans, Second Edition. Edited by David L. Kirchman
Copyright © 2008 John Wiley & Sons, Inc.

In the marine environment, the availability of N is a key factor limiting primary and secondary production (Carpenter and Capone 1983; Dugdale 1967; Ryther and Dunstan 1971). In all marine environments (estuaries, coastal, pelagic), biological demands for N often exceed availability. On ecosystem, regional, and global scales, the availability of N is ultimately controlled by the relative activities of N_2 fixation and the processes (e.g., denitrification) that form N_2 (Codispoti et al. 2001). Historically, geochemists have argued that N is not the nutrient that ultimately limits production in the oceans, since the presence of the genetic potential for obtaining N from the large atmospheric reservoir should alleviate N deficiencies over long time scales (Redfield 1958; Smith 1984; Tyrrell 1999). However, N_2 fixation rates in the ocean are controlled by a myriad of interacting environmental factors, including oxygen tension, turbulence, availability of non-N nutrients and energy needed to sustain this process, and even grazing on short and long time scales (Falkowski 1997; Howarth and Marino 1988; Paerl 1990).

For many years, rates of N_2 fixation in the surface waters of the open ocean were assumed to be negligible relative to the other inputs of N, such as upwelling and advection from nitrate-rich deep ocean waters. However, the results of biogeochemical analyses, and the application of molecular biology and high-sensitivity analytical techniques, have led to a new appreciation of the role of N_2 fixation in the oceans (Capone 2001; Capone et al. 2005; Codispoti et al. 2001; Karl et al. 2002; Lipschultz and Owens 1996). The goal of this chapter is to review nitrogen fixation from fundamental biochemistry and physiology to the biological and molecular biological aspects, focusing on recent discoveries with respect to the open ocean. Other reviews of N_2 fixation, including the biogeochemical perspective, are available (Capone 2000; Herbert 1999; Mahaffey et al. 2005).

CHEMISTRY, BIOCHEMISTRY, AND GENETICS OF N_2 FIXATION

N_2 fixation requires large amounts of energy, since there is a high activation energy for breaking the triple bond of the N_2 ($N{\equiv}N$) molecule (Postgate 1982). Abiological N_2 fixation can occur in the upper atmosphere, catalyzed by lightning discharges that result in the production of oxidized forms of N and ammonium (Galloway et al. 2004). Anthropogenic abiological N_2 fixation occurs in internal combustion engines (Galloway et al. 2004) and the Haber–Bosch industrial synthesis of nitrogenous fertilizers. Anthropogenic N_2 fixation has resulted in global-level perturbations of the N cycle (Galloway et al. 2004; Galloway and Cowling 2002; Howarth et al. 1996; Paerl 1997).

The stoichiometry of biological N_2 fixation is

$$N_2 + 8H^+ + 8e^- + 16ATP \rightarrow 2NH_3 + H_2 + 16ADP + 16P_i$$

where ATP is adenosine triphosphate, ADP is adenosine diphosphate, and P_i represents inorganic phosphates. This reaction is catalyzed by the enzyme nitrogenase, which is composed of two multisubunit metalloproteins. N_2 fixation requires large supplies of ATP and reductant (NADH or NADPH: the reduced forms of nicotinamide adesive dinucleotide and its phosphate, respectively). Hydrogen (H_2) is evolved by nitrogenase, which is recovered by uptake hydrogenases in some microorganisms (Postgate 1998). This reduction of H^+ to H_2, which consumes ATP, links H_2 production and N_2 fixation, and is not well understood (Rees and Howard 2000). H_2 production has been measured in diazotrophic cyanobacteria (Paerl 1982a), and in the ocean during a *Trichodesmium* bloom (Scranton 1983). H_2 production during N_2 fixation can be an energy loss for the diazotrophs, but may be an energy source for the microbial community (Karl et al. 2002), since many microorganisms can obtain energy through H_2 oxidation. The ATP stoichiometry of the reaction shown above is not necessarily fixed, and ratios of ATP per electron transfer can exceed two (Rees and Howard 2000). Differences in nitrogenase efficiencies may affect competition between microorganisms, but the ecological significance of varying efficiencies in the environment is not known.

Genetics and Enzymology

The molecular biology and genetics of the N_2 fixation apparatus is beyond the scope of this chapter, but the structural and functional complexity of the apparatus is important for understanding the evolution, adaptation, and ecology of diazotrophic microorganisms. The two nitrogenase proteins are called component I (MoFe (molybdenum–iron) protein, or dinitrogenase) and component II (the Fe protein or dinitrogenase reductase) (Fig. 13.1). The best-described nitrogenase requires Mo in

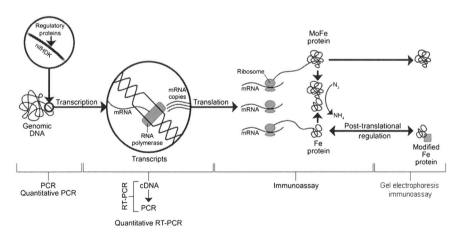

Figure 13.1 Regulation of N_2 fixation at transcriptional, translational and post-translational levels. Different levels of regulation can be assayed using molecular genetic, immunological, and proteomic methods.

a Mo–Fe cofactor (FeMo-co). A discussion of the characteristics of the proteins can be found in Dixon and Kahn (2004) and Rees and Howard (2000).

> The **Fe protein** (about 60 kDa) is composed of a pair of identical subunits encoded by the *nifH* gene. The **MoFe protein** (about 200 kDa) is composed of two sets of heterodimers, composed of the α and β subunits encoded by the *nifD* and *nifK* genes. The multisubunit structure coordinates FeS clusters, called the P clusters, and the FeMo cofactor (FeMo-co), which is presumably the site for substrate (N_2) binding (Howard and Rees 1996; Rees and Howard 2000). The primary function of the Fe protein is to reduce the MoFe protein through a series of single electron transfers. Each electron transfer requires docking and undocking of the two proteins and hydrolysis of ATP (Howard and Rees 1996). The Fe protein plays several roles, including maturation and proper insertion of FeMo-co into the MoFe protein (Rubio and Ludden 2005). The substrate-binding site for N_2 reduction is the FeMo-co in the MoFe protein (Howard and Rees 1996; Rees and Howard 2000).

The structural genes *nifHDK* are often present as a single copy in the genome. However, *Clostridium* has multiple copies of *nifH*, as well as having the conventional and alternative nitrogenases. "Alternative" nitrogenases are often present as second and/or third copies of the *nif* operon that encode non-Mo-containing component I proteins (Bishop and Premakumar 1992; Eady 1996). One of these (the "first alternative") contains Vanadium (V) in place of Mo in the cofactor; the other (the "second alternative") contains only Fe. The alternative nitrogenases differ from the Mo-containing nitrogenase since they contain a third subunit in component I encoded by *vnfG* or *anfG* (Bishop and Premakumar 1992; Eady 1996). The ecological significance of the alternative nitrogenase genes is not known (Eady 1996). *Gammaproteobacteria* with Mo-independent nitrogenases have been isolated from salt marshes (Loveless et al. 1999).

There is a poorly characterized enzyme isolated from a thermophilic *Streptomyces* that catalyzes the reduction of N_2 via a mechanism that is substantially different from nitrogenases (Ribbe et al. 1997; Rubio and Ludden 2005). Unlike other N_2 fixation reactions, the one catalyzed by the *Streptomyces* nitrogenase involves carbon monoxide (CO). The environmental significance is not yet known, but this *Streptomyces* was isolated from a very unusual environment, and it is unlikely that CO levels in the surface ocean could support activity of this nitrogenase.

The Fe protein of the conventional and alternative nitrogenases is highly conserved among microorganisms (*nifH*, *vnfH*, *anfH*). The structural genes for nitrogenase, *nifHDK*, are often contiguous and expressed as a single operon (Figs. 13.1 and 13.2). The abundance of each of the structural gene transcripts can differ, due to multiple transcript start sites or RNA processing. This has significance for assaying the activities of diazotrophs by *nifH* or *nifD* gene expression in the oceans and other natural environments.

> Multiple genes are involved in N_2 fixation. Some gene products are involved in cofactor synthesis and insertion, and others in Mo uptake (Dean and Jacobsen 1992). At least 20 nitrogenase (*nif*) genes have been characterized in *Klebsiella pneumoniae*. The *nif* genes include *nifJ*, *nifH*, *nifD*, *nifK*, *nifT*, *nifY*, *nifE*, *nifN*, *nifX*, *nifU*, *nifS*, *nifV*, *nifZ*, *nifM*, *nifF*, *nifL*, *nifA*, *nifB*, and *nifQ* (Postgate 1998), although not all of these are found in all organisms. The structural genes are sometimes on plasmids, for example, the *Desulfovibrio vulgaris* megaplasmid (200 kbp) and the *Sinorhizobium meliloti* plasmid (1.35 Mbp) (http://www.ncbi.nlm.nih.gov/) (Sobecky et al. 1997).

N_2 fixation activity is directly regulated by the expression (transcription and translation) of the *nifHDK* structural genes (Fig. 13.1). Regulation of the N_2 fixation apparatus is important, since nitrogenase is rapidly inactivated by oxygen, and the energy expenditure involved in N_2 fixation makes it advantageous to use other sources of fixed N when available. In organisms with multiple gene copies, the different nitrogenases can be regulated by Mo and V availability or levels of oxygen (Eady 1996). *Anabaena variabilis* is a heterocyst-forming cyanobacterium containing a nitrogenase that is synthesized in the heterocyst (a specialized cell that fixes N_2 and has reduced photosystem II activity) and another that is expressed in vegetative cells under anaerobic or microaerophilic conditions (Thiel et al. 1995). *A. variabilis* also contains a V nitrogenase (Thiel 1993).

Evolution of N_2 Fixation

The diversity of N_2-fixing microorganisms and the distribution in different habitats is partially a legacy of the evolution of life on Earth. The evolution of N_2 fixation has been reviewed elsewhere (Berman-Frank et al. 2003; Raven and Yin 1998; Zehr et al. 2006). Although the Earth's atmosphere has always been dominated by N_2 (Galloway 2003), ammonium may have been present in the early Earth's atmosphere, and this has been argued to preclude the need for N_2 fixation (Towe 2002). It is sometimes debated whether nitrogenase evolved early, or was laterally transferred later in evolution (Postgate and Eady 1988; Raymond et al. 2004; Young 1992, 2005). Coincident with changes in oxygen concentrations in the atmosphere and oceans were changes in the chemistry and availability of metals and ions required for N_2 fixation. The early oceans are believed to have been anoxic, and hence reduced, with high concentrations of biologically available ferrous ions (Fe^{2+}) (Raven and Yin 1998). Thus, the Fe nitrogenase may have been the first nitrogenase, with the Mo and V nitrogenases arising later (Berman-Frank et al. 2003; Raymond et al. 2004). Most phylogenetic analyses conclude that nitrogenase evolved during the early stages of prokaryotic evolution, although there are examples of lateral transfer, and lateral transfer is required to explain the current distribution of Mo and V nitrogenases (Fani et al. 2000; Leigh 2000; Raymond et al. 2004; Young 1992; Zehr et al. 1995).

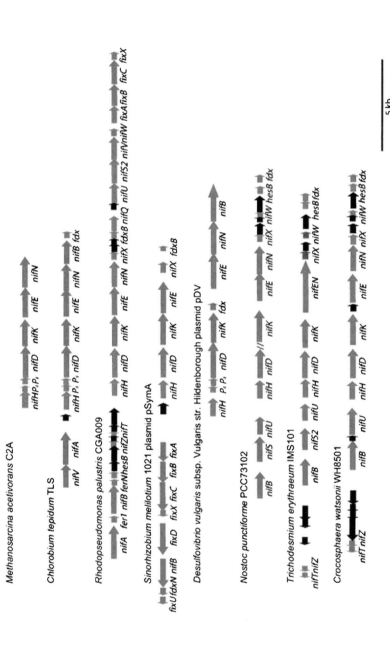

Figure 13.2 Nitrogenase gene arrangements from genomes of a number of microorganisms. Multiple genes are required for assembly and activity of nitrogenase. There is conservation of gene organization in related organisms, but also differences between major taxonomic groups. Gene arrangements and other *nif* genes could be targets for studying genetic diversity in natural populations.

> Nitrogenase, or some form of prenitrogenase genes, could have initially served a function other than for N_2 fixation. It could have played a role in cyanide reduction early in the Earth's history when cyanide was produced by photooxidation of atmospheric methane (Kasting and Siefert 2001), and then later became functional in N_2 fixation (Fani et al. 2000; Kasting and Siefert 2001; Raymond et al. 2004). It is also conceivable that nitrogenase served to make N biologically available early on, if ammonium in the Archean was depleted by early life, including methanogens, other anaerobic chemoheterotrophs, and autotrophs (Ehrlich 1990; Kasting and Siefert 2002; Nealson and Rye 2004; Raven and Yin 1998; Rosing 1999; Sprent and Raven 1992). These organisms would have been present about 3.8 Ga ago (Nisbet and Fowler 2003). Extant methanogens and sulfate reducers have *nif* genes (Leigh 2000), and were probably present over 3.5 Ga ago (Canfield 1998; Shen et al. 2001).

Phylogeny of Nitrogenase

Phylogenetically, the nitrogenase Fe protein *(nifH)* genes form four or five deeply branching clusters (Fig. 13.3) (Zehr et al. 2003). One of these comprises the evolutionarily related protochlorophyllide reductases (see below). Another deep cluster includes *nif*-like genes from *Archaea* that are not involved in N_2 fixation. The three major clusters, Clusters I–III, all contain *nifH* genes that encode active nitrogenases. Cluster II comprises the second alternative *nifH* genes, the Fe-only nitrogenases. The first alternative V-containing nitrogenases do not form a distinct cluster in the *nifH* tree, but can usually be distinguished from their conventional nitrogenase within the same organism (e.g., *Azotobacter vinelandii* or *Anabaena variabilis*).

Within Cluster I, *nifH* and 16S rRNA phylogenies are largely congruent. The phylogeny of *nifH* provides a way to characterize or identify uncultivated N_2-fixing microorganisms (Zehr et al. 2003). Ultimately, confirming the classification of marine diazotrophs is dependent upon cultivation or by linking *nifH* genes to rRNA genes by genomic approaches, since there are some examples of possible lateral gene transfer of *nifH* (Cantera et al. 2004; Raymond et al. 2004).

Genomics of N_2 Fixation

The genomes of relatively few N_2-fixing microorganisms have been sequenced. Nitrogenase genes have not been found in the genomes of abundant marine organisms, including *Pelagibacter* (Giovannoni et al. 2005), which is a cultured representative of the abundant SAR11 clade, *Silicibacter* (Moran et al. 2004), *Prochlorococcus* (Rocap et al. 2003), and marine *Synechococcus* (Palenik et al. 2003). Chapter 4 discusses genomics in marine microbial ecology.

There are draft genome sequences of two important oceanic diazotrophs: *Trichodesmium* and *Crocosphaera* (www.jgi.doe.gov). The nitrogenase gene organizations of these two cyanobacteria are highly conserved along with the genes of other cyanobacteria (Fig. 13.2). The interesting contribution of the

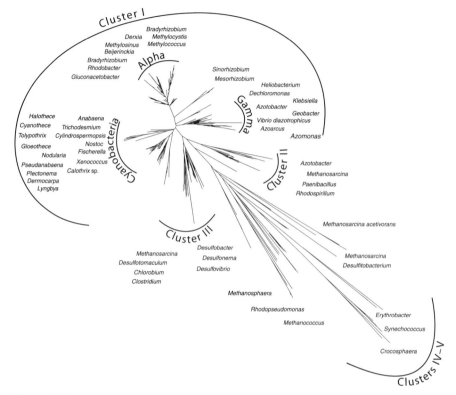

Figure 13.3 Phylogenetic distribution of nitrogenase (*nifH*) genes, showing representative cultivated microorganisms. Diazotrophs include cyanobacteria, *Proteobacteria*, and gram-positive microorganisms with diverse physiologies.

genome sequences is not the nitrogenase genes themselves, but the genomic information that can provide hypotheses as to how these microorganisms compete in the marine environment. For example, genomic analysis has suggested that there may be differences in the phosphorus-utilization pathways of *Trichodesmium* and *Crocosphaera* (Dyhrman et al. 2006; Dyhrman and Haley 2006). The *dps* gene that encodes an Fe-binding protein was found and subsequently expressed in *Escherichia coli* (Castruita et al. 2006). The genomes of several more marine diazotrophic cyanobacteria are currently being sequenced (http://www.moore.org/microgenome, http://www.wust.edu).

There may be differences in regulatory pathways as well. Cells sense and respond to the environment using signal transduction systems composed of sensor and response regulator protein domains. One-component and two-component signal transduction systems have been identified in *Trichodesmium* (Ulrich et al. 2005), but the number of one-component systems is small for its genome size (Ulrich et al. 2005). In the draft sequences, there were many more two component systems in *Crocosphaera* than in *Trichodesmium* (Zehr et al. 2005).

CHEMISTRY, BIOCHEMISTRY, AND GENETICS OF N_2 FIXATION

Diversity of N_2-Fixing Microorganisms

There are diverse N_2-fixing microorganisms in terrestrial and aquatic environments. N_2 fixation has not yet been found in Eukarya, but is found in the *Bacteria* and *Archaea* domains (Young 1992). N_2-fixing *Bacteria* and *Archaea* are genetically, physiologically, and ecologically diverse, and N_2 fixation appears to be widely scattered throughout prokaryotic lineages (Fig. 13.3). However, the nitrogenase genes are not necessarily conserved within lineages, so that closely related microorganisms may differ in their ability to fix N_2. For example, *Vibrio diazotrophicus* fixes N_2, but *V. cholerae* does not.

N_2 fixation and the nitrogenase genes are found in chemoheterotrophs (aerobic, anaerobic, and facultative), chemolithotrophs, photolithotrophs, and photoheterotrophs (see Chapter 1 for definitions of these terms). Diazotrophic representatives include (1) phototrophic bacteria (other than cyanobacteria, such as *Chlorobium*, *Chromatium*, and *Rhodospirillum*), (2) cyanobacteria, including all heterocystous filamentous (*Aphanizomenon*, *Calothrix*, and *Nodularia*), some nonheterocystous filamentous (*Oscillatoria*, *Lyngbya*, and *Trichodesmium*) and unicellular (*Gloeothece* and *Synechococcus*) genera, (3) strict anaerobes (e.g., *Clostridium* and *Desulfovibrio*), (4) heterotrophs (e.g., *Klebsiella*, *Vibrio*, and *Azotobacter*), (5) Fe oxidizers (e.g., *Thiobacillus*), and (6) archaeal methanogens.

Examples from all of these groups are found in marine environments, although different groups are most abundant or active only in specific habitats consistent with requirements for oxygen, energy (light and organic matter), and macro- and micronutrients (Capone 1983; Paerl 1990; Potts 1980).

Regulation in Diazotrophs

N_2 fixation provides a means of obtaining N for growth at the expense of ATP and reductant. The nitrogenase enzyme is sensitive to oxygen, yet for some organisms oxygen is required as an electron sink to support the respiration needed to supply energy. The N fixed must balance growth requirements, or excess N will be lost to the environment and to other organisms. The nitrogenase reaction regenerates electron acceptors, and is an electron sink. In general, the major factors that regulate nitrogenase gene expression and activity are energy (e.g., light in the case of some phototrophs and photoheterotrophs), O_2 concentrations, and fixed inorganic N availability. Trace-metal availability for the metallocenters (Mo and V) can also control expression of the *nif* and alternative nitrogenase (*vnf* and *anf*) genes. Presumably, Fe availability, believed to limit productivity and N_2 fixation in some areas of the oceans (Berman-Frank et al. 2001a; Falkowski 1997), may also regulate N_2 fixation even at the gene expression level. The specifics of the regulatory networks (e.g., *nifLA* genes) differ among organisms, but the important point is that complex networks have evolved to balance N and carbon metabolism. An in-depth review of the general physiology and regulation of N metabolism and N_2 fixation is beyond the scope of this chapter, and is provided elsewhere (Dixon and Kahn 2004; Karl et al. 2002; Martinez-Argudo et al. 2005; Ninfa and Jiang 2005).

Methods for Assessing Diazotroph Diversity, Gene Expression, and N_2 Fixation Activity

N_2 fixation activity of enzymes or cells can be measured directly by using N_2 enriched in the stable isotope ^{15}N or indirectly using acetylene reduction (Stewart et al. 1967). Rates of N_2 fixation are most directly measured using the stable isotope ^{15}N. Acetylene (CH≡CH) is an analogue of N_2 and is reduced by conventional nitrogenases to ethylene, which is easily measured by gas chromatography or, more recently, by photoacoustics (Zuckermann et al. 1997). These methods are reviewed elsewhere (Capone 1983; Capone and Montoya 2001; Paerl 1998; Zehr and Montoya 2007). Relatively recently, stable-isotope techniques have been developed that can assay N_2 incorporation at the cell level. Secondary ion mass spectrometry (SIMS) has been used to visualize isotope incorporation of ^{13}C at cell-level resolution (Orphan et al. 2001), and applications for ^{15}N are being developed (D. G. Capone, personal communication). The natural abundance of ^{15}N in particulate material can also be used to estimate the importance of N_2 fixation rates in natural samples (Zehr and Montoya 2007). Particulate N that has been fixed from atmospheric N_2 has a ratio of $^{15}N : ^{14}N$ slightly lower than atmospheric N_2.

Historically, N_2 fixers have been identified, enumerated, and characterized by classical microbiological enrichment techniques (Guerinot and Colwell 1985; Wynn-Williams and Rhodes 1974). Cultivation techniques can yield N scavengers (surviving on trace concentrations of fixed N contaminants in media) rather than diazotrophs (Wynn-Williams and Rhodes 1974). For years, the important diazotroph *Trichodesmium* (Fig. 13.4) eluded cultivation, but within the last two decades a few strains have been retained in many laboratories. One of these strains, *Trichodesmium* sp. IMS 101, originally isolated from the Gulf Stream off the North Carolina Atlantic coast (Prufert-Bebout et al. 1993), is now available in culture collections, and a draft genome sequence is also available. *Crocosphaera*, a unicellular cyanobacterium (Fig. 13.4), was cultivated from Atlantic waters in the mid-1980s (Waterbury and Rippka 1989), and related strains have been cultivated from both the Atlantic and Pacific (Falcón et al. 2004b; Zehr et al. 2001a, b). Recently, the symbiont of *Chaetoceros*, *Calothrix* sp. (Fig. 13.4), was isolated and is now maintained asymbiotically in N-free media (Foster et al. 2007).

The high degree of similarity of the *nifH* gene among diverse taxa has made it attractive for use in cultivation-independent molecular studies, since *nifH* can be amplified with "universal" degenerate polymerase chain reaction (PCR) primers (Langlois et al. 2005; Lovell et al. 2000; Mehta et al. 2003; Olson et al. 1999; Steppe et al. 1996; Zehr and McReynolds 1989; Zehr et al. 1995, 2003; Zehr and Capone 1996; Zehr and Paerl 1998). Phylogenetic analysis of amplified *nifH* sequences can help to identify diazotrophs (Fig. 13.3), since *nifH* phylogeny largely mirrors rRNA phylogeny (Zehr et al. 2003). Functional gene cultivation-independent approaches are useful, since gene transcription into messenger RNA (mRNA) can be directly assessed (Omoregie et al. 2004b; Zani et al. 2000). Gene transcription does not necessarily lead to N_2 fixation, since subsequent protein translation, modification, and transcript degradation can regulate N_2 fixation activity

CHEMISTRY, BIOCHEMISTRY, AND GENETICS OF N_2 FIXATION

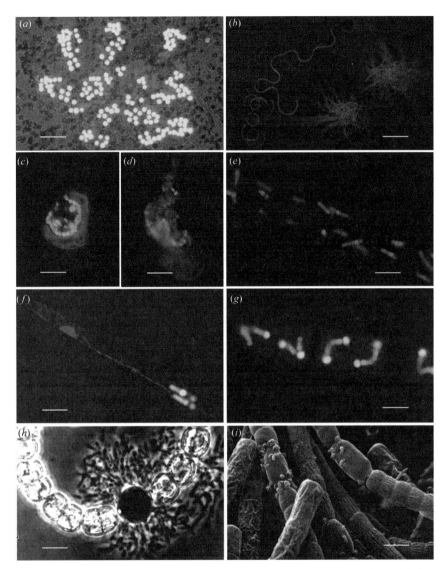

Figure 13.4 Photomicrographs of cultivated and uncultivated marine cyanobacterial diazotrophs. (*a*) Uncultivated unicellular cyanobacteria from the open ocean that are likely diazotrophs. (*b*) Diversity of *Trichodesmium* morphology in natural populations. (*c, d*) Symbiotic unicellular cyanobacteria. (*e*) Heterocystous symbiont of diatoms. (*f*) Heterocystous symbiont of diatoms. (*g*) Heterocystous symbiont of diatoms. (*h, i*) Association of bacteria with heterocysts of heterocyst-forming cyanobacteria. (See insert for color representation.)

(Fig. 13.1). The detection of *nifH* mRNA potentially identifies the organisms likely to be involved in N_2 fixation. The large database of *nifH* gene sequences facilitates the application of other techniques to quantify nitrogenase genes and to assess diversity. Methods such as terminal restriction fragment length polymorphism (T-RFLP) and denaturing gradient gel electrophoresis (DGGE) have been used for *nifH* gene characterization. Quantitative PCR methods have made it possible to target specific groups of microorganisms to determine the distribution and autecology of specific *nifH* phylotypes (Short et al. 2004). This approach can be combined with reverse transcription to identify actively transcribed genes (Church et al. 2005b; Short and Zehr 2007).

DNA macroarrays and microarrays have been used to assay whole-genome expression, or to examine diversity using "phylochips" (Jenkins et al. 2004; Moisander et al. 2006; Steward et al. 2004; Taroncher-Oldenburg et al. 2003; Tiquia et al. 2004). In this approach, different *nifH* gene probes are spotted on the DNA microarray, and the array is then hybridized to *nifH* amplification products (Fig. 13.5).

Estimates of N_2 fixation rates in the ocean obtained from biogeochemical analysis have recently been reviewed (Mahaffey et al. 2005). N_2 fixation can be measured by incorporation of N over time during incubation. This method is only applicable when both rates and biomass are high, which occurs in blooms in some estuarine and coastal systems (e.g., the Baltic Sea) (Elmgren and Larsson 2001). Over large spatial scales, the ratios of regeneration of inorganic N and phosphorous (P) have been used to identify regions of N_2 fixation. N_2 fixation rates are estimated by gradients in N^* (see the text box) integrated along isopycnal surfaces over appropriate spatial scales if rates of mixing along the isopycnals are estimated.

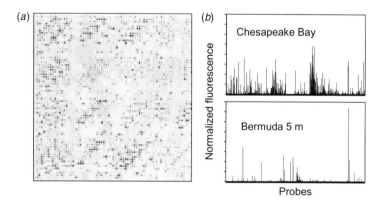

Figure 13.5 (*a*) A *nifH* microarray hybridization to an oligonucleotide microarray containing 768 *nifH* probes. *nifH* was amplified from a Chesapeake Bay North Bay surface water sample collected in October 2002. (*b*) Response of individual probe spots in hybridizations to samples from the Chesapeake Bay and Sargasso Sea (Bermuda Hydrostation, 5 m depth). The target concentration was the same in each of the hybridizations, and the data were normalized using internal standards.

CHEMISTRY, BIOCHEMISTRY, AND GENETICS OF N_2 FIXATION

> Elevated dissolved inorganic N:P ratios are produced from regeneration of particulate material with non-Redfield high N:P elemental ratios that result from diazotrophy. N^* is a variable that is a linear function of regenerated N and P (forced to global Redfield ratios of N:P and global $N^* = 0$). Integration of N^* has been used to estimate basin scale N_2 fixation and denitrification rates and identify regions of activity (Deutsch et al. 2001; Gruber and Sarmiento 1997). These calculations require multiple assumptions, including the global balance of denitrification and N_2 fixation, an estimate of the elemental composition of diazotrophs, and extrapolation along pycnoclines over large spatial scales. Furthermore, N^* does not necessarily reflect the composition of local sinking material due to the time scales of remineralization (Mahaffey et al. 2005). Chapter 14 also discusses N^*.

Remote sensing and modeling approaches are needed to scale rate information to regional and ocean basin scales. In the open ocean, blooms of *Trichodesmium* and symbiont-containing diatoms can be detected by remote sensing (Fig. 13.6) (Capone et al. 1998; Carpenter et al. 1999; Dupouy et al. 1998; Hood et al. 2002; Subramaniam et al. 1999, 2002). There are constraints on remote sensing, but blooms of highly reflective microorganisms (such as *Trichodesmium* gas vacuoles) concentrated in the surface waters can be detected by satellite or aircraft sensors. Detection of surface-dwelling blooms of the heterocystous, aggregate-forming cyanobacteria *Nodularia*, *Aphanizomenon*, and *Anabaena*, which are a dominant feature of the surface waters of the Baltic Sea and are a significant source of N (Elmgren and Larsson 2001; Kononen et al. 1996), could also be approached using remote sensing techniques.

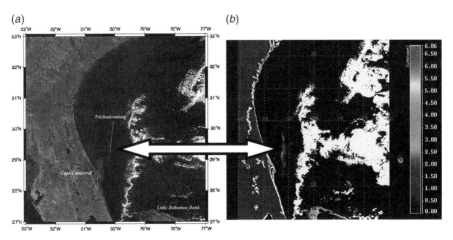

Figure 13.6 Remote sensing of blooms of *Trichodesmium*. (*a*) True-color image of surface ocean, indicating position of *Trichodesmium* bloom detected by remote sensing. (*b*) Remote sensing image of *Trichodesmium* using the algorithm of Subramaniam et al. (2002). Modified from Subramaniam et al. (2002) and used with permission from Elsevier. (See insert for color representation.)

Models have focused primarily on *Trichodesmium* (Hood et al. 2001). The model of Hood et al. (2001) assumes that diazotrophic growth is a function of light, and predicts the distribution of *Trichodesmium* in the North Atlantic Ocean. Interestingly, this model predicts high abundances and N_2 fixation rates of *Trichodesmium* as part of successional events from upwelling regions (Hood et al. 2004). Coles et al. (2004) estimated the contribution of N_2 fixation to phytoplankton in the North Atlantic on the basis of relationships between chlorophyll, sea-surface height (SSH), and sea-surface temperature (SST), which were observed by remote sensing. A model for the North Pacific incorporated light, temperature, P, and wind stress, and the N_2 fixation rate was predicted from growth rate and elemental composition (Fennel et al. 2005). N_2 fixation can also be estimated from CO_2 drawdown (Lee et al. 2002).

Models of N_2 fixation cellular processes have begun to be devised to predict daily N_2 fixation patterns in individual populations or species (Rabouille et al. 2006; Stephens et al. 2003). These models are useful for generating hypotheses for how and why diazotrophs fix N_2 during certain periods and to understand the energetics of N_2 fixation and photosynthesis. Ultimately, models at this cellular scale will provide a necessary link between ecosystem models and organismal biology.

ECOPHYSIOLOGICAL ASPECTS OF N_2 FIXATION

N_2 fixation rates in the environment are controlled by microbial community structure, grazing, and environmental factors that regulate gene expression and activity of populations. Environmental factors that have been hypothesized or shown to regulate and control the activity, abundance and distribution of marine diazotrophs include (1) energy in the form of light or organic matter; (2) oxygen inhibition; (3) nutrient availability, including silica, N and P, N : P ratios, chemical forms and sources; (3) Fe and other trace elements (e.g., Mo); (4) organic matter composition and concentration; (5) salinity; (6) fixed N availability; (7) temperature; (8) water column turbulence, stratification and stability; and (9) biotic interactions (symbiosis, synergism, and grazing) (cf. Paerl 1990).

Cyanobacterial N_2 fixation is supported directly, in the light, by photosynthesis, or indirectly through the metabolism of fixed carbon compounds. Cyanobacteria that fix N_2 in the light are primarily filamentous cyanobacteria that form heterocysts—specialized cells for N_2 fixation that lack photosynthetic O_2 evolution. *Trichodesmium* also fixes N_2 primarily during the day. Most unicellular or filamentous non-heterocyst-forming cyanobacteria fix N_2 during the night, using respiration of carbon storage compounds for energy production. A recent study in hot spring mats indicated that fermentation may support dark N_2 fixation in a unicellular cyanobacterium (Steunou et al. 2006). Since increasing light intensity increases photosynthetic rates up to a point, increasing light intensity increases N_2 fixation rates in cyanobacteria that fix N_2 during the light. Both photosynthesis-versus-irradiance curves (P vs. I) and photosynthesis versus N_2 fixation curves exhibit saturating behavior, with inhibition at higher light intensities. Presumably, energy does not limit phototrophic diazotrophs in the upper euphotic zone or the surfaces of microbial mats. In the case of N_2 fixation, inhibition at higher light intensities is partially due to the inhibition of nitrogenase activity by oxygen evolved in photosynthesis.

Heterotrophic N_2 fixation can be supported by a wide variety of organic substrates, metabolized aerobically (e.g., *Azotobacter*) or anaerobically (e.g., clostridia, methanogens, and sulfate reducers). Energy is probably an important factor regulating heterotrophic diazotrophs in the oligotrophic ocean, but perhaps is not as important in microbial mats. Interestingly, the expression of heterotrophic *nifH* genes exhibited a slight diel cycle (Church et al. 2005b), which may reflect the link between heterotrophic metabolism and organic matter substrates provided by photoautotrophs.

Several different metabolic strategies are used by diazotrophic microorganisms to obtain energy while avoiding inactivation by oxygen. Aerobic heterotrophs (*Azotobacter*) use polysaccharide capsules to restrict influx of O_2 from the environment, while maintaining high respiratory rates to maintain low intracellular O_2 concentrations and generate ATP. Strict anaerobes, such as *Clostridium*, support N_2 fixation with fermentative metabolism in anoxic habitats. Purple bacteria, including *Rhodospirillum* and *Rhodobacter*, can support N_2 fixation by photoautotrophic (anaerobic in the light) or photoheterotrophic (microaerophilic) metabolism (Postgate 1998). Under some circumstances, nitrogenase can act as an electron sink rather than a N source, such as in *Rhodobacter*, when other electron acceptors are unavailable (Tichi and Tabita 2001).

The cyanobacteria have a number of strategies for fixing N_2 while depending upon a phototrophic metabolism that generates oxygen (Bergman et al. 1997; Berman-Frank et al. 2003; Gallon 1992). Some cyanobacteria spatially separate N_2 fixation from photosynthesis by forming heterocysts. Many cyanobacteria separate the two processes in time, and they fix N_2 primarily in the dark. Organisms such as *Trichodesmium* fix N_2 during the day, yet do not have heterocysts. This paradox has been the subject of many studies.

Patterns of localization of the nitrogenase protein in different cells along filaments (termed diazocytes: El-Shehawy et al. 2003; Fredriksson and Bergman 1995) indicate that there may be some differences in distribution of nitrogenase in cells along the filaments, but the expected inverse relationship between nitrogenase activity and photosystem II oxygen evolution in individual cells has not yet been demonstrated. Paerl et al. (1989) used immunolocalization to demonstrate that nitrogenase was widely distributed along and among individual trichomes of *Trichodesmium*. Microelectrode and cellular $^{14}CO_2$ microautoradiographic studies of photosynthetic activity indicated that localized differences in oxygen tension and reductant supplies, may be regulating enzyme activity in aggregates (Paerl and Bebout 1988b). Berman-Frank et al. (2001b) suggested that the Mehler reaction is elevated during the middle part of the light period to alleviate O_2 inactivation of nitrogenase during the day. This mechanism was hypothesized to work in concert with a complex mechanism involving temporal (short time interval) and spatial (different cells) separation of photosynthesis and N_2 fixation (Berman-Frank et al. 2001b; Küpper et al. 2004). Clearly, the mechanisms involved are still only partially understood, and, as a result, remain unclear.

Since N_2 fixation is generally more energetically expensive than the metabolism of fixed dissolved inorganic N, the presence of dissolved inorganic N should affect the selection for diazotrophs and the expression of nitrogenase activity. In general, *nif* gene transcription is repressed in the presence of ammonium, although there are exceptions. Nitrate and ammonium inhibit nitrogenase activity and gene expression in a number of cyanobacteria, including heterocystous and unicellular strains (Flores and Herrero 1994). Interestingly, the reduction of nitrate appears to be thermodynamically less favorable than N_2 fixation (Karl et al. 2002): it is the high activation energy required to break the triple bond of N_2 that makes N_2 fixation energetically expensive. Holl and Montoya (2005) showed that *Trichodesmium* rapidly shifts to use of nitrate even at relatively low concentrations (micromolar), and shifts back to diazotrophy when nitrate concentrations fall below micromolar levels. N_2 fixation by *Nodularia* in the presence of low concentrations of nitrate has been observed in culture (Sanz-Alférez and Campo 1994). Since some microorganisms continue to fix N_2 in the presence of nitrate, it is perhaps not surprising that cyanobacterial *nifH* expression was detected in the Chesapeake Bay in the freshwater reaches where nitrate concentrations were micromolar (Short and Zehr 2007). Intriguingly, the diversity of the genes and organisms is not necessarily correlated with the degree of N limitation (Fig. 13.7) (Zehr et al. 2003). Environments that are not extremely N-limited, such as sediments with high concentrations and gradients of ammonium (see Chapter 14), can have a high diversity of *nifH* genes (Burns et al. 2002). Mats have diverse *nifH* genes. Here, N_2 fixation is largely carried out by phototrophs, which provide the bulk of the biomass (Bebout et al. 1993).

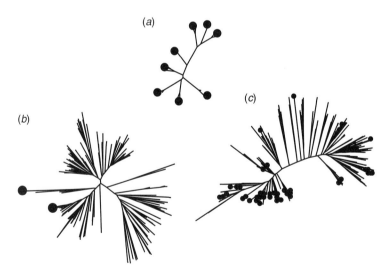

Figure 13.7 Diversity of *nifH* genes and expression in open-ocean (*a*), Chesapeake Bay Estuary (*b*), and Guerrero Negro cyanobacterial mat (*c*) habitats. Circles indicate genes that are expressed, highlighting differences in genetic potential and the number of genes that are expressed in different habitats.

> The Chesapeake Bay (Maryland and Virginia) and Neuse River (North Carolina) Estuaries have a very high diversity of *nifH* genes (Affourtit et al. 2001; Jenkins et al. 2004). Manipulations of salt marsh sediments with additions of fixed N resulted in little perturbation of the diazotroph community (Bagwell and Lovell 2000), which suggests that the diazotrophs are not selected or retained because of the function of their *nif* genes. *nifH* phylotypes have their own spatial and temporal distributions (Short et al. 2004), which would suggest that microorganisms containing the different *nifH* gene types (*nifH* phylotypes) are selected by the environmental factors. However, since most of the genes are not expressed (Fig. 13.7), the distribution is not due to the selective advantage of the *nif* genes themselves.

Specific nutrient deficiencies may restrict the activities and geographic extent of N_2 fixation. The trace metals Mo and Fe (and potentially V) are important components of nitrogenase. Mo has been investigated as a nutrient potentially limiting N_2 fixation (Cole et al. 1993; Howarth and Cole 1985). Howarth and Cole (1985) proposed that the relatively high (>20 mmol/L) concentrations of sulfate (SO_4^{2-}), which is a structural analogue of the most common form of molybdenum found in seawater, namely, molybdate (MoO_4^{2-}), could competitively (via the uptake process) inhibit N_2 fixation. Competitive inhibition of MoO_4^{2-} uptake by high SO_4^{2-} concentrations was shown by Cole et al. (1993). However, MoO_4^{2-} is highly soluble in seawater, with concentrations on the order of 100 µmol/L (Collier 1985). Ter Steeg et al. (1986) and Paulsen et al. (1991) showed that, despite the potential for SO_4^{2-} competition, Mo availability was ensured at concentrations much lower than 100 µmol/L. The presence of "alternative" non-Mo-requiring nitrogenases in bacterial and cyanobacterial diazotrophs (Bishop and Premakumar 1992) could provide a mechanism by which Mo limitation can be circumvented.

Fe is required for the metallocenters of nitrogenase, and diazotrophic microorganisms exhibit a relatively (compared with Mo) high demand for this metal. Fe was initially suggested to be important in controlling N_2 fixation in *Trichodesmium* (Rueter 1988). It appears to be generally assumed that Fe controls N_2 fixation in the oceans (Berman-Frank et al. 2001a). The chemistry of Fe in seawater is complex. Interestingly, Fe limitation appears to be far less commonplace in the benthos than in the planktonic environment (Paerl and Prufert 1987; Paerl et al. 1994; Paerl 2000). This is likely because Fe is sequestered from the water column as either precipitated or particle-associated Fe, which is then solubilized in periodically reduced sediments.

Dissolved organic matter (DOM) content has been suggested to be a possible modulator of N_2-fixing cyanobacterial growth and bloom potential. The hypothesized mechanism for DOM-stimulated cyanobacterial growth is that DOM induces nutrient assimilatory enzymes and heterotrophy (Antia et al. 1991) and/or provides energy and nutrition for associated heterotrophic bacteria, which are known to optimize the growth of "host" cyanobacteria (Paerl and Pinckney 1996a, b). Heterotrophic N_2-fixing microorganisms have been discovered in the open ocean, based on the presence and expression of *nifH* genes (Bird and Karl 1991; Church et al. 2005a; Falcón et al. 2004a; Langlois et al. 2005; Zehr et al. 1998). Although these organisms have not been cultivated, these bacteria are likely to be heterotrophic *Gamma-* or *Alphaproteobacteria* that rely at least partially on DOM for energy. DOM availability in the surface ocean may limit heterotrophic N_2 fixation.

Salinity, per se, is not a strong modulator of either the establishment or activities of diazotrophs. A wide variety of active N_2-fixers has been observed in the plankton and benthos of estuarine, coastal, and open-ocean environments, and even hypersaline lakes and lagoons (Paerl et al. 2000, 2003; Potts 1980). However, hypersaline conditions, where salinity greatly exceeds (by more than fourfold) seawater levels (35 psu), have been shown to inhibit N_2 fixation (Paerl et al. 2003; Pinckney and Paerl 1997). Factors other than salinity, including nutrient limitation, turbulence, organic matter supply, and grazing (Marino et al. 2006), appear to control N_2 fixation activity along the freshwater to open ocean salinity continuum.

Temperature is an important factor regulating enzymes and the distribution of organisms. N_2 fixation is found in a wide spectrum of temperatures, from Antarctic lakes to hot springs (Paerl and Priscu 1998; Postgate 1998; Steunou et al. 2006). It is interesting that the most abundant cyanobacterial diazotrophs in the oceans are either filamentous nonheterocystous cyanobacteria or unicellular cyanobacteria. It is not clear why free-living heterocystous cyanobacteria are not more abundant in the tropical open ocean, although heterocystous strains are often found in symbiotic relationships in the plankton (Villareal 1992). An interesting hypothesis implicates a possible role of temperature in selecting for filamentous nonheterocystous cyanobacteria. Staal et al. (2003) showed that the diffusion of oxygen into N_2-fixing cells is constrained by the solubility of oxygen (which is a function of temperature and salinity) and that there was no advantage for a heterocyst cell wall in the tropical open ocean, since oxygen diffusion would limit the respiration required to support N_2 fixation.

Turbulence exerts a strong impact on phytoplankton growth and structural integrity (Fogg et al. 1973; Thomas and Gibson 1990). Increased levels of turbulence may inhibit growth of diazotrophs (Fogg et al. 1973; Fogg 1982). Aquatic environments with persistent elevated turbulence may have a lower abundance of active N_2-fixing heterocystous cyanobacteria. In laboratory experiments where shear rates representative of surface wind-mixed conditions were applied to bloom-forming cyanobacteria (*Anabaena* and *Nodularia*), Kucera (1996) and Moisander (2002a) showed that rates of N_2 fixation and photosynthesis can be suppressed by strong turbulence. However, other studies did not find a negative relationship between turbulence and N_2 fixation (Burford 2006; Gervais et al. 1997; Howarth et al. 1993; Keto et al. 1992; Nakano et al. 2001). The negative impacts of elevated shear could be due to (1) breakage or weakening of cyanobacterial filaments, specifically at the delicate heterocyst–vegetative cell junction, causing O_2 inactivation of nitrogenase in heterocysts (Fogg 1969); (2) disruption of bacterial–cyanobacterial associations (Paerl 1990).

Diazotrophs form many symbiotic relationships with organisms in terrestrial and aquatic environments. These can be mutualistic interactions or loose associations. There are a number of symbiotic relationships between heterocystous cyanobacteria and the diatom genera *Rhizosolenia*, *Hemiaulus*, *Bacteriastrum*, and *Chaetoceros* in the marine environment (Villareal 1992) (Fig. 13.4). Numerous other symbioses have been reported (Carpenter and Foster 2002; Foster et al. 2006a, b), including unicellular cyanobacterial associations with tintinnids, dinoflagellates, and radiolarians (Foster et al. 2006a, b). Heterotrophic bacterial symbionts in the diatoms *Rhizosolenia* have also been reported (Martinez et al. 1983). Evidence of N_2 fixation was found in symbionts of a dinoflagellate (*Histioneis* sp.) host using immunolabeling with nitrogenase coupled to transmission electron microscopy (TEM) (Foster et al. 2006a). In addition, 16S rRNA sequences obtained from another *Histioneis* sp. host were similar to the diazotroph *Cyanothece* sp. ATCC, and provided evidence for N_2-fixing potential in the cyanobiont (Foster et al. 2006b).

Cyanobacterial–microbial associations are commonly observed among naturally occurring and cultured bloom-forming genera, including *Anabaena*, *Aphanizomenon*, *Nodularia*, and *Trichodesmium* (Paerl and Kellar 1978; Paerl 1982b, 2000; Paerl and Millie 1996). Examples of consortial N_2 fixation are the aggregates of the common heterocystous cyanobacterium *Nodularia* in the Baltic Sea, and species of the nonheterocystous aggregate-forming *Lyngbya* as planktonic and benthic aggregates (Paerl and Kuparinen 2002). Associated microorganisms include bacteria, fungi, protozoans, and eukaryotic algae. Many associations appear to be intimate, occurring within and around colonies and aggregates of filaments, and within fibrillar–mucilaginous sheaths, capsules, and exuded slimes (Paerl 1982b) (Fig. 13.4).

Specific associations, where certain microbial populations exclusively attach to specific types of cyanobacterial host cells (i.e., akinetes and heterocysts), have also been observed (Fig. 13.4). The mechanistic basis for cyanobacterial–bacterial synergism is poorly understood, and remains a subject of intense scrutiny. Proposed mutually beneficial mechanisms include exchange of metabolites and growth factors, as well as detoxifying roles of associated bacteria (Paerl 1982b; Paerl et al. 2000).

Ecology of Diazotrophs in the Open Ocean

Biological evidence on the rates and distribution of diazotrophs points to higher diversity, and wider geographical distributions and activity, than previously appreciated (Capone et al. 2005; Church et al. 2005a; Langlois et al. 2005; Montoya et al. 2004; Zehr et al. 2001b). Several diazotrophic cyanobacterial genera have long been recognized even without cultivation-independent approaches. These organisms include *Trichodesmium*, which has been mentioned several times already. Long before molecular studies and its cultivation in the laboratory, this microbe was well recognized to be a filamentous, non-heterocyst-forming cyanobacterium that forms aggregates or colonies visible to the unaided eye (Capone et al. 1997; LaRoche and Breitbarth 2005) (Fig. 13.4). Key characteristics of *Trichodesmium* are its colonial growth (although it also occurs as free filaments), the ability to fix N_2 simultaneously with O_2 evolution from oxygenic photosynthesis, the formation of dense surface blooms, and buoyancy regulation by gas vesicles (LaRoche and Breitbarth 2005). There appear to be several species of *Trichodesmium*, based on morphology of the filaments (trichomes) and genetic analyses (Janson et al. 1999a; Orcutt et al. 2002).

Blooms of *Trichodesmium* have been observed frequently in calm conditions, suggesting that water column stability and stratification selects for *Trichodesmium* (Carpenter and Capone 1992; Karl et al. 1995, 1999). *Trichodesmium* is most abundant during summer stratified conditions in the North Pacific Gyre at Station ALOHA (Karl et al. 1995). The reason for a correlation between N_2 fixation and *Trichodesmium* and lack of turbulence or water column stability was assumed to be due to the effect of turbulence in mixing O_2 within the colony (Paerl and Bebout 1988a), but it is now known that *Trichodesmium* colony formation is not absolutely necessary for N_2 fixation, although it may affect the efficacy of this process.

Many studies have focused on the ability of *Trichodesmium* to fix N_2 in the light without heterocysts. One hypothesis is that there is combined temporal and spatial separation of activities (Berman-Frank et al. 2001b), with reduced photosynthetic activities from uncoupling of photosystem II (PSII) (which produces O_2) from photosytem I (PSI) on short time scales that would allow N_2 fixation to occur (Küpper et al. 2004). Direct evidence that cells are active in N_2 fixation but not in photosynthesis is lacking, as there is not yet a way to measure N_2 fixation at the cellular level. This may become possible in the near future using SIMS technology (D. Capone, personal communication). Other mechanisms that have been proposed to be involved in facilitating N_2 fixation are the Mehler reaction and the high respiration rate in *Trichodesmium* (Berman-Frank et al. 2001b; Kana 1993), reviewed by LaRoche and Breitbarth (2005). Intriguingly, it has been ignored that unicellular cyanobacterial cultures will fix N_2 under constant, albeit low levels of illumination, even though it is not possible to differentiate cells even in a unicellular culture. This has been observed in a *Synechococcus* sp. (really a *Cyanothece* sp.) (Mitsui et al. 1987), a *Cyanothece* sp. (Schneegurt et al. 1994), and a *Gloeothece* sp. (Ortega-Calvo and Stal 1991).

Filamentous heterocyst-forming cyanobacteria are symbiotic with some genera of diatoms, but are not well characterized, since they have not been brought into culture (Gómez et al. 2005; Janson et al. 1999b). By comparing microscopic observations with PCR results, Foster and Zehr (2006) were able to show that the different symbiotic associations harbor phylogenetically distinct populations. Quantitative PCR data demonstrate that they are widely distributed (Church et al. 2005a; Foster et al. 2007). The diatom symbionts are important diazotrophs, since diatoms can rapidly sink out of the water column and export C and N to deep water.

A number of other genera of diazotrophic cyanobacteria have been reported, including the heterocystous cyanobacterium *Anabaena gerdii* (Carpenter and Janson 2001). These cyanobacteria are not reported frequently, although a heterocystous cyanobacterial *nifH* was previously amplified from Station BATS in the North Atlantic (Zehr et al. 1998). Undoubtedly, more diazotrophs will be discovered to be widely distributed as quantitative methods are used for mapping nitrogenase genes in the sea.

PCR amplification of *nifH* genes from ocean samples has yielded a number of noncyanobacterial phylotypes (Bird et al. 2005; Langlois et al. 2005; Zehr et al. 1998). Recent work has shown that some of these phylotypes are present and expressing the *nifH* gene in the mesopelagic (Hewson et al. 2007). One group of *nifH* sequences has repeatedly been found in ocean surveys—the group called UMB by Bird et al. (2005). Phylogenetically, these bacteria appear to be *Gammaproteobacteria*, and the *nifH* gene is expressed (Bird et al. 2005; Church et al. 2005a). However, the modes of metabolism of these bacteria, and how they can support N_2 fixation are unknown.

In the early 1990s, there was considerable debate as to whether members of the common cyanobacterial genus *Synechococcus* could fix N_2 in the plankton. This was partially due to the reports of a marine cyanobacterium, called *Synechococcus* (now believed to be a *Cyanothece* sp.), that fixed N_2 (Mitsui et al. 1987). A marine unicellular cyanobacterium, then called *Erythrosphaera* (also called marine *Synechocystis*, or *Crocosphaera watsonii*), had been isolated from the Atlantic Ocean (Waterbury and Rippka 1989). This unicellular cyanobacterium, however, was not a *Synechococcus*, but rather a larger cell, 2–8 μm in diameter, that contained phycoerythrin. A similarly sized unicellular cyanobacterium, *Cyanothece* sp. ATCC 51142, was isolated from the Gulf of Mexico (http://www.atcc.org) (Reddy et al. 1993). It was not generally believed at that time that the unicellular cyanobacteria of this morphology were abundant in the open ocean plankton. However, reports of large (larger than marine *Synechococcus* spp.) phycoerythrin-containing cells were made by Neveux et al. (1999) in the South Pacific and by Campbell et al. (1997) at Station ALOHA in the North Pacific.

When molecular approaches targeting the *nifH* gene were used to investigate planktonic N_2 fixation, nanoplanktonic cyanobacterial and picoplanktonic noncyanobacterial *nifH* genes were found in the Atlantic and Pacific Ocean (Zehr et al. 1998). Cyanobacterial *nifH* sequences were repeatedly detected in the Pacific Ocean from Station ALOHA (Zehr et al. 1998). These gene sequences were of two types, called Groups A and B, and were shown to be expressed using reverse-transcriptase (RT)-PCR (Zehr et al. 2001b). Some of the cyanobacterial *nifH* sequences were closely related to sequences from *Crocosphaera*.

Quantitative PCR assays show that unicellular cyanobacteria and diatom-associated symbionts can be at least as abundant as *Trichodesmium* in the North Pacific Ocean (Church et al. 2005b). Similarly, analysis of PCR amplification products in the North Atlantic Ocean indicate that unicellular cyanobacterial diazotrophs are present, and may have a greater temperature range than *Trichodesmium* (Langlois et al. 2005). The diazotrophs have different daily patterns of *nifH* gene expression and activity (Church et al. 2005b). Intriguingly, the Group A unicellular cyanobacterial phylotype has maximum *nifH* gene expression during the day. Although *nifH* gene expression can be temporally offset from N_2 fixation, the pattern suggests N_2 fixation during the day. This would be the first unicellular cyanobacterium to fix N_2 during the day of a light–dark cycle. It could be that different daily patterns of N_2 fixation have implications for competition, since the turnover of nitrogenase involves cycling of Mo and Fe on a daily basis (Tuit et al. 2004). The fact that *Trichodesmium* fixes N_2 during the day, along with the proposed Group A unicellular cyanobacteria, may suggest that there is an ecological advantage to daytime N_2 fixation in warm oligotrophic oceans.

N_2 Fixation Rates: From Cells to Basins Perhaps one of the most important and challenging aspects of marine N_2 fixation is to integrate N_2 fixation inputs from scales of microorganisms to ocean basins with coincident temporal variability in biological and physical processes. Rates of N_2 fixation by *Trichodesmium* and the diatom symbionts have been estimated for cultured and natural populations (Capone et al. 2005; Carpenter et al. 1999; Letelier and Karl 1998). Estimates of N_2 fixation in *Crocosphaera* have been obtained in culture (Tuit et al. 2004), but N_2 fixation rates of natural populations of smaller unicellular cyanobacterial cells (*Crocosphaera* and the uncultivated Group A presumed unicellular cyanobacteria) can only be obtained indirectly by size fractionation (Montoya et al. 2004). Rates of N_2 fixation in *Trichodesmium* can be on the order of 10–30 fmol N/cell/h, which may be higher than cellular rates of unicellular cyanobacteria (Mahaffey et al. 2005). In experiments with *Trichodesmium* and *Crocosphaera*, daily maximum per cell rates in *Trichodesmium* ranged from approximately 4- to 15-fold higher in *Trichodesmium* (Tuit et al. 2004). The variability in rates highlights the difficulty in extrapolating rates obtained in culture to the environment, where gradients in light and nutrients, as well as hydrographic factors, affect the spatial and temporal distribution of N_2 fixation rates.

Areal and basin scale N_2 fixation rates have been estimated by a variety of means, including tracer methods, These estimates have been comprehensively summarized by Mahaffey et al. (2005). N_2 fixation by *Trichodesmium* ranges from 1.4 to 898 μmol $N/m^2/d$, with the highest rates in the tropics (Mahaffey et al. 2005). Rates up to 4000 μmol $N/m^2/d$ for the small-size fraction were reported by Montoya et al. (2004) near the Australian coast. Diatom symbionts can also be responsible for high rates, up to 3500 μmol $N/m^2/d$ (Mahaffey et al. 2005). Biogeochemical estimates, based for example on N^* (Gruber and Sarmiento 1997; Michaels et al. 1996), can exceed these rates (Mahaffey et al. 2005).

In the past decade, the conceptual framework of N_2 fixation in the open ocean has changed, and it is now generally believed that N_2 fixation is an important process in the N cycle that is involved in supporting carbon export (Gruber and Sarmiento 1997; Karl et al. 2001b; Michaels et al. 1996). Several independent lines of evidence have been interpreted to be indications of N_2 fixation, including low $^{15}N:^{14}N$ ratios in surface waters, increasing concentrations of dissolved organic nitrogen at Station ALOHA, attempts to balance carbon budgets (Lee et al. 2002), and high N^* ratios in some regions (Capone 2001; Mahaffey et al. 2005). Modeling studies indicate that N_2 fixation and surface ocean nitrification are adequate to explain observed productivity at the BATS site in the North Atlantic (Bissett et al. 1999).

Estimates of N_2 fixation rates from biogeochemical calculations initially appeared to be much larger than could be explained by N_2 fixation by *Trichodesmium* alone.

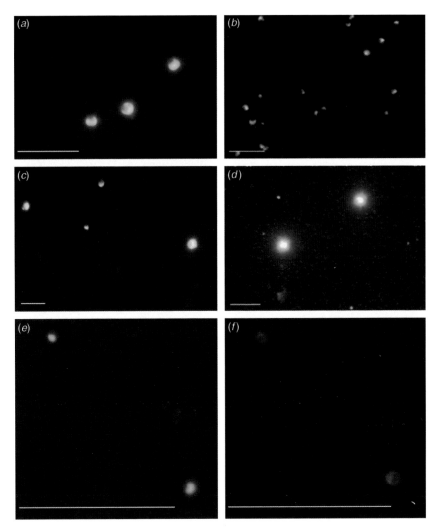

Figure 6.8 Epifluorescence images of picoeukaryotes from environmental samples illuminated via FISH. (*a, b*) Surface samples from the English Channel hybridized with a Chlorophyta-specific probe (CHLO02) and a *Micromonas* genus-specific probe (MICRO01), respectively, using TSA-FISH. The cytoplasm of hybridized cells has green fluorescence (FITC dye). (*c, d*) Florida Straits samples from 5 m and 85 m depth, respectively, hybridized with the Haptophyta-specific probe (PRYM02) using the TSA-FISH technique and FITC. The natural red/orange fluorescence of the cyanobacteria can be seen. The sample shown in (*d*) has been counterstained with the DNA-specific dye DAPI (blue) to allow visualization of heterotrophic bacteria. (*e, f*) A surface sample from the Blanes Bay (Mediterranean Sea) hybridized with the probe NS4 (Massana et al. 2004), specific for MAST-4 cells. In contrast to other images shown, the FISH method used to illuminate MAST cells in (*f*) did not involve signal amplification, but rather used a monolabeled probe linked to the fluorescent dye CY3 (red). The DAPI counterstaining from the same microscopic field is shown in (*e*). Scale bars represent 10 μm, except in (*e, f*), for which they represent 20 μm. (*a*) Courtesy of F. Not. (*b*) Courtesy of E. Foulon. (*c, d*) Courtesy of J. A. Hilton, F. Not, and A. Z. Worden. (*e, f*) Courtesy of R. Massana.

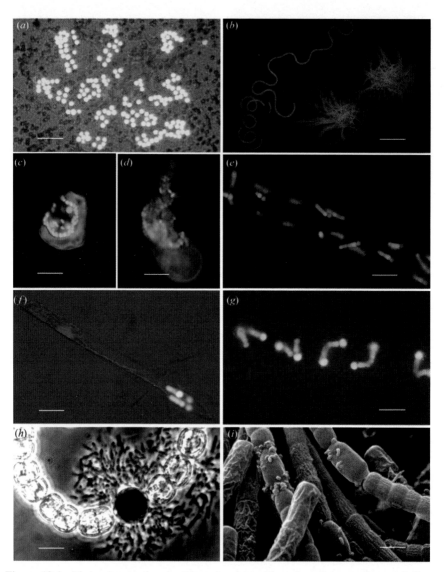

Figure 13.4 Photomicrographs of cultivated and uncultivated marine cyanobacterial diazotrophs. (*a*) Uncultivated unicellular cyanobacteria from the open ocean that are likely diazotrophs. (*b*) Diversity of *Trichodesmium* morphology in natural populations. (*c*, *d*) Symbiotic unicellular cyanobacteria. (*e*) Heterocystous symbiont of diatoms. (*f*) Heterocystous symbiont of diatoms. (*g*) Heterocystous symbiont of diatoms. (*h*, *i*) Association of bacteria with heterocysts of heterocyst-forming cyanobacteria.

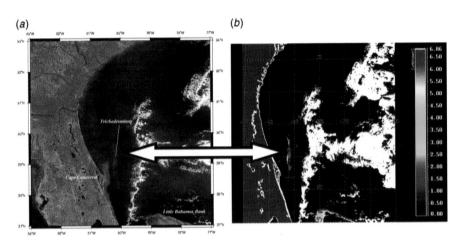

Figure 13.6 Remote sensing of blooms of *Trichodesmium*. (*a*) True-color image of surface ocean, indicating position of *Trichodesmium* bloom detected by remote sensing. (*b*) Remote sensing image of *Trichodesmium* using the algorithm of Subramaniam et al. (2002). Modified from Subramaniam et al. (2002) and used with permission from Elsevier.

Analysis of extensive datasets of *Trichodesmium* indicates that the two estimates are closer than originally thought, and that N_2 fixation by *Trichodesmium* can be as great or greater than nitrate advection from deep water (Capone et al. 2005). *Trichodesmium* rates may still be underestimated, since rates are usually measured in *Trichodesmium* aggregates, and, in some cases, free trichomes can be a significant fraction of the biomass (Chang 2000; Letelier and Karl 1998; Orcutt et al. 2001). Including N_2 fixation by free trichomes tripled the estimates of the annual input in the Sargasso Sea (Orcutt et al. 2001). There has also been recognition of the presence (Church et al. 2005b; Zehr et al. 2001b), gene expression (Church et al. 2005b), and activity (Montoya et al. 2004) of unicellular cyanobacteria. Although there are several problems in obtaining geographical information on N_2 fixation by *Trichodesmium* (LaRoche and Breitbarth 2005), the contribution of the other diazotrophs is even harder to study, because of their small size. The geochemically-derived rates of N_2 fixation based on N^* have been questioned, as there are many assumptions including the region over which to integrate N^* values (Hansell et al. 2004). If the estimates of N_2 fixation by *Trichodesmium* and from biogeochemical budgets converge (Mahaffey et al. 2005), then a new enigma will emerge, since the diatom symbionts (Carpenter et al. 1999) and unicellular diazotrophic cyanobacteria (Montoya et al. 2004) have been shown to contribute substantially to N_2 fixation rates.

It is now clear that N_2 fixation plays an important role in "new" production in the oligotrophic gyres. Studies of N_2 fixation at the two oceanic long-term monitoring sites, using different approaches, reached the conclusion that N_2 fixation is important in the North Atlantic and North Pacific gyres (Karl et al. 1997; Michaels et al. 2000; Orcutt et al. 2001).

Dore et al. (2002) used natural abundance of ^{15}N and enriched $^{15}N_2$ tracer experiments to assess the significance of N_2 fixation at Station ALOHA. N_2 fixation was strongly seasonal, but there was a large discrepancy between the rates determined from sinking particles and from tracer rate measurements. However, the rates determined from sinking particles agreed well with biogeochemically derived rates (Deutsch et al. 2001) and accounted for a mean of 48 percent of the annual N export from the surface layer. Sources of P and Fe (P from remineralization in advected deep water, and Fe from terrestrially derived aeolian dust) have implications for the ultimate limiting factor and how N_2 fixation has responded, and will respond, to climate changes (Karl 2002). This may partially explain why N_2 fixation can be affected by El Niño events, for example (Karl 1999, 2002).

Controls on N_2 Fixation Most likely, oceanic N_2 fixation is controlled by a complex interplay of physical, chemical, and biological factors. Correlation studies indicate that N_2 fixation in the open ocean appears to be in roughly 20 or 25 °C waters (Capone et al. 1997; LaRoche and Breitbarth 2005). The unicellular diazotrophic cyanobacterium *Crocosphaera* appears also to have a high temperature optimum (J. Waterbury, personal communication). However, diazotroph genes and transcripts have been detected in waters at least as low as 20 °C, and perhaps occur at even lower temperatures. Temperature was not correlated with *Trichodesmium* abundance in low latitudes north and south of the equator in the eastern tropical North Atlantic (Tyrrell et al. 2003).

Inorganic N and P concentrations are low in the North Atlantic and Pacific gyres (Wu et al. 2000), and soluble reactive P concentrations appear to be decreasing in the North Pacific (Karl et al. 2001b). There is evidence for P limitation in the North Pacific (Björkman et al. 2000; Karl et al. 2001b), and diazotrophy may cause the decreasing P (Karl et al. 2001a). P stress was shown in natural populations of *Trichodesmium* using a cell-level assay (Dyhrman et al. 2002). There was a correlation between P and N_2 fixation, indicating that P, and light, may be major factors controlling N_2 fixation in the central North Atlantic (Sañudo-Wilhelmy et al. 2001). This may be due to the fact that P is in much lower concentrations in the North Atlantic than in the Pacific (Wu et al. 2000). Although P additions stimulated N_2 fixation by *Trichodesmium* in the Great Barrier Reef lagoon (Fu and Bell 2003), additions of inorganic P to North Pacific Subtropical Gyre water did not consistently stimulate unicellular cyanobacterial gene expression, abundance, or activity (Zehr et al. 2007).

Different diazotrophs may have different strategies for obtaining P. Genomics studies indicate that *Trichodesmium* may be able to utilize a previously little studied group of P compounds, the phosphonates (Dyhrman et al. 2006). It has also been proposed that *Trichodesmium* could migrate to the phosphocline (Karl et al. 1992; Villareal and Carpenter 1990). Experimental and modeling evidence indicates that this is feasible (White et al. 2006), but direct evidence is lacking. *Crocosphaera* has high- and low-affinity P transporters (Dyhrman and Haley 2006). The relative abilities to compete for sources, in concert with the forms and availability of P, may determine the spatial distribution and succession of diazotroph species.

Fe limits productivity of nondiazotrophs in HNLC regions, and should be important in controlling diazotrophs, since the nitrogenase proteins contain Fe (Kustka et al. 2002, 2003). On a theoretical basis, it has been predicted that *Trichodesmium* Fe requirements are roughly 10-fold greater than non-N_2-fixing phytoplankton (Kustka et al. 2003). The effect of Fe on *Trichodesmium* was tested experimentally (Paerl et al. 1994; Rueter et al. 1992). Rueter et al. (1992) suggested that *Trichodesmium* colonies were able to intercept Fe derived from aeolian sources (volcanic emissions, dust from desertification, and air pollution from industrial and automotive emissions) by trapping particulate Fe in the weblike matrix of trichomes. Although the estimates of Fe needs for N_2 fixation in *Trichodesmium* have been reduced from those originally reported (see Kustka et al. 2003; Sañudo-Wilhelmy et al. 2001), there is still a high Fe requirement, presumably to support the Fe metallocenters in nitrogenase. Highest rates in the eastern tropical Atlantic Ocean and in the Amazon River plume correlated with Fe concentrations (Galloway et al. 2004; Mahaffey et al. 2005; Voss et al. 2004). Analysis of the distribution of atmospheric Fe deposition suggested that Fe could limit N_2 fixation in 75 percent of the world's oceans (Berman-Frank et al. 2001a). However, in the North Atlantic Ocean, there was no correlation between Fe and N_2 fixation activity (Sañudo-Wilhelmy et al. 2001).

The requirements and quotas of Mo in *Trichodesmium* and *Crocosphaera* have been evaluated in cultures and natural populations (Tuit et al. 2004). Field populations had Mo:C ratios far in excess of that needed for N_2 fixation as determined in culture experiments, indicating that Mo does not limit N_2 fixation in the open ocean (Tuit et al. 2004).

Role of N_2 Fixation in Food Webs N_2 fixation is important ecologically in several respects. The fixed N can enter the food chain, is exported to deep waters, or is released into dissolved pools. We probably know the most about the fate of N in *Trichodesmium* biomass, but this is still unclear. *Trichodesmium* has been implicated in the release of ammonium (Mulholland and Capone 2000) and organic N (Bronk et al. 1994). Release of N (and other nutrients) from *Trichodesmium* (and diatom) blooms has been suggested to support secondary blooms of eukaryotic algae (Lenes et al. 2001; Wilson 2003). *Trichodesmium* is rarely found in sediment traps, but symbiotic diatoms are. Thus, there may be large differences in the contribution of individual diazotrophs to carbon and N export. Similarly, there are likely to be substantial differences in transmission through the food chain. Unicellular cyanobacteria could conceivably be very actively grazed by a number of organisms, whereas the large, sometimes toxic (Hawser et al. 1992; Preston et al. 1998), aggregates of *Trichodesmium* appear to be principally grazed by planktonic harpacticoid copepods such as *Macrosetella*.

There are still major gaps in our understanding of the role of N_2 fixation in the open ocean. Estimates of abundances and rates of N_2 fixation are severely limited by the number of samples. Research cruises to study diazotrophy usually sample regions of known diazotroph occurrences, rather than employing random sampling to map distributions. Another complication is that the highly seasonal and bloom-forming nature of some diazotrophs means that they are often missed even in the monthly sampling programs at Stations ALOHA and the Bermuda Atlantic Time Series (BATS). The factors affecting N_2 fixation, even Fe, are still poorly understood. Finally, the pathways of fixed N_2 are not well known—for example, whether the different diazotrophs contribute unequally to export or the food chain.

Estuarine and Coastal Waters

Estuaries are characterized by hydrological, salinity, and biogeochemical gradients. Typically, phosphorus availability tends to limit primary production in the upstream, freshwater segments of estuaries, while N assumes a more dominant role as a limiting nutrient in the more saline, downstream marine waters (cf. Fisher et al. 1988). Some estuaries also have brackish regions where N and P co-limitation occur, and there can be seasonal shifts between N and P limitation (Elmgren and Larsson 2001; Fisher et al. 1988; Paerl et al. 1995a). Coastal waters draining estuaries or isolated from estuaries tend to be strongly N-limited (Nixon 1995; Paerl 1997). One would therefore expect these more meso- to euhaline lower estuarine and coastal waters to be potential habitats for N_2-fixers. Indeed, in some systems, including coastal and open water regions of the brackish Baltic Sea, as well as lagoonal estuarine and coastal ecosystems, N_2-fixers can account for a significant and at times dominant fraction of phytoplankton and benthic microalgal biomass. In addition, bacterial N_2-fixers may be found in these systems, most often in the benthos (Paerl and Zehr 2000). Molecular studies, based on *nifH* analyses, indicate that a diverse taxonomic potential exists for N_2 fixation in these waters (Affourtit et al. 2001; Burns et al. 2002; Jenkins et al. 2004). Despite the genetic potential, N_2 fixation activity is frequently absent or present at ecologically insignificant rates, and, if it is present, it is confined to sedimentary or biofilm habitats. Recent studies indicate that there may be physical barriers to the establishment and dominance of N_2-fixers in N-limited estuaries, especially in the water column. The relatively turbulent properties of estuarine waters, which include strong wind and tidal mixing, and high rates of small-scale shear, may restrict the establishment and proliferation of diazotrophic cyanobacterial and bacterial communities (Moisander et al. 2002b; Paerl et al. 1995b, 1996). In particular, strong and persistent vertical mixing of near-surface waters prevents dominance by bouyant filamentous diazotrophic cyanobacterial bloom genera (e.g., *Anabaena, Aphanizomenon, Nodularia*, and *Trichodesmium*). Lastly, periodic hypoxia in stratified estuaries can result in large fluxes of ammonium from the sediments into the water column. These ammonium pulses may act to repress nitrogenase activity and thus negate potential competitive advantages of N_2-fixing cyanobacteria and bacteria.

In summary, estuarine and coastal waters have a diverse genetic potential for N_2 fixation, which, under favorable conditions (e.g., midsummer stratified conditions in the Baltic Sea), can be readily expressed. However, more often, persistently wind-mixed surface waters and periodically flushed and nutrient-pulsed conditions in these environments represent physical and chemical barriers to N_2-fixers, thus restricting their dominance and bloom potentials. This, combined with the fact that estuarine and coastal systems are frequent sites of active denitrification and P sufficiency, helps explain why these systems exhibit chronic N deficiency.

Benthic Habitats, Including Microbial Mats and Reefs

Because of their strong and extensive biogeochemical zonation, microbial mats support metabolically and taxonomically diverse diazotrophic communities (Paerl et al. 2000). These include cyanobacteria (nonheterocystous filamentous, heterocystous, and coccoid), anaerobic photosynthetic bacteria, microheterotrophs, and a range of chemolithotrophs (Omoregie et al. 2004a; Steppe et al. 1996; Steppe and Paerl 2002; Zehr et al. 1995). Microbial mats are widespread in shallow marine environments, including estuaries, coastal lagoons, salt marshes, mudflats, and reefs. In deeper waters, mats inhabit specialized environments where organic matter and energy sources for sustaining N_2 fixation may abound. These include hydrothermal vents and seeps (see below). Chapter 14 also discusses N_2 fixation in the benthos.

In benthic sediment and mat environments, N_2 fixation can be a large and sometimes dominant source of "new" N supporting primary and secondary production (Paerl et al. 2000). Very high rates in *Spartina* salt marshes can be due to cyanobacterial mats or epiphytes (Carpenter et al. 1978; Currin and Paerl 1998; Moisander et al. 2005). Diazotroph diversity is high in the rhizosphere (Lovell et al. 2000). Sulfate-reducing bacteria have been implicated in N_2 fixation of salt marshes (Nielsen et al. 2001). Interestingly, diazotroph diversity remains high even when inorganic N is added, indicating that the diazotrophs are competitive with nondiazotrophs (Piceno and Lovell 2000a), and reverse sample genome probing showed that individual populations were not displaced by fertilization (Bagwell and Lovell 2000). Acetylene reduction activity was detectable in control treatments where interstitial ammonium concentrations are approximately 5 $\mu mol/L$ even in the control plots (Bagwell et al. 1998). Diazotroph populations appear to be different between different grasses (Lovell et al. 2000). The results of a variety of perturbations indicate that the diazotroph assemblages are stable (Bagwell and Lovell 2000; Lovell et al. 2001; Piceno and Lovell 2000a, b).

Coral reefs are relatively productive ecosystems that are commonly found in N-depleted, oligotrophic tropical and subtropical marine waters. It has been suggested that the relatively productive nature of these ecosystems is likely due to N_2 fixation (D'Elia and Wiebe 1990). Coral reefs are sites of both cyanobacterial and heterotrophic bacterial N_2 fixation (Capone 1983), although measured rates are few and environmental regulation of this process is not well understood (Charpy-Roubaud et al. 2001). Nitrogenase activity has been measured on surfaces of benthic organisms (Charpy-Rouband and Larkum 2005), in sediments (Capone et al. 1992; Charpy-Roubaud et al. 2001; Koop et al. 2001; Mayajima et al. 2001), and in filamentous cyanobacteria endemic to coral reefs (Kayanne et al. 2005). Rohwer et al (2002) observed several cyanobacterial ribotypes associated with coral surfaces, of which at least three are known to fix N_2. Phycoerythin was found in a scleractinian coral that was due to a symbiotic unicellular cyanobacterium (Lesser et al. 2004). Cyanobacteria of the genus *Oscillatoria* have been observed in benthic microalgal assemblages of coral reefs (Heil et al. 2004). Unpublished data suggest that water column N_2 fixation in coral reef lagoons is mostly attributable to the size fraction smaller than 5 μm (see Charpy 2005), indicating that unicellular cyanobacteria may be an ecologically important component of the coral reef N cycle. While the phylogeny of N_2-fixing microorganisms on coral reefs has not previously been studied at the *nifH* level, a recent fingerprinting study of sediment diazotrophic assemblages, utilizing terminal restriction fragment length polymorphism (TRFLP) of the *nifH* gene, indicated that diverse and spatially variable assemblages persist within reef sediments (Hewson and Fuhrman 2006). Thus, diazotrophic prokaryotes may be an important source of new N in nutrient-impoverished coral reef ecosystems.

Deep Water and Hydrothermal Vents

Deep-water environments typically have high concentrations of inorganic and organic N, and thus traditionally N_2 fixation was not believed to occur in these environments. However, recent studies have found N_2-fixing microorganisms in deep-water environments. Some hydrothermal vent fluids contain relatively low concentrations of combined N (Karl et al. 1988; Lilley et al. 1993), and have been the target for studies of *nifH*. Mehta et al. (2003) reported a wide suite of diazotrophic phylotypes within hydrothermal vent fluids, where the majority of phylotypes were in Clusters II and III, which contain sequences from alternative *nifH*, anaerobic bacteria, and archaea. Mehta and Baross (2006) recently described a new thermophilic N_2-fixing isolate. Additional studies of deep-sea bacteria in seafloor rift valleys have found the presence of more Cluster I, II, and IV sequences, and another group of divergent, coldwater (nonthermophilic) archaeal-like genes (Mehta et al. 2005). A recent study of meso- to abyssopelagic diazotroph assemblages found that endemic, low-abundance, and actively transcribed phylotypes were present in waters of the Sargasso Sea, beneath the Costa Rica Upwelling dome, near O'ahu, and in eastern North Pacific waters (Hewson et al. 2007). Anomalously light $\delta^{15}N$ values suggested that N_2 fixation may occur in sediments and biomass near cold methane seeps in the Gulf of Mexico (Brooks et al. 1987; Joye et al. 2004). This is consistent with the observation of type II methanotrophs, which can contain nitrogenase genes (Auman et al. 2001). Thus, N_2 fixation may be common in deep-water environments, despite the presence of high concentrations of combined N.

SUMMARY

1. Biological N_2 fixation is an energetically expensive process (requiring ATP and reductant), and is sensitive to oxygen.

2. Nitrogenase is an ancient enzyme, and the evolution of the enzyme and the organisms that contain it reflect the anoxic–oxic transitions in the ocean in Earth's history.

3. N_2-fixing microorganisms are diverse, represented by taxa throughout the *Archaea* and *Bacteria*, and by many physiological groups. Most cultivated marine N_2-fixing *Bacteria* have come from sediments or mats. Three open-ocean cyanobacteria have been cultivated: unicellular (*Crocosphaera*), filamentous nonheterocystous (*Trichodesmium*), and a heterocystous symbiont of *Chaetoceros*.

4. N_2 fixation and the nitrogenase genes are highly regulated in response to fixed N. Regulation, which involves many genes and gene products, can include a circadian rhythm in cyanobacteria such as *Trichodesmium*.

5. Methods for studying N_2 fixation include tracer and natural-abundance approaches, but now include a wide variety of genetic and genomic techniques, such as DNA microarrays and gene expression assays, and remote sensing technology.

6. Recent N_2 fixation estimates for the open ocean are higher than previously believed. Revised estimates for *Trichodesmium* approach these higher estimates, and unicellular cyanobacteria and diatom symbionts may contribute substantial to global N_2 fixation. New organisms continue to be discovered (Zehr et al. 2001b).

7. A variety of factors control N_2 fixation, ranging from turbulence and grazing to nutrient and trace element availability. Major knowledge gaps exist on the spatial and temporal distributions of N_2-fixing microorganisms in the sea, and the factors that control their activity. The roles of P and Fe in constraining N_2 fixation are still unclear, as are the conditions that select for different diazotroph groups.

ACKNOWLEDGMENTS

J. Zehr was supported by the National Science Foundation (OCE-9981437, OCE-0131762 and OCE-0425363), the Gordon and Betty Moore Foundation, and the Center for Microbial Oceanography Research and Education (C-MORE). H. Paerl was supported by the National Science Foundation (MCB 0132528 and OCE 0327056), the U.S. Department of Agriculture NRI Project 00-35101-9981, U.S. EPA STAR Project R82867701. The technical and editorial support of A. Joyner, J. Dyble, K. Pennebaker, and A. Michels is much appreciated. We thank R. Foster, I. Hewson, P. Moisander, Tuo Shi, and K. Achilles for their contributions in preparing this review.

REFERENCES

Affourtit, J., J. P. Zehr, and H. W. Paerl. 2001. Distribution of nitrogen-fixing microorganisms along the Neuse River Estuary, North Carolina. *Microb. Ecol.* 41: 114–123.

Antia, N. J., P. J. Harrison, and L. Oliveira. 1991. The role of dissolved organic nitrogen in phytoplankton nutrition, cell biology and ecology. *Phycologia* 30: 1–89.

Auman, A. J., C. C. Speake, and M. E. Lidstrom. 2001. *nifH* sequences and nitrogen fixation in type I and type II methanotrophs. *Appl. Environ. Microbiol.* 67: 4009–4016.

Bagwell, C. E., and C. R. Lovell. 2000. Persistence of selected *Spartina alterniflora* rhizoplane diazotrophs exposed to natural and manipulated environmental variability. *Appl. Environ. Microbiol.* 66: 4625–4633.

Bagwell, C. E., Y. M. Piceno, A. Ashburne-Lucas, and C. R. Lovell. 1998. Physiological diversity of the rhizosphere diazotroph assemblages of selected salt marsh grasses. *Appl. Environ. Microbiol.* 64: 4276–4282.

Bebout, B. M., M. W. Fitzpatrick, and H. W. Paerl. 1993. Identification of the sources of energy for nitrogen fixation and physiological characterization of nitrogen-fixing members of a marine microbial mat community. *Appl. Environ. Microbiol.* 59: 1495–1503.

Bergman, B. B., J. R. Gallon, A. N. Rai, and L. J. Stal. 1997. N_2 fixation by non-heterocystous cyanobacteria. *FEMS Microbiol. Rev.* 19: 139–185.

Berman-Frank, I., J. T. Cullen, Y. Shaked, R. M. Sherrell, and P. G. Falkowski. 2001a. Iron availability, cellular iron quotas, and nitrogen fixation in *Trichodesmium*. *Limnol. Oceanogr.* 46: 1249–1260.

Berman-Frank, I., P. Lundgren, Y. B. Chen, H. Küpper, Z. Kolber, B. Bergman, and P. Falkowski. 2001b. Segregation of nitrogen fixation and oxygenic photosynthesis in the marine cyanobacterium *Trichodesmium*. *Science* 294: 1534–1537.

Berman-Frank, I., P. Lundgren, and P. Falkowski. 2003. Nitrogen fixation and photosynthetic oxygen evolution in cyanobacteria. *Res. Microbiol.* 154: 157–164.

Bird, C., J. Martinez Martinez, A. G. O'Donnell, and M. Wyman. 2005. Spatial distribution and transcriptional activity of an uncultured clade of planktonic diazotrophic γ-Proteobacteria in the Arabian Sea. *Appl. Environ. Microbiol.* 71: 2079–2085.

Bird, D. F., and D. M. Karl. 1991. Spatial patterns of glutamate and thymidine assimilation in Bransfield Strait, Antarctica during and following the austral spring bloom. *Deep-Sea Res.* I 38: 1057–1075.

Bishop, P. E., and R. Premakumar. 1992. Alternative nitrogen fixation systems. In G. Stacey, R. H. Burris, and H. J. Evans (eds.), *Biological Nitrogen Fixation*. Chapman and Hall, pp. 736–762.

Bissett, W. P., J. J. Walsh, D. A. Dieterle, and K. L. Carder. 1999. Carbon cycling in the upper waters of the Sargasso Sea: I. Numerical simulation of differential carbon and nitrogen fluxes. *Deep-Sea Res.* I 46: 205–269.

Björkman, K., A. L. Thomson-Bulldis, and D. M. Karl. 2000. Phosphorus dynamics in the North Pacific subtropical gyre. *Aquat. Microb. Ecol.* 22: 185–198.

Bronk, D. A., P. M. Glibert, and B. B. Ward. 1994. Nitrogen uptake, dissolved organic nitrogen release, and new production. *Science* 265: 1843–1846.

Brooks, J. M., M. C. Kennicutt, C. R. Fisher, S. A. Macko, K. Cole, R. R. Bidigare, and R. D. Vetter. 1987. Deep sea hydrocarbon seep communities: Evidence for energy and nutritional carbon sources. *Science* 238: 1138–1142.

Burford, M. A. 2006. A comparison of phytoplankton community assemblages in artificially and naturally mixed subtropical water reservoirs. *Freshwater Biol.* 51: 973–982.

Burns, J. A., J. P. Zehr, and D. G. Capone. 2002. Nitrogen-fixing phylotypes of Chesapeake Bay and Neuse River Estuary sediments. *Microb. Ecol.* 44: 336–343.

Campbell, L., H. B. Liu, H. A. Nolla, and D. Vaulot. 1997. Annual variability of phytoplankton and bacteria in the subtropical North Pacific Ocean at Station ALOHA during the 1991–1994 ENSO event. *Deep-Sea Res.* I 44: 167–192.

Canfield, D. E. 1998. A new model for Proterozoic ocean chemistry. *Nature* 396: 450–453.

Cantera, J. J. L., H. Kawasaki, and T. Seki. 2004. The nitrogen-fixing gene (*nifH*) of *Rhodopseudomonas palustris*: a case of lateral gene transfer? *Microbiology* 150: 2237–2246.

Capone, D. G. 1983. Benthic nitrogen fixation. In E. J. Carpenter and D. G. Capone (eds.), *Nitrogen in the Marine Environment*. Academic Press, pp. 105–137.

Capone, D. G. 2000. The marine microbial nitrogen cycle. In David L. Kirchman (ed.), *Microbial Ecology of the Oceans*, 1st edn. Wiley-Liss, pp. 455–494.

Capone, D. G. 2001. Marine nitrogen fixation: What's the fuss? *Curr. Opin. Microbiol.* 4: 341–348.

Capone, D. G., and J. P. Montoya. 2001. Nitrogen fixation and denitrification. In J. Paul (ed.), *Marine Microbiology*. Academic Press, pp. 501–515.

Capone, D. G., S. E. Dunham, S. G. Horrigan, and L. E. Duguay. 1992. Microbial nitrogen transformations in unconsolidated coral reef sediments. *Mar. Ecol. Prog. Ser.* 80: 75–88.

Capone, D. G., J. P. Zehr, H. W. Paerl, B. Bergman, and E. J. Carpenter. 1997. *Trichodesmium*: a globally significant marine cyanobacterium. *Science* 276: 1221–1229.

REFERENCES

Capone, D. G., A. Subramaniam, and E. J. Carpenter. 1998. An extensive bloom of the N_2-fixing cyanobacterium *Trichodesmium erythraeum* in the central Arabian Sea. *Mar. Ecol. Prog. Ser.* 172: 281–292.

Capone, D. G., J. A. Burns, J. P. Montoya, A. Subramaniam, C. Mahaffey, T. Gunderson, A. F. Michaels, and E. J. Carpenter. 2005. Nitrogen fixation by *Trichodesmium* spp.: An important source of new nitrogen to the tropical and subtropical North Atlantic Ocean. *Global Biogeochem. Cycles* 19:GB2024.

Carpenter, E. J., and D. G. Capone (eds.). 1983. *Nitrogen in the Marine Environment*. Academic Press.

Carpenter, E. J., and D. G. Capone. 1992. Nitrogen fixation in *Trichodesmium* blooms. In E. J. Carpenter, D. G. Capone, and J. G. Rueter (eds.), *Marine Pelagic Cyanobacteria*: Trichodesmium *and Other Diazotrophs*. Kluwer, pp. 211–218.

Carpenter, E. J., and R. A. Foster. 2002. Marine symbioses. In A. N. Rai, B. Bergman, and U. Rasmussen (eds.), *Cyanobacteria in Symbiosis*, Kluwer, pp. 11–18.

Carpenter, E. J., and S. Janson. 2001. *Anabaena gerdii* sp nov., a new planktonic filamentous cyanobacterium from the South Pacific Ocean and Arabian Sea. *Phycologia* 40: 105–110.

Carpenter, E. J., C. D. Van Raalte, and I. Valiela. 1978. Nitrogen fixation by algae in a Massachusetts salt marsh. *Limnol. Oceanogr.* 23: 318–327.

Carpenter, E. J., J. P. Montoya, J. Burns, M. R. Mulholland, A. Subramaniam, and D. G. Capone. 1999. Extensive bloom of a N_2 fixing diatom/cyanobacterial association in the tropical Atlantic Ocean. *Mar. Ecol. Prog. Ser.* 185: 273–283.

Castruita, M., M. Saito, P. C. Schottel, L. A. Elmegreen, S. Myneni, E. I. Stiefel, and F. M. M. Morel. 2006. Overexpression and characterization of an iron storage and DNA-binding Dps protein from *Trichodesmium erythraeum*. *Appl. Environ. Microbiol.* 72: 2918–2924.

Chang, J. 2000. Precision of different methods used for estimating the abundance of the nitrogen-fixing marine cyanobacterium, *Trichodesmium* Ehrenberg. *J. Exp. Mar. Biol. Ecol.* 245: 215–224.

Charpy, L. 2005. Importance of photosynthetic picoplankton in coral reef ecosystems. *Vie Milieu–Life Environ.* 55: 217–223.

Charpy-Rouband, C., and A. W. D. Larkum. 2005. Dinitrogen fixation by exposed communities on the rum of Tikehau atoll (Tuamotu Archipelago, French Polynesia). *Coral Reefs* 24: 622–628.

Charpy-Roubaud, C., L. Charpy, and A. W. D. Larkum. 2001. Atmospheric dinitrogen fixation by benthic communities of Tikehau Lagoon (Tuamotu Archipelago, French Polynesia) and its contribution to benthic primary production. *Mar. Biol.* 139: 991–997.

Church, M. J., B. D. Jenkins, D. M. Karl, and J. P. Zehr. 2005a. Vertical distributions of nitrogen-fixing phylotypes at Station ALOHA in the oligotrophic North Pacific Ocean. *Aquat. Microb. Ecol.* 38: 3–14.

Church, M. J., C. M. Short, B. D. Jenkins, D. M. Karl, and J. P. Zehr. 2005b. Temporal patterns of nitrogenase gene (*nifH*) expression in the oligotrophic North Pacific Ocean. *Appl. Environ. Microbiol.* 71: 5362–5370.

Codispoti, L. A., J. A. Brandes, J. P. Christensen, A. H. Devol, S. W. A. Naqvi, H. W. Paerl, and T. Yoshinari. 2001. The oceanic fixed nitrogen and nitrous oxide budgets: Moving targets as we enter the anthropocene? *Sci. Mar.* 65: 85–105.

Cole, J., J. Lane, R. Marino, and R. Howarth. 1993. Molybdenum assimilation by cyanobacteria and phytoplankton in freshwater and salt water. *Limnol. Oceanogr.* 38: 25–35.

Coles, V. J., C. Wilson, and R. R. Hood. 2004. Remote sensing of new production fuelled by nitrogen fixation. *Geophys. Res. Lett.* 31: Art. No. C06007. doi: 10.1029/2003G1019018.

Collier, R. W. 1985. Molybdenum in the Northeast Pacific Ocean. *Limnol. Oceanogr.* 30: 1351–1354.

Currin, C. A., and H. W. Pearl. 1998. Epiphytic nitrogen fixation associated with standing dead shoots of smooth cordgrass, *Spartina alterniflora*. *Estuaries* 21: 108–117.

D'Elia, C. F., and W. J. Wiebe. 1990. Biochemical nutrient cycles in coral reef ecosystems. In Z. Dubinsky (ed.), *Ecosystems of the World 25: Coral Reefs*. Elsevier, pp. 49–74.

Dean, D. R., and M. R. Jacobsen. 1992. Biochemical genetics of nitrogenase. In G. Stacey, R. H. Burris, and H. J. Evans (eds.), *Biological Nitrogen Fixation*. Chapman and Hall, pp. 763–834.

Deutsch, C., N. Gruber, R. M. Key, J. L. Sarmiento, and A. Ganachaud. 2001. Denitrification and N2 fixation in the Pacific Ocean. *Global Biogeochem. Cycles* 15: 483–506.

Dixon, R., and D. Kahn. 2004. Genetic regulation of biological nitrogen fixation. *Nat. Rev. Microbiol.* 2: 621–631.

Dore, J. E., J. R. Brum, L. M. Tupas, and D. M. Karl. 2002. Seasonal and interannual variability in sources of nitrogen supporting export in the oligotrophic subtropical North Pacific Ocean. *Limnol. Oceanogr.* 47: 1595–1607.

Dugdale, R. C. 1967. Nutrient limitation in the sea: Dynamics, identification and significance. *Limnol. Oceanogr.* 12: 685–695.

Dupouy, C., M. Petit, and Y. Dandonneau. 1988. Satellite detected cyanobacteria bloom in the southwestern tropical Pacific—Implication for oceanic nitrogen-fixation. *Int. J. Remote Sens.* 9: 389–396.

Dyhrman, S. T., and S. T. Haley. 2006. Phosphorus scavenging in the unicellular marine diazotroph *Crocosphaera watsonii*. *Appl. Environ. Microbiol.* 72: 1452–1458.

Dyhrman, S. T., E. A. Webb, D. M. Anderson, J. W. Moffett, and J. B. Waterbury. 2002. Cell-specific detection of phosphorus stress in *Trichodesmium* from the western north Atlantic. *Limnol. Oceanogr.* 47: 1832–1836.

Dyhrman, S. T., P. D. Chappell, S. T. Haley, J. W. Moffett, E. D. Orchard, J. B. Waterbury, and E. A. Webb. 2006. Phosphonate utilization by the globally important marine diazotroph *Trichodesmium*. *Nature* 439: 68–71.

Eady, R. R. 1996. Structure–function relationships of alternative nitrogenases. *Chem. Rev.* 96: 3013–3030.

Ehrlich, H. L. 1990. *Geomicrobiology*. Marcel Dekker.

El-Shehawy, R., C. Lugomela, A. Ernst, and B. Bergman. 2003. Diurnal expression of *hetR* and diazocyte development in the filamentous non-heterocystous cyanobacterium *Trichodesmium erythraeum*. *Microbiology* 149: 1139–1146.

Elmgren, R., and U. Larsson. 2001. Nitrogen and the Baltic Sea: Managing nitrogen in relation to phosphorus. *The Scientific World* 1: 371–377.

Falcón, L. I., E. J. Carpenter, F. Cipriano, B. Bergman, and D. G. Capone. 2004a. N_2 fixation by unicellular bacterioplankton from the Atlantic and Pacific Oceans: Phylogeny and in situ rates. *Appl. Environ. Microbiol.* 70: 765–770.

REFERENCES

Falcón, L. I., S. Lindvall, K. Bauer, B. Bergman, and E. J. Carpenter. 2004b. Ultrastructure of unicellular N_2 fixing cyanobacteria from the tropical North Atlantic and subtropical North Pacific Oceans. *J. Phycol.* 40: 1074–1078.

Falkowski, P. G. 1997. Evolution of the nitrogen cycle and its influence on the biological sequestration of CO_2 in the ocean. *Nature* 387: 272–275.

Fani, R., R. Gallo and P. Lio. 2000. Molecular evolution of nitrogen fixation: The evolutionary history of the *nifD*, *nifK*, *nifE*, and *nifN* genes. *J. Mol. Evol.* 51: 1–11.

Fennel, K., M. Follows, and P. G. Falkowski. 2005. The co-evolution of the nitrogen, carbon and oxygen cycles in the Proterozoic ocean. *Am. J. Sci.* 305: 526–545.

Fisher, T. R., L. W. Harding, D. W. Stanley, and L. G. Ward. 1988. Phytoplankton, nutrients, and turbidity in the Chesapeake, Delaware, and Hudson estuaries. *Estuar. Coast. Shelf Sci.* 27: 61–93.

Flores, E., and A. Herrero. 1994. Assimilatory nitrogen metabolism and its regulation. In D. A. Bryant (ed.), *The Molecular Biology of Cyanobacteria*, Kluwer, pp. 487–517.

Fogg, G. E. 1969. The physiology of an algal nuisance. *Proc. Roy. Soc. Lond., Ser. B, Biol. Sci.* 173: 175–189.

Fogg, G. E. 1982. Marine plankton. In N. G. Carr and B. A. Whitton (eds.), *The Biology of Cyanobacteria*. University of California Press, pp. 491–513.

Fogg, G. E., W. D. P. Stewart, P. Fay, and A. E. Walsby. 1973. *The Blue-Green Algae*. Academic Press.

Foster, R. A., and J. P. Zehr. 2006. Characterization of diatom-cyanobacteria symbioses on the basis of *nifH*, *hetR*, and 16S rRNA sequences. *Environ. Microbiol.* 8: 1913–1925.

Foster, R. A., E. J. Carpenter, and B. Bergman. 2006a. Unicellular cyanobionts in open ocean dinoflagellates, radiolarians, and tintinnids: Ultrastructural characterization and immunolocalization of phycoerythrin and nitrogenase. *J. Phycol.* 42: 453–463.

Foster, R. A., J. L. Collier, and E. J. Carpenter. 2006b. Reverse transcription PCR amplification of cyanobacterial symbiont 16S rRNA sequences from single non-photosynthetic eukaryotic marine planktonic host cells. *J. Phycol.* 42: 243–250.

Foster, R. A., A. Subramaniam, C. Mahaffey, E. J. Carpenter, D. G. Capone, and J. P. Zehr. 2007. Influence of the Amazon River plume on distributions of free-living and symbiotic cyanobacteria in the western tropical North Atlantic Ocean. *Limnol. Oceanogr.* 52: 517–532.

Fredriksson, C., and B. Bergman. 1995. Nitrogenase quantity varies diurnally in a subset of cells within colonies of the non-heterocystous cyanobacteria *Trichodesmium* spp. *Microbiology* 141: 2471–2478.

Fu, F. X., and P. R. F. Bell. 2003. Growth, N_2 fixation and photosynthesis in a cyanobacterium, *Trichodesmium* sp., under Fe stress. *Biotechnol. Lett.* 25: 645–649.

Gallon, J. R. 1992. Tansley Review No. 44/Reconciling the incompatible: N_2 fixation and O_2. *New Phytol.* 122: 571–609.

Galloway, J. N. 2003. The global nitrogen cycle. In W. Schlesinger (ed.), *Treatise on Geochemistry*, Vol. 8, Elsevier, pp. 557–583.

Galloway, J. N., and E. B. Cowling. 2002. Reactive nitrogen and the world: 200 years of change. *Ambio* 31: 64–71.

Galloway, J., F. J. Dentener, D. G. Capone, et al. 2004. Nitrogen cycles: Past, present and future. *Biogeochemistry* 70: 153–226.

Gervais, F., D. Opitz, and H. Behrendt. 1997. Influence of small-scale turbulence and large-scale mixing on phytoplankton primary production. *Hydrobiologia* 342: 95–101.

Giovannoni, S. J., H. J. Tripp, S. Givan, et al. 2005. Genome streamlining in a cosmopolitan oceanic bacterium. *Science* 309: 1242–1245.

Gómez, F., K. Furuya, and S. Takeda. 2005. Distribution of the cyanobacterium *Richelia intracellularis* as an epiphyte of the diatom *Chaetoceros compressus* in the western Pacific Ocean. *J. Plankton Res.* 27: 323–330.

Gruber, N., and J. L. Sarmiento. 1997. Global patterns of marine nitrogen fixation and denitrification. *Global Biogeochem. Cycles* 11: 235–266.

Guerinot, M. L., and R. R. Colwell. 1985. Enumeration, isolation and characterization of N_2-fixing bacteria from seawater. *Appl. Environ. Microbiol.* 50: 350–355.

Hansell, D. A., N. R. Bates, and D. B. Olson. 2004. Excess nitrate and nitrogen fixation in the North Atlantic Ocean. *Mar. Chem.* 84: 243–265.

Hawser, S. P., J. M. O'Neil, M. R. Romans, and G. A. Codd. 1992. Toxicity of blooms of the cyanobacterium *Trichodesmium* to zooplankton. *J. Appl. Phycol.* 4: 79–86.

Heil, C. A., K. A. Chaston, A. Jones, P. Bird, B. Longstaff, S. Costanzo, and W. C. Dennison. 2004. Benthic microalgae in coral reef sediments of the southern Great Barrier Reef, Australia. *Coral Reefs* 23: 336–343.

Herbert, R. A. 1999. Nitrogen cycling in coastal marine ecosystems. *FEMS Microbiol. Rev.* 23: 563–590.

Hewson, I., and J. A. Fuhrman. 2006. Spatial and vertical biogeography of coral reef sediment bacterial and diazotroph communities. *Mar. Ecol. Prog. Ser.* 306: 79–86.

Hewson, I., P. H. Moisander, K. M. Achilles, C. A. Carlson, B. D. Jenkins, E. A. Mondragon, A. E. Morrison, and J. P. Zehr. 2007. Characteristics of diazotrophs in surface to abyssopelagic waters of the Sargasso Sea. *Aquat. Microb. Ecol.* 46: 15–30.

Holl, C. M., and J. P. Montoya. 2005. Interactions between nitrate uptake and nitrogen fixation in continuous cultures of the marine diazotroph, *Trichodesmium* (cyanobacteria). *J. Phycol.* 41: 1178–1183.

Hood, R. R., N. R. Bates, D. G. Capone, and D. B. Olson. 2001. Modeling the effect of nitrogen fixation on carbon and nitrogen fluxes at BATS. *Deep-Sea Res.* II 48: 1609–1648.

Hood, R. R., A. Subramaniam, L. R. May, E. J. Carpenter, and D. G. Capone. 2002. Remote estimation of nitrogen fixation by *Trichodesmium*. *Deep-Sea Res.* II 49: 123–147.

Hood, R. R., V. J. Coles, and D. G. Capone. 2004. Modeling the distribution of *Trichodesmium* and nitrogen fixation in the Atlantic Ocean. *J. Geophys. Res.* 109: C06006.

Howard, J. B., and D. C. Rees. 1996. Structural basis of biological nitrogen fixation. *Chem. Rev.* 96: 2965–2982.

Howarth, R. W., and J. J. Cole. 1985. Molybdenum availability, nitrogen limitation, and phytoplankton growth in natural waters. *Science* 229: 653–655.

Howarth, R. W., and R. Marino. 1988. Nitrogen fixation in freshwater, estuarine, and marine ecosystems. 2. Biogeochemical controls. *Limnol. Oceanogr.* 33: 688–701.

Howarth, R. W., T. Butler, K. Lunde, D. Swaney, and C. R. Chu. 1993. Turbulence and planktonic nitrogen fixation: A mesocosm experiment. *Limnol. Oceanogr.* 38: 1696–1711.

Howarth, R. W., G. Billen, D. Swaney, et al. 1996. Regional nitrogen budgets and riverine N and P fluxes for the drainages to the North Atlantic Ocean: Natural and human influences. *Biogeochemistry* 35: 75–139.

REFERENCES

Janson, S., B. Bergman, E. J. Carpenter, S. J. Giovannoni, and K. Vergin. 1999a. Genetic analysis of natural populations of the marine diazotrophic cyanobacterium *Trichodesmium*. *FEMS Microbiol. Ecol.* 30: 57–65.

Janson, S., J. Wouters, and B. C. E. J. Bergman. 1999b. Host specificity in the *Richelia*–diatom symbiosis revealed by *hetR* gene sequence analysis. *Environ. Microbiol.* 1: 431–438.

Jenkins, B. D., G. F. Steward, S. M. Short, B. B. Ward, and J. P. Zehr. 2004. Fingerprinting diazotroph communities in the Chesapeake Bay by using a DNA macroarray. *Appl. Environ. Microbiol.* 70: 1767–1776.

Joye, S. B., A. Boetius, B. N. Orcutt, J. P. Montoya, H. N. Schulz, M. J. Erickson, and S. K. Lugo. 2004. The anaerobic oxidation of methane and sulfate reduction in sediments from Gulf of Mexico cold seeps. *Chem. Geol.* 205: 219–238.

Kana, T. M. 1993. Rapid oxygen cycling in *Trichodesmium thiebautii*. *Limnol. Oceanogr.* 38: 18–24.

Karl, D. M. 1999. A sea of change: Biogeochemical variability in the North Pacific Subtropical Gyre. *Ecosystems* 2: 181–214.

Karl, D. M. 2002. Nutrient dynamics in the deep blue sea. *Trends Microbiol.* 10: 410–418.

Karl, D. M., G. T. Taylor, T. J. Novitsky, H. W. Jannasch, C. O. Wirsen, N. R. Pace, D. J. Lane, G. J. Olsen, and S. J. Giovannoni. 1988. A microbiological study of Guaymas Basin high temperature hydrothermal vents. *Deep-Sea Res.* 35: 777–791.

Karl, D. M., R. Letelier, D. V. Hebel, D. F. Bir, and C. D. Winn. 1992. *Trichodesmium* blooms and new nitrogen in the North Pacific Gyre. In E. J. Carpenter, D. G. Capone, and J. G. Rueter (eds.), *Marine Pelagic Cyanobacteria:* Trichodesmium *and Other Diazotrophs*. Kluwer, pp. 219–238.

Karl, D., R. Letelier, D. Hebel, L. Tupas, J. Dore, J. Christian, and C. Winn. 1995. Ecosystem changes in the North Pacific subtropical gyre attributed to the 1991–92 El Nino. *Nature* 373: 230–234.

Karl, D., R. Letelier, L. Tupas, J. Dore, J. Christian, and D. Hebel. 1997. The role of nitrogen fixation in biogeochemical cycling in the subtropical North Pacific Ocean. *Nature* 388: 533–538.

Karl, D. M., D. F. Bird, K. Björkman, T. Houlihan, R. Shackelford, and L. Tupas. 1999. Microorganisms in the accreted ice of Lake Vostok, Antarctica. *Science* 286: 2144–2147.

Karl, D. M., R. R. Bidigare, and R. M. Letelier. 2001a. Long-term changes in plankton community structure and productivity in the North Pacific Subtropical Gyre: The domain shift hypothesis. *Deep-Sea Res.* II 48: 1449–1470.

Karl, D. M., K. M. Björkman, J. E. Dore, L. Fujieki, D. V. Hebel, T. Houlihan, R. M. Letelier, and L. M. Tupas. 2001b. Ecological nitrogen-to-phosphorus stoichiometry at station ALOHA. *Deep-Sea Res.* II 48: 1529–1566.

Karl, D., A. Michaels, B. Bergman, et al. 2002. Dinitrogen fixation in the world's oceans. *Biogeochemistry* 57/58: 47–98.

Kasting, J. F., and J. L. Siefert. 2001. Biogeochemistry—The nitrogen fix. *Nature* 412: 26–27.

Kasting, J. F., and J. L. Siefert. 2002. Life and the evolution of Earth's atmosphere. *Science* 296: 1066–1068.

Kayanne, H., M. Hirota, M. Yamamuro, and I. Koike. 2005. Nitrogen fixation of filamentous cyanobacteria in a coral reef measured using three different methods. *Coral Reefs* 24: 197–200.

Keto, J., J. Horppila, and T. Kairesalo. 1992. Regulation of the development and species dominance of summer phytoplankton in Lake Vesijarvi—predictability of enclosure experiments. *Hydrobiologia* 243: 303–310.

Kononen, K., J. Kuparinen, K. Mäkelä, J. Laanemets, J. Pavelson, and S. Nõmmann. 1996. Initiation of cyanobacterial blooms in a frontal region at the entrance to the Gulf of Finland, Baltic Sea. *Limnol. Oceanogr.* 41: 98–112.

Koop, K., D. Booth, A. Broadbent, et al. 2001. ENCORE: The effect of nutrient enrichment on coral reefs. Synthesis of results and conclusions. *Mar. Pollution Bull.* 42: 91–120.

Kucera, S. A. 1996. The influence of small-scale turbulence on N_2 fixation and growth in heterocystous cyanobacteria. MS Thesis, University of North Carolina, Chapel Hill.

Küpper, H., N. Ferimazova, I. Setlik, and I. Berman-Frank. 2004. Traffic lights in *Trichodesmium*. Regulation of photosynthesis for nitrogen fixation studied by chlorophyll fluorescence kinetic microscopy. *Plant Physiol.* 135: 2120–2133.

Kustka, A., E. J. Carpenter, and S. A. Sañudo-Wilhelmy. 2002. Iron and marine nitrogen fixation: progress and future directions. *Res. Microbiol.* 153: 255–262.

Kustka, A., S. Sañudo-Wilhelmy, E. J. Carpenter, D. G. Capone, and J. A. Raven. 2003. A revised estimate of the iron use efficiency of nitrogen fixation, with special reference to the marine cyanobacterium *Trichodesmium* spp. (Cyanophyta). *J. Phycol.* 39: 12–25.

Langlois, R. J., J. LaRoche, and P. A. Raab. 2005. Diazotrophic diversity and distribution in the tropical and subtropical Atlantic Ocean. *Appl. Environ. Microbiol.* 71: 7910–7919.

LaRoche, J., and E. Breitbarth. 2005. Importance of the diazotrophs as a source of new nitrogen in the ocean. *J. Sea Res.* 53: 67–91.

Lee, K., D. M. Karl, R. Wanninkhof, and J. Z. Zhang. 2002. Global estimates of net carbon production in the nitrate-depleted tropical and subtropical oceans. *Geophys. Res. Lett.* 29: 1907.

Leigh, J. 2000. Nitrogen fixation in methanogens—the Archaeal perspective. In E. W. Triplett (ed.), *Prokaryotic Nitrogen Fixation; a Model System for the Analysis of a Biological Process.* Horizon Scientific Press, pp. 657–669.

Lenes, J. M., B. P. Darrow, C. Cattrall, et al. 2001. Iron fertilization and the *Trichodesmium* response on the West Florida shelf. *Limnol. Oceanogr.* 46: 1261–1277.

Lesser, M. P., C. H. Mazel, M. Y. Gorbunov, and P. G. Falkowski. 2004. Discovery of symbiotic nitrogen-fixing cyanobacteria in corals. *Science* 305: 997–1000.

Letelier, R., and D. Karl. 1998. *Trichodesmium* spp. physiology and nutrient fluxes in the North Pacific subtropical gyre. *Aquat. Microb. Ecol.* 15: 265–276.

Lilley, M. D., D. A. Butterfield, E. J. Olsen, J. E. Lupton, S. A. Macko, and R. E. McDuff. 1993. Anomalous CH_4 and NH_4^+ concentrations at an unsedimented mid-ocean-ridge hydrothermal system. *Nature* 364: 45–47.

Lipschultz, F., and N. J. P. Owens. 1996. An assessment of nitrogen fixation as a source of nitrogen to the North Atlantic Ocean. *Biogeochemistry* 35: 261–274.

REFERENCES

Loveless, T. M., J. R. Saah, and P. E. Bishop. 1999. Isolation of nitrogen-fixing bacteria containing molybdenum-independent nitrogenases from natural environments. *Appl. Environ. Microbiol.* 65: 4223–4226.

Lovell, C. R., Y. M. Piceno, J. M. Quattro, and C. E. Bagwell. 2000. Molecular analysis of diazotroph diversity in the rhizosphere of the smooth cordgrass, *Spartina alterniflora*. *Appl. Environ. Microbiol.* 66: 3814–3822.

Lovell, C. R., M. J. Friez, J. W. Longshore, and C. E. Bagwell. 2001. Recovery and phylogenetic analysis of *nifH* sequences from diazotrophic bacteria associated with dead aboveground biomass of *Spartina alterniflora*. *Appl. Environ. Microbiol.* 67: 5308–5314.

Mahaffey, C., A. F. Michaels, and D. G. Capone. 2005. The conundrum of marine N_2 fixation. *Am. J. Sci.* 305: 546–595.

Marino, R., F. Chan, R. W. Howarth, M. L. Pace, and G. E. Likens. 2006. Ecological constraints on planktonic nitrogen fixation in saline estuaries. I. Nutrient and trophic controls. *Mar. Ecol. Prog. Ser.* 309: 25–39.

Martinez, L., M. W. Silver, J. M. King, and A. L. Alldredge. 1983. Nitrogen fixation by floating diatom mats: a source of new nitrogen to oligotrophic ocean waters. *Science* 221: 152–154.

Martinez-Argudo, I., R. Little, N. Shearer, P. Johnson, and R. Dixon. 2005. Nitrogen fixation: key genetic regulatory mechanisms. *Biochem. Soc. Trans.* 33: 152–156.

Mayajima, T., M. Suzumura, Y. Umezawa, and I. Koike. 2001. Microbiological nitrogen transformation in carbonate sediments of a coral reef lagoon and associated seagrass beds. *Mar. Ecol. Prog. Ser.* 217: 273–286.

Mehta, M. P., and J. A. Baross. 2006. Nitrogen fixation at 92 °C by a hydrothermal vent archaeon. *Science* 314: 1783–1786.

Mehta, M. P., D. A. Butterfield, and J. A. Baross. 2003. Phylogenetic diversity of nitrogenase (*nifH*) genes in deep-sea and hydrothermal vent environments of the Juan de Fuca Ridge. *Appl. Environ. Microbiol.* 69: 960–970.

Mehta, M. P., J. A. Huber, and J. A. Baross. 2005. Incidence of novel and potentially archaeal nitrogenase genes in the deep Northeast Pacific Ocean. *Environ. Microbiol.* 7: 1525–1534.

Michaels, A. F., D. Olson, J. L. Sarmiento, J. W. Ammerman, K. Fanning, R. Jahnke, A. H. Knap, F. Lipschultz, and J. M. Prospero. 1996. Inputs, losses and transformations of nitrogen and phosphorus in the pelagic North Atlantic Ocean. *Biogeochemistry* 35: 181–226.

Michaels, A. F., D. M. Karl, and A. H. Knap. 2000. Temporal studies of biogeochemical dynamics in oligotrophic oceans. In R. B. Hanson, H. W. Ducklow, and J. G. Field (eds.), *The Changing Ocean Carbon Cycle: A Midterm Synthesis of the Joint Global Ocean Flux Study*. Cambridge University Press, pp. 392–413.

Mitsui, A., S. Cao, A. Takahashi, and T. Arai. 1987. Growth synchrony and cellular parameters of the unicellular nitrogen-fixing marine cyanobacterium, *Synechococcus* sp. strain Miami Bg 043511 under continuous illumination. *Physiol. Plant.* 69: 1–8.

Moisander, P. H., J. L. Hench, K. Kononen, and H. W. Paerl. 2002a. Small-scale shear effects on heterocystous cyanobacteria. *Limnol. Oceanogr.* 47: 109–119.

Moisander, P. H., E. McClinton, and H. W. Paerl. 2002b. Salinity effects on growth, photosynthetic parameters, and nitrogenase activity in estuarine planktonic cyanobacteria. *Microb. Ecol.* 43: 432–442.

Moisander, P. H., M. F. Piehler, and H. W. Paerl. 2005. Diversity and activity of epiphytic nitrogen-fixers on standing dead stems of the salt marsh grass *Spartina alterniflora*. *Aquat. Microb. Ecol.* 39: 271–279.

Moisander, P. H., L. Shiue, G. F. Steward, B. D. Jenkins, B. M. Bebout, and J. P. Zehr. 2006. Application of a *nifH* oligonucleotide microarray for profiling diversity of N_2-fixing microorganisms in marine microbial mats. *Environ. Microbiol.* 8: 1721–1735.

Montoya, J. P., C. M. Holl, J. P. Zehr, A. Hansen, T. A. Villareal, and D. G. Capone. 2004. High rates of N_2 fixation by unicellular diazotrophs in the oligotrophic Pacific Ocean. *Nature* 430: 1027–1032.

Moran, M. A., A. Buchan, J. M. Gonzalez, et al. 2004. Genome sequence of *Silicibacter pomeroyi* reveals adaptations to the marine environment. *Nature* 432: 910–913.

Mulholland, M., and D. G. Capone. 2000. The nitrogen physiology of the marine N_2-fixing cyanobacteria *Trichodesmium* spp. *Trend Plant Sci.* 5: 148–153.

Nakano, S., K. Hayakawa, J. J. Frenette, T. Nakajima, C. M. Jiao, S. Tsujimura, and M. Kumagai. 2001. Cyanobacterial blooms in a shallow lake: A large-scale enclosure assay to test the importance of diurnal stratification. *Arch. Hydrobiol.* 150: 491–509.

Nealson, K. H., and R. Rye. 2004. Evolution of metabolic pathways. In W. Schesinger (ed.), *Treatise on Geochemistry*, Vol. 8. Elsevier, pp. 41–61.

Neveux, J., F. Lantoine, D. Vaulot, D. Marie, and J. Blanchot. 1999. Phycoerythrins in the southern tropical and equatorial Pacific Ocean: Evidence for new cyanobacterial types. *J. Geophys. Res.* 104: 3311–3321.

Nielsen, L. B., K. Finster, D. T. Welsh, A. Donelly, R. A. Herbert, R. de Wit, and B. A. Lomstein. 2001. Sulphate reduction and nitrogen fixation rates associated with roots, rhizomes and sediments from *Zostera noltii* and *Spartina maritima* meadows. *Environ. Microbiol.* 3: 63–71.

Ninfa, A. J., and P. Jiang. 2005. PII signal transduction proteins: Sensors of α-ketoglutarate that regulate nitrogen metabolism. *Curr. Opin. Microbiol.* 8: 168–173.

Nisbet, E. G., and C. M. R. Fowler. 2003. The early history of life. In W. Schlesinger (ed.), *Treatise on Geochemistry*, Vol. 8. Elsevier, pp. 1–39.

Nixon, S. W. 1995. Coastal marine eutrophication—a definition, social causes, and future concerns. *Ophelia* 41: 199–219.

Olson, J. B., R. W. Litaker, and H. W. Paerl. 1999. Ubiquity of heterotrophic diazotrophs in marine microbial mats. *Aquat. Microb. Ecol.* 19: 29–36.

Omoregie, E. O., L. L. Crumbliss, B. M. Bebout, and J. P. Zehr. 2004a. Comparison of diazotroph community structure in *Lyngbya* sp. and *Microcoleus chthonoplastes* dominated microbial mats from Guerrero Negro, Baja, Mexico. *FEMS Microbiol. Ecol.* 47: 305–318.

Omoregie, E. O., L. L. Crumbliss, B. M. Bebout, and J. P. Zehr. 2004b. Determination of nitrogen-fixing phylotypes in *Lyngbya* sp. and *Microcoleus chthonoplastes* cyanobacterial mats from Guerrero Negro, Baja California, Mexico. *Appl. Environ. Microbiol.* 70: 2119–2128.

Orcutt, K. M., F. Lipschultz, K. Gundersen, R. Arimoto, A. F. Michaels, A. H. Knap, and J. R. Gallon. 2001. A seasonal study of the significance of N_2 fixation by *Trichodesmium* spp. at the Bermuda Atlantic Time-series Study (BATS) site. *Deep-Sea Res. II* 48: 1583–1608.

Orcutt, K. M., U. Rasmussen, E. A. Webb, J. B. Waterbury, K. Gundersen, and B. Bergman. 2002. Characterization of *Trichodesmium* spp. by genetic techniques. *Appl. Environ. Microbiol.* 68: 2236–2245.

REFERENCES

Orphan, V. J., C. H. House, K. U. Hinrichs, K. D. McKeegan, and E. F. DeLong. 2001. Methane-consuming archaea revealed by directly coupled isotopic and phylogenetic analysis. *Science* 293: 484–487.

Ortega-Calvo, J. J., and L. J. Stal. 1991. Diazotrophic growth of the unicellular cyanobacterium *Gloeothece* sp PCC 6909 in continuous culture. *J. Gen. Microbiol.* 137: 1789–1797.

Paerl, H. W. 1982a. In situ H_2 production and utilization by natural populations of N_2 fixing blue–green algae. *Can. J. Bot.* 60: 2542–2546.

Paerl, H. W. 1982b. Chpt. 17. Interactions with bacteria. In N. G. Carr and B. A. Whitton (eds.), *The Biology of Cyanobacteria*. Blackwell, pp. 441–461.

Paerl, H. W. 1990. Physiological ecology and regulation of N_2 fixation in natural waters. *Adv. Microb. Ecol.* 8: 305–344.

Paerl, H. W. 1997. Coastal eutrophication and harmful algal blooms: Importance of atmospheric deposition and groundwater as "new" nitrogen and other nutrient sources. *Limnol. Oceanogr.* 42: 1154–1165.

Paerl, H. W. 1998. Microbially mediated nitrogen cycling. In R. Burlage (ed.), *Techniques in Microbial Ecology*. Oxford University Press, pp. 3–30.

Paerl, H. W. 2000. Marine plankton. In M. Potts and B. A. Whitton (eds.), *The Biology and Ecology of Cyanobacteria*. Blackwell, pp. 121–148.

Paerl, H., and B. Bebout. 1988a. Microelectrode determinations of O_2-depleted microzones in marine *Oscillatoria* spp. (*Trichodesmium*): A mechanism promoting contemporaneous N_2 fixation and oxygenic photosynthesis. *Science* 241: 442–445.

Paerl, H. W., and B. M. Bebout. 1988b. Direct measurement of O_2-depleted microzones in marine *Oscillatoria:* Relation to nitrogen fixation. *Science* 241: 442–445.

Paerl, H. W., and P. E. Kellar. 1978. Significance of bacterial *Anabaena* (Cyanophyceae) associations with respect to N_2 fixation in freshwater. *J. Phycol.* 14: 254–260.

Paerl, H. W., and J. Kuparinen. 2002. Microbial aggregates and consortia. In G. Bitton (ed.), *Encyclopedia of Environmental Microbiology*. Wiley, pp. 160–181.

Paerl, H. W., and D. F. Millie. 1996. Physiological ecology of toxic cyanobacteria. *Phycologia.* 35: 160–167.

Paerl, H. W., and J. L. Pinckney. 1996a. Microbial consortia: Their roles in aquatic production and biogeochemical cycling. *Microb. Ecol.* 31: 225–247.

Paerl, H. W., and J. L. Pinckney. 1996b. A mini-review of microbial consortia: their roles in aquatic production and biogeochemical cycling. *Microb. Ecol.* 31: 225–247.

Paerl, H. W., and L. E. Prufert. 1987. Oxygen-poor microzones as potential sites of microbial N_2 fixation in nitrogen-depleted aerobic marine waters. *Appl. Environ. Microbiol.* 53: 1078–1087.

Paerl, H. W., and J. C. Priscu. 1998. Microbial phototrophic, heterotrophic and diazotrophic activities associated with aggregates in the permanent ice cover of Lake Bonney, Antarctica. *Microb. Ecol.* 36: 221–230.

Paerl, H. W., and J. P. Zehr 2000. Marine nitrogen fixation. In D. L. Kirchman (ed.), *Microbial Ecology of the Oceans*, 1st edn. Wiley-Liss, pp. 387–426.

Paerl, H. W., J. C. Priscu, and D. L. Brawner. 1989. Immunochemical localization of nitrogenase in marine *Trichodesmium* aggregates: Relationship to N_2 fixation potential. *Appl. Environ. Microbiol.* 55: 2965–2975.

Paerl, H. W., L. E. Prufert-Bebout, and C. Guo. 1994. Iron-stimulated N_2 fixation and growth in natural and cultured populations of the planktonic marine cyanobacteria *Trichodesmium* spp. *Appl. Environ. Microbiol.* 60: 1044–1047.

Paerl, H. W., M. A. Mallin, C. A. Donahue, M. Go, and B. L. Peierls. 1995a. Nitrogen loading sources and eutrophication of the Neuse River Estuary, NC: Direct and indirect roles of atmospheric deposition. UNC Water Resources Research Institute, Raleigh, NC.

Paerl, H. W., J. L. Pinckney, and S. A. Kucera. 1995b. Clarification of the structural and functional roles of heterocysts and anoxic microzones in the control of pelagic nitrogen-fixation. *Limnol. Oceanogr.* 40: 634–638.

Paerl, H. W., M. Fitzpatrick, and B. M. Bebout. 1996. Seasonal nitrogen fixation dynamics in a marine microbial mat—potential roles of cyanobacteria and microheterotrophs. *Limnol. Oceanogr.* 41: 419–427.

Paerl, H. W., J. L. Pinckney, and T. F. Steppe. 2000. Cyanobacterial-bacterial mat consortia: examining the functional unit of microbial survival and growth in extreme environments. *Environ. Microbiol.* 2: 11–26.

Paerl, H. W., T. F. Steppe, K. C. Buchan, and M. Potts. 2003. Hypersaline cyanobacterial mats as indicators of elevated tropical hurricane activity and associated climate change. *Ambio* 32: 87–90.

Palenik, B., B. Brahamsha, F. W. Larimer, et al. 2003. The genome of a motile marine *Synechococcus*. *Nature* 424: 1037–1042.

Paulsen, D. M., H. W. Paerl, and P. E. Bishop. 1991. Evidence that molybdenum-dependent nitrogen-fixation is not limited by high sulfate concentrations in marine environments. *Limnol. Oceanogr.* 36: 1325–1334.

Piceno, Y. M., and C. R. Lovell. 2000a. Stability in natural bacterial communities: I. Nutrient addition effects on rhizosphere diazotroph assemblage composition. *Microb. Ecol.* 39: 32–40.

Piceno, Y. M., and C. R. Lovell. 2000b. Stability in natural bacterial communities: II. Plant resource allocation effects on rhizosphere diazotroph assemblage composition. *Microb. Ecol.* 39: 41–48.

Pinckney, J. L., and H. W. Paerl. 1997. Anoxic photosynthesis and N_2 fixation by a microbial mat community in a Bahamian hypersaline lagoon. *Appl. Environ. Microbiol.* 63: 420–426.

Postgate, J. R. 1982. *The Fundamentals of Nitrogen Fixation*. Cambridge University Press.

Postgate, J. R. 1998. *Nitrogen Fixation*. Cambridge University Press.

Postgate, J. R., and R. R. Eady. 1988. The evolution of biological nitrogen fixation. In H. Bothe, F. J. de Bruijn, and W. E. Newton (eds.), *Nitrogen Fixation: Hundred Years After*. Gustav Fischer, pp. 31–40.

Potts, M. 1980. Blue–green algae (cyanophyta) in marine coastal environment of the Sinai Peninsula: distribution, zonation, stratification and taxonomic diversity. *Phycologia* 19: 60–73.

Preston, N. P., M. A. Burford, and D. J. Stenzel. 1998. Effects of *Trichodesmium* spp. blooms on penaeid prawn larvae. *Mar. Biol.* 131: 671–679.

Prufert-Bebout, L., H. W. Paerl, and C. Lassen. 1993. Growth, nitrogen fixation, and spectral attenuation in cultivated *Trichodesmium* species. *Appl. Environ. Microbiol.* 59: 1367–1375.

Rabouille, S., M. Staal, L. J. Stal, and K. Soetaert. 2006. Modeling the dynamic regulation of nitrogen fixation in the cyanobacterium *Trichodesmium* sp. *Appl. Environ. Microbiol.* 72: 3217–3227.

Raven, J. A., and Z. H. Yin. 1998. The past, present and future of nitrogenous compounds in the atmosphere, and their interactions with plants. *New Phytol.* 139: 205–219.

Raymond, J., J. L. Siefert, C. R. Staples, and R. E. Blankenship. 2004. The natural history of nitrogen fixation. *Mol. Biol. Evol.* 21: 541–554.

Reddy, K. J., J. B. Haskell, D. M. Sherman, and L. A. Sherman. 1993. Unicellular, aerobic nitrogen-fixing cyanobacteria of the genus *Cyanothece*. *J. Bacteriol.* 175: 1284–1292.

Redfield, A. C. 1958. The biological control of chemical factors in the environment. *Am. Sci.* 46: 205–222.

Rees, D. C., and J. B. Howard. 2000. Nitrogenase: Standing at the crossroads. *Curr. Opin. Chem. Biol.* 4: 559–566.

Ribbe, M., D. Gadkari, and O. Meyer. 1997. N_2 fixation by *Streptomyces thermoautotrophicus* involves a molybdenum-dinitrogenase and a manganese-superoxide oxidoreductase that couple N_2 reduction to the oxidation of superoxide produced from O_2 by a molybdenum–CO dehydrogenase. *J. Biol. Chem.* 272: 26627–26633.

Rocap, G., F. W. Larimer, J. Lamerdin, et al. 2003. Genome divergence in two *Prochlorococcus* ecotypes reflects oceanic niche differentiation. *Nature* 424: 1042–1047.

Rohwer, F., V. Seguritan, F. Azam, and N. Knowlton. 2002. Diversity and distribution of coral-associated bacteria. *Mar. Ecol. Prog. Ser.* 243: 1–10.

Rosing, M. T. 1999. ^{13}C-depleted carbon microparticles in $>$ 3700-Ma sea-floor sedimentary rocks from west Greenland. *Science* 283: 674–676.

Rubio, L. M., and P. W. Ludden. 2005. Maturation of nitrogenase: a biochemical puzzle. *J. Bacteriol.* 187: 405–414.

Rueter, J. G. 1988. Iron stimulation of photosynthesis and nitrogen fixation in Anabaena 7120 and Trichodesmium (Cyanophyceae). *J. Phycol.* 24: 249–254.

Rueter, J., D. A. Hutchins, R. W. Smith, and N. L. Unsworth. 1992. Iron nutrition of Trichodesmium: establishment of culture and characteristics of N_2-fixation. In E. J. Carpenter, D. G. Capone and J. G. Rueter (eds.), *Marine Pelagic Cyanobacteria: Trichodesmium and Other Diazotrophs*. Kluwer, pp. 289–306.

Ryther, J. H., and W. M. Dunstan. 1971. Nitrogen, phosphorus, and eutrophication in the coastal marine environment. *Science* 171: 1008–1013.

Sañudo-Wilhelmy, S. A., A. B. Kustka, C. J. Gobler, et al. 2001. Phosphorus limitation of nitrogen fixation by *Trichodesmium* in the central Atlantic Ocean. *Nature* 411: 66–69.

Sanz-Alférez, S., and F. F.d. Campo. 1994. Relationship between nitrogen fixation and nitrate metabolism in the *Nodularia* strains M1 and M2. *Planta* 194: 339–345.

Schneegurt, M. A., D. M. Sherman, S. Nayar, and L. A. Sherman. 1994. Oscillating behavior of carbohydrate granule formation and dinitrogen fixation in the cyanobacterium *Cyanothece* sp. strain ATCC 51142. *J. Bacteriol.* 176: 1586–1597.

Scranton, M. I. 1983. Gaseous nitrogen compounds in the marine environment. In E. J. Carpenter and D. G. Capone (eds.), *Nitrogen in the Marine Environment*. Academic Press, pp. 37–64.

Sharp, J. H. 1983. The distribution of inorganic nitrogen and dissolved and particulate organic nitrogen in the sea. In E. J. Carpenter and D. G. Capone (eds.), *Nitrogen in the Marine Environment*. Academic Press, pp. 1–36.

Shen, Y. A., R. Buick, and D. E. Canfield. 2001. Isotopic evidence for microbial sulphate reduction in the early Archaean era. *Nature* 410: 77–81.

Short, S. M., and J. P. Zehr. 2007. Nitrogenase gene expression in the Chesapeake Bay Estuary. *Environ. Microbiol.* 9: 1591–1596.

Short, S. M., B. D. Jenkins, and J. P. Zehr. 2004. Spatial and temporal distribution of two diazotrophic bacteria in the Chesapeake Bay. *Appl. Environ. Microbiol.* 70: 2186–2192.

Smith, S. V. 1984. Phosphorous versus nitrogen limitation in the marine environment. *Limnol. Oceanogr.* 29: 1149–1160.

Sobecky, P. A., T. J. Mincer, M. C. Chang, and D. R. Helinski. 1997. Plasmids isolated from marine sediment microbial communities contain replication and incompatibility regions unrelated to those of known plasmid groups. *Appl. Environ. Microbiol.* 63: 888–895.

Sprent, J. I., and J. A. Raven. 1992. Evolution of nitrogen-fixing symbioses. In G. Stacey, R. H. Burris and H. J. Evans (eds.), *Biological Nitrogen Fixation*. Chapman and Hall, pp. 461–496.

Staal, M., F. J. R. Meysman, and L. J. Stal. 2003. Temperature excludes N_2-fixing heterocystous cyanobacteria in the tropical oceans. *Nature* 425: 504–507.

Stephens, N., K. J. Flynn, and J. R. Gallon. 2003. Interrelationships between the pathways of inorganic nitrogen assimilation in the cyanobacterium *Gloeothece* can be described using a mechanistic mathematical model. *New Phytol.* 160: 545–555.

Steppe, T. F., and H. W. Paerl. 2002. Potential N_2 fixation by sulfate-reducing bacteria in a marine intertidal microbial mat. *Aquat. Microb. Ecol.* 28: 1–12.

Steppe, T. F., J. B. Olson, H. W. Paerl, R. W. Litaker, and J. Belnap. 1996. Consortial N_2 fixation: A strategy for meeting nitrogen requirements of marine and terrestrial cyanobacterial mats. *FEMS Microbiol. Ecol.* 21: 149–156.

Steunou, A. S., D. Bhaya, M. M. Bateson, M. C. Melendrez, D. M. Ward, E. Brecht, J. W. Peters, M. Kuhl, and A. R. Grossman. 2006. In situ analysis of nitrogen fixation and metabolic switching in unicellular thermophilic cyanobacteria inhabiting hot spring microbial mats. *Proc. Natl. Acad. Sci. USA* 103: 2398–2403.

Steward, G. F., B. D. Jenkins, B. B. Ward, and J. P. Zehr. 2004. Development and testing of a DNA macroarray to assess nitrogenase (*nifH*) gene diversity. *Appl. Environ. Microbiol.* 70: 1455–1465.

Stewart, W. D. P., G. P. Fitzgerald, and R. H. Burris. 1967. In situ studies on nitrogen fixation using the acetylene reduction technique. *Proc. Natl. Acad. Sci. USA* 58: 2071–2078.

Subramaniam, A., E. J. Carpenter, D. Karentz, and P. G. Falkowski. 1999. Bio-optical properties of the marine diazotrophic cyanobacteria *Trichodesmium* spp. I. Absorption and photosynthetic action spectra. *Limnol. Oceanogr.* 44: 608–617.

Subramaniam, A., C. W. Brown, R. R. Hood, E. J. Carpenter, and D. G. Capone. 2002. Detecting *Trichodesmium* blooms in SeaWiFS imagery. *Deep-Sea Res.* II 49: 107–121.

Taroncher-Oldenburg, G., E. M. Griner, C. A. Francis, and B. B. Ward. 2003. Oligonucleotide microarray for the study of functional gene diversity in the nitrogen cycle in the environment. *Appl. Environ. Microbiol.* 69: 1159–1171.

REFERENCES

Ter Steeg, P. F., P. J. Hanson, and H. W. Paerl. 1986. Growth-limiting quantities and accumulation of molybdenum in *Anabaena oscillarioides* (cyanobacteria). *Hydrobiologia* 140: 143–147.

Thiel, T. 1993. Characterization of genes for an alternative nitrogenase in the cyanobacterium *Anabaena variabilis*. *J. Bacteriol.* 175: 6276–6286.

Thiel, T., E. M. Lyons, J. C. Erker, and A. Ernst. 1995. A second nitrogenase in vegetative cells of a heterocyst-forming cyanobacterium. *Proc. Natl. Acad. Sci.* 92: 9358–9362.

Thomas, W. H., and C. H. Gibson. 1990. Quantified small-scale turbulence inhibits a red tide dinoflagellate, *Gonyaulax polyedra* Stein. *Deep-Sea Res.* 37: 1583–1593.

Tichi, M. A., and F. R. Tabita. 2001. Interactive control of *Rhodobacter capsulatus* redox-balancing systems during phototrophic metabolism. *J. Bacteriol.* 183: 6344–6354.

Tiquia, S. M., L. Wu, S. C. Chong, S. Passovets, D. Xu, Y. Xu, and J. Zhou. 2004. Evaluation of 50-mer oligonucleotide arrays for detecting microbial populations in environmental samples. *BioTechniques* 36: 664–675.

Towe, K. M. 2002. Evolution of nitrogen fixation. *Science* 295: 798–799.

Tuit, C., J. Waterbury, and G. Ravizza. 2004. Diel variation of molybdenum and iron in marine diazotrophic cyanobacteria. *Limnol. Oceanogr.* 49: 978–990.

Tyrrell, T. 1999. The relative influences of nitrogen and phosphorus on oceanic primary production. *Nature* 400: 525–531.

Tyrrell, T., E. Maranon, A. J. Poulton, A. R. Bowie, D. S. Harbour, and E. M. S. Woodward. 2003. Large-scale latitudinal distribution of *Trichodesmium* spp. in the Atlantic Ocean. *J. Plankton Res.* 25: 405–416.

Ulrich, L. E., E. V. Koonin, and I. B. Zhulin. 2005. One-component systems dominate signal transduction in prokaryotes. *Trends Microbiol.* 13: 52–56.

Villareal, T. A. 1992. Marine nitrogen-fixing diatom–cyanobacteria symbioses. In E. J. Carpenter, D. G. Capone, and J. G. Rueter (eds.), *Marine Pelagic Cyanobacteria: Trichodesmium and other Diazotrophs*. Kluwer, pp. 163–175.

Villareal, T. A., and E. J. Carpenter. 1990. Diel buoyancy regulation in the marine diazotrophic cyanobacterium *Trichodesmium thiebautii*. *Limnol. Oceanogr.* 35: 1832–1837.

Villareal, T. A., and E. J. Carpenter. 2003. Buoyancy regulation and potential for vertical migration in the oceanic cyanobacterium *Trichodesmium*. *Microb. Ecol.* 45: 1–10.

Voss, M., P. Croot, K. Lochte, M. Mills and I. Peeken. 2004. Patterns of nitrogen fixation along 10 °N in the tropical Atlantic. *Geophys. Res. Lett.* 31: Art. No. L23S09. doi: 10.1029/2004GL020127.

Waterbury, J. B., and R. Rippka. 1989. Cyanobacteria. Subsection I. Order Chroococcales Wettstien 1924, Emend. Rippka. et al., 1979. In J. T. Staley, M. P. Bryant, N. Pfenning, and J. G. Holt (eds.), *Bergey's Manual of Systematic Bacteriology*. Williams & Wilkins, pp. 1728–1729.

White, A. E., Y. H. Spitz, and R. M. Letelier. 2006. Modeling carbohydrate ballasting by *Trichodesmium* spp. at Station ALOHA. *Mar. Ecol. Prog. Ser.* 323: 35–45.

Wilson, C. 2003. Late summer chlorophyll blooms in the oligotrophic North Pacific Subtropical Gyre. *Geophys. Res. Lett.* 30. doi: 10.1029/2003GL01770.

Wu, J., W. Sunda, E. A. Boyle and D. M. Karl. 2000. Phosphate depletion in the western North Atlantic Ocean. *Science* 289: 759–762.

Wynn-Williams, D. D., and M. E. Rhodes. 1974. Nitrogen fixation in seawater. *J. Appl. Bacteriol.* 37: 203–216.

Young, J. P. W. 1992. Phylogenetic classification of nitrogen-fixing organisms. In G. Stacey, H. J. Evans and R. H. Burris (eds.), *Biological Nitrogen Fixation*. Chapman and Hall, pp. 43–86.

Young, J. P. W. 2005. The phylogeny and evolution of nitrogenases. In R. Palacios and W. E. Newton (eds.), *Genomes and Genomics of Nitrogen-Fixing Organisms*. Springer, pp. 221–241.

Zani, S., M. T. Mellon, J. L. Collier, and J. P. Zehr. 2000. Expression of *nifH* genes in natural microbial assemblages in Lake George, NY detected with RT-PCR. *Appl. Environ. Microbiol.* 66: 3119–3124.

Zehr, J. P., and D. G. Capone. 1996. Problems and promises of assaying the genetic potential for nitrogen fixation in the marine environment. *Microb. Ecol.* 32: 263–281.

Zehr, J. P., and L. A. McReynolds. 1989. Use of degenerate oligonucleotides for amplification of the *nifH* gene from the marine cyanobacterium *Trichodesmium* spp. *Appl. Environ. Microbiol.* 55: 2522–2526.

Zehr, J. P., and J. P. Montoya. 2007. Measuring N_2 fixation in the field. In H. Bothe, S. Ferguson, and W. E. Newton (eds.), *Biology of the Nitrogen Cycle*. Elsevier, pp. 193–205.

Zehr, J. P., and H. W. Paerl. 1998. Nitrogen fixation in the marine environment: Genetic potential and nitrogenase expression. In K. E. Cooksey (ed.), *Molecular Approaches to the Study of the Ocean*. Chapman and Hall, pp. 285–302.

Zehr, J. P., M. Mellon, S. Braun, W. Litaker, T. Steppe, and H. W. Paerl. 1995. Diversity of heterotrophic nitrogen fixation genes in a marine cyanobacterial mat. *Appl. Environ. Microbiol.* 61: 2527–2532.

Zehr, J. P., M. T. Mellon, and S. Zani. 1998. New nitrogen fixing microorganisms detected in oligotrophic oceans by the amplification of nitrogenase (*nifH*) genes. *Appl. Environ. Microbiol.* 64: 3444–3450.

Zehr, J. P., P. J. Turner, E. Omoregie, A. Hansen, G. F. Steward, J. P. Montoya, L. Tupas, and D. M. Karl. 2001a. Nitrogenase gene expression in the North Pacific Gyre. In *Proceedings of Aquatic Sciences Meeting, American Society for Limnology and Oceanography*, Albuquerque, New Mexico.

Zehr, J. P., J. B. Waterbury, P. J. Turner, J. P. Montoya, E. Omoregie, G. F. Steward, A. Hansen, and D. M. Karl. 2001b. Unicellular cyanobacteria fix N_2 in the subtropical North Pacific Ocean. *Nature* 412: 635–638.

Zehr, J. P., B. D. Jenkins, S. M. Short, and G. F. Steward. 2003. Nitrogenase gene diversity and microbial community structure: A cross-system comparison. *Environ. Microbiol.* 5: 539–554.

Zehr, J. P., B. Methé, and R. Foster. 2005. New nitrogen-fixing microorganisms from the oceans: Biological aspects and global implications. In Y.-P. Wang, M. Lin, Z.-X. Tian, C. Elmerich, and W. E. Newton (eds.), *Biological Nitrogen Fixation, Sustainable Agriculture and the Environment*. Springer-Verlag, pp. 361–365.

Zehr, J. P., M. J. Church, and P. H. Moisander. 2006. Diversity, distribution and biogeochemical significance of nitrogen-fixing microorganisms in anoxic and suboxic ocean environments. In L. N. Neritin (ed.), *Past and Present Water Column Anoxia*. Springer-Verlag, pp. 337–371.

Zehr, J. P., J. P. Montoya, B. D. Jenkins, I. Hewson, E. Mondragon, C. M. Short, M. J. Church, A. Hansen, and D. M. Karl. 2007. Experiments linking nitrogenase gene expression to nitrogen fixation in the North Pacific subtropical gyre. *Limnol. Oceanogr.* 52: 169–183.

Zuckermann, H., M. Staal, L. J. Stal, J. Reuss, S. T. L. Hekkert, F. Harren, and D. Parker. 1997. On-line monitoring of nitrogenase activity in cyanobacteria by sensitive laser photoacoustic detection of ethylene. *Appl. Environ. Microbiol.* 63: 4243–4251.

NITROGEN CYCLING IN SEDIMENTS

BO THAMDRUP

Institute of Biology, University of Southern Denmark, DK-5230 Odense M, Denmark

TAGE DALSGAARD

National Environmental Research Institute, University of Aarhus, DK-8600 Silkeborg, Denmark

INTRODUCTION

Fixed nitrogen makes up less than 0.1 percent of the nitrogen in the biosphere, and is often a limiting nutrient for algae that require nitrogen in fixed, or combined, form, typically as ammonium, nitrate, or some nitrogen-containing organic compounds (Falkowski et al. 1998; Capone 2000). The largest pool of nitrogen on Earth is ammonium bound in rocks and sediments (1.8×10^{22} g), while atmospheric N_2 accounts for 0.39×10^{22} g. Fixed nitrogen in the oceans amounts to less than 0.01 percent of the atmospheric pool, split between inorganic forms (5.7×10^{17} g), living biomass (0.005×10^{17} g) and nitrogen in organic detritus (5.3×10^{17} g) (Delwiche 1970; Söderlund and Roswall 1982). The small size of the fixed nitrogen pool may seem surprising considering that nitrogen fixation ($N_2 \rightarrow NH_4^+$), a process that is quite widespread in the prokaryotic domains, is believed to have been active through most of Earth's history (Falkowski 1997), and that nitrate is the thermodynamically stable form of nitrogen under oxic conditions. The reason is that two other microbial processes return the fixed nitrogen to the N_2 pool: denitrification, a respiratory reduction of nitrate or nitrite, and anaerobic ammonium oxidation

---

Microbial Ecology of the Oceans, Second Edition. Edited by David L. Kirchman
Copyright © 2008 John Wiley & Sons, Inc.

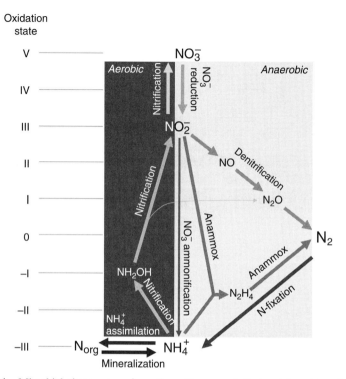

Figure 14.1 Microbial nitrogen transformations. Compounds that mainly exist as short-lived intermediates are shown in smaller type.

(anammox), a newly discovered metabolism that couples nitrite reduction and ammonium oxidation (Fig. 14.1). Both of these processes are obligately anaerobic. Thus, since the size of the fixed nitrogen pool is determined by the balance between nitrogen fixation and the loss of fixed nitrogen to N_2, the persistence of anoxic environments profoundly influences the conditions of life in our otherwise oxic world.

Nitrogen fixation is a reductive process that generates ammonium, which is incorporated into organic matter, and released again when organic nitrogen compounds are degraded (Fig. 14.1). Denitrification and anammox (see the text box) require nitrate or nitrite. The production of these substrates is accomplished by two additional groups of specialized prokaryotes: ammonium oxidizers that produce nitrite, and nitrite oxidizers that produce nitrate. Collectively these organisms, which close the nitrogen cycle between fixation and production of N_2, are known as nitrifiers. The nitrification processes require molecular oxygen. Thus, specialized microbes play key roles in all parts of the biogeochemical nitrogen cycle, and, to complete the cycle, nitrogen must pass through both oxic and anoxic environments, which makes oxic/anoxic interfaces sites of particular importance.

INTRODUCTION

> **What is Denitrification?**
>
> Two different definitions of denitrification (at least) are used in the biogeochemical and microbiological literature. In the wider sense, denitrification refers to the conversion of fixed nitrogen forms to nitrogen gases, mainly nitrous oxide (N_2O) or dinitrogen (N_2), regardless of the mechanism, while denitrification in the stricter sense refers specifically to N_2O or N_2 production through the well-defined dissimilatory microbial respiratory pathway, which involves the enzymes nitrite reductase, nitric oxide reductase, and nitrous oxide reductase. This pathway is also called canonical denitrification (see, e.g., Codispoti 2007). In the quantitative sense, the coverage of the two definitions has been almost the same, as microbial respiration was thought to be the prevailing mechanism for the removal of fixed nitrogen in most environments. However, the discovery of anammox as an alternative pathway, which involves different enzymes and accounts for most of the N_2 formation in some environments, has accentuated the difference between the definitions. We here adopt the more common, strict definition.

The sea floor is an important compartment in the marine nitrogen cycle, particularly because it harbors the most widely distributed anoxic niche in the marine realm, often within millimeters of the sediment surface. In the sea, anoxia is otherwise restricted to the cores of the major oxygen minimum zones, anoxic basins, the centers of larger marine snow aggregates with high microbial activity, and possibly the guts of some animals. Sediments have three major functions in the nitrogen cycle (Fig. 14.2):

1. The removal of fixed nitrogen through permanent burial of nitrogen-containing organic compounds and particle-bound ammonium.
2. The regeneration of dissolved inorganic nitrogen, mainly as ammonium and nitrate, through ammonification coupled to the remineralization of organic detritus, in combination with nitrification in the oxic part of the sediment.
3. The loss of fixed nitrogen through denitrification and anammox in the anoxic part of the sediment.

While these functions are related to the input and degradation of organic detritus, it should be noted that one-third of the global shelf sediment area is within the photic zone, and when given sufficient light and little physical disturbance, dense populations of microalgae may develop on the sediment surface (Gattuso et al. 2006). The benthic microalgal community (or microphytobenthos) assimilates nutrients and thus affects the exchange of nitrogen between the sediment and the overlying water. The algae may in some cases prevent release of soluble nitrogen compounds from the sediment and may even take up nitrogen from the water column (see e.g., Rysgaard et al. 1993; Tyler et al. 2003; Sundbäck et al. 2004). Through their production of oxygen and assimilation of nitrate and ammonium, the algae can

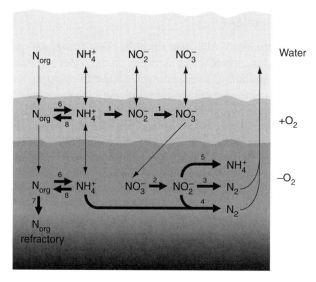

Figure 14.2 The nitrogen cycle of marine sediments. Thin arrows indicate transport processes and thick arrows indicate microbial transformations.

also influence both nitrification and denitrification (Rysgaard et al. 1995; Risgaard-Petersen 2003).

Benthic denitrification and anammox together constitute the largest sink term in the marine nitrogen cycle, accounting for about 70 percent of the fixed nitrogen loss in current budgets (Codispoti 2007). The importance of this function is emphasized by a discrepancy in the best-estimate nitrogen budget, with the loss terms substantially exceeding the inputs, which implies that the oceans may currently be loosing fixed nitrogen at a rate of 10 percent per 100 years (Codispoti 1995, 2007). Such a loss should influence primary production, with all the consequence that this may have for the marine carbon cycle.

Nitrogen fixation, nitrification, nitrate reduction, denitrification, and anammox have been shown to be important players in the redox cycle of nitrogen, but other processes may also be active. Dissimilatory reduction of nitrite to ammonium (Fig. 14.1) is generally a minor sink for nitrate and nitrite in marine sediments, but it may be significant under special circumstances, for example, at locations where nitrate-accumulating giant sulfur bacteria are abundant. As shown by the discovery of anammox and, even more recently, of anaerobic nitrite oxidation by phototrophic bacteria (Griffin et al. 2007), other pathways in the nitrogen cycle may remain to be discovered. Several reactions involving nitrogen compounds are thermodynamically favorable under conditions often found in nature, including reactions with manganese, which have been proposed to affect nitrogen cycling in some sediments, through either the reduction of nitrate with Mn^{2+} or the oxidation of ammonium with manganese oxide (Luther et al. 1997; Hulth et al. 1999), but the importance of these processes has not been demonstrated.

This chapter examines the microbial ecology and biogeochemistry of benthic nitrogen cycling. We follow nitrogen as it enters the sediment, through different transformations, until it leaves the sediment again or is buried. We discuss the—quite limited—knowledge of the organisms involved, as well as the regulation of the individual processes by environmental factors. In particular, we focus on novel processes and insights from new methodological developments. The interactions of the different organisms and processes are presented through case studies, and we discuss the contribution of the benthic nitrogen cycle to marine nitrogen budgets.

INPUTS

The main source of nitrogen to the sediment is generally the depositional flux of organic detritus, with its associated microflora. The bulk nitrogen content of this organic matter varies somewhat, depending mainly on its source and age. Typical C:N ratios in organic matter of planktonic origin vary from around 6–7 in fresh phytoplankton to about 10 in material that reaches the bottom of the deep sea, as a result of the preferential degradation of nitrogen-rich compounds (see, e.g., Antia 2005). Detritus from macrophytes, either terrestrial or seagrasses and seaweeds, has a lower nitrogen content, with C:N ratios of about 20 (Duarte 1992). Despite these variations, the organic nitrogen flux is roughly proportional to the carbon flux, and thus varies similarly with location and oceanic regime (see, e.g., Jahnke 1996).

Organic nitrogen has an oxidation state of -3 and occurs mainly as amide and amino groups or in heterocyclic structures in compounds such as amino acids and nucleotides and their polymers. Protein constitutes 30–60 percent of the organic carbon in sinking particulates and pelagic sediments (Hedges et al. 2001; Nelson and Baldock 2005), and similarly accounts for 30–70 percent of organic nitrogen when quantified as total hydrolyzable amino acids, a fraction that largely consists of protein amino acids (e.g., Henrichs et al. 1984; Burdige and Martens 1988; Dauwe and Middelburg 1998). The remaining fraction of the organic nitrogen is poorly defined, but includes amino sugars and DNA as minor contributors (Dauwe and Middelburg 1998; dell'Anno et al. 2002; Niggemann and Schubert 2006).

Sediments may also take up dissolved nitrogen compounds, mainly in the form of nitrate, from the overlying water. While nitrate uptake is typically small relative to the flux of organic nitrogen, and only delivers a minor part of the nitrate consumed by benthic denitrification, it may be significant in removing nitrate from the water column. This is particularly important in coastal waters with high nitrogen loading, where benthic denitrification contributes to the removal of nitrate from the water, and where nitrate uptake may even exceed the flux of detrital nitrogen. The influx of nitrate can also be important for benthic metabolism in regions with nitrate-rich, oxygen-depleted waters, where nitrate may be a major electron acceptor in benthic carbon oxidation (Canfield 1993). An extreme example of this occurs in sediments populated by giant filamentous sulfur bacteria of the genera *Thioploca* and *Beggiatoa*, which can stretch out from the sediment and actively take up nitrate from the bottom water. These bacteria concentrate nitrate to levels of 0.5 mmol/L in a large central vacuole (Fossing et al. 1995; Huettel et al. 1996). After gliding back into the sediment, bacteria in these genera couple the reduction of nitrate to ammonium with the oxidation of hydrogen sulfide to sulfate (Otte et al. 1999).

The nitrogen requirements of benthic organisms for biomass production are generally met by the supply of fixed nitrogen liberated during the degradation of organic matter. Nitrogen fixation is therefore not expected to be important, because it would be an unnecessary expenditure of energy. In fact, expression of nitrogenase genes is repressed by ammonium availability (see Chapter 13). Exceptions include systems with benthic photosynthesis, where cyanobacteria and anoxygenic phototrophs may fix nitrogen (Capone 1983; Herbert 1999), with cyanobacterial mats as the extreme example (see, e.g., Bebout et al. 1987). In many cases, however, benthic algae obtain much of their nitrogen from the mineralization processes in the underlying sediment and act as a barrier towards the release of fixed nitrogen to the water column (see, e.g., Rysgaard et al. 1993; Tyler et al. 2003). Nitrogen fixation may also be important in sediments with rooted macrophytes such as *Spartina* marshes and seagrass meadows, where diazotrophs thriving on root exudates in the rhizosphere may supply up to 50 percent of the nitrogen demand of the plants in extreme cases (Herbert 1999; Nielsen et al. 2001). Consistent with the wide distribution of nitrogen fixation among many different groups of heterotrophic and chemolithoauotrophic prokaryotes (Postgate 1998; Zehr et al. 2003; and see Chapter 13), the diversity of diazotrophs in such sediments is large. But sulfate-reducing bacteria seem to be particularly important contributors to the process (Herbert 1999; Lovell et al. 2000; Bagwell et al. 2002; Burns et al. 2002).

Nitrogen fixation is also measurable in unvegetated sediments, although it is generally unimportant as a source of nitrogen there, with depth-integrated rates corresponding to less than 5 percent of typical organic nitrogen sedimentation rates (Capone 1983; Herbert 1999; Burns et al. 2002). The process is detected even in the presence of ammonium, but it is stimulated by the addition of polysaccharides and the removal of ammonium (Tibbles et al. 1994; Capone and Carpenter 1982), suggesting that some organisms maintain the genetic potential and a slight activity in order to meet fluctuations in the relative availability of organic carbon and fixed nitrogen (Burns et al. 2002).

TRANSFORMATIONS

Microbes and Microbial Processes

Ammonification: $N_{org} \rightarrow NH_4^+$ With organic matter as the main source of nitrogen in sediments, it is the release of nitrogen from these compounds that feeds dissimilatory redox processes of the benthic nitrogen cycle. The microbial degradation of organic detritus is initiated by extracellular enzymatic hydrolysis, which solubilizes the organic matter to mono- or oligomers that can be taken up by cells. As a result, dissolved organic nitrogen (DON) builds up in pore waters, and may contribute to the benthic nitrogen exchange (Lomstein et al. 1998; Burdige and Zheng 1998).

Microbial Metabolism

All living organisms have two basic types of requirements: (1) matter—the molecular building blocks—from which they can produce biomass for growth and reproduction through anabolic processes; (2) energy, needed for growth, maintenance, active transport, and motility. The acquisition of matter, such as nitrogen, is called assimilation, while dissimilatory processes are those associated with extracting energy from the environment. There are two sources of energy for life: (1) light, which is used by phototrophic organisms; (2) the energy available from the forming and breaking of molecular bonds in chemical reactions of compounds available in the environment, used by chemotrophic organisms. With the exception of the newly discovered anaerobic nitrite-oxidizing phototrophic bacteria, all known microbes that carry out dissimilatory transformations of nitrogen compounds are all chemotrophs and obtain their energy from redox reactions, that is, processes involving oxidation and reduction, the transfer of electrons from one atom to another. For example, aerobic nitrite oxidizers extract energy from oxidation of the nitrogen atom in nitrite coupled to the reduction of oxygen, with the transfer of two electrons (see Reaction (5) below). In order to conserve energy from the reaction, the electrons are not transferred directly from the electron donor (here nitrite) to the acceptor (oxygen), but rather pass through a series of membrane-bound electron carriers. The passage of electrons through this electron transport chain drives a translocation of protons across the membrane, and the resulting electrochemical gradient in turn drives the production of adenosine triphosphate (ATP), the universal energy currency of living cells.

Ammonium is the inorganic end-product of organic nitrogen mineralization, and is generated through several different biochemical pathways, depending on the nitrogen compound, the organism, and the presence or absence of oxygen (Fuchs 1999; Buckel 1999). The release of ammonium through these pathways is collectively referred to as ammonification. In addition to direct deamination, ammonium is also generated from the hydrolysis of urea [$CO(NH_2)_2$], which results from the degradation of heterocyclic compounds such as nucleic acids (Lomstein et al. 1989; Therkildsen and Lomstein 1994; Therkildsen et al. 1996). Protein amino acids are an important source of ammonium in sediments, although estimates of their contribution vary widely, from about 25 to about 80 percent of ammonium production (Burdige and Martens 1988; Mayer and Rice 1992; Pantoja and Lee 2003). The rapid microbial turnover of urea suggests that RNA may also be an important source of ammonium in some sediments (Lomstein et al. 1989; Pedersen et al.

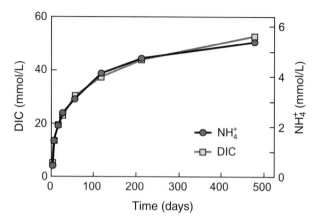

Figure 14.3 Coupled accumulation of dissolved inorganic carbon (DIC) and ammonium resulting from carbon oxidation and ammonification during a 1.3-year-long anoxic batch incubation of a marine sediment. Freeze-dried phytoplankton was added as a source of highly reactive detritus at $t = 0$, and rates decreased 30-fold during the incubation. Note that, due to reversible adsorption, the soluble ammonium concentration shown here represents only approximately half of the ammonium content of the sediment at any given time. Redrawn after Jensen et al. (2005).

1993; Therkildsen et al. 1996), while extracellular DNA has been estimated to account for 7 percent of nitrogen mineralization in deep-sea sediments (Dell'Anno and Danovaro 2005).

Rates of ammonification are tightly linked to the mineralization of organic carbon (Fig. 14.3), the rate of which depends on the concentration and quality of the organic matter (Berner 1980). Thus, at steady state, the ratio of the rates of carbon dioxide and ammonium production in the sediment corresponds to the bulk C:N ratio of the complex organic detritus being mineralized. The tight coupling of carbon and nitrogen mineralization is illustrated by the nearly constant ratio of dissolved inorganic carbon (DIC) and ammonium accumulation typically observed in anoxic sediment incubations for as long as 1 year (see, e.g., Kristensen et al. 1999; Jensen et al. 2005). An uncoupling of carbon and nitrogen mineralization may be associated with pulses of organic detritus to the sediment, because ammonification precedes the terminal oxidation of organic carbon. This is particularly pronounced under anoxic conditions, where the two processes are carried out by different organisms: ammonification is primarily associated with the initial hydrolysis and fermentation, while carbon mineralization is associated with terminal electron accepting processes, such as sulfate reduction (see, e.g., Kristensen and Hansen 1995; Canfield et al. 2005).

Assimilation: $NH_4^+ \rightarrow N_{org}$ To the extent that carbon oxidation sustains growth of organotrophic prokaryotes, mineralization is also tightly linked to the assimilation of nitrogen, with ammonium as the primary inorganic source. Thus, the net release of ammonium from the mineralization processes is the difference between gross mineralization and assimilation (Blackburn 1979). Based on $^{15}\text{NH}_4^+$ dilution experiments, assimilation is estimated to consume about 30 percent of the gross ammonium production in coastal sediments (Blackburn and Henriksen 1983). The C:N ratio of bacterial biomass is relatively low, about 5 (Fagerbakke et al. 1996), compared with the bulk C:N ratio of organic detritus in sediments, which is typically 7 or more. Net microbial growth will therefore increase the net C:N mineralization ratio. The magnitude of this increase depends on the growth yields of the microbes, which to a first approximation scale with the free energy yield of the metabolic reaction and thus are higher for aerobes than for anaerobes (Thauer et al. 1977; VanBriesen 2002). Thus, the most pronounced effects of bacterial growth on net nitrogen mineralization are associated short-term sedimentation events, where the population size increases rapidly (see, e.g., van Duyl et al. 1993; Pedersen et al. 1999).

Nitrification

$NH_4^+ \rightarrow NO_2^-$ Under anoxic conditions and in the absence of nitrate and nitrite, ammonium is the stable end-product of nitrogen mineralization (Fig. 14.1). The further transformation of ammonium is initiated by the enzyme ammonium monooxygenase (AMO) in the ammonium-oxidizing nitrifiers. AMO is thought to depend on the presence of molecular oxygen in the oxidation of ammonium to hydroxylamine:

$$\text{NH}_4^+ + \text{O}_2 + \text{H}^+ + 2\text{e}^- \rightarrow \text{NH}_2\text{OH} + \text{H}_2\text{O} \quad (1)$$

The electrons required for this reaction are delivered from the further oxidation of hydroxylamine to nitrite by hydroxylamine oxidoreductase (HAO):

$$\text{NH}_2\text{OH} + \text{H}_2\text{O} \rightarrow \text{NO}_2^- + 5\text{H}^+ + 4\text{e}^- \quad (2)$$

The excess electrons liberated during this step are delivered to oxygen through an electron transport chain, or used for autotrophic carbon fixation (Wood 1986). The claimed oxygen dependence of AMO is now challenged by studies of *Nitrosomonas europaea*, the model organism for ammonium oxidation. In this bacterium, dinitrogen tetroxide (N_2O_4), the spontaneously formed dimer of nitrogen dioxide, can substitute for oxygen under anoxic conditions, in an AMO-catalyzed reaction that generates hydroxylamine and nitric oxide (Schmidt and Bock 1997; Kampschreur et al. 2006):

$$\text{NH}_4^+ + \text{N}_2\text{O}_4 + \text{H}^+ + 2\text{e}^- \rightarrow \text{NH}_2\text{OH} + 2\text{NO} + \text{H}_2\text{O} \quad (3)$$

Hydroxylamine is oxidized to nitrite via Reaction (2). The excess electrons are in this case used to reduce approximately half of the nitrite to N_2 through denitrification. Aerobic ammonium oxidation is inhibited by acetylene, but Schmidt et al. (2001) found that this inhibition was relieved by the addition of NO_2, and concluded that NO_2/N_2O_4 generated intracellularly is the obligatory substrate of AMO in *N. europaea*. According to this hypothesis, the role of O_2 under oxic conditions is to regenerate N_2O_4 from NO produced through Reaction (3), in an acetylene-sensitive step:

$$2NO + O_2 \rightarrow N_2O_4 \qquad (4)$$

Note that the sum of (3) and (4) is identical to (1).

The finding that AMO is not obligately aerobic opens the possibility of anaerobic ammonium oxidation by organisms like *N. europaea*, with potential significance for the marine nitrogen cycle. However, nitrogen dioxide is not abundant in aquatic systems and anoxic sources are not known, so the possibilities for substituting oxygen in Reaction (4) with other oxidants need to be explored. Finally, it must be determined whether the important marine ammonium oxidizers function similarly as *N. europaea*.

In addition to the nitrogen species mentioned above, ammonium oxidizers can also produce nitrous oxide (N_2O). This is in part explained by a spontaneous oxidation of hydroxylamine, but it now seems that the gas is primarily produced through nitrite reduction (Poth and Focht 1985; Schmidt et al. 2004), and we therefore discuss this nitrifier denitrification in the section on denitrification.

Ammonium oxidizers in culture are phylogenetically clustered within a few taxa of the prokaryotic domains. Most known species are found in the genera *Nitrosomonas* and *Nitrosospira* in the *Betaproteobacteria*, while the species *Nitrosococcus oceani* and *N. halophilus* belong to the *Gammaproteobacteria* (Head et al. 1993; Teske et al. 1994). Recently, an ammonium-oxidizing, nitrite-producing archaeon has been isolated with the proposed name *Nitrosopumilus maritimus*, as the first representative of what will likely develop into a large group of ammonium-oxidizing species within the *Crenarchaeota* (Könneke et al. 2005). The crenarchaea possess AMO genes that are homologous to those found in bacterial nitrifiers, suggesting some similarity in the enzymatic mechanism (Schleper et al. 2005). In addition to these organisms, which are all autotrophs, ammonium oxidation is found in several heterotrophic bacteria such as *Alcaligenes* sp. and *Paracoccus denitrificans* (Focht and Verstraete 1977). The enzymes for heterotrophic nitrification appear distinct from those of the autotrophic pathway, and the organisms seem to dissipate rather than conserve energy through the process, which shows highest activity when coupled to aerobic denitrification (Robertson and Kuenen 1990; Richardson et al. 1998; Jetten 2001). Heterotrophic nitrification is believed to be of minor importance in aquatic systems, but its physiological and biogeochemical roles are poorly understood (Ward 2000).

Both betaproteobacterial and crenarchaeotal ammonium oxidizers are very diverse in marine sediments, based on AMO gene analysis (Nold et al. 2000; Nicolaisen and Ramsing 2002; Francis et al. 2003, 2005), whereas the gammaproteobacterial *Nitrosococcus oceani*, which is widespread in seawater (Ward and O'Mullan 2002), has not yet been detected in sediments (Nold et al. 2000). The betaproteobacterial clones tend to cluster in *Nitrosomonas*- and *Nitrobacter*-like groups, which suggests that organisms in culture are not greatly dissimilar to those active in nature. There is, however, no quantitative information on the relative abundance and activity of the different groups in sediments, although, based on the relative success of AMO gene amplification, Francis et al. (2005) suggested that crenarchaeal ammonium oxidizers may be more widespread than their bacterial analogs.

$NO_2^- \rightarrow NO_3^-$ Surprisingly, no known autotrophic ammonium oxidizer can continue the oxidative path to form nitrate from nitrite. Instead, this process is catalyzed by a second, polyphyletic group of bacteria by means of the enzyme nitrite oxidoreductase, which delivers electrons to oxygen through a short electron transport chain (Bock and Wagner 2001):

$$2NO_2^- + O_2 \rightarrow 2NO_3^- \qquad (5)$$

Nitrite oxidizers may grow chemolithoautotrophically, but some are also capable of chemolithoheterotrophic, or, in the case of *Nitrobacter*, chemoorganoheterotrophic growth through oxygen or nitrate respiration of simple organic compounds (Bock and Wagner 2001; Lipski et al. 2001). Nitrite oxidizers in culture include the genera *Nitrobacter*, *Nitrospina*, and *Nitrococcus*, affiliated with the *Alpha-*, *Delta-*, and *Gammaproteobacteria*, respectively, and *Nitrospira* which forms a separate phylum within the *Bacteria* (Teske et al. 1994, Ehrich et al. 1995). Our understanding of nitrite oxidation comes mainly from *Nitrobacter* cultures, but cultivation-independent methods have shown that *Nitrospira* species dominate nitrite-oxidizing communities in wastewater treatment systems and natural environments, including freshwater sediments (see, e.g., Wagner et al. 1996; Bartosch et al. 2002; Altmann et al. 2003). This may well also be true for marine environments, where little is known about the microbes behind the process.

Most recently, an alternative mechanism for the oxidation has been discovered in an anoxygenic phototrophic purple bacterium that uses nitrite as an electron donor in photosynthesis under anaerobic conditions (Griffin et al. 2007):

$$2NO_2^- + CO_2 + H_2O \rightarrow 2NO_3^- + [CH_2O] \qquad (6)$$

The process has not yet been studied in natural environments, and so far is the only phototrophic metabolism in the nitrogen cycle.

$NH_4^+ \rightarrow NO_3^-$ The nitrification processes are affected by many factors, including salinity, light, pH, and hydrogen sulfide (Herbert 1999; Canfield et al. 2005), but the most important control is the availability of the primary substrates: ammonium and oxygen. Although it requires two separate groups of organisms, nitrification in marine sediments is often treated only as the sum of the two reactions. This is reasonable, because ammonium oxidation is typically the rate-limiting step, so that nitrite oxidation keeps nitrite levels low, and because ammonium oxidation is probably the only significant source of nitrite for nitrite oxidizers. This view is supported by determinations of microscale distributions of nitrate and nitrite in oxic surface sediments (Nielsen et al. 2004). Thus, even in highly active mangrove sediments where oxygen penetrates only 1–2 mm, nitrite oxidation accounted for 87–104 percent of

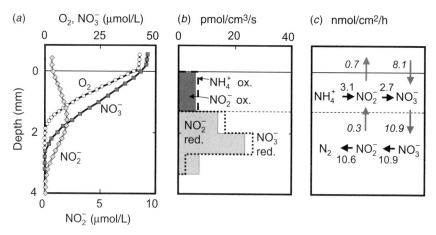

Figure 14.4 Depth distributions of oxygen, nitrite, and nitrate (*a*), and of ammonium oxidation, nitrite oxidation, nitrate reduction, and nitrite reduction (*b*), and depth-integrated rates and fluxes (*c*) in a tidal river mangrove sediment. Concentrations were determined with microsensors, and the depth distribution of production and consumption were derived by inverse modeling, assuming transport by molecular diffusion. Redrawn after Meyer et al. (2005).

ammonium oxidation (Fig. 14.4) (Meyer et al. 2005). Still, there is some evidence that ammonium oxidation can substantially exceed nitrite oxidation at low oxygen concentrations (Henriksen and Kemp 1988). An extreme example of this is a wastewater treatment process known as CANON (complete autotrophic nitrogen removal over nitrite), where fixed nitrogen is removed through a close coupling between microaerobic ammonium oxidation and anammox, while nitrite oxidizers lose the competition for oxygen as long as ammonium is abundant (Third et al. 2001; Sliekers et al. 2002). Such an interaction has also been observed in the suboxic zone of the Black Sea water column (Lam et al. 2007), and could possibly also function in some sediments.

Anaerobic Processes

Nitrate Reduction: $NO_3^- \rightarrow NO_2^-$ In the presence of oxygen, nitrate is the thermodynamically stable form of nitrogen, whereas in the anoxic parts of the sediment it is a reactant in dissimilatory redox reactions that lead to the formation of N_2 or ammonium. The first step in these reductive pathways is the reduction of nitrate to nitrite, which requires catalysis because of the high chemical stability of the nitrate ion at environmentally relevant temperatures (see, e.g., Fanning 2000). This process is often treated as part of the denitrification pathway, but is really a process in its own right, since it is found in many organisms that are not denitrifiers, including anammox and nitrite-ammonifying bacteria (see below) as well as some that do not posses enzymes for further reduction of nitrite (Zumft 1997). Nitrate reduction is catalyzed by nitrate reductases, a group of related enzymes, which contain a molybdenum cofactor (Table 14.1) (Zumft 1997; Stolz and Basu 2002). It is not known which types of organisms are generally responsible for nitrate reduction in the sediment, but the first detailed investigation in an estuary indicates a high diversity of nitrate-reducers with little similarity to those from other environments or from pure cultures (Smith et al. 2007). Transient accumulation of nitrite is often observed during nitrate consumption (Fig. 14.4), emphasizing the partial uncoupling of the $NO_3^- \rightarrow NO_2^-$ step from further transformations, either because the different steps are performed by different organisms, or related to the separate regulation of the different enzymes in one type of organism.

From nitrite, three pathways of further reduction branch out: denitrification and anammox, which both lead to N_2, and dissimilatory reduction to ammonium. The balance of these processes is important in determining whether the sediment acts as a sink or source for nitrogen.

Denitrification: $NO_2^- \rightarrow N_2$ Denitrification is the respiratory reduction of nitrite via the free radical nitric oxide to gaseous nitrous oxide and N_2,

$$NO_2^- \rightarrow NO \rightarrow N_2O \rightarrow N_2 \qquad (7)$$

with these three steps often preceded by $NO_3^- \rightarrow NO_2^-$ in the same organism (Zumft 1997). The denitrifying organisms are highly versatile with respect to electron donor, and almost any sort of organic matter may be utilized (Beauchamp et al. 1989; Payne 1981). Moreover, several reduced inorganic compounds may also serve as electron donor. These include hydrogen (Smith et al. 1994), hydrogen sulfide (Aminuddin and Nicholas 1973), thiosulfate (Ishaque and Aleem 1973), ferrous iron (Benz et al. 1998; Straub et al. 1996; Weber et al. 2006a), carbon monoxide (King 2006), and methane (Raghoebarsing et al. 2006).

The ability to denitrify is widely distributed among the prokaryotes, and does not follow a recognizable phylogenetic pattern (Knowles 1982; Zumft 1997). Members of both *Bacteria* and *Archaea* can denitrify (Cabello et al. 2004), and even a fungus can carry out this reaction (Takaya 2002). Recently, denitrification has been shown in foraminifera that accumulate nitrate intracellularly to concentrations of hundreds of millimolar (Risgaard-Petersen et al. 2006). Many denitrifiers are found in the *Alpha-*, *Beta-*, and *Gammaproteobacteria*, including well-studied isolates such as *Paracoccus denitrificans, Alcaligenes faecalis*, and *Pseudomonas* spp. Most denitrifiers appear to be facultative anaerobes. The lack of phylogenetic clustering complicates the identification of environmentally important denitrifiers (see the text box). Functional gene analysis targeting mainly the nitrite and nitrous oxide reductase genes *nirS*, *nirK*, and *nosZ* has shown that denitrifiers in marine sediments are a highly diverse group and that most members do not seem to be closely related to cultivated species, although some relatives of *Alcaligenes* and *Pseudomonas* spp. are present (Scala and Kerkhoff 1999; Braker et al. 2000; Liu et al. 2003; Nogales et al. 2002; Smith et al. 2007). These studies also indicate a substantial geographical variation in the community structure of denitrifiers.

Targets for Identification of Nitrogen-Metabolizing Bacteria in Sediments

One of the greatest current challenges in studies of benthic nitrogen cycling lies in determining which organisms are involved in the different transformations. This task is complicated by the lack of phylogenetic coherency among the functional groups of microorganisms involved. As discussed in this chapter, only anammox bacteria seem to form a single distinct phylogenetic clade, lending themselves to detection with phylogenetic markers, whereas organisms performing the reductive pathways of the nitrogen cycle in particular are spread over many prokaryotic phyla and even eukaryotes in the case of denitrification. In this situation, it is particularly useful to identify the organisms by probing for the genes that encode the enzymes, or subunits of these, directly involved in the nitrogen transformations. The most relevant genes for such a functional gene analysis are given in Table 14.1. The characteristic enzymes of the anammox process have not yet been identified with certainty. However, anammox bacteria possess both *narG* and *nirS* genes known from nitrate reducers and denitrifiers (Strous et al. 2006), and possibly also *nrfA* known from nitrite ammonifiers (Kartal et al. 2007a).

Nitrite reduction is carried out by two entirely different types of nitrite reductases, which contain either cytochrome cd_1 or copper at the reactive site. The presence of these types in the cell appears to be mutually exclusive, and the cytochrome cd_1 type is most common in cultivated strains (about 75 percent: Zumft 1997). Two

TABLE 14.1 Processes in the Nitrogen Cycle, their Associated Enzymes and Examples of Genes that Encode the Enzymes or Selected Subunits[a]

Process	Enzyme	Gene(s)
N_2 fixation	Nitrogenase	*nifH*
NH_4^+ oxidation	NH_4^+ monooxygenase	*amoA*
	NH_2OH oxidoreductase	*hao*
NO_2^- oxidation	NO_2^- oxidoreductase	*nxrA (norA)*
NO_3^- reduction	NO_3^- reductase, membrane-bound	*narG*
	NO_3^- reductase, periplasmic	*napA*
Denitrification	NO_2^- reductase, cytochrome cd_1	*nirS*
	NO_2^- reductase, copper type	*nirK*
	NO reductase	*norB*
	N_2O reductase	*nosZ*
Nitrite ammonification	NO_2^- reductase, cytochrome c	*nrfA*
Anammox[b]	NO_2^- reductase	*nirS*
	N_2H_4 hydrolase	*hh*
	NH_2OH oxidoreductase	*hao*

[a] See text box "Targets for Identification of Nitrogen-Metabolizing Bacteria in Sediments."
[b] The enzymatics of the anammox process are not yet fully described. Enzymes and genes listed here are based on a model arising from analysis of the genome of "*Candidatus* Kuenenia stuttgartiensis" (Strous et al. 2006).

somewhat different types of respiratory nitric oxide reductase have been described (Hendriks et al. 2000), while only one type of nitrous oxide reductase is known (Zumft 1997). In terms of regulation, the denitrification pathway consists of two modules: $NO_2^- \to N_2O$ and $N_2O \to N_2$. The reaction $NO_3^- \to NO_2^-$ constitutes a separate third module in organisms that carry out the complete reduction of nitrate to N_2 (Zumft 1997).

This stepwise reduction may lead to extracellular accumulation of one or more of the intermediates, and hence opens up the possibility of transfer between different organisms and exchange with the overlying water. Different organisms regulate the enzymes of the nitrate reduction/denitrification pathway differently, and thus accumulate the intermediates to different degrees (Betlach and Tiedje 1981; Carlson and Ingraham 1983). Maxima of nitrite and nitrous oxide concentrations are commonly found in the zone of nitrate reduction in marine sediments. In anoxic incubations of homogenized sediment with nitrate, nitrite accumulated during the first part of the incubation up to a concentration of 70 percent of the initial nitrate concentration, followed by reduction to N_2 (Dalsgaard and Thamdrup 2002). In intact sediment, nitrite does not reach such high concentrations, but it still accumulates (Fig. 14.4) (Meyer et al. 2005; Sørensen 1978; Tuominen et al. 1999; Usui et al. 1998). Nitrous oxide reaches submicromolar levels in the pore water of intact sediment, where it may be produced both by nitrification and by denitrification (Jørgensen et al. 1984; Senga et al. 2006; Sørensen 1978; Usui et al. 1998, 2001). Sediments can act as both sources and sinks of bottom-water nitrite and nitrous oxide, but the fluxes of these species are generally at least an order of magnitude lower than those of nitrate and N_2 (Fig. 14.4) (Hall et al. 1996; Laursen and Seitzinger 2002; Dong et al. 2002; Robinson et al. 1998). We have no general understanding of how the accumulation of the intermediates is controlled in natural communities, but there are special cases when a bottleneck may be predicted. For instance the presence of sulfide may inhibit the last step in the sequence and lead to accumulation of nitrous oxide both in pure culture (Sørensen et al. 1980) and in marine sediments (Senga et al. 2006). Also, low levels of oxygen cause the accumulation of nitrous oxide (Jørgensen et al. 1984).

Denitrification is generally inhibited by oxygen, which affects all steps of the pathway (Tiedje et al. 1982; Zumft 1997). Aerobic denitrification has, however, been reported by pure culture studies, in which oxygen and nitrate were consumed simultaneously (Robertson et al. 1995; Robertson and Kuenen 1984). So far, however, denitrification at higher oxygen concentrations has not been reported to be significant in natural environments. For instance, microsensor studies in freshwater sediments have revealed denitrification in the oxic part of the sediment, but only up to oxygen concentrations of about 20 µmol/L and at reduced rates (Lorenzen et al. 1998). In suboxic waters, denitrification does not occur before oxygen is close to typical detection limits of about 4 µmol/L (Codispoti et al. 2001). Aerobically grown ammonium-oxidizing bacteria can produce nitrous oxide through nitrite reduction via nitric oxide, and mainly at low oxygen levels (Poth and Focht 1985; Schmidt et al. 2004; Wrage et al. 2001; Shaw et al. 2006). This capability is widespread within the ammonium oxidizers (Shaw et al. 2006; Casciotti and Ward

2005), but its physiological role is not clear. Under anoxic conditions, the ammonium oxidizer *Nitrosomonas europaea* can carry out complete denitrification of nitrite to N_2 coupled to hydrogen oxidation (Bock et al. 1995). Nitrous oxide production has been linked to nitrifier denitrification in some marine sediments (see, e.g., Usui et al. 1998), but in general little is known about the role of nitrifiers in benthic denitrification.

The availability of nitrate is the main factor limiting the depth-integrated rate of denitrification in most marine sediments, with nitrate reduction typically restricted to a narrow zone at the oxic/anoxic interface (Fig. 14.4, and see Fig. 14.7 below) (see, e.g., Canfield et al. 1993; Brandes and Devol 1995). Nitrate is supplied by two sources: the bottom water and nitrification in the oxic part of the sediment (Fig. 14.4). The relative importance of these sources depends on the nitrate concentration in the bottom water and the thickness of the oxic surface layer. A thick oxic zone favors nitrification over ammonium loss, while it slows down the diffusional flux of nitrate from the bottom water to the zone of denitrification, and vice versa (Nielsen et al. 1990). Nitrate reduction and, thereby, denitrification is often tightly coupled to nitrification (Jenkins and Kemp 1984; Seitzinger 1988), and even with an oxygen penetration of just 1 mm, 30 percent of the nitrate from nitrification can be denitrified (Rysgaard et al. 1994). Bottom-water nitrate can be the main nitrate source in sediments of eutrophic coastal waters and oxygen-minimum zones with high nitrate and/or low oxygen concentrations, and to a first approximation the importance of this source varies inversely with the oxygen penetration depth (see, e.g., Christensen et al. 1990; Jensen et al. 1994; Middelburg et al. 1996).

While benthic denitrification is an important sink in the global nitrogen cycle (see below) and in some ecosystems (see, e.g., Seitzinger 1988; Nixon et al. 1996), its contribution to benthic carbon oxidation is in most cases low: less than 10 percent of the total (Thamdrup 2000; Canfield et al. 2005). Larger contributions are possible particularly when the process is fuelled by nitrate from the bottom water. Most likely, however, much of the denitrification is actually not directly coupled to carbon oxidation, but rather to the reoxidation of reduced inorganic iron and sulfur compounds produced deeper in the sediment (Thamdrup 2000; Canfield et al. 2005). Denitrification rates would also be overestimated if anammox is active, since this process has only been considered in the most recent quantitative studies (see, e.g., Risgaard-Petersen et al. 2003).

Nitrite Ammonification: $NO_2^- \rightarrow NH_4^+$ Several microbes can reduce nitrite to ammonium rather than to N_2 (Cole 1987). Like denitrification, this process is often coupled to nitrate reduction, and it is known as dissimilatory nitrate/nitrite reduction to ammonium (DNRA) or nitrite ammonification. Some fermenting bacteria use the reaction as an electron sink, which allows them to oxidize organic substrates further and thereby generate more adenosine triphosphate (ATP) through substrate-level phosphorylation (Cole and Brown 1980; Keith et al. 1982; Macfarlane and Herbert 1982). The possibly more widespread use is in electron transport phosphorylation during respiration. Here, nitrite is reduced without free intermediates to ammonium by a cytochrome *c*-bearing nitrite reductase, which is entirely different from the nitrite reductases of denitrification (Table 14.1) (Simon 2002). This type of metabolism is known from a range of different organotrophic organisms, including facultative anaerobes and anaerobic dissimilatory iron, sulfur, and sulfate reducers, such as *Escherichia coli*, *Geobacter metallireducens*, and *Desulfovibrio desulfuricans* (Simon 2002), as well as from anammox bacteria (see below). It can also be coupled to the oxidation of ferrous iron or hydrogen sulfide, for example in giant sulfur bacteria such as *Thioploca* sp. and *Beggiatoa* sp. (Finneran 2002; Weber et al. 2006b; Otte et al. 1999). The majority of nitrite ammonifiers in estuarine sediments, identified in nitrite reductase clone libraries, were related to *Deltaproteobacteria* such as sulfate- and iron-reducers (Smith et al. 2007), suggesting that they may be able to use a selection of anaerobic electron acceptors.

Nitrite ammonifiers must compete with denitrifiers for both electron donors and acceptors. In this competition, denitrifiers should be favored by the higher energy yield from the reduction of nitrite to N_2 than to ammonium, but experiments with pure cultures indicate that denitrifiers are less efficient than nitrite ammonifiers in conserving energy from the respiration, resulting in similar growth yields for the two groups (Strohm et al. 2007). Consistent with this, the relative importance of the two processes in sediments varies. Denitrifiers seem to win the competition in sediments with low to moderate organic loading and in some coastal sediments (see, e.g., Binnerup et al. 1992; Rysgaard et al. 1993; Dalsgaard and Thamdrup 2002), while nitrite ammonification is mainly important in relatively reducing sediments as found in the nutrient-rich coastal zone, where it may consume more nitrite than does denitrification (Rysgaard et al. 1996; Christensen et al. 2000; Tobias et al. 2001; An and Gardner 2002). In such sediments, much of the nitrite ammonification may be linked to sulfide oxidation, for example, by nitrate-accumulating giant sulfur bacteria that transport nitrate into the sediment. When nitrate-accumulating *Beggiatoa* filaments were added to a sulfidic sediment, their coupled sulfide oxidation and nitrate reduction/nitrite ammonification created a more than 1 cm thick suboxic zone separating the fronts of oxygen and hydrogen sulfide, and shifted the dominating product of benthic nitrate consumption from N_2 to ammonium (Fig. 14.5) (Sayama et al. 2005). Similarly, dense *Thioploca* populations are estimated to oxidize 20–30 percent of the hydrogen sulfide produced in sediments of the Chilean oxygen minimum zone, and may, through their active uptake of nitrate from the bottom water, account for most of the nitrate consumption in those sediments (Fossing et al. 1995; Zopfi et al. 2001). Quite likely, sulfide or iron oxidation is also important for nitrite ammonification in other sediments (see, e.g., Weber et al. 2006b).

Anammox: $NH_4^+ + NO_2^- \rightarrow N_2$ From its discovery in the 1860's (Payne 1981) until the mid 1990s denitrification was the only known microbial process that could return fixed nitrogen to the large pool of N_2 in the atmosphere, although there were several good indications of anaerobic ammonium oxidation in marine water columns (Cline and Richards 1972; Richards 2008, b) and sediments (Bender et al. 1989; Schulz et al. 1994). The first direct proof of anaerobic ammonium oxidation came from wastewater treatment, where it was discovered that ammonium

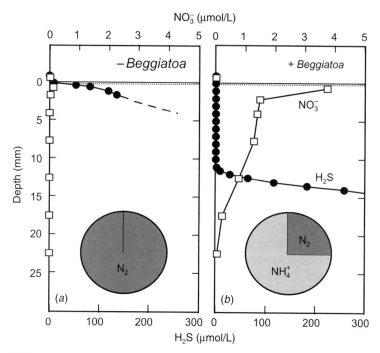

Figure 14.5 Experimental study of the effect of *Beggiatoa* on nitrogen cycling in a sulfidic sediment. (*a*) and (*b*) show the depth distributions of nitrate and soluble sulfide in the same sediment without and with *Beggiatoa*. The oxygen penetration depth of 0.2 mm is indicated by the dotted lines. The inserted pie charts show the relative importance of denitrification and nitrite ammonification as sinks for nitrate in the two treatments. Nitrate concentrations represent the whole sediment nitrate pool, including nitrate accumulated intracellularly by *Beggiatoa*. Redrawn after Sayama et al. (2005).

was oxidized anaerobically with nitrite in a dissimilatory microbial process (Mulder et al. 1995; van de Graaf et al. 1995):

$$NH_4^+ + NO_2^- \rightarrow N_2 + 2H_2O \qquad (8)$$

Later, anaerobic ammonium oxidation coupled to nitrite reduction was demonstrated experimentally in marine sediments (Thamdrup and Dalsgaard 2002; Dalsgaard and Thamdrup 2002). The sediment process is very similar to anammox in wastewater reactors, as evident by isotope pairing in ^{15}N studies and the sensitivity to acetylene and methanol as inhibitors, but not to the nitrification inhibitor allylthiourea (Dalsgaard and Thamdrup 2002; Jensen et al. 2007). It is technically difficult to link anaerobic ammonium oxidation in a sediment to the specific activity of anammox bacteria (but see Risgaard-Petersen et al. 2004; Tal et al. 2005; Schmid et al. 2007), but in suboxic marine waters, there is a close correlation between anammox activity and the abundance of close relatives to wastewater anammox bacteria, which substantiates that the same microbial process takes place in wastewater and marine environments (Kuypers et al. 2003, 2005; Schmid et al. 2007).

> ## Use of ^{15}N to Study Microbial Nitrogen Cycling
>
> The rarer stable nitrogen isotope ^{15}N, with an abundance of 0.3663 percent, is very useful for tracing nitrogen metabolisms in natural systems. Incubations with ^{15}N-labeled nitrogen compounds and analysis by mass spectrometry have been used to quantify all the major pathways of nitrogen metabolism (see, e.g., Blackburn 1979; Nishio et al. 1982, 1983; Risgaard-Petersen and Rysgaard 1995; Ward 2007). ^{15}N-based techniques are particularly useful in studies of N_2 formation because of distinct patterns in the pairing of N atoms during denitrification and anammox. In denitrification, the formation of N_2O and N_2 occurs through random pairing of nitrogen atoms during NO reduction. Thus, the reduction of nitrite with a ^{15}N mole fraction of $F_{nitrite}$ results in the formation in N_2 with relative abundances for masses 28, 29, and 30 ($^{14}N^{14}N$, $^{14}N^{15}N$, $^{15}N^{15}N$) of $(1 - F_{nitrite})^2$, $(1 - F_{nitrite}) \times F_{nitrite}$, and $F_{nitrite}^2$, respectively. In contrast, anammox produces N_2 by one-to-one pairing of nitrogen from NO_2^- and NH_4^+, and the two processes can therefore be quantified simultaneously in incubations of homogenized sediment where [^{15}N]nitrite is added (or generated through the reduction of [^{15}N]nitrate), provided that the fraction of ^{15}N in nitrite is known (Thamdrup and Dalsgaard 2002). In a common assay for denitrification in intact sediment cores, [^{15}N]nitrate is added to the bottom water and, after initial equilibration, denitrification rates are determined from the rates of $^{14}N^{15}N$ and $^{15}N^{15}N$ production (Nielsen 1992). This technique assumes that added [^{15}N]nitrate and [^{14}N]nitrate formed through nitrification in the sediment are homogeneously mixed in the denitrification zone, and that denitrification is the only source of N_2. Modifications of the whole-core technique for determination of both denitrification and anammox have been proposed, but have not yet been extensively tested (Risgaard-Petersen et al. 2003; Trimmer et al. 2006).
>
> Stable nitrogen isotopes can also be used at their natural abundances to study the nitrogen cycle, because many of the processes are associated with distinct kinetic isotope effects, which impart small deviations in the natural abundance (Canfield et al. 2005, and references therein). In incubations with addition of ^{15}N-labeled compounds, the isotopic enrichment is typically so large that the kinetic effects are of marginal importance.

Although anammox bacteria have so far not been isolated in pure culture and only slow-growing enrichment cultures based on wastewater inocula have been described, a great deal of phylogenetic and biochemical information is available. Known anammox organisms belong to the bacterial phylum *Planctomycetes*, and four candidate genera have been identified: Brocadia, Kuenenia, Scalindua and Anammoxoglobus (Strous et al. 1999; Jetten et al. 2003; Kartal et al. 2007b).

Interestingly, all anammox bacteria that have been identified so far in marine environments belong to the "*Candidatus* Scalindua" genus (Kuypers et al. 2003, 2005; Risgaard-Petersen et al. 2004; Rich et al. 2008). This contrasts strongly with the high diversity of other nitrate and nitrate reducers (see above). A report of an anammox-like process in a strain of the denitrifier *Pseudomonas mendocina* (*Gammaproteobacteria*) isolated from a wastewater reactor, however, challenges the monophyletic distribution of the process (Hu et al. 2006).

The members of the *Planctomycetes* have unusual characteristics such as a cytoplasm divided into compartments by membranes, one of which, the nucleoid, contains DNA (Fuerst 2005). The anammox bacteria also contain a membrane structure known as the anammoxosome, which contains unique lipids, the ladderanes, that may be used as biomarkers to identify the anammox bacteria (Damsté et al. 2002; Van Niftrik et al. 2004; Kuypers et al. 2003, 2005). The anammoxosome is thought to contain hydrazine, a unique intermediate of the anammox process, and to carry many of the enzymes involved in anammox (Schalk et al. 1998; Strous et al. 2006). The proposed pathway also involves nitric oxide as an intermediate formed through nitrite reduction by a cd_1 nitrite reductase, with hydrazine formation through a combination of nitric oxide and ammonium in a putative hydrazine hydrolase (Strous et al. 2006).

Anammox bacteria originally appeared to be obligate chemolithoautotrophs (Mulder et al. 1995; Strous et al. 1999), but it is now clear that they are metabolically versatile. Some strains oxidize short-chain fatty acids, with the reduction of nitrate via nitrite to ammonium, and may assimilate carbon from propionate (Güven et al. 2005; Kartal et al. 2007a,b). Furthermore, Kuenenia stuttgartiensis can couple the oxidation of formate to the reduction of manganese or iron oxide (Strous et al. 2006). This versatility has implications for the potential role of anammox bacteria in natural environments, where their assumed slow and low-yielding chemolithoautotrophic growth, which is known from enrichment experiments, would suggest that they are not important.

Anammox may also be more flexible with respect to oxygen sensitivity than originally thought. Whereas the process is obligately anaerobic in wastewater reactors, where it is reversibly inhibited by just 1 µmol/L of oxygen (Strous et al. 1997), marine anammox bacteria appear to be microaerotolerant, since they are present and active at oxygen concentrations up to about 10 µmol/L (Kuypers et al. 2005, Jensen et al. 2008). This may give them an advantage over denitrifiers and contribute to their dominance in oxygen-depleted waters (Kuypers et al. 2005, Thamdrup et al. 2006, Jensen et al. 2008).

The contribution of anammox to N_2 production in the marine sediments investigated so far varies from less than 1 to about 70 percent, with a reasonable coverage from estuarine and coastal sites to the continental shelf, but our understanding of the factors controlling this variation is rudimentary (Dalsgaard et al. 2005; Kuypers et al. 2006). There is a clear trend for an increasing importance for anammox in N_2 production with increasing water depth (Fig. 14.6). Denitrification is favored in sediments with high rates of carbon mineralization, and hence anammox is of little importance in shallow estuarine and coastal sediments. The highest contributions

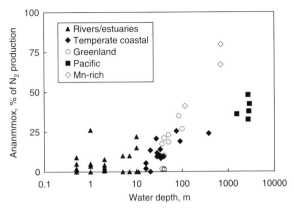

Figure 14.6 Relative contribution of the anammox process to N_2 production (anammox plus denitrification) in marine sediments as a function of water depth. Data were compiled by Dalsgaard et al. (2005), supplemented with data from Meyer et al. (2005), Engström et al. (2006), Rich et al. (2008), Hietanen and Kuparinen (2008), and B. Thamdrup (unpublished results). Data from shallow estuaries were plotted arbitrarily at 0.5, 1, and 2 m depth.

for anammox in sediments is found at manganese-rich sites (with Mn(IV) content of about 5 percent by weight: Thamdrup and Dalsgaard 2002; Engström et al. 2005). The two types of organisms seem to have similar, low half-saturation constants for nitrite (and nitrate) in the sediment (≤ 3 μmol/L: Dalsgaard and Thamdrup 2002), suggesting that their relative importance reflects the availability of their respective electron donors rather than a competition for nitrite and nitrate. When manganese oxide is present, denitrifiers suffer from competition with dissimilatory manganese reducers, while anammox bacteria remain the main anaerobic sink for ammonium (Dalsgaard and Thamdrup 2002), but manganese reduction is generally not important for carbon oxidation in marine sediments (Thamdrup 2000). If phytoplankton detritus with a typical composition of $(CH_2O)_{106}(NH_3)_{16}H_3PO_4$ is mineralized completely through nitrate reduction, denitrification, and anammox, the contribution of anammox to total N_2 production is 29 percent (Dalsgaard et al. 2003). This fraction seems to be a realistic average for the contribution of anammox for N_2 production in shelf sediments. Recent measurements suggest a somewhat higher contribution in the deep sea (Fig. 14.6) (Dalsgaard et al. 2005).

Studies of anammox in marine sediments have so far focused mainly on its relation to denitrification and less on the relative importance of anammox and nitrification in the oxidation of ammonium. In deep-sea sediments with an extended redox zonation, anammox bacteria are probably responsible for the efficient removal of the ammonium generated in the anoxic part of the sediment before it reaches the aerobic ammonium oxidizers (see, e.g., Bender et al. 1989; Schulz et al. 1994). In such sediments, however, most of the organic matter is oxidized in the oxic surface layer, and ammonium oxidation is therefore also mainly aerobic. Conversely, in sediments dominated by anaerobic mineralization, the zone of nitrate reduction is thin, and a large fraction of the ammonium reaches oxic depths

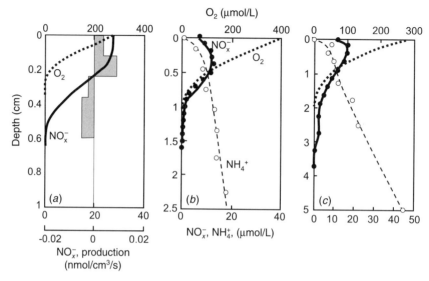

Figure 14.7 Examples of porewater distributions of nitrate plus nitrite, nitrite, and ammonium in relation to oxygen in marine sediments: (*a*) Randers Fjord; (*b*) Young Sound; (*c*) Skagerrak. Oxygen and, for Randers Fjord, nitrate plus nitrite distributions are shown as continuous traces based on microsensor measurements at high spatial resolution. For Randers Fjord, net rates of nitrate plus nitrite production and consumption derived from reaction-transport modelling are shown as grey bars. Data are from Risgaard-Petersen et al. (2004), Rysgaard et al. (1998), Canfield et al. (1993), and B. Thamdrup and T. Dalsgaard (unpublished results). Note the different depth scales.

(see Fig. 14.7 below). Thus, in the examples in Figures 14.7 and 14.8, described in more detail below, anammox accounts for 3, 2, and 16 percent of the estimated ammonium oxidation (Randers Fjord, Young Sound, and Skagerrak Basin, respectively; estimated as net ammonification − ammonium efflux). The most promising sites for a major importance of anammox as an ammonium sink are sediments that underlie oxygen-depleted, nitrate-rich waters. A symbiotic relationship between anammox bacteria and *Thioploca* has been suggested to explain the distributions of ammonium and nitrogen isotopes in such sediments (Prokopenko et al. 2006), but the presence and activity of the anammox bacteria there await direct quantification.

Processes Involving Mn and Fe

The redox potentials of couples such as NO_3^-/N_2 and NO_3^-/NH_4^+ are such that several interactions between nitrogen and manganese and iron cycling are thermodynamically favorable. The oxidation by bacteria of ferrous iron coupled to the reduction of nitrate to N_2 or NH_4^+ was mentioned above, but other reactions are also possible. Some mineral surfaces catalyze an abiotic reduction of nitrate and

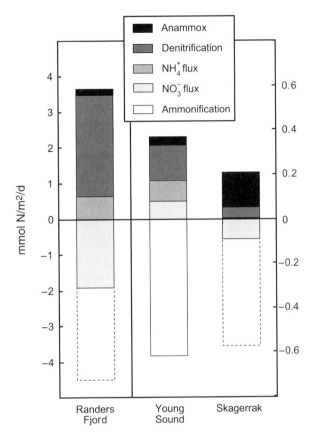

Figure 14.8 Dissolved inorganic nitrogen budgets for three sediments. Negative and postive values are sources and sinks, respectively. Note the different scales for Randers Fjord and Young Sound–Skagerrak Basin. Data from Randers Fjord represent an annual average of values from seven stations. The anammox contribution was only determined at one station at a separate occasion. Data from Young Sound are an annual average from one site, with anammox contributions measured on a separate occasion. Data from Skagerrak combine results from several samplings. Stippled borders of the bars for ammonification at Randers Fjord and Skagerrak indicate that net ammonification was estimated from the flux of dissolved inorganic carbon assuming a C : N ratio for mineralization of 10 : 1, similar to that determined for Young Sound. Data are from Nielsen et al. (2001), Risgaard-Petersen et al. (2003, 2005), Rysgaard et al. (1998, 2001), Hall et al. (1996), and Thamdrup and Dalsgaard (2002).

nitrite by ferrous iron to nitrous oxide or N_2 (Ottley et al. 1987; Sørensen and Thorling 1991), but this type of chemodenitrification is thought to be slow compared with the rates of microbial processes (Canfield et al. 2005). There is as yet no direct evidence of Mn^{2+} serving as electron donor for nitrate reduction, as proposed for some marine sediments (Luther et al. 1997).

Manganese oxides have been suggested to function as oxidants for ammonium in marine sediments (Luther et al. 1997; Aller et al. 1998; Hulth et al. 1999; Anschutz et al. 2005). The process could be either a direct oxidation to N_2 or a nitrification-like oxidation to nitrate/nitrite followed by denitrification. Attempts to detect such a process directly in a wide range of sediments using anoxic incubations with $^{15}NH_4^+$ have so far been unsuccessful, suggesting that nitrogen/manganese coupling is not of quantitative importance (Dalsgaard et al. 2005), but the role of highly reactive Mn(III), which forms as an intermediate during manganese oxidation, may require further investigation (Luther et al. 1997; Anschutz et al. 2005; Tebo et al. 2005).

NITROGEN BUDGETS

Benthic Budgets

During benthic transformations of nitrogen, many different nitrogen compounds accumulate either transiently or permanently in the pore water. These compounds, which include DON, urea, ammonium, nitrous oxide, nitrite, nitrate, and N_2, are all to a greater or lesser extent subject to exchange with the bottom water. Of least quantitative importance in terms of benthic exchange are nitrous oxide and nitrite, the fluxes of which are generally at least one order of magnitude lower than those of nitrate and N_2 (Hall et al. 1996; Robinson et al. 1998; Laursen and Seitzinger 2002; Dong et al. 2002). Conversely, each of the remaining species, as well as the DON pool, may dominate benthic exchange on specific occasions. The contribution of DON to benthic exchange is particularly important in shallow coastal sediments, where DON fluxes are comparable in magnitude to the fluxes of dissolved inorganic nitrogen, and where benthic microalgae may be an important DON source at the sediment surface, although at other times the algae may also take up DON (Eyre and Ferguson 2002; Tyler et al. 2003; Sundbäck et al. 2004).

The contribution of ammonium to benthic nitrogen exchange depends mainly on the efficiency of nitrification. With air-saturated bottom water, an oxic surface layer of just 1–2 mm thickness—typical of active coastal sediments—may act as an efficient barrier against the escape of ammonium (see, e.g., Jensen et al. 1994; Rysgaard et al. 1994). Decreasing oxygen levels in the bottom water will decrease the penetration depth and thereby favor the efflux of ammonium. Likewise, the ammonium flux can be stimulated through the input of organic matter, which will act both as a source of ammonium and as a sink for oxygen (Caffrey et al. 1993). As discussed above, sulfidic sediments represent an extreme in this respect where nitrite ammonification may contribute further to the ammonium flux (Fig. 14.5). Furthermore, sulfide inhibits nitrification (Joye and Hollibaugh 1995). Infauna will often stimulate nitrification by extending the area of the oxic/anoxic interface through their burrowing (see, e.g., Pelegri and Blackburn 1994), but porewater irrigation can also contribute to the loss of ammonium from the sediment (see, e.g., Kristensen and Hansen 1999; Berg et al. 2003). Thus, ammonium is often an important component in benthic nitrogen exchange, and the contribution of the ammonium fluxes to the benthic nitrogen budget is highly variable in both space and time (see, e.g., Blackburn and Henriksen 1983; Rysgaard et al. 1998; Laursen and Seitzinger 2002).

While ammonium fluxes are normally directed out of the sediment, the direction of nitrate fluxes is variable, depending on bottom-water nitrate concentrations and on the relative importance of nitrification and the pathways of nitrate reduction. In coastal sediments where benthic metabolism and bottom-water oxygen and nitrate levels vary seasonally, the sediment can act interchangeably as a source and a sink for nitrate (see, e.g., Rysgaard et al. 1998). Nitrate fluxes are generally directed into more reducing sediments with shallow oxygen penetration, where nitrate is reduced close to the sediment surface and nitrification is limited (Fig. 14.7). As the oxic zone widens and nitrification becomes more efficient, nitrate may start to escape the sediment, but, due to the tight coupling of nitrification and denitrification, and possibly anammox, N_2 production often exceeds the nitrate efflux. Nitrate first becomes the dominant species released from the seafloor in oligotrophic sediments of the deep sea (Middelburg et al. 1996).

Some of the variability in benthic nitrogen cycling is illustrated in three examples ranging from a shallow estuary to a 700 m deep continental basin (Figs. 14.7 and 14.8). The examples were chosen to represent different organic loadings, and are all extensively studied and part of the relatively small set of locations where both denitrification and anammox rates have been determined. The locations include (1) Randers Fjord, a shallow eutrophic Danish estuary opening into the Kattegat between Denmark and Sweden, and subject to high nitrate concentrations throughout the year (Nielsen et al. 2001; Meyer et al. 2001; Risgaard-Petersen et al. 2004); (2) a 35 m-deep station in the fjord Young Sound in Northeast Greenland, which is covered by ice for about 9 months of the year, with a bloom of phytoplankton during the summer thaw (Rysgaard et al. 1998, 1999); and (3) a 700 m-deep station in the Skagerrak Basin between Denmark and Norway, which is the record holder in terms of the contribution of anammox to N_2 production and of dissimilatory manganese reduction to carbon oxidation (Canfield et al. 1993; Thamdrup and Dalsgaard 2000, 2002). The bottom water is well oxygenated at all three sites, but varies with respect to nitrate from $5-10$ μmol/L in Young Sound and Skagerrak to $30-300$ μmol/L in Randers Fjord, where the highest concentrations are experienced in winter.

The porewater distributions of oxygen, nitrate plus nitrite, and ammonium exemplified in Figure 14.7 are qualitatively similar at the three sites, but plot on different depth scales. Thus, oxygen penetration depth increases with water depth from about 2 mm in Randers Fjord to almost 2 cm in Skagerrak, and nitrate is consumed in a narrow zone below the oxic/anoxic interface. At all three sites, convex nitrate profiles indicate nitrification throughout the oxic zone, and ammonium determinations in Young Sound and Skagerrak indicate that ammonium reaches the sediment surface.

When averaged over the entire year, rates of benthic metabolism and exchange are approximately one order of magnitude higher in Randers Fjord than in Young Sound and Skagerrak (Fig. 14.8). There is a pronounced seasonal variation in the magnitude of benthic fluxes in Randers Fjord, with nitrate uptake peaking in winter when nitrate concentrations are highest, and ammonium efflux peaking in summer (Nielsen et al. 2001). In Young Sound, rates are relatively stable over the year, except for a small peak in activity following the summer bloom (Rysgaard et al. 1998), while no pronounced seasonality has been observed in the Skagerrak Basin (Canfield et al. 1993; Thamdrup and Dalsgaard 2000). At all three sites, ammonification is the main source of inorganic nitrogen to the sediment, although in Randers Fjord this source is nearly matched by the uptake of nitrate from the nitrate-rich bottom water (Fig. 14.8). The sediment is also a sink for nitrate in the Skagerrak Basin (Hall et al. 1996), whereas it is a net source in Young Sound, although nitrate uptake is occasionally observed there too (Rysgaard et al. 1998). Both in Randers Fjord and Skagerrak, the influx of nitrate seems to conflict with the nitrate gradient at the sediment surface (Fig. 14.7), most likely because nitrate uptake at both sites is stimulated by porewater advection by the infauna (Nielsen et al. 2001; Munksby et al. 2002; Risgaard-Petersen et al. 2004). Porewater mixing and advection is also important in Young Sound (Berg et al. 2001, 2003).

Ammonium accounts for a minor part of the nitrogen liberated from the sediments, which is dominated by N_2 (Fig. 14.8). DON fluxes were not measured at these sites, while the flux of urea was measured in Young Sound, where it corresponded to 30 percent of the ammonium flux (Rysgaard et al. 1998). The small fluxes of ammonium compared with ammonification rates indicate that most ammonium is consumed by nitrification. In Randers Fjord, however, assimilation by benthic algae may also play a role (Risgaard-Petersen 2003). Denitrification dominates N_2 production in Randers Fjord and Young Sound, whereas anammox accounts for about 70 percent of N_2 production in the Skagerrak Basin (Thamdrup and Dalsgaard 2002; Risgaard-Petersen et al. 2004). The 18 percent contribution of anammox to N_2 production in Young Sound is quite typical for shelf sediments, while the low but persistent contribution in Randers Fjord (4–24 percent at different times of the year), is high for shallow sediments (Fig. 14.6), but may be explained by availability of nitrate throughout the year. At all three sites, denitrification accounts for less than 10 percent of the carbon oxidation in the sediment (Canfield et al. 1993; Rysgaard et al. 1998; Nielsen et al. 2001).

The examples discussed above cover only a part of the wide range of benthic environments in the oceans. Important types missing are sediments from beyond the continental shelf and from oxygen-minimum zones. Details of the nitrogen cycling in such environments are not well studied, however, but should be an important topic for future research.

Oceanic Budgets

Because nitrogen is a limiting nutrient in oceanic primary production, the global dynamics of marine fixed nitrogen are an important element in our understanding of the oceanic carbon cycle. Sediments are the site of two of the three major sinks in current budgets of marine fixed nitrogen, including benthic N_2 production, by

NITROGEN BUDGETS

TABLE 14.2 The Marine Fixed Nitrogen Budget (rates in Tg N/year)[a]

Sources		Sinks	
Rivers	25–43	Pelagic denitrification	60–150
Atmosphere	40–50	Benthic denitrification	200–300
N_2 fixation	40–135	Burial	10–27
Total sources	105–228	Total sinks	270–477

[a]After Middelburg et al. (1996), Brandes and Devol (2002), and Codispoti (2007).

denitrification and anammox, which is the largest sink overall, and burial of organic nitrogen (Table 14.2). The third important sink is pelagic N_2 production, which takes place in the oxygen-deficient cores of the oceanic oxygen minimum zones. On the source side, nitrogen fixation is the largest term, followed by transport by rivers and through the atmosphere.

Nitrogen : Phosphate Ratios in the Ocean

The biogeochemical cycles of carbon, nitrogen, and phosphorus in the ocean are coupled through a quite invariable stoichiometric relationship between these elements in marine biomass in bulk. Thus, the average composition of marine biomass is, to a good approximation, described by the Redfield formula: $(CH_2O)_{106}(NH_3)_{16}H_3PO_4$ (Redfield et al. 1963). As a result of this stoichiometry, concentrations of inorganic fixed nitrogen (which is essentially all nitrate) and phosphate in ocean waters are strongly correlated, co-varying in response to assimilation and mineralization at a ratio of 16 N to 1 P. Sources and sinks such as nitrogen fixation and denitrification that affect the two elements independently will cause deviations from this ratio, which is therefore useful for localizing sources and sinks in the marine nitrogen budget (see, e.g., Codispoti and Richards 1976). The deviation of the nitrate concentrations from the expected relationship is expressed by the parameter $N^* = [NO_3^-] - 16 \times [PO_4^{3-}] + 2.9$ μmol/L (Gruber and Sarmiento 1997); the constant term (2.9 μmol/L) is included because a regression of nitrate versus phosphate concentrations does not pass through the origin. Mapping of the distribution of N^* within the water column, in combination with water residence times, allows estimation of net N_2 fixation and production, and application of this calculation to the oceanic oxygen minimum zones forms the basis of the estimates of pelagic N_2 production (Table 14.2) (Codispoti and Richards 1976; Gruber and Sarmiento 1997).

Interestingly, the budget as it stands is markedly out of balance, with sinks exceeding sources even when ranges in estimates of the individual terms are considered (Middelburg et al. 1996; Codispoti et al. 2001; Codispoti 2007). The residence time of fixed nitrogen in the oceans is only about 2000 years with respect to the sinks, and if the imbalance is real and persists, it would cause depletion of the pool in about 3000 years, with serious consequences for carbon dioxide uptake by the oceans (Brandes and Devol 2002). Thus, while the imbalance could be real at present, it cannot be representative of the Holocene. Since the nitrogen isotope record in marine sediments does not support major changes during the Holocene, it can be argued that the imbalance must be due to faults in the budget. The estimates of burial and pelagic denitrification appear quite robust, and although the contribution of benthic denitrification rests on a small experimental database, similar numbers have been reached repeatedly through widely different approaches (Middelburg et al. 1996; Codispoti et al. 2001; Brandes and Devol 2002; Codispoti 2007). Thus, nitrogen fixation seems to be the most likely term to need reevaluation (see Chapter 13).

With a reevaluation pending, it would seem that the discovery of anammox as a new marine sink for fixed nitrogen could worsen the imbalance of the budget. More likely, however, the current denitrification estimates mainly need to be partitioned between anammox and dentrification in the strict sense used throughout this chapter (see the text box on "What is Denitrification?"). Thus, anaerobic ammonium oxidation only increased N_2 production by 12 percent when the global benthic contribution was estimated with a prognostic diagenetic model (Middelburg et al. 1996). Unless the nitrogen isotope fractionation associated with anammox in sediments is markedly different from that of denitrification, denitrification estimates based on isotope balances (Brandes and Devol 2002) should not be sensitive to the pathway of benthic N_2 formation. Experimental determinations of benthic denitrification include the efflux of N_2, which is insensitive to the pathway (see, e.g., Devol et al. 1997), and the isotope pairing technique, which will overestimate N_2 production if anammox is active but neglected (Risgaard-Petersen et al. 2003). Nonetheless, the short residence time of fixed nitrogen emphasizes that the availability of fixed nitrogen is sensitive to small changes in sources and sinks. To understand and predict such changes, it is therefore important to know the distribution and regulatory characteristics of all the processes at play.

SUMMARY

1. Sediments have three main functions in the marine nitrogen cycle: recycling of fixed nitrogen to the water column; burial of detrital organic nitrogen; and removal of fixed nitrogen through denitrification and the anammox process.

2. The main input of nitrogen to marine sediments is in sinking organic detritus, while nitrate uptake may dominate in sediments underlying nitrate-rich waters. Nitrogen fixation by prokaryotes in the sediment is a minor source. Sediments within the photic zone may harbor microalgal communities that assimilate inorganic nitrogen from the water column.

3. The benthic nitrogen cycle is dominated by a diverse set of dissimilatory microbial processes during which nitrogen is oxidized or reduced and intermediates accumulate. The processes are performed by different groups of

specialized bacteria, and by newly discovered ammonium-oxidizing archaea and denitrifying foraminifera. We know little about the specific organisms involved.
4. Nitrate is typically not important as an electron acceptor for benthic carbon oxidation, but may contribute substantially to the oxidation of hydrogen sulfide by nitrate-accumulating giant sulfur bacteria.
5. Nitrogen is exported from the sediment in several different organic and inorganic forms. The inorganic forms, which generally constitute most of the benthic efflux, include nitrate, N_2, ammonium, nitrite, and nitrous oxide. In most cases, nitrate and N_2 dominate the nitrogen flux from the sediment.
6. Benthic N_2 production by denitrifiers and anammox bacteria together accounts for approximately two-thirds of the sink term in global marine nitrogen budgets. These budgets are currently unbalanced, possibly due to underestimation of marine nitrogen fixation.
7. Important next steps in studies of benthic nitrogen cycling include identification of the organisms involved, ecophysiological investigations of important players, and experimental investigations of nitrogen cycling in deep-sea sediments.

REFERENCES

Aller, R. C., P. O. J. Hall, P. D. Rude, and J. Y. Aller. 1998. Biogeochemical heterogeneity and suboxic diagenesis in hemipelagic sediments of the Panama Basin. *Deep-Sea Res.* 45: 133–165.

Altmann, D., P. Stief, R. Amann, D. de Beer, and A. Schramm. 2003. In situ distribution and activity of nitrifying bacteria in freshwater sediment. *Environ. Microbiol.* 5: 798–803.

Aminuddin, M., and D. J. D. Nicholas. 1973. Sulphide oxidation linked to the reduction of nitrate and nitrite in *Thiobacillus denitrificans*. *Biochim. Biophys. Acta* 325: 81–93.

An, S. M., and W. S. Gardner. 2002. Dissimilatory nitrate reduction to ammonium (DNRA) as a nitrogen link, versus denitrification as a sink in a shallow estuary (Laguna Madre/Baffin Bay, Texas). *Mar. Ecol. Prog. Ser.* 237: 41–50.

Anschutz, P., K. Dedieu, F. Desmazes, and G. Chaillou. 2005. Speciation, oxidation state, and reactivity of particulate manganese in marine sediments. *Chem. Geol.* 218: 265–279.

Antia, A. N. 2005. Solubilization of particles in sediment traps: Revising the stoichiometry of mixed layer export. *Biogeosciences* 2: 189–204.

Bagwell, C. E., J. R. La Rocque, G. W. Smith, S. W. Polson, M. J. Friez, J. W. Longshore, and C. R. Lovell. 2002. Molecular diversity of diazotrophs in oligotrophic tropical seagrass bed communities. *FEMS Microbiol. Ecol.* 39: 113–119.

Bartosch, S., C. Hartwig, E. Spieck, and E. Bock. 2002. Immunological detection of nitrospira-like bacteria in various soils. *Microb. Ecol.* 43: 26–33.

Beauchamp, E. G., J. T. Trevors, and J. W. Paul. 1989. Carbon sources for bacterial denitrification. *Adv. Soil Sci.* 10: 113–142.

Bebout, B. M., H. W. Paerl, K. M. Crocker, and L. E. Prufert. 1987. Diel interactions of oxygenic photosynthesis and N_2 fixation (acetylene-reduction) in a marine microbial mat community. *Appl. Environ. Microbiol.* 53: 2353–2362.

Bender, M., R. A. Jahnke, R. Weiss, W. Martin, D. T. Heggie, J. Orchardo, and T. Sowers. 1989. Organic carbon oxidation and benthic nitrogen and silica dynamics in San Clemente Basin, a continental borderland site. *Geochim. Cosmochim. Acta* 53: 685–697.

Benz, M., A. Brune, and B. Schink. 1998. Anaerobic and aerobic oxidation of ferrous iron at neutral pH by chemoheterotrophic nitrate-reducing bacteria. *Arch. Microbiol.* 169: 159–165.

Berg, P., S. Rysgaard, P. Funch, and M. K. Sejr. 2001. Effects of bioturbation on solutes and solids in marine sediments. *Aquat. Microb. Ecol.* 26: 81–94.

Berg, P., S. Rysgaard, and B. Thamdrup. 2003. Dynamic modeling of early diagenesis and nutrient cycling. A case study in an Arctic marine sediment. *Am. J. Sci.* 303: 905–955.

Berner, R. A. 1980. *Early Diagenesis.* Princeton University Press.

Betlach, M. R., and J. M. Tiedje. 1981. Kinetic explanation for accumulation of nitrite, nitric oxide, and nitrous oxide during bacterial denitrification. *Appl. Environ. Microbiol.* 42: 1074–1084.

Binnerup, S. J., K. Jensen, N. P. Revsbech, M. H. Jensen, and J. Sørensen. 1992. Denitrification, dissimilatory reduction of nitrate to ammonium, and nitrification in a bioturbated estuarine sediment as measured with N-15 and microsensor techniques. *Appl. Environ. Microbiol.* 58: 303–313.

Blackburn, T. H. 1979. Method for measuring rates of NH_4^+ turnover in anoxic marine-sediments, using a N-15–NH_4^+ dilution technique. *Appl. Environ. Microbiol.* 37: 760–765.

Blackburn, T. H., and K. Henriksen. 1983. Nitrogen cycling in different types of sediments from Danish waters. *Limnol. Oceanogr.* 28: 477–493.

Bock, E., and M. Wagner. 2001. Oxidation of inorganic nitrogen compounds as energy source. In M. Dworkin, et al. (eds.), *The Prokaryotes: An Evolving Electronic Resource for the Microbiological Community*, 3rd edn. Springer-Verlag, pp. 414–430.

Bock, E., I. Schmidt, R. Stuven, and D. Zart. 1995. Nitrogen loss caused by denitrifying *Nitrosomonas* cells using ammonium or hydrogen as electron-donors and nitrite as electron-acceptor. *Arch. Microbiol.* 163: 16–20.

Braker, G., J. Z. Zhou, L. Y. Wu, A. H. Devol, and J. M. Tiedje. 2000. Nitrite reductase genes (*nirK* and *nirS*) as functional markers to investigate diversity of denitrifying bacteria in Pacific northwest marine sediment communities. *Appl. Environ. Microbiol.* 66: 2096–2104.

Brandes, J. A., and A. H. Devol. 1995. Simultaneous nitrate and oxygen respiration in coastal sediments: evidence for discrete diagenesis. *J. Mar. Res.* 53: 771–797.

Brandes, J. A., and A. H. Devol. 2002. A global marine-fixed nitrogen isotopic budget: Implications for Holocene nitrogen cycling. *Global Biogeochem. Cycles* 16: doi: 10.1029/2001GB001856.

Buckel, W. 1999. Anaerobic energy metabolism. In J. W. Lengeler, G. Drews, and H. G. Schlegel (eds.), *Biology of the Prokaryotes.* Blackwell, pp. 278–326.

Burdige, D. J., and C. S. Martens. 1988. Biogeochemical cycling in an organic-rich coastal marine basin, 10: The role of amino-acids in sedimentary carbon and nitrogen cycling. *Geochim. Cosmochim. Acta* 52: 1571–1584.

REFERENCES

Burdige, D. J., and S. Zheng. 1998. The biogeochemical cycling of dissolved organic nitrogen in estuarine sediments. *Limnol. Oceanogr.* 43: 1796–1813.

Burns, J. A., J. P. Zehr, and D. G. Capone. 2002. Nitrogen-fixing phylotypes of Chesapeake Bay and Neuse River estuary sediments. *Microb. Ecol.* 44: 336–343.

Cabello, P., M. D. Roldan, and C. Moreno-Vivian. 2004. Nitrate reduction and the nitrogen cycle in archaea. *Microbiol. SGM* 150: 3527–3546.

Caffrey, J. M., N. P. Sloth, H. F. Kaspar, and T. H. Blackburn. 1993. Effect of organic loading on nitrification and denitrification in a marine sediment microcosm. *FEMS Microbiol. Ecol.* 12, 159–167.

Canfield, D. E. 1993. Organic matter oxidation in marine sediments. In R. Wollast, F. T. Mackenzie, and L. Chou (eds.), *Interactions of C, N, P, and S Biogeochemical Cycles and Global Change*. Springer-Verlag, pp. 333–363.

Canfield, D. E., B. B. Jørgensen, H. Fossing, et al. 1993. Pathways of organic carbon oxidation in three continental margin sediments. *Mar. Geol.* 113: 27–40.

Canfield, D. E., E. Kristensen, and B. Thamdrup. 2005. *Aquatic Geomicrobiology*. Elsevier Academic Press.

Capone, D. G. 1983. N_2 fixation in seagrass communities. *Mar. Technol. Soc. J.* 17: 32–37.

Capone, D. G. 2000. The marine microbial nitrogen cycle. In D. L. Kirchman (ed.), *Microbial Ecology of the Oceans*, 1st edn. Wiley-Liss, pp. 455–494.

Capone, D. G., and E. J. Carpenter. 1982. Nitrogen-fixation in the marine environment. *Science* 217: 1140–1142.

Carlson, C. A., and J. L. Ingraham. 1983. Comparison of denitrification by *Pseudomonas stutzeri*, *Pseudomonas aeruginosa*, and *Paracoccus denitrificans*. *Appl. Env. Microbiol.* 45: 1247–1253.

Casciotti, K. L., and B. B. Ward. 2005. Phylogenetic analysis of nitric oxide reductase gene homologues from aerobic ammonia-oxidizing bacteria. *FEMS Microbiol. Ecol.* 52, 197–205.

Christensen, P. B., L. P. Nielsen, J. Sorensen, and N. P. Revsbech. 1990. Denitrification in nitrate-rich streams—diurnal and seasonal-variation related to benthic oxygen-metabolism. *Limnol. Oceanogr.* 35: 640–651.

Christensen, P. B., S. Rysgaard, N. P. Sloth, T. Dalsgaard, and S. Schwærter. 2000. Sediment mineralization, nutrient fluxes, denitrification and dissimilatory nitrate reduction to ammonium in an estuarine fjord with sea cage trout farms. *Aquat. Microb. Ecol.* 21: 73–84.

Cline, J. D., and F. A. Richards. 1972. Oxygen deficient conditions and nitrate reduction in the eastern tropical northern Pacific Ocean. *Limnol. Oceanogr.* 17: 885–900.

Codispoti, L. A. 1995. Biogeochemical cycles—Is the ocean losing nitrate? *Nature* 376: 724.

Codispoti, L. A. 2007. An oceanic fixed nitrogen sink exceeding 400 Tg N a^{-1} vs. the concept of homeostasis in the fixed-nitrogen inventory. *Biogeosciences* 4: 233–253.

Codispoti, L. A., and F. A. Richards. 1976. Analysis of horizontal regime of denitrification in Eastern Tropical North Pacific. *Limnol. Oceanogr.* 21: 379–388.

Codispoti, L. A., J. A. Brandes, J. P. Christensen, A. H. Devol, S. W. A. Naqvi, H. W. Paerl, and T. Yoshinary. 2001. The oceanic fixed nitrogen and nitrous oxide budgets: moving targets as we enter the anthropocene? *Sci. Mar.* 65 (Suppl. 2): 85–105.

Cole, J. A. 1987. Assimilatory and dissimilatory reduction of nitrate to ammonia. In J. A. Cole and J. Ferguson (eds.), *The Nitrogen and Sulphur Cycles*. Cambridge University Press, pp. 281–329.

Cole, J. A., and C. M. Brown. 1980. Nitrite reduction to ammonia by fermentative bacteria: A short circuit in the biological nitrogen cycle. *FEMS Microbiol. Lett* 7: 65–72.

Dalsgaard, T., and B. Thamdrup. 2002. Factors controlling anaerobic ammonium oxidation with nitrite in marine sediments. *Appl. Environ. Microbiol.* 68: 3802–3808.

Dalsgaard, T., D. E. Canfield, J. Petersen, B. Thamdrup, and J. Acuña-Gonzalez. 2003. N_2 production by the anammox reaction in the anoxic water column of Golfo Dulce, Costa Rica. *Nature* 422: 606–608.

Dalsgaard, T., B. Thamdrup, and D. E. Canfield. 2005. Anaerobic ammonium oxidation (anammox) in the marine environment. *Res. Microbiol.* 156: 457–464.

Damsté, J. S. S., M. Strous, W. I. C. Rijpstra, E. C. Hopmans, J. A. Geenevasen, A. C. T. Van Duin, L. A. Van Niftrik, and M. S. M. Jetten. 2002. Linearly concatenated cyclobutane lipids form a dense bacterial membrane. *Nature* 419: 708–712.

Dauwe, B., and J. J. Middelburg. 1998. Amino acids and hexosamines as indicators of organic matter degradation state in North Sea sediments. *Limnol. Oceanogr.* 43: 782–798.

Dell'Anno, A., and R. Danovaro. 2005. Extracellular DNA plays a key role in deep-sea ecosystem functioning. *Science* 309: 2179.

Dell'Anno, A., S. Bompadre, and R. Danovaro. 2002. Quantification, base composition, and fate of extracellular DNA in marine sediments. *Limnol. Oceanogr.* 47: 899–905.

Delwiche, C. C. 1970. The nitrogen cycle. *Sci. Am.* 223: 136–146.

Devol, A. H., L. A. Codispoti, and J. P. Christensen. 1997. Summer and winter denitrification rates in western Arctic shelf sediments. *Continent. Shelf Res.* 17: 1029.

Dong, L. F., D. B. Nedwell, G. J. C. Underwood, D. C. O. Thornton, and I. Rusmana. 2002. Nitrous oxide formation in the Colne estuary, England: The central role of nitrite. *Appl. Env. Microbiol.* 68: 1240–1249.

Duarte, C. M. 1992. Nutrient concentration of aquatic plants—patterns across species. *Limnol. Oceanogr.* 37: 882–889.

Ehrlich, S., D. Behrens, E. Lebedeva, W. Ludwig, and E. Bock. 1995. A new obligately chemolithoautotrophic, nitrite-oxidizing bacterium, *Nitrospira moscoviensis* sp. nov. and its phylogenetic relationship. *Arch. Microbiol.* 164: 16–23.

Engström, P., T. Dalsgaard, S. Hulth, and R. C. Aller. 2005. Anaerobic ammonium oxidation by nitrite (anammox): Implications for N_2 production in coastal marine sediments. *Geochim. Cosmochim. Acta* 69: 2057–2065.

Eyre, B. D., and A. J. P. Ferguson. 2002. Comparison of carbon production and decomposition, benthic nutrient fluxes and denitrification in seagrass, phytoplankton, benthic microalgae- and macroalgae-dominated warm-temperate Australian lagoons. *Mar. Ecol. Prog. Ser.* 229: 43–59.

Fagerbakke, K. M., M. Heldal, and S. Norland. 1996. Content of carbon, nitrogen, oxygen, sulfur and phosphorus in native aquatic and cultured bacteria. *Aquatic Microb. Ecol.* 10: 15–27.

Falkowski, P. G. 1997. Evolution of the nitrogen cycle and its influence on the biological sequestration of CO_2 in the ocean. *Nature* 387: 272–275.

Falkowski, P. G., R. T. Barber, and V. Smetacek. 1998. Biogeochemical controls and feedbacks on ocean primary production. *Science* 281: 200–206.

REFERENCES

Fanning, J. C. 2000. The chemical reduction of nitrate in aqueous solution. *Coord. Chem. Rev.* 199: 159–179.

Finneran, K. T., H. M. Forbush, C. V. G. VanPraagh, and D. R. Lovley. 2002. *Desulfitobacterium metallireducens* sp. nov., an anaerobic bacterium that couples growth to the reduction of metals and humic acids as well as chlorinated compounds. *Int. J. Syst. Evol. Microbiol.* 52: 1929–1935.

Focht, D. D., and W. Verstraete. 1977. Biochemical ecology of nitrification and denitrification. *Adv. Microb. Ecol.* 1: 135–214.

Fossing, H., V. A. Gallardo, B. B. Jørgensen, et al. 1995. Concentration and transport of nitrate by the mat-forming sulphur bacterium *Thioploca. Nature* 374: 713–715.

Francis, C. A., G. D. O'Mullan, and B. B. Ward. 2003. Diversity of ammonia monooxygenase (*amoA*) genes across environmental gradients in Chesapeake Bay sediments. *Geobiology* 1: 129–140.

Francis, C. A., K. J. Roberts, J. M. Beman, A. E. Santoro, and B. B. Oakley. 2005. Ubiquity and diversity of ammonia-oxidizing archaea in water columns and sediments of the ocean. *Proc. Natl. Acad. Sci. USA* 102: 14683–14688.

Fuchs, G. 1999. Oxidation of organic compounds. In J. W. Lengeler, G. Drews, and H. G. Schlegel (eds.), *Biology of the Prokaryotes*. Blackwell, pp. 187–233.

Fuerst, J. A. 2005. Intracellular compartmentation in planctomycetes. *Annu. Rev. Microbiol.* 59: 299–328.

Gattuso, J. P., B. Gentili, C. M. Duarte, J. A. Kleypas, J. J. Middelburg, and D. Antoine. 2006. Light availability in the coastal ocean: Impact on the distribution of benthic photosynthetic organisms and their contribution to primary production. *Biogeosciences* 3: 489–513.

Griffin, B. M., J. Schott, and B. Schink. 2007. Nitrite, an electron donor for anoxygenic photosynthesis. *Science* 316: 1870.

Gruber, N., and J. L. Sarmiento. 1997. Global patterns of marine nitrogen fixation and denitrification. *Global Biogeochem. Cycles* 11: 235–266.

Güven, D., A. Dapena, B. Kartal, et al. 2005. Propionate oxidation by and methanol inhibition of anaerobic ammonium-oxidizing bacteria. *Appl. Environ. Microbiol.* 71, 1066–1071.

Hall, P. O. J., S. Hulth, G. Hulthe, A. Landen, and A. Tengberg. 1996. Benthic nutrient fluxes on a basin-wide scale in the Skagerrak (north-eastern North Sea). *J. Sea Res.* 35: 123–137.

Head, I. M., W. D. Hiorns, T. M. Embley, A. J. McCarthy, and J. R. Saunders. 1993. The phylogeny of autotrophic ammonia-oxidizing bacteria as determined by analysis of 16S ribosomal-RNA gene-sequences. *J. Gen. Microbiol.* 139: 1147–1153.

Hedges, J. I., J. A. Baldock, Y. Gelinas, C. Lee, M. Peterson, and S. G. Wakeham. 2001. Evidence for non-selective preservation of organic matter in sinking marine particles. *Nature* 409: 801–804.

Hendriks, J., A. Oubrie, J. Castresana, A. Urbani, S. Gemeinhardt, and M. Saraste. 2000. Nitric oxide reductases in bacteria. *Biochim. Biophys. Acta Bioenergetics* 1459: 266–273.

Henrichs, S. M., J. W. Farrington, and C. Lee. 1984. Peru upwelling region sediments near 15° S. 2: Dissolved free and total hydrolyzable amino-acids. *Limnol. Oceanogr.* 29: 20–34.

Henriksen, K., and W. M. Kemp. 1988. Nitrification in estuarine and coastal marine sediments. In T. H. Blackburn and J. Sørensen (eds.), *Nitrogen Cycling in Coastal Marine Environments*. Wiley, pp. 207–249.

Herbert, R. A. 1999. Nitrogen cycling in coastal marine ecosystems. *FEMS Microbiol. Rev.* 23: 563–590.

Hietanen S., and J. Kuparinen. 2008. Seasonal and short-term variation in denitrification and anammox at a coastal station on the Gulf of Finland, Baltic Sea. *Hydrobiologia* 596: 67–77.

Hu, B. L., P. Zheng, J. Y. Li, X. Y. Xu, and R. C. Jin. 2006. Identification of a denitrifying bacterium and verification of its anaerobic ammonium oxidation ability. *Sci. China, Ser. C, Life Sci.* 49: 460–466.

Huettel, M., S. Forster, S. Klöser, and H. Fossing. 1996. Vertical migration in the sediment-dwelling sulfur bacteria *Thioploca* spp. in overcoming diffusion limitations. *Appl. Environ. Microbiol.* 62: 1863–1872.

Hulth, S., R. C. Aller, F. Gilbert. 1999. Coupled anoxic nitrification/manganese reduction in marine sediments. *Geochim. Cosmochim. Acta* 63: 49–66.

Ishaque, M., and M. I. H. Aleem. 1973. Intermediates of denitrification in the chemoautotroph *Thiobacillus denitrificans*. *Arch. Microbiol.* 94: 269–282.

Jahnke, R. A. 1996. The global ocean flux of particulate organic carbon: Areal distribution and magnitude. *Global Biogeochem. Cycles* 10: 71–88.

Jenkins, M. C., and W. M. Kemp. 1984. The coupling of nitrification and denitrification in two estuarine sediments. *Limnol. Oceanogr.* 29: 609–619.

Jensen, K., N. P. Sloth, N. Risgaard-Petersen, S. Rysgaard, and N. P. Revsbech. 1994. Estimation of nitrification and denitrification from microprofiles of oxygen and nitrate in model sediment systems. *Appl. Environ. Microbiol.* 60: 2094–2100.

Jensen, M. M., M. Holmer, and B. Thamdrup. 2005. Composition and diagenesis of neutral carbohydrates in sediments of the Baltic–North Sea transition. *Geochim. Cosmochim. Acta* 69: 4085–4099.

Jensen, M. M., B. Thamdrup, and T. Dalsgaard. 2007. Effects of specific inhibitors on anammox and denitrification in a marine sediment. *Appl. Environ. Microbiol.* 73: 3151–3158.

Jensen, M. M., M. M. M. Kuypers, G. Lavik, and B. Thamdrup. 2008. Rates and regulation of anaerobic ammonium oxidation and denitrification in the Black Sea. *Limnol. Oceanogr.* 53: 23–36.

Jetten, M. S. M. 2001. New pathways for ammonia conversion in soil and aquatic systems. *Plant Soil* 230: 9–19.

Jetten, M. S. M., O. Sliekers, M. Kuypers, et al. 2003. Anaerobic ammonium oxidation by marine and freshwater planctomycete-like bacteria. *Appl. Microbiol. Biotechnol.* 63: 107–114.

Jørgensen, K. S., H. B. Jensen, and J. Sørensen. 1984. Nitrous oxide production from nitrification and denitrification in marine sediment at low oxygen concentrations. *Can. J. Microbiol.* 30: 1073–1078.

Joye, S. B., and J. T. Hollibaugh. 1995. Influence of sulfide inhibition of nitrification on nitrogen regeneration in sediments. *Science* 270: 623–625.

Kampschreur, M. J., N. C. G. Tan, C. Picioreanu, M. S. M. Jetten, I. Schmidt, and M. C. M. van Loosdrecht. 2006. Role of nitrogen oxides in the metabolism of ammonia-oxidizing bacteria. *Biochem. Soc. Trans.* 34: 179–181.

Kartal, B., M. M. M. Kuypers, G. Lavik, J. Schalk, H. Op den Camp, M. S. M. Jetten, and M. Strous. 2007a. Anammox bacteria disguised as denitrifiers: Nitrate reduction to dinitrogen gas via nitrite and ammonium. *Environ. Microbiol.* 9: 635–642.

Kartal, B., J. Rattray, L. A. van Niftrik, et al. 2007b. Candidatus "*Anammoxoglobus propionicus*" a new propionate oxidizing species of anaerobic ammonium oxidizing bacteria. *Syst. Appl. Microbiol.* 30: 39–49.

Keith, S. M., G. T. Macfarlane, and R. A. Herbert. 1982. Dissimilatory nitrate reduction by a strain of *Clostridium butyricum* isolated from estuarine sediments. *Arch. Microbiol.* 132: 62–66.

King, G. M. 2006. Nitrate-dependent anaerobic carbon monoxide oxidation by aerobic CO-oxidizing bacteria. *FEMS Microbiol. Ecol.* 56: 1–7.

Knowles, R. 1982. Denitrification. *Microbiol. Rev.* 46: 43–70.

Könneke, M., A. E. Bernhar, J. R. de la Torre, C. B. Walker, J. B. Waterbury, and D. A. Stahl. 2005. Isolation of an autotrophic ammonia-oxidizing marine archaeon. *Nature* 437: 543–546.

Kristensen, E., and K. Hansen. 1995. Decay of plant detritus in organic-poor marine sediment—production-rates and stoichiometry of dissolved C-compounds and N-compounds. *J. Mar. Res.* 53: 675–702.

Kristensen, E., A. H. Devol, and H. E. Hartnett. 1999. Organic matter diagenesis in sediments on the continental shelf and slope of the Eastern Tropical and temperate North Pacific. *Continent. Shelf Res.* 19: 1331–1351.

Kuypers, M. M. M., A. O. Sliekers, G. Lavik, M. Schmid, B. B. Jørgensen, J. G. Kuenen, J. S. S. Damsté, M. Strous, and M. S. M. Jetten. 2003. Anaerobic ammonium oxidation by anammox bacteria in the Black Sea. *Nature* 422: 608–611.

Kuypers, M. M. M., G. Lavik, D. Woebken, M. Schmid, B. M. Fuchs, R. Amann, B. B. Jørgensen, and M. S. M. Jetten. 2005. Massive nitrogen loss from the Benguela upwelling system through anaerobic ammonium oxidation. *Proc. Natl. Acad. Sci. USA* 102: 6478–6483.

Kuypers, M. M. M., G. Lavik, and B. Thamdrup. 2006. Anaerobic ammonium oxidation in the marine environment. In L. N. Neretin (ed.), *Past and Present Water Column Anoxia*. Springer-Verlag, pp. 311–335.

Lam, P., M. M. Jensen, G. Lavik, D. F. McGinnis, B. Müller, C. J. Schubert, R. Amann, B. Thamdrup, and M. M. M. Kuypers. 2007. Linking crenarchaeal and bacterial nitrification to anammox in the Black Sea. *Proc. Natl. Acad. Sci. USA* 14: 7104–7109.

Laursen, A. E., and S. P. Seitzinger. 2002. The role of denitrification in nitrogen removal and carbon mineralization in Mid-Atlantic Bight sediments. *Continent. Shelf Res.* 22: 1397–1416.

Lipski, A., E. Spieck, A. Makolla, and K. Altendorf. 2001. Fatty acid profiles of nitrite-oxidizing bacteria reflect their phylogenetic heterogeneity. *Syst. Appl. Microbiol.* 24: 377–384.

Liu, X. D., S. M. Tiquia, G. Holguin, et al. 2003. Molecular diversity of denitrifying genes in continental margin sediments within the oxygen-deficient zone off the Pacific coast of Mexico. *Appl. Environ. Microbiol.* 69: 3549–3560.

Lomstein, B. A., T. H. Blackburn, and K. Henriksen. 1989. Aspects of nitrogen and carbon cycling in the northern Bering Shelf sediment. 1: The significance of urea turnover in the Mineralization of NH_4^+. *Mar. Ecol. Prog. Ser.* 57: 237–247.

Lomstein, B. A., A. G. U. Jensen, J. W. Hansen, J. B. Andreasen, L. S. Hansen, J. Berntsen, and H. Kunzendorf. 1998. Budgets of sediment nitrogen and carbon cycling in the shallow water of Knebel Vig, Denmark. *Aquat. Microb. Ecol.* 14: 69–80.

Lorenzen, J., L. H. Larsen, T. Kjær, and N. P. Revsbech. 1998. Biosensor determination of the microscale distribution of nitrate, nitrate assimilation, nitrification, and denitrification in a diatom-inhabited freshwater sediment. *Appl. Env. Microbiol.* 64: 3264–3269.

Lovell, C. R., Y. M. Piceno, J. M. Quattro, and C. E. Bagwell. 2000. Molecular analysis of diazotroph diversity in the rhizosphere of the smooth cordgrass, *Spartina alterniflora*. *Appl. Environ. Microbiol.* 66: 3814–3822.

Luther, G. W., III, B. Sundby, B. L. Lewis, and P. J. Brendel. 1997. Interactions of manganese with the nitrogen cycle: alternative pathways to dinitrogen. *Geochim. Cosmochim. Acta* 61: 4043–4052.

Macfarlane, G. T., and R. A. Herbert. 1982. Nitrate dissimilation by *Vibrio* spp. isolated from estuarine sediments. *J. Gen. Microbiol.* 128: 2463–2468.

Mayer, L. M., and D. L. Rice. 1992. Early diagenesis of protein—a seasonal study. *Limnol. Oceanogr.* 37: 280–295.

Meyer, R. L., T. Kjaer, and N. P. Revsbech. 2001. Use of NO_x^- microsensors to estimate the activity of sediment nitrification and NO_x^- consumption along an estuarine salinity, nitrate, and light gradient. *Aquatic Microb. Ecol.* 26: 181–193.

Meyer, R. L., N. Risgaard-Petersen, and D. E. Allen. 2005. Correlation between anammox activity and microscale distribution of nitrite in a subtropical mangrove sediment. *Appl. Environ. Microbiol.* 71: 6142–6149.

Middelburg, J. J., K. Soetaert, P. M. J. Herman, and C. H. R. Heip. 1996. Denitrification in marine sediments: A model study. *Global Biogeochem. Cycles* 10: 661–673.

Mulder, A., A. A. Van De Graaf, L. A. Robertson, and J. G. Kuenen. 1995. Anaerobic ammonium oxidation discovered in a denitrifying fluidized bed reactor. *FEMS Microbiol. Ecol.* 16: 177–184.

Munksby, N., M. Benthien, and R. N. Glud. 2002. Flow-induced flushing of relict tube structures in the central Skagerrak (Norway). *Mar. Biol.* 141: 939–945.

Nelson, P. N., and J. A. Baldock. 2005. Estimating the molecular composition of a diverse range of natural organic materials from solid-state C-13 NMR and elemental analyses. *Biogeochemistry* 72: 1–34.

Nicolaisen, M. H., and N. B. Ramsing. 2002. Denaturing gradient gel electrophoresis (DGGE) approaches to study the diversity of ammonia-oxidizing bacteria. *J. Microbiol. Meth.* 50: 189–203.

Nielsen, K., N. Risgaard-Petersen, B. Sømod, S. Rysgaard, and T. Bergo. 2001. Nitrogen and phosphorus retention estimated independently by flux measurements and dynamic modelling in the estuary, Randers Fjord, Denmark. *Mar. Ecol. Prog. Ser.* 219: 25–40.

Nielsen, L. P. 1992. Denitrification in sediment determined from nitrogen isotope pairing. *FEMS Microbiol. Ecol.* 86: 357–362.

Nielsen, L. P., P. B. Christensen, N. P. Revsbech, and J. Sorensen. 1990. Denitrification and oxygen respiration in biofilms studied with a microsensor for nitrous oxide and oxygen. *Microb. Ecol.* 19: 63–72.

Nielsen, M., L. H. Larse, M. S. M. Jetten, and N. P. Revsbech. 2004. Bacterium-based NO_2^- biosensor for environmental applications. *Appl. Environ. Microbiol.* 70: 6551–6558.

Niggemann, J., and C. J. Schubert. 2006. Sources and fate of amino sugars in coastal Peruvian sediments. *Geochim. Cosmochim. Acta* 70: 2229–2237.

Nishio, T., I. Koike, and A. Hattori. 1982. Denitrification, nitrate reduction, and oxygen consumption in coastal and estuarine sediments. *Appl. Environ. Microbiol.* 43: 648–653.

Nishio, T., I. Koike, and A. Hattori. 1983. Estimates of denitrification and nitrification in coastal and estuarine sediments. *Appl. Environ. Microbiol.* 45: 444–450.

REFERENCES

Nixon, S. W., J. W. Ammerman, L. P. Atkinson, et al. 1996. The fate of nitrogen and phosphorus at the land sea margin of the North Atlantic Ocean. *Biogeochemistry* 35: 141–180.

Nogales, B., K. N. Timmis, D. B. Nedwell, and A. M. Osborn. 2002. Detection and diversity of expressed denitrification genes in estuarine sediments after reverse transcription–PCR amplification from mRNA. *Appl. Environ. Microbiol.* 68: 5017–5025.

Nold, S. C., J. Z. Zhou, A. H. Devol, and J. M. Tiedje. 2000. Pacific northwest marine sediments contain ammonia-oxidizing bacteria in the beta subdivision of the Proteobacteria. *Appl. Environ. Microbiol.* 66: 4532–4535.

Otte, S., J. G. Kuenen, L. P. Nielsen, et al. 1999. Nitrogen, carbon, and sulfur metabolism in natural *Thioploca* samples. *Appl. Environ. Microbiol.* 65: 3148–3157.

Ottley, C. J., W. Davison, and W. M. Edmunds. 1997. Chemical catalysis of nitrate reduction by iron(II). *Geochim. Cosmochim. Acta* 61: 1819–1828.

Pantoja, S., and C. Lee. 2003. Amino acid remineralization and organic matter lability in Chilean coastal sediments. *Org. Geochem.* 34: 1047–1056.

Payne, W. J. 1981. *Denitrification.* Wiley, New York.

Pedersen, A. G. U., J. Berntsen, and B. A. Lomstein. 1999. The effect of eelgrass decomposition on sediment carbon and nitrogen cycling: A controlled laboratory experiment. *Limnol. Oceanogr.* 44: 1978–1992.

Pedersen, H., B. A. Lomstein, and T. H. Blackburn. 1993. Evidence for bacterial urea production in marine ssediments. *FEMS Microbiol. Ecol.* 12: 51–59.

Pelegri, S. P., and T. H. Blackburn. 1994. Bioturbation effects of the amphipod *Corophium volutator* on microbial nitrogen transformations in marine sediments. *Mar. Biol.* 121: 253–258.

Postgate, J. R. 1998. *Nitrogen Fixation*, 3rd. edn. Cambridge University Press.

Poth, M., and D. D. Focht. 1985. N-15 kinetic-analysis of N_2O production by *Nitrosomonas europaea*—an examination of nitrifier denitrification. *Appl. Environ. Microbiol.* 49: 1134–1141.

Prokopenko, M. G., D. E. Hammond, W. M. Berelson, J. M. Bernhard, L. Stott, and R. Douglas. 2006. Nitrogen cycling in the sediments of Santa Barbara basin and Eastern Subtropical North Pacific: Nitrogen isotopes, diagenesis and possible chemosymbiosis between two lithotrophs (*Thioploca* and Anammox)—"riding on a glider." *Earth Planet. Sci. Lett.* 242: 186–204.

Raghoebarsing, A. A., A. Pol, K. T. Van De Pas-Schoonen, et al. 2006. A microbial consortium couples anaerobic methane oxidation to denitrification. *Nature* 440: 918–921.

Redfield, A. C., B. H. Ketchum, and F. A. Richards. 1963. The influence of organisms on the composition of sea-water. In M. N. Hill (ed.), *The Sea*, Vol. 2. Interscience, pp. 26–77.

Rich, J. J., O. R. Dale, B. Song, and B. B. Ward. 2008. Anaerobic ammonium oxidation (Anammox) in Chesapeake Bay sediments. *Microb. Ecol.* 55: 311–320.

Richards, F. A. 1965a. Anoxic basins and fjords. In J. P. Riley and G. Skirrow (eds.), *Chemical Oceanography*. Academic Press, pp. 611–645.

Richards, F. A. 1965b. Chemical observations in some anoxic, sulfide-bearing basins and fjords. In E. A. Pearson (ed.), *Advances in Water Pollution Research*, Pergamon Press, pp. 215–232.

Richardson, D. J., J. M. Wehrfritz, A. Keech, et al. 1998. The diversity of redox proteins involved in bacterial heterotrophic nitrification and aerobic denitrification. *Biochem. Soc. Trans.* 26: 401–408.

Risgaard-Petersen, N. 2003. Coupled nitrification–denitrification in autotrophic and heterotrophic estuarine sediments: On the influence of benthic microalgae. *Limnol. Oceanogr.* 48: 93–105.

Risgaard-Petersen, N., and S. Rysgaard. 1995. Nitrate reduction in sediments and waterlogged soil measured by ^{15}N techniques. In K. Alef and P. Nannipieri (eds.), *Anaerobic Microbial Activities in Soil*. Academic Press, pp. 287–310.

Risgaard-Petersen, N., L. P. Nielsen, S. Rysgaard, T. Dalsgaard, and R. L. Meyer. 2003. Application of the isotope pairing technique in sediments where anammox and denitrification coexist. *Limnol. Oceanogr. Meth.* 1: 63–73.

Risgaard-Petersen, N., R. L. Meyer, M. Schmid, M. S. M. Jetten, A. Enrich-Prast, S. Rysgaard, and N. P. Revsbech. 2004. Anaerobic ammonium oxidation in an estuarine sediment. *Aquat. Microb. Ecol.* 36: 293–304.

Risgaard-Petersen, N., R. L. Meyer, and N. P. Revsbech. 2005. Denitrification and anaerobic ammonium oxidation in sediments: effects of microphytobenthos and NO_3^-. *Aquat. Microb. Ecol.* 40: 67–76.

Risgaard-Petersen, N., A. M. Langezaal, S. Ingvardsen, et al. 2006. Evidence for complete denitrification in a benthic foraminifer. *Nature* 443: 93–96.

Robertson, L. A., and J. G. Kuenen. 1984. Aerobic denitrification: A controversy revived. *Arch. Microbiol.* 139: 351–354.

Robertson, L. A., and J. G. Kuenen. 1990. Combined heterotrophic nitrification and aerobic denitrification in *Thiosphaera pantotropha* and other bacteria. *Antonie van Leeuwenhoek* 57: 139–152.

Robertson, L. A., T. Dalsgaard, N. P. Revsbech, and J. G. Kuenen. 1995. Confirmation of aerobic denitrification in batch cultures, using gas-chromatography and N-15 mass-spectrometry. *FEMS Microbiol. Ecol.* 18: 113–119.

Robinson, A. D., D. B. Nedwell, R. M. Harrison, and B. G. Ogilvie. 1998. Hypernutrified estuaries as sources of N_2O emission to the atmosphere: the estuary of the River Colne, Essex, UK. *Mar. Ecol. Prog. Ser.* 164: 59–71.

Rysgaard, S., N. Risgaard-Petersen, L. P. Nielsen, and N. P. Revsbech. 1993. Nitrification and denitrification in lake and estuarine sediments measured by the ^{15}N dilution technique and isotope pairing. *Appl. Environ. Microbiol.* 59: 2093–2098.

Rysgaard, S., N. Risgaard-Petersen, N. P. Sloth, K. Jensen, and L. P. Nielsen. 1994. Oxygen regulation of nitrification and denitrification in sediments. *Limnol. Oceanogr.* 39: 1643–1652.

Rysgaard, S., P. B. Christensen, and L. P. Nielsen. 1995. Seasonal-variation in nitrification and denitrification in estuarine sediment colonized by benthic microalgae and bioturbating infauna. *Mar. Ecol. Prog. Ser.* 126: 111–121.

Rysgaard, S., N. Risgaard-Petersen, and N. P. Sloth. 1996. Nitrification, denitrification, and nitrate ammonification in sediments of two coastal lagoons in southern France. *Hydrobiologia* 329: 133–141.

Rysgaard, S., B. Thamdrup, N. Risgaard-Petersen, H. Fossing, P. Berg, P. B. Christensen, and T. Dalsgaard. 1998. Seasonal carbon and nutrient mineralization in a high-Arctic coastal marine sediment, Young Sound, Northeast Greenland. *Mar. Ecol. Prog. Ser.* 175: 261–276.

Rysgaard, S., T. G. Nielsen, and B. W. Hansen. 1999. Seasonal variation in nutrients, pelagic primary production and grazing in a high-Arctic coastal marine ecosystem, Young Sound, Northeast Greenland. *Mar. Ecol. Prog. Ser.* 179: 13–25.

Rysgaard, S., H. Fossing, and M. M. Jensen. 2001. Organic matter degradation through oxygen respiration, denitrification, and manganese, iron, and sulfate reduction in marine sediments (the Kattegat and the Skagerrak). *Ophelia* 55: 77–91.

Rysgaard, S., R. N. Glud, N. Risgaard-Peterssen, and T. Dalsgaard. 2004. Denitrification and anammox activity in Arctic marine sediments. *Limnol. Oceanogr.* 49: 1493–1502.

Sayama, M., N. Risgaard-Petersen, L. P. Nielsen, H. Fossing, and P. B. Christensen. 2005. Impact of bacterial NO_3^- transport on sediment biogeochemistry. *Appl. Environ. Microbiol.* 71: 7575–7577.

Scala, D. J., and L. J. Kerkhof. 1999. Diversity of nitrous oxide reductase (*nosZ*) genes in continental shelf sediments. *Appl. Environ. Microbiol.* 65: 1681–1687.

Schalk, J., H. Oustad, J. G. Kuenen, and M. S. M. Jetten. 1998. The anaerobic oxidation of hydrazine: a novel reaction in microbial nitrogen metabolism. *FEMS Microbiol. Lett.* 158: 61–67.

Schleper, C., G. Jurgens, and M. Jonuscheit. 2005. Genomic studies of uncultivated archaea. *Nature Rev. Microbiol.* 3: 479–488.

Schmid, M. C., N. Risgaard-Petersen, J. van de Vossenberg, et al. 2007. Anaerobic ammonium-oxidizing bacteria in marine environments: Widespread occurrence but low diversity. *Environ. Microbiol.* 9: 1476–1484.

Schmidt, I., and E. Bock. 1997. Anaerobic ammonia oxidation with nitrogen dioxide by *Nitrosomonas eutropha*. *Arch. Microbiol.* 167: 106–111.

Schmidt, I., E. Bock, and M. S. M. Jetten. 2001. Ammonia oxidation by *Nitrosomonas eutropha* with NO_2 as oxidant is not inhibited by acetylene. *Microbiol. SGM* 147: 2247–2253.

Schmidt, I., R. J. M. Van Spanning, and M. S. M. Jetten. 2004. Denitrification and ammonia oxidation by *Nitrosomonas europaea* wild-type, and NirK- and NorB-deficient mutants. *Microbiol. SGM* 150: 4107–4114.

Schulz, H. D., A. Dahmke, U. Schinzel, K. Wallmann, and M. Zabel. 1994. Early diagenetic processes, fluxes, and reaction rates in sediments of the South Atlantic. *Geochim. Cosmochim. Acta* 58: 2041–2060.

Seitzinger, S. P. 1988. Denitrification in freshwater and coastal marine ecosystems: ecological and geochemical significance. *Limnol. Oceanogr.* 33: 702–724.

Senga, Y., K. Mochida, R. Fukumori, N. Okamoto, and Y. Seike. 2006. N_2O accumulation in estuarine and coastal sediments: The influence of H_2S on dissimilatory nitrate reduction. *Estuar. Coast. Shelf Sci.* 67: 231–238.

Shaw, L. J., G. W. Nicol, Z. Smith, J. Fear, J. I. Prosser, and E. M. Baggs. 2006. *Nitrosospira* spp. can produce nitrous oxide via a nitrifier denitrification pathway. *Environ. Microbiol.* 8: 214–222.

Simon, J. 2002. Enzymology and bioenergetics of respiratory nitrite ammonification. *FEMS Microbiol. Rev.* 26: 285–309.

Sliekers, A. O., N. Derwort, J. L. C. Gomez, M. Strous, J. G. Kuenen, and M. S. M. Jetten. 2002. Completely autotrophic nitrogen removal over nitrite in one single reactor. *Water Res.* 36: 2475–2482.

Smith, C. J., D. B. Nedwell, L. F. Dong, and A. M. Osborn. 2007. Diversity and abundance of nitrate reductase genes (*narG* and *napA*), nitrite reductase genes (*nirS* and *nrfA*), and their transcripts in estuarine sediments. *Appl. Environ. Microbiol.* 73: 3612–3622.

Smith, R. L., M. L. Ceazan, and M. H. Brooks. 1994. Autotrophic, hydrogen-oxidizing, denitrifying bacteria in groundwater, potential agents for bioremediation of nitrate contamination. *Appl. Env. Microbiol.* 60: 1949–1955.

Söderlund, R., and T. Rosswall. 1982. The nitrogen cycles. In O. Huntziger (ed.), *The Natural Environment and the Biogeochemical Cycles*. Springer-Verlag, p. 70.

Sørensen, J. 1978. Occurence of nitric and nitrous oxides in coastal marine sediment. *Appl. Environ. Microbiol.* 36: 809–813.

Sørensen, J., and L. Thorling. 1991. Stimulation by lepidocrocite (γ-FeOOH) of Fe(II)-dependent nitrite reduction. *Geochim. Cosmochim. Acta* 55: 1289–1294.

Sørensen, J., J. M. Tiedje, and R. B. Firestone. 1980. Inhibition by sulfide of nitric and nitrous oxide reduction by denitrifying *Pseudomonas flourescens*. *Appl. Environ. Microbiol.* 39: 105–108.

Stolz, J. F., and P. Basu. 2002. Evolution of nitrate reductase: Molecular and structural variations on a common function. *ChemBioChem* 3: 198–206.

Straub, K. L., M. Benz, B. Schink, and F. Widdel. 1996. Anaerobic, nitrate-dependent microbial oxidation of ferrous iron. *Appl. Environ. Microbiol.* 62: 1458–1460.

Strohm, T. O., B. Griffin, W. G. Zumft, and B. Schink. 2007. Growth yields in bacterial denitrification and nitrate ammonification. *Appl. Environ. Microbiol.* 73: 1420–1424.

Strous, M., E. Vangerven, J. G. Kuenen, and M. Jetten. 1997. Effects of aerobic and microaerobic conditions on anaerobic ammonium-oxidizing (anammox) sludge. *Appl. Env. Microbiol.* 63: 2446–2448.

Strous, M., J. A. Fuerst, E. H. M. Kramer, et al. 1999. Missing lithotroph identified as new planctomycete. *Nature* 400: 446–449.

Strous, M., E. Pelletier, S. Mangenot, et al. 2006. Deciphering the evolution and metabolism of an anammox bacterium from a community genome. *Nature* 440: 790–794.

Sundbäck, K., F. Linares, F. Larson, A. Wulff, and A. Engelsen. 2004. Benthic nitrogen fluxes along a depth gradient in a microtidal fjord: The role of denitrification and microphytobenthos. *Limnol. Oceanogr.* 49: 1095–1107.

Takaya, N. 2002. Dissimilatory nitrate reduction metabolisms and their control in fungi. *J. Biosci. Bioeng.* 94: 506–510.

Tal, Y., J. E. M. Watts, and H. J. Schreier. 2005. Anaerobic ammonia-oxidizing bacteria and related activity in Baltimore inner Harbor sediment. *Appl. Environ. Microbiol.* 71: 1816–1821.

Tebo, B. M., H. A. Johnson, J. K. McCarthy, and A. S. Templeton. 2005. Geomicrobiology of manganese(II) oxidation. *Trends Microbiol.* 13: 421–428.

Teske, A., E. Alm, J. M. Regan, S. Toze, B. E. Rittmann, and D. A. Stahl. 1994. Evolutionary relationships among ammonia-oxidizing and nitrite-oxidizing bacteria. *J. Bacteriol.* 176: 6623–6630.

Thamdrup, B. 2000. Microbial manganese and iron reduction in aquatic sediments. *Adv. Microb. Ecol.* 16: 41–84.

Thamdrup, B., and T. Dalsgaard. 2000. The fate of ammonium in anoxic manganese oxide-rich marine sediment. *Geochim. Cosmochim. Acta* 64: 4157–4164.

Thamdrup, B., and T. Dalsgaard. 2002. Production of N_2 through anaerobic ammonium oxidation coupled to nitrate reduction in marine sediments. *Appl. Environ. Microbiol.* 68: 1312–1318.

Thamdrup, B., T. Dalsgaard, M. M. Jensen, O. Ulloa, L. Farias, and R. Escribano. 2006. Anaerobic ammonium oxidation in the oxygen-deficient waters off northern Chile. *Limnol. Oceanogr.* 51: 2145–2156.

Thauer, R. K., K. Jungermann, and K. Decker. 1977. Energy-conservation in chemotropic anaerobic bacteria. *Bacteriol. Rev.* 41: 100–180.

Therkildsen, M. S., and B. A. Lomstein. 1994. Seasonal variation in sediment urea turnover in a shallow estuary. *Mar. Ecol. Prog. Ser.* 109: 77–82.

Therkildsen, M. S., G. M. King, and B. A. Lomstein. 1996. Urea production and turnover following the addition of AMP, CMP, RNA and a protein mixture to a marine sediment. *Aquat. Microb. Ecol.* 10: 173–179.

Third, K. A., A. O. Sliekers, J. G. Kuenen, and M. S. M. Jetten. 2001. The CANON system (completely autotrophic nitrogen-removal over nitrite) under ammonium limitation: Interaction and competition between three groups of bacteria. *Syst. Appl. Microbiol.* 24: 588–596.

Tibbles, B. J., M. I. Lucas, V. E. Coyne, and S. T. Newton. 1994. Nitrogenase activity in marine-sediments from a temperate salt-marsh lagoon—modulation by complex polysaccharides, ammonium and oxygen. *J. Exp. Mar. Biol. Ecol.* 184: 1–20.

Tiedje, J. M., A. J. Sexstone, D. D. Myrold, and J. A. Robinson. 1982. Denitrification: Ecological niches, competition and survival. *Antonie van Leeuwenhoek* 48: 569–583.

Tobias, C. R., I. C. Anderson, E. A. Canuel, and S. A. Macko. 2001. Nitrogen cycling through a fringing marsh-aquifer ecotone. *Mar. Ecol. Prog. Ser.* 210: 25–39.

Trimmer, M., N. Risgaard-Petersen, J. C. Nicholls, and P. Engström. 2006. Direct measurement of anaerobic ammonium oxidation (anammox) and denitrification in intact sediment cores. *Mar. Ecol. Prog. Ser.* 326: 37–47.

Tuominen, L., K. Makela, K. K. Lehtonen, H. Haahti, S. Hietanen, and J. Kuparinen. 1999. Nutrient fluxes, porewater profiles and denitrification in sediment influenced by algal sedimentation and bioturbation by *Monoporeia affinis*. *Estuar. Coast. Shelf Sci.* 49: 83–97.

Tyler, A. C., K. J. Mcglathery, and I. C. Anderson. 2003. Benthic algae control sediment-water column fluxes of organic and inorganic nitrogen compounds in a temperate lagoon. *Limnol. Oceanogr.* 48: 2125–2137.

Usui, T., I. Koike, and N. Ogura. 1998. Vertical profiles of nitrous oxide and dissolved oxygen in marine sediments. *Mar. Chem.* 59: 253–270.

Usui, T., I. Koike, and N. Ogura. 2001. N_2O production, nitrification and denitrification in an estuarine sediment. *Estuar. Coast. Shelf Sci.* 52: 769–781.

VanBriesen, J. M. 2002. Evaluation of methods to predict bacterial yield using thermodynamics. *Biodegradation* 13: 171–190.

van de Graaf, A., A. Mulder, P. De Brujin, M. S. M. Jetten, L. A. Robertson, and J. G. Kuenen. 1995. Anaerobic oxidation of ammonium is a biologically mediated process. *Appl. Env. Microbiol.* 61: 1246–1251.

van Duyl, F. C., W. van Raaphorst, and A. J. Kop. 1993. Benthic bacterial production and nutrient sediment–water exchange in sandy North-Sea sediments. *Mar. Ecol. Prog. Ser.* 100: 85–95.

van Niftrik, L. A., J. A. Fuerst, J. S. S. Damste, J. G. Kuenen, M. S. M. Jetten, and M. Strous. 2004. The anammoxosome: An intracytoplasmic compartment in anammox bacteria. *FEMS Microbiol. Lett.* 233: 7–13.

Wagner, M., G. Rath, H. P. Koops, J. Flood, and R. Amann. 1996. In situ analysis of nitrifying bacteria in sewage treatment plants. *Water Sci. Technol.* 34: 237–244.

Ward, B. B. 2000. Nitrification in the marine nitrogen cycle. In D. L. Kirchman (ed.), *Microbial Ecology of the Oceans*, 1st edn. Wiley-Liss, pp. 427–453.

Ward, B. B. 2007. Nitrogen cycling in aquatic environments. In C. J. Hurst, R. L. Crawford, J. L. Garland, D. A. Lipson, A. L. Mills, and L. D. Stetzenbach (eds.), *Manual of Environmental Microbiology*, 3rd edn. ASM Press, pp. 511–522.

Ward, B. B., and G. D. O'Mullan. 2002. Worldwide distribution of *Nitrosococcus oceani*, a marine ammonia-oxidizing gamma-proteobacterium, detected by PCR and sequencing of 16S rRNA and *amo*A genes. *Appl. Environ. Microbiol.* 68: 4153–4157.

Weber, K. A., J. Pollock, K. A. Cole, S. M. O'Connor, L. A. Achenbach, and J. D. Coates. 2006a. Anaerobic nitrate-dependent iron(II) bio-oxidation by a novel lithoautotrophic betaproteobacterium, strain 2002. *Appl. Env. Microbiol.* 72: 686–694.

Weber, K. A., M. M. Urrutia, P. F. Churchill, R. K. Kukkadapu, and E. E. Roden. 2006b. Anaerobic redox cycling of iron by freshwater sediment microorganisms. *Environ. Microbiol.* 8: 100–113.

Wood, P. M. 1986. Nitrification as a bacterial energy source. In J. I. Prosser (ed.), *Nitrification*. IRL Press, pp. 39–62.

Wrage, N., G. L. Velthof, M. L. Van Beusichem, and O. Oenema. 2001. Role of nitrifier denitrification in the production of nitrous oxide. *Soil Biol. Biochem.* 33: 1723–1732.

Zehr, J. P., B. D. Jenkins, S. M. Short, and G. F. Steward. 2003. Nitrogenase gene diversity and microbial community structure: A cross-system comparison. *Environ. Microbiol.* 5: 539–554.

Zopfi, J., T. Kjaer, L. P. Nielsen, and B. B. Jorgensen. 2001. Ecology of *Thioploca* spp.: Nitrate and sulfur storage in relation to chemical microgradients and influence of *Thioploca* spp. on the sedimentary nitrogen cycle. *Appl. Environ. Microbiol.* 67: 5530–5537.

Zumft, W. G. 1997. Cell biology and molecular basis of denitrification. *Microbiol. Mol. Biol. Rev.* 61: 533–616.

INDEX

ABC. *See* Adenosine triphosphate binding cassette (ABC)
Actinobacteria
  bacterial structure and patterns, 58
  comparative genomic analysis, 110–118
  genomes, 108
  geographical distribution, 58
  nitrogen sources, 356
  size, 410
Actinomycetes, 58
Activity probes, 251–253
  cellular components, 250
Activity protocols, 267–269
Adenosine triphosphate (ATP), 5, 92, 143, 211, 339, 482
Adenosine triphosphate binding cassette (ABC), 339
Aerobic anoxygenic phototrophic bacteria, 5, 131, 132
  abundance, 142
  diversity, 139–141
  ecological significance, 142
  energy production pigment, 6
  habitat, 139–141
  Pacific Ocean, 368
  physiology, 140
  rediscovery, 139
Aeropyrum pernix K1
  genomes, 94
Agreia sp. PHSC20c1
  comparative genomic analysis, 110–118

Alcaligenes faecalis
  denitrification, 539
Allochromatium vinosum, 150
Allopatry, 422
Aloha. *See* North Pacific Subtropical Gyre (Station Aloha)
Alphaproteobacteria, 55, 59, 217
  comparative genomic analysis, 110–118
  DFAA, 231
  DOM, 498
  MAR-FISH, 275
  nitrogen fixation, 362
  proteorhodopsins, 144
  resource control, 369
  seasonal variation, 73
Alternative lifestyles and heterotrophs, 170–171
Alteromonadales
  comparative genomic analysis, 110–118
Alteromonas, 51
  comparative genomic analysis, 110–118
  macleodii, 52, 64
Amino acids. *See also* Dissolved free amino acids (DFAA)
  photoheterotrophic marine prokaryotes, 134
  uptake, 134
Ammonia, 363–364

Microbial Ecology of the Oceans, Second Edition. Edited by David L. Kirchman
Copyright © 2008 John Wiley & Sons, Inc.

Ammonium, 533–534
　ammonification, 534
　benthic nitrogen exchange, 550–551
　depth distribution, 538
　oxidation, 536, 538
　porewater distribution, 548
Ammonium monooxygenase (AMO), 535–536
AMO. *See* Ammonium monooxygenase (AMO)
Amoeboids, 388–389
Amoebophrya, 16
Anabaena
　detection, 493
　gerdii, 500
　microbial associations, 499
　nitrogenase, 485
　nitrogen fixation, 498
　variabilis, 485
Anaerobic Ammonium Oxidation (anammox), 20
　benthic denitrification, 530
　nitrogen production, 546–548, 554
Anaerobic anoxygenic phototrophs, 132
Anammox. *See* Anaerobic Ammonium Oxidation (anammox)
Anaplerotic reactions, 142
Anoxygenic phototrophs, 132
Antarctica flagellate abundance, 392
Aphanizomenon, 55
　detection, 493
　nitrogen fixation, 489
Apusomonads, 389
Apusomonas, 395
Arabian Sea
　BCD and PP, 348
　denitrifiers, 20
Arafura Sea, 78
Archaea. *See also* Bacterial community structure and patterns
　vs. bacteria, 13
　community structure and patterns, 45–79
　oceanic discovery, 13–14, 60–61
Archaea
　denitrification, 539
　DOM, 231
　nitrification, 103
　nitrogen fixation, 489
　phylogeny, 48

Archaebacteria, 13
Archaeoglobus fulgidus DSM4304
　genomes, 94
Arctic Ocean
　nutrient-dose experiments, 359
　picobiliphytes, 177
ARISA. *See* Automated rRNA Intergenic Spacer Analysis (ARISA)
Atlantic Ocean. *See also* North Atlantic
　bacterial respiration, 312
　BGE, 315
　plankton measured and predicted activity, 323
ATP. *See* Adenosine triphosphate (ATP)
Attached bacteria
　deep waters, 225
　total bacterial abundance and production, 226–227
Aurantimonas sp. SI85-9A1
　comparative genomic analysis, 110–118
Aureococcus
　anophagefferens, 169, 193
　HMW, 169
Automated rRNA Intergenic Spacer Analysis (ARISA), 66, 77
　North Atlantic, 78
Autoradiography, 259
Autotrophic bacteria
　abundance, 446
Autotrophic organisms, 52
Azam, Farooq, 33, 34, 36
Azotobacter
　nitrogen fixation, 489, 495

Bacillus sp. NRRL B-14911
　comparative genomic analysis, 110–118
Bacteria. *See also* Cyanobacteria; Dissolved organic matter (DOM) bacteria interactions; Heterotrophic bacteria; Marine bacteria; Particulate organic matter (POM), bacteria interactions; Seawater organic matter-bacteria interactions
　abundance, 226–227, 343–346
　assemblages, 281, 408–410
　attached, 225–227
　autotrophic, 446
　BGE, 281
　bound nutrients, 407

bulk growth rate, 281
cell abundance and size, 10
cell physiological state, 261–262
cell size, 409–410
chlorophyll, 343–346
classically culturable, 49–54
deep waters, 225
denitrification, 539
DOM, 231
flagellate interaction, 413
fluorescently labeled, 401–402
generation times, 13
gene role category distribution, 99
heat-killed bacteria, 401–402
marine bacterioplankton physiological structure, 261–262
motility, 410–411
nitrification, 103
nitrogen fixation, 489
nitrogen-metabolizing identification, 540
organic matter utilization, 230
photoheterotrophic, 5–6, 52
phylogeny, 48
production, 226–227
protective mechanism, 409–410
proteorhodopsin-bearing, 5, 143–150
remineralize, 407
resistance mechanisms, 409
shaping, 408–410
single-cell activity, 261–262
virus predation, 453
Bacterial carbon demand (BCD), 300, 348
  Arabian Sea, 348
  euphotic zone rates, 348
Bacterial community composition (BCC), 414
  schematic diagram, 70
  top-down control, 451
  viruses controlling, 467
Bacterial community structure and patterns, 45–79, 230
  Actinobacteria, 58
  bacterioplankton diversity, 63–65
  Bacteroidetes, 52
  Betaproteobacteria, 59
  bottom-up control, 68
  classically culturable *Bacteria*, 49–54
  cyanobacteria, 52–54
  description and factors, 67–71
  Gammaproteobacteria, 51

global distribution, 75
kill the winner hypothesis, 71
latitudinal gradient and degree of endemism, 76
marine gammaproteobacterial clusters, 57
marine group A, 59
marine group B, 59
microdiversity, 64
microscale patterns, 74
not-yet-cultured bacteria, 57–59
patchiness and large eddies, 77–78
richness and evenness, 65
Roseobacter Clade of Marine *Alphaproteobacteria*, 50
SAR11 cluster, 55–56
SAR116 cluster, 59
seasonal variation, 72–73
sea water culturable bacteria, 55–56
short-term variation, 72
sideways control, 69
spatial variation, 74–78
species concept, 63
temporal variation, 72–73
top-down control, 70
Bacterial diversity and community dynamics impact on microbial processes, 452–456
Bacterial growth efficiency (BGE), 12, 300, 315, 348
  bacterial assemblage, 281
  CTC positive cells, 283
  equations, 316
Bacterial production (BP), 11, 348, 406–407
  bacterial respiration, 314
  measurement scale, 314
  vs. primary production, 9
  vs. protistan grazing, 404
Bacterial respiration (BR), 321, 406–407. *See also* Heterotrophic bacterial respiration
  bacterial abundance, 312
  bacterial production, 314
  BGE, 306–307
  community respiration, 316, 321
  deep waters, 309–310
  DOC, 11, 312
  equations, 316–318
  measurement scale, 314
  seasonal variation, 308
  temperature, 311

Bacteriastrum
  symbiotic relationships, 499
Bacteriochlorophylls, 140
Bacterioplankton. *See also* Marine
    bacterioplankton
  ALOHA, 338
  assemblages physiologic structure, 245
  biomass, 347
  biomass *vs.* living surface area, 34
  Chesapeake Bay, 122
  distribution, 76, 245
  diversity, 63–65
  mass and production, 344
  physiological structure, 246
  population size, 350
  rank-abundance distributions, 68
  starvation, dormancy and viability, 246
  taxonomic-trophic compartments, 35
Bacterivorous flagellates
  functional ecology, 394–396
  HNF culture experiments, 397–400
  marine bacterioplankton protistan grazing, 394–396
Bacterivorous nanoflagellates
  feeding types, 395
Bacteroidetes, 48, 52, 57, 58, 69, 76, 106
  bacterial and archaeal community structure and patterns, 52
  comparative genomic analysis, 110–118
  environmental reductionism, 106
  leucine incorporation, 137
  MAR-FISH, 275
  resource control, 369
  seasonal variation, 73
Baltic Sea, 79
  nitrogen fixation, 54–55, 505–506
Bardach, John, 31
Barents Sea
  flagellate abundance, 392
  picoeukaryotes, 192
Bathycoccus, 186
  prasinos, 167
  Sargasso Sea, 186
BATS. *See* Bermuda Atlantic Times Series (BATS)
BCC. *See* Bacterial community composition (BCC)
BCD. *See* Bacterial carbon demand (BCD)

Beggiatoa
  nitrogen, 532, 544
Benthic budgets
  nitrogen budgets, 550–551
Benthic denitrification, 542
Benthic habitats
  ecophysiological aspects, 506
Bermuda Atlantic Times Series (BATS), 73, 309
  bacterioplankton mass and production, 344
  BCD and PP, 348
  chlorophyll, 344
  nitrogen fixation, 502, 505
  nutrient-dose experiments, 357
  nutrients, 344
  surface-water temperature, 344
Betaproteobacteria
  ammonium oxidation, 536
  bacterial and archaeal community structure and patterns, 59
  genomes, 108
  resource control, 369
BGE. *See* Bacterial growth efficiency (BGE)
Bicosoeca, 389, 395
Bicosoecids, 395
  published autecological laboratory studies, 398
Biogeochemical cycling
  impact on microbial processes, 450–451
Biogeographic barriers, 422
Bioinformatics, 92
Biological guilds
  abundance, 446
Blackman Limitation, 336
Black Sea
  denitrifiers, 20
Blanes Bay
  Alphaproteobacteria and *Bacteroidetes*
    seasonal variation, 73
  flagellate abundance, 392
BLAST, 98, 100, 463, 465
Blastopirellula marina DSM3645T
  comparative genomic analysis, 110–118
Blue-green algae. *See Cyanobacteria*
Blue light-absorbing proteorhodopsin (BPR), 145

Bodo
  designis, 399
  published autecological laboratory studies, 398
  saltans, 395
Bodonids, 388, 389
  published autecological laboratory studies, 398
Bolidomonas, 169
Bottom-up control, 343
  bacterial and archaeal community structure and patterns, 68
BP. *See* Bacterial production (BP)
BPR. *See* Blue light-absorbing proteorhodopsin (BPR)
BR. *See* Bacterial respiration (BR)

Caecitellus, 419
  parvulus, 398
  published autecological laboratory studies, 398
Cafeteria, 419
  published autecological laboratory studies, 398
  roenbergensis, 15
Calothrix
  cultivation, 490
  nitrogen fixation, 489
Candidatus Scalindua, 546
CANON. *See* Complete autotrophic nitrogen removal over nitrite (CANON)
Capillary electrophoresis–single-strand conformation polymorphism (CE–SSCP), 73
Carbohydrates
  DOM, 354–355
Carbon
  flow into bacterioplankton, 406–407
Carbon dioxide, 21
  cycle, 29
Carboxyl-rich alicyclic molecules (CRAM), 222
CARD-FISH, 175, 257
Cariaco Basin
  denitrifiers, 20
Caribbean
  nutrient-dose experiments, 357
Cellulophaga sp. MED134
  comparative genomic analysis, 110–118

Cenarchaeum symbiosum, 63, 103
Cercomonas, 395
CE-SSCP. *See* Capillary electrophoresis–single-strand conformation polymorphism (CE-SSCP)
Chaetoceros
  cultivation, 490
  symbiotic relationships, 499
Chemoautotrophs, 52
Chemoheterotrophs, 52
Chemolithoautotrophs, 14
Chemolithotrophs, 19
Chesapeake Bay
  bacterioplankton, 122
  nifH, 496, 497
Chlorobium
  nitrogen fixation, 489
Chlorophyll, 344
  bacterial abundance, 343–346
  picoeukaryotes taxa, 165
Choanoeca perplexa, 395
Choanoflagellates, 388, 389–390, 394
  published autecological laboratory studies, 398
Chromatium
  nitrogen fixation, 489
Chromosomes, 93
Chrysochromulina, 172
Chrysomonads
  published autecological laboratory studies, 398
Ciliates, 388
Ciliophrys, 395
  published autecological laboratory studies, 398
Clade, 174
Clone libraries, 415
  protistan assemblages, 420
Clostridium, 484
  nitrogenase gene arrangements, 486
  nitrogen fixation, 489, 495
Coastal waters
  ecophysiological aspects, 505
Cobalt, 365
Coccolithophorids, 5
Codosiga sp.
  published autecological laboratory studies, 398
COG, 98

Colorless microflagellates, 15
Colwellia, 51
   environmental reductionism, 106
   genomes, 94
   psychrerythraea, 94, 106
Community cell growth and metabolic rates
   vs. individual, 280–281
Community growth rates, 11
   vs. population growth estimates, 281
   vs. single-cell based growth estimates, 281
Community manipulation techniques, 402
Community respiration (CR), 300
   bacterial respiration, 316, 321
   seasonal variation, 308
Comparative genomic analysis, 108, 110–118
   aromatic monomers, 110–118
   carbon monoxide oxidation, 110–118
   chitin, 110–118
   DMSP, 110–118
   glycine betaine, 110–118
   marine prokaryote genomics, 107–121
   metagenomics, 107–121
   methylotrophy, 110–118
   nitrogen cycling, 110–118
   sulfur oxidation, 110–118
   taurine, 110–118
Complete autotrophic nitrogen removal over nitrite (CANON), 538
Congregibacter litoralis KT71
   leucine incorporation, 137
Conserved genes, 462
Coral reefs
   nitrogen fixation, 507
Coral Sea, 78
CR. *See* Community respiration (CR)
CRAM. *See* Carboxyl-rich alicyclic molecules (CRAM)
Crenarchaea nitrification, 363
Crenarchaeota, 60–61, 103
   ammonium oxidation, 536
   total prokaryotic abundance, 62
Croceibacter atlanticus
   comparative genomic analysis, 110–118
   nitrogen sources, 356
Crocosphaera, 486
   nitrogen fixation, 487, 501, 503
   phosphorus-utilization pathways, 488
   watsonii, 501

Cryothecomonas, 419
Cryptic species, 420
CTC. *See* Cyano-ditolyltetrazoliumchloride (CTC)
Culture-independent genetic analysis, 46
Cyanobacteria, 4, 5
   bacterial and archaeal community structure and patterns, 52–54
   cell abundance and size, 10
   comparative genomic analysis, 110–118
   energy production pigment, 6
   as facultative heterotrophs, 132
   genomes, 108
   nitrogen fixation, 494–495
Cyano-ditolyltetrazoliumchloride (CTC), 251, 253, 255, 256, 259
BGE, 283
Cyanophages
   Prochlorococcus, 459
Cyanothece
   nitrogen fixation, 501
Cytophaga-Flavobacteria-Bacteroides, 73. *See also* Bacteroidetes

DAPI. *See* Diamidino phenylindole (DAPI)
DCAA. *See* Dissolved combined amino acids (DCAA)
DCMU. *See* Dichlorophenyl-1-dimethylurea (DCMU)
Deep waters
   attached bacteria, 225
   bacterial respiration, 309–310
Degree of uncoupling, 406
Deltaproteobacteria, 20, 59
   denitrification, 543
   genomes, 108
Denaturing gradient gel electrophoresis (DGGE), 67, 492
Denitrification, 529, 539
   oxygen inhibition, 541–542
Denitrifiers, 20, 538, 539, 540, 543
   function and type, 2
Desulfotalea psychrophila LSv54
   genomes, 94
Desulfovibrio desulfuricans
   denitrification, 543
Detritus food web, 30
DFA. *See* Discriminant function analysis (DFA)

DFAA. *See* Dissolved free amino acids (DFAA)
DGGE. *See* Denaturing gradient gel electrophoresis (DGGE)
Diamidino phenylindole (DAPI), 32, 62, 187, 224, 252, 259
  positive particles, 260
Diaphanoeca sp.
  published autecological laboratory studies, 398
Diatoms, 5
Diazotrophs, 481
  defined, 19
  diversity, gene expression, and nitrogen fixation activity, 490–493
  regulation, 489
DIC. *See* Dissolved inorganic carbon (DIC)
Dichlorophenyl-1-dimethylurea (DCMU), 132
Dilution technique, 402
Dimethylsulfide (DMS), 150, 231, 370
Dimethylsulfoniopropionate (DMSP), 98, 231, 369
  demethylation, 103
  genes, 103
  Prochlorococcus, 133–134
Dinitrogen, 362
  fixers, 2, 18
  production, 21
Dinoflagellates, 16, 390
Direct count assays based on epifluorescence microscopy, 32
Discriminant function analysis (DFA), 74
Dispersal range, 422
Dissimilatory nitrate/nitrite reduction to ammonium (DNRA), 21, 542
Dissolved combined amino acids (DCAA), 219
Dissolved free amino acids (DFAA), 212
  Alphaproteobacteria, 231
  bacterial growth, 355
  concentrations, 214
Dissolved inorganic carbon (DIC)
  ammonification, 534
Dissolved organic carbon (DOC), 7
  bacterial respiration, 11
  bacterial respiration (BR), 312
  characteristics, 209
  composite profile, 209
  concentrations, 352
  global fluxes, 209
Dissolved organic matter (DOM), 7–9, 207, 299–300, 338–339
  Alphaproteobacteria, 498
  Archaea, 231
  carbohydrates, 354–355
  concentrations, 352
  fluxes, 7–8
  heterotrophic bacteria, 5
  labile, 352
  limitation, 353, 354–360
  nitrogen fixation, 497
  trend reversal, 220
Dissolved organic matter (DOM) bacteria interactions, 211–223
  LMW, 211–214
  polymeric DOM protein model, 217–219
  refractory DOM, 220–222
Dissolved organic nitrogen (DON), 7, 169
  concentrations, 352
Dissolved organic phosphorus (DOP), 7
  concentrations, 352
DMS. *See* Dimethylsulfide (DMS)
DMSP. *See* Dimethylsulfoniopropionate (DMSP)
DNA
  fingerprinting methods, 66–67
  sequencing future directions, 122
  synthesis and replication, 258
  viral, 464
DNA–DNA hybridization, 72
DNRA. *See* Dissimilatory nitrate/nitrite reduction to ammonium (DNRA)
DOC. *See* Dissolved organic carbon (DOC)
Dokodinoa, 57
DOM. *See* Dissolved organic matter (DOM)
DON. *See* Dissolved organic nitrogen (DON)
DOP. *See* Dissolved organic phosphorus (DOP)
Dormancy, 249
Doubling time, 11
Ducklow, Hugh, 36, 39

Ecological niche, 421–422
Ecosystems
  energy and matter flow, 337
Ecotypes, 420

EGT. *See* Environmental genome tags (EGT)
18S rRNA, 170–172, 175, 191, 387, 420
Electron acceptors, 21
Electron transport system activity (ETS), 310
Emiliania huxleyi, 170, 321
Endosymbiotic events, 160
Energy production pigments, 6
English Channel
  picobiliphytes, 177
  picoeukaryotes, 188
Environmental arrays, 123–124
Environmental genome tags (EGT), 101
Environmental polymerase chain reaction (envPCR) surveys, 170
Environmental transcriptomics, 123, 124
Enzymatic hydrolysis
  POM, 228
Epifluorescent techniques, 446–447
  picoeukaryotes, 187
Erythrobacter
  comparative genomic analysis, 110–118
  litoralis HTCC2594, 110–118
  NAP1, 110–118, 356
  nitrogen sources, 356
Escherichia coli
  denitrification, 543
  phosphorus-utilization pathways, 488
  proteorhodopsins, 147
  starvation, 247
EST. *See* Expressed sequence tag (EST) libraries
Estuary. *See* specific name
ETS. *See* Electron transport system activity (ETS)
Eukarya, 60–61
Eukaryotes
  cell evolution, 387
  cyanobacteria, 3–4
  defined, 386
  endosymbiotic events, 160
  phytoplankton, 3–4, 5, 6
Euphotic zone rates
  bacterial carbon demand, 348
Euryarchaeota, 60
  total prokaryotic abundance, 62
Export, 9
Expressed sequence tag (EST) libraries, 193–194

Facultative heterotrophs, 132–133
Facultative photoheterotrophy, 138
FALS. *See* Forward-angle light scatter (FALS)
Fasham, Michael, 39
FCM. *See* Flow cytometry (FCM)
Fenchel, Tom, 35
Firmicutes
  comparative genomic analysis, 110–118
FISH. *See* Fluorescence in situ hybridization (FISH)
Fixed carbon repackaging and recovery pathways, 38
Fixed nitrogen budget
  marine, 553
Flagellates, 15, 388–389, 392. *See also* Microflagellates
  bacteria interaction, 413
Flavobacteria
  bacterium BBFL7, 110–118
  bacterium HTCC2170, 110–118
  comparative genomic analysis, 110–118
  MED0217, 110–118
FLB. *See* Fluorescently labeled, heat-killed bacteria (FLB)
Flow cytometry (FCM), 180, 182
  Gulf Stream DCM, 181
Flow into bacterioplankton
  carbon, 406–407
Flow sorting of specific cell fractions, 259
Fluorescence in situ hybridization (FISH), 52, 139, 175, 180, 257, 369, 402, 415
  microautoradiography, 61
  Micromonas, 190
  probes, 416–417
Fluorescently labeled, heat-killed bacteria (FLB), 401–402, 403
Fluorescently labeled virus (FLV), 448–449
FLV. *See* Fluorescently labeled virus (FLV)
Food webs. *See also* Marine pelagic food webs
  detritus, 30
  diversity, 444–446
  impact on microbial processes, 444–446
  marine viruses community dynamics, 444–446
Forward-angle light scatter (FALS), 180
Fox, George, 39

Fractionation
  microbial activity, 302
  size, 301
F-ratio, 9
Functional genes, 105
Functional genomics, 122
Functional screen, 107
Fundamentals of Ecology, 29
Fungi, 16

Gammaproteobacteria, 51
  activity distribution, 274
  ammonium oxidation, 536
  archaeal structure and patterns, 51
  bacterial structure and patterns, 51
  BCC, 414
  comparative genomic analysis, 110–118
  denitrification, 539
  DOM, 498
  Gammaproteobacterium HTCC2207, 110–118
  Gammaproteobacterium KT71, 110–118
  genomes, 93
  MAR-FISH, 275
  Mo-independent nitrogenases, 484
  nitrification, 363
  nitrogen fixation, 362
  North Sea, 370
  resource control, 369
GenBank, 75, 98, 190, 459, 463, 465
Genes
  annotation starting points, 100
  annotation tools, 98
  conserved, 462
  defined, 96
  DMSP demethylation, 103
  horizontal annotation, 100
  nitrogen fixation, 485
  photoheterotrophy, 105
  role category distribution, 99
  sulfatase, 106
Genetic material
  phosphorus, 360–361
Gene transfer agents (GTA), 456
Genomes, 108
  Gammaproteobacteria, 93
  marine bacteria, 94
  marine prokaryote genomics and metagenomics, 92–94

  Nanoarchaeum equitans, 96
  Prochlorococcus, 92
  sequence and assembly, 92–94
Genomic islands, 108
Genomic sequencing
  marine viral diversity, 457
  Ostreococcus tauri, 193
Geobacillus kaustophilus HTA426
  genomes, 94
Geobacter metallireducens
  denitrification, 543
Giovannoni, Stephen, 40
Glaciecola Oceanospirillum, 51
Global distribution, 421
  bacterial and archaeal structure, 75
  and diversity, 420–421
  protistan ecology molecular tools, 420–421
Global Ocean Sampling Project, 74, 101
  marine metagenomic library, 103
GNS. *See* Green non-sulfur (GNS) bacterial division
Goldman, Joel, 38
Gonyaulax polyhedra, 16
Gordon and Betty Moore Foundation
  microbial genome sequencing project initiative, 140
GPR. *See* Green light-absorbing proteorhodopsins (GPRs)
Gramella forsetii
  environmental reductionism, 106
  genomes, 94
Grazers
  bacterioplankton population size, 350
  function and type, 2
  removal experiments, 350
Grazing rates, 402–403
Great plate count anomaly, 32, 39, 46, 414
Great white sharks
  abundance, 446
Green light-absorbing proteorhodopsins (GPRs), 145
Green non-sulfur (GNS) bacterial division, 59
GTA. *See* Gene transfer agents (GTA)
Gulf of Mexico
  marine metagenomic library, 103
Gulf Stream DCM
  FCM, 181
Gymnodinium, 16

Halobacterium sp. NRC1
  transcriptional regulation, 147
Hawaii Ocean Time-series (HOT), 309
  bacterioplankton biomass, 347
  bacterioplankton mass and production, 344
  BCD, 348
  chlorophyll, 344
  nutrients, 344
  PP, 348
  surface-water temperature, 344
Hemiaulus
  symbiotic relationships, 499
Heterotroph
  and alternative lifestyles, 170–171
Heterotrophic bacteria, 10–12
  abundance, 446
  DOM, 5
  growth rates and biomass levels, 12
Heterotrophic bacterial respiration, 299–325.
  See also Bacterial respiration
  changing environment, 324–325.
  community respiration, 315–316
  environmental and ecological factors, 311–314
  measurements, 301–304, 319–320
  predicting, 317–320
  primary production, 321–323
  spatial variability, 309
  temporal variability, 308
  variability, 304–309
Heterotrophic bacterivorous protists
  marine genera, 388
Heterotrophic flagellates, 404
  abundance, 392
  defined, 386
  FISH probes, 416–417
  PCR primers, 416–417
  size-selectivity feeding, 410
  total bacterivory, 404
Heterotrophic microplankton, 391
Heterotrophic nanoflagellates (HNF), 391
  assemblages, 391–393
  bacterial consumption, 397
  batch culture, 399
  chemostat, 399
  continuous culture, 399
  culture experiments, 397–400
  culturing, 399
  ecological relevance, 397
  grazing, 407
  growth on bacteria, 397
  images, 15, 164, 393
  prey size spectrum, 397
  published autecological laboratory studies, 398
  resistance to starvation, 397
  taxa, 398
Heterotrophic nanoplankton (HNAN), 15
Heterotrophic nitrogen fixation, 495
Heterotrophic prokaryotes
  function and type, 2
Heterotrophic protists, 14–16
  dinoflagellates, 16
  illustration, 15, 164, 393
  marine fungi, 16
  marine pelagic food webs, 34–35
  microzooplanktonic protists (20–200 mm), 16
  nanoflagellates (2–20 mm), 16
Hexanoyloxyfucoxanthin, 166
High-molecular-weight (HMW), 169
High nutrient low chlorophyll (HNLC)
  ecosystems, 365–366
  iron, 19
  oceans, 19
High-performance liquid chromatography (HPLC), 167, 180, 182
Hillea marina, 170
HNAN. *See* Heterotrophic nanoplankton (HNAN)
HNF. *See* Heterotrophic nanoflagellates (HNF)
HNLC. *See* High nutrient low chlorophyll (HNLC)
Homologs, 97
Host-range studies
  marine viral diversity, 458
HOT. *See* Hawaii Ocean Time-series (HOT)
HPLC. *See* High-performance liquid chromatography (HPLC)

ICTV. *See* International Committee on Taxonomy of Viruses (ICTV)
Idiomarina
  baltica, 110–118
  comparative genomic analysis, 110–118
  loihiensis, 110–118
  loihiensis L2TR, 94

Imantonia rotunda, 170
Inactivity, 249
Indian Ocean
 flagellate abundance, 392
 HNF, 391
 picoeukaryotes, 192
Individual cell growth and metabolic rates
 vs. community, 280–281
Infrared fast-repetition-rate (IRFRR)
 fluorometry, 139
Inorganic nitrogen budgets
 dissolved, 549
Inorganic nutrient limitation
 nitrogen, 361–363
 phosphorus, 364
 sea bacterial dynamics resource control, 361–363, 364, 365
 trace nutrients, 365
Internal enzymatic capacity, 258
International Committee on Taxonomy of Viruses (ICTV), 463
IRFRR. *See* Infrared fast-repetition-rate (IRFRR) fluorometry
Iron, 19, 21, 365–366
 cycling, 548–549
 nitrogen fixation, 484, 497, 504
 protein, 484
Isolated marine heterotrophic nanoflagellates taxa, 398

Jakoba libera
 published autecological laboratory studies, 398
Jakobids
 flagellates, 388, 389
 published autecological laboratory studies, 398
Janibacter HTCC2649
 comparative genomic analysis, 110–118
Jannasch, Holgar, 34
Jannaschia sp. CCS1
 comparative genomic analysis, 110–118
 genomes, 94
JGOFS. *See* Joint Ocean Flux Study (JGOFS)
Johannes, Robert E., 30, 31
Joint Ocean Flux Study (JGOFS), 39

Karenia brevis, 176
Katablepharids, 390

KEGG, 98
Kill the winner, 454–455
 bacterial and archaeal patterns, 71
Klebsiella
 nitrogen fixation, 489

Labile dissolved organic matter, 352
Labile low-molecular weight
 DOM bacteria interactions, 211–214
Laboea spiralis, 16
Land plants, 3
 vs. phytoplankton, 3–4
LASL. *See* Linker-amplified shotgun library (LASL)
Latency, 249
Law of the Minimum, 335–336
Legendre, Louis, 39
Length heterogeneity PCR (LHPCR), 69
Leucine
 incorporation, 136
 light effect on bacterial production, 6, 135–137
 photostimulation, 368–369
LHPCR. *See* Length heterogeneity PCR (LHPCR)
Light
 leucine incorporation, 6, 137
 Pelagibacter ubique, 368
 sea bacterial dynamics resource control, 368
Lingulodinium polyhedron, 176
Linker-amplified shotgun library (LASL), 465
LMW. *See* Low-molecular weight (LMW)
Loktanella vestfoldensis SKA53
 comparative genomic analysis, 110–118
Low-molecular weight (LMW)
 DOM bacteria interactions, 211–214
Lyngbya
 nitrogen fixation, 489
Lysogenic phage, 456
 Vibrio cholerae, 456
Lysogeny, 457

Macromolecular composition, 258
Macrosetella
 nitrogen fixation, 505
Magnetococcus MC1
 comparative genomic analysis, 110–118

Manganese, 21, 365
  cycling, 548–549
  oxides, 549–550
MAR. *See* Microautoradiography (MAR)
MAR-FISH. *See* Fluorescence in situ hybridization (FISH), microautoradiography
Marine Archaea, 13
  community structure and patterns, 60–62
  genomes, 94
Marine bacteria
  gene role category distribution, 99
Marine bacterioplankton, 243–284
  aromatic monomers, 110–118
  assemblages physiological states distributions, 245
  assemblages physiological structure, 245
  assemblages single-cell characteristics distribution, 265–275
  bacterial abundance, 279
  bacterial assemblages shaping, 408–413
  bacterial cell physiological state, 261–262
  bacterial cell size determining vulnerability, 408–410
  bacterial community composition, 413
  bacterial production *vs.* protistan grazing, 404
  bacterial size classes, 273
  bacterivorous flagellates functional ecology, 394–400
  bottom-up *vs.* top-down control, 405
  carbon monoxide oxidation, 110–118
  chitin, 110–118
  community *vs.* individual cell growth, 280–281
  comparative genomic analysis, 110–118
  culturing bias, 414–419
  description, 250–262
  dilute environment, 394–396
  DMSP, 110–118
  ecological functions, 406–407
  ecological implications, 279–283
  function, 422
  global distribution and diversity, 420–421
  glycine betaine, 110–118
  HNF culture experiments, 397–400
  marine gradients, 271–272
  marine HNF natural assemblages, 391–393
  metabolic rates, 280–281
  nitrogen cycling and methylotrophy, 110–118
  nutrient acquisition, 118–121
  phylogenetic groups, 274–275
  phylogenetic organization, 386–389
  physiological fractions ecological role, 283
  physiological fractions loss and persistence, 263–264
  physiological state dynamics, 276–278
  physiological states, 265–269
  prokaryote antipredator traits, 411–412
  protistan bacterivory impact, 401–407
  protistan bacterivory quantification, 401–402
  protistan bacterivory rates in sea, 403
  protistan ecology molecular tools, 414–422
  protistan grazing, 383–422
  protists functional size classes, 390
  regulation, 260
  simultaneous determination, 270
  single-cell activity operational categories, 259
  single-cell parameters and bulk assemblage response, 282
  single-cell properties and methodological approaches, 250–258
  sulfur oxidation, 110–118
  taurine, 110–118
  uncultured heterotrophic flagellate diversity, 422
  viability, 246
Marine cyanobacterial diazotrophs photomicrographs, 491
Marine environment nitrogen fixation molecular and ecological aspects, 481–507
  benthic habitats, 506
  biochemistry, 482–493
  chemistry, 482–493
  deep water and hydrothermal vents, 507
  diazotroph diversity, 490–493
  diazotrophs regulation, 489
  ecophysiological aspects, 494–507
  enzymology, 483–484
  estuarine and coastal waters, 505
  gene expression, 490–493
  genetics, 482–493
  genomics, 487

microorganism diversity, 489
nitrogenase phylogeny, 487
nitrogen fixation activity, 490–493
open ocean diazotroph ecology, 499–504
Marine fixed nitrogen budget, 553
Marine food web, 445
Marine fungi, 16
Marine gammaproteobacterial clusters
  community structure and patterns, 57
Marine group A, 59
Marine group B, 59
Marine heterotrophic bacterivorous
    protists, 388
Marine heterotrophic nanoflagellates
    assemblages, 391–393
Marine heterotrophic protists
  marine pelagic food webs, 34–35
Marine metagenomic datasets
  Sargasso Sea, 108
Marine metagenomic library
  Gulf of Mexico, 103
Marine microbial ecology
  defined, 1
  prokaryote genomics and metagenomics, 103–121
Marine microbial genomics
  approaches, 101
  metagenomic techniques, 101
Marine microbiology
  development, 28
Marine Microbiology: A Monograph on Hydrobacteriology, 28
Marine pelagic food webs, 27–42
  bacterial abundance, 32
  bacterial activity, 33
  marine heterotrophic protists, 34–35
  methods improvement, 32–33
  microbial loop, 36–38
  molecular revolution, 39
  1950 and before, 28
  1950–1974, 29–31
  1970s–1980s, 32
  1990-present, 39
Marine phototrophic flagellates, 390
Marine plankton communities
  reactive cells, 266
Marine prokaryote genomics and
    metagenomics, 91–125
  comparative genomics and metagenomics, 107–121

environmental reductionism, 106
finding genes, 95
finding operons, 96
functional annotation, 96–99
future directions, 122–124
genome composition ecology, 103
genome sequence and assembly, 92–94
marine microbial ecology, 103–121
pure-culture genomics *vs*. metagenomics, 100–102
reverse biogeochemistry, 104–105
tame or wild, 100–102
Marine sediments
  nitrogen cycle, 530
Marine stramenopiles (MAST), 176
  characteristics, 419
Marine viral diversity
  examination methods, 457–458
  genomic sequencing, 457
  host-range studies, 458
  metagenomic sequencing, 458
  PFGE, 458
  signature gene amplification, 458
  TEM, 458
Marine viruses, 17
  bacterial diversity, 452–456
  biogeochemical cycling, 450–451
  community dynamics, 443–466
  culture-based studies, 458
  culture-independent methods, 459
  defining characteristics, 466–467
  direct counts and viral numbers, 444–446
  diversity, 457–465
  food web, 444–446
  general production rates, 449
  host-specificity, 454
  impact on microbial processes, 443–466
  marine sediments, 449
  metagenomic studies, 462
  pelagic systems, 447–448
  production and decay rates, 447–449
  signature genes, 461
  transmission electron microscopy, 460
  whole-genome profiling, 461
Marinomonas, 51
  comparative genomic analysis, 110–118
  MED121, 110–118
  nitrogen sources, 356

MAST. *See* Marine stramenopiles (MAST)
MDA. *See* Multiple-displacement amplification (MDA)
Mediterranean Sea
  Alphaproteobacteria and *Bacteroidetes* seasonal variation, 73
  bacterial abundance, 277–278
  bacterial respiration, 309–312
  bacterioplankton mass and production, 344
  BGE, 315–316
  chlorophyll, 344
  HNF, 391
  leucine incorporation, 136
  mesocosm experiment, 69
  nutrient-dose experiments, 357
  nutrients, 344
  phosphorous, 19
  picobiliphytes, 177
  picoeukaryotes, 192
  surface-water temperature, 344
Mesocosm, 33
  Mediterranean Sea, 69
Mesodinium rubrum, 16
Mesozooplankton
  taxonomic-trophic compartments, 35
Metagenomics, 122
  datasets, 108
  Gulf of Mexico library, 103
  marine microbial genomics, 101
  Sargasso Sea, 108, 122–123
  sequencing, 458
  techniques, 101
Metatranscriptomics, 123
Methanocaldococcus jannaschii DSM 2661, 94
Methanococcus maripaludis S2, 94
Methanopyrus kandleri AV19, 94
Methanosarcina acetivorans C2A, 94
Methylotrophy genes, 105
Microautoradiography (MAR), 134, 257, 275
Microautoradiography FISH. *See* Fluorescence in situ hybridization (FISH), microautoradiography
Microbial diversity, 422
Microbial ecology
  defined, 1
  marine, 1, 103–121
  prokaryote genomics and metagenomics, 103–121

Microbial food web
  integration, 194
  mortality and contributions, 191–192
Microbial genome sequencing project initiative
  Gordon and Betty Moore Foundation, 140
Microbial genomics
  approaches, 101
  marine, 101
  metagenomic techniques, 101
Microbial loop
  box model diagram, 37
  marine pelagic food webs, 36–38
Microbial nitrogen cycling
  ^{15}N, 545
Microbial nitrogen transformation, 528
Microbial Seascapes, 35
Microbiology
  marine development, 28
Microcystis aeruginosa, 134
MICRO-FISH. *See* Fluorescence in situ hybridization (FISH), microautoradiography
Microflagellates
  colorless, 15
Micromonas, 387, 396
  FISH, 190
  mortality, 192
  pusilla, 167
  Q-PCR, 190–191
  Sargasso Sea, 186
  sex-related genes, 174
Microscopy, 175
Microzooplanktonic protists (20–200 mm), 16
Miller, Charles B., 38
Mineralization, 8
Mineral protection
  organic matter, 230
Mixed approaches, 258
Mixotrophic flagellates, 390
MLST. *See* Multilocus sequence typing (MLST)
Molecular phylogenetics
  picoeukaryotes ecology and diversity, 172–177
Molybdenum
  gammaproteobacteria, 484

independent nitrogenases, 484
nitrogen fixation, 505
protein, 484
Monograph on Hydrobacteriology, 28
Monosiga, 395
Monterey Bay
  SAR86-II, 147
  SAR86-II proteorhodopsin genes, 149
  SAR86 16S rRNA mapping, 148
Morita, Richard, 28
Morphological integrity, 258
Multilocus sequence typing (MLST), 63–64
Multiple-displacement amplification (MDA), 122, 123

NABE. *See* North Atlantic Bloom Experiment (NABE)
Nannochloropsis, 169
Nanoarchaeum equitans, 96
Nanoarchaeum equitans Kin4-M, 94
Nanochlorum eucaryotum, 167
Nanoflagellates (2–20 μm), 14–16
Nanopore sequencing, 122
NAST-E. *See* North Atlantic Subtropical Gyre–East (NAST-E)
NATO. *See* North America Treaty Organization (NATO)
Neighbor-joining distance tree
  picoeukaryotes, 168
Nekton
  taxonomic-trophic compartments, 35
New production, 9
Nickel, 365
Nitrate, 21
  depth distribution, 538
  porewater distribution, 548
  production, 21
  reduction to ammonium, 21, 542
Nitrate reduction
  depth distribution, 538
  dissimilatory, 21
Nitric oxidation, 537
Nitrification, 103, 363
Nitrifiers, 19
  function and type, 2
Nitrite, 362–363
  ammonifiers, 543
  depth distribution, 538
  oxidation, 538

porewater distribution, 548
reduction, 538
reduction to ammonium, 21, 542
Nitrobacter
  comparative genomic analysis, 110–118
  Nb 311A, 110–118
  nitric oxidation, 537
Nitrococcus
  comparative genomic analysis, 110–118
  mobilis Nb231, 110–118
  nitric oxidation, 537
Nitrogen. *See also* Dinitrogen
  enzymes and energetic expense, 362
  inorganic nutrient limitation, 361–363
  phosphate ratios, 553
  phosphorus ratios, 493
  production, 554
  reduction, 362
  remineralize, 407
  sources, 356
Nitrogenase, 485
  gene arrangements, 486, 487
  gene phylogenetic distribution, 488
Nitrogen budgets
  benthic budgets, 550–551
  dissolved, 549
  inorganic, 549
  oceanic budgets, 552–553
Nitrogen cycle, 544. *See also* Sediment nitrogen cycling
  genes, 105
  oceanic, 18
  processes, 540
Nitrogen fixation, 54–55, 362, 484, 487, 489, 494–495, 501, 507
  genes, 485
  regulation, 483
  temperature, 498
  turbulence, 498
Nitrogen-metabolizing bacteria identification
  sediments, 540
Nitrosococcus
  ammonium oxidation, 536
  halophilus, 536
  oceani, 536
Nitrosomonas
  ammonium oxidation, 536
  europaea, 535–536

Nitrosospira
  ammonium oxidation, 536
Nitrospina
  nitric oxidation, 537
Nitrospira
  nitric oxidation, 537
Noctiluca, 16
Nodularia
  detection, 493
  microbial associations, 499
  nitrogen fixation, 489, 496, 498
Nodular spumigena, 55
North America Treaty Organization (NATO)
  Advanced Research Institute Flows of Energy and Material in Marine Ecosystems, 36
North Atlantic
  ARISA, 78
  picoeukaryotes, 184
North Atlantic Bloom Experiment (NABE)
  bacterioplankton biomass, 347
  bacterioplankton biomass and production, 344
  BCD, 348
  chlorophyll, 344
  nutrients, 344
  PP, 348
  surface-water temperature, 344
North Atlantic Subtropical Gyre–East (NAST-E), 322–323
North Pacific
  marine metagenomic library, 103
North Pacific Gyre, 19
  leucine incorporation, 136
North Pacific Subtropical Gyre (Station ALOHA), 108
  bacterioplankton abundance, 338
  leucine incorporation, 136
  leucine photostimulation, 368–369
  marine metagenomic library, 103
  metagenomic dataset, 101
  nitrogen fixation, 502, 505
  pelagophytes, 184
  PP, 338
  proteorhodopsin genes, 148
  TOC, 338
North Sea
  bacterial respiration, 312, 313, 321
  beta-glucosidase, 218
  BGE, 315

Gammaproteobacteria, 370
  picobiliphytes, 177
Norwegian Sea
  HNF, 391
Nostoc
  nitrogenase gene arrangements, 486, 488
Nucleosides uptake
  photoheterotrophic marine prokaryotes, 134
Nutrient, 344, 355
  acquisition genes, 105
Nutrient cycling
  viral lysis, 466
Nutrient-dose experiments
  bacterioplankton population size, 350

Ocean ecosystems
  energy and matter flow, 337
Oceanicaulis alexandrii
  comparative genomic analysis, 110–118
Oceanic budgets
  nitrogen budgets, 552–553
Oceanic carbon cycle
  microbial role, 3
Oceanic coccoid cyanobacterial genera
  comparison, 4
Oceanic microbes
  functional groups, 2
Oceanic nitrogen cycle, 18
Oceanicola
  batensis, 110–118
  comparative genomic analysis, 110–118
  granulosus, 110–118
Oceanic phototrophic microbes, 6
Oceanobacillus
  comparative genomic analysis, 110–118
  iheyensis, 110–118
  iheyensis HTE831, 94
Oceanobacter sp. RED65
  comparative genomic analysis, 110–118
Oceanospirilliaceae
  comparative genomic analysis, 110–118
Oceanospirillum MED92
  comparative genomic analysis, 110–118
Odum, Eugene, 29–30
Odum, Howard, 29
Oikomonas sp.
  published autecological laboratory studies, 398

INDEX

Oikopleura, 384
Oligonucleotide microarray hybridization, 492
Oligotrophic marine *Gammaproteobacteria* (OMG), 57
OM43 clade, 59
OMG. *See* Oligotrophic marine *Gammaproteobacteria* (OMG)
Open reading frame (ORF), 96
Operational heterotrophs, 7
Operational taxonomic unit (OTU), 64, 74, 77, 174
ORF. *See* Open reading frame (ORF)
Organic Chemistry in Its Applications to Agriculture and Physiology, 335
Organic matter
  mineral protection, 230
  size continuum, 224
Organic matter-bacteria
  interactions, 207–231
  bacterial community structure, 230
  DOM bacteria interactions, 211–223
  extracellular hydrolytic enzymes, 215–216
  future challenges, 231
  labile LMW DOM, 211–214
  organic matter inventory and fluxes, 208–210
  organic matter utilization, 230
  polymeric DOM protein model, 217–219
  POM-bacteria interactions, 223–229
  POM continuum, 223
  POM fluxes, 223–228
  POM-mineral interactions, 229
  refractory DOM, 220–222
Orthologs, 97
Oscillatoria
  nitrogen fixation, 489, 507
Osmotic shock, 263
Osmotroph, 339
Ostreococcus, 396, 421
  flow cytometry, 182
  genomic sequencing, 193
  HMW, 169
  Sargasso Sea, 174
  sex-related genes, 174
  tauri, 167, 193
OTU. *See* Operational taxonomic unit (OTU)

Oxidative genes, 105
Oxygen, 21
  depth distribution, 538
  inhibition denitrification, 541–542

Pace, Norman, 40
Pacific Ocean
  AAnP, 368
  bacterioplankton mass and production, 344
  BCD and PP, 348
  chlorophyll, 344
  choanoflagellates, 394
  denitrifiers, 19
  iron, 365
  nutrient-dose experiments, 357, 358
  nutrients, 344, 355
  phytoplankton, 38
  picoeukaryotes, 185
Paracoccus denitrificans
  denitrification, 539
Paralogs, 97
Paraphysomonas, 388, 389, 419
  abundance, 414–415
  imperforata, 399, 414–415
  published autecological laboratory studies, 398
Particulate organic matter (POM), 207
  bacteria interactions, 223, 229
  continuum, 223
  enzymatic hydrolysis, 228
  mineral interactions, 229
  organic matter-bacteria interactions in seawater, 223
Parvularcula bermudensis
  comparative genomic analysis, 110–118
PCR. *See* Polymerase chain reaction (PCR)
Pedinellids
  published autecological laboratory studies, 398
Pelagibacter, 262
  nitrogen fixation, 487
Pelagibacter ubique, 6, 55, 75
  comparative genomic analysis, 110–118
  genomes, 94, 95, 96
  HTCC106, 94
  leucine incorporation, 137
  light, 368
  nitrogen sources, 356
  proteorhodopsins, 144
  size, 409
  transporters, 355

Pelagic systems, 408
 viral production and decay rates, 447–448
Pelagococcus, 169
Pelagomonas
 calceolata, 169
 flow cytometry, 182
PF. See Phototrophic flagellates (PF)
Pfam, 98
PFGE. See Pulsed-field gel electrophoresis (PFGE)
Pfiesteria piscicida, 16
PHA. See Polyhydroxyalkanoate (PHA)
Phaeocystis cordata, 170
Phage Proteomic Tree, 462
Phagotrophy
 eukaryotic cell evolution, 387
Phosphate, 459
Phosphorus, 19
 Crocosphaera, 488
 genetic material, 360–361
 inorganic nutrient limitation, 364
 remineralize, 407
 Sargasso Sea, 364
 utilization pathways, 488
Photoautotrophs, 52
Photobacterium
 comparative genomic analysis, 110–118
 environmental reductionism, 106
 genomes, 94, 95
 profundum, 95, 106
 profundum SS9, 94, 110–118
 profundum 3TCK, 110–118
 proteorhodopsins, 144
 SKA34, 110–118
Photoheterotroph
 function and type, 2
Photoheterotrophic bacteria, 5–6, 52, 131–151
Photoheterotrophic marine prokaryotes, 131–158
 abundance, 142, 146–149, 150
 activity, 146–149
 cyanobacteria as facultative heterotrophs, 132
 diversity, 139
 ecological significance, 142, 150
 implications, 138
 light and dark incubations field studies, 135

marine AAnP bacteria habitats and diversity, 139–141
nucleosides and amino acids uptake, 134
physiology, 140
proteorhodopsin-containing prokaryotes, 143–150
proteorhodopsin genotypes, 144
proteorhodopsin spectral tuning, 145
rediscovery, 139
taxonomic distributions, 144
unicellular cyanobacteria facultative photoheterotrophy, 132–138
urea and DMSP uptake, 133
Photoheterotrophy genes, 105
Photosynthesis
 Prochlorococcus, 459
Photosynthetic reaction center, 140
Phototroph, 132
Phototrophic flagellates (PF)
 abundance, 392
 marine, 390
 total bacterivory, 404
Phototrophic microbes
 oceanic, 6
Phylochips, 492
Phylogenetic organization
 marine bacterioplankton protistan grazing, 386–389
Phylogenetic relationships
 Prochlorococcus, 50
 Synechococcus, 50
Phylogenetic trees, 47, 387
 proteorhodopsins, 145
 PufM, 141
 stramenopiles, 418
Phylogeny
 picoeukaryotes, 173
 prokaryotes, 47
Phylotype, 174
Phytoplankton, 38
 biomass vs. living surface area, 34
 cell abundance and size, 10
 generation times, 13
 growth rates and biomass levels, 12
 vs. land plants, 3–4
 taxonomic-trophic compartments, 35
PI. See Propidium iodide (PI)
Picobiliphytes, 177
Picochlorum atomus, 167

Picochlorum eucaryotum, 167
Picocyanobacteria, 407
Picoeukaryotes, 161–162, 184, 192
  abundance and activity, 182–184
  alternative lifestyles, 170–171
  A-PCR, 186
  biological traits, 162–171
  carotenoids, 166
  chlorophylls, 165
  classification, 162–171
  ecology and diversity, 159–194
  English Channel, 188
  environmental diversity and molecular phylogenetics, 172–177
  epifluorescence, 187
  FISH, 186
  functional roles, 162–171
  genomic approaches, 193
  heterotrophs, 170–171
  images, 15, 164, 393
  microbial food web integration, 194
  microbial food web mortality and contributions, 191–192
  mixed picophytoplankton assemblages, 182–184
  neighbor-joining distance tree, 168
  photoautotrophs, 163–169
  phylogeny, 173
  population quantifying, 186–189
  population quantifying methodological challenges, 190
  quantifying mixed assemblages methods, 180–181
Picophagus flagellatus, 171
Picophytoeukaryotic taxa
  abundance, 189
Picophytoplankton
  abundance relationships, 183
Picoplankton, 5
Pinguiochrysis pyriformis, 169
Planctomycete, 545
  comparative genomic analysis, 110–118
  Rhodopirellula baltica genome, 96
Plankton
  Atlantic Ocean, 323
  biomass *vs.* living surface area, 34
  reactive cells, 266
  respiratory activity, 33
  taxonomic-trophic compartments, 35

Plate count anomaly, 32, 39, 46, 414
Polaribacter irgensii 23-P
  comparative genomic analysis, 110–118
Polyhydroxyalkanoate (PHA), 106
Polymerase chain reaction (PCR), 142, 415, 461–463
  primers, 416–417
Polymeric DOM protein model
  bacteria interactions, 217–219
Polynucleobacter, 401
  size, 409–410
POM. *See* Particulate organic matter (POM)
Pomeroy, Lawrence, 30, 31
Population growth estimates
  vs. community growth estimates, 281
Porewater distribution, 548
Poterioochromonas malhamensis, 399
PP. *See* Primary production (PP)
Prasinophyceae, 163
Predation viruses, 453
Primary production (PP), 300, 322
  ALOHA, 338
  Arabian Sea, 348
  vs. bacterial production, 9
  euphotic zone rates, 348
  function, 2
  type, 2
Principle of Competitive Exclusion, 341
Prochlorococcus, 4–5, 47, 64
  amino acids uptake, 134
  comparative genomic analysis, 110–118
  cyanophages, 459
  discovery, 53
  distribution, 54
  DMSP uptake, 133–134
  environmental reductionism, 106
  facultative heterotrophs, 132–133
  facultative photoheterotrophy, 138
  flow cytometry, 182
  gene role category distribution, 99
  genomes, 92, 94
  genomic islands, 65
  HNF grazing, 407
  light and dark incubations, 135–136
  light effect on bacterial production, 6
  marinus, 99
  marinus CCMP1375, 94
  marinus MED4, 94
  marinus MIT9211, 110–118

Prochlorococcus (Continued)
　marinus MIT9313, 94
　nucleoside uptake, 134
　photosynthesis, 459
　phylogenetic relationships, 50
　vs. Synechococcus, 4
　urea uptake, 133–134
Prokaryotes. *See also* Marine prokaryote genomics and metagenomics
　membrane and cell wall components, 221–222
　physicochemical surface properties, 412
　seawater, 47–48
Prokaryotic genes, 95
Propidium iodide (PI), 258
Proteins
　bacterial growth, 355
　iron, 484
　molybdenum, 484
　polymeric DOM model, 217–219
　Sargasso Sea, 101
Proteorhodopsin-containing prokaryotes
　abundance and activity ecological significance, 150
　photoheterotrophic marine prokaryotes, 143–150
　proteorhodopsin genotypes, 144
　taxonomic distributions, 144
Proteorhodopsin genes
　Sargasso Sea, 148
　Station Aloha, 148
Proteorhodopsins
　Escherichia coli, 147
　genotypes and taxonomic distributions, 144
　Photobacterium, 144
　phylogenetic tree, 145
Protistan assemblages
　clone libraries, 420
Protistan bacterivory impact on marine bacterioplankton
　bacterial production *vs.* protistan grazing, 404
　marine bacterioplankton protistan grazing, 401–402, 404
　protistan bacterivory quantification, 401–402
Protistan bacterivory quantification
　impact on marine bacterioplankton, 401–402

Protistan ecology molecular tools
　culturing bias, 414–419
　global distribution and diversity, 420–421
　marine bacterioplankton protistan grazing, 414–421
Protistan grazing
　bacterial assemblages shaping, 408–413
　marine bacterioplankton, 383–422
Protists
　biogeography and diversity, 421
　cell abundance and size, 10
　defined, 386
　functional size classes, 390
　marine bacterioplankton protistan grazing, 390
　metabolisms, 161
Protozoa
　abundance, 446
　biomass *vs.* living surface area, 34
　defined, 386
Protozooplankton
　taxonomic-trophic compartments, 35
Prymnesiophytes
　FISH, 188
　quartet puzzling tree, 171
Pseudoalteromonas, 51
　comparative genomic analysis, 110–118
　genomes, 94
　haloplanktis, 110–118
　haloplanktis TAC125, 93, 94
　nitrification, 363
　tunicata D2, 110–118
Pseudobodo
　published autecological laboratory studies, 398
Pseudomonas
　aeruginosa, 221
　denitrification, 539
　mendocina, 546
Psychromonas CNPT3
　comparative genomic analysis, 110–118
Pteridomonas
　danica, 399
　published autecological laboratory studies, 398
PufM, 140
　phylogenetic tree, 141

Pulsed-field gel electrophoresis (PFGE), 458, 461
  marine viral diversity, 458
Pure-culture genomics, 122
  marine prokaryote genomics and metagenomics, 100–102
  vs. metagenomics, 100–102
Pyramimonas, 387
Pyrobaculum aerophilum IM2, 94
Pyrococcus
  abyssi GE5, 94
  furiosus DSM3638, 94
  horikoshii OT3, 94
Pyrosequencing, 122, 463

Quantitative real-time PCR (qPCR), 147, 180
  Micromonas, 190–191
Quartet puzzling tree
  prymnesiophytes, 171

Radiolabeled bacteria, 402
Radiotracers, 212
RBH. *See* Reciprocal best hits (RBH)
Reactive bacteria
  total bacterial abundance, 272
Reactive cells
  marine plankton communities, 266
Reciprocal best hits (RBH), 100
Refractory dissolved organic matter, 352
  bacteria interactions, 220–222
Reinekea MED297
  comparative genomic analysis, 110–118
Remote sensing
  Trichodesmium, 493–494
Resource control, 335–370
  comparative approaches, 343–348
  compounds, 354–360
  defining limitation, 349–350
  DOM limitation, 351–352
  experimental approaches, 349–350
  inorganic nutrient limitation, 361–365
  light, 368
  nitrogen, 361–363
  nutrient uptake kinetics, 339–342
  phosphorus, 364
  populations, 369–370
  temperature-DOM interactions, 366–367
  trace nutrients, 365
Resources
  competing for limited, 342

Respiration rates units, 305
Respiratory activity, 258
Respiratory quotient (RQ), 303
Reverse biogeochemistry, 92
  discovery, 103
  marine prokaryote genomics and metagenomics, 104–105
Reverse-transcriptase polymerase chain reaction (RT-PCR), 148, 501
Rhizobacterales
  comparative genomic analysis, 110–118
Rhizosolenia
  symbiotic relationships, 499
Rhodobacterales bacterium HTCC2654
  comparative genomic analysis, 110–118
Rhodopirellula baltica
  comparative genomic analysis, 110–118
  gene role category distribution, 99
  SH1 genomes, 94
  sulfatase genes, 106
Rhodopseudomonas
  nitrogenase gene arrangements, 486
Rhodopsins, 144
Rhynchomonas
  published autecological laboratory studies, 398
Ribosomal RNA (rRNA), 46–47, 50, 52, 56, 59, 65, 124. *See also* 16S rRNA
  BAC library, 144
  18S, 170–172, 175, 191, 387, 420
  gene clone libraries, 67
  global distribution, 75
  operons, 94
  vs. rDNA libraries, 423
  SAR86 mapping, 148
Richelia, 54
Robiginitalea biformata HTCC2501
  comparative genomic analysis, 110–118
Rohdospirillum
  nitrogen fixation, 489
Roseobacter
  activity distribution, 274
  comparative genomic analysis, 110–118
  MED193, 110–118
Roseobacter Clade of Marine *Alphaproteobacteria*, 50–51, 108–109
Roseovarius
  comparative genomic analysis, 110–118
  nubinhibens, 110–118

Ross Sea
  BCD and PP, 348
RQ. *See* Respiratory quotient (RQ)
rRNA. *See* Ribosomal RNA (rRNA)
RT-PCR. *See* Reverse-transcriptase polymerase chain reaction (RT-PCR)

Salpingoeca, 15
  infusorium, 15
San Pedro Ocean Time Series, 74
Sapelo Island, 30
SAR11, 40. *See also Pelagibacter*
  clade, 56
  cluster, 55–56
  comparative genomic analysis, 110–118
  diversity, 56
SAR116 cluster, 59
SAR202 cluster, 59
Sargasso Sea, 40, 59, 66
  Bathycoccus, 186
  dataset, 103
  DMSP demethylation genes, 103
  marine metagenomic datasets, 108
  metagenomic database, 122–123
  Micromonas, 186
  nutrient-dose experiments, 357
  nutrients, 355
  Ostreococcus, 174
  phosphorous, 364
  picobiliphytes, 177
  picoeukaryotes, 184
  proteins, 101
  proteorhodopsin genes, 148
SAR86-II
  Monterey Bay, 147
  proteorhodopsin genes, 149
SAR86 16S rRNA
  mapping, 148
  Monterey Bay, 147
SATL. *See* South Atlantic Subtropical Gyre (SATL)
Scripps Institution of Oceanography, 28
Sea Microbes, 35
Seasonal variation
  community respiration, 308
  Synechococcus, 72
Sea-surface height (SSH), 494
Sea-surface microlayer (SML), 309

Sea-surface temperature (SST), 494
Seawater
  culturable bacteria, 55–56
  direct count, 447
  organic matter size continuum, 224
  prokaryotes, 47–48
Secondary ion mass spectrometry (SIMS), 490
Sediment
  marine, 530
  nitrogen-metabolizing bacteria identification, 540
Sediment nitrogen cycling, 527–552
  benthic budgets, 550–551
  inputs, 531
  manganese and iron, 548–549
  microbes, 532–547
  nitrogen budgets, 550–553
  oceanic budgets, 552–553
  transformations, 532–549
SEED, 98
Semilabile dissolved organic matter, 352
Sequencing
  complete *vs.* draft, 92–93
Sex-related genes, 174
Shewanella, 51
  comparative genomic analysis, 110–118
  frigidimarina, 110–118
Sieburth, John, 35
Signature genes, 462
  amplification, 458
  marine viral diversity, 458
Silicibacter
  comparative genomic analysis, 110–118
  genomes, 93, 94, 95
  nitrogen fixation, 487
  nitrogen sources, 356
  pomeroyi, 51, 95, 110–118, 356
  pomeroyi DSS-3, 94
  TM1040, 93, 110–118
SIMS. *See* Secondary ion mass spectrometry (SIMS)
Single-cell based growth estimates
  vs. community growth estimates, 281
Single-cell sequencing, 122
Sinorhizobium
  nitrogenase gene arrangements, 486

16S rRNA, 46–47, 59, 65, 250
  bacteria, 467
  gene clone libraries, 67
  global distribution, 75
  link between, 230
  nitrogenase, 487
  phylogeny, 217
  probes, 253
Size fractionation, 301
  microbial activity, 302
Slow growth, 249
Small scale heterogeneity, 446
SML. *See* Sea-surface microlayer (SML)
Sorokin, Yuri, 34
Southampton Water
  choanoflagellates, 394
South Atlantic Subtropical Gyre (SATL), 322–323
Southern Ocean
  choanoflagellates, 394
  iron, 365–366
South Pacific
  pelagophytes, 184
  picoeukaryotes, 184
Spartina
  alterniflora, 30
  nitrogen fixation, 506–507, 532
Spatial variation
  bacterial and archaeal structure, 74–78
Speciation mechanisms, 422
Species, 422. *See also* specific genus
Sphingomonas, 262
  comparative genomic analysis, 110–118
  SKA58, 110–118
  starvation, 247
Sphingopyxis alaskensis
  comparative genomic analysis, 110–118
Spumella, 395
  growth rates, 400
  published autecological laboratory studies, 398
SSH. *See* Sea-surface height (SSH)
SST. *See* Sea-surface temperature (SST)
STARFISH, 257. *See also* Fluorescence in situ hybridization (FISH), microautoradiography
Starvation, 247
  HNF, 397
  survival, 249

Status of cell membrane, 258
Steele, John, 30–31
Stephanoeca sp.
  published autecological laboratory studies, 398
Stoeckeria, 15
  algicida, 15
Stramenopiles
  flagellates, 389
  phylogenetic tree, 418
Streptomyces
  nitrogen fixation, 484
Strombidium capitatum, 15
Structure of Marine Ecosystems, 30
Subarctic Pacific Ecosystem Research (SUPER) program, 38
Substrate tracking autoradiography (STAR) FISH, 257. *See also* Fluorescence in situ hybridization (FISH), microautoradiography
Substrate uptake, 258
Sulfatase genes
  Rhodopirellula baltica, 106
Sulfate, 21
Sulfitobacter
  comparative genomic analysis, 110–118
  EE36, 110–118
  NAS4.1, 110–118
SUPER. *See* Subarctic Pacific Ecosystem Research (SUPER) program
Surface area of cells (SA), 10, 34
Surface-water temperature, 344
Survival starvation, 249
Symbiomonas scintillans, 164, 171–172, 174
Symbiotic relationships, 499
Sympatry, 422
Synechococcus, 4–5, 18
  amino acids uptake, 134
  CC9902, 94
  comparative genomic analysis, 110–118
  cyanophages, 459
  discovery, 53
  DMSP uptake, 133–134
  facultative heterotrophs, 132–133
  facultative photoheterotrophy, 138
  flow cytometry, 182
  genomes, 93, 94, 95
  HNF grazing, 407
  light and dark incubations, 135–136

Synechococcus (Continued)
  nitrogen fixation, 487, 501
  nucleoside uptake, 134
  phages, 456
  photosynthesis, 459
  phylogenetic relationships, 50
  vs. Prochlorococcus, 4
  RS9917, 110–118
  seasonal variation, 72
  urea uptake, 133–134
  WH8102, 94
Synechocystis
  nitrogen fixation, 501

TdR. *See* Tritium-labeled thymidine (TdR) incorporation
Teal, John, 30
Telonema, 177–178
TEM. *See* Transmission electron microscopy (TEM)
Temperature
  bacterial respiration, 311
  nitrogen fixation, 498
Temporal variation
  bacterial and archaeal community structure and patterns, 72–73
Tenacibaculum sp. MED152
  comparative genomic analysis, 110–118
Terminal restriction fragment length polymorphism (T-RFLP), 66, 77, 313, 492, 507
Terrestrial plants. *See* Land plants
Tetraparma pelagica, 169
Thermotoga maritima MSB8
  genomes, 94
Thiobacillus
  nitrogen fixation, 489
Thioploca, 543, 548
  nitrogen, 532
TIGRFAM, 98
TOC. *See* Total organic carbon (TOC)
Top-down population control, 343
Torres Strait, 79
Total bacterivory
  heterotrophic flagellates, 404
Total organic carbon (TOC)
  ALOHA, 338
Total prokaryotic abundance, 62
Toxins, 410

Trace nutrients
  inorganic nutrient limitation, 365
Transcriptional regulation
  Halobacterium sp. NRC1, 147
Transmission electron microscopy (TEM), 254, 445, 460
  marine viral diversity, 458
T-RFLP. *See* Terminal restriction fragment length polymorphism (T-RFLP)
Trichodesmium, 53, 483
  characteristics, 499
  cultivation, 490
  ecology, 499–500
  microbial associations, 499
  nitrogenase gene arrangements, 486
  nitrogen fixation, 487, 489, 495–497, 502–505
  phosphorus-utilization pathways, 488
  remote sensing, 493–494
Tritium-labeled thymidine (TdR) incorporation, 33
Trophic interactions, 8
TSA. *See* Tyramide signal amplification (TSA) FISH
Tyramide signal amplification (TSA) FISH, 175

Underlying water (ULW), 309
Unicellular cyanobacteria facultative photoheterotrophy
  cyanobacteria as facultative heterotrophs, 132
  implications, 138
  nucleosides and amino acids uptake, 134
  photoheterotrophic marine prokaryotes, 132, 134, 138
University of Georgia Marine Institute, 30
Unrooted Bayesian tree of marine *Gammaproteobacteria*, 49

Viable but not culturable (VBNC), 247–248, 249
Vibrio, 71
  alginolyticus 12G01, 110–118
  angularium, 51, 73
  angustum, 110–118
  cholerae, 51, 456
  comparative genomic analysis, 110–118
  diazotrophicus, 489

fischeri ES114, 94
lysogenic phage, 456
MED222, 110–118
nitrification, 363
nitrogen fixation, 489
parahaemolyticus RIMD 2210633, 94
splendidus, 64
splendidus 12B01, 110–118
starvation, 247
vulnificus CMCP6, 94
vulnificus YJ016, 94
Vibrionales
  comparative genomic analysis, 110–118
Viral DNA
  purifying, 464
Viral lysis
  nutrient cycling, 466
Viral production and decay rates
  diversity, 447–448
  impact on microbial processes, 447–448
  marine viruses community dynamics, 447–448
  pelagic systems, 447–448
Viral shunt, 445
Virioplankton
  taxonomic-trophic compartments, 35

Viruses. *See also* Marine viruses
  abundance, 446
  cell abundance and size, 10
  controlling, 467
  function and type, 2
  predation, 453
Virus-like particles (VLP), 445

Waksman, Selman, 28
Whole-genome shotgun (WGS)
  metagenomic library, 101
  sequencing, 92, 101
Williams, Peter, 33
Woese, Carl, 39
Wood, E.J. Ferguson, 32
Woods Hole Oceanographic Institution, 28

Yayanos, Aristides, 34

Zinc, 365
ZoBell, Claude, 28
ZoBell's box diagram of carbon flows, 29
Zooplankton
  abundance, 446
  biomass *vs.* living surface area, 34
  cell abundance and size, 10